D1334397

Abstract Regular Polytopes

Abstract regular polytopes stand at the end of more than two millennia of geometrical research, which began with regular polygons and polyhedra. They are highly symmetric combinatorial structures with distinctive geometric, algebraic, or topological properties, in many ways more fascinating than traditional regular polytopes and tessellations. The rapid development of the subject in the past 20 years has resulted in a rich new theory, featuring an attractive interplay of mathematical areas, including geometry, combinatorics, group theory, and topology. Abstract regular polytopes and their groups provide an appealing new approach to understanding geometric and combinatorial symmetry.

This is the first comprehensive up-to-date account of the subject and its ramifications, and meets a critical need for such a text, because no book has been published in this area of classical and modern discrete geometry since Coxeter's *Regular Polytopes* (1948) and *Regular Complex Polytopes* (1974). The book should be of interest to researchers and graduate students in discrete geometry, combinatorics, and group theory.

Peter McMullen is Professor of Mathematics at University College London.

Egon Schulte is Professor of Mathematics at Northeastern University in Boston.

ENCYCLOPEDIA OF MATHEMATICS AND ITS APPLICATIONS

Volume 92
Abstract Regular Polytopes

27 N. H. Bingham, C. M. Goldie, and J. L. Teugels *Regular Variation*
28 P. P. Petrushev and V. A. Popov *Rational Approximation of Real Functions*
29 N. White (ed.) *Combinatorial Geometries*
30 M. Pohst and H. Zassenhaus *Algorithmic Algebraic Number Theory*
31 J. Aczel and J. Dhombres *Functional Equations in Several Variables*
32 M. Kuczma, B. Choczewski, and R. Ger *Iterative Functional Equations*
33 R. V. Ambartzumian *Factorization Calculus and Geometric Probability*
34 G. Gripenberg, S.-O. Londen, and O. Staffans *Volterra Integral and Functional Equations*
35 G. Gasper and M. Rahman *Basic Hypergeometric Series*
36 E. Torgersen *Comparison of Statistical Experiments*
37 A. Neumaier *Interval Methods for Systems of Equations*
38 N. Korneichuk *Exact Constants in Approximation Theory*
39 R. Brualdi and H. Ryser *Combinatorial Matrix Theory*
40 N. White (ed.) *Matroid Applications*
41 S. Sakai *Operator Algebras in Dynamical Systems*
42 W. Hodges *Basic Model Theory*
43 H. Stahl and V. Totik *General Orthogonal Polynomials*
45 G. Da Prato and J. Zabczyk *Stochastic Equations in Infinite Dimensions*
46 A. Björner et al. *Oriented Matroids*
47 G. Edgar and L. Sucheston *Stopping Times and Directed Processes*
48 C. Sims *Computation with Finitely Presented Groups*
49 T. Palmer *Banach Algebras and the General Theory of *-Algebras*
50 F. Borceux *Handbook of Categorical Algebra I*
51 F. Borceux *Handbook of Categorical Algebra II*
52 F. Borceux *Handbook of Categorical Algebra III*
54 A. Katok and B. Hasselblatt *Introduction to the Modern Theory of Dynamical Systems*
55 V. N. Sachkov *Combinatorial Methods in Discrete Mathematics*
56 V. N. Sachkov *Probabilistic Methods in Discrete Mathematics*
57 P. M. Cohn *Skew Fields*
58 R. Gardner *Geometric Tomography*
59 G. A. Baker Jr. and P. Graves-Morris *Padé Approximants*
60 J. Krajicek *Bounded Arithmetic, Propositional Logic, and Complexity Theory*
61 H. Groemer *Geometric Applications of Fourier Series and Spherical Harmonics*
62 H. O. Fattorini *Infinite Dimensional Optimization and Control Theory*
63 A. C. Thompson *Minkowski Geometry*
64 R. B. Bapat and T. E. S. Raghavan *Nonnegative Matrices with Applications*
65 K. Engel *Sperner Theory*
66 D. Cvetkovic, P. Rowlinson, and S. Simic *Eigenspaces of Graphs*
67 F. Bergeron, G. Labelle, and P. Leroux *Combinatorial Species and Tree-Like Structures*
68 R. Goodman and N. Wallach *Representations and Invariants of the Classical Groups*
69 T. Beth, D. Jungnickel, and H. Lenz *Design Theory I, 2ed*
70 A. Pietsch and J. Wenzel *Orthonormal Systems for Banach Space Geometry*
71 G. E. Andrews, R. Askey, and R. Roy *Special Functions*
72 R. Ticciati *Quantum Field Theory for Mathematicians*
73 M. Stern *Semimodular Lattices*
74 I. Lasiecka and R. Triggiani *Control Theory for Partial Differential Equations I*
75 I. Lasiecka and R. Triggiani *Control Theory for Partial Differential Equations II*
76 A. A. Ivanov *Geometry of Sporadic Groups 1*
77 A. Schinzel *Polynomials with Special Regard to Reducibility*
78 H. Lenz, T. Beth, and D. Jungnickel *Design Theory II, 2ed*
79 T. Palmer *Banach Algebras and the General Theory of *-Algebras II*
80 O. Stormark *Lie's Structural Approach to PDE Systems*
81 C. F. Dunkl, Y. Xu *Orthogonal Polynomials of Several Variables*
82 J. P. Mayberry *The Foundations of Mathematics in the Theory of Sets*

ENCYCLOPEDIA OF MATHEMATICS AND ITS APPLICATIONS

Abstract Regular Polytopes

PETER McMULLEN
University College London

EGON SCHULTE
Northeastern University

PUBLISHED BY THE PRESS SYNDICATE OF THE UNIVERSITY OF CAMBRIDGE
The Pitt Building, Trumpington Street, Cambridge, United Kingdom

CAMBRIDGE UNIVERSITY PRESS
The Edinburgh Building, Cambridge CB2 2RU, UK
40 West 20th Street, New York, NY 10011-4211, USA
477 Williamstown Road, Port Melbourne, VIC 3207, Australia
Ruiz de Alarcón 13, 28014 Madrid, Spain
Dock House, The Waterfront, Cape Town 8001, South Africa

http://www.cambridge.org

First published 2002

Printed in the United States of America

Typeface Times New Roman 10/12.5 pt. *System* LaTeX 2_ε [TB]

A catalog record for this book is available from the British Library.

Library of Congress Cataloging in Publication Data

McMullen, Peter, 1955–

Abstract regular polytopes / Peter McMullen, Egon Schulte.

p. cm. – (Encyclopedia of mathematics and its applications)

Includes bibliographical references and index.

ISBN 0-521-81496-0

1. Polytopes. I. Schulte, Egon, 1955– II. Title. III. Series.

QA691 .M395 2002

516.3′5 – dc21 2002017391

ISBN 0 521 81496 0 hardback

To Donald Coxeter, a constant inspiration

Contents

Preface

Regular polyhedra have been with us since before recorded history; the appeal of the beauty of geometric figures to the artistic senses well predates any mathematical investigation of them. However, it also seems to be the case that formal mathematics begins with such regular figures as an important topic, and a strong strain of mathematics since classical times has centred on them. Indeed, the subject of regular polyhedra has shown an enormous potential for revival. Before the present time, the most recent renaissance began in the early nineteenth century. The modern abstract theory started as an offshoot of this, but with the parallel (but separate) growth of the idea of geometries has taken on a new and vigorous life.

When we embarked in 1988 on the project of composing a coherent account of this modern theory, we had heard the old adage to the effect that, if one tries to write a paper, then one believes that everything is known, but when one starts to write a book, then one realizes that nothing is known. At the time, neither of us fully appreciated the truth of the saying. We began with a number of particular results, and an initial focus on a problem of classifying a certain class of abstract regular polytopes raised by Branko Grünbaum, but with only the merest outlines of a general theory. The foundations of the latter had been laid by Danzer and Schulte, although earlier intimations of what would be needed had appeared elsewhere (we elaborate on this at the beginning of Chapter 2). But the full development of the theory and its ramifications has occurred during the writing of this book, and has been largely inspired by it. Indeed, one major problem for us was that the subject seemed to be expanding faster than we could put it down on paper!

We have received much support from friends and colleagues during the preparation of this volume; as well, we have probably sorely tried their patience, because they had quite reasonably expected the book to have been published long before now. We wish particularly to thank Asia Ivič Weiss and Barry Monson for their encouragement and (we are sure) forebearance, but especially for providing various opportunities for us to collaborate. We are also very grateful to the NSF and NSA, which have supported Schulte's research, and have enabled McMullen to visit him. Above all, Donald Coxeter, the doyen of the classical subject, has been a constant inspiration to us.

1

Classical Regular Polytopes

Our purpose in this introductory chapter is to set the scene for the rest of the book. We shall do this by briefly tracing the historical development of the subject. There are two main reasons for this. First, we wish to recall the historical traditions which lie behind current research. It is all too easy to lose track of the past, and it is as true in mathematics as in anything else that those who forget history may be compelled to repeat it. But perhaps more important is the need to base what we do firmly in the historical tradition. A tendency in mathematics to greater and greater abstractness should not lead us to abandon our roots. In studying abstract regular polytopes, we shall always bear in mind the geometric origins of the subject. We hope that this introductory survey will help the reader to find a firm basis from which to view the modern subject.

The chapter has four sections. In the first, we provide an historical sketch, leading up to the point at which the formal material of Chapter 2 begins. The second is devoted to an outline of the theory of regular convex polytopes, which provide so much of the motivation for the abstract definitions which we subsequently adopt. In the third, we treat various generalizations of regular polytopes, mainly in ordinary euclidean space, including the classical regular star-polytopes. In the fourth, we introduce regular maps, which are the first examples of abstract regular polytopes, although the examples considered here occur before the general theory was formulated.

1A The Historical Background

Regular polytopes emerge only gradually out of the mists of history. Apart from certain planar figures, such as squares and triangles, the cube, in the form of a die, was probably the earliest known to man. Gamblers would have used dice from the earliest days, and a labelled example helped linguists to work out the Etruscan words for "one" up to "six". The Etruscans also had dodecahedral dice; examples date from before 500BCE, and may even be much earlier. The other three "platonic" solids appear not to have been employed in gambling; two out of the three do not roll well in any case.

However, it is only somewhat later that the regular solids were studied for their own sakes, and the leap from them to the regular star-polyhedra, analogous to that from pentagon to pentagram, had to await the later middle ages. The nineteenth century first saw regular polytopes of higher dimension, but the real flowering of what is, in origin, one of the oldest branches of mathematics occurred only in the twentieth century.

In this section, we shall give a brief outline of the historical background to the theory of regular polytopes. This is not intended to be totally comprehensive, although we have attempted to give the salient features of more than two millenia of investigations in the subject.

The Classical Period

Before the Greeks

As we have already said, the cube was probably the first known regular polyhedron; certainly it was well known before the ideas of geometry and symmetry had themselves been formulated. Curiously, though, stone balls incised in ways that illustrate all the symmetry groups of the regular polyhedra were discovered in Scotland in the nineteenth century; they appear to date from the first half of the third millenium BCE (see [137, Chapter 7]).

The Egyptians were also aware of the regular tetrahedron and octahedron. As an aside, we pose the following question. Many attempts have been made to explain why the pyramids are the shape they are, or, more specifically, why the ratio of height to base of a square pyramid is always roughly the same particular number. In particular, such explanations often manage to involve π in some practical way, such as measurements by rolling barrels – the Egyptians' theoretical value $\frac{256}{81} \approx 3.16$ for π was fairly accurate. Is it possible, though, that a pyramid was intended to be half an octahedron? The actual angles of slopes of the four proper pyramids at Giza vary between $50°47'$ and $54°14'$; the last is only a little short of half the dihedral angle of the octahedron, namely, $\arccos(1/\sqrt{3}) \approx 54°44'$.

The Early Greeks

Despite various recent claims to the contrary, it seems clear that the Greeks were the first to conceive of mathematics as we now understand it. (The mere listing of, for example, Pythagorean triples does not constitute mathematics; a proof of Pythagoras's theorem obviously does.) According to Proclus (412–485CE), the discovery of the five regular solids was attributed by Eudemus to Pythagoras (c582–c500BCE) himself; the fact that a point of the plane is exactly surrounded by six equilateral triangles, four squares or three regular hexagons (these giving rise to the three regular tilings of the plane) was also known to his followers. They knew as well of the regular pentagram, apparently regarding it as a symbol of health; it has been suggested that this also gave them their first example of incommensurability. (For a fuller account of the origin of the regular solids, consult [438].)

The regular solids in ordinary space were named after Plato (Aristocles son of Ariston, 427–347BCE) by Heron; this seems to be one of the earliest mathematical misattributions. Indeed, their first rigorous mathematical treatment was by Theaetetus (c415–369BCE, when he was killed in battle). In his *Timaeus*, Plato does discuss the regular solids, but while his enthusiasm for and appreciation of the figures are obvious, it is also evident that his discussion falls short of a full mathematical investigation. However, one very perceptive idea does appear there. An equilateral triangle is regarded by him as formed from the six right-angled subtriangles into which it is split by its altitudes. The three solids with triangular faces (tetrahedron, octahedron, and icosahedron) are then built up from these subdivided triangles. This anticipates the construction of the Coxeter kaleidoscope of their reflexion planes by more than two millennia. But the general principle was not fully recognized by Plato; this is exhibited by his splitting of the square faces of the cube into four isosceles (instead of eight) triangles. Moreover, the dodecahedron is not seen in this way at all. In the *Timaeus*, one has the impression that the existence of the dodecahedron (identified with the universe) almost embarrasses Plato. The other four regular solids are identified with the four basic elements – tetrahedron = fire, octahedron = air, cube = earth, and icosahedron = water – in a preassumed scheme which is not at all scientific. (In the *Phaido*, amusingly, Plato also describes dodecahedra; they appear as stuffed leather balls made out of twelve multicoloured pentagonal pieces, an interesting near anticipation of some modern association footballs.)

To Plato's pupil Aristotle (384–322BCE) is attributed the mistaken assertion that the regular tetrahedron tiles ordinary space. Unfortunately, such was the high regard in which Aristotle was held in later times that his opinion was not challenged until comparatively recently, although its falsity could have been established at the time it was made.

Euclid

Euclid's *Elements* ($\Sigma\tau o\iota\chi\epsilon\iota\alpha$) is undoubtedly the earliest surviving true mathematics book, in the sense that it fully recognizes the characteristic mathematical paradigm of axiom–definition–theorem–proof. Until early in the twentieth century, parts of it, notably the first six Books ($\Sigma\chi o\lambda\iota\alpha$), provided, essentially unchanged, a fine introduction to basic geometry. It is unclear to what extent Euclid discovered his material or merely compiled it; our ignorance of Euclid himself extends to our being uncertain of more than that, as we are told by Proclus, he lived and worked in Alexandria at the time of Ptolemy I Soter (reigned 323–283BCE).

Of course, it is to Euclid that we look for the first rigorous account of the five regular solids; Proclus even claimed that *Elements* is designed to lead up to the discussion of them. Whether or not that contention can be justified, Book XIII is devoted to the regular solids. (Incidentally, this book and Book X are less than thoroughly integrated into the rest of the text, suggesting that they were incorporated from an already existing work, which may well have been written by Theaetetus himself.) The scholium (theorem) of that book demonstrates that there are indeed only five regular solids. The proof is straightforward, and (in essence) remains that still used: the angle at a vertex of a regular p-gon is $(1 - \frac{2}{p})\pi$, and so for q of them to fit around a vertex of a regular solid, one requires that $q(1 - \frac{2}{p})\pi < 2\pi$, or, in other words,

1A1 $$\frac{1}{p} + \frac{1}{q} > \frac{1}{2}.$$

Euclid did not phrase the result in quite this way, but this is what the proof amounts to.

Further results about the regular solids occur in Books XIV and XV (which were written around 300CE, and so were not by Euclid), such as their metrical properties, and in particular some anticipation of duality. The details, and modern explanations of the results, are described in [120].

Archimedes

We should also briefly mention here a contribution of Archimedes (c287–212BCE). The Archimedean polyhedra themselves are beyond the scope of this book. However, Archimedes did use regular polygons – actually, the 96-gon – to find his famous bounds $3\frac{10}{71} < \pi < 3\frac{1}{7}$. The remainder of the many mathematical results of Archimedes are not relevant to the topic of this book, but their significance cannot be allowed to pass altogether unnoticed.

The Romans

The Romans were fine architects and engineers, but contributed less to mathematics. However, among the various works attributed to the great pagan philosopher Anicius Manlius Severinus Boetius (c480–524CE) (his name is usually miswritten as "Boethius") is a translation of most of the first three books of Euclid. Since he certainly translated Plato and Aristotle (with a view to reconciling them), this is a possibility which cannot lightly be dismissed.

The Mediaeval Period

The Early Middle Ages

The Christian Roman Empire got off to a poor start in its treatment of learning; under a decree of Emperor Theodosius I ("the Great") concerning pagan monuments, in around 389–391 Bishop Theophilus ordered that the great library of Alexandria (or, at least, that part in the Serapeum) be pillaged. (It is uncertain how much of the original library had survived to this time; it is said – though the event is disputed – that the larger part, the Brucheum, was burnt around 47BCE when Caesar set fire to the Egyptian fleet during the Roman civil wars. The later story of the Muslim destruction under 'Amr is of much more dubious provenance.) A little later, the last of the Alexandrine philosophers, the talented and beautiful Hypatia (c375–415), was flayed with oyster shells by a Christian mob at the instigation of Bishop Cyril (who was later canonized).

The attitude of the Byzantine (Eastern Roman) Empire to mathematics (and the other sciences) was distinctly ambiguous, alternating between encouragement and suppression. Justinian I (reigned 527–565) initially seemed supportive, but soon closed the

Academy at Athens in 529, although there was probably little resulting loss to mathematics. It is due to a few people in the ninth (particularly Leo "the mathematician") and tenth centuries that we have the Greek manuscripts of Euclid which survive; the earliest dates from 888. Similarly, a tenth-century manuscript of Archimedes was re-used in the twelfth century (a palimpsest) for religious texts; fortunately, the original mathematics can be recovered by modern techniques. It must be concluded that the Byzantines preserved rather than added to the corpus of knowledge.

The mathematical torch was also carried on by the Arabs, but they too seem to have added little to geometry, although their translation of Euclid helped to preserve it. (The contributions of the Islamic world to algebra are a quite different matter.) They had a good empirical knowledge of symmetry; in the Alhambra there are patterns which exemplify many of the seventeen possible planar symmetry groups (and the rest can be produced by slight modifications of some of the others).

The Later Middle Ages

From about the twelfth century, mathematical knowledge began seeping back into western Europe. Around the 1120s, Aethelard (Adelard) of Bath, known as "Philosophus Anglorum", produced a translation of Euclid; while he knew Greek, this is more likely to have been from Arabic. (It was first printed in Venice in 1482 under the name of Campanus of Novara, with an unhelpful commentary, but the attribution to Aethelard is universally accepted.)

Rather later, Thomas Bradwardine "the Profound Doctor" (c1290–1349), Archbishop of Canterbury for just forty days after his consecration (he died of plague), systematically investigated star-polygons, obtaining $\{\frac{n}{d}\}$ by stellating $\{\frac{n}{d-1}\}$. (The notation will be explained later in the chapter.)

Although Kepler and Poinsot (see the following subsection) are credited with discovering the regular star-polyhedra in three dimensions, the polyhedron $\{\frac{5}{2}, 5\}$ was depicted in 1420 by Paolo Uccello (1397–1475), while $\{5, \frac{5}{2}\}$ occurs in an engraving of 1568 by Wenzel Jamnitzer (1508–85); however, it is unlikely that they fully appreciated the differences between these figures and others that they drew.

The Modern Period

Before Schläfli

Johannes Kepler (1571–1630) began the modern investigation of regular polytopes by his discovery of the two star-polyhedra $\{\frac{5}{2}, 5\}$ (strictly, perhaps, a rediscovery) and $\{\frac{5}{2}, 3\}$ (see [248, p. 122]). He also investigated various regular star-polygons, particularly the heptagons; for the latter, he showed that the side lengths of the three heptagons $\{7\}$, $\{\frac{7}{2}\}$ and $\{\frac{7}{3}\}$ inscribed in the unit circle are the roots of the equation

$$\lambda^6 - 7\lambda^4 + 14\lambda^2 - 7 = 0. \qquad \textbf{1A2}$$

In a sense, Kepler stands on a cusp. The lingering effect of mediaeval (or perhaps even classical) thought on him shows in his attempt to relate the relative sizes of the orbits of the planets to the ratios of in- and circumradii of the regular polyhedra; later in his life he demonstrated that these orbits were ellipses.

The Greeks had proved that certain regular polygons, notably the pentagon, were constructible using ruler and compass alone. In 1796, the young Carl Friedrich Gauss (1777–1855) showed that, if a regular n-gon $\{n\}$ can be so constructed, then n is a power of 2 times a product of distinct Fermat primes, of the form

1A3
$$p = 2^{2^k} + 1$$

for some k; in fact, the condition is sufficient as well as necessary. The only known Fermat primes are those for $k = 0, 1, 2, 3, 4$; if there are no others, then an odd such n is a divisor of $2^{32} - 1$. In 1809, Louis Poinsot (1777–1859) rediscovered the first two regular star-polyhedra, and found their duals $\{5, \frac{5}{2}\}$ (again, perhaps really a rediscovery) and $\{3, \frac{5}{2}\}$ (see [342]); very soon afterwards, in 1811, Augustin Louis Cauchy (1789–1857) proved that the list of such regular star-polyhedra was now complete (see [76]).

Schläfli

At a time when very few mathematicians had any concept of working in higher dimensional spaces, Ludwig Schläfli (1814–95) discovered regular polytopes and honeycombs in four and more dimensions around 1850 (see [355, §17, 18]). In fact, he found all the groups of the regular polytopes whose symmetry groups are generated by reflexions in hyperplanes in euclidean spaces. But against all his evidence he refused to recognize the dual pair $\{\frac{5}{2}, 5\}$ and $\{5, \frac{5}{2}\}$ as "genuine" polyhedra (because they have non-zero genus), and so would not accept either the regular 4-polytopes which have these as facets or vertex-figures, even though calculating the spherical volumes of corresponding tetrahedra on the 3-sphere was a central part of his treatment.

From 1880 onwards, the regular polytopes in higher dimensions were rediscovered many times, beginning with Stringham [405]. We refer to [120, p. 144] for the relevant details. Edmund Hess [215] found the remaining regular star-polytopes, and S. L. van Oss [337] proved that the enumeration was complete. (For an argument avoiding consideration of each separate case, see [280] and Section 7D in this work.)

Coxeter

The subject of regular polytopes had gone into somewhat of a decline when it was taken up by H. S. M. (Donald) Coxeter (born 1907). His investigations and consolidation of the theory culminated in his famous book *Regular Polytopes* [120], whose first edition was published in 1948. His contributions are too numerous to list here individually, but we should at least mention Coxeter diagrams and the complete classification of the discrete euclidean reflexion groups among all Coxeter groups. We shall mention this latter material in Section 1B.

But Coxeter also pointed towards later developments of the theory. In particular, when J. F. Petrie (1907–72) (the inventor of the skew polygon which bears his name) found the two regular skew apeirohedra $\{4, 6\,|\,4\}$ and $\{6, 4\,|\,4\}$, he immediately found the third $\{6, 6\,|\,3\}$, and set the whole theory in a general context (see [105]). He also looked at regular maps and their automorphism groups, regarding the star-polyhedra as particular examples; he first observed that the Petrie polygons of a regular map themselves (usually) form another regular map (see [131, p. 112]). We shall provide an introduction to this area in Section 1D.

In 1975, Grünbaum (see [198]) gave the theory a further impetus. He generalized the regular skew polyhedra, by allowing skew polygons as faces as well as vertex-figures. He found eight more individual examples and twelve infinite families (with non-congruent realizations of isomorphic apeirohedra), and Dress [148, 150] completed the classification by finding the final case and establishing the completeness of the list. Again, we shall consider this work later, in the appropriate place (see Section 7E).

Finally, regular polytopes also formed the cradle of Tits's work on buildings (see [415–417]). Buildings of spherical type are the natural geometric counterparts of simple Lie groups of Chevally type. Regular polytopes, or, more generally, Coxeter complexes (see Sections 2C and 3A), occur here as fundamental structural components, namely, as the "apartments" of buildings. In a further generalization, Buekenhout [55, 58] introduced the notion of a diagram geometry to find a geometric interpretation for the twenty-six sporadic groups (see [14, 143, 244]). Although we shall not discuss buildings and diagram geometries in detail, the present book has nevertheless been considerably influenced by these developments.

History teaches us that the subject of regular polyhedra has shown an enormous potential for revival. One natural explanation is that the beauty of the geometric figures appeals to the artistic senses [20, 384].

1B Regular Convex Polytopes

We begin this section with a short discussion of convexity, which we shall need again in Chapter 5. For fuller details, we refer the reader to any one of a number of standard texts, for example [197, 357].

A subset K of n-dimensional euclidean space $\mathbb{E}^n$ is *convex* if, with each two of its points x and y, it contains the *line segment*

$$[xy] := \{(1-\lambda)x + \lambda y \mid 0 \leqslant \lambda \leqslant 1\}.$$

The intersection of convex sets is again convex, and so the *convex hull* $\operatorname{conv} S$ of a set $S \subseteq \mathbb{E}^n$ is well defined as the smallest convex set which contains S. The convex hull of a finite set of points is a *convex polytope*; in this section, we shall frequently drop the qualifying term "convex" and talk simply about a polytope. A polytope P is *k-dimensional*, or a *k-polytope*, if its affine hull is k-dimensional. Here, an *affine subspace* of $\mathbb{E}^n$ is a subset A which contains each line

$$xy := \{(1-\lambda)x + \lambda y \mid \lambda \in \mathbb{R}\}$$

between two points $x, y \in A$; the *affine hull* aff S of a subset S is similarly the smallest affine subspace of $\mathbb{E}^n$ which contains S.

Bear in mind that a non-empty affine subspace A is a translate of a unique linear subspace

$$L := A - A = A - x$$

for any $x \in A$; by definition $\dim A := \dim L$. The empty set $\emptyset$ is the affine subspace of dimension -1; it is also a polytope. We further refer to 2-polytopes as *polygons* and to 3-polytopes as *polyhedra*.

The simplest example of an n-polytope is an n-*simplex*, which is the convex hull of an affinely independent set of $n+1$ points. Here, a set $\{a_0, a_1, \ldots, a_n\}$ is *affinely independent* if, whenever $\lambda_0, \lambda_1, \ldots, \lambda_n \in \mathbb{R}$ are such that

$$\sum_{i=0}^{n} \lambda_i a_i = o, \quad \sum_{i=0}^{n} \lambda_i = 0,$$

then $\lambda_0 = \lambda_1 = \cdots = \lambda_n = 0$; this is the natural extension of the notion of linear independence. (We use "o" to denote the zero vector.)

A hyperplane

$$H(u, \alpha) := \{x \in \mathbb{E}^n \mid \langle x, u \rangle = \alpha\}$$

supports a convex set K, with *outer normal* u, if

$$\alpha = \sup\{\langle x, u \rangle \mid x \in K\}.$$

The intersection $H(u, \alpha) \cap K$ is then an (*exposed*) *face* of K. An n-polytope P has faces of each dimension $0, \ldots, n-1$, which are themselves polytopes. Often, $\emptyset$ and P itself are counted as faces of P, called the *improper* faces; the other faces are *proper*. We write $\mathcal{P}(P) = \mathcal{P}$ for the family of all faces of P. The faces of dimensions 0, 1, $n-2$ and $n-1$ are also referred to as its *vertices, edges, ridges* and *facets*, respectively; more generally, a face of dimension j is called a j-*face*.

The notation vert P is usual for the set of vertices of a polytope P; then $P =$ conv(vert P). If $v \in$ vert P, and if H is a hyperplane which strictly separates v from vert $P \setminus \{v\}$, then $H \cap P$ is called a *vertex-figure* of P at v. In the cases we shall consider in the following, we may usually choose the vertices of the vertex-figure at v in some special way; traditionally, they are the midpoints of the edges through v, although we shall frequently violate the strict terms of the definition, and take the other vertices of the edges through v instead.

Before we proceed further, we list various properties of a convex n-polytope P, which will motivate many of the definitions we adopt in Chapter 2.

- $\mathcal{P}$ is a lattice, under the partial ordering $F \leqslant G$ if and only if $F \subseteq G$. The *meet* of two faces F and G is then $F \wedge G := F \cap G$, and the *join* $F \vee G$ is the (unique) smallest face of P which contains F and G.
- If $F < G$ are two faces of P with $\dim G - \dim F = 2$, then there are exactly two faces J of P such that $F < J < G$.

- For every two faces F, G of P with $F \leqslant G$, the *section*

$$G/F := \{J \in \mathcal{P} \mid F \leqslant J \leqslant G\}$$

of $\mathcal{P}$ is isomorphic to the face-lattice of a polytope of dimension $\dim G - \dim F - 1$. (For $F = \emptyset$, we have $G/F = G$ by a minor abuse of notation; when $\dim F \geqslant 0$, proceed by induction, namely, by successive construction of vertex-figures.)

Two faces F and G of P are called *incident* if $F \leqslant G$ or $G \leqslant F$.

- If $\dim P \geqslant 2$, then $\mathcal{P}$ is *connected*, in the sense that any two proper faces F and G of P can be joined by a chain $F =: F_0, F_1, \ldots, F_k := G$ of proper faces of P, such that F_{i-1} and F_i are incident for $i = 1, \ldots, k$. Hence, $\mathcal{P}$ is *strongly connected*, in that the same is true for every section G/F of $\mathcal{P}$ such that $\dim G \geqslant \dim F + 3$.
- The boundary $\operatorname{bd} P$ of P is homeomorphic to an $(n-1)$-sphere; in particular, if $n \geqslant 3$, then $\operatorname{bd} P$ is simply connected.

We call two polytopes P and Q (*combinatorially*) *isomorphic* if their face-lattices $\mathcal{P}(P)$ and $\mathcal{P}(Q)$ are isomorphic, so that there is a one-to-one inclusion preserving correspondence between the faces of P and those of Q. Similarly, P and Q are *dual* if $\mathcal{P}(P)$ and $\mathcal{P}(Q)$ are anti-isomorphic, giving a one-to-one inclusion reversing correspondence between the faces of P and those of Q. The notation P^* for a dual of P will occur quite often.

A *flag* of an n-polytope P is a maximal subset of pairwise incident faces of P; thus, it is of the form $\{F_{-1}, F_0, \ldots, F_{n-1}, F_n\}$, with

$$F_{-1} \subset F_0 \subset \cdots \subset F_{n-1} \subset F_n.$$

Here we introduce the conventions $F_{-1} := \emptyset$ and $F_n := P$ for an n-polytope P; the inclusions are strict, so that $\dim F_j = j$ for each $j = 0, \ldots, n-1$. The improper faces $\emptyset$ and P are often omitted from the specification of a flag, since they belong to all of them. The family of flags of P is denoted $\mathcal{F}(P)$. Flags thus have the following properties.

- For each $j = 0, \ldots, n-1$, there is a unique flag $\Phi^j \in \mathcal{F}(P)$ which differs from a given flag Φ in its j-face alone. Two such flags Φ and Φ^j are called *adjacent*, or, more exactly, *j-adjacent*.
- P is *strongly flag-connected*, in that for each two flags Φ and Ψ of P, there exists a chain $\Phi =: \Phi_0, \Phi_1, \ldots, \Phi_k := \Psi$, such that Φ_{i-1} and Φ_i are adjacent for each $i = 1, \ldots, k$, and $\Phi \cap \Psi \subseteq \Phi_i$ for each $i = 1, \ldots, k-1$.

The *symmetry group* $G(P)$ of P consists of the isometries g of $\mathbb{E}^n$ such that $Pg = P$.[†] Then P is called *regular* if $G(P)$ is transitive on the family $\mathcal{F}(P)$ of flags of P; this form of the definition seems to have been given first by Du Val in [156, p. 63].

Alternative definitions of regularity of an n-polytope are common in the literature. We list some of them here; a comprehensive discussion of this topic occurs in [279].

[†] In such algebraic contexts, we write maps after their arguments throughout the book. Compositions of maps thus occur in their natural order; that is, they are read from left to right. Note that these conventions are a change from those in some of our earlier publications.

- A polygon is regular if its edges have the same length, and the angles at its vertices are equal (or, its vertices lie on a circle).
- For $n \geqslant 3$, an n-polytope is regular if its facets are regular and congruent (or isomorphic), and its vertex-figures are isomorphic. (This formulation depends on Cauchy's rigidity theorem; see [242, p. 335].)
- For every n, an n-polytope P is regular if, for each $j = 0, \ldots, n-1$, its symmetry group $G(P)$ is transitive on the j-faces of P.

A *reflexion* R in $\mathbb{E}^n$ is an involutory isometry; it has a *mirror*

$$\{x \in \mathbb{E}^n \mid xR = x\}$$

of fixed points with which it is identified, so that the same notation R is employed for it. A *hyperplane reflexion* has a hyperplane as its mirror.

A *Coxeter group* is one of the form $G := \langle R_0, \ldots, R_{n-1} \rangle$, the group generated by $R_0, \ldots, R_{n-1}$, which satisfies relations solely of the form

$$(R_i R_j)^{p_{ij}} = E,$$

the identity, where the $p_{ij} = p_{ji}$ are positive integers (or infinity) satisfying $p_{jj} = 1$ for each $j = 0, \ldots, n-1$. In addition, we call G a *string* (Coxeter) group if $p_{ij} = 2$ whenever $0 \leqslant i < j - 1 \leqslant n - 2$; this group is then denoted $[p_1, \ldots, p_{n-1}]$. We shall discuss Coxeter groups in full generality in Chapter 3.

1B1 Theorem *The symmetry group $G(P)$ of a regular convex n-polytope P is a finite string Coxeter group, with generators R_j for $j = 0, \ldots, n-1$ which are hyperplane reflexions, and $p_j := p_{j-1,j} \geqslant 3$ for $j = 1, \ldots, n-1$ (in the previous notation). Conversely, any finite string Coxeter group for which $p_j \geqslant 3$ for $j = 1, \ldots, n-1$ is the symmetry group of a regular convex polytope.*

Proof. Let us explain how this result arises. Fix a flag $\Phi = \{F_{-1}, F_0, \ldots, F_{n-1}, F_n\}$ of a regular n-polytope P, with the conventions introduced previously. Denote by q_j the centroid of F_j for $j = 0, \ldots, n$ (by this, we mean the centroid of its vertices), and, for each $j = 0, \ldots, n-1$, let

1B2 $$H_j := \operatorname{aff}\{q_i \mid i \neq j\}.$$

It is not hard to see that $\{q_0, \ldots, q_n\}$ is affinely independent, so that each H_j is a hyperplane. If R_j is the (hyperplane) reflexion whose mirror is H_j, then $G(P) = \langle R_0, \ldots, R_{n-1} \rangle$.

We see this as follows. In any n-polytope P, and for any flag Φ of P, for each $j = 0, \ldots, n-1$, let Φ^j (as before) be the unique flag of P which is j-adjacent to Φ. Then R_j is the unique symmetry of P which interchanges Φ and Φ^j. The simple-connectedness of the boundary of P (for $n \geqslant 3$ – the case $n = 2$ is trivial) then leads directly to the first assertion of the theorem. Many of the details of the proof are exactly as in that of Theorem 1B3, and so we shall postpone them until then.

We shall leave the converse of Theorem 1B1 until we have discussed Coxeter groups in more detail. However, the essence of the argument lies in the fact that a finite

Coxeter group always admits a faithful representation as a euclidean reflexion group (see Theorem 3B1). □

With the regular n-polytope P, we can associate its *Schläfli symbol* $\{p_1, \dots, p_{n-1}\}$, where the p_j are given by Theorem 1B1. We may observe already that the mirrors H_j and their images under $G(P)$ split $\mathbb{E}^n$ up into convex (actually simplicial) cones. The H_j themselves bound one of these, the *fundamental region* for $G(P)$, which is that generated by $q_1, \dots, q_{n-1}$, with q_0 as apex. For $i \neq j$, the dihedral angle between H_i and H_j is π/p_{ij}. Moreover, for $j = 0, \dots, n$, we have

$$F_j = \operatorname{conv} q_0 \langle R_0, \dots, R_{j-1} \rangle,$$

with $F_n = P$ as before. This procedure is known as *Wythoff's construction* (see [120, §11.6; 466]); we shall meet it again in Chapter 5. It exhibits each face F_j as a regular j-polytope.

The Wythoff construction can be applied to q_{n-1}, using the subgroup $\langle R_{n-k}, \dots, R_{n-1} \rangle$ to give the face $\hat{F}_k$ of a base flag $\{\hat{F}_0, \dots, \hat{F}_{n-1}\}$ of a dual polytope P^* of P. We see that P^* is also regular; its Schläfli symbol is $\{p_{n-1}, \dots, p_1\}$, reversing that of P.

The number "3" occurs frequently in Schläfli symbols, and so we adopt a suitable brief notation; more generally, a string $p, \dots, p$ of length k is abbreviated to p^k. With this convention, we can list all the regular (convex) n-polytopes (see [120, Chapter 7; 167, Chapter I.4, I.5; 217, §23]). They are:

- for $n = 0$, the point;
- for $n = 1$, the (line-) segment: $\{\,\}$;
- for $n = 2$, for each $p \geqslant 3$, a polygon: $\{p\}$;
- for $n = 3$, five polyhedra: $\{3, 3\}$, $\{3, 4\}$, $\{4, 3\}$, $\{3, 5\}$, $\{5, 3\}$;
- for $n = 4$, six polytopes: $\{3, 3, 3\}$, $\{3, 3, 4\}$, $\{4, 3, 3\}$, $\{3, 4, 3\}$, $\{3, 3, 5\}$, $\{5, 3, 3\}$;
- for $n \geqslant 5$, three polytopes: $\{3^{n-1}\}$, $\{3^{n-2}, 4\}$, $\{4, 3^{n-2}\}$.

We shall see why this list is complete in Chapter 3 (compare Table 3B1). The regular polyhedra are the five Platonic solids, namely, the tetrahedron, octahedron, cube, icosahedron and dodecahedron, respectively. For $n = 4$, the regular polytopes are the 4-simplex, 4-cross-polytope, 4-cube, 24-cell, 600-cell and 120-cell, respectively (see also [403] for historical comments about the 120-cell). For $n \geqslant 5$ we only have the n-simplex, n-cross-polytope and n-cube, respectively.

The concept of regular convex polytope in $\mathbb{E}^n$ can be generalized in many different ways. In this section, we stay within the context of convex polytopes; in the following sections, we shall generalize the concept in various other directions.

An *automorphism* of a polytope P is a permutation γ of its face-lattice $\mathcal{P}$ which preserves inclusion; that is, γ is an automorphism of $\mathcal{P}$ in the usual sense. The automorphisms of P form a group $\Gamma(P)$, called the *automorphism group* of P. Following [277] (see also [278]), we say that P is *combinatorially regular* if $\Gamma(P)$ is transitive on $\mathcal{F}(P)$. As examples, every n-polytope with $n \leqslant 2$ is combinatorially regular, as is every simplex.

The main result of [277] is

1B3 Theorem *A combinatorially regular polytope is isomorphic to an ordinary regular polytope.*

Proof. We shall sketch the main details of the proof here. The core idea is to show that, if P is combinatorially regular, then $\Gamma(P)$ is a finite Coxeter group. We may then appeal to two facts: first, a finite Coxeter group is always the automorphism group of some regular convex polytope (this uses Wythoff's construction and the fact that a finite Coxeter group is isomorphic to a reflexion group, both mentioned previously); second, two combinatorially regular polytopes are isomorphic if and only if their automorphism groups are isomorphic.

So, with the same conventions as before, we fix a *base* flag $\Phi := \{F_{-1}, F_0, \ldots, F_n\}$ of the combinatorially regular n-polytope P. For each $j = 1, \ldots, n-1$, there is a $p_j \geqslant 3$, such that the section F_{j+1}/F_{j-2} is a p_j-gon; by the combinatorial regularity of P, the same is true of each flag of P, with the same numbers p_j. We shall show that

1B4 $$\Gamma(P) \cong [p_1, \ldots, p_{n-1}],$$

the string Coxeter group defined previously.

If Ψ is any flag of P, then, as before, for $j = 0, \ldots, n-1$ we write Ψ^j for the j-adjacent flag of P to Ψ. It is immediately clear that, if $\gamma \in \Gamma(P)$, then $\Psi^j\gamma = (\Psi\gamma)^j$. From this, we deduce that, if $\gamma \in \Gamma(P)$ is such that $\Phi\gamma = \Phi$, then $\gamma = \varepsilon$, the identity, since γ will also fix each flag adjacent to Φ, and thus all flags of P, using the flag-connectedness of P. As a further consequence, there is a one-to-one correspondence between $\mathcal{F}(P)$ and $\Gamma(P)$, so that $|\Gamma(P)| = \operatorname{card} \mathcal{F}(P)$.

For each $j = 0, \ldots, n-1$, there is thus a unique $\rho_j \in \Gamma(P)$, such that $\Phi\rho_j := \Phi^j$. We claim that $\Gamma(P) = \langle \rho_0, \ldots, \rho_{n-1} \rangle$, and then establish the required isomorphism. The reason that the ρ_j generate $\Gamma(P)$ is straightforward. If $\gamma \in \Gamma(P)$, write $\Psi := \Phi\gamma$. Since $\mathcal{F}(P)$ is connected, we may find a sequence $\Phi = \Phi_0, \ldots, \Phi_r = \Psi$ of flags, such that Φ_{s-1} and Φ_s are adjacent for each $s = 1, \ldots, r$. Thus for each s, we have $\Phi_s = \Phi_{s-1}^{j(s)}$ for some $j(s) \in \{0, \ldots, n-1\}$. There is now an easy induction argument on r, using the observation we made before, whose general step for $s \geqslant 1$ is

1B5 $$\Phi_s = \Phi_{s-1}^{j(s)} = \left(\Phi\rho_{j(s-1)} \cdots \rho_{j(1)}\right)^{j(s)} = \Phi\rho_{j(s)} \cdots \rho_{j(1)}.$$

With $s = r$, appealing to the uniqueness of γ leads to

1B6 $$\gamma = \rho_{j(r)} \cdots \rho_{j(1)} \in \langle \rho_0, \ldots, \rho_{n-1} \rangle,$$

as we wished to show.

We next need to establish that the relations explicitly given by the definition of $[p_1, \ldots, p_{n-1}]$ in terms of its generators $\rho_0, \ldots, \rho_{n-1}$ suffice to determine $\Gamma(P)$. We proceed as follows. If

$$\rho_{j(r)} \cdots \rho_{j(1)} = \varepsilon$$

is a relation in $\Gamma(P)$ involving certain of its generators, then

$$\Phi^{j(1)\cdots j(r)} = \Phi,$$

by reversing the earlier discussion. We can now show where the various kinds of relations arise; indeed, they come from

$$\Phi^{(ij)^{p_{ij}}} = \Phi,$$

which just describe the structure of various sections of P. (The sections of dimensions 1 and 2 give $p_{jj} = 1$ and the values of the $p_j = p_{j-1,j}$; the other relations arise from $\Phi^{ij} = \Phi^{ji}$ if $i < j - 1$.)

We now look at a general relationship of this kind in a more geometric way. With a flag $\Psi = \{G_0, \ldots, G_{n-1}\}$ (we suppress the improper faces), we associate an $(n-1)$-simplex

1B7 $$T(\Psi) := \operatorname{conv}\{q(G_j) \mid j = 0, \ldots, n-1\},$$

where $q(G)$ denotes the centroid of (the vertices of) the polytope G. (In effect, we are constructing the order complex of the face-lattice of P; see Section 2C; as is often the case, we identify a simplicial complex with its underlying point-set – its "polyhedron".) Two such $(n-1)$-simplices share a common $(n-2)$-face if and only if the corresponding flags are adjacent. We associate with this sequence of adjacent flags an *interior loop* in the boundary $\operatorname{bd} P$ of P; this is a continuous loop in $\operatorname{bd} P$ which passes successively from one simplex $T(\Psi)$ to the next adjacent one, and does not meet an $(n-3)$-face of any of them. If $n \leqslant 2$, there is nothing to be said, since we have already identified such n-polytopes as combinatorially regular (and then their groups $\Gamma(P)$ are as described). When $n \geqslant 3$, we can appeal to the fact that $\operatorname{bd} P$ is simply connected. This means that any loop in $\operatorname{bd} P$ is contractable to a point within $\operatorname{bd} P$. We thus contract the interior loop associated with a relation in $\Gamma(P)$ to a point in $T(\Phi)$; clearly, we can do this in such a way that it never meets a face of any of the simplices $T(\Psi)$ of dimension less than $n-3$. The resulting operations on the sequences of flags are of two types:

- moving to an adjacent flag, and then returning, corresponding to a relation $\Psi^{jj} = \Psi$;
- moving the loop over an $(n-3)$-face of some simplex $T(\Psi)$, corresponding to a relation $\Psi^{(ij)^{p_{ij}}} = \Psi$.

Each gives a relation in $\Gamma(P)$ conjugate to one of those defining the Coxeter group $[p_1, \ldots, p_{n-1}]$; we conclude that $\Gamma(P) \cong [p_1, \ldots, p_{n-1}]$, as was claimed. This ends our discussion of the proof, for reasons that we have already indicated. □

In fact, a careful analysis of this proof, specifically the inductive step in (1B5), shows that something considerably short of the full force of combinatorial regularity will yield the same consequences.

1B8 Corollary *An n-polytope P is combinatorially regular if and only if there exists a flag Φ of P, such that, for each $j = 0, \ldots, n-1$, there is a $\rho_j \in \Gamma(P)$, such that $\Phi^j = \Phi\rho_j$.*

In [279], various other criteria for combinatorial regularity were established. The most important is [279, Theorem 4A1], which was rediscovered in [155]; we introduce it in a more general context here. Call an n-polytope P *equivelar* if, for each $j = 1, \ldots, n-1$, there exists a number p_j, such that, for each flag $\Psi = \{G_{-1}, G_0, \ldots, G_n\}$ of P, the section G_{j+1}/G_{j-2} is a p_j-gon. (The concept of equivelarity was introduced in a somewhat different context in [310] for polyhedral manifolds; see Section 1D.) This is now a purely combinatorial notion; that is, it does not explicitly mention the automorphism group $\Gamma(P)$. Nevertheless, we have

1B9 Theorem *An equivelar convex polytope is combinatorially regular.*

Proof. We may clearly take $n \geqslant 3$ here. The essence of the proof is the following. If Φ is a given flag of P, any other flag is of the form

$$\Phi^{j(1)\cdots j(r)},$$

for some $j(1), \ldots, j(r)$. Given another flag Ψ of P, we define a mapping $\gamma := \gamma_\Psi : \mathcal{F}(P) \to \mathcal{F}(P)$ by

$$\Phi^{j(1)\cdots j(r)}\gamma := \Psi^{j(1)\cdots j(r)}.$$

For a general polytope, this mapping would not be well defined. However, since P is equivelar and bd P is simply connected, any alternative expression

$$\Phi^{k(1)\cdots k(s)} = \Phi^{j(1)\cdots j(r)}$$

leads to an interior loop associated with

$$\Phi^{k(1)\cdots k(s)j(r)\cdots j(1)} = \Phi,$$

and reversing the contraction of the loop, but now basing it on Ψ, leads to

$$\Psi^{k(1)\cdots k(s)j(r)\cdots j(1)} = \Psi,$$

or

$$\Psi^{k(1)\cdots k(s)} = \Psi^{j(1)\cdots j(r)},$$

showing that γ is well defined. We have therefore explicitly constructed $\gamma \in \Gamma(P)$ such that $\Phi\gamma = \Psi$; hence P is combinatorially regular. □

Various other criteria for combinatorial regularity of a convex n-polytope P can be deduced from Theorem 1B9. These sit between the original definition and its reformulation in terms of equivelarity; they are all taken from [279], and we list them without proof.

- $\Gamma(P)$ is transitive on the facets of P, one facet F of P is combinatorially regular, and its automorphism group $\Gamma(F)$ is a subgroup of $\Gamma(P)$. (There is an obvious dual criterion in terms of vertex-figures.)
- For each $j = 1, \ldots, n-1$, the group $\Gamma(P)$ is transitive on the incident pairs of $(j-2)$- and $(j+1)$-faces of P.

The last two criteria involve a common condition on a polytope P.

1B10 For each face F of P, and each $\gamma \in \Gamma(P)$ such that $F\gamma = F$, there exist (commuting) $\gamma_-, \gamma_+ \in \Gamma(P)$, such that $\gamma = \gamma_-\gamma_+$, and γ_- (γ_+) fixes each face G of P with $G \subset F$ ($F \subset G$).

- $\Gamma(P)$ satisfies (1B10), and, for each $j = 1, \ldots, n-1$, is transitive on the incident pairs of $(j-1)$- and j-faces of P.
- $\Gamma(P)$ satisfies (1B10), and there exists a flag $\Phi = \{F_{-1}, F_0, \ldots, F_n\}$ of P such that, for each $j = 0, \ldots, n-1$, there is a $\sigma_j \in \Gamma(P)$ which interchanges F_j and the unique other j-face F_j^* of P which satisfies $F_{j-1} \subset F_j^* \subset F_{j+1}$.

The very stringent requirements in the definition of regularity of polytopes can be relaxed in many different ways, yielding a great variety of weaker "regularity" notions. For example, a convex polytope is called *semi-regular* if its facets are regular and its symmetry group is transitive on the vertices (see [32, 103, 108, 123, 124, 235]). There are also other related concepts, but we shall not employ any of them; for a survey, see [273, 371].

1C Extensions of Regularity

We now treat some different notions of regular polytope; we shall gradually increase the level of generality. First, there are the infinite analogues in ordinary euclidean space, the regular honeycombs. A *honeycomb* or *tessellation* is a collection P of n-polytopes, called *cells*, which tiles $\mathbb{E}^n$ face-to-face; that is, two of its cells have disjoint interiors and meet on a common face of each (which may be empty), and these cells cover $\mathbb{E}^n$. A *flag* of a honeycomb P is defined as for an $(n+1)$-polytope, of which it is an infinite analogue; then F_n will be a cell of P, which is why we also use the alternative designation *facet*.

As might be expected, a honeycomb P will be regular if its symmetry group $G(P)$ (the group of isometries of $\mathbb{E}^n$ which preserves P) is transitive on the flags of P. Once again, $G(P)$ will be a Coxeter group generated by hyperplane reflexions; for $j = 0, \ldots, n$, the mirror H_j of the jth generating reflexion R_j is defined exactly as (1B2) (there is no centre q_{n+1}, of course). This time, though, the fundamental region for $G(P)$ is an n-simplex $\text{conv}\{q_0, \ldots, q_n\}$.

The regular honeycombs in $\mathbb{E}^n$ can be classified; again, we refer forward to Chapter 3 for details (see Table 3B2). The list is

- for $n = 1$, the single *apeirogon*: $\{\infty\}$;
- for $n = 2$, three tessellations: $\{4, 4\}$, $\{3, 6\}$, $\{6, 3\}$;
- for $n = 3$, a single honeycomb: $\{4, 3, 4\}$;
- for $n = 4$, three honeycombs: $\{4, 3, 3, 4\}$; $\{3, 3, 4, 3\}$, $\{3, 4, 3, 3\}$;
- for $n \geqslant 5$, a single honeycomb: $\{4, 3^{n-2}, 4\}$.

For each n, we have the tessellation by cubes; this is the apeirogon if $n = 1$, or $\{4, 3^{n-2}, 4\}$ if $n \geqslant 2$, giving the square tessellation $\{4, 4\}$ if $n = 2$. In the plane, there are also the triangular and hexagonal tessellations $\{3, 6\}$ and $\{6, 3\}$, respectively. Finally, in dimension 4, we also have the two exceptional tessellations $\{3, 3, 4, 3\}$ and $\{3, 4, 3, 3\}$, with 4-cross-polytopes or 24-cells as tiles, respectively.

Having permitted honeycombs in euclidean space, the next natural step is to generalize yet further to hyperbolic space. Again, we shall consider such honeycombs in detail in Chapter 3, where a complete classification of them will be given (see Table 3C1).

Next, one can allow non-convex polytopes, such as pentagrams and their higher dimensional analogues. At first, we proceed in an elementary way, constructing such polytopes by building them up, facet by facet. We begin with the planar regular polygons; non-planar polygons will only be talked about a little later. A *polygon* (more exactly, a *p-gon*) is now defined to be a finite set of line-segments $[a_{i-1}a_i]$ $(i = 1, \dots, p)$ in a (2-dimensional) plane, with $a_p = a_0$. Thus a polygon is thought of as a closed path formed by line-segments which are its *edges* (facets); the end points of these segments are its *vertices*.

As before, a polygon P has a symmetry group $G(P)$, and P is regular if $G(P)$ is transitive on the set $\mathcal{F}(P)$ of flags of P. Since $G(P)$ permutes the vertices of P, their centroid

$$c := \frac{1}{p} \sum_{i=0}^{p-1} a_i$$

is invariant under $G(P)$. If we choose c to be the origin of the coordinate system in the plane, then $G(P)$ will be an orthogonal group. The vertices a_i will lie on a circle centred at c, the edges of P will have the same length, and the angles of P at its vertices will be the same. Thus we have a less abstract concept of regularity than in the formal definition.

An edge $[a_{i-1}a_i]$ of P will subtend an angle $2d\pi/p$ at c, for some integer d satisfying $1 \leqslant d < \frac{1}{2}p$, with the greatest common divisor $(p, d) = 1$. If $d > 1$, then P will be non-convex, and is called a *regular star-polygon*. The Schläfli symbol for P is then $\{\frac{p}{d}\}$.

For $n \geqslant 3$, we define a *regular star-polytope* P as follows. This will be formed from finitely many $(n-1)$-polytopes, meeting in pairs on their $(n-2)$-faces. These facets are congruent and regular, and the vertex-figures of P (which must exist) are also congruent regular $(n-1)$-polytopes. One can also make a more abstract definition, regarding P as a configuration of affine subspaces of various dimensions from 0 to $n-1$ (these substitute for the faces of the same dimensions) satisfying the same conditions as those we listed for a convex polytope; the symmetry group $G(P)$ will be obliged to be transitive on the flags of P.

Whichever definition we adopt, we conclude, much as in the convex case, that $G(P)$ is a finite group generated by hyperplane reflexions. From this point, one may proceed in one of several ways; we briefly trace the historical development. As we remarked in Section 1A, the regular pentagram $\{\frac{5}{2}\}$ was already known to the Greeks, and the regular star-polygons were systematically investigated by Bradwardine. The completeness of the enumeration of the regular star-polyhedra (Kepler–Poinsot polyhedra) was established by Cauchy, and the remaining regular star-polytopes (which are confined to $\mathbb{E}^4$) were found by Schläfli and Hess.

These constructions were all more or less recursive, and hence synthetic. It really had to await Coxeter's interest in them for group theory to begin to play the pre-eminent role it does now. Coxeter also laid the foundations for a number of systematic constructions which relate different regular polytopes. Thus, for example, while Hess had shown, by enumeration of all the cases, that a regular star-polytope must have the same vertices as those of some regular convex polytope, a complete explanation of this result was not provided until [280, p. 592]. We shall discuss this result in an appropriate context in Section 7D.

A further generalization of the notion of regular polyhedron arose from an observation by Petrie in 1926. One still insists on convex polygonal faces, but now allows the vertex-figures to be regular skew (non-planar) polygons. He and Coxeter did not take the next step of allowing skew faces as well; this was done by Grünbaum in 1965. Finally, there are abstract generalizations, such as regular maps, which we shall begin to discuss in Section 1D. Each concept leads to a greater level of generalization. The definitions here will anticipate those in Chapters 2 and 5, to which reference should be made.

1D Regular Maps

It is no part of our overall plan to give the whole theory of regular maps at this stage. What we wish to do, rather, is to describe a number of examples which we can then use to illustrate the next few chapters.

Informally, a *map* $\mathcal{P}$ is a family of polygons (which for our present purposes may be apeirogons, that is, infinite), such that

- any two polygons meet in a common edge or vertex, or do not meet at all;
- each edge belongs to precisely two polygons;
- the polygons containing a given vertex form a single cycle of adjacent polygons (sharing a common edge);
- between any two polygons is a chain of adjacent polygons.

The map $\mathcal{P}$ will have an automorphism group $\Gamma(\mathcal{P})$, and $\mathcal{P}$ will be *regular* if $\Gamma(\mathcal{P})$ is transitive on the set $\mathcal{F}(\mathcal{P})$ of flags of $\mathcal{P}$. If the regular map $\mathcal{P}$ has p-gonal faces and q-gonal vertex-figures, then it is said to be of *(Schläfli) type* $\{p, q\}$.

Each regular polyhedron or apeirohedron gives rise to a regular map, in the natural way. We wish to describe several classes of such regular maps, which admit more or less concrete definitions; these are in terms of various edge-paths on the map.

First among these is the *Petrie property*. Locally, at least, a map has a definite orientation, relative to one of its polygons, say. (Globally, a map need not be orientable, even if it is finite. If a map has infinitely many polygons through each vertex, it is often hard to discuss its underlying topology in any sensible way.) An edge-path which uses two successive edges of a polygonal face, but not three, is called a *Petrie polygon* of $\mathcal{P}$. (This notion is traditionally attributed to J. F. Petrie; but only recently did we learn from Professor Coxeter that such polygons were investigated before Petrie in Reinhardt [345, p. 11], where the concept was introduced for general convex polyhedra. However, it seems that its significance for regular polytopes was indeed first pointed out by Petrie.) All Petrie polygons are equivalent under $\Gamma(\mathcal{P})$. A regular map of type $\{p, q\}$ which is determined by the length r of its Petrie polygons is denoted by $\{p, q\}_r$. A famous example of such a regular map is Klein's map $\{3, 7\}_8$, connected with the solution of polynomial equations of degree 7 (see [250, p. 461; 251, p. 260; 252, p. 109]); another is Dyck's map $\{3, 8\}_6$ (see [157, p. 488; 158]).

The path in a regular map $\mathcal{P}$ formed by the edges which successively take the second exit on the left (in a local orientation), rather than the first, at each vertex, is called a *hole*. Again, all (left and right) holes of $\mathcal{P}$ are equivalent under $\Gamma(\mathcal{P})$. A regular map of type $\{p, q\}$ which is determined by the length h of its holes is denoted by $\{p, q \mid h\}$. We give just a few examples here. The Poinsot polyhedron $\{5, \frac{5}{2}\}$ is isomorphic to $\{5, 5 \mid 3\}$. The map $\{4, 4 \mid h\}$, for $h \geqslant 3$, is obtained from an $h \times h$ array of squares by directly identifying opposite sides to form a torus (later, we shall permit the case $h = 2$ as well). Finally, the Petrie–Coxeter regular skew polyhedra mentioned in Section 1C are $\{4, 6 \mid 4\}$ and its dual $\{6, 4 \mid 4\}$, discovered by Petrie, and $\{6, 6 \mid 3\}$, found immediately afterwards by Coxeter.

The last class we wish to consider at this stage comprises the regular *toroidal polyhedra* or, more briefly, the *regular tori*. The symmetry groups of the three planar regular tessellations or tilings $\{3, 6\}$, $\{4, 4\}$ and $\{6, 3\}$, by triangles, squares and hexagons, respectively, contain (normal) subgroups of translations. For the first two, the subgroups are transitive on the vertices of the tilings, and for the last two, they are transitive on the faces. We may then identify vertices, edges and faces of such a tiling $\mathcal{P}$ by a subgroup T of the translations. If T is a normal subgroup of the whole symmetry group, then (with a few exceptions) the map $\mathcal{P}/T$ obtained by this identification will be regular. The exceptional cases arise when T is too large a subgroup of the whole translation group.

Such a regular map is alternatively designated by a symbol $\mathcal{P}_{\mathbf{s}}$, where $\mathbf{s} := (s, t)$ is a certain non-negative integer vector. Let us begin with the easiest case $\mathcal{P} = \{4, 4\}$. We can take the vertices of $\{4, 4\}$ to be those points of $\mathbb{E}^2$ with integer cartesian coordinates. If (s, t) is any non-zero integer vector, then (s, t) and the point $(t, -s)$ obtained by rotating it by $\pi/2$ about the origin $o := (0, 0)$ generate a subgroup $T = \Lambda_{(s,t)}$ of translations (by integer linear combinations). The map resulting from $\{4, 4\}$ by identification under $\Lambda_{(s,t)}$ is denoted by $\{4, 4\}_{\mathbf{s}} = \{4, 4\}_{(s,t)} := \{4, 4\}/\Lambda_{(s,t)}$. This map will have the topological type of a torus, whence the general descriptive name.

Now, as we have defined it, the lattice $\Lambda_{(s,t)}$ has the full rotational symmetry of the whole tiling $\{4, 4\}$. This has two consequences. First, we may replace (s, t), if necessary, by one of its images under rotation by multiples of $\pi/2$, to make $s,\ t \geqslant 0$; this is unique, except when one of the coordinates is zero, in which case we take $(s, 0)$ in preference to $(0, s)$. Second, the identified map $\{4, 4\}_{(s,t)}$ will also have full rotational symmetry. However, it will only have full reflexional symmetry, so that it is regular, if $(t, s) \in \Lambda_{(s,t)}$ also; bearing in mind that (s, t) and its rotational images generate $\Lambda_{(s,t)}$, this means that $t = 0$ or $t = s$. Restoring the symmetry between s and t enables us to write this condition in the form $st(s - t) = 0$. If $st(s - t) \neq 0$, then $\{4, 4\}_{(s,t)}$ will only have rotational symmetry; the technical term to describe such a map is *chiral*,

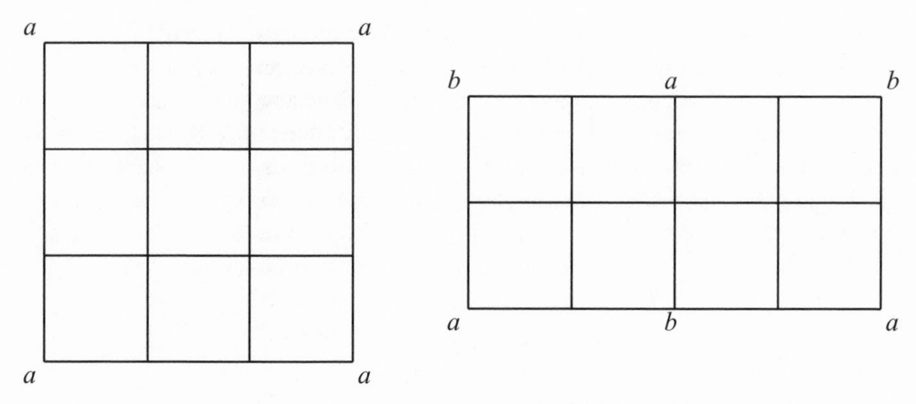

Figure 1D1. The tori $\{4, 4\}_{(3,0)}$ and $\{4, 4\}_{(2,2)}$.

which will be defined generally at the end of Section 2B. Incidentally, the "small" vectors $(s, t) = (1, 0)$ and $(1, 1)$ (which generate "large" translation subgroups) must be excluded here, because the resulting maps degenerate. Two small examples, namely, $\{4, 4\}_{(3,0)}$ and $\{4, 4\}_{(2,2)}$, are illustrated in Figure 1D1; points with the same label are identified by translation.

We may treat the other two tilings $\{3, 6\}$ and $\{6, 3\}$ similarly, and, in fact, together. The vertices of $\{3, 6\}$ (or the centres of the faces of $\{6, 3\}$) may also be taken to be the points with integer coordinates, but this time with respect to an oblique pair of axes, at an angle $\pi/3$. Again, a non-zero integer vector (s, t) generates a translation subgroup $\Lambda_{(s,t)}$, with the aid of its images by rotations through multiples of $\pi/3$, namely, $(t, -s)$ and $(t - s, s + t)$ (we may ignore their negatives). Thus, we may suppose that (s, t) is a non-negative vector. As before, we have maps $\{3, 6\}_{\mathbf{s}} = \{3, 6\}_{(s,t)}$ and $\{6, 3\}_{\mathbf{s}} = \{6, 3\}_{(s,t)}$ on the torus. They will generally be chiral: the condition for regularity is again $st(s - t) = 0$. Observe that, in this case, only the "small" vector $(s, t) = (1, 0)$ must be excluded. As with the other tori, two small examples, namely, $\{3, 6\}_{(3,0)}$ and $\{6, 3\}_{(1,1)}$, illustrate this in Figure 1D2.

These notions can be generalized in many ways, but we shall postpone further discussion until we can do it in terms of the groups. However, let us briefly mention

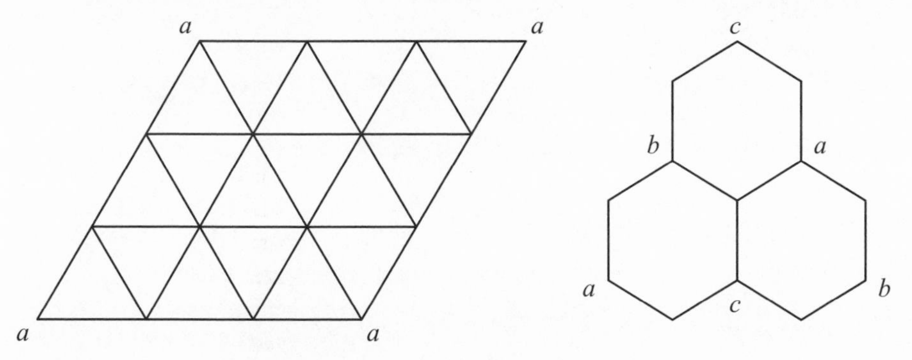

Figure 1D2. The tori $\{3, 6\}_{(3,0)}$ and $\{6, 3\}_{(1,1)}$.

(if only to dismiss) a generalization of regularity for maps formulated in [310]. A map is called *equivelar* of *type* $\{p, q\}$ if each of its faces is a p-gon, and these faces meet q at each vertex. A regular map must, of course, be equivelar, but the converse is far from true. Large classes of equivelar maps were constructed in [310, 311], where the interest was in whether they admitted polyhedral embeddings in $\mathbb{E}^3$; some of these maps are regular, particularly the simplest cases of those of type $\{4, q\}$ and $\{p, 4\}$ (see [50, 51, 309]).

2

Regular Polytopes

It is not an easy task to trace precisely the history of the idea of abstract regular polytopes. While it is clear that the concept has its roots in the classical theory, and notably in Coxeter's work [120], more recently there have been several parallel developments which have influenced the theory of abstract regular polytopes.

From the point of view of discrete geometry, it appears that combinatorial regularity was first studied in McMullen [277] in the context of combinatorially regular convex polytopes (see Section 1B). In its generality, the notion of an abstract regular polytope was largely anticipated in Grünbaum's paper [199] on structures which he called regular polystromata. Then, in 1977, Danzer introduced the more restrictive concept, based on Grünbaum's work, of a regular incidence complex; see Danzer and Schulte [141], although the definitions adopted were anticipated by McMullen (in a geometric context) in [280, p. 578]. Among these regular incidence complexes, the abstract regular polytopes, or regular incidence polytopes as they were first called, are particularly close to the traditional polytopes, and form a special class of polytope-like structures with a distinctive geometric and topological appeal. It seems that a more systematic study of these objects was begun by Schulte [362–364].

Although the points of view are quite different, the theory of abstract regular polytopes has gained considerably from its interaction with the theory of Tits buildings (Tits [417], Brown [54], Ronan [351], Scharlau [354]) and diagram geometries (Buekenhout [55, 58], Buekenhout & Cohen [59], Kantor [243], Pasini [340]). Apart from the fact that buildings and geometries need not be equipped with an incidence relation given by a partial order, the basic difference of the subjects results from the property (P4) defined in Section 2A. Formally, this property still makes each abstract polytope a geometry, or more precisely a thin geometry with a string diagram ([340, p. 424]), but this geometry is degenerate in the same sense in which a triangle is a degenerate projective plane. It is therefore more natural to study these structures from the point of view of polytope theory, and to regard them as abstract polytopes.

As another influence, we mention the work of Dress and others on classification problems for tilings in spaces of two or more dimensions (see [149, 151–154, 317]).

In this chapter, we introduce the basic notation and concepts of the theory of abstract regular polytopes. Our notation will generally be patterned after the traditional theory, and will provide a convenient framework to study combinatorially regular structures resembling the classical regular polytopes.

The chapter is organized as follows. In Section 2A, we discuss the concept of an abstract polytope without reference to regularity. Regular abstract polytopes, or as we shall say *abstract regular polytopes*, are then the subject of Section 2B. In Sections 2C and 2D, we deal with order complexes and quotients of abstract polytopes. In Section 2E, we study the groups of abstract regular polytopes and characterize them as string C-groups. Finally, in Section 2F, we shall discuss presentations of automorphism groups, and introduce some appropriate notation.

2A Abstract Polytopes

In the literature, the term "polytope" has been used in various contexts and with quite different meanings; examples of this are mentioned in Chapter 1. Here we shall use the term in an abstract sense to specify combinatorial or geometric structures with properties analogous to those of convex polytopes in euclidean spaces. Our polytopes will be precisely those structures which elsewhere have been called *incidence polytopes*. We do not choose the shorter term *polytope* merely for the sake of brevity. In fact, extensive study of these structures over the past few years has lead to a theory very much along the lines of the classical theory, and often with more exciting ramifications. We feel that this completely justifies our terminology.

An *abstract polytope* $\mathcal{P}$ *of* (*finite*) *rank* n ($\geqslant -1$), or, more briefly, an *abstract* n*-polytope*, is a partially ordered set (or *poset* for short) with properties (P1), ..., (P4) below. (These conditions should be compared with those satisfied by convex polytopes; see Section 1B.) The elements of $\mathcal{P}$ are called the *faces* of $\mathcal{P}$. Sometimes we use the term *face-set* of $\mathcal{P}$ to denote the underlying set of $\mathcal{P}$ (without reference to the partial order).

Two faces F and G of $\mathcal{P}$ are said to be *incident* if $F \leqslant G$ or $F \geqslant G$. A *chain* of $\mathcal{P}$ is a totally ordered subset of $\mathcal{P}$. A chain has *length* i ($\geqslant -1$) if it contains exactly $i+1$ faces. Note that, by definition, the empty set is a chain (of length -1). The maximal chains are called the *flags* of $\mathcal{P}$. We denote the set of all flags of $\mathcal{P}$ by $\mathcal{F}(\mathcal{P})$. It is clear that each chain is contained in a flag of $\mathcal{P}$.

We begin with the first two defining properties for abstract polytopes, of which (P1) is purely technical.

(P1) $\mathcal{P}$ contains a least face and a greatest face; they are denoted by F_{-1} and F_n, respectively.

(P2) Each flag of $\mathcal{P}$ has length $n+1$ (that is, contains exactly $n+2$ faces including F_{-1} and F_n).

For any two faces F and G of $\mathcal{P}$ with $F \leqslant G$, we call

$$G/F := \{H \mid H \in \mathcal{P}, F \leqslant H \leqslant G\}$$

a *section* of $\mathcal{P}$. In general, there is little possibility of confusion if we identify a face F and the section F/F_{-1}; the context should make clear if F is considered as an element of $\mathcal{P}$ or as a section of $\mathcal{P}$. Note that each section itself is a poset with properties (P1) and (P2), with the rank chosen appropriately; in fact, the subsequent conditions ensure that each section will indeed be an abstract polytope of this rank. Each section of $\mathcal{P}$ distinct from $\mathcal{P}$ itself is called a *proper* section of $\mathcal{P}$.

The properties (P1) and (P2) imply that $\mathcal{P}$ has a natural *rank function*, which we denote by "rank". More precisely, if F is a face of $\mathcal{P}$ and the rank of F/F_{-1} is i, then we set rank $F := i$ and call F a *face of $\mathcal{P}$ of rank i*, or, more briefly, an *i-face* of $\mathcal{P}$. It follows that rank $F_{-1} = -1$ and rank $F_n = n$, so that F_{-1} and F_n are the only faces of $\mathcal{P}$ of these ranks. The faces F_{-1} and F_n are called the *improper* faces of $\mathcal{P}$; all other faces of $\mathcal{P}$ are *proper*. By our convention, $\mathcal{P}$ and its n-face F_n are identified. We also write rank $\mathcal{P} := n$ to indicate that $\mathcal{P}$ has rank n. Finally, we denote by $\mathcal{P}_i$ the set of all i-faces of $\mathcal{P}$, for $i = -1, 0, \ldots, n$.

Further to emphasize the analogy with the theory of convex polytopes, we also call the faces of $\mathcal{P}$ of ranks $0, 1, n-2$ or $n-1$ its *vertices, edges, subfacets* (or *ridges*) and *facets*, respectively. If F is a face of $\mathcal{P}$, then the section F_n/F is said to be the *co-face at F*, or the *vertex-figure at F* if F is a vertex. If F is an i-face, then we also call the co-face at F a *co-i-face*. If F and G are faces of $\mathcal{P}$ such that $F \leqslant G$, then the rank of the section G/F is rank G − rank $F - 1$. In particular, the ranks of a face F/F_{-1} (or F) and its co-face F_n/F are rank F and $n - 1 -$ rank F, respectively; the latter is also called the *co-rank of F*. It follows that vertex-figures have rank $n - 1$ and co-rank 1. We shall use the term *k-section* to mean a section of $\mathcal{P}$ of rank k.

If Ω is a chain of $\mathcal{P}$, then we call the set of ranks of its elements the *type* of Ω. In dealing with chains, and specifically with flags, we shall agree on the following conventions. If we write a chain Ω in the form $\Omega = \{F, G, H, \ldots\}$, then the faces in Ω occur in their natural order; that is, $F < G < H < \cdots$. Further, in the notation $\Omega = \{F_{i_1}, F_{i_2}, \ldots, F_{i_m}\}$, the indices will specify the ranks of the faces; that is, rank $F_{i_k} = i_k$ for $k = 1, \ldots, m$. Similarly, if the type of Ω is $I = \{i_1, i_2, \ldots, i_m\}$, then it will be taken for granted that $i_1 < i_2 < \cdots < i_m$.

Occasionally, we shall find it convenient to omit the improper faces from chains; for example, we may write flags in the form $\{F_0, \ldots, F_{n-1}\}$.

Our next defining property concerns the connectedness of $\mathcal{P}$. A poset $\mathcal{P}$ of rank n with properties (P1) and (P2) is called *connected* if either $n \leqslant 1$, or $n \geqslant 2$ and for any two proper faces F and G of $\mathcal{P}$ there exists a finite sequence of proper faces $F = H_0, H_1, \ldots, H_{k-1}, H_k = G$ of $\mathcal{P}$ such that H_{i-1} and H_i are incident for $i = 1, \ldots, k$. We say that $\mathcal{P}$ is *strongly connected* if each section of $\mathcal{P}$ (including $\mathcal{P}$ itself) is connected. Note that, in general, connectedness of each proper section of $\mathcal{P}$ does not imply connectedness of $\mathcal{P}$ itself. Our next defining property is

(P3) $\mathcal{P}$ is strongly connected.

There are various ways to weaken the requirements of (P3) and obtain more general kinds of posets (for example, see [400–402]). We have intentionally chosen the strongest possible condition.

Though the definition of global and local connectedness is satisfactory from an intuitive point of view, it turns out that an equivalent definition in terms of flag-connectedness is more useful.

Given a poset with properties (P1) and (P2), we call two flags of $\mathcal{P}$ *adjacent* if one differs from the other in exactly one face; if this face has rank i, then the two flags are said to be i*-adjacent*. (Note that, in contrast to some treatments elsewhere, a flag is *not* considered to be adjacent to itself.) Then $\mathcal{P}$ is *flag-connected* if any two distinct flags Φ and Ψ of $\mathcal{P}$ can be *joined* by a sequence

$$\Phi = \Phi_0, \Phi_1, \ldots, \Phi_{k-1}, \Phi_k = \Psi$$

of flags, such that Φ_{j-1} and Φ_j are adjacent for $j = 1, \ldots, k$. Note that flag-connectedness implies connectedness. Further, $\mathcal{P}$ is *strongly flag-connected* if each section of $\mathcal{P}$ is flag-connected, or equivalently, if any two distinct flags Φ and Ψ of $\mathcal{P}$ can be joined by a sequence $\Phi = \Phi_0, \Phi_1, \ldots, \Phi_{k-1}, \Phi_k = \Psi$ of flags, all containing $\Phi \cap \Psi$, such that Φ_{j-1} and Φ_j are adjacent for $j = 1, \ldots, k$. The proof of the following proposition is straightforward, but because of its importance we give it here.

2A1 Proposition *Let $\mathcal{P}$ be a poset with properties (P1) and (P2). Then $\mathcal{P}$ is strongly connected if and only if it is strongly flag-connected.*

Proof. The proof proceeds by induction on $n := \operatorname{rank} \mathcal{P}$, there being nothing to prove if $n \leqslant 1$. So take $n \geqslant 2$. First, suppose that $\mathcal{P}$ is strongly connected, and let Φ and Ψ be two flags of $\mathcal{P}$. If $\Omega := \Phi \cap \Psi$ is non-empty, say $\Omega = \{F_{j_1}, \ldots, F_{j_k}\}$ with $k \geqslant 1$ (we ignore the improper faces, and follow the conventions adopted before), then each section $F_{ji}/F_{j_{i-1}}$ is strongly connected for $i = 1, \ldots, k+1$, where we write $F_{j_0} := F_{-1}$ and $F_{j_{k+1}} := F_n$. Each section has rank smaller than n, and so we may successively move the part of Φ in that section to that of Ψ by a sequence of adjacent flags. Putting these sequences together then gives a sequence of adjacent flags from Φ to Ψ, all containing Ω.

If $\Omega = \emptyset$, take any sequence $G_0, \ldots, G_k$ of successively incident faces, with $G_0 \in \Phi$ and $G_k \in \Psi$; such a sequence exists since $\mathcal{P}$ is (strongly) connected. For each $i = 1, \ldots, k-1$, there is a flag $\Phi_i \supseteq \{G_{i-1}, G_i\}$; set $\Phi_0 := \Phi$ and $\Phi_k := \Psi$. We now appeal to the foregoing part of the proof to move Φ_{i-1} to Φ_i by a sequence of adjacent flags (containing F_i) for each $i = 1, \ldots, k$; concatenation then gives such a sequence from Φ to Ψ, as required.

The converse is rather easier. If F and G are any two proper faces of $\mathcal{P}$ (and we may assume that they are not already incident), then choose flags Φ and Ψ of $\mathcal{P}$ with $F \in \Phi$ and $G \in \Psi$. Let $\Phi = \Phi_0, \ldots, \Phi_k = \Psi$ be any sequence of successively adjacent flags of $\mathcal{P}$. We immediately find a sequence of vertices and edges of $\mathcal{P}$, each incident with its neighbours, as follows. Take the first vertex and edge in Φ, and change vertex or edge only according as Φ_i is 0- or 1-adjacent to Φ_{i-1}; the last vertex will lie in Ψ. The first vertex is incident with F (or is F itself), and similarly for the last vertex and G; this gives the required sequence of successively incident proper faces from F to G. □

It follows from this proposition that (P3) takes the equivalent form

(P3′) $\mathcal{P}$ is strongly flag-connected.

Our last defining property is a certain homogeneity requirement for the sections of rank 1. Roughly speaking, this property says that the poset is basically "real". It is this property which is responsible for the close connexion with traditional polytope theory.

(P4) For each $i = 0, 1, \dots, n-1$, if F and G are incident faces of $\mathcal{P}$, of ranks $i-1$ and $i+1$, respectively, then there are precisely *two* i-faces H of $\mathcal{P}$ such that $F < H < G$.

For more general kinds of posets or geometries, the homogeneity parameter 2 in (P4) must be replaced by other values (if it exists at all). However, for abstract polytopes it is crucial that this value is 2. Note that (P4) can be rephrased by saying that all 1-sections of $\mathcal{P}$ are of diamond shape; see Figure 2A1(b).

For later use, we introduce the following notation. By (P4), if $n \geqslant 1$ then, for each $j = 0, \dots, n-1$ and each flag Φ of $\mathcal{P}$, there exists precisely one adjacent flag differing from Φ in the j-face; we shall denote this flag by Φ^j. We extend this notation by writing

$$\Phi^{j(1)\cdots j(r-1)j(r)} := \left(\Phi^{j(1)\cdots j(r-1)}\right)^{j(r)}$$

for $r \geqslant 2$. Note the following basic facts about adjacent flags.

2A2 Lemma *Let Φ be a flag of an abstract n-polytope $\mathcal{P}$, and let $0 \leqslant j \leqslant k \leqslant n-1$. Then*

(a) $\Phi^{jj} = \Phi$;
(b) $\Phi^{jk} = \Phi^{kj}$ *for* $k \geqslant j+2$.

To summarize our definitions, an *abstract n-polytope* $\mathcal{P}$ (that is, an *abstract polytope of rank n*) is a poset which satsifies properties (P1), ..., (P4). If the rank is not specified, we shall simply refer to $\mathcal{P}$ as an *abstract polytope*. Further, we shall usually drop the qualification *abstract* and allow it to be understood; thus, for example, *n-polytope* will generally mean *abstract* n-polytope. To avoid confusion, we shall then refer to the traditional (regular) polytopes in euclidean spaces by such terms as "convex (regular) polytopes" or "star-polytopes". Similarly, if we discuss geometric realizations of abstract polytopes, either an additional qualification or the context will make it clear whether the term "polytope" is being used in an abstract or a geometric sense.

We have special terms for polytopes of small rank; we often call a 2-polytope a *polygon*, and a 3-polytope a *polyhedron*.

The underlying face-set of a polytope $\mathcal{P}$ can be finite or infinite. Accordingly, we shall call $\mathcal{P}$ *finite* or *infinite*. An infinite n-polytope is also called an *n-apeirotope*; when $n = 2$, we also refer to it as an *apeirogon*, and when $n = 3$ as an *apeirohedron*. We say that $\mathcal{P}$ is *locally finite* if all its proper sections are finite.

Before we proceed, it may be helpful to provide an alternative, recursive, definition of an abstract polytope. The n-polytopes for $n \leqslant 1$ are as in Figure 2A1. For $n \geqslant 2$, an n-polytope is a connected poset satisfying (P1), such that each $(n-1)$-section is an

$(n-1)$-polytope. The picture we should have here is of building a polytope by sticking its facets together along its ridges, in such a way that they fit polytopally around each vertex.

As can be expected from the use of our terminology, convex polytopes of dimension n (or, rather, their face-lattices) are indeed finite (abstract) n-polytopes. For $n \geqslant 1$, the (face-to-face) tessellations of euclidean n-space provide examples of $(n+1)$-apeirotopes.

Of course, there are many more examples of polytopes. For instance, if we begin with the tessellation $\mathcal{T}$ of the euclidean plane by squares and identify points in the plane which are equivalent under a suitable translation subgroup Λ of the symmetry group $G(\mathcal{T})$ of $\mathcal{T}$, then we obtain a decomposition of the 2-torus into topological squares, 4 meeting at each "vertex". The set of all "vertices", "edges" and "squares" in this decomposition (together with suitably adjoined improper faces), ordered by inclusion, is a finite polyhedron (3-polytope). This is an example of a map of type $\{4, 4\}$ on the torus; see Sections 1D and 6D for more details. It is constructed from an apeirohedron by making suitable identifications.

Note that the decomposition of the torus will fail to give a polytope if Λ is the full translation subgroup of $G(\mathcal{T})$. In fact, in this case the decomposition consists of only one "square", two "edges" and one "vertex"; in other words, there are not enough "faces" for the decomposition to form a 3-polytope.

A map on a surface is *polytopal* if its face-set is a 3-polytope. The example of the torus map with only one square face is not polytopal. More generally, a member of a family of posets is called *polytopal* if it is an abstract polytope.

Returning to the general discussion, we observe that polytopes of rank at most 2 are essentially trivial. A polytope of rank -1 consists of the single face $F_{-1}(= F_n)$, together with the trivial partial order. A polytope of rank 0 has only two faces – the improper faces F_{-1} and F_0 $(= F_n)$. The Hasse diagram of a 1-polytope is diamond-shaped; there are exactly two vertices, each incident with F_{-1} and the (improper) edge F_1. Note that all 1-sections of polytopes of higher rank are of this kind. The Hasse diagrams of the polytopes of ranks 0 and 1 are illustrated in Figure 2A1. (In a Hasse diagram for $\mathcal{P}$, faces of $\mathcal{P}$ of the same rank are depicted by nodes on the same diagram level, and two nodes on adjacent diagram levels are joined by a branch if and only if the corresponding faces of $\mathcal{P}$ are incident.)

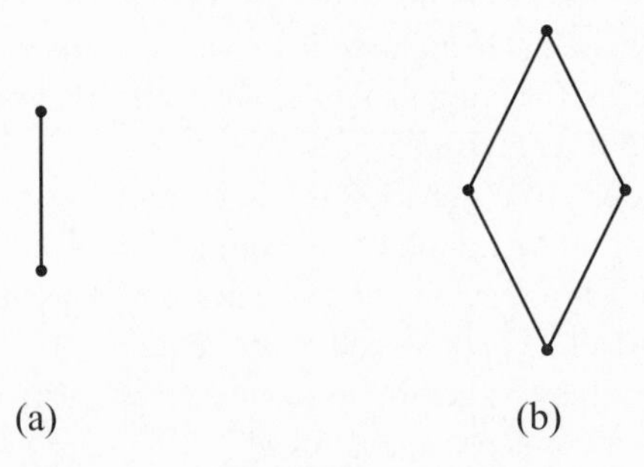

Figure 2A1. The polytopes of ranks 0 or 1.

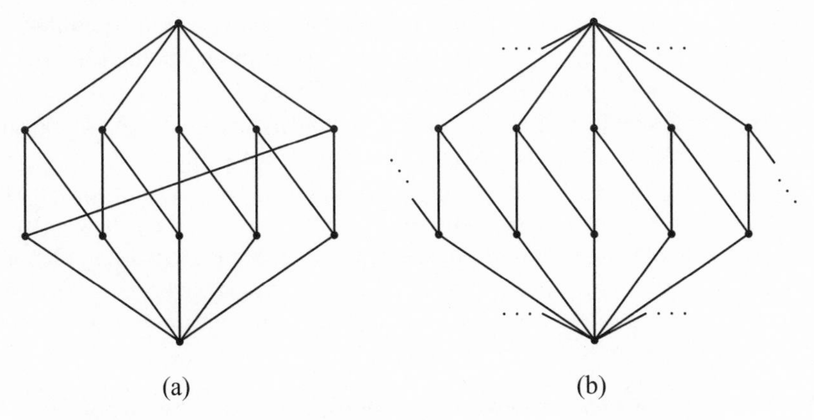

Figure 2A2. The Hasse diagrams for the pentagon and the apeirogon.

It follows easily from the defining properties (P1), ..., (P4) that the Hasse diagram of a 2-polytope $\mathcal{P}$ is necessarily of one of the two kinds illustrated in Figure 2A2. According as $\mathcal{P}$ is finite or infinite, it is identical to the diagram of a convex p-gon $\{p\}$, with $2 \leqslant p < \infty$, or the apeirogon $\{\infty\}$; see Sections 1B and 1C. Part (a) of the figure shows the pentagon (that is, $p = 5$); part (b), the apeirogon.

General notions for incidence structures like homomorphism, isomorphism, automorphism or duality have a corresponding analogue in the theory of polytopes. A map $\varphi\colon \mathcal{P} \to \mathcal{Q}$ between (the face-sets of) two polytopes $\mathcal{P}$ and $\mathcal{Q}$ is called a *homomorphism* (*between polytopes*) if it *preserves incidence*; that is, $F \leqslant G$ in $\mathcal{P}$ implies that $F\varphi \leqslant G\varphi$ in $\mathcal{Q}$. An *isomorphism* $\varphi\colon \mathcal{P} \to \mathcal{Q}$ between $\mathcal{P}$ and $\mathcal{Q}$ is a bijection for which both φ and φ^{-1} are homomorphisms; that is, $F \leqslant G$ in $\mathcal{P}$ if and only if $F\varphi \leqslant G\varphi$ in $\mathcal{Q}$. Clearly, the inverse of an isomorphism is again an isomorphism. Two polytopes $\mathcal{P}$ and $\mathcal{Q}$ are said to be *isomorphic* if there exists an isomorphism of $\mathcal{P}$ onto $\mathcal{Q}$; in this situation we write $\mathcal{P} \cong \mathcal{Q}$, or, by abuse of notation, $\mathcal{P} = \mathcal{Q}$. Note that isomorphisms preserve ranks, while this need not be true for homomorphisms.

For example, as we saw earlier, each 2-polytope is isomorphic to a finite polygon $\{p\}$ or the infinite apeirogon $\{\infty\}$. Since the term "polygon" has been used in a somewhat more general meaning in the context of Tits buildings (see [341, 417, p. 56; 422]), we should point out here that we are adopting the traditional terminology according to which a polygon is isomorphic to a convex polygon or to an apeirogon. (Thus our polygons are "generalized polygons" which are "thin".)

An *automorphism* of a polytope $\mathcal{P}$ is an isomorphism of $\mathcal{P}$ onto itself. The set of all automorphisms of a polytope $\mathcal{P}$ forms a group, the *automorphism group* $\Gamma(\mathcal{P})$ of $\mathcal{P}$; we also refer to it simply as the *group* of $\mathcal{P}$. Clearly, if $\mathcal{P}$ is finite, then $\Gamma(\mathcal{P})$ is also finite.

When discussing abstract polytopes, we are usually interested in the *isomorphism class* or *type* of a polytope rather than in the actual "representation" of the polytope. Hence we shall usually not distinguish an abstract polytope from another polytope which is isomorphic to it; in other words, isomorphic polytopes are considered to be the same. This point of view changes drastically in the theory of geometric realizations of abstract polytopes in Chapter 5. There we shall be concerned with the various realizations which a given abstract polytope can have, and then the actual representation is essential.

A bijection $\varphi: \mathcal{P} \to \mathcal{Q}$ between two polytopes $\mathcal{P}$ and $\mathcal{Q}$ is a *duality* if both φ and φ^{-1} reverse incidence; that is, $F \leqslant G$ in $\mathcal{P}$ if and only if $F\varphi \geqslant G\varphi$ in $\mathcal{Q}$. We call $\mathcal{P}$ and $\mathcal{Q}$ *duals* of each other if there exists a duality of $\mathcal{P}$ onto $\mathcal{Q}$.

Given an n-polytope $\mathcal{P}$, there exists up to isomorphism precisely one polytope which is dual to $\mathcal{P}$; this is again an n-polytope, and is denoted by $\mathcal{P}^*$. It can be constructed from $\mathcal{P}$ by leaving the face-set of $\mathcal{P}$ unchanged, but reversing the partial order on $\mathcal{P}$. Then the i-faces of $\mathcal{P}$ are in one-to-one correspondence with the $(n-1-i)$-faces of $\mathcal{P}^*$. It is immediate that $\mathcal{P}$ and $\mathcal{P}^*$ have isomorphic automorphism groups. Moreover, $(\mathcal{P}^*)^* \cong \mathcal{P}$. We note the following simple facts.

2A3 Lemma *Let $\mathcal{P}$, $\mathcal{Q}$ be n-polytopes, $\varphi: \mathcal{P} \to \mathcal{Q}$ a map, and Φ a flag of $\mathcal{P}$.*

(a) If φ is an isomorphism, then φ preserves adjacency; that is, $\Phi^j\varphi = (\Phi\varphi)^j$ for $j = 0, \ldots, n-1$.
(b) If φ is a duality, then $\Phi^j\varphi = (\Phi\varphi)^{n-1-j}$ for $j = 0, \ldots, n-1$.

A polytope $\mathcal{P}$ is *self-dual* if it is isomorphic to its dual $\mathcal{P}^*$. By a *polarity* of a self-dual polytope $\mathcal{P}$, we mean an involutory duality of $\mathcal{P}$ onto itself. Self-dual polytopes need not possess a polarity (see [10, 15, 205, 385]). However, self-dual regular polytopes necessarily do; see Proposition 2B17. For a self-dual polytope $\mathcal{P}$, the set of all automorphisms and dualities of $\mathcal{P}$ forms a group, the *extended group* $\bar{\Gamma}(\mathcal{P})$ of $\mathcal{P}$. This group contains $\Gamma(\mathcal{P})$ as a subgroup of index 2. If $\mathcal{P}$ admits a polarity, then

$$\bar{\Gamma}(\mathcal{P}) = \Gamma(\mathcal{P}) \rtimes C_2,$$

a semi-direct product of $\Gamma(\mathcal{P})$ by C_2 (see also Proposition 2B17). The semi-direct product $G := W \rtimes \Lambda$ means that Λ acts as a group of automorphisms on W on the right; thus, $W \trianglelefteq G$, a normal subgroup, with quotient $G/W \cong \Lambda$. An alternative notation for the semi-direct product would be $\Lambda \ltimes W$.)

According to our definition, the automorphisms of a polytope $\mathcal{P}$ are permutations of the face-set of $\mathcal{P}$. We shall often consider automorphisms as operating on certain subsets of $\mathcal{P}$, or on sets of subsets of $\mathcal{P}$. For example, the group $\Gamma(\mathcal{P})$ admits a representation as a permutation group on the set $\mathcal{P}_i$ of all i-faces of $\mathcal{P}$, for each fixed $i = 0, \ldots, n-1$. In general, however, this representation will not be faithful. The fact that $\mathcal{P}$ is a lattice is sufficient for faithfulness, but weaker conditions will also serve. (Recall that a poset is a *lattice* if any two of its elements have a (unique) infimum and a supremum in the poset.)

More important to us is the action of $\Gamma(\mathcal{P})$ on the set $\mathcal{F}(\mathcal{P})$ of all flags of $\mathcal{P}$. This is trivially faithful.

2A4 Proposition *For each polytope $\mathcal{P}$, the group $\Gamma(\mathcal{P})$ acts freely on $\mathcal{F}(\mathcal{P})$.*

Proof. Assume that $\varphi \in \Gamma(\mathcal{P})$ fixes one flag Φ (say). By Lemma 2A3, we have $\Phi^j\varphi = (\Phi\varphi)^j = \Phi^j$ for $j = 0, \ldots, n-1$. It follows that φ fixes each flag adjacent to Φ. Then, by (P3) and Proposition 2A1, φ fixes each flag of $\mathcal{P}$, so that $\varphi = \varepsilon$. □

2A5 Proposition *If $\mathcal{P}$ is a finite polytope, then the order of $\Gamma(\mathcal{P})$ divides $|\mathcal{F}(\mathcal{P})|$. If $|\Gamma(\mathcal{P})| = |\mathcal{F}(\mathcal{P})|$, then $\Gamma(\mathcal{P})$ is simply flag-transitive (and thus $\mathcal{P}$ is regular in the sense defined in Section 2B).*

Proof. By Proposition 2A4, each orbit of $\Gamma(\mathcal{P})$ on $\mathcal{F}(\mathcal{P})$ has size $|\Gamma(\mathcal{P})|$; in fact, if $\Phi \in \mathcal{F}(\mathcal{P})$ and $\varphi, \psi \in \Gamma(\mathcal{P})$, then $\Phi\varphi = \Phi\psi$ if and only if $\varphi = \psi$. This implies the first part. For the second part, if $|\Gamma(\mathcal{P})| = |\mathcal{F}(\mathcal{P})|$, then there is only one orbit, so that $\Gamma(\mathcal{P})$ must be flag-transitive; by Proposition 2A4 it must indeed be simply flag-transitive. □

If Ω is a set of faces of $\mathcal{P}$, then we denote by $\Gamma(\mathcal{P}, \Omega)$ the stabilizer of Ω in $\Gamma(\mathcal{P})$. In our applications, Ω will typically be a chain of $\mathcal{P}$, so that stabilizing Ω as a set is equivalent to stabilizing each of its elements. Note that, by Proposition 2A4, $\Gamma(\mathcal{P}, \Omega)$ is trivial if Ω is a flag of $\mathcal{P}$. If F is a face of $\mathcal{P}$, we write $\Gamma(\mathcal{P}, F)$ instead of $\Gamma(\mathcal{P}, \{F\})$. For the next proposition, recall our convention for writing chains.

2A6 Proposition *Let $\mathcal{P}$ be an n-polytope.*

(a) *If $-1 \leqslant i < j \leqslant n$ and $\Omega = \{F_{-1}, F_0, \ldots, F_{i-1}, F_i, F_j, F_{j+1}, \ldots, F_n\}$ is a chain of $\mathcal{P}$, then $\Gamma(\mathcal{P}, \Omega)$ is (isomorphic to) a subgroup of $\Gamma(F_j/F_i)$.*
(b) *If F is a vertex of $\mathcal{P}$, then $\Gamma(\mathcal{P}, F)$ is a subgroup of the group of the vertex-figure of $\mathcal{P}$ at F. Similarly, if F is a facet, then $\Gamma(\mathcal{P}, F)$ is a subgroup of the group of F (or more precisely, of F/F_{-1}).*

Proof. For (a), observe that $\Gamma(\mathcal{P}, \Omega)$ clearly acts on the section F_j/F_i. To see that this action is faithful, note that Ω is contained in a flag Φ (say) of $\mathcal{P}$. Hence, if φ in $\Gamma(\mathcal{P}, \Omega)$ acts trivially on F_j/F_i, then it must keep Φ fixed, so that $\varphi = \varepsilon$ by Proposition 2A4. This proves (a), and part (b) consists of special cases of (a). □

Let $\mathcal{P}$ be an n-polytope, and let $0 \leqslant k \leqslant n-1$. The *k-skeleton* $\mathrm{skel}_k(\mathcal{P})$ of $\mathcal{P}$ is the poset consisting of all proper faces of $\mathcal{P}$ of rank at most k (together with the induced partial order). Note that, if two "improper" faces are adjoined to $\mathrm{skel}_k(\mathcal{P})$, then one obtains a poset of rank $k+1$ with properties (P1), (P2) and (P3). The 1-skeleton of $\mathcal{P}$ is also called the *edge-graph* of $\mathcal{P}$. Note that this is a connected graph with possibly multiple edges. The edge-graph of the dual $\mathcal{P}^*$ is also referred to as the *dual 1-skeleton* or *dual edge-graph*. If an n-polytope $\mathcal{P}$ is a lattice, then its edge-graph is n-connected, meaning that, for every pair of vertices of $\mathcal{P}$, there exist n pairwise disjoint edge-paths having these vertices as end points (see [21, 92, p. 10; 184, p. 266]).

By definition, the 1-sections of an n-polytope $\mathcal{P}$ are all of the same kind. Generally this will not remain true for sections of higher rank, indeed not even for the 2-sections determined by faces of the same rank.

For $n \geqslant 2$ and $i = 1, \ldots, n-1$, if F is an $(i-2)$-face and G an $(i+1)$-face of $\mathcal{P}$ incident with F, then we write $p_i(F, G)$ for the number of i-faces (or $(i-1)$-faces) of $\mathcal{P}$ in the section G/F; then G/F is isomorphic to the 2-polytope with Schläfli symbol $\{p_i(F, G)\}$. If these numbers depend only on i but not on the choice of F and G, then we set $p_i := p_i(F, G)$ for each i, and call $\mathcal{P}$ *equivelar of (combinatorial Schläfli) type*

$\{p_1, \ldots, p_{n-1}\}$, or, simply, *of type* $\{p_1, \ldots, p_{n-1}\}$; the latter symbol is also called the *Schläfli symbol* of $\mathcal{P}$. (This generalizes the definition of equivelarity for polyhedral manifolds, introduced in [310]; see also Section 1D.) Note that any of the numbers p_i can take the value ∞.

When $n = 1$, we shall assign to $\mathcal{P}$ the "empty" symbol $\{\,\}$, and call this the *Schläfli symbol* of $\mathcal{P}$.

It follows immediately that each section of rank at least 1 of an equivelar polytope $\mathcal{P}$ is also equivelar (and thus possesses a Schläfli symbol). In fact, if $\mathcal{P}$ is of type $\{p_1, \ldots, p_{n-1}\}$, and if F is an $(i-2)$-face and G a $(j+1)$-face with $i \leqslant j$ and $F \leqslant G$, then G/F is of type $\{p_i, p_{i+1}, \ldots, p_j\}$. In particular, if $n \geqslant 3$, then the facets are of type $\{p_1, \ldots, p_{n-2}\}$ and the vertex-figures are of type $\{p_2, \ldots, p_{n-1}\}$. Note that each polygon (2-polytope) indeed has a Schläfli symbol; this coincides with its original symbol. If $\mathcal{P}$ is self-dual, then necessarily $p_i = p_{n-i}$ for all i, so that the symbol is symmetrical.

For convex regular polytopes in euclidean spaces, the symbol coincides with the classical Schläfli symbol [120, p. 129]. In fact, within the class of convex polytopes, the Schläfli symbol determines the polytope up to isomorphism (see Section 1B). However, for abstract polytopes this is no longer true. A simple example is the icosahedron and the hemi-icosahedron (constructed from the former by identifying antipodes); both are of type $\{3, 5\}$.

In using the Schläfli symbol $\{p_1, \ldots, p_{n-1}\}$, we repeat the convention introduced in Section 1B. If a symbol contains a string of k (say) equal integers p (say), then we shorten the symbol by replacing the string by a single p with a superscript k. For example, the symbol for the ordinary n-cube would be $\{4, 3^{n-2}\}$, standing for $\{4, 3, 3, \ldots, 3\}$ with $n-2$ entries 3.

In Chapter 5, which is about realizations of abstract polytopes, we shall introduce a finer version of the Schläfli symbol. There the numbers p_i are allowed to be fractions as in the case of regular star-polytopes in euclidean spaces. However, in the combinatorial Schläfli symbol, all entries are integers, even if classically a polytope is denoted by a symbol involving fractions. For example, the *great icosahedron* $\{3, \frac{5}{2}\}$ is of type $\{3, 5\}$; it is indeed (combinatorially) isomorphic to the icosahedron (see [120, pp. 106, 107]).

For an n-polytope $\mathcal{P}$, we write $f_i(\mathcal{P})$, or simply f_i, for the number of its i-faces. The vector $f(\mathcal{P}) := (f_0(\mathcal{P}), f_1(\mathcal{P}), \ldots, f_{n-1}(\mathcal{P}))$ is then called the *face-vector* of $\mathcal{P}$.

We shall occasionally use certain *incidence parameters* of polytopes. Assume for simplicity that $\mathcal{P}$ is a finite polytope of rank $n \geqslant 2$, with the property that any two sections which are defined by faces of the same ranks are isomorphic. This condition will be satisfied for regular polytopes; see Proposition 2B3.

If $-1 \leqslant i \leqslant j \leqslant n$, then we write k_{ij} $(= k_{ij}(\mathcal{P}))$ for the number of flags of any section G/F, where F is an i-face and G an j-face of $\mathcal{P}$ with $F \leqslant G$. Then $k_{-1,n} = |\mathcal{F}(\mathcal{P})|$, and $k_{i-1,i+1} = 2$ for all i. There are various obvious relationships between the numbers f_i and k_{ij} of $\mathcal{P}$. For example, if $n \geqslant 2$, then, for $i = 1, \ldots, n-2$,

2A7 $$k_{-1,n} = f_0 k_{0n} = f_{n-1} k_{-1,n-1} = k_{-1,i} f_i k_{in}.$$

Furthermore, if $-1 \leqslant i, j \leqslant n$, then we denote by N_{ij} the number of j-faces of $\mathcal{P}$

which are incident with a fixed i-face of $\mathcal{P}$. Clearly, if F is an i-face of $\mathcal{P}$, then $N_{ij} = f_{j-i-1}(F_n/F)$ if $j \geqslant i$, or $N_{ij} = f_j(F/F_{-1})$ if $j \leqslant i$. Then, for $n \geqslant 2$, the number of flags of $\mathcal{P}$ is given by

2A8 $|\mathcal{F}(\mathcal{P})| = k_{-1,n} = f_0 N_{01} N_{12} \cdots N_{n-2,n-1} = f_{n-1} N_{n-1,n-2} N_{n-2,n-3} \cdots N_{10}.$

2B Regular Polytopes

In the traditional theory of regular polytopes, there are many equivalent ways of defining regularity (see Section 1B). The strongest and, at the same time, most flexible definition is in terms of the flag-transitivity of the symmetry group of the polytope [156, p. 63]. In the theory of abstract polytopes, the analogous definition of regularity is again appropriate for various reasons. On the one hand, it is restrictive enough to distinguish polytopes with a high degree of (combinatorial) symmetry from polytopes with less symmetry, while on the other it allows one to develop a mathematically satisfying theory.

Before we begin the general discussion, the reader should be warned that similar generalizations of other definitions of traditional regularity will generally fail. For example, if we took the inductive definition and said that a polytope is regular if its facets and vertex-figures are regular, then *all* abstract polytopes would be regular; hence the definition would be useless. In fact, to begin the induction we would call each polytope of rank at most 2 regular, and in turn, this would make each 3-polytope regular, and so on.

Similarly, we could also try to define regularity of a polytope by requiring transitivity of its automorphism group on the faces of each rank j. This would indeed imply strong restrictions for the polytopes, but still make the class of "regular" polytopes too large. For example, polytopes which we shall call chiral (see the end of the section) also have this property.

Let $\mathcal{P}$ be an (abstract) n-polytope. We call $\mathcal{P}$ *(combinatorially) regular* if its automorphism group $\Gamma(\mathcal{P})$ is transitive on its flags. Before we derive simple properties of regular polytopes, we state the following fact without proof (compare Section 1B, and see Section 7D).

2B1 Theorem *All convex regular polytopes and all regular star-polytopes in euclidean space are abstract regular polytopes.*

The following is a restatement of part of Proposition 2A5. Recall that $\mathcal{F}(\mathcal{P})$ denotes the set of flags of $\mathcal{P}$.

2B2 Proposition *The group $\Gamma(\mathcal{P})$ of a regular polytope $\mathcal{P}$ is simply flag-transitive. In particular, $|\Gamma(\mathcal{P})| = |\mathcal{F}(\mathcal{P})|$ if $\mathcal{P}$ is finite.*

The flag-transitivity of the group of a regular polytope $\mathcal{P}$ immediately implies the transitivity of $\Gamma(\mathcal{P})$ on the faces of each rank. In particular, $\Gamma(\mathcal{P})$ acts transitively on the vertices and facets of $\mathcal{P}$. More generally, for any type of chain of $\mathcal{P}$, the group $\Gamma(\mathcal{P})$ acts transitively on the set of all chains of this type. However, no such transitivity property for chains of any type other than flags themselves will actually imply flag-transitivity of $\Gamma(\mathcal{P})$. For example, the groups of chiral (or irreflexible) maps on surfaces act transitively on the chains of each type which are not flags. However, these maps are not regular in our sense; compare the definition of chirality, or see Coxeter and Moser [131, p. 102].

For the next proposition, recall that $\Gamma(\mathcal{P}, \Omega)$ denotes the stabilizer of a chain Ω of $\mathcal{P}$.

2B3 Proposition *Let $\mathcal{P}$ be a regular n-polytope.*

(a) *All sections of $\mathcal{P}$ are regular polytopes, and any two sections which are defined by faces of the same ranks are isomorphic. In particular, $\mathcal{P}$ has isomorphic facets and isomorphic vertex-figures. Furthermore, $\mathcal{P}$ is equivelar (that is, possesses a Schläfli symbol).*
(b) *The group of each section of $\mathcal{P}$ is a subgroup of $\Gamma(\mathcal{P})$. More precisely, if F_j is a j-face and F_k a k-face with $-1 \leqslant j < k \leqslant n$ and $F_j < F_k$, then $\Gamma(F_k/F_j)$ is isomorphic to $\Gamma(\mathcal{P}, \Omega)$, where Ω is any chain of type $\{-1, 0, \ldots, j-1, j, k, k+1, \ldots, n\}$ which includes both F_j and F_k.*

Proof. Part (a) is obvious from what was said earlier. To prove (b), recall that, by Proposition 2A6, the group $\Gamma(\mathcal{P}, \Omega)$ is a subgroup of $\Gamma(F_k/F_j)$. Since $\Gamma(\mathcal{P}, \Omega)$ is flag-transitive on F_k/F_j, it follows from Proposition 2B2 that the groups actually coincide. □

The following result is crucial for an intuitive understanding of the concept of regularity; in a sense, it "localizes" the definition. Recall that, for $j = 0, \ldots, n-1$, the j-adjacent flag of a flag Φ is denoted by Φ^j. Further, bear in mind Lemma 2A3(a); since automorphisms of a polytope $\mathcal{P}$ are isomorphisms (of $\mathcal{P}$ into itself), it follows additionally that $(\Phi\varphi)^j = \Phi^j\varphi$ for each j and each $\varphi \in \Gamma(\mathcal{P})$.

2B4 Proposition *An n-polytope $\mathcal{P}$ is regular if and only if, for some flag Φ of $\mathcal{P}$ and each $j = 0, \ldots, n-1$, there exists a (unique) involutory automorphism ρ_j of $\mathcal{P}$ such that $\Phi\rho_j = \Phi^j$.*

Proof. The condition (with any choice of flag Φ) is clearly necessary for regularity. Note that $\Phi^{jj} = \Phi$ implies that $\Phi(\rho_j^2) = (\Phi\rho_j)\rho_j = \Phi^j\rho_j = (\Phi\rho_j)^j = \Phi^{jj} = \Phi$, yielding $\rho_j^2 = \varepsilon$; that is, ρ_j is an involution.

To prove sufficiency, let $\Phi \in \mathcal{F}(\mathcal{P})$ be such that the condition is satisfied. Then we have to show that, for any given flag Ψ of $\mathcal{P}$, there exists $\varphi \in \Gamma(\mathcal{P})$ such that $\Phi\varphi = \Psi$. By (P3) and Proposition 2A1, there exists a sequence $\Phi = \Phi_0, \Phi_1, \ldots, \Phi_{m-1}, \Phi_m = \Psi$ of flags such that Φ_{i-1} and Φ_i are adjacent for $i = 1, \ldots, m$. We prove the existence of φ by induction on m, the case $m = 0$ being trivial. When $m \geqslant 1$, then, by the inductive hypothesis, there exists $\psi \in \Gamma(\mathcal{P})$ such that $\Phi\psi = \Phi_{m-1}$. But Φ_{m-1} and $\Psi = \Phi_m$ are j-adjacent for some j, so that $\Psi = \Phi_{m-1}^j = (\Phi\psi)^j = \Phi^j\psi = \Phi\rho_j\psi$, by Lemma 2A3(a) and the given assumption on Φ. In other words, $\Psi = \Phi_m = \Phi\varphi$ with $\varphi := \rho_j\psi$, completing the inductive step. □

The proof of Proposition 2B4 makes essential use of the flag-connectedness of $\mathcal{P}$. For later use, it is interesting to note that the multiplication of an element of $\Gamma(\mathcal{P})$ from the left by ρ_j corresponds to passing from a suitable flag to its j-adjacent flag. Note further that, by Proposition 2B4, such automorphisms ρ_j exist for one flag Φ if and only if they exist for each flag of $\mathcal{P}$. More precisely, if Φ and Ψ are flags and $\Phi\varphi = \Psi$ with $\varphi \in \Gamma(\mathcal{P})$, then for each $j = 0, \ldots, n-1$, we have $\Phi\rho_j = \Phi^j$ if and only if $\Psi\varphi^{-1}\rho_j\varphi = \Psi^i$. That is, the involutory generators of $\Gamma(\mathcal{P})$ corresponding to Ψ are the conjugates $\varphi^{-1}\rho_j\varphi$ of those corresponding to Φ.

From now on, let $\mathcal{P}$ be a *regular* n-polytope. We choose one fixed flag

$$\Phi := \{F_{-1}, F_0, \ldots, F_n\}$$

of $\mathcal{P}$, and call it the *base flag*. Throughout, we shall reserve the notation F_j for the *base* j-face of the base flag Φ. For $j \in N := \{0, \ldots, n-1\}$, let ρ_j denote the unique involutory automorphism of $\mathcal{P}$ such that

$$\Phi\rho_j = \Phi^j;$$

that is, $F_i\rho_j = F_i$ if and only if $i \neq j$.

In view of the following results, the elements $\rho_0, \rho_1, \ldots, \rho_{n-1}$ are called the *distinguished generators* of $\Gamma := \Gamma(\mathcal{P})$ (*with respect to the base flag* Φ). Whenever generators $\rho_0, \ldots, \rho_{n-1}$ of Γ occur, then we are (explicitly or implicitly) assuming that a base flag has been chosen to which these generators belong. We also call the subgroups of Γ generated by subsets of $\{\rho_0, \ldots, \rho_{n-1}\}$ the *distinguished subgroups* of Γ. For technical reasons, we also define $\rho_j := \varepsilon$ if $j < 0$ or $j \geqslant n$.

Let us introduce some further notation, which we shall employ frequently in what follows. Let $J \subseteq N$ $(= \{0, \ldots, n-1\}$, as before). We write

2B5 $$\Phi_J := \{F_j \in \Phi \mid j \in J\},$$

and

2B6 $$\Gamma_J := \langle \rho_j \mid j \notin J \rangle$$

for the distinguished subgroup of Γ generated by the *complementary* set of ρ_j. In the latter case, we shall further use the shorthand $\Gamma_j := \Gamma_{\{j\}}$. Observe that $\Gamma_\emptyset = \Gamma$ and $\Gamma_N = \{\varepsilon\}$, the trivial subgroup.

2B7 Proposition *Let Φ_J be the chain of type J with $\Phi_J \subseteq \Phi$, where Φ is the base flag of $\mathcal{P}$. Then $\Gamma(\mathcal{P}, \Phi_J) = \Gamma_J$.*

Proof. It is clear that $\Gamma_J \leqslant \Gamma(\mathcal{P}, \Phi_J)$. To prove equality of the two groups, note first that $\Gamma(\mathcal{P}, \Phi_J)$ is simply transitive on the set of all flags Ψ with $\Phi_J \subset \Psi$. Hence, it suffices to show that Γ_J is also transitive on these flags.

Let Ψ be a flag with $\Phi_J \subset \Psi$. Choose a sequence

$$\Phi = \Phi_0, \Phi_1, \ldots, \Phi_{m-1}, \Phi_m = \Psi$$

of flags, all containing Φ_J, such that Φ_{i-1} and Φ_i are adjacent for all i. As in the proof of Proposition 2B4, we proceed by induction on m, the case $m = 0$ being trivial. By the inductive hypothesis, there exists $\psi \in \Gamma_J$ such that $\Phi\psi = \Phi_{m-1}$. As in that proof, Φ_{m-1} and $\Psi = \Phi_m$ are j-adjacent for some j. But now $j \notin J$, since $\psi \in \Gamma_J$ and $\Phi_J \subset \Phi_i$ for all i. Again, $\Psi = \Phi\rho_j\psi = \Phi\varphi$ with $\varphi := \rho_j\psi$, and since $\psi, \rho_j \in \Gamma_J$, we have $\varphi \in \Gamma_J$, as required. □

2B8 Proposition *Let $\mathcal{P}$ be a regular n-polytope, and let $\rho_0, \ldots, \rho_{n-1}$ be the distinguished generators of its group with respect to some flag. Then $\Gamma(\mathcal{P}) = \langle \rho_0, \ldots, \rho_{n-1} \rangle$.*

Proof. Apply Proposition 2B7 with $\Phi_J = \emptyset$ (that is, $J = \emptyset$). □

2B9 Proposition *Let $\mathcal{P}$ be a regular n-polytope, and let $\Gamma = \langle \rho_0, \ldots, \rho_{n-1} \rangle$ be its group.*

(a) If $-1 \leqslant j \leqslant k \leqslant n$, then

$$\Gamma(F_k/F_j) \cong \langle \rho_{j+1}, \rho_{j+2}, \ldots, \rho_{k-1} \rangle.$$

In particular,

$$\Gamma(F_j/F_{-1}) \cong \langle \rho_0, \ldots, \rho_{j-1} \rangle,$$
$$\Gamma(F_n/F_j) \cong \langle \rho_{j+1}, \ldots, \rho_{n-1} \rangle.$$

(b) If $\mathcal{P}$ is of type $\{p_1, \ldots, p_{n-1}\}$, then, for $1 \leqslant k \leqslant n-1$,

$$\Gamma(F_{k+1}/F_{k-2}) \cong \langle \rho_{k-1}, \rho_k \rangle \cong D_{p_k},$$

the dihedral group of order $2p_k$.

Proof. To prove (a), let $J = \{-1, 0, \ldots, j-1, j, k, k+1, \ldots, n\}$, and apply Propositions 2B3 and 2B7 with type J. Note that the group in (a) is trivial if $j = k$ or $j = k-1$. Part (b) is a special case of (a). □

2B10 Proposition *If $I, J \subseteq N$, then*

$$\langle \rho_i \mid i \in I \rangle \cap \langle \rho_i \mid i \in J \rangle = \langle \rho_i \mid i \in I \cap J \rangle.$$

Proof. Proposition 2B7 says that, for each $K \subseteq N$, if Φ_K is the chain of type K contained in Φ and $\Gamma_K = \langle \rho_i \mid i \notin K \rangle$, then Γ_K is the stabilizer of Φ_K. The present proposition thus claims that

$$\Gamma_{N\setminus I} \cap \Gamma_{N\setminus J} = \Gamma_{(N\setminus I)\cup(N\setminus J)} = \Gamma_{N\setminus(I\cap J)}.$$

Now one inclusion is obvious. For the other, we just note that an element γ of the intersection $\Gamma_{N\setminus I} \cap \Gamma_{N\setminus J}$ stabilizes $\Phi_{N\setminus I}$ and $\Phi_{N\setminus J}$; hence $\Phi_{N\setminus I} \cup \Phi_{N\setminus J} = \Phi_{N\setminus(I\cap J)}$. Thus $\gamma \in \Gamma_{N\setminus(I\cap J)} = \langle \rho_i \mid i \in I \cap J \rangle$, as required. □

We shall refer to the condition of Proposition 2B10 as the *intersection property* of $\Gamma(\mathcal{P})$ (*with respect to the generators* $\rho_0, \ldots, \rho_{n-1}$).

We next have a consequence of Lemma 2A2(b).

2B11 Proposition *If $|j-k| \geqslant 2$, then $(\rho_j\rho_k)^2 = \varepsilon$.*

Proof. This just restates the property $\Phi^{jk} = \Phi^{kj}$. □

By Proposition 2B8, the automorphism group $\Gamma(\mathcal{P})$ is a group generated by involutions, a term which we shall often abbreviate to *ggi*. Proposition 2B11 says the generators ρ_j and ρ_k commute if $|j-k| \geqslant 2$. This commutation rule implies that the ggi $\Gamma(\mathcal{P})$ is of a special kind, called a *string* ggi, or *sggi* for short. The rule is crucial for the theory of regular polytopes. As we shall see, it implies that the group belongs to a string diagram. Note that, by Proposition 2B9(b), we have $(\rho_{j-1}\rho_j)^2 = \varepsilon$ only if $p_j = 2$; we shall see that this implies a certain degeneracy of the group (in a sense that we shall make plain later).

2B12 Proposition *If* $0 \leqslant k \leqslant n-1$, *then*

$$\begin{aligned}\Gamma(\mathcal{P}, F_k) = \Gamma_k &= \langle \rho_i \mid i \neq k \rangle \\ &= \langle \rho_0, \ldots, \rho_{k-1} \rangle \times \langle \rho_{k+1}, \ldots, \rho_{n-1} \rangle \\ &= \Gamma(F_k/F_{-1}) \times \Gamma(F_n/F_k).\end{aligned}$$

Proof. The first equations follow from Propositions 2B7 and 2B9. Further, by Proposition 2B11, if $i < k < j$, then $\rho_i\rho_j = \rho_j\rho_i$. It follows that each element of $\langle \rho_0, \ldots, \rho_{k-1} \rangle$ commutes with each of $\langle \rho_{k+1}, \ldots, \rho_{n-1} \rangle$. Hence

$$\Gamma_k = \langle \rho_0, \ldots, \rho_{k-1} \rangle \langle \rho_{k+1}, \ldots, \rho_{n-1} \rangle = \langle \rho_{k+1}, \ldots, \rho_{n-1} \rangle \langle \rho_0, \ldots, \rho_{k-1} \rangle,$$

a product of normal subgroups of Γ_k (note that one or other subgroup is trivial if $k = 0$ or $n-1$). But, by Proposition 2B10, the intersection of the subgroups is trivial, so that we have an internal direct product. □

The next theorem is the core result in the characterization of the structure of a regular polytope $\mathcal{P}$ in terms of the distinguished generators $\rho_0, \ldots, \rho_{n-1}$ of its group. By the transitivity properties of the group, we can write each j-face of $\mathcal{P}$ in the form $F_j\varphi$, where $\varphi \in \Gamma(\mathcal{P})$ and F_j is the base j-face of $\mathcal{P}$ (in the base flag Φ). We begin with a lemma.

2B13 Lemma *Let* $0 \leqslant j, k \leqslant n-1$, *and let* G_j *be a* j*-face of* $\mathcal{P}$. *Then* G_j *is incident with* F_k *if and only if* $G_j = F_j\gamma$ *for some* $\gamma \in \Gamma_k$.

Proof. Suppose that $j \leqslant k$; the proof with $j > k$ is similar. If $G_j = F_j\gamma$ with $\gamma \in \Gamma_k$, then $G_j \leqslant F_k\gamma = F_k$, as claimed. For the converse, let Ψ be any flag of $\mathcal{P}$ such that $\{G_j, F_k\} \subseteq \Psi$. Then, by Proposition 2B7, $F_k \in \Phi \cap \Psi$ implies that $\Psi = \Phi\gamma$ for some $\gamma \in \Gamma(\mathcal{P}, F_k) = \Gamma_k$. Thus $G_j = F_j\gamma$, as required. □

2B14 Theorem *Let* $0 \leqslant j \leqslant k \leqslant n-1$, *and let* $\varphi, \psi \in \Gamma$. *Then the following conditions are equivalent:*

(a) $F_j\varphi \leqslant F_k\psi$;
(b) $\varphi\psi^{-1} \in \langle \rho_{j+1}, \ldots, \rho_{n-1} \rangle \langle \rho_0, \ldots, \rho_{k-1} \rangle$;
(c) $\Gamma_j\varphi \cap \Gamma_k\psi \neq \emptyset$.

Proof. We shall prove the equivalence in the form (a) ⇒ (c) ⇒ (b) ⇒ (a).

If (a) holds, then $F_j\varphi\psi^{-1} \leqslant F_k$; thus $\varphi\psi^{-1} \in \Gamma_k$ by Lemma 2B13. In turn, this says that $\Gamma_j\varphi\psi^{-1} \cap \Gamma_k \neq \emptyset$; hence $\Gamma_j\varphi \cap \Gamma_k\psi \neq \emptyset$, which is (c).

Next, suppose that (c) holds. It follows, using the commuting properties of the ρ_i (compare the proof of Proposition 2B12), that

$$\begin{aligned}\varphi\psi^{-1} \in \Gamma_j\Gamma_k &= \langle\rho_0,\ldots,\rho_{j-1}\rangle\langle\rho_{j+1},\ldots,\rho_{n-1}\rangle\langle\rho_0,\ldots,\rho_{k-1}\rangle\langle\rho_{k+1},\ldots,\rho_{n-1}\rangle \\ &= \langle\rho_{j+1},\ldots,\rho_{n-1}\rangle\langle\rho_{k+1},\ldots,\rho_{n-1}\rangle\langle\rho_0,\ldots,\rho_{j-1}\rangle\langle\rho_0,\ldots,\rho_{k-1}\rangle \\ &= \langle\rho_{j+1},\ldots,\rho_{n-1}\rangle\langle\rho_0,\ldots,\rho_{k-1}\rangle,\end{aligned}$$

as required for (b).

Finally, assume that (b) holds. Then we can write $\varphi\psi^{-1} = \alpha\beta$, for some $\alpha \in \langle\rho_{j+1},\ldots,\rho_{n-1}\rangle$ and $\beta \in \langle\rho_0,\ldots,\rho_{k-1}\rangle$. We deduce that

$$F_j\varphi\psi^{-1} = F_j\alpha\beta = F_j\beta \leqslant F_k\beta = F_k,$$

so that $F_j\varphi \leqslant F_k\psi$, which is (a). This completes the proof. □

Theorem 2B14 has important consequences. In effect, it says that, as is familiar from similar situations in the theory of transitive permutation groups, we may identify a face $F_j\varphi$ of $\mathcal{P}$ with the right coset $\Gamma_j\varphi$ of the stabilizer $\Gamma_j = \Gamma(\mathcal{P}, F_j) = \langle\rho_i \mid i \neq j\rangle$ of F_j in $\Gamma(\mathcal{P})$. Then Theorem 2B14 tells us when two such cosets must be regarded as "incident". This "incidence" of cosets has analogues in the theory of Tits buildings and diagram geometries (see [340, §10.1; 417, §1.4]); it was first discovered by Tits [416]. In Section 2E, this approach will be explored further.

By Propositions 2B9 and 2B11, if $\mathcal{P}$ is a regular n-polytope of type $\{p_1,\ldots,p_{n-1}\}$, then the distinguished generators of $\Gamma(\mathcal{P})$ satisfy relations $(\rho_j\rho_k)^{p_{jk}} = \varepsilon$ for $0 \leqslant j \leqslant k \leqslant n-1$, where

2B15
$$p_{jk} = \begin{cases} 1, & \text{if } j = k; \\ p_k, & \text{if } j = k-1; \\ 2, & \text{if } j \leqslant k-2. \end{cases}$$

Generally, however, there will be further independent relations too. In any case, anticipating Chapter 3, if $W = \langle r_0,\ldots,r_{n-1}\rangle$ denotes the Coxeter group with the diagram

then $r_j \mapsto \rho_j$ for $j = 0,\ldots,n-1$ induces a surjective homomorphism of W onto $\Gamma(\mathcal{P})$, which is one-to-one on the subgroups $\langle r_{k-1}, r_k\rangle$ for $k = 1,\ldots,n-1$. In particular, $\Gamma(\mathcal{P})$ is a quotient of W. In Section 3D, we shall see that W itself is the group of a regular n-polytope.

The next proposition discusses the case where one of the numbers p_k is 2.

2B16 Proposition *Let $\mathcal{P}$ be a regular polytope of rank $n \geqslant 2$ and type $\{p_1,\ldots,p_{n-1}\}$, and let $1 \leqslant k \leqslant n-1$. Then $p_k = 2$ if and only if each $(k-1)$-face of $\mathcal{P}$ is incident* *with each k-face. In this case* $\Gamma(\mathcal{P}) = \langle \rho_0, \ldots, \rho_{k-1} \rangle \times \langle \rho_k, \ldots, \rho_{n-1} \rangle$ *is an internal direct product.*

Proof. If $p_k = 2$, then $\rho_{k-1}\rho_k = \rho_k\rho_{k-1}$, and it follows that $\rho_i\rho_j = \rho_j\rho_i$ whenever $i < k \leqslant j$. Thus

$$\Gamma(\mathcal{P}) = \langle \rho_k, \ldots, \rho_{n-1} \rangle \langle \rho_0, \ldots, \rho_{k-1} \rangle;$$

hence, by Proposition 2B14(b), $F_{k-1}\varphi \leqslant F_k\psi$ for all $\varphi, \psi \in \Gamma(\mathcal{P})$. Furthermore, the two subgroups $\langle \rho_0, \ldots, \rho_{k-1} \rangle$ and $\langle \rho_k, \ldots, \rho_{n-1} \rangle$ of $\Gamma(\mathcal{P})$ commute elementwise and have a trivial intersection; that is, $\Gamma(\mathcal{P})$ is the (internal) direct product of these groups.

Conversely, if each $(k-1)$-face of $\mathcal{P}$ is incident with each k-face, then this is also true for the faces in the section F_{k+1}/F_{k-2} of $\mathcal{P}$; hence this section must be a digon, with $p_k = 2$. □

We next consider a self-dual regular n-polytope $\mathcal{P}$, with group $\Gamma(\mathcal{P})$ and extended group $\bar{\Gamma}(\mathcal{P})$. Recall that an involutory duality of $\mathcal{P}$ is called a polarity of $\mathcal{P}$.

2B17 Proposition *Let $\mathcal{P}$ be a self-dual regular n-polytope.*

(a) There exists a unique polarity ω of $\mathcal{P}$ which fixes the base flag Φ (but reverses the order of its faces).

(b) For each $j = 0, \ldots, n-1$,

$$\omega\rho_j\omega = \rho_{n-1-j},$$

so that ω induces a group automorphism of $\Gamma(\mathcal{P})$. In particular,

$$\bar{\Gamma}(\mathcal{P}) = \Gamma(\mathcal{P}) \rtimes C_2,$$

a semi-direct product of $\Gamma(\mathcal{P})$ by C_2.

(c) If the group automorphism in (b) is an inner automorphism of $\Gamma(\mathcal{P})$, given by conjugation with an involution in $\Gamma(\mathcal{P})$, then $\bar{\Gamma}(\mathcal{P}) = \Gamma(\mathcal{P}) \times C_2$.

Proof. To prove (a), let ψ be any duality of $\mathcal{P}$, and define $\Psi := \Phi\psi$. Since $\mathcal{P}$ is regular, there exists $\varphi \in \Gamma(\mathcal{P})$ such that $\Psi\varphi = \Phi$. But then $\omega := \psi\varphi$ is a duality of $\mathcal{P}$ which maps Φ onto itself. It follows that ω^2 is an automorphism of $\mathcal{P}$ which fixes Φ. Hence $\omega^2 = \varepsilon$.

For (b), note that both $\omega\rho_j\omega$ and ρ_{n-1-j} are automorphisms of $\mathcal{P}$ which map Φ onto Φ^{n-1-j}. Hence $\omega\rho_j\omega = \rho_{n-1-j}$ for each j, and so

$$\bar{\Gamma}(\mathcal{P}) = \Gamma(\mathcal{P}) \rtimes \langle \omega \rangle = \Gamma(\mathcal{P}) \rtimes C_2.$$

Finally, for (c), let $\tau \in \Gamma(\mathcal{P})$ be such that $\tau^2 = \varepsilon$, and $\tau\rho_j\tau^{-1} = \omega\rho_j\omega$ for all j. Since $\rho_0, \ldots, \rho_{n-1}$ generate $\Gamma(\mathcal{P})$, this implies that $\tau\varphi\tau^{-1} = \omega\varphi\omega$ for all $\varphi \in \Gamma(\mathcal{P})$. In particular, with $\varphi = \tau$ we have $\omega\tau\omega = \tau = \tau^{-1}$; thus $(\omega\tau)^2 = \varepsilon$. Define $\omega_0 := \omega\tau$. Then, $\omega_0 \notin \Gamma(\mathcal{P})$, $\omega_0^2 = \varepsilon$, and $\omega_0\rho_j\omega_0^{-1} = \rho_j$ for each j. It follows that ω_0 commutes with each element of $\Gamma(\mathcal{P})$; thus $\bar{\Gamma}(\mathcal{P}) = \Gamma(\mathcal{P}) \times C_2$, as required. □

As a general rule, the extended group $\bar{\Gamma}(\mathcal{P})$ of a self-dual regular polytope $\mathcal{P}$ will not be a *direct* product of $\Gamma(\mathcal{P})$ and C_2. One example is the regular 24-cell $\{3, 4, 3\}$, whose extended group has order 2304. On the other hand, for the n-simplex $\{3^{n-1}\}$, the extended group is the direct product $S_{n+1} \times C_2$ (see [123, p. 574] or Lemma 8A2).

To conclude this section, we briefly discuss chirality; for more detailed accounts, see Schulte and Weiss [372, 373]. Chiral polytopes form an important class of nearly regular polytopes. Intuitively, they have complete rotational symmetry, but not the full symmetry by reflexion. To motivate the discussion, we begin with some remarks about polytopes which are regular.

Let $\mathcal{P}$ be a regular n-polytope with group $\Gamma(\mathcal{P}) = \langle \rho_0, \ldots, \rho_{n-1} \rangle$, and let $\Phi = \{F_{-1}, F_0, \ldots, F_n\}$ be the corresponding base flag. For $j = 1, \ldots, n-1$, define

2B18
$$\sigma_j := \rho_{j-1}\rho_j.$$

Then σ_j fixes each face in $\Phi \setminus \{F_{j-1}, F_j\}$, and cyclically permutes ("rotates") consecutive j-faces (and consecutive $(j-1)$-faces) in the polygonal 2-section F_{j+1}/F_{j-2} of $\mathcal{P}$. The subgroup

$$\Gamma^+(\mathcal{P}) := \langle \sigma_1, \ldots, \sigma_{n-1} \rangle$$

of $\Gamma(\mathcal{P})$ is of index at most 2. It is called the *rotation subgroup* of $\Gamma(\mathcal{P})$, or the *rotation group* of $\mathcal{P}$; its elements are the (*combinatorial*) *rotations* of $\mathcal{P}$. If the index is 2, we also say that $\mathcal{P}$ is *directly regular*. If $\mathcal{P}$ is of type $\{p_1, \ldots, p_{n-1}\}$, then the generators of $\Gamma^+(\mathcal{P})$ satisfy the relations

2B19
$$\begin{aligned} \sigma_j^{p_j} &= \varepsilon \quad \text{for } 1 \leqslant j \leqslant n-1, \\ (\sigma_j\sigma_{j+1}\cdots\sigma_k)^2 &= \varepsilon \quad \text{for } 1 \leqslant j < k \leqslant n-1, \end{aligned}$$

but in general some further independent relations too.

An n-polytope $\mathcal{P}$ is called *chiral* if its group $\Gamma(\mathcal{P})$ has precisely two orbits on the flags, such that adjacent flags belong to distinct orbits. Then the following is a local characterization of chirality.

An n-polytope $\mathcal{P}$ is chiral if and only if $\mathcal{P}$ is not regular, but for some *base flag* $\Phi = \{F_{-1}, F_0, \ldots, F_n\}$ of $\mathcal{P}$ there exist automorphisms $\sigma_1, \ldots, \sigma_{n-1}$ of $\mathcal{P}$ with the following property: if $j = 1, \ldots, n-1$, then σ_j fixes each face in $\Phi \setminus \{F_{j-1}, F_j\}$, and cyclically permutes consecutive j-faces in the 2-section F_{j+1}/F_{j-2} of $\mathcal{P}$. For a chiral polytope $\mathcal{P}$, we can choose the orientation of the σ_j for $j = 1, \ldots, n-1$ in such a way that, if F_j' denotes the j-face of $\mathcal{P}$ with $F_{j-1} < F_j' < F_{j+1}$ and $F_j' \neq F_j$, then

$$F_j\sigma_j = F_j';$$

thus $F_{j-1}'\sigma_j = F_{j-1}$. Then $\sigma_1, \ldots, \sigma_{n-1}$ satisfy the relations (2B19), and generate the group $\Gamma(\mathcal{P})$. They are called the *distinguished generators* (*with respect to* Φ) of $\Gamma(\mathcal{P})$.

In the present book we shall mainly deal with regular polytopes, and so we shall not fully develop the theory of chirality. For an introduction to chiral polytopes, the reader is referred to [119, 372, 373]. Interesting classes of chiral polytopes are also discussed in [237, 322–324, 333–335, 374].

2C Order Complexes

In this section, we discuss basic properties of order complexes of abstract polytopes. We shall not deal with the subject in full generality, but instead restrict ourselves to results which we shall use later.

It is a standard technique to associate with any partially ordered set $\mathcal{P}$ its order complex $\mathcal{C}(\mathcal{P})$ [30, Chapter 34; 398, p. 120]. Let $\mathcal{P}$ be any polytope of rank $n \geqslant 1$, which is not necessarily regular. The *order complex* $\mathcal{C} = \mathcal{C}(\mathcal{P})$ of $\mathcal{P}$ is the (abstract) $(n-1)$-dimensional simplicial complex, whose vertices are the proper faces of $\mathcal{P}$, and whose simplices are the chains of $\mathcal{P}$ which do not contain the improper faces F_{-1} and F_n of $\mathcal{P}$. Then the maximal simplices in $\mathcal{C}$ are $(n-1)$-dimensional and are in one-to-one correspondence with the flags of $\mathcal{P}$. We follow the usual convention of identifying a simplex in a simplicial complex with the set of its vertices.

The natural convention is thus that the order complex of the point (0-polytope) is empty. However, this is sometimes inconvenient. One way of getting round the problem is to consider instead the *augmented* order complex $\bar{\mathcal{C}} := \bar{\mathcal{C}}(\mathcal{P})$. We adjoin an extra vertex F_n corresponding to the maximal face of $\mathcal{P}$, and take as the simplices of $\bar{\mathcal{C}}$ the chains of $\mathcal{P}$ which omit F_{-1}. Thus

$$\bar{\mathcal{C}} = \mathcal{C} * \{F_n\},$$

the join of $\mathcal{C}$ and the point-complex $\{F_n\}$, where the *join* of two simplicial complexes $\mathcal{C}$ and $\mathcal{D}$ is as usual

$$\mathcal{C} * \mathcal{D} := \{S \cup T \mid S \in \mathcal{C} \text{ and } T \in \mathcal{D}\},$$

using the conventional description of a simplex. (Note that the point complex $\{F_n\}$ contains $\emptyset$ as well as F_n.) The maximal simplices of $\bar{\mathcal{C}}$ are now n-dimensional, and each contains F_n. Recalling that the *link* of a simplex Ω in a simplicial complex $\mathcal{D}$ is

$$\mathrm{lk}(\Omega, \mathcal{D}) := \{\Omega' \in \mathcal{D} \mid \Omega \cup \Omega' \in \mathcal{D} \text{ and } \Omega \cap \Omega' = \emptyset\},$$

we observe that we recover the original complex $\mathcal{C}$ as

$$\mathcal{C} = \mathrm{lk}(F_n, \bar{\mathcal{C}}).$$

We shall see how to use the augmented complex in the proof of Proposition 2C1.

In discussing the order complex $\mathcal{C}$, we shall not distinguish between chains of $\mathcal{P}$ which only differ in improper faces. In particular, we shall denote a chain of $\mathcal{P}$ and the corresponding simplex of $\mathcal{C}$ by the same symbol, always with the understanding that any improper faces have been removed from the chain. A similar convention applies to the augmented complex $\bar{\mathcal{C}}$, except that now only F_{-1} is ignored.

The 1-skeleton of $\mathcal{C}$ is also called the *comparability graph* of $\mathcal{P}$ (or more exactly, of $\mathcal{P} \setminus \{F_{-1}, F_n\}$). Its vertices are the *proper* faces of $\mathcal{P}$, and two distinct vertices are joined by an edge if and only if the corresponding faces of $\mathcal{P}$ are incident.

For $j = 0, \ldots, n-1$, two $(n-1)$-simplices of $\mathcal{C}$ are called (j-)*adjacent* if the corresponding flags of $\mathcal{P}$ are (j-)adjacent. A simplex Ω of $\mathcal{C}$ is said to be of *type* $J \subseteq N := \{0, \ldots, n-1\}$ if the corresponding chain of $\mathcal{P}$ is of type J. In particular, the chain of type J in the base flag of $\mathcal{P}$ gives rise to the simplex $\{F_j \mid j \in J\}$ in $\mathcal{C}$. We call a vertex of $\mathcal{C}$ a *vertex of type* $j = 0, \ldots, n-1$ if its type as a simplex of $\mathcal{C}$ is $\{j\}$. For each $(n-1)$-simplex Φ, there is a one-to-one correspondence between its faces and the subsets of N, the correspondence being provided by the type function. Note that two $(n-1)$-simplices of $\mathcal{C}$ are j-adjacent if and only if they differ precisely in their vertices of type j. An $(n-2)$-face of an $(n-1)$-simplex Φ of $\mathcal{C}$ is said to be a *wall* of Φ, or the *j-wall* of Φ if its type is $N \setminus \{j\}$.

In the case of the augmented complex $\bar{\mathcal{C}}$, it is n-simplices which have walls, and all walls contain F_n.

The defining properties of $\mathcal{P}$ translate into strong topological properties of $\mathcal{C}$. In particular, by property (P4) in Section 2A, each $(n-1)$-simplex of $\mathcal{C}$ is j-adjacent to exactly one other, for each $j = 0, \ldots, n-1$. When rephrased for $\mathcal{C}$, the flag-connectedness of $\mathcal{P}$ says that, for any two $(n-1)$-simplices Φ and Ψ of $\mathcal{C}$, there exists a sequence $\Phi = \Phi_0, \Phi_1, \ldots, \Phi_{k-1}, \Phi_k = \Psi$ of $(n-1)$-simplices of $\mathcal{C}$, such that Φ_{i-1} and Φ_i are adjacent for $i = 1, \ldots, k$. The strong flag-connectedness of $\mathcal{P}$ translates into a similar property for each link of $\mathcal{C}$. Note that $\mathrm{lk}(\{F\}, \mathcal{C})$ can be identified with the order complex $\mathcal{C}(F/F_{-1})$ for each facet F of $\mathcal{P}$. A similar remark applies to the vertices and corresponding vertex-figures of $\mathcal{P}$.

For a convex polytope P, the (geometric realization of the) order complex is known as the *barycentric subdivision* of the boundary bd P of P. Similarly, the augmented complex is the barycentric subdivision of P itself. For abstract polytopes $\mathcal{P}$, the order complex $\mathcal{C}$ provides a natural way of associating with $\mathcal{P}$ a topological space $|\mathcal{C}|$, the underlying point-set of $\mathcal{C}$ (somewhat confusingly for our context, this is usually called the *polyhedron* of $\mathcal{C}$). In Chapter 6, we shall investigate the case when this is a spherical, euclidean, or hyperbolic space-form.

The action of the group $\Gamma(\mathcal{P})$ of a polytope $\mathcal{P}$ induces in an obvious way a faithful action on $\mathcal{C}$ (or on $\bar{\mathcal{C}}$). If $\mathcal{P}$ is regular, Φ the base flag of $\mathcal{P}$, and $\rho_0, \ldots, \rho_{n-1}$ the distinguished generators of $\Gamma(\mathcal{P})$, then each ρ_j maps the *base* $(n-1)$*-simplex* Φ of $\mathcal{C}$ onto the j-adjacent simplex Φ^j, leaving the wall $\Phi \cap \Phi^j$ of Φ fixed.

Occasionally, if we consider the group $\Gamma(\mathcal{P})$ of a regular polytope $\mathcal{P}$ as acting on $\mathcal{C}$, we shall think of $\Gamma(\mathcal{P})$ as a *combinatorial reflexion group* generated by the *combinatorial reflexions* $\rho_0, \ldots, \rho_{n-1}$ in the *reflexion walls* of the base simplex Φ of $\mathcal{C}$. This terminology is motivated by the corresponding geometric counterpart in the traditional theory, where ρ_j occurs as an actual reflexion in a hyperplane cutting across $|\mathcal{C}|$.

It is easy to recover an n-polytope $\mathcal{P}$ from its order complex $\mathcal{C}$. The following proposition will be used in Section 3D, and describes the reconstruction in terms of the group $\Gamma(\mathcal{P})$ of a regular polytope $\mathcal{P}$ and its action on $\mathcal{C}$. In practice, we work with the augmented complex $\bar{\mathcal{C}}$, for reasons which will become clear in the proof. If Ω is a chain of proper faces of $\mathcal{P}$, then we write $\bar{\Omega} := \Omega \cup \{F_n\}$; similarly, $\bar{J} := J \cup \{n\}$ for $J \subseteq N \ (= \{0, \ldots, n-1\})$.

2C1 Proposition *Let $\mathcal{P}$ be a regular polytope of rank $n \geqslant 1$, and let $\bar{\mathcal{C}}$ be its augmented order complex. Let $\Phi = \{F_{-1}, F_0, \ldots, F_n\}$ be the base flag of $\mathcal{P}$, and let $\rho_0, \ldots, \rho_{n-1}$ be the corresponding distinguished generators of $\Gamma(\mathcal{P})$. For $j = 0, \ldots, n$, define the* *subcomplex* $\bar{\mathcal{C}}_j$ *of* $\bar{\mathcal{C}}$ *by*

$$\bar{\mathcal{C}}_j := \{\Phi_J\chi \mid J \subseteq \{0, \ldots, j\} \text{ and } \chi \in \langle\rho_0, \ldots, \rho_{j-1}\rangle\};$$

here, $\Phi(\bar{N}) := \bar{\Phi}$. *Define* $F'_{-1} := \emptyset$, *and for* $j = 0, \ldots, n$ *set*

$$F'_j := |\bar{\mathcal{C}}_j|.$$

Finally, let

$$\mathcal{P}' := \{F'_j\varphi \mid -1 \leqslant j \leqslant n \text{ and } \varphi \in \Gamma(\mathcal{P})\},$$

and order $\mathcal{P}'$ *by (set) inclusion. Then* $\mathcal{P}'$ *and* $\mathcal{P}$ *are isomorphic regular polytopes with the same groups. Furthermore, the generators* $\rho_0, \ldots, \rho_{n-1}$ *are the distinguished generators of* $\Gamma(\mathcal{P}') = \Gamma(\mathcal{P})$ *belonging to the base flag* $\Phi' = \{F'_{-1}, F'_0, \ldots, F'_n\}$ *of* $\mathcal{P}'$.

Proof. What we have done here is identify a face F of $\mathcal{P}$ with the underlying point-set of the associated augmented order complex $\bar{\mathcal{C}}(F/F_{-1})$. (We could have worked with the order complex $\mathcal{C}$, except that the vertices disappear!) As we remarked earlier, when P is a convex polytope and the augmented complex is expressed by genuine barycentric subdivision, this recovers the faces of P geometrically.

Since our complexes are *abstract*, there is no harm in identifying a subcomplex of $\bar{\mathcal{C}}$ with its underlying point-set. (Remember that a subcomplex contains every sub-simplex of each of its simplices.) We first note that, in this sense, the faces of $\mathcal{P}'$ are subcomplexes of $\bar{\mathcal{C}}$; indeed, F'_j is the augmented complex of F_j/F_{-1} (augmented, of course, by F_j rather than F_n). In particular, $F'_{-1} = \emptyset$ is the natural definition, and $F'_n = \bar{\mathcal{C}}$ itself.

All the properties of $\mathcal{P}'$ can easily be derived from the equivalence

$$F'_j\varphi \subseteq F'_k\psi \iff j \leqslant k \quad \text{and} \quad \varphi\psi^{-1} \in \langle\rho_{j+1}, \ldots, \rho_{n-1}\rangle\langle\rho_0, \ldots, \rho_{k-1}\rangle$$

(compare Theorem 2B14(b)). To prove this equivalence, we may clearly assume that $\psi = \varepsilon$.

First, let $F'_j\varphi \subseteq F'_k$. Since $\Phi_{\{0,\ldots,j\}} \in F'_j$, by definition there exist $\chi \in \langle\rho_0, \ldots, \rho_{k-1}\rangle$ and $K \subseteq \{0, \ldots, k\}$, such that $\Phi_{\{0,\ldots,j\}}\varphi = \Phi_K\chi$. But then $K = \{0, \ldots, j\}$ and $\varphi\chi^{-1} \in \Gamma_{\{0,\ldots,j\}} = \langle\rho_{j+1}, \ldots, \rho_{n-1}\rangle$; hence, $j \leqslant k$ and $\varphi \in \langle\rho_{j+1}, \ldots, \rho_{n-1}\rangle\langle\rho_0, \ldots, \rho_{k-1}\rangle$.

Conversely, let $j \leqslant k$, and let $\varphi = \alpha\beta$ with $\alpha \in \langle\rho_{j+1}, \ldots, \rho_{n-1}\rangle$, and $\beta \in \langle\rho_0, \ldots, \rho_{k-1}\rangle$. Let $\chi \in \langle\rho_0, \ldots, \rho_{j-1}\rangle$, and suppose that $J \subseteq \{0, \ldots, j\}$. Then, $\alpha\chi = \chi\alpha$ and $\chi\beta \in \langle\rho_0, \ldots, \rho_{k-1}\rangle$. It follows that

$$\Phi_J\chi\varphi = \Phi_J\chi\alpha\beta = \Phi_J\alpha\chi\beta = \Phi_J\chi\beta \in F'_k,$$

so that, by definition, $F'_j\varphi \subseteq F'_k$. □

We conclude this section by briefly discussing how the order complex $\mathcal{C} = \mathcal{C}(\mathcal{P})$ of a regular polytope $\mathcal{P}$ can be completely described in terms of the group $\Gamma(\mathcal{P})$.

To begin with, we set everything in a more general context (compare [54, Chapter III]). For $k \geqslant 1$, let $\Gamma = \langle\sigma_1, \ldots, \sigma_k\rangle$ be a ggi, that is, a group with a system of involutory generators $\sigma_1, \ldots, \sigma_k$. For $\emptyset \neq J \subseteq K := \{1, \ldots, k\}$, define $\Gamma_J := \langle\sigma_i \mid i \notin J\rangle$;

in particular, $\Gamma_K = \{\varepsilon\}$. (The reader should recall from time to time that Γ_J gets smaller as J gets bigger. If we were to include $J = \emptyset$, we would obtain instead the augmented complex $\bar{\mathcal{C}}(\Gamma)$, with $\Gamma_\emptyset = \Gamma$; in this case, we could allow $k = 0$ as well.) Assume that Γ and its generators satisfy the *intersection property*

$$\Gamma_I \cap \Gamma_J = \Gamma_{I \cup J} \quad \text{for all } I, J \subseteq K.$$

We now consider the partial order $\leqslant_{\mathcal{C}}$ defined on the set $\mathcal{C}_\Gamma = \mathcal{C}(\Gamma; \sigma_1, \ldots, \sigma_k)$ of all right cosets $\Gamma_J \varphi$, with $\varphi \in \Gamma$ and $\emptyset \neq J \subseteq K$, by

2C2
$$\Gamma_I \varphi \leqslant_{\mathcal{C}} \Gamma_J \psi \iff \Gamma_I \varphi \subseteq \Gamma_J \psi.$$

In other words, $\leqslant_{\mathcal{C}}$ is just inclusion of subsets of group elements. Then the intersection property for Γ implies that, in $\mathcal{C}_\Gamma$, we have

$$\Gamma_I \varphi \leqslant_{\mathcal{C}} \Gamma_J \psi \iff I \subseteq J \quad \text{and} \quad \varphi \psi^{-1} \in \Gamma_I.$$

In particular, $\mathcal{C}_\Gamma$ carries the structure of an abstract simplicial $(k-1)$-complex whose simplices are (in one-to-one correspondence with) the elements in $\mathcal{C}_\Gamma$, with inclusion of simplices corresponding to incidence in the poset $\mathcal{C}_\Gamma$. More precisely, the maximal simplices (of dimension $k-1$) are given by the cosets of $\{\varepsilon\} = \Gamma_K$, or, equivalently, by the elements of Γ. The faces (of dimension $|J| - 1$) of a maximal simplex $\Gamma_K \varphi$ are the cosets $\Gamma_J \varphi$ with $J \subseteq K$. By construction, Γ acts faithfully on $\mathcal{C}_\Gamma$ as a group of automorphisms of an abstract simplicial complex, and the "closure" $\{\Gamma_I \mid I \subseteq K\}$ of the maximal simplex $\{\varepsilon\}$ is a "fundamental region" for the action of Γ on $\mathcal{C}_\Gamma$.

A particularly interesting case arises if Γ is a Coxeter group. Then the simplicial complex $\mathcal{C}_\Gamma$ is called the *Coxeter complex* of Γ [259, Chapter 5; 417, Chapter 2]. For a geometric realization of $\mathcal{C}_\Gamma$ in this case, see also Section 3A.

If $\mathcal{P}$ is a regular n-polytope, then the previous construction applies with $\Gamma := \Gamma(\mathcal{P}) = \langle \rho_0, \ldots, \rho_{n-1} \rangle$ (and $\sigma_i = \rho_i$), and gives an abstract simplicial $(n-1)$-complex $\mathcal{C}_\Gamma = \mathcal{C}_{\Gamma(\mathcal{P})}$ which is isomorphic to the order complex $\mathcal{C}$ of $\mathcal{P}$. In particular, if $\Phi = \{F_{-1}, F_0, \ldots, F_n\}$ denotes the base flag of $\mathcal{P}$, then an isomorphism $\mathcal{C}_\Gamma \to \mathcal{C}$ is given by

2C3
$$\Gamma_J \varphi \mapsto \Phi_J \varphi$$

for $\varphi \in \Gamma = \Gamma(\mathcal{P})$ and $J \subseteq N$, with Φ_J defined as in Proposition 2C1.

It follows that the order complex of a regular polytope $\mathcal{P}$ can be completely described in terms of $\Gamma(\mathcal{P})$. Note that, in the light of this observation, Proposition 2C1 can be rewritten in terms of $\mathcal{C}_{\Gamma(\mathcal{P})}$.

2D Quotients

A standard method of deriving new structures from a given combinatorial structure is by making suitable identifications. This technique also applies to abstract polytopes. Although similar considerations hold more generally for other kinds of posets, we shall restrict ourselves here to abstract polytopes. We shall largely follow [301].

By a *pre-polytope* we mean a poset satisfying (P1), (P2) and (P4) in Section 2A, but not necessarily (P3); that is, we drop the requirement of strong connectedness and, in fact, do not even assume connectedness. Note that terms like "face", "rank", "flag", "homomorphism", "automorphism" and similar terms carry over naturally to pre-polytopes.

Let $\mathcal{P}$ and $\mathcal{Q}$ be two pre-polytopes of the same rank. A homomorphism $\varphi\colon \mathcal{P} \to \mathcal{Q}$ is called a *rap-map* if φ is rank (preserving) and adjacency preserving, the latter meaning that adjacent flags of $\mathcal{P}$ are mapped onto (distinct) adjacent flags of $\mathcal{Q}$. Note that φ is surjective if $\mathcal{Q}$ is flag-connected. A surjective rap-map φ is called a *covering*. A covering φ is called a *k-covering* if it maps sections of $\mathcal{P}$ of rank at most k isomorphically onto corresponding sections of $\mathcal{Q}$, so that, if $F, G \in \mathcal{P}$ are such that $F < G$ and $\operatorname{rank} G - \operatorname{rank} F = k+1$, then φ induces an isomorphism of G/F onto $G\varphi/F\varphi$. If there exists a covering $\varphi\colon \mathcal{P} \to \mathcal{Q}$, then we also say that $\mathcal{P}$ is a *covering of* $\mathcal{Q}$, or that $\mathcal{P}$ *covers* $\mathcal{Q}$, or that $\mathcal{Q}$ *is covered by* $\mathcal{P}$; in this situation, we occasionally write $\mathcal{P} \searrow \mathcal{Q}$. We also use similar terminology for k-coverings. A particularly interesting case is that of $(n-1)$-coverings; here, φ preserves the structure of the facets and vertex-figures of $\mathcal{P}$.

2D1 Proposition *Let $\mathcal{P}$ be a polytope, $\mathcal{Q}$ a pre-polytope and $\varphi\colon \mathcal{P} \to \mathcal{Q}$ a rap-map. Then $\mathcal{Q}$ is a polytope (and φ a covering) if and only if the image of each section of $\mathcal{P}$ is a section of $\mathcal{Q}$.*

Proof. Suppose first that the image of every section of $\mathcal{P}$ is a section of $\mathcal{Q}$. Then φ is surjective (we can take the section to be $\mathcal{P}$ itself). Let $\hat{F}, \hat{G}$ be faces of $\mathcal{Q}$ with $\hat{F} < \hat{G}$. Then $\hat{G} = G\varphi$ for some face G of $\mathcal{P}$. Since, by assumption, $\hat{F} \in \hat{G}/\hat{F}_{-1} = (G/F_{-1})\varphi$, we have $\hat{F} = F\varphi$ for some face F with $F < G$. By assumption again, $\hat{G}/\hat{F} = (G/F)\varphi$. Since G/F is connected, it follows that $\hat{G}/\hat{F}$ is connected. Since $\hat{F}, \hat{G}$ are arbitrary, we see that $\mathcal{Q}$ is strongly connected, and hence a polytope.

Conversely, suppose that the image of some section of $\mathcal{P}$ is not a section of $\mathcal{Q}$. Let G/F be a minimal such section of $\mathcal{P}$. If $\hat{F} := F\varphi$, $\hat{G} := G\varphi$, we then have $(G/F)\varphi \subset \hat{G}/\hat{F}$. Hence there is a face $\hat{H}$ of $\mathcal{Q}$, such that $\hat{F} < \hat{H} < \hat{G}$, but $\hat{H} \notin (G/F)\varphi$. But if $\hat{H}$ is any such face, then no flag of $\hat{G}/\hat{F}$ containing $\hat{H}$ can meet $(G/F)\varphi$ (except at $\hat{F}$ and $\hat{G}$), otherwise G/F would fail to be minimal. It follows that $\hat{G}/\hat{F}$ is not flag-connected, so that $\mathcal{Q}$ is not a polytope. □

2D2 Corollary *Let $\mathcal{P}$ be a polytope, $\mathcal{Q}$ a pre-polytope, and $\varphi\colon \mathcal{P} \to \mathcal{Q}$ a rap-map. Then $\mathcal{Q}$ is a polytope if and only if, whenever $\hat{F}, \hat{G}, \hat{H}$ are faces of $\mathcal{Q}$ with $\hat{F} < \hat{H} < \hat{G}$, and $\hat{F} = F\varphi$, $\hat{G} = G\varphi$ for some faces F, G of $\mathcal{P}$, then $\hat{H} = H\varphi$ for some $H \in G/F$.*

Proof. This really restates Proposition 2D1, since the condition $\hat{F} = F\varphi$, $\hat{G} = G\varphi$ with $F < G$ defines $\hat{G}/\hat{F}$. □

A common way to obtain coverings is by the construction of quotients. Let $\mathcal{P}$ be an n-polytope and Σ a subgroup of $\Gamma(\mathcal{P})$. In most of our applications, $\mathcal{P}$ will be regular and Σ will be a normal subgroup of $\Gamma(\mathcal{P})$. However, to begin with at least, we do not generally impose these restrictions.

The set of orbits of Σ in $\mathcal{P}$ is denoted $\mathcal{P}/\Sigma$. If F is a face of $\mathcal{P}$, then we write its orbit under Σ in the form $F \cdot \Sigma$. We introduce a partial order on $\mathcal{P}/\Sigma$ as follows: if $\hat{F}, \hat{G} \in \mathcal{P}/\Sigma$, then $\hat{F} \leqslant \hat{G}$ if and only if $\hat{F} = F \cdot \Sigma$ and $\hat{G} = G \cdot \Sigma$ for some faces F and G of $\mathcal{P}$ with $F \leqslant G$. The set $\mathcal{P}/\Sigma$ together with this partial order is called the *quotient of* $\mathcal{P}$ *with respect to* Σ. The mapping $\pi : \mathcal{P} \to \mathcal{P}/\Sigma$ given by $F\pi := F \cdot \Sigma$ is called the *canonical projection.* A *quotient polytope* of $\mathcal{P}$ is a quotient which is again a polytope.

2D3 Proposition *Let* $\mathcal{P}$ *be an n-polytope and* Σ *a subgroup of* $\Gamma(\mathcal{P})$. *Then* $\mathcal{P}/\Sigma$ *is a flag-connected poset of rank n which has the properties (P1) and (P2) in Section 2A.*

Proof. Properties (P1) and (P2) in Section 2A are obvious from the construction. In particular, $F_{-1} \cdot \Sigma = \{F_{-1}\}$ and $F_n \cdot \Sigma = \{F_n\}$. The rank function of $\mathcal{P}/\Sigma$ is defined by $\operatorname{rank}(F \cdot \Sigma) := \operatorname{rank} F$ for $F \in \mathcal{P}$. The canonical projection $\pi : \mathcal{P} \to \mathcal{P}/\Sigma$ induces a surjective mapping $\pi : \mathcal{F}(\mathcal{P}) \to \mathcal{F}(\mathcal{P}/\Sigma)$ between the sets of flags of $\mathcal{P}$ and $\mathcal{P}/\Sigma$ (we denote it by the same symbol); thus, if $\Psi = \{G_{-1}, G_0, \ldots, G_n\} \in \mathcal{F}(\mathcal{P})$, then

$$\Psi\pi := \{G_{-1} \cdot \Sigma, G_0 \cdot \Sigma, \ldots, G_n \cdot \Sigma\}.$$

The surjectivity of π on $\mathcal{F}(\mathcal{P}/\Sigma)$ is an immediate consequence of the definition of the partial order on $\mathcal{P}/\Sigma$. But then the flag-connectedness of $\mathcal{P}/\Sigma$ follows from the fact that π maps adjacent flags of $\mathcal{P}$ onto equal or adjacent flags of $\mathcal{P}/\Sigma$. □

Similar arguments to those in the proof of Proposition 2D3 show that all sections of $\mathcal{P}/\Sigma$ of rank $n-1$ are connected. More generally, if F is a face of $\mathcal{P}$, then any two flags of $\mathcal{P}/\Sigma$ which contain the face $F \cdot \Sigma$ can be joined by a suitable sequence of flags, all of which contain $F \cdot \Sigma$. However, further connectivity properties of $\mathcal{P}/\Sigma$ will depend upon the choice of Σ.

To illustrate the fact that $\mathcal{P}/\Sigma$ need not be a pre-polytope, consider the square tessellation $\mathcal{P} = \{4, 4\}$ with vertex-set $\mathbb{Z}^2$, and take for Σ the group generated by the translation τ with translation vector $(1, 1)$. Then the fundamental region of Σ is an infinite strip, and $\mathcal{P}/\Sigma$ is a "tessellation" of this strip. If $\Phi = \{F_0, F_1, F_2\}$ is the base flag of $\mathcal{P}$ with $F_0 = (0, 0)$, $F_1 = \{(0, 0), (0, 1)\}$ and $F_2 = \{(0, 0), (0, 1), (1, 0), (1, 1)\}$ (we list faces by their vertices), then there are four faces of rank 1 in the 1-section $(F_2 \cdot \Sigma)/(F_0 \cdot \Sigma)$ of $\mathcal{P}/\Sigma$, namely, $H \cdot \Sigma$ with H an edge of F_2.

In general, we are interested in quotients $\mathcal{P}/\Sigma$ which are again polytopes, and have particular kinds of sections isomorphic to those of the original polytope $\mathcal{P}$. This property can be achieved by imposing certain conditions on Σ. For the next two results, we do not insist on polytopes being regular; hence, until further notice, Φ may be *any* flag.

Let $-1 \leqslant f < g \leqslant n$, and let F be an f-face and G a g-face of $\mathcal{P}$ such that $F < G$. Then $F\pi$ is an f-face and $G\pi$ a g-face of $\mathcal{P}/\Sigma$ such that $F\pi < G\pi$. Let Ω be a chain of $\mathcal{P}$ of type $\{-1, 0, \ldots, f-1, f, g, g+1, \ldots, n\}$ which contains F and G. By Σ_Ω we denote the stabilizer of Ω in Σ. Then we have the canonical projection $\beta : G/F \to (G/F)/\Sigma_\Omega$, as well as the homomorphism $\gamma : (G/F)/\Sigma_\Omega \to G\pi/F\pi$, defined for

$H \in G/F$ by

2D4 $$(H \cdot \Sigma_\Omega)\gamma := H \cdot \Sigma.$$

In particular, $\beta\gamma = \pi$ (or, more exactly, $\beta\gamma$ is the restriction $\pi|_{G/F}: G/F \to G\pi/F\pi$ of π to G/F). Further, γ is injective if and only if two faces of G/F are equivalent modulo Σ_Ω whenever they are equivalent modulo Σ. Clearly, if γ is an isomorphism, then it induces a bijection $\mathcal{F}((G/F)/\Sigma_\Omega) \to \mathcal{F}(G\pi/F\pi)$.

2D5 Proposition *Let $\mathcal{P}$ be an n-polytope and Σ a subgroup of $\Gamma(\mathcal{P})$. Then $\mathcal{P}/\Sigma$ is an n-polytope if and only if the following two conditions hold.*

- *(a) The map γ of (2D4) is surjective for all F, G, and Ω as before.*
- *(b) The maps β, γ are injective if* $\operatorname{rank} G - \operatorname{rank} F = 2$.

Proof. First, note that γ is surjective if and only if π (restricted to G/F) is surjective. Also, $\mathcal{P}/\Sigma$ satisfies (P4) if and only if β, γ are isomorphisms when $\operatorname{rank} G - \operatorname{rank} F = 2$. But then the statement follows from Propositions 2D1 and 2D3. □

We denote by $\mathcal{F}(\mathcal{P})/\Sigma$ the set of orbits of $\mathcal{F}(\mathcal{P})$ under the action of Σ. Then we have the map $\mu: \mathcal{F}(\mathcal{P})/\Sigma \to \mathcal{F}(\mathcal{P}/\Sigma)$, given for $\Psi \in \mathcal{F}(\mathcal{P})$ by

2D6 $$(\Psi \cdot \Sigma)\mu := \Psi\pi.$$

It follows from the proof of Proposition 2D3 that μ is surjective. Our next lemma tells us when $\mathcal{F}(\mathcal{P})/\Sigma$ and $\mathcal{F}(\mathcal{P}/\Sigma)$ can be identified naturally under μ.

2D7 Lemma *Let $\mathcal{P}$ be an n-polytope and let Σ be a subgroup of $\Gamma(\mathcal{P})$ such that $\mathcal{P}/\Sigma$ is a polytope. Then the map μ of (2D6) is a bijection if and only if, for all F, G and Ω as before, the map γ of (2D4) is an isomorphism.*

Proof. By Proposition 2D5, the maps γ are surjective. Now suppose that μ is a bijection, and let F, G and Ω be as before. To prove that $\gamma: (G/F)/\Sigma_\Omega \to G\pi/F\pi$ is injective, let H and K be faces in G/F such that $H \cdot \Sigma = K \cdot \Sigma$. Choose flags Φ and Ψ of $\mathcal{P}$ such that $\Omega \subset \Phi, \Psi$ and $H \in \Phi$, $K \in \Psi$. Then $\hat{\Phi} := \{J \cdot \Sigma \mid J \in \Phi\}$ and $\hat{\Psi} := \{J \cdot \Sigma \mid J \in \Psi\}$ are flags of $\mathcal{P}/\Sigma$, both of which contain the face $H \cdot \Sigma = K \cdot \Sigma$ as well as the chain $\hat{\Omega} := \{J \cdot \Sigma \mid J \in \Omega\}$. Let $h := \operatorname{rank} H$ $(= \operatorname{rank} K)$. Since $\mathcal{P}/\Sigma$ is a polytope, there exists a sequence

$$\hat{\Phi} = \hat{\Phi}_0, \hat{\Phi}_1, \ldots, \hat{\Phi}_m = \hat{\Psi}$$

of flags of $\mathcal{P}/\Sigma$, all containing $H \cdot \Sigma$ and $\hat{\Omega}$, such that $\hat{\Phi}_{i-1}$ and $\hat{\Phi}_i$ are adjacent for $i = 1, \ldots, m$. Then, in fact, $\hat{\Phi}_{i-1}$ and $\hat{\Phi}_i$ are $j(i)$-adjacent for some $j(i)$ with $\operatorname{rank} F = f < j(i) < g = \operatorname{rank} G$ and $j(i) \neq h$. But μ is a bijection, so that, for $i = 0, \ldots, m$, there exists a flag Φ_i of $\mathcal{P}$ such that $\hat{\Phi}_i = (\Phi_i \cdot \Sigma)\mu$. Here we may assume that $\Phi_0 = \Phi$. But now, by Proposition 2D5(b), adjacent flags of $\mathcal{P}$ are never in the same orbit of Σ, so that the canonical mapping $\mathcal{F}(\mathcal{P}) \to \mathcal{F}(\mathcal{P})/\Sigma$ followed by μ sends adjacent flags of $\mathcal{P}$ to adjacent flags of $\mathcal{P}/\Sigma$. It follows that, starting with $\Phi_0 = \Phi$, we can choose the flags Φ_i of $\mathcal{P}$ in such a way that Φ_{i-1} and Φ_i are $j(i)$-adjacent for $i = 1, \ldots, m$;

here, $j(i)$ is as before. This gives us a sequence

$$\Phi \cdot \Sigma = \Phi_0 \cdot \Sigma,\ \Phi_1 \cdot \Sigma,\ \ldots,\ \Phi_m \cdot \Sigma = \Psi \cdot \Sigma$$

of orbits of Σ, where the sequence $\Phi = \Phi_0, \Phi_1, \ldots, \Phi_m$ in $\mathcal{F}(\mathcal{P})$ has the same adjacency properties as the corresponding sequence in $\mathcal{F}(\mathcal{P}/\Sigma)$. Note that we cannot conclude that $\Phi_m = \Psi$. However, since $f < j(i) < g$ and $j(i) \neq h$ for all i, each flag Φ_i of $\mathcal{P}$ must contain both the face H and the chain Ω. In addition, $\Phi_m \cdot \Sigma = \Psi \cdot \Sigma$, so that there exists $\varphi \in \Sigma$ such that $\Phi_m\varphi = \Psi$; thus $H\varphi = K$. But $\Omega \subseteq \Psi$, so that $\varphi \in \Sigma_\Omega$. It follows that $H \cdot \Sigma_\Omega = K \cdot \Sigma_\Omega$, which proves that γ is injective.

Conversely, let all maps γ be isomorphisms. To prove that μ is injective, let

$$\Phi = \{F_{-1}, F_0, \ldots, F_{n-1}, F_n\}, \quad \Psi = \{F_{-1}, G_0, \ldots, G_{n-1}, F_n\}$$

be flags of $\mathcal{P}$ such that $\Phi\pi = \Psi\pi$. Then $\Phi \cdot \Sigma = \Psi \cdot \Sigma$, if we can prove by induction on j that there exists $\varphi \in \Sigma$ such that $F_i\varphi = G_i$ for $0 \leqslant i \leqslant j$. For $j = 0$, this is true by assumption. To prove it for $j \geqslant 1$, assume that $\varphi \in \Sigma$ has been chosen so that $F_i\varphi = G_i$ for $i < j$. Then $G_{j-1} \leqslant G_j, F_j\varphi$. But $F_j \cdot \Sigma = G_j \cdot \Sigma$, so that $F_j\varphi$ and G_j are faces of the section F_n/G_{j-1}, which are in the same orbit of Σ. On the other hand, the map γ, defined for the case $G := F_n$, $F := G_{j-1}$ and $\Omega := \{F_{-1}, G_0, \ldots, G_{j-1}, F_n\}$, is an isomorphism, so that $F_j\varphi$ and G_j are also in the same orbit of Σ_Ω; that is, $F_j\varphi\tau = G_j$ for some $\tau \in \Sigma_\Omega$.

But if $i < j$, then we also have $F_i\varphi\tau = G_i\tau = G_i$, so that $\varphi\tau$ has the required property with respect to j. This completes the proof. □

Lemma 2D7 explains why, in most applications, it is natural to require that the map γ of (2D4) be an isomorphism. We shall usually impose this condition. For regular polytopes, we also give a further justification in Proposition 2E18.

The following proposition identifies the groups of quotients of regular polytopes. See also Proposition 2E24 for a generalization.

2D8 Proposition *Let $\mathcal{P}$ be a regular n-polytope and Σ a subgroup of $\Gamma(\mathcal{P})$. Suppose that, for all F, G and Ω as before, the map γ of (2D4) is an isomorphism. Then $\mathcal{P}/\Sigma$ is an n-polytope with group $\Gamma(\mathcal{P}/\Sigma) \cong N_{\Gamma(\mathcal{P})}(\Sigma)/\Sigma$, where $N_{\Gamma(\mathcal{P})}(\Sigma)$ is the normalizer of Σ in $\Gamma(\mathcal{P})$. In particular, $\mathcal{P}/\Sigma$ is regular if and only if Σ is normal in $\Gamma(\mathcal{P})$, in which case $\Gamma(\mathcal{P}/\Sigma) \cong \Gamma(\mathcal{P})/\Sigma$.*

Proof. By Lemma 2D7, the map μ of (2D6) is a bijection. Now, to prove the statement about $\Gamma(\mathcal{P}/\Sigma)$, observe that, for each $\varphi \in N_{\Gamma(\mathcal{P})}(\Sigma)$, the map $\hat{\varphi}$ given by $(F \cdot \Sigma)\hat{\varphi} := F\varphi \cdot \Sigma$ induces an automorphism of $\mathcal{P}/\Sigma$. Thus, there is a homomorphism $\alpha\colon N_{\Gamma(\mathcal{P})}(\Sigma) \to \Gamma(\mathcal{P}/\Sigma)$, given by $\varphi\alpha = \hat{\varphi}$. Let $\tau \in \ker(\alpha)$. Then, for any $\Phi \in \mathcal{F}(\mathcal{P})$, we have $\Phi\pi = (\Phi\tau)\pi$. Hence, since μ is injective, we see that $\Phi\tau = \Phi\varphi$ for some $\varphi \in \Sigma$; thus $\tau = \varphi \in \Sigma$. It follows that $\ker(\alpha) = \Sigma$. It remains to be shown that α is surjective.

Let $\chi \in \Gamma(\mathcal{P}/\Sigma)$, and let $\Phi = \{F_{-1}, F_0, \ldots, F_n\}$ be the base flag of $\mathcal{P}$. Since $\mathcal{P}$ is regular (by assumption) and χ preserves flags of $\mathcal{P}/\Sigma$, we have $(\Phi\pi)\chi = (\Phi\varphi)\pi$ for some $\varphi \in \Gamma(\mathcal{P})$. We shall prove that this implies that $((\Phi\psi)\pi)\chi = (\Phi\psi\varphi)\pi$ for each $\psi \in \Gamma(\mathcal{P})$.

To see this, join the flags Φ and $\Phi\psi$ of $\mathcal{P}$ by a sequence

$$\Phi = \Phi_0, \Phi_1, \ldots, \Phi_{m-1}, \Phi_m = \Phi\psi,$$

in which any two consecutive flags are adjacent. Then

$$\Phi\varphi = \Phi_0\varphi, \Phi_1\varphi, \ldots, \Phi_{m-1}\varphi, \Phi_m\varphi = \Phi\psi\varphi$$

is a similar such sequence joining $\Phi\varphi$ and $\Phi\psi\varphi$, in which the flags $\Phi_{k-1}\varphi$ and $\Phi_k\varphi$ differ in a face of the same rank as do Φ_{k-1} and Φ_k, for each $k = 1, \ldots, m$. But the images of adjacent flags in $\mathcal{P}$ under the canonical mapping $\pi\colon \mathcal{F}(\mathcal{P}) \to \mathcal{F}(\mathcal{P}/\Sigma)$ cannot coincide, so that the sequences

$$\Phi\pi = \Phi_0\pi, \Phi_1\pi, \ldots, \Phi_{m-1}\pi, \Phi_m\pi = (\Phi\psi)\pi$$

and

$$(\Phi\varphi)\pi = (\Phi_0\varphi)\pi, (\Phi_1\varphi)\pi, \ldots, (\Phi_{m-1}\varphi)\pi, (\Phi_m\varphi)\pi = (\Phi\psi\varphi)\pi$$

have the same adjacency properties as their pre-images. Hence, since $(\Phi\pi)\chi = (\Phi\varphi)\pi$ and χ preserves the type of adjacency, we must have $(\Phi_k\pi)\chi = (\Phi_k\varphi)\pi$ for each $k = 0, \ldots, m$. For $k = m$, this proves that $((\Phi\psi)\pi)\chi = (\Phi\psi\varphi)\pi$, as required.

Now, for $\psi \in \Sigma$, this equation implies that

$$(\Phi\psi\varphi)\pi = ((\Phi\psi)\pi)\chi = (\Phi\pi)\chi = (\Phi\varphi)\pi.$$

But this in turn shows that $\Phi\psi\varphi = \Phi\varphi\tau$, and hence that $\psi\varphi = \varphi\tau$ for some $\tau \in \Sigma$. It follows that $\varphi \in N_{\Gamma(\mathcal{P})}(\Sigma)$. Further, since $((\Phi\psi)\pi)\chi = ((\Phi\psi)\varphi)\pi$ for each $\psi \in \Gamma(\mathcal{P})$, we have

$$(F \cdot \Sigma)\chi = (F\pi)\chi = (F\varphi)\pi = F\varphi \cdot \Sigma = (F \cdot \Sigma)\hat{\varphi}$$

for each F in $\mathcal{P}$; thus $\chi = \hat{\varphi}$. This proves that α is surjective, and hence that $\Gamma(\mathcal{P}/\Sigma) = N_{\Gamma(\mathcal{P})}(\Sigma)/\Sigma$.

Finally, consider the base flag $\Phi\pi$ of $\mathcal{P}/\Sigma$. Then any flag of $\mathcal{P}/\Sigma$ is of the form $(\Phi\varphi)\pi$, with $\varphi \in \Gamma(\mathcal{P})$. Hence, $\mathcal{P}/\Sigma$ is regular if and only if, for each $\varphi \in \Gamma(\mathcal{P})$, there exists $\tau \in N_{\Gamma(\mathcal{P})}(\Sigma)$ such that $(\Phi\varphi)\pi = (\Phi\pi)\hat{\tau} = (\Phi\tau)\pi$; thus, $(\Phi\varphi) \cdot \Sigma = (\Phi\tau) \cdot \Sigma$, or, equivalently, $\varphi \in \tau\Sigma \subset N_{\Gamma(\mathcal{P})}(\Sigma)$. It follows that $\mathcal{P}/\Sigma$ is regular if and only if $N_{\Gamma(\mathcal{P})}(\Sigma) = \Gamma(\mathcal{P})$, in which case $\Sigma_{\Gamma(\mathcal{P})}(\Sigma)/\Sigma = \Gamma(\mathcal{P})/\Sigma$. This completes the proof. □

The following Proposition 2D9 deals with the interesting special case in which all the maps γ are isomorphisms. See also Proposition 2E19 for an equivalent version of this proposition, and Proposition 2E22 for necessary and sufficient conditions under which it applies.

2D9 Proposition *Let $\mathcal{P}$ be an n-polytope, and let Σ be a subgroup of $\Gamma(\mathcal{P})$ such that each orbit of Σ meets each proper section of $\mathcal{P}$ in at most one face. Then, for* *all F, G and Ω as before, the map γ of (2D4) is an isomorphism (with $\Sigma_\Omega = \{\varepsilon\}$ if G/F is proper), and the sections $G\pi/F\pi$ of $\mathcal{P}/\Sigma$ and G/F of $\mathcal{P}$ are isomorphic. In particular, $\pi\colon \mathcal{P} \to \mathcal{P}/\Sigma$ is an $(n-1)$-covering of n-polytopes.*

Proof. Let G/F be a proper section of $\mathcal{P}$. By our assumptions, $\Sigma_\Omega = \{\varepsilon\}$, so that in (2D4) we may identify $H \cdot \Sigma_\Omega$ with H. Since $G/F \cap (H \cdot \Sigma) = \{H\}$ for all $H \in G/F$, the map γ is clearly injective. To prove surjectivity, let $H \cdot \Sigma \in G\pi/F\pi = (G \cdot \Sigma)/(F \cdot \Sigma)$. Here we may assume that $F \leqslant H$ and also that there exists $\varphi \in \Sigma$ such that $H \leqslant G\varphi$. But if $F \neq F_{-1}$, then G and $G\varphi$ are faces in the proper section F_n/F of $\mathcal{P}$ which are in the same orbit; hence we must have $G = G\varphi$ and $H \in G/F$. If $F = F_{-1}$, then $G \neq F_n$, and similar arguments work with the rôles of F and G interchanged. It follows that γ is surjective. □

To conclude this section, we discuss analogues of the intersection property for the groups of regular polytopes.

2D10 Proposition *Let $\mathcal{P}$ be a regular n-polytope with group $\Gamma(\mathcal{P}) = \langle \rho_0, \ldots, \rho_{n-1} \rangle$, and let Σ be a subgroup of $\Gamma(\mathcal{P})$ such that, for all F, G and Ω as before, the map γ of (2D4) is an isomorphism. Then, for each f, g with $-1 \leqslant f < g \leqslant n$ and each $\varphi \in \Gamma(\mathcal{P})$,*

$$\langle \rho_k \mid k > f \rangle \varphi\Sigma \cap \langle \rho_k \mid k < g \rangle \varphi\Sigma = \langle \rho_k \mid f < k < g \rangle \varphi\Sigma.$$

Proof. By assumption, $\mathcal{P}/\Sigma$ is a polytope, and the map μ of (2D6) is a bijection. Let $\Phi = \{F_{-1}, F_0, \ldots, F_n\}$ be the base flag of $\mathcal{P}$. Write $\hat{F}_i := F_i\varphi \cdot \Sigma$ for $i = -1, 0, \ldots, n$. Consider the orbits $(\Phi\gamma\varphi) \cdot \Sigma$, with

$$\gamma\varphi\Sigma \subseteq \langle \rho_{f+1}, \ldots, \rho_{n-1} \rangle \varphi\Sigma \cap \langle \rho_0, \ldots, \rho_{g-1} \rangle \varphi\Sigma.$$

Each of their images under μ contains $\{\hat{F}_{-1}, \hat{F}_0, \ldots, \hat{F}_f, \hat{F}_g, \hat{F}_{g+1}, \ldots, \hat{F}_n\}$, so that their restrictions to sections of $\mathcal{P}/\Sigma$ of type $\{f, \ldots, g\}$ form flags of $\hat{F}_g/\hat{F}_f$. But when

$$F = F_f\varphi,\ G = F_g\varphi,\ \Omega = \{F_{-1}, F_0\varphi, \ldots, F_f\varphi, F_g\varphi, F_{g+1}\varphi, \ldots, F_{n-1}\varphi, F_n\},$$

the map γ is an isomorphism, so that these flags are the images under μ of orbits of the form $(\Phi\gamma\varphi) \cdot \Sigma$, with $\gamma \in \langle \rho_{f+1}, \ldots, \rho_{g-1} \rangle$. But μ is bijective, so that the orbits must actually coincide. Thus $\gamma\varphi\Sigma \subseteq \langle \rho_{f+1}, \ldots, \rho_{g-1} \rangle \varphi\Sigma$. This proves one inclusion, and the opposite inclusion is trivial. □

Note that, if Σ is normal in $\Gamma(\mathcal{P})$, then Proposition 2D10 just gives the intersection property for $\Gamma(\mathcal{P}/\Sigma) = \Gamma(\mathcal{P})/\Sigma$, the group of the regular polytope $\mathcal{P}/\Sigma$. For a general subgroup Σ, the condition of Proposition 2D10 is really a condition for orbits of flags of $\mathcal{P}$ and thus cannot be expected to characterize polytopality of $\mathcal{P}/\Sigma$ completely. In contrast, the condition of our next proposition can be seen as a condition for flags of $\mathcal{P}/\Sigma$, and does indeed characterize polytopality.

2D11 Proposition *Let $\mathcal{P}$ be a regular n-polytope with group $\Gamma(\mathcal{P}) = \langle \rho_0, \ldots, \rho_{n-1} \rangle$, and let Σ be a subgroup of $\Gamma(\mathcal{P})$. Then $\mathcal{P}/\Sigma$ is an n-polytope if and only if the following two conditions hold.*

(a) *For each* f, g, k *with* $-1 \leqslant f < k < g \leqslant n$, *and each* $\varphi \in \Gamma(\mathcal{P})$,

$$\begin{aligned}\langle \rho_i \mid i \neq k\rangle\langle \rho_i \mid i > f\rangle\varphi\Sigma \cap \langle \rho_i \mid i \neq k\rangle\langle \rho_i \mid i < g\rangle\varphi\Sigma \\ = \langle \rho_i \mid i \neq k\rangle\langle \rho_i \mid f < i < g\rangle\varphi\Sigma.\end{aligned}$$

(b) *For each* $k = 0, \ldots, n-1$ *and each* $\varphi \in \Gamma(\mathcal{P})$,

$$\langle \rho_i \mid i \neq k\rangle\rho_k \cap \varphi^{-1}\Sigma\varphi = \emptyset.$$

Proof. We use Proposition 2D5. First, we prove that (a) of that proposition and (a) of the present one are equivalent. Let $\Phi = \{F_{-1}, F_0, \ldots, F_n\}$ be the base flag of $\mathcal{P}$. Then general chains of $\mathcal{P}$ are equivalent under $\Gamma(\mathcal{P})$ to subsets of Φ, so that for Proposition 2D5 we may assume that $F = F_f\varphi$, $G = F_g\varphi$ and $\Omega = \{F_i\varphi \mid i \neq f+1, \ldots, g-1\}$ for some $\varphi \in \Gamma(\mathcal{P})$; thus $\gamma\colon (F_g\varphi/F_f\varphi)/\Sigma_\Omega \to (F_g\varphi \cdot \Sigma)/(F_f\varphi \cdot \Sigma)$. Let $f < k < g$. The k-faces of $\mathcal{P}/\Sigma$ incident with $F_g\varphi \cdot \Sigma$ are just those of the form $F_k\alpha_1\varphi \cdot \Sigma$ with $\alpha_1 \in \langle \rho_0, \ldots, \rho_{g-1}\rangle$, while those incident with $F_f\varphi \cdot \Sigma$ are of the form $F_k\alpha_2\varphi \cdot \Sigma$ with $\alpha_2 \in \langle \rho_{f+1}, \ldots, \rho_{n-1}\rangle$. In particular, the k-faces in $F_g\varphi \cdot \Sigma/F_f\varphi \cdot \Sigma$ can be expressed in both ways, and if $F_k\alpha_1\varphi \cdot \Sigma = F_k\alpha_2\varphi \cdot \Sigma$, then $\alpha_1\varphi \in \langle \rho_i \mid i \neq k\rangle\alpha_2\varphi\Sigma$.

Hence, if the condition of (a) holds, then $\alpha_1\varphi \in \langle \rho_i \mid i \neq k\rangle\alpha_3\varphi\Sigma$ for some $\alpha_3 \in \langle \rho_i \mid f < i < g\rangle$; thus $F_k\alpha_1\varphi \cdot \Sigma = F_k\alpha_3\varphi \cdot \Sigma$; but $F_k\alpha_3\varphi \in F_g\varphi/F_f\varphi$, so that γ is surjective. Conversely, if γ is surjective, then a k-face $F_k\alpha_1\varphi \cdot \Sigma = F_k\alpha_2\varphi \cdot \Sigma$ in $F_g\varphi \cdot \Sigma/F_f\varphi \cdot \Sigma$ is of the form $F_k\alpha_3\varphi \cdot \Sigma$ with $\alpha_3 \in \langle \rho_l \mid f < l < g\rangle$. Hence, if $\beta_1, \beta_2 \in \langle \rho_i \mid i \neq k\rangle$, $\alpha_1 \in \langle \rho_0, \ldots, \rho_{g-1}\rangle$ and $\alpha_2 \in \langle \rho_{f+1}, \ldots, \rho_{n-1}\rangle$ are such that $\beta_1\alpha_1\varphi\Sigma = \beta_2\alpha_2\varphi\Sigma$, then $F_k\alpha_1\varphi \cdot \Sigma = F_k\alpha_2\varphi \cdot \Sigma = F_k\alpha_3\varphi \cdot \Sigma$ for some $\alpha_3 \in \langle \rho_i \mid f < i < g\rangle$; thus $\beta_1\alpha_1\varphi\Sigma \subseteq \langle \rho_i \mid i \neq k\rangle\langle \rho_i \mid f < i < g\rangle\varphi\Sigma$. This proves one inclusion of (a), and the other is trivial.

Next we need to show that parts (b) of the two propositions also correspond to each other. First note that Σ does not contain a conjugate of any ρ_i (this is a weaker requirement than (b)) if and only if orbits under Σ of adjacent flags of $\mathcal{P}$ are distinct; in fact, if $\varphi \in \Gamma(\mathcal{P})$, then $(\Phi\varphi)^i \cdot \Sigma = (\Phi\varphi)(\varphi^{-1}\rho_i\varphi) \cdot \Sigma$. Hence, if (b) of Proposition 2D11 holds, then the orbits of adjacent flags of $\mathcal{P}$ must be distinct, so that the maps β, γ of (b) in Proposition 2D5 must be injective. Conversely, suppose that (b) of Proposition 2D11 does not hold for some k and φ; that is, that $\rho_k\varphi^{-1} \in \langle \rho_i \mid i \neq k\rangle\varphi^{-1}\Sigma$. Then $F_k\rho_k\varphi^{-1} \cdot \Sigma = F_k\varphi^{-1} \cdot \Sigma$, so that the two faces $F_k\rho_k\varphi^{-1}$ and $F_k\varphi^{-1}$ of the section $F_{k+1}\varphi^{-1}/F_{k-1}\varphi^{-1}$ are identified under $\pi = \beta\gamma$. Hence β, γ cannot both be injective. This completes the proof. □

2E C-Groups

In Section 2B, we derived various properties of the groups of regular polytopes. In Theorem 2B14, we proved that the combinatorial structure of a regular n-polytope can be completely described in terms of the distinguished generators $\rho_0, \ldots, \rho_{n-1}$ of the group $\Gamma(\mathcal{P})$. In this section, we shall characterize the groups of regular polytopes as what are called string C-groups. In particular, given a string C-group Γ, we shall construct a regular polytope $\mathcal{P} = \mathcal{P}(\Gamma)$ whose group is Γ. We use this correspondence between regular polytopes and string C-groups to prove further results on the groups of regular polytopes. As one of the most important consequences of this approach, we may think of regular polytopes and corresponding string C-groups as being essentially the same objects.

Let Γ be a group generated by involutions $\rho_0, \ldots, \rho_{n-1}$; we shall usually assume that $n \geqslant 1$. During this section, these generators will be fixed; as in Section 2B, we then refer to them as the *distinguished generators* of Γ. Further, the subgroups $\langle \rho_j \mid j \in J \rangle$ with $J \subseteq N := \{0, \ldots, n-1\}$ are called the *distinguished subgroups* of Γ. Then Γ is called a *C-group* if Γ has the *intersection property with respect to its distinguished generators*; that is, for each $J,\ K \subseteq N$,

2E1 $$\langle \rho_j \mid j \in J \rangle \cap \langle \rho_j \mid j \in K \rangle = \langle \rho_j \mid j \in J \cap K \rangle.$$

Here, "C" stands for "Coxeter", though clearly not every C-group is a Coxeter group; see Chapter 3. It is immediate from the definition that the subgroups $\langle \rho_j \mid j \in J \rangle$ are themselves C-groups, with distinguished generators those ρ_j with $j \in J$. In this context, we also call the trivial group a C-group; here we set $\rho_j = \varepsilon$ for each integer j.

Note that the concept of a C-group involves a pair consisting of both a group and a system of distinguished generators. If we simply speak about a C-group Γ, then we implicitly assume that a system of generators has been specified. However, note that a group Γ can be a C-group with respect to quite different (non-conjugate) systems of generators; we shall see examples in Chapter 7.

A C-group Γ is called a *string C-group* if its generators satisfy the relations

2E2 $$(\rho_j \rho_k)^2 = \varepsilon \quad \text{if } j, k = 0, \ldots, n-1 \quad \text{and} \quad |j-k| \geqslant 2.$$

In this case, the underlying Coxeter diagram for Γ is a string diagram; see Section 3A. The consequences of the commutation rule (2E2) for the generators of Γ are similar to those for the generators of the group $\Gamma(\mathcal{P})$ in Section 2B. In particular,

$$\Gamma_k := \langle \rho_j \mid j \neq k \rangle \cong \langle \rho_0, \ldots, \rho_{k-1} \rangle \times \langle \rho_{k+1}, \ldots, \rho_{n-1} \rangle,$$

an internal direct product, for all $k = 1, \ldots, n-2$.

While there is still a huge variety of groups with property (2E2), the number of possibilities is cut down considerably by requiring the intersection property (2E1). By Propositions 2B10 and 2B11, if a group is the group of a regular polytope $\mathcal{P}$, then it is a string C-group. In this case, the intersection property depends essentially on the fact that the subgroups $\langle \rho_j \mid j \in J \rangle$ are stabilizers of chains of $\mathcal{P}$. However, for an arbitrary group Γ with property (2E2), we do not know of any general method to decide whether (2E1) holds or not. It seems that this property has to be verified separately for each group, or, at least, each class of groups. We shall see that, in some instances, the intersection property for Γ can be related to the corresponding property for another group, for example, a Coxeter group.

It follows from the definition that, in a non-trivial C-group Γ, the distinguished subgroups $\langle \rho_j \mid j \in J \rangle$ with $J \subset N$ are pairwise distinct. To see this, observe first that, by (2E1), we have

$$\rho_i \notin \langle \rho_j \mid j \in J \rangle \quad \text{if } i \notin J.$$

Now, if $J, K \subseteq \{0, \dots, n-1\}$ and $\langle \rho_j \mid j \in J \rangle = \langle \rho_j \mid j \in K \rangle$, then (2E1) implies that $\langle \rho_j \mid j \in J \rangle = \langle \rho_j \mid j \in J \cap K \rangle = \langle \rho_j \mid j \in K \rangle$; hence, it follows from what was said earlier that $J = J \cap K = K$, as required.

We now construct a regular n-polytope $\mathcal{P}$ from a string C-group $\Gamma = \langle \rho_0, \dots, \rho_{n-1} \rangle$, with $n \geqslant 1$. For $i = -1, 0, \dots, n$, we define

$$\Gamma_j := \langle \rho_i \mid i \neq j \rangle;$$

further, we set

$$\Gamma_{-1} = \Gamma_n = \Gamma.$$

Observe that the subgroups $\Gamma_0, \dots, \Gamma_{n-1}$ are mutually distinct, and distinct from Γ.

For $j \in N$, we take as the set of j-faces of $\mathcal{P}$ (that is, its faces of rank j) the set of all right cosets $\Gamma_j \varphi$ in Γ, with $\varphi \in \Gamma$. As improper faces of $\mathcal{P}$, we chose two copies of Γ: one denoted by Γ_{-1}, and the other by Γ_n; in this context, they are regarded as distinct. This formulation is motivated by the considerations following Proposition 2B14. Then, for the right cosets of Γ_{-1} and Γ_n, we have $\Gamma_{-1}\varphi = \Gamma_{-1}$ and $\Gamma_n \varphi = \Gamma_n$ for all $\varphi \in \Gamma$. On (the set of all proper and improper faces of) $\mathcal{P}$, we define $\Gamma_j \varphi \leqslant \Gamma_k \psi$ to mean

2E3 $\quad -1 \leqslant j \leqslant k \leqslant n \quad \text{and} \quad \varphi\psi^{-1} \in \langle \rho_{j+1}, \dots, \rho_{n-1} \rangle \langle \rho_0, \dots, \rho_{k-1} \rangle.$

Our first remark is trivial.

2E4 Lemma *The group Γ acts on $\mathcal{P}$ as a family of order preserving automorphisms.*

An equivalent condition to (2E3) is given by

2E5 Proposition *Let Γ be a string C-group, and let $\mathcal{P}$ be as before. Then $\Gamma_j \varphi \leqslant \Gamma_k \psi$ if and only if $-1 \leqslant j \leqslant k \leqslant n$ and $\Gamma_j \varphi \cap \Gamma_j \psi \neq \emptyset$.*

Proof. The equivalence of the two definitions can be checked as for the groups of regular polytopes; see the proof of Theorem 2B14. Indeed, that (2E3) implies the condition of the proposition was proved in the theorem; the opposite implication is even easier. □

The condition $\Gamma_j \varphi \cap \Gamma_k \psi \neq \emptyset$ of Proposition 2E5 (without assuming that $j \leqslant k$) defines a notion of incidence for a general C-group, and yields what are called *thin diagram geometries*; for details, see [65, pp. 1165, 1187]. We shall comment on this again a little later.

Occasionally, if the dependence on Γ and $\rho_0, \dots, \rho_{n-1}$ is to be emphasized, we write $\mathcal{P}(\Gamma)$ or $\mathcal{P}(\Gamma; \rho_0, \dots, \rho_{n-1})$ for $\mathcal{P}$. We also make appropriate definitions of $\mathcal{P}(\Gamma)$ for the case where Γ is trivial and $n = -1$ or 0. If $n = -1$ or 0, there is only one polytope of rank n. The appropriate definition is that $\mathcal{P}(\Gamma) = \mathcal{P}(\{\varepsilon\})$ be this polytope, except that for $n = -1$ we just take one copy of Γ rather than two.

We first show that $\mathcal{P}$ is a poset.

2E6 Lemma *The condition (2E3) induces a partial order on $\mathcal{P}$.*

Proof. For reflexivity and antisymmetry of $\leqslant$ we can appeal to Proposition 2E5. Certainly, a coset $\Gamma_j\varphi$ is incident with itself, which is reflexivity. If two distinct cosets $\Gamma_j\varphi$ and $\Gamma_k\psi$ are incident, then $j \neq k$, since distinct cosets of the same subgroup are disjoint. Then $\Gamma_j\varphi < \Gamma_k\psi$ or $\Gamma_j\varphi > \Gamma_k\psi$ as $j < k$ or $j > k$, which implies antisymmetry.

Finally, if $-1 \leqslant j \leqslant i \leqslant k \leqslant n$, we have

$$\textbf{2E7} \qquad \begin{cases} \langle \rho_{j+1}, \ldots, \rho_{n-1}\rangle\langle \rho_0, \ldots, \rho_{i-1}\rangle \cdot \langle \rho_{i+1}, \ldots, \rho_{n-1}\rangle\langle \rho_0, \ldots, \rho_{k-1}\rangle \\ \quad = \langle \rho_{j+1}, \ldots, \rho_{n-1}\rangle\langle \rho_0, \ldots, \rho_{k-1}\rangle. \end{cases}$$

Transitivity of $\leqslant$ is then an immediate consequence if we appeal to the original definition of $\leqslant$ in (2E3). It follows that $\leqslant$ is indeed a partial order. □

We remark that the definition of the partial order extends more generally to groups which have property (2E2) alone; in fact, we have not really used (2E1) so far.

Before we embark on the main theorem, whose proof we shall split into several lemmas, we extend the conventions of Section 2B to the poset $\mathcal{P}$. Clearly, $\Phi := \{\Gamma_0, \ldots, \Gamma_{n-1}\}$ is a flag of $\mathcal{P}$ (we shall frequently ignore the improper faces in this context), which we naturally call the *base* flag; its faces are also called the *base* faces of $\mathcal{P}$. Since Φ is a flag, so is its image $\Phi\varphi = \{\Gamma_0\varphi, \ldots, \Gamma_{n-1}\varphi\}$ for any $\varphi \in \Gamma$. The set of flags of $\mathcal{P}$ is denoted $\mathcal{F} = \mathcal{F}(\mathcal{P})$ ($= \mathcal{F}(\Gamma)$, if we wish to emphasize Γ instead). Last, let $K \subseteq N$. If $\Psi \in \mathcal{F}$ then Ψ_K denotes the chain of type, K contained in Ψ. We also write $\Gamma_K := \langle \rho_j \mid j \notin K\rangle$.

We next establish

2E8 Lemma *Γ is transitive on all chains of $\mathcal{P}$ of each given type $K \subseteq N$.*

Proof. Let $\{\Gamma_i\varphi_i \mid i \in K\}$ be a chain of type K. We proceed by induction. Suppose that, for some $k \in K$, we have already shown that there exists a $\psi \in \Gamma$ such that $\Gamma_i\varphi_i = \Gamma_i\psi$ for $i \geqslant k$. Let $j \in K$ be the next smaller number than k (assuming that there is one). Then $\Gamma_j\varphi_j \leqslant \Gamma_k\psi$ implies by (2E3) that $\varphi_j\psi^{-1} \in \langle \rho_{k+1}, \ldots, \rho_{n-1}\rangle\langle \rho_0, \ldots, \rho_{j-1}\rangle$, say $\varphi_j\psi^{-1} = \alpha\beta$, with $\alpha \in \langle \rho_{k+1}, \ldots, \rho_{n-1}\rangle$ and $\beta \in \langle \rho_0, \ldots, \rho_{j-1}\rangle$. It follows that $\alpha^{-1}\varphi_j = \beta\psi =: \chi$, say, and hence that

$$\begin{aligned} \Gamma_i\chi &= \Gamma_i\beta\psi = \Gamma_i\psi, \quad \text{for } i \geqslant k, \\ \Gamma_j\chi &= \Gamma_j\alpha^{-1}\varphi_j = \Gamma_j\varphi_j, \end{aligned}$$

giving the same property with j instead of k (and ψ replaced by χ). This is the inductive step, and the lemma follows. □

2E9 Lemma *If $K \subseteq N$, then the stabilizer of the chain Φ_K of type K in the base flag Φ is Γ_K.*

Proof. Let $\varphi \in \Gamma$. Then φ stabilizes Φ_K if and only if $\Gamma_j\varphi = \Gamma_j$; hence $\varphi \in \Gamma_j$, for each $j \in K$. Equivalently,

$$\varphi \in \bigcap\{\Gamma_j \mid j \in K\} = \bigcap_{j \in K}\langle \rho_i \mid i \neq j\rangle = \Gamma_K,$$

by the intersection property (2E1). This is the lemma. □

2E10 Corollary *Γ is simply transitive on $\mathcal{F}$. Hence the action of Γ on $\mathcal{P}$ is faithful.*

Proof. Indeed, by Lemma 2E9, the stabilizer of the base flag $\Phi = \Phi_N$ is $\Gamma_N = \{\varepsilon\}$. The assertions of the corollary follow at once. □

We can now state the main characterization result.

2E11 Theorem *Let $n \geqslant 1$, and let $\Gamma = \langle \rho_0, \ldots, \rho_{n-1} \rangle$ be a string C-group and $\mathcal{P} := \mathcal{P}(\Gamma)$ the corresponding poset. Then $\mathcal{P}$ is a regular n-polytope such that $\Gamma(\mathcal{P}) = \Gamma$.*

Proof. For $\mathcal{P}$ we need to check the defining properties (P1), ..., (P4) of polytopes; see Section 2A. The property (P1) is trivially satisfied with Γ_{-1} and Γ_n as the least and greatest face, respectively. In fact, by (2E3), $\Gamma_{-1} \leqslant \Gamma_j \varphi \leqslant \Gamma_n$ for all φ and all j.

Next, we deduce at once from Lemma 2E8 that every chain Ω in $\mathcal{P}$ of type K can be expressed in the form $\Omega = \Phi_K \varphi$, for some $\varphi \in \Gamma$. In particular, Ω is contained in the flag $\Phi\varphi$, which gives (P2).

We then prove (P4). If, in Lemma 2E8, we take $K = N \setminus \{j\}$ for any $j \in N$, we see that the stabilizer of $\Phi_{N\setminus\{j\}} = \{\Gamma_0, \ldots, \Gamma_{j-1}, \Gamma_{j+1}, \ldots, \Gamma_{n-1}\}$ is $\Gamma_{N\setminus\{j\}} = \langle \rho_j \rangle = \{\varepsilon, \rho_j\}$. Hence there is exactly one flag apart from Φ which contains $\Phi_{N\setminus\{j\}}$, namely,

$$\Phi^j = \{\Gamma_0, \ldots, \Gamma_{j-1}, \Gamma_j \rho_j, \Gamma_{j+1}, \ldots, \Gamma_{n-1}\} = \Phi \rho_j.$$

The transitivity of Γ on $\mathcal{F}$ (Corollary 2E10 then implies that a general flag $\Psi = \Phi\varphi$ (for some suitable $\varphi \in \Gamma$) possesses a unique j-adjacent flag $\Psi^j = \Phi^j \varphi = (\Phi\varphi)^j$, as required.

Finally, we demonstrate (P3), in the alternative form (P3′) of strong flag-connectedness. Corollary 2E10 shows that Γ is (simply) transitive on $\mathcal{F}$; hence, to prove (P3′), it suffices to consider the special case where one flag is the base flag Φ. If $\Psi \in \mathcal{F}$ is another flag, let $K \subseteq N$ be such that $\Phi \cap \Psi = \Phi_K$. Since $\Phi_K \subseteq \Psi$, Lemma 2E9 and Corollary 2E10 show that $\Psi = \Phi\varphi$ for some (unique) $\varphi \in \Gamma_K$. This says that

$$\Psi = \Phi \rho_{k(1)} \cdots \rho_{k(m)} = \Phi^{k(m)\cdots k(1)},$$

for some $k(1), \ldots, k(m) \in N \setminus K$, giving an adjacency sequence of flags which each contain Φ_K; here, we have used the previous part of the proof. This establishes the theorem. □

Let us collect together various other properties of the regular polytope $\mathcal{P}(\Gamma)$ which we have proved (implicitly or explicitly) earlier.

2E12 Proposition *Let $\mathcal{P}$ be the regular n-polytope associated with the string C-group $\Gamma = \langle \rho_0, \ldots, \rho_{n-1} \rangle$.*

(a) *Let $-1 \leqslant j < k \leqslant n$, and let F be a j-face and G a k-face of $\mathcal{P}$ with $F \leqslant G$. Then the section G/F of $\mathcal{P}$ is isomorphic to $\mathcal{P}(\langle \rho_{j+1}, \ldots, \rho_{k-1} \rangle)$. In particular, the facets and vertex-figures of $\mathcal{P}$ are isomorphic to the regular $(n-1)$-polytopes $\mathcal{P}(\langle \rho_0, \ldots, \rho_{n-2} \rangle)$ and $\mathcal{P}(\langle \rho_1, \ldots, \rho_{n-1} \rangle)$, respectively.*

(b) $\mathcal{P}$ *is of type* $\{p_1, \ldots, p_{n-1}\}$, *where* p_j *is the period of* $\rho_{j-1}\rho_j$ *in* Γ *for* $j = 1, \ldots, n-1$.
(c) $\mathcal{P}$ *is finite if and only if* Γ *is finite.*
(d) $\mathcal{P}$ *is self-dual if and only if there exists an involutory group automorphism* ω *of* Γ *such that* $\rho_j\omega = \rho_{n-1-j}$ *for each* $j = 0, \ldots, n-1$.

Proof. To prove part (a), note that the case $j = k-1$ is trivial by definition. Hence, we can assume that $-1 \leqslant j < k-1 \leqslant n-1$. The transitivity of $\mathcal{P}$ on chains of a given type (Lemma 2E8) implies that it suffices to prove the result for the section $\mathcal{P}(j,k) := \Gamma_k/\Gamma_j$. Let $K := \{0, \ldots, j, k, \ldots, n-1\} \subset N$. There is a one-to-one correspondence between chains of $\mathcal{P}(j,k)$ and chains of $\mathcal{P}$ which contain Φ_K. In particular, using Lemma 2E8 again, we deduce that each face $\Gamma_i\varphi \in \mathcal{P}(j,k)$ (with $j \leqslant i \leqslant k$) admits a representation with φ in the stabilizer of Φ_K, namely, $\Gamma_K = \langle \rho_{j+1}, \ldots, \rho_{k-1} \rangle$. Hence, the group of $\mathcal{P}(j,k)$ is isomorphic to Γ_K, as claimed. Observe that there is an induced bijection $\Gamma_i\varphi \leftrightarrow (\Gamma_K)_i\varphi$ for $j \leqslant i \leqslant k$ and $\varphi \in \Gamma_K$.

Next, part (b) is a special case of part (a), because $\langle \rho_{j-1}, \rho_j \rangle \cong D_{p_j}$ for $j = 1, \ldots, n-1$, and part (c) is trivial.

Finally we prove (d). By Proposition 2B17, if $\mathcal{P}$ is self-dual, then we can take for ω the restriction to Γ of the inner automorphism of $\bar{\Gamma}$ which is induced by the polarity which fixes Φ. Conversely, let ω be as in (e). Then, for $-1 \leqslant j \leqslant k \leqslant n$,

$$(\langle \rho_{j+1}, \ldots, \rho_{n-1} \rangle \langle \rho_0, \ldots, \rho_{k-1} \rangle)\omega = \langle \rho_0, \ldots, \rho_{n-j-2} \rangle \langle \rho_{n-k}, \ldots, \rho_{n-1} \rangle.$$

By (2E3), this implies that the mapping $\delta\colon \mathcal{P} \to \mathcal{P}$ given, for $-1 \leqslant i \leqslant n$ and $\varphi \in \Gamma$, by

$$(\Gamma_i\varphi)\delta := \Gamma_{n-1-i}(\varphi\omega)$$

is a (well-defined) duality of $\mathcal{P}$ onto itself. Hence, $\mathcal{P}$ is self-dual. The mapping δ is indeed the unique polarity which fixes Φ. This completes the proof. □

We note here two immediate consequences of the earlier results and the underlying construction of $\mathcal{P}(\Gamma)$. First,

2E13 Corollary *The string C-groups are precisely the groups of regular polytopes.*

The one-to-one correspondence between string C-groups and regular polytopes will be crucial for all further considerations. Second, we have

2E14 Corollary *Let* $n \geqslant -1$, *let* $\mathcal{P}$ *be a regular n-polytope, and let* $\rho_0, \ldots, \rho_{n-1}$ *be the distinguished generators of* $\Gamma(\mathcal{P})$ *associated with the base flag* $\Phi = \{F_{-1}, F_0, \ldots, F_n\}$ *of* $\mathcal{P}$. *Then the regular polytopes* $\mathcal{P}$ *and* $\mathcal{P}(\Gamma(\mathcal{P}))$ *(or more exactly,* $\mathcal{P}(\Gamma(\mathcal{P}); \rho_0, \ldots, \rho_{n-1})$*) are isomorphic. In particular, the mapping* $\pi\colon \mathcal{P} \to \mathcal{P}(\Gamma(\mathcal{P}))$ *given by*

$$(F_j\varphi)\pi := \langle \rho_i \mid i \neq j \rangle \varphi,$$

for $-1 \leqslant j \leqslant n$ *and* $\varphi \in \Gamma(\mathcal{P})$, *is an isomorphism.*

In constructing self-dual regular polytopes, Proposition 2E12(d) will usually be applied in the following form.

2E15 Proposition *Let $n \geqslant 1$, and let $\Gamma = \langle \rho_0, \dots, \rho_{n-1} \rangle$ be a string C-group. Assume that Γ has a presentation in terms of the generators $\rho_0, \dots, \rho_{n-1}$ which is symmetric, in the sense that it is invariant under the exchange*

$$\rho_j \leftrightarrow \rho_{n-1-j}$$

for $i = 0, \dots, n-1$. Then $\mathcal{P} = \mathcal{P}(\Gamma)$ is self-dual.

Proof. The exchange induces an involutory group automorphism ω of Γ. Then Proposition 2E12(d) applies. □

In actual applications, the proof of the intersection property (2E1) for string C-groups can often be reduced to the consideration of only a few cases. Here, the following proposition is quite helpful.

2E16 Proposition *Let $\Gamma = \langle \rho_0, \dots, \rho_{n-1} \rangle$ be an sggi (that is, Γ is a group satisfying (2E2)), and suppose that its subgroup $\Gamma_{n-1} := \langle \rho_0, \dots, \rho_{n-2} \rangle$ is a string C-group (with respect to its generators).*

(a) If $\Gamma_0 := \langle \rho_1, \dots, \rho_{n-1} \rangle$ is also a string C-group, and $\Gamma_{n-1} \cap \Gamma_0 = \langle \rho_1, \dots, \rho_{n-2} \rangle$, then Γ itself is a string C-group.
(b) If $\Gamma_{n-1} \cap \langle \rho_k, \dots, \rho_{n-1} \rangle = \langle \rho_k, \dots, \rho_{n-2} \rangle$ for each $k = 1, \dots, n-1$, then Γ is also a string C-group.

Proof. We first prove part (b). As usual, let $\Gamma_j := \langle \rho_i \mid i \neq j \rangle$ for $j = 0, \dots, n-1$. Define $\Delta := \langle \rho_0, \dots, \rho_{n-2} \rangle = \Gamma_{n-1}$, and let $\Delta_j := \langle \rho_i \mid 0 \leqslant i \leqslant n-2$ and $i \neq j \rangle$ for $j = 0, \dots, n-2$. First note that, for $j = 0, \dots, n-2$,

$$\begin{aligned} \Gamma_i \cap \Delta &= \Delta \cap \langle \rho_{j+1}, \dots, \rho_{n-1} \rangle \langle \rho_0, \dots, \rho_{j-1} \rangle \\ &= (\Delta \cap \langle \rho_{j+1}, \dots, \rho_{n-1} \rangle) \langle \rho_0, \dots, \rho_{j-1} \rangle \\ &= \langle \rho_{j+1}, \dots, \rho_{n-2} \rangle \langle \rho_0, \dots, \rho_{j-1} \rangle \\ &= \Delta_j. \end{aligned}$$

To prove (2E1), it suffices to check that, for each $K \subseteq N := \{0, \dots, n-1\}$,

$$\Gamma_K := \langle \rho_j \mid j \notin K \rangle = \bigcap \{\Gamma_j \mid j \in K\}.$$

Now $\Delta = \Gamma_{n-1}$ is a C-group, so that, if $n-1 \in K$, then

$$\begin{aligned} \bigcap \{\Gamma_j \mid j \in K\} &= \bigcap \{\Gamma_j \cap \Delta \mid j \in K \setminus \{n-1\}\} \\ &= \bigcap \{\Delta_j \mid j \in K \setminus \{n-1\}\} \\ &= \Delta_{K \setminus \{n-1\}} \\ &= \Gamma_K, \end{aligned}$$

as required. On the other hand, if $n-1 \notin K$, let $k \in K$ be its largest member. Then

$$\begin{aligned}\bigcap\{\Gamma_j \mid j \in K\} &= \bigcap\{\Gamma_j \mid j \in K \setminus \{k\}\} \cap \langle \rho_{k+1}, \ldots, \rho_{n-1}\rangle\langle \rho_0, \ldots, \rho_{k-1}\rangle \\ &= \langle \rho_{k+1}, \ldots, \rho_{n-1}\rangle\Big(\bigcap\{\Gamma_j \mid j \in K \setminus \{k\}\} \cap \langle \rho_0, \ldots, \rho_{k-1}\rangle \cap \Gamma_{n-1}\Big) \\ &= \langle \rho_{k+1}, \ldots, \rho_{n-1}\rangle\Big(\bigcap\{\Gamma_j \mid j \in K \setminus \{k\}\} \cap \Gamma_k \cap \Gamma_{n-1}\Big) \\ &= \langle \rho_{k+1}, \ldots, \rho_{n-1}\rangle\langle \rho_j \mid j \notin K \cup \{n-1\}\rangle \\ &= \Gamma_K,\end{aligned}$$

where we have used the previous part of the proof. This gives part (b).

For part (a), note that, for $j = 1, \ldots, n-1$,

$$\begin{aligned}\Gamma_{n-1} \cap \langle \rho_j, \ldots, \rho_{n-1}\rangle &= (\Gamma_{n-1} \cap \Gamma_0) \cap \langle \rho_j, \ldots, \rho_{n-1}\rangle \\ &= \langle \rho_1, \ldots, \rho_{n-2}\rangle \cap \langle \rho_j, \ldots, \rho_{n-1}\rangle \\ &= \langle \rho_j, \ldots, \rho_{n-2}\rangle,\end{aligned}$$

by property (2E1) for Γ_0. Now we can appeal to part (b). □

In many cases we shall encounter, quotient relations play a vital role. In verifying that a given group is a string C-group, the following result is frequently useful. We shall refer to its condition as the *quotient criterion* (previously, we have often used the term "quotient lemma").

2E17 Theorem *Let $\Gamma = \langle \rho_0, \ldots, \rho_{n-1}\rangle$ be an sggi (a group satisfying (2E2)), and let $\Delta = \langle \sigma_0, \ldots, \sigma_{n-1}\rangle$ be a string C-group (with respect to the distinguished generators σ_i). If the mapping $\rho_j \to \sigma_j$ for $j = 0, \ldots, n-1$ induces a homomorphism $\pi : \Gamma \to \Delta$, which is one-to-one on $\Gamma_{n-1} := \langle \rho_0, \ldots, \rho_{n-2}\rangle$ or on $\Gamma_0 := \langle \rho_1, \ldots, \rho_{n-1}\rangle$, then Γ is also a string C-group, and π induces a covering $\mathcal{P}(\Gamma) \searrow \mathcal{P}(\Delta)$ of the corresponding polytopes.*

Proof. Assume that π is one-to-one on Γ_{n-1}. By Proposition 2E16, it suffices to check that

$$\Gamma_{n-1} \cap \langle \rho_j, \ldots, \rho_{n-1}\rangle = \langle \rho_j, \ldots, \rho_{n-2}\rangle$$

for each $j = 1, \ldots, n-1$. Let $\varphi \in \Gamma_{n-1} \cap \langle \rho_j, \ldots, \rho_{n-1}\rangle$. Then,

$$\varphi\pi \in \langle \sigma_0, \ldots, \sigma_{n-2}\rangle \cap \langle \sigma_j, \ldots, \sigma_{n-1}\rangle = \langle \sigma_j, \ldots, \sigma_{n-2}\rangle,$$

by our assumption that Δ is a C-group. Hence $\varphi\pi$ has a pre-image in $\langle \rho_j, \ldots, \rho_{n-2}\rangle$. But π is one-to-one on Γ_{n-1}, so that φ is the only pre-image of $\varphi\pi$ in Γ_{n-1}. It follows that $\varphi \in \langle \rho_j, \ldots, \rho_{n-2}\rangle$. This proves one inclusion, and the other is trivial. Therefore Γ is a string C-group. Finally, it is immediate that

$$\langle \rho_j \mid j \neq k\rangle\varphi \mapsto \langle \sigma_j \mid j \neq k\rangle(\varphi\pi),$$

for $k = -1, 0, \ldots, n$ and $\varphi \in \Gamma$, gives a covering $\mathcal{P}(\Gamma) \searrow \mathcal{P}(\Delta)$. □

Our next two results relate quotients of string C-groups to quotients of polytopes. In particular, in Proposition 2E18 it is explained why, in our discussion in Section 2D, it was usually assumed that the maps γ of (2D4) are isomorphisms. Finally, Proposition 2E19 is equivalent to Proposition 2D9.

2E18 Proposition *Let $\mathcal{P}$ be a regular n-polytope with group $\Gamma(\mathcal{P}) = \langle \rho_0, \ldots, \rho_{n-1} \rangle$, and let Σ be a normal subgroup of $\Gamma(\mathcal{P})$. If $\Gamma(\mathcal{P})/\Sigma$ is a string C-group (with distinguished generators $\Sigma\rho_0, \ldots, \Sigma\rho_{n-1}$), then the two polytopes $\mathcal{P}/\Sigma$ and $\mathcal{P}(\Gamma(\mathcal{P})/\Sigma)$ are isomorphic, with group $\Gamma(\mathcal{P}/\Sigma) = \Gamma(\mathcal{P})/\Sigma$, and all maps γ of (2D4) are isomorphisms.*

Proof. Let $\Phi = \{F_{-1}, F_0, \ldots, F_n\}$ be the base flag of $\mathcal{P}$. Write $\Delta := \Gamma(\mathcal{P})/\Sigma$, and let $\pi\colon \Gamma(\mathcal{P}) \to \Delta$ be the canonical projection. For $k = -1, 0, \ldots, n$, define $\Gamma_k := \langle \rho_j \mid j \neq k \rangle$, and $\Delta_k := \langle \Sigma\rho_j \mid j \neq k \rangle$. Note that $\Delta_k\pi^{-1} = \Gamma_k\Sigma$ for each k. By assumption, Δ is a string C-group, so that $\mathcal{P}(\Delta)$ is defined, with base flag $\{\Delta_{-1}, \Delta_0, \ldots, \Delta_n\}$. Consider the mapping $\kappa\colon \mathcal{P}/\Sigma \to \mathcal{P}(\Delta)$ given by

$$(F_j\varphi)\kappa := \Delta_j(\varphi\pi)$$

for $j = -1, 0, \ldots, n$ and $\varphi \in \Gamma(\mathcal{P})$. For each j and $\varphi, \psi \in \Gamma(\mathcal{P})$, we have $F_j\varphi \cdot \Sigma = F_j\psi \cdot \Sigma$ in $\mathcal{P}/\Sigma$ if and only if $\Gamma_j\Sigma\varphi = \Gamma_j\Sigma\psi$; that is, if and only if $\Delta_j(\varphi\pi) = \Delta_j(\psi\pi)$ in $\mathcal{P}(\Delta)$. It follows that κ is a bijection. But κ also preserves incidence in both directions. In fact, in $\mathcal{P}/\Sigma$, a pair of incident faces is of the form $F_j\varphi \cdot \Sigma$ and $F_k\tau\varphi \cdot \Sigma$ with $j \leqslant k$, $\varphi \in \Gamma(\mathcal{P})$ and $\tau \in \langle \rho_{k+1}, \ldots, \rho_{n-1} \rangle$. Under κ, they are mapped onto the faces $\Delta_j(\varphi\pi)$ and $\Delta_k(\tau\pi)(\varphi\pi)$ of $\mathcal{P}(\Delta)$, which are again incident because $\tau\pi \in \langle \Sigma\rho_{j+1}, \ldots, \Sigma\rho_{n-1} \rangle$. Conversely, if $\Delta_j(\varphi\pi)$ and $\Delta_k(\tau\pi)(\varphi\pi)$ are incident in $\mathcal{P}(\Delta)$, with $\tau\pi \in \langle \Sigma\rho_{j+1}, \ldots, \Sigma\rho_{n-1} \rangle$, then $\tau \in \langle \rho_{k+1}, \ldots, \rho_{n-1} \rangle \Sigma$; thus $F_j\varphi \cdot \Sigma$ and $F_k\tau\varphi \cdot \Sigma$ are incident in $\mathcal{P}/\Sigma$.

It follows that $\mathcal{P}/\Sigma$ and $\mathcal{P}(\Delta)$ are isomorphic polytopes and thus have the same group Δ. By Lemma 2D7, all maps γ are isomorphisms if the mapping $\mu\colon \mathcal{F}(\mathcal{P})/\Sigma \to \mathcal{F}(\mathcal{P}/\Sigma)$ of (2D6) is a bijection. To prove this, consider the chain of maps

$$\begin{array}{ccccccc} \mathcal{F}(\mathcal{P}(\Delta)) & \stackrel{\lambda}{\longrightarrow} & \Delta & \stackrel{\nu}{\longrightarrow} & \mathcal{F}(\mathcal{P})/\Sigma & \stackrel{\mu}{\longrightarrow} & \mathcal{F}(\mathcal{P}/\Sigma) \\ \{\Delta_j(\varphi\pi)\}_j & \longrightarrow & \varphi\pi & \longrightarrow & \Phi\varphi \cdot \Sigma & \longrightarrow & \{F_j\varphi \cdot \Sigma\}_j \quad (= (\Phi\varphi)\pi). \end{array}$$

Here, λ is a bijection, because $\mathcal{P}(\Delta)$ is a regular polytope with (the simply flag-transitive) group Δ, and ν is a bijection, because $\Phi\varphi \cdot \Sigma = \Phi\psi \cdot \Sigma$ in $\mathcal{F}(\mathcal{P})/\Sigma$ if and only if $\varphi\pi = \Sigma\varphi = \Sigma\psi = \psi\pi$. But $\lambda\nu\mu$ is also a bijection, because it is the map induced by the isomorphism $\kappa^{-1}\colon \mathcal{P}(\Delta) \to \mathcal{P}/\Sigma$. Hence μ must be a bijection, and the proof is complete. □

2E19 Proposition *Let $\mathcal{P}$ be a regular n-polytope with group $\Gamma(\mathcal{P}) = \langle \rho_0, \ldots, \rho_{n-1} \rangle$, and let Σ be a normal subgroup of $\Gamma(\mathcal{P})$ such that*

$$\Sigma \cap \langle \rho_1, \ldots, \rho_{n-1} \rangle \langle \rho_0, \ldots, \rho_{n-2} \rangle = \{\varepsilon\}.$$

(a) $\Gamma(\mathcal{P})/\Sigma$ *is a string C-group, and* $\mathcal{P}/\Sigma$ *and* $\mathcal{P}(\Gamma(\mathcal{P})/\Sigma)$ *are isomorphic regular polytopes with group* $\Gamma(\mathcal{P}/\Sigma) = \Gamma(\mathcal{P})/\Sigma$.

(b) The facets and vertex-figures of $\mathcal{P}/\Sigma$ *are isomorphic to the facets and vertex-figures of* $\mathcal{P}$*, respectively, so that* $\mathcal{P}$ *is an* $(n-1)$*-covering of* $\mathcal{P}/\Sigma$.

Proof. We use the same notation as that in the proof of Proposition 2E18. First, note that $\Sigma \cap \Gamma_{n-1} = \{\varepsilon\} = \Sigma \cap \Gamma_0$, so that the restrictions of π to the subgroups Γ_{n-1} and Γ_0 are isomorphisms. In particular, $\Gamma_{n-1}\pi = \langle \Sigma\rho_0, \ldots, \Sigma\rho_{n-2}\rangle$ and $\Gamma_0\pi = \langle \Sigma\rho_1, \ldots, \Sigma\rho_{n-1}\rangle$ are string C-groups.

To prove that Δ is a string C-group, observe that (2E2) holds trivially. By Proposition 2E16, for (2E1) it suffices to check that

$$\Gamma_{n-1}\pi \cap \Gamma_0\pi = \langle \rho_1, \ldots, \rho_{n-2}\rangle\pi.$$

Let $\varphi \in \Gamma_{n-1}$ and $\psi \in \Gamma_0$, and suppose that $\varphi\pi = \psi\pi$. Then $\psi\varphi^{-1} \in \ker(\pi) \cap \Gamma_0\Gamma_{n-1} = \Sigma \cap \Gamma_0\Gamma_{n-1} = \{\varepsilon\}$; thus $\varphi = \psi \in \Gamma_{n-1} \cap \Gamma_0 = \langle \rho_1, \ldots, \rho_{n-2}\rangle$. But this proves (2E1). Now we can apply Proposition 2E18 to complete the proof of part (a). Part (b) follows from $\Delta_{n-1} \cong \Gamma_{n-1}$ and $\Delta_0 \cong \Gamma_0$. □

We conclude this section by discussing some results on quotient polytopes of regular polytopes, which preserve the facets and vertex-figures but are not necessarily themselves regular. These are defined by an interesting class of subgroups Σ called sparse. The name "sparse" for such groups was introduced in Hartley [208, Chapter 5; 209, 210], but the groups themselves occur earlier, for example, in [145, 245, 301]. The definition builds on a generalization of the hypothesis of Proposition 2E19.

Let $\mathcal{P}$ be a regular n-polytope with group $\Gamma(\mathcal{P}) = \langle \rho_0, \ldots, \rho_{n-1}\rangle$. A subgroup Σ of $\Gamma(\mathcal{P})$ is called *sparse* if

2E20 $$\varphi^{-1}\Sigma\varphi \cap \langle \rho_1, \ldots, \rho_{n-1}\rangle\langle \rho_0, \ldots, \rho_{n-2}\rangle = \{\varepsilon\}$$

for each $\varphi \in \Gamma(\mathcal{P})$. Equivalently, Σ is sparse if and only if

2E21 $$\Sigma \cap \langle \varphi^{-1}\rho_1\varphi, \ldots, \varphi^{-1}\rho_{n-1}\varphi\rangle\langle \varphi^{-1}\rho_0\varphi, \ldots, \varphi^{-1}\rho_{n-2}\varphi\rangle = \{\varepsilon\}$$

for each $\varphi \in \Gamma(\mathcal{P})$. Note that $\varphi^{-1}\rho_0\varphi, \ldots, \varphi^{-1}\rho_{n-1}\varphi$ are the distinguished generators of $\Gamma(\mathcal{P})$ associated with the flag $\Phi\varphi$, where Φ is the base flag of $\mathcal{P}$.

We begin with the observation that the sparse subgroups are precisely the groups Σ which satisfy the hypothesis of Proposition 2D9.

2E22 Lemma *Let* $\mathcal{P}$ *be a regular n-polytope with group* $\Gamma(\mathcal{P})$*, and let* Σ *be a subgroup of* $\Gamma(\mathcal{P})$*. Then* Σ *is sparse if and only if each orbit of* Σ *meets each proper section of* $\mathcal{P}$ *in at most one face.*

Proof. First, assume that the condition for the orbits holds. Let $\varphi \in \Gamma(\mathcal{P})$, and let $\psi \in \Sigma$ be such that $\varphi\psi\varphi^{-1} \in \langle \rho_1, \ldots, \rho_{n-1}\rangle\langle \rho_0, \ldots, \rho_{n-2}\rangle$. Then, by Proposition 2B14, $F_0\varphi\psi \leqslant F_{n-1}\varphi$ in $\mathcal{P}$. Since the vertex $F_0\varphi$ is the only element in its orbit which is contained in the facet $F_{n-1}\varphi/F_{-1}$ of $\mathcal{P}$, we must have $F_0\varphi\psi = F_0\varphi$. It follows that ψ maps the whole vertex-figure $F_n/F_0\varphi$ of $\mathcal{P}$ onto itself. Since each orbit meets this vertex-figure in at most one face, ψ must fix each face of $F_n/F_0\varphi$. It follows that $\psi = \varepsilon$, as required.

It is sufficient to prove the converse for facets and vertex-figures, because every proper section of $\mathcal{P}$ is a section of a facet or a vertex-figure. By duality we may consider only a facet, $F_{n-1}\varphi/F_{-1}$ (say) with $\varphi \in \Gamma(\mathcal{P})$. A generic j-face of this facet has the form $F_j\alpha\varphi$ with $\alpha \in \langle \rho_0, \ldots, \rho_{n-2}\rangle$. Now, if an orbit of Σ meets $F_{n-1}\varphi/F_{-1}$ in two faces of rank j (say), then there are $\alpha, \beta \in \langle \rho_0, \ldots, \rho_{n-2}\rangle$ and $\tau \in \Sigma$ such that $F_j\alpha\varphi = F_j\beta\varphi\tau$. Replacing φ by $\beta\varphi$ (and α by $\alpha\beta^{-1}$), we may assume that $\beta = \varepsilon$. Then $F_j\alpha\varphi = F_j\varphi\tau$, and therefore

$$\varphi\tau\varphi^{-1}\alpha^{-1} \in \langle \rho_i \mid i \neq j\rangle = \langle \rho_i \mid i > j\rangle\langle \rho_i \mid i < j\rangle \subseteq \langle \rho_1, \ldots, \rho_{n-1}\rangle\langle \rho_0, \ldots, \rho_{n-2}\rangle.$$

But $\alpha \in \langle \rho_0, \ldots, \rho_{n-2}\rangle$, and so we also have $\varphi\tau\varphi^{-1} \in \langle \rho_1, \ldots, \rho_{n-1}\rangle\langle \rho_0, \ldots, \rho_{n-2}\rangle$. Since Σ is sparse, this implies that $\varphi\tau\varphi^{-1} = \varepsilon$, and hence that $\tau = \varepsilon$. It follows that the two faces in the same orbit must actually coincide, which completes the proof. □

The following result is now a direct consequence of Lemma 2E22 and Propositions 2D9 and 2D8. It generalizes Proposition 2E19.

2E23 Proposition *Let $\mathcal{P}$ be a regular n-polytope, and let Σ be a sparse subgroup of $\Gamma(\mathcal{P})$.*

(a) $\mathcal{P}/\Sigma$ is an n-polytope with group $\Gamma(\mathcal{P}/\Sigma) \cong N_{\Gamma(\mathcal{P})}(\Sigma)/\Sigma$, where $N_{\Gamma(\mathcal{P})}(\Sigma)$ is the normalizer of Σ in $\Gamma(\mathcal{P})$. In particular, $\mathcal{P}/\Sigma$ is regular if and only if Σ is normal in $\Gamma(\mathcal{P})$.

(b) The facets and vertex-figures of $\mathcal{P}/\Sigma$ are isomorphic to the facets and vertex-figures of $\mathcal{P}$, respectively.

Our next proposition determines when two quotients by sparse subgroups are isomorphic. In particular, this proves again the statement of Proposition 2D8 about the group of a quotient polytope. See also Kato [245, Theorem 4], Davis [145, Corollary 2.15], and Hartley [208, Theorem 5.2.46; 209, Theorem 5.3] for different proofs of Proposition 2E24 (for the universal polytopes $\mathcal{P} = \{p_1, \ldots, p_{n-1}\}$).

2E24 Proposition *Let $\mathcal{P}$ be a regular n-polytope, and let Σ and Ξ be sparse subgroups of $\Gamma(\mathcal{P})$. Then the quotients $\mathcal{P}/\Sigma$ and $\mathcal{P}/\Xi$ are isomorphic if and only if Σ and Ξ are conjugate subgroups of $\Gamma(\mathcal{P})$.*

Proof. The idea is to generalize Proposition 2D8 and its proof. One direction is straightforward. If Σ and Ξ are conjugate and $\Xi = \varphi^{-1}\Sigma\varphi$ with $\varphi \in \Gamma(\mathcal{P})$, then the mapping $\hat{\varphi}: \mathcal{P}/\Sigma \to \mathcal{P}/\Xi$, defined for $F \in \mathcal{P}$ by $(F \cdot \Sigma)\hat{\varphi} := F\varphi \cdot \Xi$, is an isomorphism between the two quotients. In fact, since $\varphi\Xi = \Sigma\varphi$, this is a well-defined bijection which preserves incidence in both directions.

For the proof of the converse we can adjust the proof of Proposition 2D8 (for the surjectivity of the homomorphism π defined there). Now, let $\mu: \mathcal{P}/\Sigma \to \mathcal{P}/\Xi$ be an isomorphism of polytopes. Let $\pi_\Sigma: \mathcal{P} \to \mathcal{P}/\Sigma$ and $\pi_\Xi: \mathcal{P} \to \mathcal{P}/\Xi$ be the corresponding canonical projections, and let Φ be the usual base flag of $\mathcal{P}$. Since $\mathcal{P}$ is regular and μ maps flags to flags, we have $(\Phi\pi_\Sigma)\mu = (\Phi\varphi)\pi_\Xi$ for some $\varphi \in \Gamma(\mathcal{P})$. By arguments similar to those of the proof of Proposition 2D8 (with appropriate changes of π to π_Σ or π_Ξ), we can now conclude that $((\Phi\psi)\pi_\Sigma)\mu = (\Phi\psi\varphi)\pi_\Xi$ for each $\psi \in \Gamma(\mathcal{P})$.

Next we make two clever choices for ψ. First, for $\psi \in \Sigma$ this gives

$$(\Phi\psi\varphi)\pi_\Xi = ((\Phi\psi)\pi_\Sigma)\mu = (\Phi\pi_\Sigma)\mu = (\Phi\varphi)\pi_\Xi.$$

By Lemma 2E22 and Proposition 2D9, as well as Lemma 2D7, we can identify the set of flags of $\mathcal{P}/\Xi$ with the set of orbits of $\mathcal{F}(\mathcal{P})$ under the action of Ξ. Therefore, if $\psi \in \Sigma$, we have $\Phi\psi\varphi = \Phi\varphi\tau$; thus, $\psi\varphi = \varphi\tau$ for some $\tau \in \Xi$. It follows that $\varphi^{-1}\Sigma\varphi \subseteq \Xi$.

Second, with $\psi := \varphi^{-1}$, the earlier equation implies that $(\Phi\pi_\Xi)\mu^{-1} = (\Phi\varphi^{-1})\pi_\Sigma$, so that the same arguments applied to the inverse μ^{-1} of μ, with φ replaced by φ^{-1}, give the other inclusion $\varphi^{-1}\Sigma\varphi \supseteq \Xi$. In particular, Σ and Ξ are conjugate subgroups. (Also, as in the proof of Proposition 2D8, we have $\mu = \hat{\varphi}$, with $\hat{\varphi}$ as defined before.) This concludes the proof. □

When applied to (locally finite) regular tesssellations $\mathcal{P}$ in spherical, euclidean or hyperbolic n-space E, Proposition 2E24 can be seen as a combinatorial analogue of the following general fact about such n-dimensional space-forms. If Σ and Ξ are two discrete subgroups of the isometry group $\mathcal{I}(E)$ of E which act freely on E, then the corresponding space-forms E/Σ and E/Ξ are isometric if and only if (their fundamental groups) Σ and Ξ are conjugate in $\mathcal{I}(E)$ (see Section 6A). Here, the condition on the action of Σ and Ξ on E corresponds to the condition of Proposition 2E24 that the subgroups be sparse. In fact, the analogy goes even further if E is hyperbolic and $n \geqslant 3$. Two hyperbolic space-forms E/Σ and E/Ξ of dimension $n \geqslant 3$ are homeomorphic if and only if Σ and Ξ are conjugate in $\mathcal{I}(E)$. This is a consequence of Mostow rigidity (see Section 6J, and Thurston [411, §5.7]). So, in a sense, Proposition 2E24 can be viewed as a combinatorial rigidity result.

We shall further explore the relations between space-forms and quotient polytopes in Chapter 6.

2F Presentations of Polytopes

In Chapter 3, we shall discuss in more detail the basic groups which underly the groups of regular polytopes, namely, the Coxeter groups. We have already informally introduced the concept of a Coxeter group; for present purposes, we can think of it as prescribed by the Schläfli type of a regular polytope, so that it is a group $\langle \rho_0, \ldots, \rho_{n-1} \rangle$, whose generators satisfy relations $(\rho_i\rho_j)^{p_{ij}} = \varepsilon$, where

$$p_{ij} \begin{cases} = 1, & \text{if } i = j, \\ = p_j \geqslant 2, & \text{if } j = i+1, \\ = 2, & \text{if } j \geqslant i+2. \end{cases}$$

(Compare here (2B15).) This group is denoted $[p_1, \ldots, p_{n-1}]$, and since (as will be shown in Section 3D) it is a C-group, it is the group of the *universal* regular n-polytope written $\{p_1, \ldots, p_{n-1}\}$.

We have also seen that the general regular polytope of Schläfli type $\{p_1, \ldots, p_{n-1}\}$ is a quotient of the universal polytope (compare Section 3D). Such a quotient (or, rather, its group) will generally be describable by means of the imposition of further relations on the group $[p_1, \ldots, p_{n-1}]$. We introduce here some notation, which enables us to avoid clumsy circumlocutions of the form "the polytope whose group is the Coxeter group $[p_1, \ldots, p_{n-1}]$, factored out by the relations ...".

In fact, the notation will merely be the combinatorial expression of these additional relations; it will also give us some extra flexibility, in that we can then apply the notation to describe a quotient of a general regular polytope, which is not necessarily universal (in the sense used earlier).

When we constructed the group $\Gamma = \Gamma(\mathcal{P})$ of a regular n-polytope $\mathcal{P}$ in Section 2B, we obtained a very direct correspondence between group elements and chains of successively adjacent flags:

$$\Phi^{j(1)\cdots j(r)} \longleftrightarrow \rho_{j(r)} \cdots \rho_{j(1)}.$$

We refer to $j(1) \cdots j(r)$ as a *flag-sequence*; since $\mathcal{P}$ is regular, we may even forget on which flag the sequence is based, since if we start from a different flag, then we obtain a conjugate group element. In particular, a relation $\rho_{j(r)} \cdots \rho_{j(1)} = \varepsilon$ just gives a closed chain relation $\Phi^{j(1)\cdots j(r)} = \Phi$, and conversely. (Note, by the way, that since the ρ_j are involutions, the group relation is equivalent to $\rho_{j(1)} \cdots \rho_{j(r)} = \varepsilon$ as well.) From a combinatorial point of view, then, the information about the relation is encapsulated in the closed flag-sequence or *flag-cycle* $s := j(1) \cdots j(r)$. This dispenses with any reliance on the notation chosen for the distinguished generators of the group Γ.

We therefore introduce the following notation. Let $\mathcal{P}$ be a regular n-polytope of Schläfli type $\{p_1, \ldots, p_{n-1}\}$. If $\mathcal{P}$ is the quotient of the universal polytope $\{p_1, \ldots, p_{n-1}\}$, obtained by imposing the extra flag-cycles $\mathcal{R} = \{s_1, \ldots, s_k\}$, with $s_1, \ldots, s_k$ sequences in $0, \ldots, n-1$, then we write

2F1 $$\mathcal{P} := \{p_1, \ldots, p_{n-1}\}/\langle\langle s_1, \ldots, s_k \rangle\rangle.$$

The understanding here is, of course, that none of $s_1, \ldots, s_k$ corresponds to the basic relations for the Schläfli type itself.

This latter restriction is only for convenience; there is no theoretical reason why we should not start from the universal regular n-polytope $\{\infty, \ldots, \infty\}$, and let s_j be the flag-cycle $((j-1)j)^{p_j}$ for $j = 1, \ldots, n-1$. But it would be silly to do this – after all, our aim is to simplify the notation.

It is too early to give many examples of how the notation works. However, we can anticipate a little. We shall see in Section 7B that the notation for the polyhedra determined by their Petrie polygons or by their holes, which we met in Section 1D, is

$$\{p, q\}_r = \{p, q\}/\langle\langle (012)^r \rangle\rangle,$$
$$\{p, q \,|\, h\} = \{p, q\}/\langle\langle (0121)^h \rangle\rangle.$$

In practice, we do not use the latter notation, since the former is more succinct. This will also frequently be the case – we adopt a brief notation which is shorter than the canonical one just defined. However, we shall find ourselves in circumstances where the new notation will prove extremely useful.

Observe that, in fact, we need not start from a universal polytope. If $\mathcal{P}$ is obtained from another, not necessarily universal, regular n-polytope $\mathcal{Q}$ by the imposition of extra relations on its group, then we may similarly write

$$\mathcal{P} := \mathcal{Q}/\langle\langle s_1, \ldots, s_k \rangle\rangle,$$

where $s_1, \ldots, s_k$ correspond to the *additional* relations imposed on the group $\Gamma(\mathcal{Q})$ to get its quotient $\Gamma(\mathcal{P})$, or, in other words, to the additional flag-cycles which occur in $\mathcal{P}$.

It is appropriate to make some further comments.

2F2 Proposition *Minimal relations determining the group of a regular polytope $\mathcal{P}$ as a quotient of a Coxeter group $[p_1, \ldots, p_{n-1}] = \langle \rho_0, \ldots, \rho_{n-1} \rangle$ involve all the ρ_i in consecutive blocks.*

Proof. This is clear, when we recall that a subgroup $\Gamma_j = \langle \rho_i \mid i \neq j \rangle$ of $\Gamma = \Gamma(\mathcal{P})$ is an internal direct product

$$\Gamma_j = \langle \rho_0, \ldots, \rho_{j-1} \rangle \times \langle \rho_{j+1}, \ldots, \rho_{n-1} \rangle,$$

if $j \neq 0$ or $n - 1$. □

What Proposition 2F2 implies is that (for instance) it would be inappropriate to impose a flag-cycle $(01542)^5$, since this would be a consequence of the simpler cycles $(012)^5$ and $(45)^5$.

So far, we have discussed the presentation of the group of a regular polytope in very general terms. We now come to a result of considerable practical importance, to which we shall frequently appeal. A general element γ of the group $\Gamma = \Gamma(\mathcal{P}) = \langle \rho_0, \ldots, \rho_{n-1} \rangle$ (say) of a regular n-polytope $\mathcal{P}$ is of the form

2F3
$$\gamma = \alpha_0 \rho_0 \alpha_1 \rho_0 \cdots \rho_0 \alpha_k,$$

with $\alpha_0, \ldots, \alpha_k \in \Gamma_0 := \langle \rho_1, \ldots, \rho_{n-1} \rangle$, the group of the vertex-figure of $\mathcal{P}$ at its base vertex $v := F_0$, for $i = 0, \ldots, k$. (It is now more convenient to work with group elements, rather than with flag-chains.) With γ, we can associate a path in $\mathcal{P}$ with k edges leading from v to $v\gamma$. If $k = 0$, the path consists of v $(= v\alpha_0)$ alone. For $k > 0$, let $(E'_1, \ldots, E'_{k-1})$ be an edge-path associated with $\gamma' := \alpha_0 \rho_0 \alpha_1 \rho_0 \cdots \alpha_{k-1}$. With γ is then associated the path $(E_1, \ldots, E_k)$, given by

$$\begin{aligned} E_1 &:= E\alpha_k \; (= E\rho_0\alpha_k), \\ E_i &:= E'_{i-1}\rho_0\alpha_k \quad \text{for } i = 2, \ldots, k, \end{aligned}$$

where $E := F_1$ is the base edge of $\mathcal{P}$. Of course, this path will not generally be unique, since it depends on the particular expression for γ.

Conversely, an edge-path $(E_1, \ldots, E_k)$ from v corresponds to such an element $\gamma \in \Gamma$, in which ρ_0 occurs k times. If $k > 0$, then there is an $\alpha_k \in \Gamma_0$ such that $E_1 = E\alpha_k$. The shorter path $(E'_1, \ldots, E'_{k-1})$, given by

$$E'_i := E_{i+1}\alpha_k^{-1}\rho_0$$

for $i = 1, \ldots, k-1$, also starts at v, and we can repeat to obtain γ as before, with a free choice of α_0.

In the context of group presentations, we deduce what we shall refer to as the *circuit criterion*:

2F4 Theorem *Let $\mathcal{P}$ be a regular polytope. Then the group $\Gamma = \Gamma(\mathcal{P})$ of $\mathcal{P}$ is determined by the group of its vertex-figure, and the relations on the distinguished generators of Γ induced by the edge-circuits of $\mathcal{P}$ which contain the initial vertex.*

Proof. A relation on Γ can be written in the form

$$\alpha_0 \rho_0 \alpha_1 \rho_0 \cdots \alpha_{k-1} \rho_0 = \varepsilon,$$

with $\alpha_i \in \Gamma_0$ for $i = 0, \ldots, k-1$, which corresponds to an edge-circuit starting and ending at v. Conversely, such an edge-circuit is equivalent under Γ_0 to one beginning with E, and this gives rise to a relation as before (now the element α_0 will be determined by the circuit). This is the result. □

Note, by the way, that among the edge-circuits which determine the group are those of the 2-faces $\{p_1\}$ of $\mathcal{P}$. Again, we shall leave until somewhat later providing instances of how the circuit criterion works in practice.

We end with a comment on the geometric interpretation of Theorem 2F4. As we saw earlier a relation of the form

$$\alpha_0 \rho_0 \alpha_1 \rho_0 \cdots \alpha_{k-1} \rho_0 = \varepsilon,$$

with $\alpha_i \in \Gamma_0$ for $i = 0, \ldots, k-1$, corresponds to an edge-circuit. However, we do not seem to proceed along the circuit to obtain the relation. In fact, though, we do, if we recall the original association between group elements and flag-chains; this relation corresponds to a sequence

$$s := 0a_{k-1}0 \cdots 0a_1 0a_0,$$

with each of $a_{k-1}, \ldots, a_0$ sequences in $1, \ldots, n-1$. Each time we encounter a ‘0” in the flag-chain, we change vertices in a common edge; thus we do directly obtain a corresponding edge-circuit.

3

Coxeter Groups

A central role in the theory of abstract regular polytopes is played by an important class of groups known as Coxeter groups. Historically, Coxeter groups made their first appearance as the symmetry groups of the classical regular polytopes and tessellations, or as related reflexion groups [120, Chapter 11]. Since then, they have occurred in many branches of mathematics.

The purpose of the first three sections of this chapter is to review some of the basic properties of Coxeter groups. We shall not give a full exposition of the subject here, but instead focus attention on those results which will be used later. In particular, we shall state most results without proof. For further notes, references, historical remarks, and proofs, the reader is referred to Cohen [81, 82], Coxeter [120], Grove and Benson [195], Hiller [219], Humphreys [222], Koszul [254], and Tits [417].

Then, in Section 3D, we discuss the class of universal regular polytopes $\{p_1, \ldots, p_{n-1}\}$, which includes the traditional convex regular polytopes as well as the euclidean or hyperbolic regular tessellations. Finally, in Section 3E, we describe how the Brianchon–Gram theorem for angle sums of convex polytopes can be employed to calculate the order of a finite Coxeter group in a purely elementary manner.

3A The Canonical Representation

We shall restrict our study to Coxeter groups with finitely many distinguished generators, although a good part of the theory goes through for groups with infinitely many distinguished generators.

Let $M = (m_{ij})_{i,j=1,\ldots,k}$ be a $k \times k$ matrix whose entries m_{ij} are positive integers or ∞. We call M a *Coxeter matrix* if

$$\begin{cases} m_{ii} = 1, & \text{for } i = 1, \ldots, k, \\ m_{ij} = m_{ji} \geqslant 2, & \text{for } 1 \leqslant i < j \leqslant k. \end{cases}$$

Let M be a Coxeter matrix. The *Coxeter group with Coxeter matrix M* is the group $W = W(M)$ with generators $\sigma_1, \ldots, \sigma_k$, and presentation

3A1 $$(\sigma_i \sigma_j)^{m_{ij}} = \varepsilon \quad \text{for all } i,\ j \text{ with } m_{ij} \neq \infty.$$

If M is not explicitly mentioned, we simply refer to W as a *Coxeter group*. Note that the notion of a Coxeter group involves both a group W and a system of *distinguished generators* $\sigma_1, \ldots, \sigma_k$. Occasionally, we shall also call the pair $(W; \{\sigma_1, \ldots, \sigma_k\})$ a *Coxeter system*. The number k is called the *rank* of W. We write

$$K := \{1, \ldots, k\}.$$

In the following, we shall describe a geometric representation for the Coxeter group W. It follows immediately from this representation that, for all i and j, the order of the product $\sigma_i\sigma_j$ in W is indeed m_{ij}. In particular, the generators $\sigma_1, \ldots, \sigma_k$ of W are involutions, and any two generators σ_i, σ_j generate a (possibly infinite) dihedral group $D_{m_{ij}}$. If $k = 1$, then $W = \langle\sigma_1\rangle \cong C_2$. If $k = 2$, then $W = \langle\sigma_1, \sigma_2\rangle \cong D_{m_{12}}$.

For $I \subseteq K$ we define the *distinguished subgroups* W_I of W by

$$W_I := \langle\sigma_i \mid i \in I\rangle.$$

Then, $W_\emptyset = \{\varepsilon\}$ and $W_K = W$. (Elsewhere, the subgroups W_I are also called special parabolic subgroups of W.)

3A2 Theorem *Let $W = \langle\sigma_1, \ldots, \sigma_k\rangle$ be a Coxeter group with Coxeter matrix $M = (m_{ij})_{i,j\in K}$. Then the distinguished subgroups have the following properties.*

(a) Each group W_I with $\emptyset \neq I \subset K$ is (isomorphic to) the Coxeter group with Coxeter matrix $(m_{ij})_{i,j\in I}$.

(b) If $I, J \subset K$, then

3A3 $$W_I \cap W_J = W_{I\cap J}.$$

(c) The subgroups W_I with $I \subset K$ are mutually distinct. Equivalently, if $j \notin I$, then $\sigma_j \notin W_I$.

From now on, we shall refer to (3A3) as the *intersection property* of W (*with respect to its generators* $\sigma_1, \ldots, \sigma_k$). We remark that this intersection property also holds for Coxeter groups with infinitely many distinguished generators σ_i (groups defined by relations as in (3A1)).

In W we have $\sigma_i\sigma_j = \sigma_j\sigma_i$ whenever $m_{ij} = 2$. In particular, if $m_{ij} = 2$ for all i and j with $i \neq j$, then W is the direct product of k copies of C_2, one for each σ_i. More generally, if K splits into two sets I and J (say) such that $m_{ij} = 2$ whenever $i \in I$ and $j \in J$, then W is the direct product of the two (Coxeter) subgroups W_I and W_J; see Proposition 3A4.

It is convenient to represent a Coxeter group $W = W(M)$ by a diagram $\mathcal{D} = \mathcal{D}(M)$, called the *Coxeter diagram* of W. We also say that W is the *Coxeter group with diagram* $\mathcal{D}$ and write $W = W(\mathcal{D})$. This diagram $\mathcal{D}$ is a labelled graph which represents the set of defining relations (3A1). More precisely, the nodes of the graph represent the generators σ_i of W, and, for each i and j, a branch with label (mark) m_{ij} joins the ith and jth nodes. Conventionally, we omit an *improper* branch labelled 2, and also, for simplicity, the label m_{ij} is omitted from a branch whenever it takes the (most prevalent) value 3. (Occasionally, though, it is preferable to retain at least some improper branches.) Sometimes, if the actual correspondence of the nodes and the generators is to be emphasized, we also label the ith node of the graph with the index i of the corresponding σ_i. (For an historical account on the evolution of the diagram notation, see Coxeter [127].)

By a *string (Coxeter) diagram* we mean a Coxeter diagram of the form

with possibly some of the marks equal to 2. Following Coxeter [120, pp. 130, 199], this group is denoted by $[p_1, \ldots, p_{k-1}]$ if $k \geqslant 2$, or $[1]$ if $k = 1$. A *string Coxeter group* is a Coxeter group with a string diagram.

The following proposition reduces many questions on Coxeter groups to problems on groups with a connected diagram.

3A4 Proposition *Let $\mathcal{D}$ be a Coxeter diagram without improper branches, and let $\mathcal{D}_1, \ldots, \mathcal{D}_m$ be its connected components. Then*

$$W(\mathcal{D}) \cong W(\mathcal{D}_1) \times \cdots \times W(\mathcal{D}_m),$$

but no component $W(\mathcal{D}_i)$ is itself a direct product of non-trivial distinguished subgroups (given by subdiagrams of $\mathcal{D}_i$).

For example, in the string diagram, if $k \geqslant 2$ and $p_i = 2$ for some i, then

$$[p_1, \ldots, p_{k-1}] \cong [p_1, \ldots, p_{i-1}] \times [p_{i+1}, \ldots, p_{k-1}];$$

here we must interpret the left or right factor of the direct product as $[1]$ ($\cong C_2$) if $i = 1$ or $i = k - 1$, respectively. In particular,

$$[2] \cong [1] \times [1] \cong C_2 \times C_2.$$

We call a Coxeter group W (or more exactly, a Coxeter system $(W; \{\sigma_1, \ldots, \sigma_k\})$) *irreducible* or *reducible* according to whether its Coxeter diagram (after omission of all improper branches) is connected or disconnected, respectively. Then Proposition 3A4 says that every Coxeter group is the direct product of its irreducible components.

As was mentioned in Sections 1B and 1C, the symmetry groups of the convex regular n-polytopes in $\mathbb{E}^n$ (or, equivalently, the regular spherical tessellations on $\mathbb{S}^{n-1}$) and of the regular tessellations in euclidean or hyperbolic $(n-1)$-space are generated by n reflexions $R_0, \ldots, R_{n-1}$ in the walls of a fundamental simplex (which is an orthoscheme); see [120, p. 137]. If $\{p_1, \ldots, p_{n-1}\}$ is the Schläfli symbol of the polytope or tessellation, then the dihedral angle between the reflexion walls of R_i and R_j is π/p_j if $j = i + 1$ or $\pi/2$ if $|j - i| \geqslant 2$. Accordingly, $(R_{j-1}R_j)^{p_j} = E$, the identity, or $(R_i R_j)^2 = E$. Figure 3A1 illustrates the situation for the 3-cube $\{4, 3\}$.

Let us introduce a useful convention. In a geometric space E (such as a euclidean or hyperbolic space), we shall identify an involutory isometry R with its mirror $\{x \in E \mid xR = x\}$ of fixed points. Thus, in the previous paragraph, we may call R_j itself a

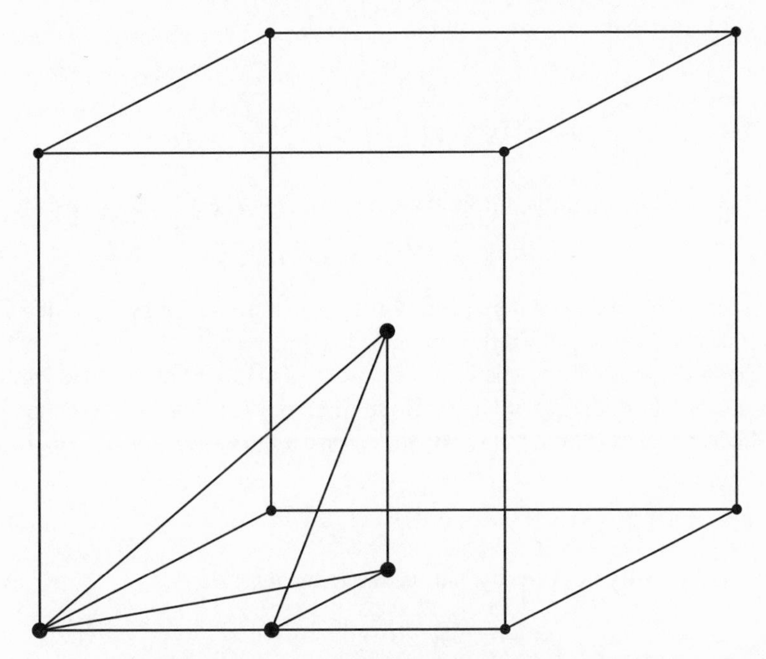

Figure 3A1. A fundamental chamber for the 3-cube.

reflexion wall. This convention is very helpful when we come to consider realizations (see Chapter 5), but we shall employ it sporadically earlier. In particular, where there is no possibility of confusion, we may also denote by E the identity isometry of E.

The following theorem is based on the simple-connectedness of the underlying spaces, and extends Theorem 1B1.

3A5 Theorem *The symmetry group of a convex regular n-polytope or a regular tessellation in euclidean or hyperbolic $(n-1)$-space is a Coxeter group. More precisely, if $\{p_1, \ldots, p_{n-1}\}$ is the Schläfli symbol of the polytope or tessellation, then this group is $[p_1, \ldots, p_{n-1}]$, with diagram*

The geometric picture for regular polytopes and tessellations suggests the construction of a representation for Coxeter groups known as the *canonical representation*. We begin by defining the *canonical bilinear form* for W.

Let $M = (m_{ij})_{i,j \in K}$ be a Coxeter matrix, and $W = W(M) = \langle \sigma_1, \ldots, \sigma_k \rangle$ the corresponding Coxeter group. Let $E := \mathbb{R}^k$, and let $\{e_1, \ldots, e_k\}$ be the standard basis of E. Now define the symmetric bilinear form $\langle \cdot, \cdot \rangle_M$ by

$$\langle e_i, e_j \rangle_M := -\cos(\pi/m_{ij}) \quad \text{for } i, j = 1, \ldots, k.$$

In particular,

$$\langle e_i, e_j \rangle_M = \begin{cases} 1, & \text{if } i = j, \\ 0, & \text{if } m_{ij} = 2, \\ -1, & \text{if } m_{ij} = \infty. \end{cases}$$

For $i = 1, \ldots, k$, consider the linear mapping $S_i \colon E \to E$ defined for $x \in E$ by

$$xS_i := x - 2\langle e_i, x \rangle_M e_i.$$

With respect to the orthogonal geometry on E, which is defined by the bilinear form $\langle \cdot, \cdot \rangle_M$, this mapping S_i is the reflexion in the hyperplane orthogonal to the vector e_i.

By $\mathcal{I}(M)$ we denote the isometry group (orthogonal group) determined by $\langle \cdot, \cdot \rangle_M$; this is the subgroup of the general linear group $\mathrm{GL}(E)$ comprising the linear transformations which preserve the form $\langle \cdot, \cdot \rangle_M$. By the construction of $\langle \cdot, \cdot \rangle_M$, for $i, j = 1, \ldots, k$, we have

$$(S_i S_j)^{m_{ij}} = E.$$

(It is tedious to verify this directly, but see the following.) Hence,

$$\sigma_i \mapsto S_i \quad (i = 1, \ldots, k)$$

defines a homomorphism

$$r \colon W \to \mathcal{I}(M) \subseteq \mathrm{GL}(E).$$

This homomorphism is called the *canonical representation* of W.

In general, the representation r will not be irreducible even if W is irreducible. In fact, the reflexion hyperplanes of $S_1, \ldots, S_k$ intersect in the radical $\operatorname{rad} M$ of $\langle \cdot, \cdot \rangle_M$, so that $\operatorname{rad} M$ is pointwise fixed under the elements of Wr. It follows that, if $\langle \cdot, \cdot \rangle_M$ is degenerate, then r is reducible, yet not completely reducible. On the other hand, if W is irreducible and $\langle \cdot, \cdot \rangle_M$ is non-degenerate, then r is an irreducible representation.

To resolve the problem that the reflexion hyperplanes of the generating reflexions may not be in general position, we pass to the *contragredient representation* r^* of r, which is obtained as follows.

Let E^* denote the dual linear space of E, namely, the space of linear functionals on E, and let $\langle \cdot, \cdot \rangle$ denote the dual pairing of E and E^*. Recall that, if $\Phi \colon E \to E$ is a linear mapping, then it induces a linear mapping $\Phi^* \colon E^* \to E^*$, given by

$$\langle x, y\Phi^* \rangle := \langle x\Phi, y \rangle,$$

for all $x \in E$ and $y \in E^*$. Moreover, for $\Phi, \Psi \colon E \to E$, we have $(\Phi\Psi)^* = \Psi^*\Phi^*$.

Let W be any Coxeter group. For $w \in W$, let $wr^* = (w^{-1}r)^*$ be the dual map of $w^{-1}r$; in other words, for all $x \in E$ and $y \in E^*$, we have

3A6
$$\langle x, y(wr^*) \rangle = \langle x(w^{-1}r), y \rangle.$$

Then the contragredient representation $r^* \colon W \to \mathrm{GL}(E^*)$ is given by $w \mapsto wr^*$. As pointed out in Corollary 3A10, it turns out that both r and r^* are injective.

From now on, we shall simply denote wr^* by w and Wr^* by W, so that we identify W with its contragredient representation on E^* (for the moment, we are anticipating Corollary 3A10). In expressing elements of W as maps on E^*, it is more convenient to use coordinates. Recall that the *dual basis* $\{e_1^*, \ldots, e_n^*\}$ of E^* satisfies $\langle e_j, e_k^* \rangle = \delta_{jk}$, the Kronecker delta. If $y = \sum_{j=1}^n \eta_j e_j^* \in E^*$, and if $z := y\sigma_i$ is given by $z = \sum_{j=1}^n \zeta_j e_j^*$, then (since $S_i = S_i^{-1}$) we have

$$\zeta_k = \langle e_k, z \rangle = \langle e_k, y\sigma_i \rangle = \langle e_k S_i, y \rangle = \langle e_k, y \rangle - 2\langle e_i, e_k \rangle_M \langle e_i, y \rangle = \eta_k - 2\langle e_i, e_k \rangle_M \eta_i.$$

In other words, the generators σ_i of W are given by

3A7
$$\begin{cases} y\sigma_i = y(\sigma_i r^*) = (\sigma_i r)y = S_i y \\ \quad\;\; = y - 2\langle e_i, y \rangle \langle e_i, \cdot \rangle_M \qquad (y \in E^*), \end{cases}$$

with $\langle e_i, \cdot \rangle_M$ the linear form on E defined by $x \mapsto \langle e_i, x \rangle_M$. Hence, for $i = 1, \ldots, k$, the generator σ_i is the affine reflexion in the hyperplane

$$H_i := \{y \in E^* \mid \langle e_i, y \rangle = 0\}$$

of E^* mapping $\langle e_i, \cdot \rangle_M$ onto $-\langle e_i, \cdot \rangle_M$.

For $i = 1, \ldots, k$, define

$$A_i := \{y \in E^* \mid \langle e_i, y \rangle > 0\}$$

and

$$\bar{A}_i := \{y \in E^* \mid \langle e_i, y \rangle \geqslant 0\}.$$

Then A_i is one of the two *open* half-spaces bounded by H_i, and $\bar{A}_i$ is its *closure*. (Here and in the following, if topological concepts like open or closed are used, they refer to the identification of E^* with real k-dimensional space $\mathbb{R}^k$.) For each $I \subseteq K$, define

$$C_I := \left(\bigcap_{i \in I} H_i \right) \cap \left(\bigcap_{i \notin I} A_i \right)$$

and

$$\bar{C}_I := \left(\bigcap_{i \in I} H_i \right) \cap \left(\bigcap_{i \notin I} \bar{A}_i \right).$$

Then C_I (resp. $\bar{C}_I$) is an *open* (resp. *closed*) simplicial cone in E^* with $(k - |I|)$-dimensional linear hull $\bigcap_{i \in I} H_i$. Also, $\bar{C}_I$ is the disjoint union of all cones C_J with $I \subseteq J \subseteq K$. In particular, $C_\emptyset$ and $\bar{C}_\emptyset$ are k-dimensional simplicial cones, and $\bar{C}_\emptyset$ is the disjoint union of all cones C_J with $J \subseteq K$.

The family of all simplicial cones $C_I w$ with $w \in W$ and $I \subset K$ is called the *chamber complex* of W. Its elements $C_\emptyset w$ are the *open chambers* of W, and the cones $\bar{C}_\emptyset w$ are the *closed chambers* of W. In particular, $C_\emptyset$ and $\bar{C}_\emptyset$ are the *open fundamental chamber* and *closed fundamental chamber* of W, respectively; see Theorem 3A11.

The following central result is a theorem of Tits and clarifies the action of W on its chamber complex. One of its immediate consequences is that the chamber complex of

W is isomorphic to the Coxeter complex $\mathcal{C}_W$ associated with the Coxeter group W as in Section 2C (see (2C2)), with the cones $C_I w$ corresponding to the right cosets $W_I w$.

3A8 Theorem *Let $I, I' \subseteq \{1, \ldots, k\}$, and let $w, w' \in W$ be such that $C_I w \cap C_{I'} w' \neq \emptyset$. Then $I = I'$, $W_I w = W_{I'} w'$, and $C_I w = C_{I'} w'$.*

By Theorem 3A8, the stabilizer in W of the cone C_I in E^* equals the subgroup W_I of W. In fact, this subgroup even stabilizes C_I pointwise. Applying these results with $I = \emptyset$ proves the next two corollaries.

3A9 Corollary *W acts simply transitively on the set of all open chambers $C_\emptyset w$ of W.*

3A10 Corollary *The representations r and r^* of W are faithful.*

3A11 Theorem *The union*

$$T := \bigcup_{w \in W} \bar{C}_\emptyset w$$

of all closed chambers in E^* has the following properties.

(a) T is a k-dimensional convex cone in E^*.
(b) The closed chamber $\bar{C}_\emptyset$ (or any other closed chamber) is a fundamental region for the action of W on T. In other words, for each point in T, its orbit under W intersects $\bar{C}_\emptyset$ in precisely one point.
(c) $T = E^*$ if and only if W is finite.

The convex cone T of Theorem 3A11 is often called the *Tits cone*. In the important special case where the bilinear form $\langle \cdot, \cdot \rangle_M$ is non-degenerate, the two representations r and r^* are equivalent in the following sense. If $\langle \cdot, \cdot \rangle_M$ is non-degenerate, the map

3A12
$$\begin{aligned} L \colon E &\to E^* \\ x &\mapsto \langle x, \cdot \rangle_M \end{aligned}$$

becomes an isomorphism, so that r and r^* are related by

$$wr^* = L^{-1}(wr)L \qquad (w \in W).$$

In particular, L^{-1} transforms the chamber complex of W in E^* into a *chamber complex* of W in E. More precisely, if

$$\bar{D}_\emptyset := \{x \in E \mid \langle x, e_i \rangle_M \geqslant 0 \text{ for } i = 1, \ldots, k\}$$

denotes the *closed fundamental chamber* of W in E, then

$$\bar{D}_\emptyset(wr) = (\bar{C}_\emptyset(wr^*))L^{-1} = (\bar{C}_\emptyset w)L^{-1} \qquad (w \in W),$$

so that W acts on the chamber complex in E as it does on the chamber complex in E^*. In particular, $\bar{D}_\emptyset$ is a fundamental region for W in the convex cone TL^{-1} of E. More generally, the results of Theorem 3A8, Corollaries 3A9 and 3A10 and Theorem 3A11 carry over with C_I replaced by certain cones D_I in E.

3B Groups of Spherical or Euclidean Type

We continue the discussion of Section 3A by considering the important case when the canonical bilinear form $\langle \cdot, \cdot \rangle_M$ of a Coxeter group $W = W(M)$ is positive definite or positive semi-definite. The following theorem of Witt [465] characterizes the case when the geometry defined by $\langle \cdot, \cdot \rangle_M$ is euclidean.

3B1 Theorem *The Coxeter group $W = W(M)$ is finite if and only if the canonical bilinear form $\langle \cdot, \cdot \rangle_M$ is positive definite.*

Note that, by Theorems 3A11(c) and 3B1, the chamber complex of W in E covers the whole space E if and only if $\langle \cdot, \cdot \rangle_M$ defines a euclidean geometry. This corresponds to the classical situation where a Coxeter group appears as a finite group generated by hyperplane reflexions in euclidean space. More exactly, the finite Coxeter groups are precisely the finite euclidean reflexion groups (generated by hyperplane reflexions).

For a finite Coxeter group W, the chamber complex in E can also be represented by a *chamber complex* of W on the unit $(k-1)$-sphere

$$\mathbb{S}_M := \{x \in E \mid \langle x, x \rangle_M = 1\}$$

defined by $\langle \cdot, \cdot \rangle_M$. In fact, each chamber in E intersects $\mathbb{S}_M$ in a spherical $(k-1)$-simplex, and W acts on these simplices in the same way as on the chambers in E. Accordingly, the finite Coxeter groups and their Coxeter diagrams are said to be of *spherical type*.

The following well-known classification result for finite Coxeter groups was obtained by Coxeter [99] (see also [96, 102]).

3B2 Theorem *The finite irreducible Coxeter groups are precisely the groups with a diagram listed in Table 3B1.*

Here and elsewhere we shall follow the original notation of Coxeter [120, §§11.4, 11.5 and Table IV] for the diagrams listed in Tables 3B1 and 3B2. In particular, the subscript at a diagram name is its number of nodes. The notation for the diagrams used in Lie Theory is added in parentheses (if it is different); see [222, §§2.4, 2.5] or the end of this section.

For a method of calculating the orders of the finite Coxeter groups (given in Table 3B1), we refer forwards to Section 3E.

We can now rephrase parts of Theorems 1B1 and 3A5 as follows.

3B3 Theorem *The finite irreducible Coxeter groups with a string diagram are precisely the symmetry groups of the convex regular polytopes, with a pair of dual polytopes corresponding to a pair of groups which are related by reversing the order of the generators.*

Note that, for a finite irreducible Coxeter group W with a string diagram, the chamber complex of W in $E = TL^{-1}$ can be realized by the cone system in E, which is obtained from the corresponding regular polytope $\mathcal{P}$ by replacing each simplex in the barycentric subdivision of $\mathcal{P}$ by a suitable simplicial cone whose apex is the centroid of $\mathcal{P}$. In

Table 3B1. *The Finite Irreducible Coxeter Groups*

Notation	Diagram	Order
A_n $(n \geq 1)$		$(n+1)!$
B_n $(=D_n)$ $(n \geq 4)$	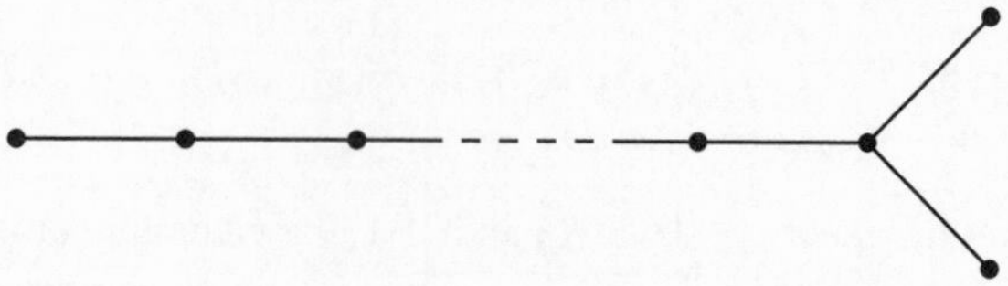	$2^{n-1}n!$
C_n $(=B_n)$ $(n \geq 2)$	 4	$2^n n!$
D_2^p $(=I_2(p))$ $(p \geq 3)$	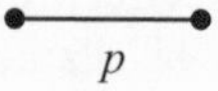 p	$2p$
E_6	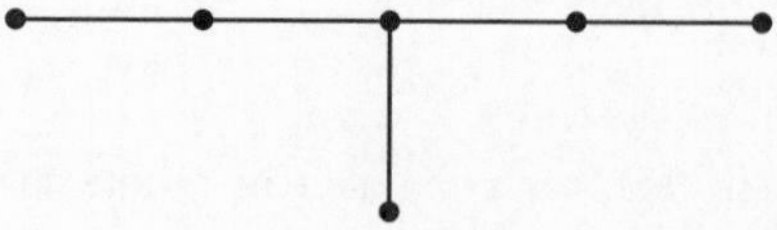	$72 \cdot 6!$
E_7	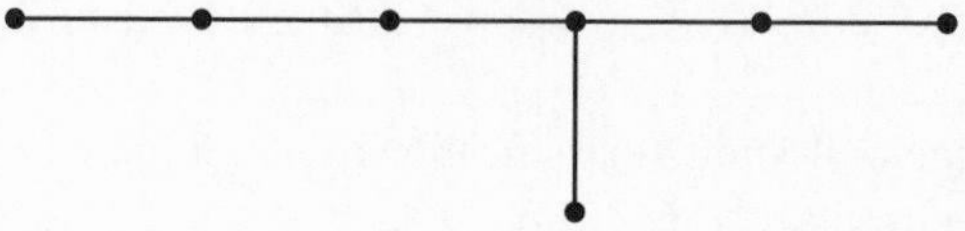	$8 \cdot 9!$
E_8	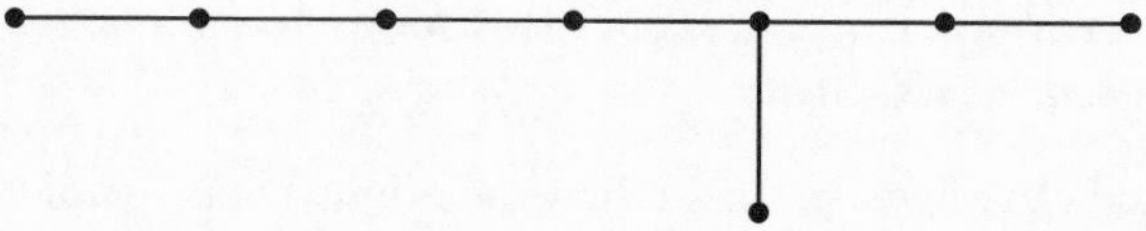	$192 \cdot 10!$
F_4	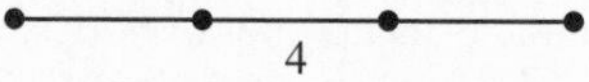 4	1152
G_3 $(= H_3)$	 5	120
G_4 $(= H_4)$	 5	14400

particular, the action of the symmetry group $G(\mathcal{P})$ of $\mathcal{P}$ on the cone system is equivalent to the action of W on its chamber complex.

A similar situation arises for the regular tessellations of euclidean spaces; here the form $\langle \cdot, \cdot \rangle_M$ becomes positive semi-definite. The general picture for an irreducible Coxeter group W with a positive semi-definite form $\langle \cdot, \cdot \rangle_M$ is as follows (see also Lemma 9B19).

Table 3B2. *The Infinite Irreducible Coxeter Groups of Euclidean Type*

Notation	Diagram
$P_n\ (=\tilde{A}_{n-1})\ (n \geq 3)$	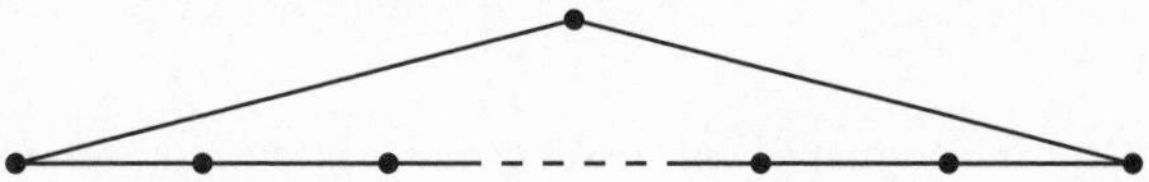
$Q_n\ (=\tilde{D}_{n-1})\ (n \geq 5)$	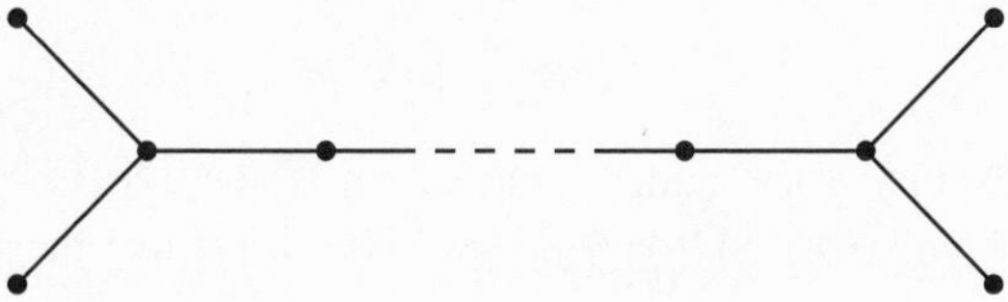
$R_n\ (=\tilde{C}_{n-1})\ (n \geq 3)$	 4 4
$S_n\ (=\tilde{B}_{n-1})\ (n \geq 4)$	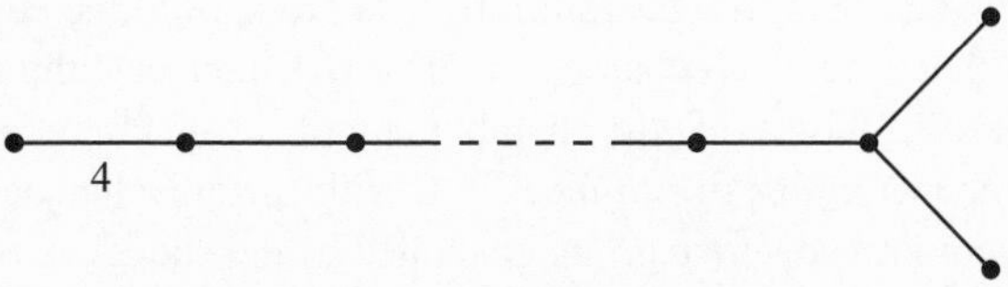 4
$T_7\ (=\tilde{E}_6)$	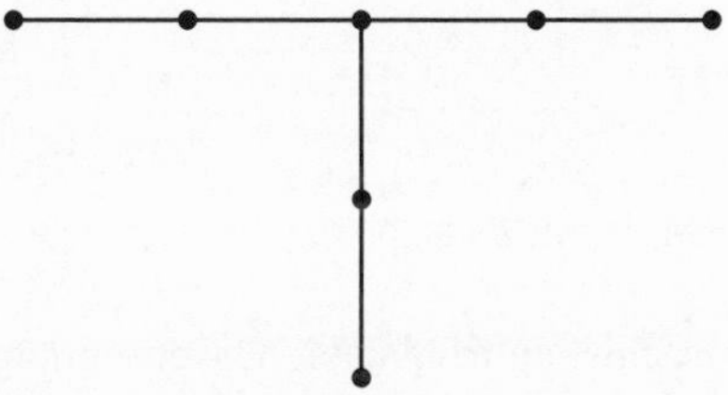
$T_8\ (=\tilde{E}_7)$	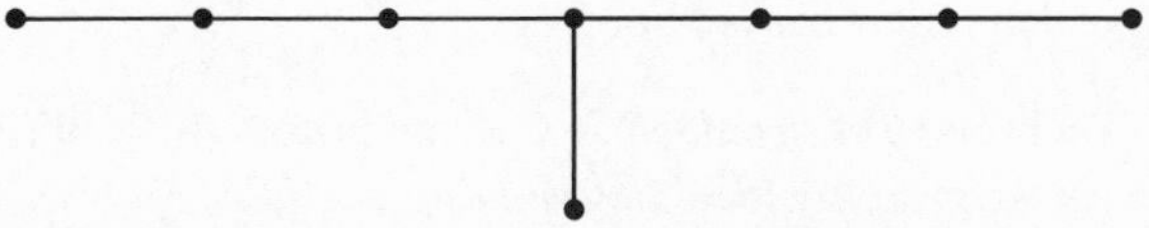
$T_9\ (=\tilde{E}_8)$	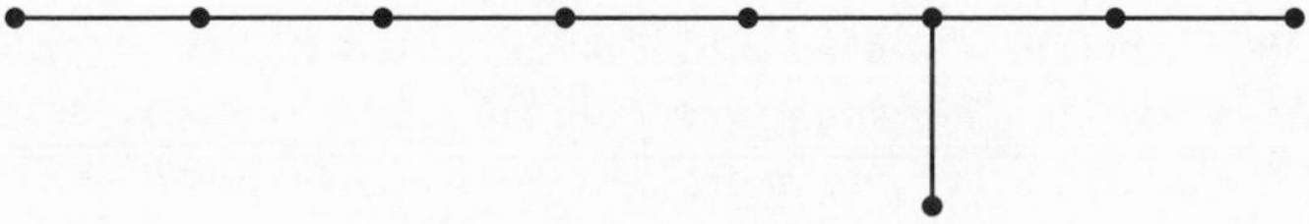
$U_5\ (=\tilde{F}_4)$	 4
$V_3\ (=\tilde{G}_2)$	 6
$W_2\ (=\tilde{A}_1)$	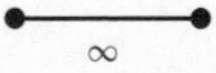 ∞

Assume that $\langle\cdot,\cdot\rangle_M$ is positive semi-definite (but not positive definite). Then the radical rad M of $\langle\cdot,\cdot\rangle_M$ in E is 1-dimensional and is spanned by a vector

$$v = \sum_{i=1}^{k} \lambda_i e_i \quad \text{with } \lambda_1, \ldots, \lambda_k > 0.$$

Then the linear map L of (3A12) induces an isomorphism of the quotient space $E/\operatorname{rad} M$ onto the hyperplane

$$H := \{y \in E^* \mid \langle v, y\rangle = 0\}$$

of E^*, the image of E under L. On $E/\operatorname{rad} M$, the form $\langle\cdot,\cdot\rangle_M$ induces a positive definite form which is carried over to H by L. It follows that the affine hyperplane

$$A := \{y \in E^* \mid \langle v, y\rangle = 1\}$$

of E^*, on which H acts by translation, inherits a euclidean structure. Finally, on this affine hyperplane, W acts faithfully as a group of affine euclidean motions.

Furthermore, each chamber of W in E^* intersects the hyperplane A in a $(k-1)$-simplex. It follows that the chamber complex (of simplicial cones) in E^* gives rise to a chamber complex of simplices in A with an equivalent action of W. Since W acts on A as an infinite discrete group generated by the euclidean reflexions in the walls of the fundamental chamber, this chamber complex covers all of A. Equivalently,

$$T \cap A = A,$$

where T is the Tits cone (see Theorem 3A11). Theorem 3B5 discusses the case when the chamber complex in A can be regarded as the barycentric subdivision of a euclidean tessellation.

Motivated by this, an irreducible Coxeter group and its diagram are said to be of *euclidean type* if the corresponding bilinear form is positive semi-definite (but not positive definite). Then we have

3B4 Theorem *The irreducible Coxeter groups $W = W(M)$ of euclidean type are precisely the groups with a diagram listed in Table 3B2.*

3B5 Theorem *The irreducible Coxeter groups $W = W(M)$ of euclidean type which have a string diagram are precisely the symmetry groups of the regular euclidean tessellations, with a pair of dual tessellations corresponding to a pair of groups which are related by reversing the orders of the generators.*

A reducible Coxeter group and its diagram $\mathcal{D}$ are of *euclidean type* if each connected component of $\mathcal{D}$ is of spherical type or euclidean type, and if at least one of the latter kind occurs. It was remarked earlier that every finite euclidean (hyperplane) reflexion group is a Coxeter group and thus is covered by Theorem 3B2 (and Proposition 3A4). A similar remark also applies to infinite discrete euclidean reflexion groups. Call a group generated by euclidean reflexions in (affine) hyperplanes *irreducible* if its set of reflexion hyperplanes cannot be decomposed into two non-empty subsets such that every two hyperplanes in different sets are perpendicular; it is *reducible* if it is not irreducible. Clearly, every reducible group is the direct product of irreducible groups. Now, if W is an infinite irreducible group generated by euclidean reflexions in (affine) hyperplanes, and if W acts discretely on the underlying euclidean space, then W is an (irreducible) Coxeter group with a diagram among the diagrams of Table 3B2, and its action on the underlying space is equivalent to its action (as a Coxeter group) on the affine hyperplane A of E^*.

To conclude this section, we draw attention to a class of finite Coxeter groups which is especially interesting in Lie Theory. These groups are defined by (crystallographic) root systems and are known as Weyl groups; see Bourbaki [40, Chapter VI] or Humphreys [222, §§1.2, 2.9].

A *root system* $\mathcal{R}$ in a euclidean k-space $\mathbb{E}^k$, with scalar product $(\cdot, \cdot)$, is a finite set of non-zero vectors, the *roots*, satisfying the conditions:

1. $\mathcal{R}$ spans $\mathbb{E}^k$, and $\mathcal{R} \cap \mathbb{R}a = \{\pm a\}$ for all $a \in \mathcal{R}$;
2. $\mathcal{R}S_a = \mathcal{R}$ for all $a \in \mathcal{R}$, where S_a is the euclidean reflexion in the (linear) hyperplane H_a orthogonal to a;
3. $2(a, b)/(b, b)$ is an integer for all $a, b \in \mathcal{R}$.

The integers in (3) are called *Cartan integers*. Every root system $\mathcal{R}$ has a *base*; that is, a subset $\mathcal{S}$ of $\mathcal{R}$ which is a basis of $\mathbb{E}^k$ such that every $a \in \mathcal{R}$ is a linear combination of vectors in $\mathcal{S}$ with integer coefficients which are all non-negative or all non-positive. The vectors in $\mathcal{S}$ are called *simple roots*.

The group W generated by the reflexions S_a $(a \in \mathcal{R})$ is a finite Coxeter group, called the *Weyl group* of $\mathcal{R}$, whose distinguished generators are given by the reflexions S_a $(a \in \mathcal{S})$ from some base $\mathcal{S}$ of $\mathcal{R}$. By Theorem 3B2, the Coxeter diagrams for the irreducible Weyl groups (those given by an "irreducible" root system) are among the diagrams listed in Table 3B1. However, not every diagram of this table belongs to a Weyl group. In fact, among the diagrams in Table 3B1, precisely the diagrams A_n, B_n, C_n, E_6, E_7, E_8, F_4 and D_2^6, namely, those whose branches carry only the labels 2, 3, 4 or 6, correspond to Weyl groups and root systems; C_n actually corresponds to a pair of dual root systems. (Compare also Theorem 9C9 for the complex analogue.)

To see how Weyl groups $W = W(M)$ fit into the earlier discussion on general Coxeter groups, we note that, by Theorem 3B1, the corresponding bilinear form $\langle \cdot, \cdot \rangle_M$ on E is positive definite, and so defines a euclidean geometry. Thus we can identify E and $\mathbb{E}^k$. Then W acts on its finite chamber complex as a euclidean reflexion group, and up to multiplication by positive scalars, the roots in $\mathcal{R}$ are precisely the normal vectors of the reflexion hyperplanes for W, with a set of simple roots given by the outer normal vectors of the walls of a fundamental chamber for W. The exclusion of G_3, G_4, and D_2^m with $m \neq 3, 4, 6$ from the list of diagrams guarantees that these normal vectors (after rescaling) meet the third requirement on the Cartan integers. Further, W stabilizes the lattice spanned by a base $\mathcal{S}$ of $\mathcal{R}$, the *root lattice* of $\mathcal{R}$. These lattices have many interesting geometric properties (see [91, Chapter 4]).

The irreducible Coxeter groups $W = W(M)$ of euclidean type, or, equivalently, the infinite discrete irreducible euclidean reflexion groups, are intimately related to Weyl groups. They are also called *affine Weyl groups*. In fact, any such group W has an isotopy group (point stabilizer) which is a Weyl group; this Weyl group is a maximal distinguished subgroup of W, so that its diagram is obtained from the diagram of W by removing a suitable node and the branches containing this node. Every Weyl group occurs as an isotropy group in an affine Weyl group. Further, affine Weyl groups are crystallographic groups (discrete groups of isometries acting with compact quotient). For more details, see Coxeter [120, §11.2].

The complexifications of the reflexion hyperplanes for a finite Coxeter group give an example of a complex "hyperplane arrangement" (see [31, §2.3; 336, §6.2]). The topology of the set-theoretic complement of these "Coxeter arrangements" in complex space has been extensively studied.

3C Groups of Hyperbolic Type

An irreducible Coxeter group $W = W(M)$ of rank k and its Coxeter diagram $\mathcal{D}$ are said to be of *hyperbolic type* if the canonical bilinear form $\langle \cdot, \cdot \rangle_M$ of W has signature $(k-1, 1)$, and if $\langle x, x \rangle_M < 0$ for each x in the open fundamental chamber

$$D_\emptyset = \{x \in E \mid \langle x, e_i \rangle_M > 0 \text{ for } i = 1, \ldots, k\}$$

in the chamber complex of W in E. Note that the latter condition says that $D_\emptyset$ lies in one, Q (say), of the two components of the (open) quadratic cone defined by the inequality

$$\langle z, z \rangle_M < 0 \quad (z \in E),$$

while the closed fundamental chamber $\bar{D}_\emptyset$ lies in the closure of Q. For a hyperbolic Coxeter group, the interior of the Tits cone T coincides with Q.

Let W be of hyperbolic type, and let $\mathbb{H}^{k-1}$ be the component of

$$\{z \in E \mid \langle z, z \rangle_M = -1\}$$

which intersects T. Then $\mathbb{H}^{k-1}$ has the natural structure of a $(k-1)$-dimensional hyperbolic space on which W acts as a hyperbolic reflexion group; here we are using the hyperboloid model of hyperbolic space (see [344, §3.2]). Now the chamber complex of W on E induces a chamber complex on $\mathbb{H}^{k-1}$ (or more exactly, on $\mathbb{H}^{k-1}$ extended by points on the absolute of $\mathbb{H}^{k-1}$), on which W acts in the same way.

The following proposition characterizes the Coxeter groups of hyperbolic type, and can be used to prove Theorem 3C4.

3C1 Proposition *An irreducible Coxeter group $W = W(M)$ with a diagram $\mathcal{D}$ and a non-degenerate bilinear form $\langle \cdot, \cdot \rangle$ is of hyperbolic type if and only if $\mathcal{D}$ is not of*

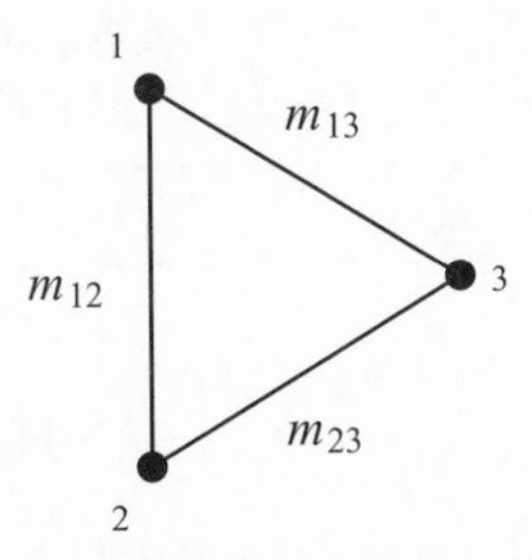

Figure 3C1. The diagram for a triangle group.

spherical type; yet, for each node i of $\mathcal{D}$, the Coxeter diagram $\mathcal{D}_i$ obtained by removing i from $\mathcal{D}$ is of spherical type or of euclidean type.

A Coxeter group W and its diagram $\mathcal{D}$ are said to be of *compact hyperbolic type* if $\mathcal{D}$ is of hyperbolic type and all the subdiagrams $\mathcal{D}_i$ of Proposition 3C1 are of spherical type. The latter condition says that the chamber complex of W in $\mathbb{H}^{k-1}$ (or in E, or in E^*) is locally finite in the sense that the link of each non-empty simplex (cone of non-zero dimension, respectively) is finite. A Coxeter group is of *non-compact hyperbolic type* if it is of hyperbolic type but not of compact hyperbolic type.

Coxeter groups of hyperbolic type exist only in ranks 3 to 10, and there are only finitely many such groups in ranks 4 to 10. Groups of compact hyperbolic type exist only in ranks 3, 4 and 5.

The irreducible Coxeter groups W of rank 3 are also called *triangle groups* and are represented by triangular diagrams as in Figure 3C1, possibly with one improper branch (with mark 2). Here, W is of spherical, euclidean or hyperbolic type according as

$$\frac{1}{m_{12}} + \frac{1}{m_{13}} + \frac{1}{m_{23}} > 1, \; = 1, \quad \text{or} \quad <1.$$

This corresponds to the fact that a (fundamental) triangle with angles π/m_{12}, π/m_{13} and π/m_{23} is necessarily a spherical, euclidean or hyperbolic triangle, respectively. If the group is irreducible and of spherical type, then up to relabelling of nodes, we must have

3C2 $$(m_{12}, m_{13}, m_{23}) = (3, 3, 2), \; (3, 4, 2), \quad \text{or} \quad (3, 5, 2)$$

giving the diagrams A_3, C_3 and G_3 of Table 3B1. The groups of euclidean type are determined by

3C3 $$(m_{12}, m_{13}, m_{23}) = (3, 6, 2), \; (4, 4, 2), \quad \text{or} \quad (3, 3, 3)$$

and correspond to the diagrams V_3, R_3, and P_3 of Table 3B2. All other triangle groups are of hyperbolic type (including those with marks ∞).

For the classification of hyperbolic Coxeter groups of rank at least 4, see Humphreys [222, §§6.8, 6.9]. We are mainly interested in groups with a string diagram. Here we have

Table 3C1. *Coxeter Groups $[p_1, \dots, p_n]$ of Hyperbolic Type, and Related Regular Honeycombs $\{p_1, \dots, p_n\}$ in $\mathbb{H}^n$*

Dimension	Group	Honeycomb
2	$[p, q]$	$\{p, q\}$
	$3 \leq p, q \leq \infty, \frac{1}{p} + \frac{1}{q} < \frac{1}{2}$	
3	$[r, 3, 5], [5, 3, r]$	$\{r, 3, 5\}, \{5, 3, r\}$
	$r = 4, 5$	
	$[3, 5, 3]$	$\{3, 5, 3\}$
	$[r, 4, 4], [4, 4, r]$	$\{r, 4, 4\}, \{4, 4, r\}$
	$r = 3, 4$	
	$[r, 3, 6], [6, 3, r]$	$\{r, 3, 6\}, \{6, 3, r\}$
	$r = 3, 4, 5, 6$	
	$[3, 6, 3]$	$\{3, 6, 3\}$
4	$[r, 3, 3, 5], [5, 3, 3, r]$	$\{r, 3, 3, 5\}, \{5, 3, 3, r\}$
	$r = 3, 4, 5$	
	$[3, 4, 3, 4], [4, 3, 4, 3]$	$\{3, 4, 3, 4\}, \{4, 3, 4, 3\}$
5	$[3, 3, 4, 3, 3]$	$\{3, 3, 4, 3, 3\}$
	$[3, 3, 3, 4, 3], [3, 4, 3, 3, 3]$	$\{3, 3, 3, 4, 3\}, \{3, 4, 3, 3, 3\}$
	$[3, 4, 3, 3, 4], [4, 3, 3, 4, 3]$	$\{3, 4, 3, 3, 4\}, \{4, 3, 3, 4, 3\}$

3C4 Theorem *The Coxeter groups of rank $n + 1$ ($n \geqslant 2$) with a string diagram of hyperbolic type are precisely the symmetry groups of the regular honeycombs in hyperbolic n-space, with a pair of dual honeycombs corresponding to a pair of groups which are related by reversing the order of the generators. In particular, the only such groups and honeycombs are those listed in Table 3C1.*

For a discussion of regular hyperbolic honeycombs see Coxeter [113] or Section 6J. If a string Coxeter group of rank $n + 1$ is of compact hyperbolic type, then the corresponding honeycomb is actually a (locally finite) regular tessellation of hyperbolic n-space by hyperbolic regular n-polytopes. Well-known examples are the symmetry groups $[p, q]$ of the regular tessellations in the hyperbolic plane.

For the general theory of hyperbolic reflexion groups, the reader is referred to Vinberg [431–433]. We remark that there are examples of discrete groups generated by hyperplane reflexions in a hyperbolic space which are Coxeter groups, but do not have a simplex as a fundamental region.

3D The Universal Polytopes $\{p_1, \dots, p_{n-1}\}$

In this section, we study Coxeter groups with string diagrams from the point of view of (abstract) regular polytopes. We shall see that, for each Coxeter group $W = [p_1, \dots, p_{n-1}]$ (with diagram 3D1), there exists a regular n-polytope $\mathcal{P}$ of type $\{p_1, \dots, p_{n-1}\}$ whose group is W. This polytope is universal among all regular polytopes $\mathcal{Q}$ of the same type $\{p_1, \dots, p_{n-1}\}$; that is, any such polytope $\mathcal{Q}$ is a quotient of $\mathcal{P}$.

Among these universal polytopes $\mathcal{P}$, we find the convex regular polytopes and the regular tessellations of euclidean or hyperbolic spaces, whose type is given by the ordinary Schläfli symbol (see Theorems 3B3, 3B5 and 3C4, as well as Theorem 3D6). We take this as motivation to denote a universal polytope $\mathcal{P}$ simply by its Schläfli symbol $\{p_1, \dots, p_{n-1}\}$. As in the classical case, these polytopes $\{p_1, \dots, p_{n-1}\}$ admit a geometric represention in real n-space, which we shall construct from the canonical representation of the corresponding Coxeter group $[p_1, \dots, p_{n-1}]$. Note that, for $n = 1$, the Schläfli symbol becomes $\{\,\}$ and represents the unique 1-polytope.

Historically, these polytopes $\mathcal{P}$ were first discovered by Tits [415]; in the notation of Tits [417, Chapter 12], they occur as the shadow geometries $Sp(\Sigma, \{1\})$, where Σ denotes the Coxeter complex for W. However, our approach and notation follow Schulte [362, Chapter 4; 363, §§5, 6], where these polytopes were rediscovered. For related work, see also Maxwell [275] and Scharlau [353].

Let $n \geqslant 2$ and $p_1, \dots, p_{n-1} \geqslant 2$, possibly with $p_i = \infty$ for some i. Consider the Coxeter group $W = [p_1, \dots, p_{n-1}]$, with generators $\sigma_0, \dots, \sigma_{n-1}$ and diagram

3D1

$$\bullet \overset{}{\underset{p_1}{\text{———}}} \bullet \underset{p_2}{\text{———}} \bullet \text{— — — —} \bullet \underset{p_{n-2}}{\text{———}} \bullet \underset{p_{n-1}}{\text{———}} \bullet$$

(Recall our convention that we regard a branch with mark 2 as missing, or, equivalently, as an improper branch.) Then the group has the presentation

3D2

$$\begin{cases} \sigma_i^2 &= \varepsilon \quad \text{for } 0 \leqslant i \leqslant n-1; \\ (\sigma_i\sigma_j)^2 &= \varepsilon \quad \text{for } 0 \leqslant i < j-1 \leqslant n-2; \\ (\sigma_{i-1}\sigma_i)^{p_i} &= \varepsilon \quad \text{for } 1 \leqslant i \leqslant n-1. \end{cases}$$

The first proposition shows that we can apply the results of Section 2E to W.

3D3 Proposition *With respect to its canonical generators $\sigma_0, \dots, \sigma_{n-1}$, the group W is a string C-group.*

Proof. This follows from Theorem 3A2. □

Then we obtain

3D4 Theorem *Let $n \geqslant 2$ and $2 \leqslant p_1, \dots, p_{n-1} \leqslant \infty$. Let $W = [p_1, \dots, p_{n-1}] = \langle \sigma_0, \dots, \sigma_{n-1} \rangle$, and let $\mathcal{P} := \mathcal{P}(W)$ be the poset associated with W as in Section 2E. Then $\mathcal{P}$ has the following properties.*

(a) $\mathcal{P}$ is a regular n-polytope of type $\{p_1, \dots, p_{n-1}\}$, which hereafter is denoted by the Schläfli symbol $\{p_1, \dots, p_{n-1}\}$.
(b) $\Gamma(\mathcal{P}) = W$, and $\sigma_0, \dots, \sigma_{n-1}$ are the distinguished generators of $\Gamma(\mathcal{P})$ with respect to some base flag of $\mathcal{P}$.
(c) The facets and vertex-figures of $\{p_1, \dots, p_{n-1}\}$ are isomorphic to $\{p_1, \dots, p_{n-2}\}$ and $\{p_2, \dots, p_{n-1}\}$, respectively. More generally, if $-1 \leqslant i \leqslant j-2 \leqslant n-2$ and F is an i-face and G a j-face of $\{p_1, \dots, p_{n-1}\}$ with $F < G$, then the section G/F is isomorphic to $\{p_{i+2}, \dots, p_{j-1}\}$.

(d) *The polytope* $\{p_1, \ldots, p_{n-1}\}$ *is self-dual if and only if* $p_i = p_{n-i}$ *for* $i = 1, \ldots, n-1$.

Proof. Apply Theorem 2E11 with $\Gamma = W$ and $\rho_i = \sigma_i$ for each i to prove parts (a) and (b). For the proof of (c), we use the fact that the subgroup $\langle \sigma_{i+1}, \ldots, \sigma_{j-1} \rangle$ of W is (isomorphic to) the Coxeter group $[p_{i+2}, \ldots, p_{j-1}]$; see Theorem 3A2(a). Then (c) follows from Theorem 2E11(b). Finally, part (d) follows from Proposition 2E15. □

From the results of Section 2B, we know that the group $\Gamma(\mathcal{Q})$ of a regular n-polytope $\mathcal{Q}$ of type $\{p_1, \ldots, p_{n-1}\}$ is necessarily a quotient of the group $W = [p_1, \ldots, p_{n-1}]$ of the n-polytope $\{p_1, \ldots, p_{n-1}\}$. The next theorem deals with the corresponding statement for polytopes. For the definition of a covering, see Section 2D.

3D5 Theorem *Let* $n \geqslant 2$ *and* $2 \leqslant p_1, \ldots, p_{n-1} \leqslant \infty$. *Then the regular* n*-polytope* $\{p_1, \ldots, p_{n-1}\}$ *is a 2-covering of every regular* n*-polytope* $\mathcal{Q}$ *of type* $\{p_1, \ldots, p_{n-1}\}$.

Proof. The proof is straightforward. Define the surjective homomorphism

$$\alpha \colon W = \langle \sigma_0, \ldots, \sigma_{n-1} \rangle \to \Gamma(\mathcal{Q}) = \langle \rho_0, \ldots, \rho_{n-1} \rangle$$

by $\sigma_i \alpha := \rho_i$ for $i = 0, \ldots, n-1$. Then $\langle \sigma_i \mid i \in I \rangle \alpha = \langle \rho_i \mid i \in I \rangle$ for each I. Let $\{F_{-1}, F_0, \ldots, F_n\}$ and $\{G_{-1}, G_0, \ldots, G_n\}$ be the base flags of $\{p_1, \ldots, p_{n-1}\}$ and $\mathcal{Q}$ which belong to the distinguished generators $\sigma_0, \ldots, \sigma_{n-1}$ and $\rho_0, \ldots, \rho_{n-1}$, respectively. Consider the mapping

$$\begin{aligned} \beta \colon \{p_1, \ldots, p_{n-1}\} &\to \mathcal{Q} \\ F_i w &\mapsto G_i(w\alpha) \quad (w \in W; \quad i = -1, 0, \ldots, n). \end{aligned}$$

By Proposition 2B14, this mapping β is a covering. For $i = 0, \ldots, n-2$, the restriction of α to $\langle \sigma_i, \sigma_{i+1} \rangle$ is one-to-one, so that the two sections F_{i+2}/F_{i-1} and G_{i+2}/G_{i-1} are isomorphic to $\{p_{i+1}\}$. Hence β is indeed a 2-covering. In particular, by Proposition 2E18,

$$\mathcal{Q} \cong \{p_1, \ldots, p_{n-1}\}/N$$

with $N = \ker \alpha$. □

Theorem 3D5 says that each regular n-polytope $\mathcal{Q}$ of type $\{p_1, \ldots, p_{n-1}\}$ can be obtained from the regular n-polytope $\{p_1, \ldots, p_{n-1}\}$ by making suitable identifications. Because of this universal property, we shall refer to the polytope $\{p_1, \ldots, p_{n-1}\}$ as the *universal polytope of type* $\{p_1, \ldots, p_{n-1}\}$.

The next theorem characterizes the regular convex polytopes and regular tessellations among the polytopes $\{p_1, \ldots, p_{n-1}\}$. We confine our attention to the (irreducible) case where $p_1, \ldots, p_{n-1} \geqslant 3$. See also Section 6B, which includes a discussion of the finite reducible case.

3D6 Theorem *The following characterizes certain families of universal regular* n*-polytopes* $\mathcal{P} := \{p_1, \ldots, p_{n-1}\}$ *and their groups* $W = [p_1, \ldots, p_{n-1}]$, *with* $n \geqslant 2$ *and* $3 \leqslant p_1, \ldots, p_{n-1} \leqslant \infty$.

(a) *The finite polytopes $\mathcal{P}$ are precisely the convex regular n-polytopes (with the same Schläfli symbol).*
(b) *The polytopes $\mathcal{P}$ with a (Coxeter) group W of euclidean type are precisely the regular tessellations of euclidean $(n-1)$-space.*
(c) *The polytopes $\mathcal{P}$ with a (Coxeter) group W of (compact) hyperbolic type are precisely the regular honeycombs (locally finite tessellations, respectively) of hyperbolic $(n-1)$-space.*

Proof. From Theorem 3B3, we know that the finite groups W are precisely the symmetry groups of the convex regular n-polytopes $\mathcal{Q}$ (with Schläfli symbol $\{p_1, \ldots, p_{n-1}\}$), with the systems of distinguished generators corresponding to each other. Part (a) now follows from the fact that $\mathcal{Q}$ and $\{p_1, \ldots, p_{n-1}\}$ are constructed from W in the same way. The proofs of (b) and (c) are similar, and use Theorems 3B5 and 3C4. For a more geometric proof, see also Theorem 3D7. □

We shall now employ the canonical representation for Coxeter groups to construct a geometric model for a universal regular polytope $\{p_1, \ldots, p_{n-1}\}$ in real n-space. In particular, this will provide a natural geometry for the ambient space on which the Coxeter group operates as the "symmetry group" of this polytope. However, note that this geometric model for $\{p_1, \ldots, p_{n-1}\}$ is not a realization in the sense of Chapter 5.

As before, let $W = [p_1, \ldots, p_{n-1}] = \langle \sigma_0, \ldots, \sigma_{n-1} \rangle$, and let $\mathcal{P} = \{p_1, \ldots, p_{n-1}\}$. In Proposition 2C1, we described how a regular polytope can be recovered from its order complex. On the other hand, as was proved at the end of Section 2C, the order complex of a regular polytope with a group Γ is isomorphic to the abstract simplicial $(n-1)$-complex $\mathcal{C}_\Gamma$ whose simplices are the right cosets of the subgroups Γ_J, with incidence given by the reverse of inclusion; see (2C2) and (2C3). It follows that Proposition 2C1 translates into a reconstruction of this regular polytope from its complex $\mathcal{C}_\Gamma$. Now, for the universal regular polytope $\mathcal{P}$ with group W, the complex $\mathcal{C}_W$ is the Coxeter complex of W. But as was pointed out in Section 3A, the Coxeter complex of W is isomorphic to the chamber complex of W which is associated with the contragredient representation $r^*\colon W \to \mathrm{GL}(E^*)$ of the canonical representation $r\colon W \to \mathrm{GL}(E)$ of W, with $E = \mathbb{R}^n$ and E^* the dual space of E. This in turn allows us to recover $\mathcal{P}$ from its chamber complex.

Equivalently, we can construct the realization of $\mathcal{P}$ in a more direct way from the chamber complex of W using Proposition 2C1 as guidance. For the notation, see Section 3A.

3D7 Theorem *Let $n \geqslant 2$ and $2 \leqslant p_1, \ldots, p_{n-1} \leqslant \infty$, and let $\mathcal{P} := \{p_1, \ldots, p_{n-1}\}$ and $W = [p_1, \ldots, p_{n-1}] = \langle \sigma_0, \ldots, \sigma_{n-1} \rangle$. For $i = -1, 0, \ldots, n$, define the subset $\hat{F}_i$ of the dual space E^* of $E = \mathbb{R}^n$ by*

$$\hat{F}_i := \bigcup_{w \in \langle \sigma_0, \ldots, \sigma_{i-1} \rangle} \bar{C}_{\{i+1, \ldots, n-1\}} w.$$

Set

$$\hat{\mathcal{P}} := \{\hat{F}_i w \mid -1 \leqslant i \leqslant n,\ w \in W\},$$

and order $\hat{\mathcal{P}}$ by inclusion. Then the following hold.

(a) $\mathcal{P}$ and $\hat{\mathcal{P}}$ are isomorphic regular polytopes (with the same groups).

(b) The generators $\sigma_0, \ldots, \sigma_{n-1}$ are the distinguished generators of $\Gamma(\hat{\mathcal{P}}) = W$ with respect to the base flag

$$\hat{\Phi} := \{\emptyset = \hat{F}_{-1}, \hat{F}_0, \ldots, \hat{F}_{n-1}, \hat{F}_n = T\}$$

of $\hat{\mathcal{P}}$, where T is the Tits cone of W (see Theorem 3A11).

(c) For $i = 0, \ldots, n-1$, the i-faces of $\hat{\mathcal{P}}$ are $(i+1)$-dimensional convex cones in E^*.

(d) As posets, $\hat{\mathcal{P}}$ (and $\mathcal{P}$) are lattices. In particular, in $\hat{\mathcal{P}}$, the (set-theoretical) intersection of any two faces is again a face.

Proof. Let $\Phi = \{F_{-1}, F_0, \ldots, F_n\}$ be the base flag of $\mathcal{P}$. By Proposition 2B14, we have $F_i w \leqslant F_j \tilde{w}$ in $\mathcal{P}$ if and only if $-1 \leqslant i \leqslant j \leqslant n$ and $w\tilde{w}^{-1} \in W_{\{i+1,\ldots,n-1\}} W_{\{0,\ldots,j-1\}}$. To prove the isomorphism between $\mathcal{P}$ and $\hat{\mathcal{P}}$, it suffices to check the corresponding property for $\hat{\mathcal{P}}$ and the $\hat{F}_i$. We may assume that $\tilde{w} = \varepsilon$.

Recall that, by Theorem 3A8, the stabilizer of a cone C_I equals the subgroup W_I of W. Now let $\hat{F}_i w \subseteq \hat{F}_j$, so that

$$C_{\{i+1,\ldots,n-1\}} w \subseteq \bar{C}_{\{i+1,\ldots,n-1\}} w \subseteq \hat{F}_i w \subseteq \hat{F}_j.$$

Then, by Theorem 3A8, we must have $i \leqslant j$ and $C_{\{i+1,\ldots,n-1\}} w = C_{\{i+1,\ldots,n-1\}} u$ for some $u \in W_{\{0,\ldots,j-1\}}$. Hence $W_{\{i+1,\ldots,n-1\}} w = W_{\{i+1,\ldots,n-1\}} u$, and so $w \in W_{\{i+1,\ldots,n-1\}} W_{\{0,\ldots,j-1\}}$.

Conversely, if $i \leqslant j$ and $w \in W_{\{i+1,\ldots,n-1\}} W_{\{0,\ldots,j-1\}}$, say $w = w_1 w_2$ with $w_1 \in W_{\{i+1,\ldots,n-1\}}$ and $w_2 \in W_{\{0,\ldots,j-1\}}$, then

$$\hat{F}_i w = \hat{F}_i w_1 w_2 = \hat{F}_i w_2 \subseteq \hat{F}_j w_2 = \hat{F}_j.$$

This proves (a).

For (b), note that the isomorphism between $\mathcal{P}$ and $\hat{\mathcal{P}}$ carries the i-face F_i in Φ onto the i-face $\hat{F}_i$ in $\hat{\Phi}$. For the proofs of (c) and (d), the reader is referred to Tits [415, 417]. In the notation of [417, Chapter 12], $\mathcal{P} = Sp(\Sigma, \{1\})$. □

By Theorem 3D7, every universal regular polytope $\mathcal{P} = \{p_1, \ldots, p_{n-1}\}$ has a realization $\hat{\mathcal{P}}$ as a system of convex cones in the dual space E^*. If the canonical bilinear form $\langle \cdot, \cdot \rangle_M$ for W is non-degenerate on E, then we can identify E with E^* *via* the isomorphism $L^{-1}\colon E^* \to E$, with L as in (3A12). Then we also obtain a realization of $\mathcal{P}$ as a system of convex cones in the original space E, namely, $\hat{\mathcal{P}} L^{-1}$.

For the remainder of this section, let $p_1, \ldots, p_{n-1} \geqslant 3$. By Theorem 3D6(a), if $\mathcal{P}$ is finite, then it is isomorphic to a convex regular n-polytope. This isomorphism can be seen geometrically. In fact, if W is finite, the form is positive definite and $\hat{\mathcal{P}} L^{-1}$ is a finite system of "polyhedral" cones whose intersections with the unit sphere

$$\mathbb{S}_M := \{z \mid \langle z, z\rangle_M = 1\}$$

determine a spherical tessellation on $\mathbb{S}_M$. The convex hull of the vertices in this tessellation is a convex regular n-polytope in E whose faces are in one-to-one correspondence with the cones in $\hat{\mathcal{P}}L^{-1}$. This polytope is indeed the ordinary convex regular polytope $\{p_1, \ldots, p_{n-1}\}$ in the euclidean geometry on E which is defined by $\langle\cdot,\cdot\rangle_M$.

Similarly, by Theorem 3D6(b, c), if $\mathcal{P}$ has a group W of euclidean or hyperbolic type, then it is isomorphic to a regular tessellation in euclidean $(n-1)$-space or a regular honeycomb in hyperbolic $(n-1)$-space, respectively. For the euclidean case, we recall from Section 3B, that the bilinear form $\langle\cdot,\cdot\rangle_M$ now induces a euclidean geometry on the affine hyperplane of E^*

$$A = \{y \in E^* \mid \langle v, y\rangle = 1\},$$

with rad $M = \langle v\rangle$, on which W acts as a euclidean reflexion group. Then the intersections of A with the cones in $\hat{\mathcal{P}}$ determine a euclidean regular tessellation in A which is isomorphic to $\mathcal{P}$. Finally, in the hyperbolic case, the component $\mathbb{H}^{n-1}$ of

$$\{z \in E \mid \langle z, z\rangle_M = -1\}$$

which intersects the Tits cone T has the structure of a hyperbolic $(n-1)$-space on which W acts as a hyperbolic reflexion group (Section 3C). Then the intersections of $\mathbb{H}^{n-1}$ with the cones in $\hat{\mathcal{P}}L^{-1}$ (possibly extended by points on the absolute) form a hyperbolic regular honeycomb which is isomorphic to $\mathcal{P}$. If W is of compact hyperbolic type, then this honeycomb is a locally finite regular tessellation of $\mathbb{H}^{n-1}$.

3E The Order of a Finite Coxeter Group

In this section, we shall describe how the Brianchon–Gram theorem for angle sums of convex polytopes can be employed to calculate the order of a finite Coxeter group in a purely elementary manner. We shall largely follow McMullen [285], but with some subsequent simplifications.

Let G be a finite group generated by reflexions in hyperplanes or *mirrors* in n-dimensional euclidean space $\mathbb{E}^n$. We shall adopt the convention of identifying a reflexion R (or, more generally, an involutory isometry of $\mathbb{E}^n$), with its mirror of fixed points $\{x \in \mathbb{E}^n \mid xR = x\}$. These hyperplanes must contain a common point; if we take this point to be the origin o of coordinates, then G is an orthogonal group. Recall that these groups are listed in Table 3B1.

The images under G of the mirrors of the generating reflexions R_j dissect the space $\mathbb{E}^n$ into congruent convex cones, which are fundamental regions for G, and whose number is obviously the order $|G|$ of G. Thus, to find $|G|$, it is only necessary to count these cones, or, equivalently, measure their normalized volumes or *angles*. However, until recently [285], if $n \geqslant 4$, no strictly elementary way of doing this was available.

Let us assume that G is *irreducible*, that is, acts irreducibly on $\mathbb{E}^n$ (with no non-trivial invariant subspaces). Then, as we saw in earlier sections (see also [120, Chapter 11]),

G is generated by precisely n reflexions, whose mirrors may be chosen to bound any one of the fundamental cones in the dissection of $\mathbb{E}^n$ just described.

If these mirrors are $R_1, \ldots, R_n$, then, for $1 \leqslant j < k \leqslant n$, the dihedral angle between R_j and R_k is π/p_{jk} for some integer $p_{jk} \geqslant 2$, and G has the presentation

$$G = \langle R_1, \ldots, R_n \mid (R_j R_k)^{p_{jk}} = E \text{ for } 1 \leqslant j \leqslant k \leqslant n \rangle,$$

where $p_{jj} = 1$ for $1 \leqslant j \leqslant n$, and E is the identity. (We begin the indexing of the reflexions with 1, for reasons which will soon become clear.) Abstract groups with such presentations are known as *Coxeter groups*; we saw in Sections 3A and 3B (see also [102]) that all finite Coxeter groups are, in fact, isomorphic to reflexion groups. (Since the groups are now geometric, we use the notation G rather than W.)

As in Section 3A, we denote G by its *Coxeter diagram* $\mathcal{G}$ (see also [120, 11.3]), which is a graph with a *node* corresponding to each reflexion R_j or mirror H_j, with nodes j and k joined by a *branch* labelled p_{jk} if the dihedral angle between H_j and H_k is π/p_{jk}. This formulation will later permit us to generalize to fractional labels $p_{jk} > 1$. We recall that it is customary to omit branches labelled 2 and, because of their frequency, to omit labels 3 on branches. The condition that G be irreducible is just that its Coxeter diagram $\mathcal{G}$ be connected; if G is reducible, then it is the direct product of the reflexion groups corresponding to the connected components of $\mathcal{G}$. Thus the connectedness condition on $\mathcal{G}$ is equivalent to G being irreducible as a Coxeter group (see Section 3A). As an example (which we shall use in the following), the Coxeter diagram of the infinite group $T_7 = [3^{2,2,2}]$ is as shown in Figure 3E1. (Refer also to Table 3B2. Recall that $[3^{k,l,m}]$ denotes the Coxeter group whose diagram is a star-like tree with a centre from which three strings with k, l and m unmarked branches emanate; see [120, p. 200]).

Let us briefly survey the earlier methods for calculating the order $|G|$ of G. First, we may associate with G a convex polytope, the numbers of whose faces of various kinds are the indices $[G : H]$ of certain subgroups H of G which are also generated by reflexions. If n is odd, then Euler's theorem (see, for example [197, Chapter 8]) and the knowledge of the orders $|H|$ will yield $|G|$ (see also the following remarks), so that the actual polytope need not be constructed. When n is even, so that Euler's theorem on its own can only yield the ratios of the numbers of faces, a suitable polytope can often be constructed by synthetic methods, and again the value of $|G|$ results. (From an historical point of view, of course, reflexion groups arise from polytopes, rather than the other way round.) In fact, various simplex dissection results (for example, in the case of the group [3, 3, 5] in the following; see also [125], which contains many

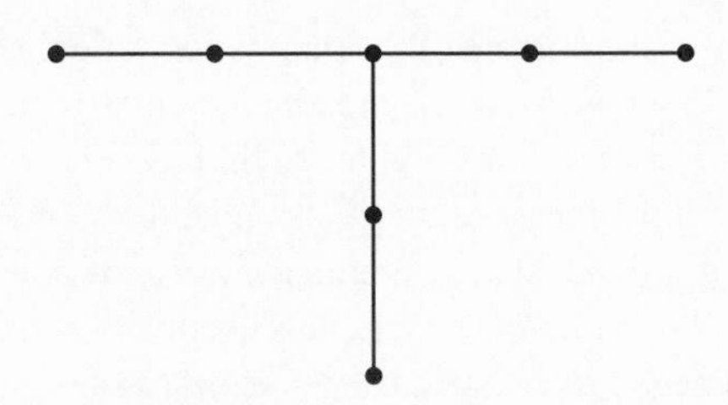

Figure 3E1. The simplex T_7.

other useful references) enable us to avoid such arguments in all but a very few cases; unfortunately, such cases are the most interesting.

Second, for $n = 4$, the order of the symmetry group of a regular polytope (which excludes only one of the five cases, and this is in any event a subgroup of index 2 in one of the others) can be calculated with the aid of a solution of a certain trigonometrical equation (see Section 7D, and the remarks on the Petrie polygon in the following). An alternative method involves the evaluation of certain integrals due to Schläfli; since the most relevant one of these cannot be evaluated directly, recourse must be had once more to the simplex dissection results (again, see [125]).

Finally, for other even $n \geqslant 6$ ($n = 6$ and $n = 8$ are the only important cases), the group G can be associated with a honeycomb, and $|G|$ can be found from the relative numbers of faces of this honeycomb. However, none of these last three methods is elementary; in particular, the last depends upon the somewhat deep result (see [120, §9.8]) that these relative numbers exist, and that the analogue of Euler's theorem holds for them. (A variant of this technique appears in [95], but it is used there with knowledge of the order of the group $[3, 3, 5]$ to calculate the densities of the regular 4-dimensional star-polytopes. It is described in [120, Chapter 14] as resting "on rather flimsy foundations"; however, see the case of $[3, 3, 5]$ in the following. The Schläfli function provides an alternative approach to calculating the densities; see [125] once more.)

The method which we shall describe here is, in a vague sense, related to the last of these approaches (we shall make the connexion more explicit later), but the result to which we shall appeal (the Brianchon–Gram theorem) relies only on the ordinary Euler theorem (in $\mathbb{E}^{n-1}$). We shall give this result, and a closely related one, in our discussion of angle-sum relations, and then apply them to our problem of determining the orders of reflexion groups in what follows.

It is worth making an additional remark at this stage. Our calculations will be purely geometric; in other words, though we often use the language of group theory, we do not really make use of the fact that the cones whose sizes we find are the fundamental regions of groups. What we do use is the fact that certain hyperplanes, with given angles between them, determine either simplicial cones or finite euclidean simplices; the criteria which must be satisfied (the Schläfli determinant condition, for which see Section 7D or [120, §7.7]; we shall take its particular applications for granted) pay no regard to whether the reflexions in these hyperplanes generate a finite or discrete group. We shall refer to this again several times later.

Angle-Sum Relations

Let K be an n-dimensional polyhedral set in $\mathbb{E}^n$ (that is, the intersection of finitely many closed half-spaces), and let F be a non-empty face of K. The (*inner*) *angle* $\alpha(F, K)$ of K at F is that proportion of a sufficiently small ball centred at a relatively interior point x of F which lies in K:

$$\alpha(F, K) = \lim_{\lambda \to 0} \frac{\mathrm{Vol}(B(x, \lambda) \cap K)}{\mathrm{Vol}(B(x, \lambda))},$$

where $B(x, \lambda)$ is the ball of radius $\lambda > 0$ with centre x, and Vol denotes volume. We shall employ two angle-sum relations.

3E1 Theorem (Brianchon–Gram) *If $n \geqslant 1$ and P is a convex n-polytope in $\mathbb{E}^n$, then*

$$\sum_F (-1)^{\dim F} \alpha(F, P) = 0.$$

3E2 Theorem (Sommerville) *If P is a polyhedral cone in $\mathbb{E}^n$, then*

$$\sum_F (-1)^{\dim F} \alpha(F, P) = (-1)^n \alpha(A, P),$$

where A is the face of apices of P.

In both theorems, the sums extend over all non-empty faces F (including P itself).

A common generalization of these theorems to arbitrary polyhedral sets is proved in [281]; it is perhaps worth noting that there it is made clear that the results hold on the level of equidissectability (in Sommerville's theorem, the cone P must be replaced by its negative on the right of the equation). Further details about the background to these results can be found there and in [197, §14.1]; for an easy proof of Theorem 3E1, see [386].

We could employ Theorems 3E2 or 3E1 directly; this is what was done in [285]. However, it is better to make a simplification of the resulting formula first. In applying Theorem 3E1, we need only consider an n-simplex T. We label the facets of T by $T_0, \ldots, T_n$, with T_0 subsequently playing a distinguished role. The *angle graph* $\mathcal{G}$ of T has node-set $N := \{0, \ldots, n\}$, and the branch-set B contains $\{j, k\}$ (for $0 \leqslant j < k \leqslant n$) precisely when the dihedral angle between T_j and T_k is *not* $\pi/2$. In the case when T is the fundamental region of some Coxeter group G, this means that $\mathcal{G}$ is just the Coxeter diagram of G, but with the labels left off (a branch corresponding to a label 2 is still omitted). We may then talk about connected subgraphs, components, and so on, of $\mathcal{G}$. Since T is an euclidean simplex, it is not hard to see that $\mathcal{G}$ itself is connected.

If $J \subseteq N$, then we write $T_J := \bigcap\{T_i \mid i \in J\}$ (thus T_j is an abbreviation for $T_{\{j\}}$). If $J \subset N$ (a proper subset), then Theorem 3E2 implies that

$$\sum_{K \subseteq J} (-1)^{\operatorname{card} K} \alpha(T_K, T) = \alpha(T_J, T),$$

since $\dim T_K = n - \operatorname{card} K$. We can now decompose the formula in Theorem 3E1 in the following way. Let $M \subset N$ be a proper subset which induces a connected subgraph $\mathcal{G}(M)$ of $\mathcal{G}$ containing node 0; formally, we also allow $M = \emptyset$ here. We write $\mathcal{M} := \mathcal{M}(N)$ for the set of such M. Define

$$\bar{M} := \{i \in N \mid \{i, j\} \notin B \text{ for any } j \in M\},$$

with $\bar{M} := N \setminus \{0\}$ if $M = \emptyset$. Thus $\bar{M}$ is the set of nodes of $\mathcal{G}$ obtained by deleting M and any neighbours of M in $\mathcal{G}$, with 0 itself taken as the sole neighbour if $M = \emptyset$. In other words, $\bar{M}$ induces the *disjoint* subgraph of $\mathcal{G}$, namely, the maximal subgraph $\mathcal{G}(\bar{M})$ of $\mathcal{G} \setminus \{0\}$ which is disconnected from $\mathcal{G}(M)$.

We now collect together the contributions $(-1)^{\dim F} \alpha(F, T)$ from faces F of T as follows. Suppose that $F = T_J$, with $J \subset N$ (note that $T_N = \emptyset$, which is not counted in Theorem 3E1; however, $T_\emptyset = T$ is counted). Write $M(F)$ for the node-set of the connected component of the induced subgraph $\mathcal{G}(J)$ of $\mathcal{G}$ which contains 0, and let $\bar{M}(F)$ be the corresponding vertex-set of the disjoint subgraph.

Writing $M := M(F)$, we see that $F = T_{M\cup K}$ for some $K \subseteq \bar{M}$, and, since our angles are scaled so that the total angle is 1, we have

$$\alpha(F, T) = \alpha(T_M, T)\alpha(T_K, T).$$

Adding together first the contributions for a fixed subset M yields

$$\begin{aligned}\sum_{M(F)=M} (-1)^{\dim F}\alpha(F, T) &= \sum_{K\subseteq\bar{M}} (-1)^{n-\operatorname{card} M-\operatorname{card} K}\alpha(T_{M\cup K}, T)\\ &= \sum_{K\subseteq\bar{M}} (-1)^{n-\operatorname{card} M-\operatorname{card} K}\alpha(T_M, T)\alpha(T_K, T)\\ &= (-1)^{\operatorname{card} M}\alpha(T_M, T)\alpha(T_{\bar{M}}, T).\end{aligned}$$

We can summarize the discussion as

3E3 Theorem *If T is a euclidean n-simplex, then*

$$\sum_{M\in\mathcal{M}} (-1)^{\operatorname{card} M}\alpha(T_M, T)\alpha(T_{\bar{M}}, T) = 0$$

Exactly similar considerations yield a simplification of Theorem 3E2. The only difference in the calculations is that we now have a simplicial n-cone T, bounded by $T_0, \ldots, T_{n-1}$ (we have changed the notation for future convenience); we thus take $N := \{0, \ldots, n-1\}$. We now allow $M = N$ to give a connected subgraph of $\mathcal{G}(T)$, and our conclusion is

3E4 Theorem *If T is a simplicial n-cone, then*

$$\sum_{M\in\mathcal{M}} (-1)^{\operatorname{card} M}\alpha(T_M, T)\alpha(T_{\bar{M}}, T) = (-1)^n\alpha(O, T),$$

where $O := \{o\}$ is the apex of T.

We now apply Theorems 3E4 and 3E3 to finite groups G generated by reflexions in $\mathbb{E}^n$. Actually, while we talk about orders of subgroups, we are really only making angle-sum calculations, and so our results sit in a wider context. Suppose that $G = \langle R_j \mid j \in N\rangle$ is a finitely generated reflexion group in $\mathbb{E}^n$; we may have to allow G to be infinite, or even non-discrete. We write $G_J := \langle R_j \mid j \in J\rangle$ for each subset $J \subseteq N$, and $g_J := |G_J|$ when this order is finite.

We begin with the easier cases. First, let G be a finite Coxeter group; we may suppose that G acts irreducibly on $\mathbb{E}^n$, so that $N = \{0, \ldots, n-1\}$. Then the mirrors R_j bound a simplicial n-cone T with facets $T_0, \ldots, T_{n-1}$. Moreover, T is a fundamental region for G, and T and its images form a dissection of $\mathbb{E}^n$ into congruent cones. Indeed, for each $J \subseteq N$, the images of T under G_J surround the face T_J of T, so that the total angle of these cones at T_J is 1. The initial discussion and the definition of angle yield at once

3E5 Lemma *For each $J \subset N$,*

$$\alpha(T_J, T) = \frac{1}{g_J}.$$

Note here that, for the case $J = \emptyset$, we have $T_\emptyset = \{E\}$, the identity subgroup, and $g_\emptyset = 1$. We use the abbreviation $g := g_N = |G|$.

Now, if n is even, the term $\alpha(T_N, T) = \alpha(O, T) = 1/g$ occurs on each side of the formula of Theorem 3E4 (or of Theorem 3E2); hence the formula yields no useful information about g. However, if n is odd, then we can move the term $1/g$ from the left to the right side of the equation, and deduce

3E6 Theorem *Let $G = \langle R_0, \ldots, R_{n-1} \rangle$ be a finite Coxeter group, with n odd. Then*

$$\frac{2}{g} = \sum_{M \in \mathcal{M} \setminus \{N\}} \frac{(-1)^{\operatorname{card} M + 1}}{g_M g_{\bar{M}}}.$$

As an example of Theorem 3E6, if $n = 3$, p, q are integers, and $G := [p, q]$ has Coxeter diagram as in Figure 3E2, then $g := g(p, q) = |G|$ is given by

$$\frac{2}{g} = \frac{1}{2q} - \frac{1}{2 \cdot 2} + \frac{1}{2p}.$$

Hence, after simplification,

$$g = \frac{8pq}{4 - (p-2)(q-2)},$$

a formula familiar from [120, 5.43]. We stress here that, if we interpret the left side of the first expression as an angle, we do not actually have to assume that p and q are integers (or even rationals).

Whether n is even or odd, in the particular case when $G = [p_1, \ldots, p_{n-1}]$ is a string group, with the usual labelling of its generators, it is the symmetry group $G(P)$ of the regular convex n-polytope $P = \{p_1, \ldots, p_{n-1}\}$. The base m-face F_m of P is stabilized by $\langle R_j \mid j \neq m \rangle$, so that the number $f_m = f_m(P)$ of m-faces of P is $g/g_{N \setminus \{m\}}$. Indeed, this also holds for the single n-face P itself of P. Since $g_{N \setminus \{m\}} = g_{\{0,\ldots,m-1\}} g_{\{m+1,\ldots,n-1\}}$, and since a subset $M \subseteq \mathcal{M}$ is of the form $M = \{0, \ldots, m-1\}$, with $\bar{M} = \{m+1, \ldots, n-1\}$, then multiplying the formula of Theorem 3E4 by g yields *Euler's Theorem*:

3E7 Theorem *Let P be a regular convex n-polytope, and let the number of m-faces of P be f_m, for $m = 0, \ldots, n$. Then*

$$\sum_{m=0}^{n} (-1)^m f_m = 1.$$

$p \qquad q$

Figure 3E2. The simplex $S(p, q)$.

In order to treat the case when n is even, we need to embed the finite Coxeter group G in an infinite group. We call G *crystallographic* if it is a subgroup of an infinite discrete group $\tilde{G}$ generated by reflexions (in the same euclidean space). Similar considerations to those of Section 3B apply; the mirrors of all the reflexions in $\tilde{G}$ dissect $\mathbb{E}^n$ into fundamental regions, which are simplices if $\tilde{G}$ (or even G) is irreducible. We apply Theorem 3E3 to these simplices.

Let T be such a simplex, bounded by the mirrors $R_0, \ldots, R_n$, which we identify with the corresponding reflexions. We shall always be able to suppose that $G = \langle R_1, \ldots, R_n \rangle$, and that any subgroups generated by other proper subsets of the R_j are isomorphic to subgroups of G (that is, that G is the *special subgroup* of $\tilde{G}$ in the sense of [120, p. 191]); some of these subgroups may actually be isomorphic to G itself. In any event, for each proper subset $J \subset N := \{0, \ldots, n\}$, we see that $G_J := \langle R_j \mid j \in J \rangle$ is a finite subgroup of $\tilde{G}$, which leaves invariant the flat $H_J := \bigcap\{R_j \mid j \in J\}$; hence the face $T_J := H_J \cap T$ of T. (The discreteness of $\tilde{G}$ ensures that all these subgroups G_J are actually finite, but we emphasize once again that we shall be performing pure angle calculations, which do not depend on this discreteness.) Theorem 3E3 and Lemma 3E5 immediately imply.

3E8 Theorem *Let $\tilde{G}$ be an irreducible discrete infinite reflexion group in $\mathbb{E}^n$, with generating reflexions $R_0, \ldots, R_n$ in the bounding hyperplanes of its fundamental region T. With $N := \{0, \ldots, n\}$ and $\mathcal{M} := \mathcal{M}(N)$ as before,*

$$\sum_{M \in \mathcal{M}} \frac{(-1)^{\operatorname{card} M}}{g_M g_{\bar{M}}} = 0.$$

We shall give the important examples of this result; in view of Theorem 3E6, of course, we are mostly interested in the cases when n is even. The interested reader will easily determine which infinite discrete reflexion group has a given finite crystallographic reflexion group as its special subgroup (comparing the lists in Tables 3B1 and 3B2 or in [120, Table IV] makes this straightforward). Let us find the orders of the groups $[3, 3, 4]$, $[3, 4, 3]$, E_6, E_7 and E_8. (We should remark that the order of $[3, 3, 4]$ can be found more simply with the aid of the generic simplex dissection results we shall discuss later; $[3, 4, 3]$ can be dealt with by another generic simplex dissection which we shall not need here, but E_6 and E_8 provide problems of a deeper kind.)

For convenience, let us denote by $S(p_1, \ldots, p_{n-1})$ the simplicial cone or simplex corresponding to the diagram of Figure 3E3, and let $\alpha(p_1, \ldots, p_{n-1})$ denote its angle, which is taken as 0 for a euclidean simplex – we shall only use the cases $n = 4$ or 5.

Figure 3E3. The simplex $S(p_1, \ldots, p_{n-1})$.

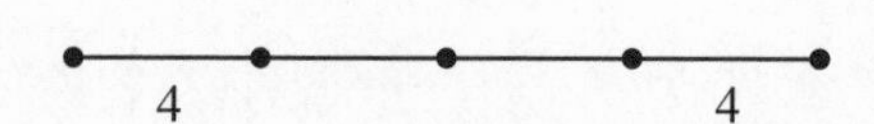

Figure 3E4. The simplex $S(4, 3, 3, 4)$.

First, $[3, 3, 4]$ is the special subgroup of the group $[4, 3, 3, 4]$. Applying Theorem 3E8 to the simplex of Figure 3E4 yields for the unknown order $g = g(3, 3, 4)$:

$$\frac{1}{g} - \frac{1}{2 \cdot 48} + \frac{1}{8 \cdot 8} - \frac{1}{48 \cdot 2} + \frac{1}{g} = 0.$$

(We have substituted the (assumed known) orders of the lower-dimensional groups.) From this easily follows $g = 384$.

In turn, $[3, 4, 3]$ is the special subgroup of $[3, 3, 4, 3]$. Using the just found order $g(3, 3, 4) = 384$, Theorem 3E8 applied to the simplex of Figure 3E5 yields for the order g:

$$\frac{1}{g} - \frac{1}{2 \cdot 48} + \frac{1}{6 \cdot 6} - \frac{1}{24 \cdot 2} + \frac{1}{384} = 0,$$

from which follows $g = 1152$.

We next come to $E_6 = [3^{2,2,1}]$. This is the special subgroup of $T_7 = [3^{2,2,2}]$ (see Figure 3E1 for its diagram), and assuming that we have already found the orders of the lower-dimensional groups ($|A_k| = (k + 1)!$ and $|B_k| = 2^{k-1}k!$ for $k = 4$ or 5 are the only extra orders we need; see [120, Table IV]), the order g satisfies

$$\frac{1}{g} + \frac{1}{2 \cdot 6!} + \frac{1}{3! \cdot 3! \cdot 3!} - \frac{1}{4! \cdot 2 \cdot 2} + \frac{2}{5! \cdot 2} - \frac{2}{6! \cdot 2} - \frac{1}{2^4 \cdot 5!} + \frac{2}{g} = 0,$$

which yields $g = 72 \cdot 6!$.

For E_7, Theorem 3E6 applies, and we have

$$\frac{2}{g} = \frac{1}{72 \cdot 6!} - \frac{1}{2 \cdot 2^4 \cdot 5!} + \frac{1}{3! \cdot 5!} - \frac{1}{4! \cdot 3! \cdot 2} + \frac{1}{7!} + \frac{1}{2^5 \cdot 6!};$$

hence $g = 8 \cdot 9!$.

Finally, we regard E_8 as the special subgroup of T_9, and Theorem 3E8 yields

$$\frac{1}{g} - \frac{1}{2 \cdot 8 \cdot 9!} + \frac{1}{3! \cdot 72 \cdot 6!} - \frac{1}{4! \cdot 2 \cdot 2^4 \cdot 5!} + \frac{1}{5! \cdot 5!} - \frac{1}{6! \cdot 3! \cdot 2} + \frac{1}{9!} + \frac{1}{2^7 \cdot 8!} = 0,$$

from which we deduce that $g = 192 \cdot 10!$.

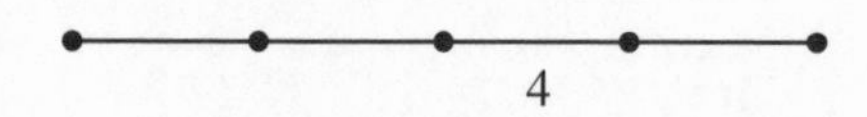

Figure 3E5. The simplex $S(3, 3, 4, 3)$.

The Group [3, 3, 5]

The non-crystallographic reflexion groups have not been dealt with thus far. However, the only group which actually escaped our treatment was [3, 3, 5], the group of the regular 600-cell (or 120-cell) in $\mathbb{E}^4$. We shall now see that even this group is amenable to our approach; as a bonus, we shall also be able to calculate the densities of the regular star-polytopes in $\mathbb{E}^4$. We shall discuss the star-polytopes in Section 7D. (The precise geometric meaning of *density* is defined in [120, p. 94], and we shall not concern ourselves with it overmuch. In any event, we are performing pure angle computations.)

Our basic result, which is just Theorem 3E3 with substitution for the angles of products of cones of dimension at most 3, is

3E9 Theorem *Let $S(p, q, r, s)$ be a euclidean 4-simplex. Then*

$$\alpha(p, q, r) + \alpha(q, r, s) = \frac{1}{8}\left(\frac{1}{p} + \frac{1}{q} + \frac{1}{r} + \frac{1}{s} - \frac{2}{ps} - 1\right).$$

(The result of [95] which Coxeter held in [120, Chapter 14] to "rest on rather flimsy foundations" is basically this, interpreted in terms of orders of groups and densities.)

We now apply Theorem 3E9 to various simplices arising in the dissection of $\mathbb{E}^4$ by the fundamental cones of [3, 3, 5]. But first we need some simplex dissection results; we refer to these as *generic*, because they do not depend on our working in any particular group (in fact, the simplices involved do not have to correspond to any group). Those we use all occur in [125].

3E10 Theorem *Let $p > 2$. Then*

(a) $\alpha(3, p, \frac{p}{2}) = 4\,\alpha(3, 3, p)$,
(b) $\alpha(p, \frac{p}{2}, p) = 6\,\alpha(3, 3, p)$.

Proof. To see this, we merely observe that the simplicial cone whose dihedral angles are all $2\pi/p$ can be dissected into 24 cones $S(3, 3, p)$, 6 cones $S(3, p, \frac{p}{2})$, or 4 cones $S(p, \frac{p}{2}, p)$. The result follows at once. □

As a useful convention, whenever $p > 1$ we shall define p' by

$$\frac{1}{p} + \frac{1}{p'} = 1.$$

Then we have

3E11 Theorem *Let C be the simplicial cone in $\mathbb{E}^4$ with dihedral angles π/p_{ij} for $1 \leqslant j < k \leqslant 4$, and let C' be the cone obtained by replacing p_{j4} by p'_{j4} for $j = 1, 2, 3$. Then*

$$\alpha(C) + \alpha(C') = \frac{1}{4}\left\{\frac{1}{p_{12}} + \frac{1}{p_{13}} + \frac{1}{p_{23}} - 1\right\}.$$

Proof. The two cones C and C' fit together along their common 4th face to form the product of a line with a 3-dimensional cone D whose dihedral angles are π/p_{ij} for

$1 \leqslant j < k \leqslant 3$. Thus $\alpha(C) + \alpha(C')$ is the angle of this product cone, which is just that of D, and so the number given in the theorem. □

We shall need two consequences of Theorem 3E11.

3E12 Corollary

$$\alpha(p, q, r') = \frac{1}{4}\left\{\frac{1}{p} + \frac{1}{q} - \frac{1}{2}\right\} - \alpha(p, q, r).$$

3E13 Corollary

$$\alpha(p, q', r) = \alpha(p, q, r) + \frac{1}{4}\left\{\frac{1}{2} - \frac{1}{q}\right\}.$$

Proof. Corollary 3E12 is a direct application of Theorem 3E11, which also yields

$$\alpha(p, q', r) = \frac{1}{4}\left\{\frac{1}{p} + \frac{1}{2} + \frac{1}{2} - 1\right\} - \alpha(p, q, r'),$$

and Corollary 3E13 follows at once from Corollary 3E12. □

We now employ these results to calculate $g := g(3, 3, 5)$ and the densities of the related regular star-polytopes. We begin by defining two numbers d_1 and d_2 by

$$\alpha\left(\tfrac{5}{2}, 3, 5\right) = d_1\, \alpha(3, 3, 5),$$
$$\alpha\left(3, 3, \tfrac{5}{2}\right) = d_2\, \alpha(3, 3, 5).$$

It is worth stressing at this point that we shall make no prior assumption that d_1 or d_2 is even rational, let alone an integer. (Of course, [280, Theorem 4.1] – see also Section 7D – shows that they must both be integers, because the regular star-polytopes $\{\frac{5}{2}, 3, 5\}$ and $\{3, 3, \frac{5}{2}\}$ have the same symmetry group $[3, 3, 5]$ as $\{3, 3, 5\}$, but that result is completely independent of the present one.) We shall take for granted [120, 14.14], which we can read as saying that certain simplices are euclidean.

Since the simplex $S(\frac{5}{2}, 3, 3, 5)$ is euclidean, we have

$$\alpha\left(3, 3, \tfrac{5}{2}\right) + \alpha(3, 3, 5) = \frac{1}{8}\left\{\frac{2}{5} + \frac{1}{3} + \frac{1}{3} + \frac{1}{5} - \frac{4}{25} - 1\right\};$$

hence

$$\frac{d_2 + 1}{g} = \frac{1}{75} = \frac{192}{14400}.$$

Next, the simplex $S(3, \frac{5}{2}, 5, 3)$ is euclidean, and so

$$\alpha\left(3, \tfrac{5}{2}, 5\right) + \alpha\left(3, 5, \tfrac{5}{2}\right) = \frac{1}{8}\left\{\frac{1}{3} + \frac{2}{5} + \frac{1}{5} + \frac{1}{3} - \frac{2}{9} - 1\right\} = \frac{1}{180}.$$

Now,

$$\alpha\left(3, 5, \tfrac{5}{2}\right) = 4\,\alpha(3, 3, 5) = \frac{4}{g}$$

from Theorem 3E10, while

$$\alpha\left(3, \tfrac{5}{2}, 5\right) = \frac{1}{4}\left\{\frac{1}{3} + \frac{2}{5} - \frac{1}{2}\right\} - \alpha\left(3, \tfrac{5}{2}, \tfrac{5}{4}\right)$$
$$= \frac{7}{120} - 4\,\alpha\left(3, 3, \tfrac{5}{2}\right)$$
$$= \frac{7}{120} - \frac{4d_2}{g},$$

from Corollary 3E12 and Theorem 3E10. Substituting, we have

$$\frac{d_2 - 1}{g} = \frac{1}{4}\left\{\frac{7}{120} - \frac{1}{180}\right\} = \frac{19}{1440} = \frac{190}{14400}.$$

We deduce immediately that $g = 14400$ and $d_2 = 191$.

If we write

$$d(p, q, r) = g\,\alpha(p, q, r) = 14400\,\alpha(p, q, r)$$

for the density of $S(p, q, r)$ (for appropriate p, q, r), we have the obvious values

$$d\left(3, 5, \tfrac{5}{2}\right) = 4,$$
$$d\left(5, \tfrac{5}{2}, 5\right) = 6$$

from Theorem 3E10, while that result and Corollaries 3E12 and 3E13 yield

$$d\left(\tfrac{5}{2}, 5, \tfrac{5}{2}\right) = 14400\left\{\alpha\left(\tfrac{5}{2}, \tfrac{5}{4}, \tfrac{5}{2}\right) + \frac{1}{4}\left[\frac{1}{2} - \frac{4}{5}\right]\right\} = 6d_2 - 1080 = 66,$$

$$d\left(3, \tfrac{5}{2}, 5\right) = 14400\,\frac{7}{120} - 4d_2 = 76,$$

where in the last equation we have not repeated our previous calculations.

It remains to find d_1. Since the simplex $S(\tfrac{5}{2}, 3, 5, \tfrac{5}{2})$ is euclidean, we have

$$\alpha\left(\tfrac{5}{2}, 3, 5\right) + \alpha\left(3, 5, \tfrac{5}{2}\right) = \frac{1}{8}\left\{\frac{2}{5} + \frac{1}{3} + \frac{1}{5} + \frac{2}{5} - \frac{8}{25} - 1\right\} = \frac{1}{600};$$

hence

$$d_1 = 14400\,\frac{1}{600} - 4 = 20.$$

This completes our calculations. We may observe that we have not used the euclidean simplices $S(\tfrac{5}{2}, 5, \tfrac{5}{2}, 5)$ or $S(5, 3, \tfrac{5}{2}, 5)$, which yield no new information.

The Petrie Polygon

We end by briefly discussing the trigonometric calculation for g when $n = 4$. Recall from [120, §12.4] that a *Petrie polygon* of a convex regular n-polytope (or an $(n-1)$-honeycomb) is a skew polygon such that any $n - 1$ consecutive edges, but no n, belong to a Petrie polygon of a facet. For this inductive definition, we declare that the Petrie polygon of a convex polygon is that polygon itself.

The length h of the Petrie polygon of the regular convex polytope $\{p, q, r\}$ is the integer solution of the equation

$$\left(\cos^2\frac{\pi}{h}-\cos^2\frac{\pi}{p}\right)\left(\cos^2\frac{\pi}{h}-\cos^2\frac{\pi}{r}\right)=\cos^2\frac{\pi}{h}\cos^2\frac{\pi}{q}$$

(see [120, 12.35]; of course, such an equation holds when p, q, r are not integers, except that there may only be rational solutions h). However, h is also related to g by

$$\frac{64h}{g}=12-p-2q-r+\frac{4}{p}+\frac{4}{r}$$

(see [120, 12.81]). In view of our independent calculations for g, this last equation yields an alternative way of finding h.

4

Amalgamation

A main thrust in the theory of regular polytopes is that of the amalgamation of polytopes of lower rank. Traditionally, the regular convex polytopes are constructed inductively, beginning with the regular polygons in the plane. The geometry of the ambient space considerably restricts the number of ways in which two regular convex n-polytopes $\mathcal{P}_1$ and $\mathcal{P}_2$ can occur as facets and vertex-figures, respectively, of a regular convex $(n+1)$-polytope $\mathcal{Q}$. Even when the simple necessary condition is satisfied that the vertex-figures of $\mathcal{P}_1$ are isomorphic to the facets of $\mathcal{P}_2$, the polytope $\mathcal{Q}$ need not exist in general. However, if we allow $\mathcal{Q}$ to be an infinite regular tessellation and the ambient space to be hyperbolic, then any two regular convex n-polytopes $\mathcal{P}_1$ and $\mathcal{P}_2$ can be "amalgamated" to form either a finite regular convex $(n+1)$-polytope or an infinite regular tessellation of euclidean or hyperbolic n-space.

This amalgamation problem generalizes readily to abstract regular polytopes. Now, in the absence of an ambient geometry, obstructions to amalgamation must necessarily come from the combinatorics of the polytopes $\mathcal{P}_1$ and $\mathcal{P}_2$. Also, as a new phenomenon, if there does exist an abstract regular $(n+1)$-polytope $\mathcal{Q}$ with facets $\mathcal{P}_1$ and vertex-figures $\mathcal{P}_2$, then in fact there can be many such polytopes, and all these are covered by a single polytope denoted $\{\mathcal{P}_1, \mathcal{P}_2\}$, and called the universal polytope. It is important to identify these universal polytopes and find criteria for their finiteness or non-finiteness.

On the level of groups, the amalgamation of polytopes translates into a problem closely related to the amalgamation of groups, which in many cases can be solved using techniques from combinatorial group theory. This approach also allows us in later chapters to study the classification of regular polytopes by local topological type.

This chapter is organized as follows. In Section 4A we introduce amalgamation of polytopes. Section 4B discusses the classification problem for universal polytopes $\{\mathcal{P}_1, \mathcal{P}_2\}$, which lies at the heart of the topological classification. Then, in Section 4C, finite quotients of infinite universal polytopes are investigated. In Section 4D we construct the "free" regular polytope with a given kind of facet. Finally, in Sections 4E and 4F, we study flat polytopes, and describe a construction of flat polytopes with pre-assigned faces and co-faces.

4A Amalgamation of Polytopes

In this section we discuss the amalgamation of polytopes, and explain how it is related to the amalgamation of groups.

Throughout the section, $\mathcal{P}_1$ and $\mathcal{P}_2$ will be two regular n-polytopes with $n \geqslant 2$. We write $\langle \mathcal{P}_1, \mathcal{P}_2 \rangle$ for the *class* of all regular $(n+1)$-polytopes whose facets are isomorphic to $\mathcal{P}_1$ and whose vertex-figures are isomorphic to $\mathcal{P}_2$.

For instance, if $\mathcal{P}_1$ is the triangle $\{3\}$ and $\mathcal{P}_2$ the pentagon $\{5\}$, the class $\langle \mathcal{P}_1, \mathcal{P}_2 \rangle$ contains exactly two regular 3-polytopes, the icosahedron $\{3, 5\}$ and the hemi-icosahedron $\{3, 5\}_5$. The hemi-icosahedron is constructed from the icosahedron by identifying antipodal vertices; see Section 1D for the notation $\{3, 5\}_5$ employed here. Similarly, in rank 4, the cube $\{4, 3, 3\}$ and the hemi-cube $\{4, 3, 3\}_4$ (obtained by identifying opposite vertices of the cube) are the only members in the class $\langle \{4, 3\}, \{3, 3\} \rangle$ of regular 4-polytopes with cubical facets $\{4, 3\}$ and tetrahedral vertex-figures $\{3, 3\}$.

We begin with a simple observation. Suppose that $\langle \mathcal{P}_1, \mathcal{P}_2 \rangle \neq \emptyset$ and that $\mathcal{Q} \in \langle \mathcal{P}_1, \mathcal{P}_2 \rangle$. Let F_{-1} and F_{n+1} be the faces of $\mathcal{Q}$ of ranks -1 and $n+1$, respectively. If F is a vertex and H a facet of $\mathcal{Q}$ such that $F < H$, then the (medial) section H/F of $\mathcal{Q}$ is a vertex-figure of the facet H/F_{-1} of $\mathcal{Q}$ as well as a facet of the vertex-figure F_{n+1}/F of $\mathcal{Q}$. Therefore, if $\langle \mathcal{P}_1, \mathcal{P}_2 \rangle \neq \emptyset$, then the vertex-figures of $\mathcal{P}_1$ must be isomorphic to the facets of $\mathcal{P}_2$. The class $\langle \mathcal{P}_1, \mathcal{P}_2 \rangle$ is empty if this condition fails to hold. However, as we shall see later, satisfying this condition alone by no means ensures that the class $\langle \mathcal{P}_1, \mathcal{P}_2 \rangle$ is non-empty.

Unless stated otherwise, we shall assume from now on that the vertex-figures $\mathcal{K}$ of $\mathcal{P}_1$ are isomorphic to the facets of $\mathcal{P}_2$. Then any regular $(n+1)$-polytope with facets $\mathcal{P}_1$ and vertex-figures $\mathcal{P}_2$ has medial sections isomorphic to the regular $(n-1)$-polytope $\mathcal{K}$.

Assume that the group of $\mathcal{P}_1$ is $\Gamma(\mathcal{P}_1) = \langle \alpha_0, \ldots, \alpha_{n-1} \rangle$, where $\alpha_0, \ldots, \alpha_{n-1}$ are the distinguished generators of $\Gamma(\mathcal{P}_1)$. As $\mathcal{P}_2$ will be the vertex-figure of the polytopes in $\langle \mathcal{P}_1, \mathcal{P}_2 \rangle$, we shift the indices of the generators for $\Gamma(\mathcal{P}_2)$ by 1, and write $\Gamma(\mathcal{P}_2) = \langle \beta_1, \ldots, \beta_n \rangle$; that is, β_i fixes all but the $(i-1)$-face of the base flag of $\mathcal{P}_2$. Note that, by our assumption on $\mathcal{P}_1$ and $\mathcal{P}_2$, the subgroups $\Gamma(\mathcal{K}) = \langle \alpha_1, \ldots, \alpha_{n-1} \rangle$ of $\Gamma(\mathcal{P}_1)$ and $\langle \beta_1, \ldots, \beta_{n-1} \rangle$ of $\Gamma(\mathcal{P}_2)$ are isomorphic under the mapping $\alpha_i \mapsto \beta_i$ for $i = 1, \ldots, n-1$.

If $\mathcal{Q}$ is any regular $(n+1)$-polytope in $\langle \mathcal{P}_1, \mathcal{P}_2 \rangle$ with group $\Gamma(\mathcal{Q}) = \langle \gamma_0, \ldots, \gamma_n \rangle$ (where $\gamma_0, \ldots, \gamma_n$ are its distinguished generators), then $\gamma_i \mapsto \alpha_i$ for $i = 0, \ldots, n-1$ and $\gamma_i \mapsto \beta_i$ for $i = 1, \ldots, n$ define isomorphisms from the subgroups $\langle \gamma_0, \ldots, \gamma_{n-1} \rangle$ and $\langle \gamma_1, \ldots, \gamma_n \rangle$ of $\Gamma(\mathcal{Q})$ onto $\Gamma(\mathcal{P}_1)$ and $\Gamma(\mathcal{P}_2)$, respectively; the former subgroup is the group of the base facet of $\mathcal{Q}$ and the latter is that of the vertex-figure at the base vertex of $\mathcal{Q}$. In particular, these isomorphisms translate into the same set of defining relations for the groups.

As a consequence, the search of the "largest" polytope in $\langle \mathcal{P}_1, \mathcal{P}_2 \rangle$ involves analysis of the group

4A1 $$\Gamma := \langle \rho_0, \ldots, \rho_n \rangle,$$

given by the following three sets of defining relations: first, for $\rho_0, \ldots, \rho_{n-1}$, the relations for $\Gamma(\mathcal{P}_1)$ in terms of $\alpha_0, \ldots, \alpha_{n-1}$ but with α_i replaced by ρ_i; second, for $\rho_1, \ldots, \rho_n$, the relations for $\Gamma(\mathcal{P}_2)$ in terms of $\beta_1, \ldots, \beta_n$ but with β_i replaced by ρ_i; third, the (necessary) commutation relation $(\rho_0\rho_n)^2 = \varepsilon$. We can now prove the following important result.

4A2 Theorem *Let $\mathcal{P}_1$ and $\mathcal{P}_2$ be regular n-polytopes such that $\langle \mathcal{P}_1, \mathcal{P}_2 \rangle \neq \emptyset$. Then $\langle \mathcal{P}_1, \mathcal{P}_2 \rangle$ contains a polytope which covers (in fact, $(n-1)$-covers) every other polytope in $\langle \mathcal{P}_1, \mathcal{P}_2 \rangle$. Its group is Γ.*

Proof. Let $\mathcal{Q} \in \langle \mathcal{P}_1, \mathcal{P}_2 \rangle$, and let $\Gamma(\mathcal{Q}) = \langle \gamma_0, \ldots, \gamma_n \rangle$, with γ_i as before. We shall apply the quotient criterion of Theorem 2E17 to $\Gamma(\mathcal{Q})$ and the group $\Gamma = \langle \rho_0, \ldots, \rho_n \rangle$. By construction of Γ, the generators γ_i of $\Gamma(\mathcal{Q})$ satisfy all the defining relations for Γ in terms of the ρ_i, but in general further independent relations too. It follows that the mappings $\rho_i \mapsto \gamma_i$ for $i = 0, \ldots, n$ induce a homomorphism $\varphi\colon \Gamma \mapsto \Gamma(\mathcal{Q})$. Further, again by construction of Γ, the mappings $(\alpha_i =)\ \gamma_i \mapsto \rho_i$ for $i = 0, \ldots, n-1$ give a homomorphism $\psi\colon \langle \gamma_0, \ldots, \gamma_{n-1} \rangle \to \langle \rho_0, \ldots, \rho_{n-1} \rangle$, which inverts the restriction of φ to $\langle \rho_0, \ldots, \rho_{n-1} \rangle$. It follows that φ is one-to-one on the subgroup $\langle \rho_0, \ldots, \rho_{n-1} \rangle$, and similarly on $\langle \rho_1, \ldots, \rho_n \rangle$. By the quotient criterion, Γ is a string C-group and φ induces a covering $\mathcal{P}(\Gamma) \mapsto \mathcal{Q}$, where $\mathcal{P}(\Gamma)$ is the regular $(n+1)$-polytope corresponding to Γ. Here we have also used Lemma 2E14 to obtain $\mathcal{Q} \cong \mathcal{P}(\Gamma(\mathcal{Q}))$.

We conclude the proof by noting that $\mathcal{P}(\Gamma) \in \langle \mathcal{P}_1, \mathcal{P}_2 \rangle$. In fact, the facets and vertex-figures of $\mathcal{P}(\Gamma)$ are the regular n-polytopes corresponding to the C-(sub)groups $\langle \rho_0, \ldots, \rho_{n-1} \rangle$ and $\langle \rho_1, \ldots, \rho_n \rangle$, and these are $\mathcal{P}_1$ and $\mathcal{P}_2$, respectively. In particular, the given covering is an $(n-1)$-covering, since it preserves the facets and vertex-figures. □

The polytope of Theorem 4A2 is denoted by the generalized Schläfli symbol $\{\mathcal{P}_1, \mathcal{P}_2\}$, and is called the *universal regular $(n+1)$-polytope with facets $\mathcal{P}_1$ and vertex-figures $\mathcal{P}_2$*, or simply, the *universal polytope in $\langle \mathcal{P}_1, \mathcal{P}_2 \rangle$*. In the following, we shall elaborate on the structure of its group $\Gamma = \Gamma(\{\mathcal{P}_1, \mathcal{P}_2\})$, using free products with amalgamation [368]. We also mention that Theorem 4A2 can be derived from more general results on coverings of "chamber complexes" [350, 418, §§2, 3, 5]. The universal polytopes provide the formal concept for the structures which Grünbaum [199, p. 196] had in mind when he investigated the "naturally generated polystromata".

The following proposition describes duality in the context of classes and universal polytopes.

4A3 Proposition *Let $\mathcal{P}_1$ and $\mathcal{P}_2$ be regular n-polytopes such that $\langle \mathcal{P}_1, \mathcal{P}_2 \rangle \neq \emptyset$. Set $\langle \mathcal{P}_1, \mathcal{P}_2 \rangle^* := \{\mathcal{P}^* \mid \mathcal{P} \in \langle \mathcal{P}_1, \mathcal{P}_2 \rangle\}$, where $\mathcal{P}^*$ denotes the dual of $\mathcal{P}$. Then $\langle \mathcal{P}_1, \mathcal{P}_2 \rangle^* = \langle \mathcal{P}_2^*, \mathcal{P}_1^* \rangle$ and $\{\mathcal{P}_1, \mathcal{P}_2\}^* = \{\mathcal{P}_2^*, \mathcal{P}_1^*\}$.*

Proof. For any polytope $\mathcal{P}$, the facets and vertex-figures of $\mathcal{P}^*$ are the duals of the vertex-figures and facets of $\mathcal{P}$, respectively. This proves the first statement. Also, $\Gamma(\mathcal{P}^*) = \Gamma(\mathcal{P})$, and the distinguished generators for $\Gamma(\mathcal{P}^*)$ are just those for $\Gamma(\mathcal{P})$ in reverse order. It follows that $\{\mathcal{P}_1, \mathcal{P}_2\}^*$ is indeed universal in $\langle \mathcal{P}_2^*, \mathcal{P}_1^* \rangle$, as required. □

We next review the concept of a free product of groups with amalgamation [269, §II.11]. We shall also use this in later sections. Each presentation of a group Λ can be written in the form $\Lambda = \langle X \mid \mathcal{R} \rangle$, where X is the set of defining generators and $\mathcal{R}$ is the set of defining relations of Λ.

Let $\{\Lambda_i \mid i \in I\}$ be a family of mutually disjoint groups indexed by a set I. Let $\Lambda_i = \langle X_i \mid \mathcal{R}_i \rangle$ for each $i \in I$, where X_i is the set of generators and $\mathcal{R}_i$ the set of relations for Λ_i. Suppose that Λ is a group and that $\{\varphi_i \mid i \in I\}$ is a family of maps such that $\varphi_i \colon \Lambda \mapsto \Lambda_i$ is an injective homomorphism for each $i \in I$. Then the *free product of the groups Λ_i amalgamating the subgroups $\Lambda\varphi_i$* is the group

$$\Pi = \left\langle \bigcup_{i \in I} X_i \mid \mathcal{R}_i \text{ for } i \in I, \text{ and } \alpha\varphi_i = \alpha\varphi_j \text{ for } \alpha \in \Lambda \text{ and } i, j \in I \right\rangle.$$

In other words, the generators of Π are all the generators of the groups Λ_i, and the defining relations for Π are all the defining relations for the Λ_i, as well as the extra relations which identify the elements $\alpha\varphi_i$ of Λ_i and $\alpha\varphi_j$ of Λ_j for each $\alpha \in \Lambda$ and $i, j \in I$. Note that the group Π depends on the groups Λ and Λ_i as well as on the homomorphisms φ_i for $i \in I$, but not on the sets X_i of generators and $\mathcal{R}_i$ of relations for each Λ_i. If the subgroup Λ is trivial, then Π is called the *free product of the groups Λ_i*; in this case, there are no extra relations.

In our applications, we usually have two groups Λ_1 and Λ_2 with a pair of isomorphic subgroups Λ and Ω, respectively. If $\varphi_1 \colon \Lambda \to \Lambda_1$ is the canonical injection and $\varphi_2 \colon \Lambda \to \Lambda_2$ is the injective homomorphism such that $\Omega = \Lambda\varphi_2$, then the extra relations take the form $\alpha\varphi_2 = \alpha$ for $\alpha \in \Lambda$. In this case, we simply write

$$\Pi = \Lambda_1 \underset{\Lambda}{*} \Lambda_2,$$

with the understanding that the corresponding map φ_2 is specified.

In the general situation, a sequence $\tau_1, \ldots, \tau_k$ of elements of Π will be called *reduced* if the following conditions hold.

4A4

1. Each τ_j is in one of the factors Λ_i of Π for $i \in I$. (More precisely, τ_j can be written as a product in the corresponding generators from X_i.)
2. Successive τ_j, τ_{j+1} come from different factors.
3. If $k > 1$, then no τ_j is in one of the subgroups $\Lambda\varphi_i$ for $i \in I$. (More informally, no τ_j comes from Λ.)
4. If $k = 1$, then $\tau_j \neq \varepsilon$.

We also define the empty sequence (with no elements) to be reduced. It is immediate that each element of Π is the product of the elements in a reduced sequence. In particular, ε is regarded as the product of the empty sequence. However, less obvious is the following converse, known as the *Normal Form Theorem for Free Products with Amalgamation*. This theorem justifies calling the groups Λ_i the *factors* of Π.

4A5 Theorem *Let Π be a free product with amalgamation as before. If $\tau_1, \ldots, \tau_k$ is a reduced sequence and $k \geqslant 1$, then $\tau_1 \cdots \tau_k \neq \varepsilon$ in Π. In particular, the groups Λ_i for $i \in I$ are embedded into Π by the maps $\tau \mapsto \tau$ for $\tau \in \Lambda_i$.*

The following result is known as the *Torsion Theorem for Free Products with Amalgamation*.

4A6 Theorem *Let Π be a free product with amalgamation as before. Then each element of finite order in Π is conjugate to an element of finite order in one of the factors Λ_i.*

Returning to the discussion of classes of polytopes, let $\mathcal{P}_1$ and $\mathcal{P}_2$ be regular n-polytopes such that the vertex-figures $\mathcal{K}$ of $\mathcal{P}_1$ are isomorphic to the facets of $\mathcal{P}_2$. As before, let $\Gamma(\mathcal{P}_1) = \langle \alpha_0, \ldots, \alpha_{n-1} \rangle$ and $\Gamma(\mathcal{P}_2) = \langle \beta_1, \ldots, \beta_n \rangle$, so that $\Gamma(\mathcal{K}) = \langle \alpha_1, \ldots, \alpha_{n-1} \rangle$. Recall the definition of $\Gamma = \langle \rho_0, \ldots, \rho_n \rangle$ in (4A1), and note that it did not assume that $\langle \mathcal{P}_1, \mathcal{P}_2 \rangle \neq \emptyset$.

Now consider the free product with amalgamation

4A7
$$\Pi := \Gamma(\mathcal{P}_1) \underset{\Gamma(\mathcal{K})}{*} \Gamma(\mathcal{P}_2),$$

which is defined by the canonical injection $\varphi_1 \colon \Gamma(\mathcal{K}) \to \Gamma(\mathcal{P}_1)$ and the injective homomorphism $\varphi_2 \colon \Gamma(\mathcal{K}) \to \Gamma(\mathcal{P}_2)$ given by $\alpha_i \mapsto \beta_i$ for $i = 1, \ldots, n-1$. Then $\Pi = \langle \alpha_0, \ldots, \alpha_{n-1}, \beta_n \rangle$, and $\alpha_i = \beta_i$ in Π for $i = 1, \ldots, n-1$. All products $\alpha_i \alpha_j$, $\beta_i \beta_j$ and $\alpha_i \beta_j$, save for $\alpha_0 \beta_n$, are contained in one of the factors $\Gamma(\mathcal{P}_1)$ or $\Gamma(\mathcal{P}_2)$, and so are determined solely by $\mathcal{P}_1$ and $\mathcal{P}_2$. However, by Theorem 4A6, the element $\alpha_0 \beta_n$ has infinite period. In particular, it does not have period 2 as is required for a string C-group. Thus, to make α_0 and β_n commute, we must factor out by the relation $(\alpha_0 \beta_n)^2 = \varepsilon$. On the other hand, while the intersection property can be shown to hold for Π (using Theorem 4A5), it is not true in general that it also passes on to a quotient.

Now let N be the normal closure of $(\alpha_0 \beta_n)^2$ in Π, and consider the quotient Π/N. Then $(N\alpha_0 N\beta_n)^2 = \varepsilon$ in Π/N. More generally, Π/N has exactly the same presentation in terms of its generators $N\alpha_0, \ldots, N\alpha_{n-1}, N\beta_n$ as Γ has in terms of $\rho_0, \ldots, \rho_n$. Therefore $\Gamma \cong \Pi/N$, and the mapping given by $\rho_i \mapsto N\alpha_i$ for $i = 0, \ldots, n-1$, and $\rho_n \mapsto N\beta_n$, induces an isomorphism $\Gamma \mapsto \Pi/N$.

4A8 Proposition *Let $\mathcal{P}_1$ and $\mathcal{P}_2$ be regular n-polytopes such that the vertex-figures $\mathcal{K}$ of $\mathcal{P}_1$ are isomorphic to the facets of $\mathcal{P}_2$. Let $\Gamma = \langle \rho_0, \ldots, \rho_n \rangle$ be as in (4A1), and let $\Pi = \Gamma(\mathcal{P}_1) *_{\Gamma(\mathcal{K})} \Gamma(\mathcal{P}_2)$ and $\langle \alpha_0, \ldots, \alpha_{n-1}, \beta_n \rangle$ be as in (4A7).*

- *(a) $\Gamma \cong \Pi/N$, where N is the normal closure of $(\alpha_0 \beta_n)^2$ in Π.*
- *(b) If $\langle \mathcal{P}_1, \mathcal{P}_2 \rangle \neq \emptyset$, then $\Gamma(\{\mathcal{P}_1, \mathcal{P}_2\}) \cong \Gamma \cong \Pi/N$.*
- *(c) The polytope $\{\mathcal{P}_1, \mathcal{P}_2\}$ exists (or, equivalently, $\langle \mathcal{P}_1, \mathcal{P}_2 \rangle \neq \emptyset$) if and only if Γ is a string C-group, $\langle \rho_0, \ldots, \rho_{n-1} \rangle \cong \Gamma(\mathcal{P}_1)$, and $\langle \rho_1, \ldots, \rho_n \rangle \cong \Gamma(\mathcal{P}_2)$.*

Proof. Part (a) has already been proved, and both part (b) and the necessity of part (c) restate Theorem 4A2. For the sufficiency of part (c), note that the conditions on Γ imply that Γ is the group of a regular $(n+1)$-polytope $\mathcal{P}$ with facets isomorphic to $\mathcal{P}_1$ and vertex-figures isomorphic to $\mathcal{P}_2$. In particular, $\langle \mathcal{P}_1, \mathcal{P}_2 \rangle \neq \emptyset$. Therefore, by Theorem 4A2, $\mathcal{P} = \{\mathcal{P}_1, \mathcal{P}_2\}$. □

This proposition explains why the polytope $\{\mathcal{P}_1, \mathcal{P}_2\}$ can be thought of as the universal amalgamation of $\mathcal{P}_1$ and $\mathcal{P}_2$ along $\mathcal{K}$. (The notation $\mathcal{P}_1 *_{\mathcal{K}} \mathcal{P}_2$ which was used in [368, p. 311] in place of $\{\mathcal{P}_1, \mathcal{P}_2\}$ emphasized this fact.) In essence, the proposition translates the existence problem for $\{\mathcal{P}_1, \mathcal{P}_2\}$ into a problem for presentation of groups. This is the form in which we shall later attempt a solution for specific $\mathcal{P}_1$ and $\mathcal{P}_2$.

We illustrate these concepts for the regular polytopes $\{p_1, \ldots, p_{n-1}\}$. Recall from Section 3D that $\{p_1, \ldots, p_{n-1}\}$ is the regular n-polytope with (Coxeter) group $[p_1, \ldots, p_{n-1}]$, and that this is universal among all regular n-polytopes of the same type $\{p_1, \ldots, p_{n-1}\}$.

4A9 Proposition *Let $n \geqslant 2$, and let $2 \leqslant p_1, \ldots, p_n \leqslant \infty$. Then*

$$\{\{p_1, \ldots, p_{n-1}\}, \{p_2, \ldots, p_n\}\} = \{p_1, \ldots, p_n\}.$$

Hence, $\langle\{p_1, \ldots, p_{n-1}\}, \{p_2, \ldots, p_n\}\rangle \neq \emptyset$.

Proof. Now we have $\mathcal{P}_1 := \{p_1, \ldots, p_{n-1}\}$, $\mathcal{P}_2 := \{p_2, \ldots, p_n\}$, and $\mathcal{K} := \{p_2, \ldots, p_{n-1}\}$; if $n = 2$, then $\mathcal{K}$ is of rank 1. Then the class $\langle \mathcal{P}_1, \mathcal{P}_2\rangle$ is non-empty, because it contains $\{p_1, \ldots, p_n\}$. Since $\Gamma(\mathcal{P}_1) = [p_1, \ldots, p_{n-1}]$ and $\Gamma(\mathcal{P}_2) = [p_2, \ldots, p_n]$, we also have $\Gamma(\{\mathcal{P}_1, \mathcal{P}_2\}) = [p_1, \ldots, p_n]$. Hence, by Theorem 3D4, $\{\mathcal{P}_1, \mathcal{P}_2\} = \{p_1, \ldots, p_n\}$. □

Proposition 4A9 says that the regular n-polytopes $\{p_1, \ldots, p_n\}$ are also universal in the class of regular polytopes with facets $\{p_1, \ldots, p_{n-1}\}$ and vertex-figures $\{p_2, \ldots, p_n\}$. For example, the 24-cell $\{3, 4, 3\}$ in euclidean 4-space is the universal regular 4-polytope with octahedral facets $\{3, 4\}$ and cubical vertex-figures $\{4, 3\}$. This is a finite polytope, in contrast to the cubical tessellation $\{4, 3, 4\}$ of euclidean 3-space, which is the universal regular 4-polytope with cubical facets and octahedral vertex-figures.

The enumeration of the finite universal polytopes $\{\mathcal{P}_1, \mathcal{P}_2\}$ will be a central topic in later chapters. We conclude this section with a result on self-dual polytopes.

4A10 Proposition *Let $\mathcal{P}_1$ and $\mathcal{P}_2$ be two regular n-polytopes such that $\langle \mathcal{P}_1, \mathcal{P}_2\rangle \neq \emptyset$. Then $\{\mathcal{P}_1, \mathcal{P}_2\}$ is self-dual if and only if $\mathcal{P}_2$ is the dual of $\mathcal{P}_1$.*

Proof. The facets of the dual of any polytope are the duals of the original vertex-figures. Thus, if $\{\mathcal{P}_1, \mathcal{P}_2\}$ is self-dual, then $\mathcal{P}_1$ and $\mathcal{P}_2$ must be duals.

For the converse, let $\Gamma(\mathcal{P}_1) = \langle \alpha_0, \ldots, \alpha_{n-1}\rangle$, $\Gamma(\mathcal{P}_2) = \langle \beta_1, \ldots, \beta_n\rangle$, and $\Gamma(\{\mathcal{P}_1, \mathcal{P}_2\}) = \Gamma = \langle \rho_0, \ldots, \rho_n\rangle$. By Theorem 2E11, the polytope $\{\mathcal{P}_1, \mathcal{P}_2\}$ is self-dual if there is an involutory group automorphism ω of Γ such that $\rho_i\omega = \rho_{n-i}$ for $i = 0, \ldots, n$. If $\mathcal{P}_2$ is the dual of $\mathcal{P}_1$, then a presentation for $\Gamma(\mathcal{P}_2)$ is obtained from one for $\Gamma(\mathcal{P}_1)$ by simply exchanging in each relation each generator α_i by β_{n-i}. This in turn gives a presentation for $\Pi := \Gamma(\mathcal{P}_1) *_{\Gamma(\mathcal{K})} \Gamma(\mathcal{P}_2)$ which is symmetric under the exchange $\alpha_i \leftrightarrow \beta_{n-i}$ for $i = 0, \ldots, n-1$. Note for this that $\mathcal{K}$ is self-dual, and that $\alpha_i \leftrightarrow \beta_i$ for $i = 1, \ldots, n-1$ gives the corresponding involutory group automorphism for $\Gamma(\mathcal{K})$. By Proposition 4A8(a), $\Gamma \cong \Pi/N$, where N is the normal closure of

$(\alpha_0\beta_n)^2$ in Π. However, if α_0 and β_n are exchanged, then $(\alpha_0\beta_n)^2$ is changed to its inverse $(\beta_n\alpha_0)^2$, and therefore N is invariant. It follows that $\alpha_i \leftrightarrow \beta_{n-i}$ for $i = 0, \ldots, n-1$ induces the required group automorphism ω on Γ. □

4B The Classification Problem

A central problem in the classical theory of regular polytopes is the complete enumeration of all regular polytopes and tessellations and their groups in spherical, euclidean or hyperbolic space. When asked within the theory of abstract polytopes, the classification problem must necessarily take a different form, because *a priori* an abstract polytope is not embedded into the geometry of an ambient space. An appropriate substitute is now the enumeration of abstract regular polytopes by their local or global topological type. In the first place this necessitates associating with abstract polytopes a natural topology, a problem which is very subtle and which in general cannot be solved uniquely. On the other hand, many polytopes admit a natural topology, and so are subject to the classification with respect to such a topology. In this context, the convex regular polytopes occur as spherical polytopes and the regular tessellations in euclidean or hyperbolic space as locally spherical polytopes.

In this section, we shall clarify the concept of combinatorial classification. This underlies the topological study which will be the subject of later chapters. Our discussion is carried out in terms of the universal polytopes $\{\mathcal{P}_1, \mathcal{P}_2\}$ which were introduced in the previous section. We begin with two simple geometric examples to illustrate several natural questions which can be asked about these polytopes.

Let us assume that we wish to construct a triangulated surface in which every vertex of the triangulation is contained in exactly 5 triangles; that is, the vertex-figures are pentagons $\{5\}$. This can be done in only two ways, both of which lead to finite triangulations. If the triangulation is "freely" generated, then the resulting surface is the 2-sphere and the triangulation is isomorphic to the icosahedron $\{3, 5\}$. However, if additional identifications are permitted, then we can also construct the hemi-icosahedron $\{3, 5\}_5$, the triangulation of the real projective plane obtained from $\{3, 5\}$ by identifying antipodal points. In the notation of the previous section, $\{3, 5\}$ and $\{3, 5\}_5$ are members of the class $\langle\{3\}, \{5\}\rangle$; in particular, they are regular. Furthermore, by Proposition 4A9, $\{\{3\}, \{5\}\} = \{3, 5\}$; that is, the icosahedron is the universal 3-polytope with triangular facets and pentagonal vertex-figures. The important point to make here is that this universal polytope is finite.

The picture changes completely if we require exactly 6 triangles around a vertex. Now there are many ways to generate triangulations, which include the infinite triangular tessellation $\{3, 6\}$ of the euclidean plane; this covers all other polytopes in the class $\langle\{3\}, \{6\}\rangle$. This class also contains infinitely many finite triangulations, such as the triangular torus map $\{3, 6\}_{(s,0)}$ which is derived from a skew $s \times s$ "chess board" of $2s^2$ triangles by identifying opposite sides; see Figure 1D2 for the case $s = 3$. In contrast to the previous example, the universal polytope $\{\{3\}, \{6\}\} = \{3, 6\}$ is now infinite.

These two examples illustrate the following general problems about universal polytopes $\{\mathcal{P}_1, \mathcal{P}_2\}$, with a given pair of regular n-polytopes $\mathcal{P}_1$ and $\mathcal{P}_2$ such that the vertex-figures $\mathcal{K}$ of $\mathcal{P}_1$ are isomorphic to the facets of $\mathcal{P}_2$.

- When is the class $\langle \mathcal{P}_1, \mathcal{P}_2 \rangle$ non-empty? Or, equivalently (by Theorem 4A2), when does the universal polytope $\{\mathcal{P}_1, \mathcal{P}_2\}$ exist?
- Given finite polytopes $\mathcal{P}_1$ and $\mathcal{P}_2$, when is $\{\mathcal{P}_1, \mathcal{P}_2\}$ finite? That is, when does this universal polytope behave like a finite convex polytope, and when like an infinite tessellation?
- Construct $\{\mathcal{P}_1, \mathcal{P}_2\}$ explicitly, and recognize its group $\Gamma(\{\mathcal{P}_1, \mathcal{P}_2\})$.

In this and later chapters, when we use the term "combinatorial classification" of regular polytopes, then in the given context we mean the classification of all the *finite universal* polytopes.

In a typical application, the facets $\mathcal{P}_1$ or the vertex-figures $\mathcal{P}_2$ will belong to a family of n-polytopes, each defined in terms of one or more parameters, and the task is then to determine exactly for which parameters the universal polytope $\{\mathcal{P}_1, \mathcal{P}_2\}$ is finite. We shall give an example in the following. Our usual approach is to translate this problem into an equivalent one for groups with presentations which depend on parameters, and then to determine the finite groups with these presentations.

As explained in the previous section, given $\mathcal{P}_1$ and $\mathcal{P}_2$, the search for the universal polytope $\{\mathcal{P}_1, \mathcal{P}_2\}$ involves analysis of the group Γ of (4A1) generated by involutions $\rho_0, \ldots, \rho_n$, subject to the relations dictated by $\Gamma(\mathcal{P}_1)$ (for $\rho_0, \ldots, \rho_{n-1}$) and $\Gamma(\mathcal{P}_2)$ (for $\rho_1, \ldots, \rho_n$), together with the commutation relation $(\rho_0\rho_n)^2 = \varepsilon$. By Proposition 4A8(a), this group is a quotient of the free product Π of $\Gamma(\mathcal{P}_1)$ and $\Gamma(\mathcal{P}_2)$ with amalgamation along their joint subgroup $\Gamma(\mathcal{K})$, the quotient being defined by the additional relation $(\rho_0\rho_n)^2 = \varepsilon$. Now, according to Proposition 4A8(c), the universal polytope $\{\mathcal{P}_1, \mathcal{P}_2\}$ exists if and only if Γ is a string C-group and its two subgroups $\langle \rho_0, \ldots, \rho_{n-1} \rangle$ and $\langle \rho_1, \ldots, \rho_n \rangle$ are isomorphic to $\Gamma(\mathcal{P}_1)$ and $\Gamma(\mathcal{P}_2)$, respectively. In practice it is usually difficult to verify these conditions.

Of course, the case $n = 2$ is easy and follows from Proposition 4A9. Here $\mathcal{P}_1 = \{p\}$ and $\mathcal{P}_2 = \{q\}$ (for some p and q), and the universal polytope $\{\mathcal{P}_1, \mathcal{P}_2\}$ of rank 3 is precisely the regular tessellation $\{p, q\}$ on the 2-sphere, in the euclidean plane, or in the hyperbolic plane. However, in higher ranks the structure of abstract regular polytopes is far less obvious, and is complicated by the lack of easily accessible non-classical examples. To illustrate this, we conclude with a typical example of a combinatorial classification problem in rank 4.

Let $\mathcal{P}_1$ be the torus map $\{6, 3\}_{(s,0)}$, which is the dual of the map $\{3, 6\}_{(s,0)}$ mentioned earlier in the section, and let $\mathcal{P}_2$ be the tetrahedron $\{3, 3\}$. Then, if it exists, $\{\mathcal{P}_1, \mathcal{P}_2\} = \{\{6, 3\}_{(s,0)}, \{3, 3\}\}$ is a regular 4-polytope with toroidal facets and spherical vertex-figures. The corresponding group Γ now has the presentation

$$\begin{aligned}
&\rho_0^2 = \rho_1^2 = \rho_2^2 = \rho_3^2 = \varepsilon \\
&(\rho_0\rho_1)^6 = (\rho_1\rho_2)^3 = (\rho_2\rho_3)^3 = (\rho_0\rho_2)^2 = (\rho_0\rho_3)^2 = (\rho_1\rho_3)^2 = \varepsilon, \\
&(\rho_0\rho_1\rho_2)^{2s} = \varepsilon.
\end{aligned}$$

There are two ways of looking at this presentation. First, we can think of it as the presentation for Γ expressed as the amalgamated free product Π, factored by the commutation relation $(\rho_0\rho_3)^2 = \varepsilon$. In other words, Γ is defined by the relations for $\Gamma(\mathcal{P}_1)$ and $\Gamma(\mathcal{P}_2)$ as well as the commutation relation. Here we use the fact that the relation in the third row is the one extra relation (in addition to those for the group [6, 3]) which defines the torus map $\{6, 3\}_{(s,0)}$ as a quotient of the euclidean plane tessellation {6, 3}; see Section 1D. Second, and equivalently, we can also observe that the relations in the first two rows are the standard relations for the Coxeter group [6, 3, 3], which is the symmetry group of the regular honeycomb {6, 3, 3} in hyperbolic 3-space; see Section 3C. The group Γ is now the quotient of [6, 3, 3] by the relation $(\rho_0\rho_1\rho_2)^{2s} = \varepsilon$; in the notation of (2F1),

$$\big\{\{6,3\}_{(s,0)}, \{3,3\}\big\} = \{6,3,6\}/\langle\langle(012)^{2s}\rangle\rangle.$$

The extra relation involves only the first three generators (but *a priori* it cannot be ruled out that it also affects the group of the vertex-figure).

In this example, only the facets $\mathcal{P}_1$ are determined by a parameter s, while the vertex-figures $\mathcal{P}_2$ remain fixed. The corresponding polytopes $\{\mathcal{P}_1, \mathcal{P}_2\}$ and their groups Γ will clearly also depend on s. We shall see in Section 11B that these polytopes $\{\mathcal{P}_1, \mathcal{P}_2\}$ exist for all $s \geqslant 2$, but that they are finite only for $s = 2, 3, 4$ (that is, when the torus map is "small").

4C Finite Quotients of Universal Polytopes

It is folklore in Riemann surface theory that each regular tessellation $\{p, q\}$ of the euclidean or hyperbolic plane is the universal 2-covering for an infinite number of finite regular maps of the same type $\{p, q\}$ on closed compact surfaces. Restated in terms of groups, this says that each infinite Coxeter group $[p, q]$ has infinitely many finite quotients which are C-groups for the same Coxeter diagram. (For example, except perhaps for finitely many values of p and q, the hyperbolic Coxeter group $[p, q]$ admits representations onto almost all symmetric and alternating groups (see [331, Theorems 1–3]).)

In this section, we follow [298] and extend this result to universal polytopes $\langle\mathcal{P}_1, \mathcal{P}_2\rangle$ which are infinite and satisfy certain conditions. Our method of proof is non-constructive and employs the concept of residual finiteness of groups. We generalize a technique used in Vince [436, Theorem 6.3] to establish a similar result for graph-theoretic structures known as "regular combinatorial maps", which also yields the statement for maps on surfaces (see also [9, 170, 188, 234, 314, 332, 435, 462, 464]). One of the consequences is that the earlier result for the plane carries over to regular tessellations in euclidean or hyperbolic spaces of any dimension. Further applications to other kinds of polytopes will be discussed in later chapters.

Let Γ be any group. Then Γ is called *residually finite* if, for any $\alpha \in \Gamma$ with $\alpha \neq \varepsilon$, there exists a homomorphism φ of Γ onto some finite group such that $\alpha\varphi \neq \varepsilon$ [271, p. 116]. The following is a useful extension of the definition.

4C1 Proposition *A group Γ is residually finite if and only if, for each finite subset $\{\alpha_1, \ldots, \alpha_m\} \subseteq \Gamma \setminus \{\varepsilon\}$, there exists a homomorphism φ of Γ onto a finite group such that $\alpha_j\varphi \neq \varepsilon$ for $j = 1, \ldots, m$.*

Proof. The sufficiency is obvious. For the necessity, let φ_i be the appropriate homomorphism associated with α_i onto some finite group, and consider the natural homomorphism of Γ into the finite group $\Gamma\varphi_1 \times \cdots \times \Gamma\varphi_m$. □

The following central result in the theory of linear groups is due to Malcev [272]; see also Wehrfritz [439, Chapter 4].

4C2 Theorem *Every finitely generated linear group over a field (or a finitely generated integral domain) is residually finite.*

4C3 Corollary *Each (finitely generated) Coxeter group is residually finite.*

Proof. By Corollary 3A10, each Coxeter group has a faithful representation over $\mathbb{R}$. Then Theorem 4C2 applies. Note that, in our use of the term, a Coxeter group is indeed finitely generated. □

The main result of this section is the following theorem. In its applications, the residually finite group will usually be a linear group.

4C4 Theorem *Let $\mathcal{P}_1$ and $\mathcal{P}_2$ be finite regular n-polytopes such that $\langle \mathcal{P}_1, \mathcal{P}_2 \rangle \neq \emptyset$. Let $\mathcal{P}$ be an infinite regular $(n+1)$-polytope in $\langle \mathcal{P}_1, \mathcal{P}_2 \rangle$ whose group $\Gamma(\mathcal{P})$ is residually finite. Then the class $\langle \mathcal{P}_1, \mathcal{P}_2 \rangle$ contains infinitely many regular $(n+1)$-polytopes which are finite and are covered (in fact, n-covered) by $\mathcal{P}$.*

Proof. Let $\Gamma(\mathcal{P}) = \langle \rho_0, \ldots, \rho_n \rangle$, with $\rho_0, \ldots, \rho_n$ the distinguished generators of $\Gamma(\mathcal{P})$. Then

$$\Gamma_1 := \langle \rho_0, \ldots, \rho_{n-1} \rangle \cong \Gamma(\mathcal{P}_1), \quad \Gamma_2 := \langle \rho_1, \ldots, \rho_n \rangle \cong \Gamma(\mathcal{P}_2).$$

We shall construct an infinite sequence of finite C-groups Δ_j for $j \geqslant 1$, such that the regular polytopes $\mathcal{Q}_j$ with groups Δ_j belong to $\langle \mathcal{P}_1, \mathcal{P}_2 \rangle$, are n-covered by $\mathcal{P}$, and are mutually non-isomorphic.

We begin with Δ_1. By our assumptions on $\mathcal{P}_1$ and $\mathcal{P}_2$, the groups Γ_1 and Γ_2 are finite, and hence so is the set $K_1 := \Gamma_1\Gamma_2 \setminus \{\varepsilon\}$. Note that $\Gamma_1 \setminus \{\varepsilon\}$, $\Gamma_2 \setminus \{\varepsilon\} \subseteq K_1$. Since $\Gamma(\mathcal{P})$ is residually finite, there exists a surjective homomorphism $\varphi_1 \colon \Gamma(\mathcal{P}) \to \Delta_1$, with Δ_1 a finite group, such that $\alpha\varphi_1 \neq \varepsilon$ for all $\alpha \in K_1$. Let N_1 denote the kernel of φ_1. Then $\Gamma_1 \cap N_1 = \{\varepsilon\} = \Gamma_2 \cap N_1$, so that φ_1 induces an isomorphism when restricted to the subgroup Γ_1 or Γ_2 of $\Gamma(\mathcal{P})$. The intersection property for $\Delta_1 = \langle \rho_0\varphi_1, \ldots, \rho_n\varphi_1 \rangle$ is now a consequence of Proposition 2E16, if we can check that $\Gamma_1\varphi_1 \cap \Gamma_2\varphi_1 = (\Gamma_1 \cap \Gamma_2)\varphi_1$. But this follows from the definition of K_1. In fact, if $\alpha_1 \in \Gamma_1$ and $\alpha_2 \in \Gamma_2$ are such that $\alpha_1\varphi_1 = \alpha_2\varphi_1$, then $\alpha_1^{-1}\alpha_2 \in \Gamma_1\Gamma_2$ and $(\alpha_1^{-1}\alpha_2)\varphi_1 = \varepsilon$, so that, by construction of K_1, we must have $\alpha_1 = \alpha_2 \in \Gamma_1 \cap \Gamma_2$, as required. This completes the proof that Δ_1 is a string C-group with distinguished generators $\rho_0\varphi_1, \ldots, \rho_n\varphi_1$.

It follows that Δ_1 is the group of a finite regular $(n+1)$-polytope $\mathcal{Q}_1 \in \langle \mathcal{P}_1, \mathcal{P}_2 \rangle$. By Proposition 2E19, $\mathcal{Q}_1$ is also n-covered by $\mathcal{P}$. The inductive step now also requires two subsets L_1 and M_1 of $\Gamma(\mathcal{P})$. First, note that N_1 is an infinite group, because $\Gamma(\mathcal{P})$ is infinite and Δ_1 is finite. Now choose $L_1 \subset N_1 \setminus \{\varepsilon\}$ such that $\mathrm{card}(K_1 \cup L_1) > |\Delta_1|$, and set $M_1 := \{\alpha^{-1}\beta \mid \alpha, \beta \in K_1 \cup L_1, \alpha \neq \beta\}$. To construct the sequence of C-groups, we can now proceed as follows.

Suppose that we have already found $\Delta_j = \Gamma(\mathcal{Q}_j)$, and finite subsets K_j, L_j, and $M_j := \{\alpha^{-1}\beta \mid \alpha, \beta \in K_j \cup L_j, \alpha \neq \beta\}$ with $K_1 \subset K_j$ and $\mathrm{card}(K_j \cup L_j) > |\Delta_j|$, so that we have a surjective homomorphism $\varphi_j \colon \Gamma(\mathcal{P}) \to \Delta_j$ whose kernel N_j satisfies $N_j \cap K_j = \emptyset$ and $L_j \subset N_j \setminus \{\varepsilon\}$. Now define $K_{j+1} := K_j \cup L_j \cup M_j$. By residual finiteness we then have a surjective homomorphism $\varphi_{j+1} \colon \Gamma(\mathcal{P}) \to \Delta_{j+1}$, with Δ_{j+1} finite, such that $\alpha\varphi_{j+1} \neq \varepsilon$ if $\alpha \in K_{j+1}$. By construction, the restriction of φ_{j+1} to $K_j \cup L_j$ is injective; thus

$$|\Delta_{j+1}| = |\Gamma(\mathcal{P})\varphi_{j+1}| \geqslant \mathrm{card}(K_j \cup L_j) > |\Delta_j|.$$

In fact, since $M_j \subset K_{j+1}$, we have $(\alpha^{-1}\beta)\varphi_{j+1} \neq \varepsilon$; thus $\alpha\varphi_{j+1} \neq \beta\varphi_{j+1}$, if $\alpha, \beta \in K_j \cup L_j$, $\alpha \neq \beta$. Furthermore, since $K_1 \subset K_{j+1}$, we can argue as for φ_1 and prove that Δ_{j+1} is a C-group, whose subgroups $\langle \rho_0\varphi_{j+1}, \ldots, \rho_{n-1}\varphi_{j+1} \rangle$ and $\langle \rho_1\varphi_{j+1}, \ldots, \rho_n\varphi_{j+1} \rangle$ are also isomorphic to Γ_1 and Γ_2, respectively.

It follows that the regular $(n+1)$-polytope $\mathcal{Q}_{j+1}$ with group Δ_{j+1} belongs to $\langle \mathcal{P}_1, \mathcal{P}_2 \rangle$ and is also n-covered by $\mathcal{P}$. Finally, since N_{j+1} is infinite and Δ_{j+1} is finite, there exists a subset L_{j+1} of $N_{j+1} \setminus \{\varepsilon\}$ such that $\mathrm{card}(K_{j+1} \cup L_{j+1}) > |\Delta_{j+1}|$. The inductive step is now complete if we set $M_{j+1} := \{\alpha^{-1}\beta \mid \alpha, \beta \in K_{j+1} \cup L_{j+1}, \alpha \neq \beta\}$.

Note that this process produces a sequence of C-groups Δ_j whose orders are strictly increasing. It follows that the regular polytopes $\mathcal{Q}_j$ with groups Δ_j are mutually non-isomorphic. □

4C5 Corollary *Let $n \geqslant 2$ and $3 \leqslant p_1, \ldots, p_{n-1} < \infty$, and let $\{p_1, \ldots, p_n\}$ be a locally finite regular tessellation of $\mathbb{E}^n$ or $\mathbb{H}^n$. Then the class $\langle \{p_1, \ldots, p_{n-1}\}, \{p_2, \ldots, p_n\} \rangle$ contains infinitely many finite regular polytopes.*

Proof. We apply Theorem 4C4, with $\mathcal{P} = \{p_1, \ldots, p_n\}$. By the assumption of local finiteness, the facets $\{p_1, \ldots, p_{n-1}\}$ and vertex-figures $\{p_2, \ldots, p_n\}$ of $\mathcal{P}$ are finite polytopes. Now $\Gamma(\mathcal{P})$ is the Coxeter group $[p_1, \ldots, p_n]$, which is residually finite. Then the result follows. □

For Corollary 4C5, note that locally finite regular tessellations in hyperbolic space occur only in dimensions 2 and 3 [113]. For $n = 2$, the necessary and sufficient condition is that $\frac{1}{p_1} + \frac{1}{p_2} < \frac{1}{2}$. If $n = 3$, the only instances are

$$\{p, q, r\} = \{3, 5, 3\},\ \{4, 3, 5\},\ \{5, 3, 4\},\ \{5, 3, 5\}.$$

For euclidean tessellations, Corollary 4C5 is subsumed by the classification of the regular toroids in Sections 6D and 6E.

Among the locally toroidal regular polytopes of a given type (class), we can often find an infinite polytope with a group isomorphic to a finitely generated real or complex linear group (see Chapters 10, 11, and 12); usually it is the universal polytope itself which has this property. In this situation, we can apply Theorem 4C4 to prove that there are infinitely many finite regular polytopes of the same locally toroidal type.

4D Free Extensions of Regular Polytopes

The regular tessellation $\{3, \infty\}$ of the hyperbolic plane by ideal hyperbolic triangles is famous for its occurrence in several areas in mathematics [54, pp. 40–46]. Its group of symmetries is the Coxeter group $[3, \infty]$, with presentation

4D1 $$\rho_0^2 = \rho_1^2 = \rho_2^2 = (\rho_0\rho_1)^3 = (\rho_0\rho_2)^2 = \varepsilon.$$

This group is isomorphic to the projective general linear group $\mathrm{PGL}_2(\mathbb{Z})$, the quotient of the general linear group $\mathrm{GL}_2(\mathbb{Z})$ of invertible 2×2 matrices over the integers $\mathbb{Z}$ by its centre $\{\pm I\}$ (where I denotes the 2×2 identity matrix); it contains the (*classical*) *modular group* $\mathrm{PSL}_2(\mathbb{Z})$ as a subgroup of index 2 (see [169, Chapter 3]). We denote a typical element of $\mathrm{PGL}_2(\mathbb{Z})$ by $[\begin{smallmatrix} a & b \\ c & d \end{smallmatrix}]$. Then, using complex coordinates, the element $[\begin{smallmatrix} a & b \\ c & d \end{smallmatrix}]$ acts on the upper half-plane model $\mathbb{H}^2$ of the hyperbolic plane by

$$z \mapsto \begin{cases} \dfrac{az+b}{cz+d} & \text{if } ad - bc = 1, \\ \dfrac{a\bar{z}+b}{c\bar{z}+d} & \text{if } ad - bc = -1, \end{cases}$$

where $\bar{z}$ is the complex conjugate of z. As generators, we can take

4D2 $$\rho_0 = \begin{bmatrix} 0 & 1 \\ 1 & 0 \end{bmatrix}, \quad \rho_1 = \begin{bmatrix} -1 & 1 \\ 0 & 1 \end{bmatrix}, \quad \rho_2 = \begin{bmatrix} -1 & 0 \\ 0 & 1 \end{bmatrix}.$$

These correspond to hyperbolic reflexions, the first to inversion in the unit circle, and the second and third to reflexions in the lines $\Re(z) = \frac{1}{2}$ and $\Re(z) = 0$, respectively. The "closed chamber"

$$\bar{C} = \left\{ z \mid \Im(z) > 0,\ 0 \leqslant \Re(z) \leqslant \tfrac{1}{2},\ |z| \geqslant 1 \right\}$$

is a fundamental region for the action of $PGL_2(\mathbb{Z})$ on $\mathbb{H}^2$. The six transforms $\bar{C}\varphi$ of $\bar{C}$, with $\varphi \in \langle \rho_0, \rho_1 \rangle$, fit together to form a triangle in the tessellation $\{3, \infty\}$, whose vertices are points at infinity. As the group $PGL_2(\mathbb{Z})$ satisfies no relations other than those of (4D1), the whole tessellation is in a sense "freely generated" by triangles. The Coxeter complex for the Coxeter group $[3, \infty]$ is now realized in $\mathbb{H}^2$ (augmented by a set corresponding to the vertex-set of $\{3, \infty\}$). Its closed chambers correspond to the transforms of $\bar{C}$ under $PGL_2(\mathbb{Z})$; see also Section 3A.

In this section, we shall prove that every regular n-polytope $\mathcal{K}$ occurs as the facet type of a "freely generated" regular $(n+1)$-polytope $\mathcal{P}$. If $\mathcal{K} = \{3\}$, then $\mathcal{P}$ is the tessellation $\{3, \infty\}$. Our construction of $\mathcal{P}$ is in terms of free products with amalgamation. For a brief introduction to such groups, see Section 4A.

Throughout this section, let $n \geqslant 2$, and let $\mathcal{K}$ be a regular n-polytope. Let $\Gamma(\mathcal{K}) = \langle \rho_0, \ldots, \rho_{n-1} \rangle$, with $\rho_0, \ldots, \rho_{n-1}$ the distinguished generators defined with respect to some base flag of $\mathcal{K}$. We write $\mathcal{F}$ for the (base) facet of $\mathcal{K}$, so that $\Gamma(\mathcal{F}) = \langle \rho_0, \ldots, \rho_{n-2} \rangle$. Let ρ_n denote the generator of the cyclic group C_2. Consider the groups

$$\Lambda_1 := \langle \rho_0, \ldots, \rho_{n-1} \rangle \cong \Gamma(\mathcal{K}),$$
$$\Lambda_2 := \langle \rho_0, \ldots, \rho_{n-2}, \rho_n \rangle \cong \Gamma(\mathcal{F}) \times C_2,$$

as well as their amalgamated free product

4D3
$$\Gamma := \Gamma(\mathcal{K}) \underset{\Gamma(\mathcal{F})}{*} (\Gamma(\mathcal{F}) \times C_2),$$

where, for $i = 0, \ldots, n-2$, the generators ρ_i of the two copies of $\Gamma(\mathcal{F})$ in $\Gamma(\mathcal{K})$ and $\Gamma(\mathcal{F}) \times C_2$ are identified in the obvious way. We shall prove that this group is a string C-group, and that the corresponding polytope has the desired properties.

4D4 Theorem *Let $n \geqslant 2$, and let $\mathcal{K}$ be a regular n-polytope of type $\{p_1, \ldots, p_{n-1}\}$, with facets $\mathcal{F}$ and vertex-figures $\mathcal{K}_0$. Then there exists a regular $(n+1)$-polytope $\mathcal{P}$ with the following properties.*

(a) $\mathcal{P}$ is of type $\{p_1, \ldots, p_{n-1}, \infty\}$, and has facets isomorphic to $\mathcal{K}$.
(b) $\mathcal{P}$ covers every regular $(n+1)$-polytope whose facets are isomorphic to $\mathcal{K}$ (that is, $\mathcal{P}$ is universal among all such polytopes).
(c) $\Gamma(\mathcal{P}) \cong \Gamma(\mathcal{K}) *_{\Gamma(\mathcal{F})} (\Gamma(\mathcal{F}) \times C_2)$, the free product of $\Gamma(\mathcal{K})$ and $\Gamma(\mathcal{F}) \times C_2$ with amalgamation of the two subgroups which are isomorphic to $\Gamma(\mathcal{F})$.
(d) $\mathcal{P} = \{\mathcal{K}, \mathcal{P}_0\}$, the universal regular $(n+1)$-polytope with facets isomorphic to $\mathcal{K}$ and with vertex-figures isomorphic to the regular n-polytope $\mathcal{P}_0$ which is constructed from $\mathcal{K}_0$ in the same way as $\mathcal{P}$ is from $\mathcal{K}$. (In other words, the construction and the properties of the polytope are hereditary in passing from it to its vertex-figures.)
(e) If $\mathcal{K}$ is a lattice, then so is $\mathcal{P}$.

Proof. First we prove that the group Γ in (4D3) is a string C-group. By construction, $\Gamma = \langle \rho_0, \ldots, \rho_n \rangle$. As defining relations for Γ, we have those of Λ_1 for $\rho_0, \ldots, \rho_{n-1}$ as well as $\rho_n^2 = (\rho_j \rho_n)^2 = \varepsilon$ for $j = 0, \ldots, n-2$. Further, by Theorem 4A5, both groups Λ_1 and Λ_2 are embedded into Γ.

By Proposition 2E16, for Γ to have the intersection property it suffices to check that

$$\Lambda_1 \cap \langle \rho_j, \ldots, \rho_n \rangle = \langle \rho_j, \ldots, \rho_{n-1} \rangle \quad \text{for } j = 1, \ldots, n.$$

To prove the non-trivial inclusion, let $\tau \in \Lambda_1 \cap \langle \rho_j, \ldots, \rho_n \rangle$ with $\tau \neq \varepsilon$. The idea is now to write τ as the product of the elements in a suitable reduced sequence; see (4A4). In particular, since $\tau \in \langle \rho_j, \ldots, \rho_n \rangle$, we can write it as

$$\tau = \tau_1 \rho_n \tau_2 \rho_n \ldots \tau_{k-1} \rho_n \tau_k \rho_n^m,$$

with $k \geqslant 1$, $\tau_1, \ldots, \tau_k \in \langle \rho_j, \ldots, \rho_{n-1} \rangle$, $\tau_2, \ldots, \tau_k \notin \langle \rho_j, \ldots, \rho_{n-2} \rangle$ $(\subseteq \Gamma(\mathcal{F}))$, and $m = 0, 1$. In fact, because $\Lambda_2 = \Gamma(\mathcal{F}) \times \langle \rho_n \rangle$ and $\Gamma(\mathcal{F}) \subseteq \Lambda_1$, it is sufficient to assume that, in a reduced sequence for τ, each element from Λ_2 is equal to ρ_n. Then

$$(\tau^{-1}\tau_1)\rho_n\tau_2\rho_n\ldots\tau_{k-1}\rho_n\tau_k\rho_n^m = \varepsilon.$$

Since $\tau \in \Lambda_1$, this expresses ε as a product of the elements in a reduced sequence (which does not include ρ_n^m if $m = 0$). By Theorem 4A5, this implies that $k = 1$ and $m = 0$. Hence, $\tau = \tau_1 \in \langle \rho_j, \ldots, \rho_{n-1} \rangle$, as required.

It follows that Γ is a string C-group. Let $\mathcal{P} := \mathcal{P}(\Gamma)$, the regular $(n+1)$-polytope with group Γ. The facets of $\mathcal{P}$ are isomorphic to $\mathcal{K}$, since $\Lambda_1 = \Gamma(\mathcal{K})$ is embedded into Γ. Also, by Theorem 4A6, the element $\rho_{n-1}\rho_n$ of Γ has infinite order, showing that $\mathcal{P}$ is of type $\{p_1, \ldots, p_{n-1}, \infty\}$. This proves part (a).

Let $\mathcal{Q}$ be any regular $(n+1)$-polytope with facets isomorphic to $\mathcal{K}$, and let $\Gamma(\mathcal{Q}) = \langle \alpha_0, \ldots, \alpha_n \rangle$. Then the distinguished generators $\alpha_0, \ldots, \alpha_n$ satisfy all the defining relations for Γ (with ρ_i replaced by α_i), but in general further independent relations as well. It follows that the mappings $\rho_i \mapsto \alpha_i$ for $i = 0, \ldots, n$ induce a homomorphism $\Gamma = \Gamma(\mathcal{P}) \mapsto \Gamma(\mathcal{Q})$, which is also one-to-one on $\langle \rho_0, \ldots, \rho_{n-1} \rangle$ $(\cong \Gamma(\mathcal{K}))$. By Theorem 2E17 (for example), $\mathcal{P}$ covers $\mathcal{Q}$, which proves part (b). Further, part (c) is true by construction.

Let $\mathcal{P}_0$ be as described in part (d), and let $\tilde{\mathcal{P}}_0$ be the vertex-figure of $\mathcal{P}$ at the base vertex of $\mathcal{P}$. Then the facets of $\tilde{\mathcal{P}}_0$ are isomorphic to $\mathcal{K}_0$. By part (b), $\tilde{\mathcal{P}}_0$ is covered by $\mathcal{P}_0$. Therefore, if the universal $(n+1)$-polytope $\{\mathcal{K}, \mathcal{P}_0\}$ exists, then it must also cover $\mathcal{P}$. To prove that the two polytopes are the same, consider the group $\tilde{\Gamma} = \langle \tilde{\rho}_0, \ldots, \tilde{\rho}_n \rangle$, which would define the polytope $\{\mathcal{K}, \mathcal{P}_0\}$ (whether it exists or not). Then, by Proposition 4A8(a) and (4A7),

$$\tilde{\Gamma} = \left(\Gamma(\mathcal{K}) \underset{\Gamma(\mathcal{K}_0)}{*} \Gamma(\mathcal{P}_0) \right) \tilde{N},$$

where $\tilde{N}$ is the normal closure of the square κ of the product of the first generator from $\Gamma(\mathcal{K})$ and the last from $\Gamma(\mathcal{P}_0)$. In particular, we can find a presentation for $\tilde{\Gamma}$ in terms of the distinguished generators $\tilde{\rho}_i$ by taking all the defining relations for $\Gamma(\mathcal{K})$ and $\Gamma(\mathcal{P}_0)$ as well as the extra relation $(\tilde{\rho}_0\tilde{\rho}_n)^2 = \varepsilon$ (corresponding to κ). However, since $\Gamma(\mathcal{P}_0)$ is itself a group of the kind described in part (c), we can break down the relations for $\Gamma(\mathcal{P}_0)$ into two kinds: the relations derived from $\Gamma(\mathcal{K}_0)$ (which are already accounted for by $\Gamma(\mathcal{K})$), and the relations $\tilde{\rho}_n^2 = (\tilde{\rho}_j\tilde{\rho}_n)^2 = \varepsilon$ for $j = 1, \ldots, n-2$. In total, this gives a presentation for $\tilde{\Gamma}$ in terms of $\tilde{\rho}_0, \ldots, \tilde{\rho}_n$ which is equivalent to the one for Γ in terms of $\rho_0, \ldots, \rho_n$. It follows that the mappings $\tilde{\rho}_i \mapsto \rho_i$ for $i = 0, \ldots, n$ induce an isomorphism $\tilde{\Gamma} \mapsto \Gamma$. Therefore $\{\mathcal{K}, \mathcal{P}_0\}$ exists, $\Gamma(\{\mathcal{K}, \mathcal{P}_0\}) = \tilde{\Gamma}$, $\mathcal{P} = \{\mathcal{K}, \mathcal{P}_0\}$, and $\mathcal{P}_0 = \tilde{\mathcal{P}}_0$. This proves part (d).

The proof of part (e) is rather elaborate, and we shall omit it. Instead, the reader is referred to Schulte [365, Theorem 3] for further details. □

We illustrate our results for the polytopes $\mathcal{K} = \{p_1, \ldots, p_{n-1}\}$; see Section 3D. In this case, $\Gamma(\mathcal{K})$ is the Coxeter group $[p_1, \ldots, p_{n-1}]$. Then $\mathcal{P} = \{p_1, \ldots, p_{n-1}, \infty\}$, and $\Gamma(\mathcal{P}) = [p_1, \ldots, p_{n-1}, \infty]$. If $n = 2$ and $p := p_1$, then $\mathcal{P}$ is the regular hyperbolic tessellation $\{p, \infty\}$, which is "freely generated" from p-gons. Now Theorem 4D4(c) takes the form

$$[p, \infty] \cong D_p \underset{C_2}{*} (C_2 \times C_2).$$

In particular, if $p = 3$, then

$$PGL_2(\mathbb{Z}) \cong [3, \infty] \cong S_3 \underset{C_2}{*} (C_2 \times C_2).$$

If $\mathcal{K}$ is the n-simplex $\{3^{n-1}\}$, then $\mathcal{P}$ is the universal $(n+1)$-polytope $\{3^{n-1}, \infty\}$, whose group is

$$[3^{n-1}, \infty] \cong S_{n+1} \underset{S_n}{*} (S_n \times C_2).$$

In general, there are many ways in which a regular n-polytope can occur as the facet of a regular $(n+1)$-polytope. In a sense, Theorem 4D4 deals with the "free construction", which leads to the "largest" polytope with a given kind of facet. In a contrasting spirit, the construction of flat polytopes described in Section 4F yields a "small" regular polytope with a given facet and vertex-figure. We conclude by remarking that the free construction can also be generalized to chiral polytopes [374, Theorem 2].

4E Flat Polytopes and the FAP

In this section, we introduce the concept of combinatorial flatness of abstract polytopes. In particular, we discuss the flat amalgamation property, or briefly, FAP, which is crucial for a construction of flat regular polytopes described in the next section.

An n-polytope $\mathcal{P}$ is called *(combinatorially) flat* if each of its vertices is incident with each of its facets. More generally, if $0 \leqslant k < m \leqslant n-1$, we say that $\mathcal{P}$ is *(combinatorially)* (k, m)*-flat* if each of its k-faces is incident with each of its m-faces. Thus a flat polytope is a $(0, n-1)$-flat polytope in this more general sense.

A 2-polytope $\mathcal{P}$ is flat if and only if it is a 2-gon; this is degenerate, with group $C_2 \times C_2$. A simple example of higher rank n is the *hemi-cube* $\{4, 3^{n-2}\}_n$, which is obtained from the n-cube $\{4, 3^{n-2}\}$ by identifying opposite vertices (the suffix n refers to the length of the Petrie polygon of $\mathcal{P}$, or, in other words, to the period of the product $\rho_0\rho_1 \cdots \rho_{n-1}$ of the distinguished generators of its group). Another example of rank 3 is the torus map $\{4, 4\}_{(2,0)}$ with four square facets (see Section 1D). On the other hand, convex polytopes can never be flat.

We begin with a number of observations which apply to all flat polytopes.

4E1 Lemma *Let $\mathcal{P}$ be an n-polytope, and let $0 \leqslant i \leqslant k < m \leqslant j \leqslant n-1$. If $\mathcal{P}$ is (k, m)-flat, then $\mathcal{P}$ is also (i, j)-flat.*

Proof. Let F_1 be an i-face and F_2 a j-face of $\mathcal{P}$. Choose a k-face G_1 with $F_1 \leqslant G_1$ and an m-face G_2 with $G_2 \leqslant F_2$. Since $G_1 \leqslant G_2$, it follows that $F_1 \leqslant F_2$, as required. □

4E2 Lemma *The dual of a (k, m)-flat n-polytope $\mathcal{P}$ is an $(n-m-1, n-k-1)$-flat polytope. In particular, the dual of a flat polytope is flat.*

Proof. The dual $\mathcal{P}^*$ is obtained from $\mathcal{P}$ by reversing the partial order on $\mathcal{P}$. Then, for each i, the i-faces of $\mathcal{P}$ are the $(n-i-1)$-faces of $\mathcal{P}^*$, and so the result follows immediately. □

We shall often appeal to Lemma 4E2, and analogous properties which directly result from duality.

4E3 Lemma *Let $\mathcal{P}$ be an n-polytope, and let $0 \leqslant k < m < i \leqslant n$. If each i-face of $\mathcal{P}$ is (k,m)-flat, then $\mathcal{P}$ is also (k,m)-flat.*

Proof. We use the connectedness of $\mathcal{P}$. Let F be a k-face and G an m-face of $\mathcal{P}$. Then there exist i-faces $H_1, \ldots, H_j$ and m-faces $G_1, \ldots, G_j$, such that

$$F < H_1 > G_1 < H_2 > G_2 < \cdots > G_{j-1} < H_j > G_j = G.$$

Since H_1 is (k,m)-flat, we have $F < G_1$ and therefore $F < H_2$. We may now proceed by induction to arrive at $F < G_j = G$. □

Of course, duality yields the same result in the case that $i < k$, and all co-i-faces of $\mathcal{P}$ are $(k-i-1, m-i-1)$-flat.

Now let $\mathcal{P}$ be a regular n-polytope, whose group is $\Gamma(\mathcal{P}) = \langle \rho_0, \ldots, \rho_{n-1} \rangle$, where $\rho_0, \ldots, \rho_{n-1}$ are (as usual in such a context) taken to be the distinguished generators of $\Gamma(\mathcal{P})$. We can tell from $\Gamma(\mathcal{P})$ whether or not $\mathcal{P}$ is flat.

4E4 Proposition *Let $\mathcal{P}$ be a regular n-polytope with group $\Gamma(\mathcal{P}) = \langle \rho_0, \ldots, \rho_{n-1} \rangle$, and let $0 \leqslant k < m \leqslant n-1$. Then $\mathcal{P}$ is (k,m)-flat if and only if*

$$\Gamma(\mathcal{P}) = \langle \rho_{k+1}, \ldots, \rho_{n-1} \rangle \langle \rho_0, \ldots, \rho_{m-1} \rangle.$$

Proof. If $\Phi = \{F_{-1}, F_0, \ldots, F_n\}$ is the base flag of $\mathcal{P}$, then $F_k\varphi < F_m\psi$ holds for all $\varphi, \psi \in \Gamma(\mathcal{P})$. But, by Theorem 2B14,

$$F_k\varphi < F_m\psi \iff \varphi\psi^{-1} \in \langle \rho_{k+1}, \ldots, \rho_{n-1} \rangle \langle \rho_0, \ldots, \rho_{m-1} \rangle,$$

and the required result follows. □

The condition on $\Gamma(\mathcal{P})$ in Proposition 4E4 is easy to check if $k = m-1$. In fact, if $\mathcal{P}$ is of type $\{p_1, \ldots, p_{n-1}\}$, then $\mathcal{P}$ is $(m-1, m)$-flat if and only if $p_m = 2$ (see Proposition 2B16); in this case, $\Gamma(\mathcal{P}) \cong \langle \rho_m, \ldots, \rho_{n-1} \rangle \times \langle \rho_0, \ldots, \rho_{m-1} \rangle$.

We next describe the class $\langle \mathcal{P}_1, \mathcal{P}_2 \rangle$, under the condition that one of the n-polytopes $\mathcal{P}_1$ or $\mathcal{P}_2$ is flat. For a proof, see [300, Proposition 1].

4E5 Proposition *Let $\mathcal{P}_1$ and $\mathcal{P}_2$ be regular n-polytopes such that $\langle \mathcal{P}_1, \mathcal{P}_2 \rangle \neq \emptyset$. If $\mathcal{P}_1$ or $\mathcal{P}_2$ is flat, then the universal polytope $\{\mathcal{P}_1, \mathcal{P}_1\}$ is flat, and is the only member of $\langle \mathcal{P}_1, \mathcal{P}_2 \rangle$.*

Flatness of polytopes is a counterintuitive phenomenon, which does not occur in the classical theory. On the other hand, many classes $\langle \mathcal{P}_1, \mathcal{P}_2 \rangle$ contain flat regular polytopes, and it is therefore desirable to find simple criteria for their existence. In the construction of such polytopes described in the next section, an important role is played by what is called the *flat amalgamation property*, or *FAP* for short.

This property is defined in the following; it relies on the following observation and the subsequent Lemma 4E7. The term FAP itself is explained by Theorem 4F9(b), which yields flat polytopes by "amalgamating" two polytopes which themselves have the FAP.

Let $\Gamma = \langle \rho_0, \ldots, \rho_{n-1} \rangle$ be any sggi, which we recall from Section 2B is a group such that $\rho_i^2 = (\rho_i \rho_j)^2 = \varepsilon$ for all i, j with $|i - j| \geqslant 2$. For convenience, we again denote the generators by $\rho_0, \ldots, \rho_{n-1}$, because Γ will usually be the group of a regular polytope. Now, if $\varphi \in \Gamma$ is such that $\varphi = \varphi_0 \rho_{j_1} \varphi_1 \rho_{j_2} \cdots \varphi_{m-1} \rho_{j_m} \varphi_m$, then clearly

4E6
$$\begin{cases} \varphi = \psi_0 (\psi_1^{-1} \rho_{j_1} \psi_1) \cdots (\psi_m^{-1} \rho_{j_m} \psi_m), \\ \text{where } \psi_r := \varphi_r \varphi_{r+1} \cdots \varphi_m \text{ for } r = 0, \ldots, m. \end{cases}$$

In particular, if $\varphi_0, \ldots, \varphi_m$ are in some specified subgroup of Γ, then $\psi_0, \psi_1, \ldots, \psi_m$ are in the same subgroup.

For $k = 0, \ldots, n-1$, we define

$$N_k^+ := \langle \varphi^{-1} \rho_i \varphi \mid i \geqslant k \text{ and } \varphi \in \Gamma \rangle,$$

$$N_k^- := \langle \varphi^{-1} \rho_i \varphi \mid i \leqslant k \text{ and } \varphi \in \Gamma \rangle,$$

which are the normal closures of $\{\rho_k, \ldots, \rho_{n-1}\}$ and $\{\rho_0, \ldots, \rho_k\}$, respectively, in $\Gamma = \langle \rho_0, \ldots, \rho_{n-1} \rangle$.

4E7 Lemma *For $0 \leqslant k \leqslant n-1$, let Γ, N_k^+ and N_k^- be as previously described. Then*

(a) $\Gamma = N_k^+ \langle \rho_0, \ldots, \rho_{k-1} \rangle = \langle \rho_0, \ldots, \rho_{k-1} \rangle N_k^+$;
(b) $N_k^+ = \langle \varphi^{-1} \rho_i \varphi \mid i \geqslant k \text{ and } \varphi \in \langle \rho_0, \ldots, \rho_{k-1} \rangle \rangle$
$\quad = \langle \varphi^{-1} \rho_k \varphi, \rho_i \mid i \geqslant k+1 \text{ and } \varphi \in \langle \rho_0, \ldots, \rho_{k-1} \rangle \rangle$;
(c) $\Gamma = N_k^- \langle \rho_{k+1}, \ldots, \rho_{n-1} \rangle = \langle \rho_{k+1}, \ldots, \rho_{n-1} \rangle N_k^-$;
(d) $N_k^- = \langle \varphi^{-1} \rho_i \varphi \mid i \leqslant k \text{ and } \varphi \in \langle \rho_{k+1}, \ldots, \rho_{n-1} \rangle \rangle$
$\quad = \langle \rho_i, \varphi^{-1} \rho_k \varphi \mid i \leqslant k-1 \text{ and } \varphi \in \langle \rho_{k+1}, \ldots, \rho_{n-1} \rangle \rangle$.

Proof. We just consider parts (a) and (b); parts (c) and (d) follow by the same arguments applied to the dual. Trivially, $N_k^+ = \Gamma$ if $k = 0$, and so all statements are true in these cases. We may thus suppose that $k > 0$. Part (a) follows from the observation preceding the lemma. In fact, in the notation of (4E6), if $\varphi \in \Gamma$, then we can take $\varphi_0, \ldots, \varphi_m \in \langle \rho_0, \ldots, \rho_{k-1} \rangle$ and $j_1, \ldots, j_m \geqslant k$, so that we have $\psi_0 \in \langle \rho_0, \ldots, \rho_{k-1} \rangle$ and $\psi_r^{-1} \rho_{j_r} \psi_r \in N_k^+$ for $r = 1, \ldots, m$. It follows that $\Gamma = \langle \rho_0, \ldots, \rho_{k-1} \rangle N_k^+$. The other equation of part (a) is clear, because N_k^+ is a normal subgroup of Γ.

For the first equation in part (b), consider a generator $\varphi^{-1} \rho_i \varphi$ of N_k^+ with $\varphi \in \Gamma$. We express φ in the form $\varphi = \varphi_0 \rho_{j_1} \varphi_1 \rho_{j_2} \cdots \varphi_{m-1} \rho_{j_m} \varphi_m$, with $\varphi_0, \ldots, \varphi_{m+1} \in \langle \rho_0, \ldots, \rho_{k-1} \rangle$ and $j_1, \ldots, j_m \geqslant k$. Noting that $i \geqslant k$, we can rewrite

$$\varphi^{-1} \rho_i \varphi = \varphi_m^{-1} \rho_{j_m} \varphi_{m-1}^{-1} \cdots \varphi_1^{-1} \rho_{j_1} \varphi_0^{-1} \rho_i \varphi_0 \rho_{j_1} \varphi_1 \cdots \varphi_{m-1} \rho_{j_m} \varphi_m,$$

as in (4E6). The corresponding element ψ_0 is now trivial, and each remaining term $\psi_r^{-1} \rho_{j_r} \psi_r$ is a conjugate of one of the generators $\rho_k, \ldots, \rho_{n-1}$ by an element of $\langle \rho_0, \ldots, \rho_{k-1} \rangle$. This proves the first equation. The second follows from the first, with the observation that $\varphi^{-1} \rho_i \varphi = \rho_i$ if $i \geqslant k+1$ and $\varphi \in \langle \rho_0, \ldots, \rho_{k-1} \rangle$. □

We can now define the flat amalgamation property (FAP). Let $\mathcal{P}$ be a regular n-polytope with group $\Gamma(\mathcal{P}) = \langle \rho_0, \ldots, \rho_{n-1} \rangle$, and let $0 \leqslant k \leqslant n-1$. We say that $\mathcal{P}$ has the *FAP with respect to its k-faces* if $\Gamma(\mathcal{P})$ is a semi-direct product of N_k^+ by $\langle \rho_0, \ldots, \rho_{k-1} \rangle$, so that the product in Lemma 4E7(a) is semi-direct:

4E8 $$\Gamma(\mathcal{P}) \cong N_k^+ \rtimes \langle \rho_0, \ldots, \rho_{k-1} \rangle,$$

or, equivalently,

$$N_k^+ \cap \langle \rho_0, \ldots, \rho_{k-1} \rangle = \{\varepsilon\}.$$

Similarly, $\mathcal{P}$ has the *FAP with respect to its co-k-faces* if and only if the product in Lemma 4E7(c) is semi-direct; that is,

4E9 $$\Gamma(\mathcal{P}) \cong N_k^- \rtimes \langle \rho_{k+1}, \ldots, \rho_{n-1} \rangle,$$

or, equivalently,

$$N_k^- \cap \langle \rho_{k+1}, \ldots, \rho_{n-1} \rangle = \{\varepsilon\}.$$

Note that in [368, §4] the term DAP (degenerate amalgamation property) was used in place of FAP for the two separate properties which we now call the FAP with respect to facets or vertex-figures. The two kinds of FAP are clearly related by duality. In fact, $\mathcal{P}$ has the FAP with respect to its co-k-faces if and only if its dual $\mathcal{P}^*$ has the FAP with respect to its $(n-k-1)$-faces. Note also that each regular polytope $\mathcal{P}$ trivially has the FAP with respect to both its vertices ($k = 0$) and co-facets ($k = n-1$), because $N_0^+ \cong \Gamma(\mathcal{P}) \cong N_{n-1}^-$.

From now on, therefore, while we shall often state results involving the FAP in full generality, we shall mainly prove them only with respect to faces. The corresponding statements for the FAP with respect to co-faces are easily obtained by duality. The following lemma describes an effective way to check whether or not a polytope has the FAP.

4E10 Lemma *Let $\mathcal{P}$ be a regular n-polytope, and let $\Gamma(\mathcal{P}) = \langle \rho_0, \ldots, \rho_{n-1} \mid \mathcal{R} \rangle$ be a presentation for its group, with $\mathcal{R}$ a set of defining relations in terms of the generators $\rho_0, \ldots, \rho_{n-1}$. Let $0 \leqslant k \leqslant n-1$, and let the k-faces of $\mathcal{P}$ be isomorphic to $\mathcal{K}$. Then $\mathcal{P}$ has the FAP with respect to its k-faces if and only if the addition of the extra relations $\rho_i = \varepsilon$ for $i \geqslant k$ gives a presentation for $\Gamma(\mathcal{K})$, or, equivalently,*

$$\langle \rho_0, \ldots, \rho_{n-1} \mid \mathcal{R} \text{ and } \rho_i = \varepsilon \text{ for } i \geqslant k \rangle \cong \Gamma(\mathcal{K}).$$

Proof. Adding the extra relations $\rho_i = \varepsilon$ for $i \geqslant k$ to those in $\mathcal{R}$ is equivalent to passing from $\Gamma(\mathcal{P})$ to its quotient by the normal closure N_k^+ of $\{\rho_i \mid i \geqslant k\}$ in $\Gamma(\mathcal{P})$. If $\mathcal{P}$ has the FAP with respect to its k-faces, this quotient is the subgroup $\langle \rho_0, \ldots, \rho_{k-1} \rangle$ of $\Gamma(\mathcal{P})$, which is isomorphic to $\Gamma(\mathcal{K})$. Conversely, if the quotient of $\Gamma(\mathcal{P})$ by N_k^+ is isomorphic to $\Gamma(\mathcal{K})$, then necessarily $N_k^+ \cap \langle \rho_0, \ldots, \rho_{k-1} \rangle = \{\varepsilon\}$ in $\Gamma(\mathcal{P})$, and therefore $\mathcal{P}$ has the FAP with respect to its k-faces. □

4E11 Lemma *The FAP is hereditary; that is, if a regular n-polytope $\mathcal{P}$ has the FAP with respect to its k-faces, then*

(a) for $j > k$, the j-faces of $\mathcal{P}$ also have the FAP with respect to their k-faces;
(b) for $j < k$, the co-j-faces of $\mathcal{P}$ have the FAP with respect to their $(k - j - 1)$-faces.

Proof. By assumption,

$$\Gamma(\mathcal{P}) = N_k^+ \langle \rho_0, \ldots, \rho_{k-1} \rangle, \quad N_k^+ \cap \langle \rho_0, \ldots, \rho_{k-1} \rangle = \{\varepsilon\}.$$

Given j, consider the normal closures L_k^+ of $\{\rho_k, \ldots, \rho_{j-1}\}$ in the group $\langle \rho_0, \ldots, \rho_{j-1} \rangle$ of the base j-face F_j of $\mathcal{P}$, as well as the normal closure M_k^+ of $\{\rho_k, \ldots, \rho_{n-1}\}$ in the group $\langle \rho_{j+1}, \ldots, \rho_{n-1} \rangle$ of the co-face at F_j. Then we have

4E12
$$\begin{cases} L_k^+ = N_k^+ \cap \langle \rho_0, \ldots, \rho_{j-1} \rangle, \\ M_k^+ = N_k^+ \cap \langle \rho_{j+1}, \ldots, \rho_{n-1} \rangle. \end{cases}$$

Clearly, the groups on the left are contained in the groups on the right. (In fact, these inclusions would already suffice to conclude the proof. However, we shall need the stronger statements later.)

To prove the opposite inclusion for the first equation, let $\varphi \in \langle \rho_0, \ldots, \rho_{j-1} \rangle$. Applying Lemma 4E7(a) to $\langle \rho_0, \ldots, \rho_{j-1} \rangle$, we have $\varphi = \eta\psi$ with $\eta \in L_k^+$ and $\psi \in \langle \rho_0, \ldots, \rho_{k-1} \rangle$. Since $L_k^+ \subset N_k^+$, this is the factorization of φ in the semi-direct product $\Gamma(\mathcal{P})$. Now, if we also have $\varphi \in N_k^+$, then $\psi = \varepsilon$ and $\varphi = \eta \in L_k^+$, as required. The opposite inclusion for the second equation is proved similarly.

To conclude the proof, we also have

$$L_k^+ \cap \langle \rho_0, \ldots, \rho_{k-1} \rangle = \{\varepsilon\} = M_k^+ \cap \langle \rho_{j+1}, \ldots, \rho_{k-1} \rangle,$$

because $L_k^+, M_k^+ \subset N_k^+$. Hence,

$$\langle \rho_0, \ldots, \rho_{j-1} \rangle \cong L_k^+ \rtimes \langle \rho_0, \ldots, \rho_{k-1} \rangle,$$
$$\langle \rho_{j+1}, \ldots, \rho_{n-1} \rangle \cong M_k^+ \rtimes \langle \rho_{j+1}, \ldots, \rho_{k-1} \rangle. \qquad \square$$

The flat amalgamation property has an interesting geometric interpretation in terms of the order complex of the polytope. Let $\mathcal{P}$ be any regular n-polytope, and let $\Phi = \{F_{-1}, F_0, \ldots, F_n\}$ be the base flag of $\mathcal{P}$. Recall from Section 2C that the *order complex* $\mathcal{C}(\mathcal{P})$ of $\mathcal{P}$ is the abstract $(n - 1)$-dimensional simplicial complex whose vertices are the *proper* faces of $\mathcal{P}$ and whose simplices are the chains Ω (totally ordered subsets) of such faces; in other words, the improper faces F_{-1} and F_n do not count in this context (see [398, p. 120]). The maximal simplices in $\mathcal{C}(\mathcal{P})$ are $(n - 1)$-dimensional, and are in one-to-one correspondence with the flags of $\mathcal{P}$. For instance, if $\mathcal{P}$ is a regular convex n-polytope or a regular tessellation of euclidean or hyperbolic $(n - 1)$-space, then $\mathcal{C}(\mathcal{P})$ corresponds to the barycentric subdivision of $\mathcal{P}$; in this context, it is helpful to regard a convex n-polytope as the tessellation of its circumscribing $(n - 1)$-sphere induced by radial projection from its centre.

In the general situation, let $0 \leqslant k \leqslant n - 1$, and define $\Omega_k := \{F_k, \ldots, F_{n-1}\}$, regarded as an $(n - k - 1)$-simplex in $\mathcal{C}(\mathcal{P})$. We write $\mathcal{D}_k := \mathrm{st}(\Omega_k, \mathcal{C}(\mathcal{P}))$ for the *star* of Ω_k in $\mathcal{C}(\mathcal{P})$, which is by definition the subcomplex induced by the simplices of $\mathcal{C}(\mathcal{P})$

which contain Ω_k. We then have

4E13 $$\mathcal{D}_k = \{\Psi\varphi \mid \Psi \subset \Phi \text{ and } \varphi \in \langle\rho_0, \ldots, \rho_{k-1}\rangle\}.$$

This subcomplex of $\mathcal{C}(\mathcal{P})$ has two kinds of vertices: those determined by the proper faces of $\mathcal{P}$ with $F < F_k$ and those given by the base faces $F_k, \ldots, F_{n-1}$. The *link* $\text{lk}(\Omega_k, \mathcal{C}(\mathcal{P})$ of Ω_k in $\mathcal{C}(\mathcal{P})$ may be identified with the order complex $\mathcal{C}(F_k/F_{-1})$ of F_k (or, rather, of the section F_k/F_{-1}); hence, as an abstract simplicial complex,

4E14 $$\mathcal{D}_k = \mathcal{C}(F_k/F_{-1}) * \mathcal{C}(\Omega_k),$$

the join of two subcomplexes, namely, $\mathcal{C}(F_k/F_{-1})$ and the complex $\mathcal{C}(\Omega_k)$ of all faces of the simplex Ω_k. (Recall that the *join* $\mathcal{C}_1 * \mathcal{C}_2$ of two abstract simplicial complexes $\mathcal{C}_1$ and $\mathcal{C}_2$ is the abstract simplicial complex whose simplices are the disjoint unions of two simplices: one from $\mathcal{C}_1$ and one from $\mathcal{C}_2$.) We repeat that, for (4E14), the improper faces of a polytope do not give vertices of its order complex.

The boundary subcomplex $\mathcal{B}_k := \text{bdst}(\Omega_k, \mathcal{C}(\mathcal{P}))$ of $\mathcal{D}_k$ is

4E15 $$\mathcal{B}_k = \{\Psi\varphi \mid \text{there exists } i \geqslant k \text{ such that } F_i \notin \Psi \text{ and } \varphi \in \langle\rho_0, \ldots, \rho_{k-1}\rangle\}.$$

This is the union of the links in $\mathcal{D}_k$ of the vertices $F_k, \ldots, F_{n-1}$ of $\tilde{F}_k$. It consists of all $(n-2)$-simplices of $\mathcal{D}_k$ which are opposite to a vertex of Ω_k in an $(n-1)$-simplex of $\mathcal{D}_k$, as well as all their lower-dimensional faces.

We claim that the $(n-2)$-simplices in $\mathcal{B}_k$ are precisely the *mirrors* or "reflexion walls" for the generators of the subgroup N_k^+ of $\Gamma(\mathcal{P})$ described in Lemma 4E7(b). To make things more precise, we consider the action of $\Gamma(\mathcal{P})$ on the topological space $|\mathcal{C}(\mathcal{P})|$ underlying $\mathcal{C}(\mathcal{P})$ (its *polyhedron* in technical terms). Recall that the generator ρ_i of $\Gamma(\mathcal{P})$ fixes all but the i-face F_i of the base flag Φ of $\mathcal{P}$, and therefore corresponds to the (combinatorial) reflexion in the wall Ψ_i of the $(n-1)$-simplex Φ of $\mathcal{C}(\mathcal{P})$ which is opposite to the vertex F_i. (We are thus identifying ρ_i with its mirror of fixed points in $|\mathcal{C}(\mathcal{P})|$.) Each generator $\varphi^{-1}\rho_i\varphi$ (with $i \geqslant k$) of N_k^+ is then the reflexion in the corresponding wall $\Psi_i\varphi$ of $\Phi\varphi$ opposite to $F_i\varphi$. But $F_i\varphi = F_i$ and $\Phi\varphi \in \mathcal{D}_k$, since $\varphi \in \langle\rho_0, \ldots, \rho_{k-1}\rangle$. Consequently, $\varphi^{-1}\rho_i\varphi$ is the reflexion in the $(n-2)$-simplex $\Phi\varphi \setminus \{F_i\}$ of $\mathcal{B}_k$, as claimed.

For instance, if $k = n-1$, then $\mathcal{D}_{n-1}$ is the star in $\mathcal{C}(\mathcal{P})$ of the vertex F_{n-1} of $\mathcal{C}(\mathcal{P})$, and

$$\mathcal{B}_{n-1} = \text{lk}(F_{n-1}, \mathcal{C}(\mathcal{P})) = \mathcal{C}(F_{n-1}/F_{-1})$$

is the link of F_{n-1} in $\mathcal{C}(\mathcal{P})$. In $|\mathcal{C}(\mathcal{P})|$, the generators $\varphi^{-1}\rho_{n-1}\varphi$ of N_{n-1}^+ are the reflexions in the $(n-2)$-simplices of this link. For the polytope $\mathcal{P}$ itself, they can be regarded as the reflexions in the $(n-2)$-faces of the base facet F_{n-1} of $\mathcal{P}$. If $\mathcal{P}$ is a regular tessellation of spherical, euclidean, or hyperbolic $(n-1)$-space, then the generators $\varphi^{-1}\rho_{n-1}\varphi$ of N_{n-1}^+ can be realized by geometric hyperplane reflexions acting on the barycentric subdivision of the underlying space given by $\mathcal{C}(\mathcal{P})$. In particular, N_{n-1}^+ is the group generated by the reflexions in the hyperplanes which contain an $(n-2)$-face of the base facet F_{n-1} of $\mathcal{P}$.

In the general situation when $0 \leqslant k \leqslant n-1$, the FAP can now be described in terms of the action of the subgroup N_k^+ of $\Gamma(\mathcal{P})$ on $|\mathcal{C}(\mathcal{P})|$. This subgroup is the group generated by the reflexions in the walls of $|\mathcal{D}_k|$ opposite to a vertex from $\{F_k, \ldots, F_{n-1}\}$. In fact,

4E16 Proposition *$\mathcal{P}$ has the FAP with respect to its k-faces if and only if $|\mathcal{D}_k|$ is a fundamental region for the action of N_k^+ on $|\mathcal{C}(\mathcal{P})|$.*

Proof. This means that the subspace $|\mathcal{D}_k|$ of $|\mathcal{C}(\mathcal{P})|$ is such that $|\mathcal{C}(\mathcal{P})|$ is the union of all transforms $|\mathcal{D}_k|\eta$ of $|\mathcal{D}_k|$ with $\eta \in N_k^+$, and no two such transforms have common interior points in $|\mathcal{C}(\mathcal{P})|$. We shall prove this in terms of $\mathcal{C}(\mathcal{P})$ itself; no problems arise from identifying a subcomplex of $\mathcal{C}(\mathcal{P})$ with its underlying polyhedron.

By Lemma 4E7(a), $\Gamma(\mathcal{P}) = \langle \rho_0, \ldots, \rho_{k-1} \rangle N_k^+$, and therefore

$$\bigcup_{\eta \in N_k^+} \mathcal{D}_k \eta = \mathcal{C}(\mathcal{P}).$$

Now suppose that $\mathcal{D}_k$ and $\mathcal{D}_k\eta$ have an $(n-1)$-simplex in common. Then there exist $\varphi_1, \varphi_2 \in \langle \rho_0, \ldots, \rho_{k-1} \rangle$ such that $\Phi\varphi_1 = \Phi\varphi_2\eta$, and therefore $\eta = \varphi_2^{-1}\varphi_1 \in N_k^+ \cap \langle \rho_0, \ldots, \rho_{k-1} \rangle$. If $\mathcal{P}$ has the FAP with respect to its k-faces, then $\eta = \varepsilon$ and $\mathcal{D}_k = \mathcal{D}_k\eta$, proving that $\mathcal{D}_k$ is a fundamental region. On the other hand, if $\eta \in N_k^+ \cap \langle \rho_0, \ldots, \rho_{k-1} \rangle$, then the definition of $\mathcal{D}_k$ immediately gives $\mathcal{D}_k = \mathcal{D}_k\eta$. Therefore, if $\mathcal{D}_k$ is a fundamental region, then $\mathcal{D}_k = \mathcal{D}_k\eta$ implies that $\eta = \varepsilon$; hence $\mathcal{P}$ has the FAP with respect to its k-faces. This establishes our claim. □

In the case $k = n-1$, with $\mathcal{P}$ a regular convex n-polytope, or a regular tessellation of euclidean or hyperbolic $(n-1)$-space, (4E16) translates into the following equivalence. The polytope $\mathcal{P}$ has the FAP with respect to its facets if and only if the base facet F_{n-1} of $\mathcal{P}$ is a fundamental region for the group N_{n-1}^+ which is generated by the hyperplane reflexions in the $(n-2)$-faces of F_{n-1}. This shows that, in the classical theory, the FAP is a rather restrictive property.

On the other hand, among the regular toroids (defined in Sections 1D, 6D or 6E) there are many examples which have the FAP with respect to facets or vertex-figure (see [305, §5; 368, §7.2]).

4F Flat Polytopes and Amalgamation

We now describe a construction of flat regular polytopes $\mathcal{P}$ with pre-assigned faces $\mathcal{P}_1$ and co-faces $\mathcal{P}_2$, which employs the flat amalgamation property (FAP) introduced in Section 4E. Our main result is Theorem 4F9, which was obtained in [305, Theorems 4.3, 4.4].

We first explain the hypotheses under which our construction works. Let $m, n \geqslant 2$, and let $\max\{0, m-n\} \leqslant k \leqslant m-2$. Let $\mathcal{P}_1$ be a regular m-polytope with group $\Gamma(\mathcal{P}_1) = \langle \alpha_0, \ldots, \alpha_{m-1} \rangle$, with $\alpha_0, \ldots, \alpha_{m-1}$ its distinguished generators. Suppose that

$\mathcal{P}_1$ has the FAP with respect to its co-k-face $\mathcal{K}$, so that

4F1
$$\begin{cases} \Gamma(\mathcal{P}_1) \cong N_k^- \rtimes \langle \alpha_{k+1}, \ldots, \alpha_{m-1} \rangle, \\ N_k^- = \langle \psi^{-1} \alpha_i \psi \mid i \leqslant k, \ \psi \in \langle \alpha_{k+1}, \ldots, \alpha_{m-1} \rangle \rangle. \end{cases}$$

If $\alpha \in \Gamma(\mathcal{P}_1)$, we write $\alpha = \eta\hat{\alpha}$ with $\eta \in N_k^-$ and $\hat{\alpha} \in \langle \alpha_{k+1}, \ldots, \alpha_{m-1} \rangle$.

The other polytope $\mathcal{P}_2$ will play the role of the co-k-face of the constructed $(k+n+1)$-polytope $\mathcal{P}$. For $\mathcal{P}_1$ and $\mathcal{P}_2$ to occur as such an m-face and co-k-face, respectively, it is necessary that $\mathcal{P}_1$ and $\mathcal{P}_2$ coincide where they overlap; hence, the $(m-k-1)$-face of $\mathcal{P}_2$ must be isomorphic to $\mathcal{K}$. We shall assume this from now on. It is therefore appropriate to change our usual convention in representing $\Gamma(\mathcal{P}_2)$; we shift the indices for its distinguished generators β_i by $k+1$, and hence write $\Gamma(\mathcal{P}_2) = \langle \beta_{k+1}, \ldots, \beta_{k+n} \rangle$. We now further suppose that $\mathcal{P}_2$ has the FAP with respect to its $(m-k-1)$-face, so that

4F2
$$\begin{cases} \Gamma(\mathcal{P}_2) \cong N_m^+ \rtimes \langle \beta_{k+1}, \ldots, \beta_{m-1} \rangle, \\ N_m^+ = \langle \psi^{-1} \beta_i \psi \mid i \geqslant m, \ \psi \in \langle \beta_{k+1}, \ldots, \beta_{m-1} \rangle \rangle. \end{cases}$$

(For notational convenience we use the same letter N for both $\mathcal{P}_1$ and $\mathcal{P}_2$.) If $\beta \in \Gamma(\mathcal{P}_2)$, we write $\beta = \zeta\hat{\beta}$, with $\zeta \in N_m^+$ and $\hat{\beta} \in \langle \beta_{k+1}, \ldots, \beta_{m-1} \rangle$. The situation is illustrated in the following diagram:

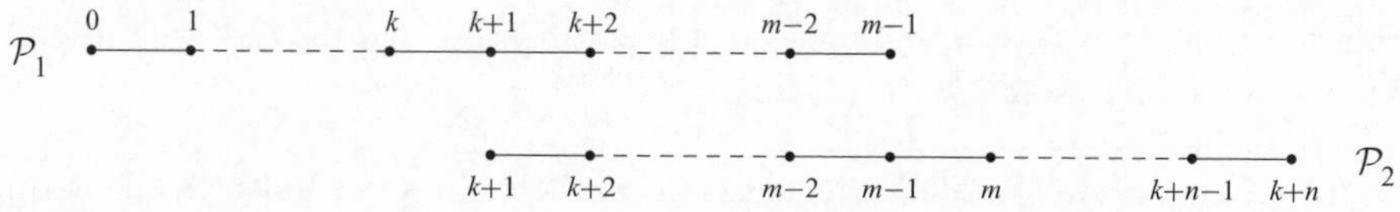

$\mathcal{P}_1$: 0, 1, k, $k+1$, $k+2$, $m-2$, $m-1$

$\mathcal{P}_2$: $k+1$, $k+2$, $m-2$, $m-1$, m, $k+n-1$, $k+n$

The mappings $\alpha_i \mapsto \beta_i$ for $i = k+1, \ldots, m-1$ induce an isomorphism

4F3
$$\kappa \colon \Gamma(\mathcal{K}) = \langle \alpha_{k+1}, \ldots, \alpha_{m-1} \rangle \to \langle \beta_{k+1}, \ldots, \beta_{m-1} \rangle.$$

In the situation described earlier consider the subgroup $\Gamma = \Gamma(\mathcal{P}_1, \mathcal{P}_2)$ of the direct product $\Gamma(\mathcal{P}_1) \times \Gamma(\mathcal{P}_2)$, which is defined by

4F4
$$\Gamma := \{(\alpha, \beta) \mid \alpha \in \Gamma(\mathcal{P}_1), \ \beta \in \Gamma(\mathcal{P}_2), \ \hat{\beta} = \hat{\alpha}\kappa\},$$

as well as its generators

4F5
$$\rho_i := \begin{cases} (\alpha_i, \varepsilon) & \text{if} \quad 0 \leqslant i \leqslant k, \\ (\alpha_i, \beta_i) & \text{if} \quad k+1 \leqslant i \leqslant m-1, \\ (\varepsilon, \beta_i) & \text{if} \quad m \leqslant i \leqslant k+n. \end{cases}$$

Then $\rho_i = (\alpha_i, \alpha_i\kappa)$ for $k+1 \leqslant i \leqslant m-1$. Further, $\rho_i^2 = (\rho_i\rho_j)^2 = \varepsilon$ for all i, j with $|i-j| \geqslant 2$.

We shall prove that Γ is a string C-group, and that the corresponding polytope, which we denote by $\mathcal{P}_1 \diamondsuit_k \mathcal{P}_2$ and call the $(k+1)$-*mix* of $\mathcal{P}_1$ and $\mathcal{P}_2$, has all the properties listed in Theorem 4F9. (Note that the operation of mixing in this way is not commutative; it is a special case of the general operation of mixing within a C-group, which we describe in Chapter 7.) The proof requires several lemmas.

4F6 Lemma *Let $\mathcal{P}_1$, $\mathcal{P}_2$, Γ, and $\rho_0, \ldots, \rho_{k+n}$ be as before. Then*

(a) $\{\varepsilon\} \times N_m^+$ *is the normal closure of* $\{\rho_m, \ldots, \rho_{k+n}\}$ *in* Γ, *and*

$$\Gamma \cong \left(\{\varepsilon\} \times N_m^+\right) \rtimes \langle \rho_0, \ldots, \rho_{m-1} \rangle \cong N_m^+ \rtimes \Gamma(\mathcal{P}_1);$$

(b) $N_k^- \times \{\varepsilon\}$ *is the normal closure of* $\{\rho_0, \ldots, \rho_k\}$ *in* Γ, *and*

$$\Gamma \cong \left(N_k^- \times \{\varepsilon\}\right) \rtimes \langle \rho_{k+1}, \ldots, \rho_{k+n} \rangle \cong N_k^- \rtimes \Gamma(\mathcal{P}_2);$$

(c) $\Gamma \cong \left(N_k^- \times N_m^+\right) \rtimes \langle \rho_{k+1}, \ldots, \rho_{m-1} \rangle \cong \left(N_k^- \times N_m^+\right) \rtimes \Gamma(\mathcal{K})$.

Proof. For part (a), if $(\alpha, \beta) \in \Gamma$ and $i \geqslant m$, then

$$(\alpha, \beta)^{-1}\rho_i(\alpha, \beta) = (\alpha, \beta)^{-1}(\varepsilon, \beta_i)(\alpha, \beta) = (\varepsilon, \beta^{-1}\beta_i\beta) \in \{\varepsilon\} \times N_m^+.$$

On the other hand, using (4F2), the same equation with $\beta \in \langle \beta_{k+1}, \ldots, \beta_{m-1} \rangle$ and $\alpha := \beta\kappa^{-1}$ shows that $\{\varepsilon\} \times N_m^+$ is contained in (and indeed equals) the normal closure of $\{\rho_m, \ldots, \rho_{n+k}\}$ in the subgroup $\langle \rho_{k+1}, \ldots, \rho_{k+n} \rangle$ of Γ; in fact, now $(\alpha, \beta) \in \langle \rho_{k+1}, \ldots, \rho_{m-1} \rangle$. This proves the statement about the normal closure. Also,

$$\left(\{\varepsilon\} \times N_m^+\right) \cap \langle \rho_0, \ldots, \rho_{m-1} \rangle = \{(\varepsilon, \varepsilon)\},$$

because the second component of an element $(\alpha, \beta) \in \langle \rho_0, \ldots, \rho_{m-1} \rangle$ is in $\langle \beta_{k+1}, \ldots, \beta_{m-1} \rangle$, and the intersection of this subgroup with N_m^+ is $\{\varepsilon\}$. Further, the canonical projection $\Gamma(\mathcal{P}_1) \to \Gamma(\mathcal{P}_1)/N_k^-$ preserves $\alpha_{k+1}, \ldots, \alpha_{m-1}$, but sends $\alpha_0, \ldots, \alpha_k$ to ε. Therefore, on $\langle \rho_0, \ldots, \rho_{m-1} \rangle$, the second component of (α, β) is just a homomorphic image of the first, proving that $\langle \rho_0, \ldots, \rho_{m-1} \rangle \cong \Gamma(\mathcal{P}_1)$. Now part (a) follows. We omit the very similar proof of part (b).

For part (c), we conclude from (4F1) and (4F2) that $\langle \rho_{k+1}, \ldots, \rho_{m-1} \rangle$ ($\cong \Gamma(\mathcal{K})$) has trivial intersection with $N_k^- \times N_m^+$. Finally, part (b) and $\langle \rho_{k+1}, \ldots, \rho_{k+n} \rangle \cong (\{\varepsilon\} \times N_m^+) \rtimes \langle \rho_{k+1}, \ldots, \rho_{m-1} \rangle$ imply that the product of these two groups is Γ, which completes the proof. □

4F7 Lemma *Let Γ be as before.*

(a) If $0 \leqslant r \leqslant k$ and $m \leqslant s \leqslant k+n$, then the subgroup $\langle \rho_r, \ldots, \rho_s \rangle$ of Γ is isomorphic to the group $\Gamma(\mathcal{Q}_1, \mathcal{Q}_2)$, which is obtained from the co-$(r-1)$-face $\mathcal{Q}_1$ of $\mathcal{P}_1$ and the $(s-k)$-face $\mathcal{Q}_2$ of $\mathcal{P}_2$ in the same way that Γ is from $\mathcal{P}_1$ and $\mathcal{P}_2$.

(b) Γ is a string C-group with distinguished generators $\rho_0, \ldots, \rho_{k+n}$.

Proof. For part (a), recall from Lemma 4E11 that the FAP is hereditary. Accordingly, $\mathcal{Q}_1$ has the FAP with respect to the co-faces at its $(k-r)$-faces, and $\mathcal{Q}_2$ has the FAP with respect to its $(m-k-1)$-faces. In particular, the group $\Gamma(\mathcal{Q}_1, \mathcal{Q}_2)$ is well defined. Also,

$$\begin{cases} \langle \rho_r, \ldots, \rho_{m-1} \rangle \cong \Gamma(\mathcal{Q}_1) \cong L_k^- \rtimes \langle \alpha_{k+1}, \ldots, \alpha_{m-1} \rangle, \\ L_k^- := \langle \psi^{-1}\alpha_i\psi \mid r \leqslant i \leqslant k,\ \psi \in \langle \alpha_{k+1}, \ldots, \alpha_{m-1} \rangle \rangle, \end{cases}$$

and

4F8
$$\begin{cases} \langle \rho_{k+1}, \ldots, \rho_s \rangle \cong \Gamma(\mathcal{Q}_2) \cong L_m^+ \rtimes \langle \beta_{k+1}, \ldots, \beta_{m-1} \rangle, \\ L_m^+ := \langle \psi^{-1}\beta_i\psi \mid m \leqslant i \leqslant s,\ \psi \in \langle \beta_{k+1}, \ldots, \beta_{m-1} \rangle \rangle. \end{cases}$$

As in (4E12), we have

$$L_k^- = N_k^- \cap \langle \alpha_r, \ldots, \alpha_{m-1} \rangle,$$
$$L_m^+ = N_m^+ \cap \langle \beta_{k+1}, \ldots, \beta_s \rangle.$$

Now, if $(\alpha, \beta) \in \langle \rho_r, \ldots, \rho_s \rangle$, then (4F5) implies that $\alpha \in \langle \alpha_r, \ldots, \alpha_{m-1} \rangle$ and $\beta \in \langle \beta_{k+1}, \ldots, \beta_s \rangle$. Write $\alpha =: \eta\hat{\alpha}$, with $\eta \in N_k^-$ and $\hat{\alpha} \in \langle \alpha_{k+1}, \ldots, \alpha_{m-1} \rangle$, and $\beta =: \zeta\hat{\beta}$ with $\zeta \in N_m^+$ and $\hat{\beta} \in \langle \beta_{k+1}, \ldots, \beta_{m-1} \rangle$. Then indeed $\eta = \alpha\hat{\alpha}^{-1} \in L_k^-$ and $\zeta = \beta\hat{\beta}^{-1} \in L_m^+$. It follows that the product decomposition for α and β relative to $\mathcal{P}_1$ and $\mathcal{P}_2$ is the same as that relative to $\mathcal{Q}_1$ and $\mathcal{Q}_2$. In particular, the isomorphism κ of (4F3) for $\mathcal{P}_1$ and $\mathcal{P}_2$ carries over to $\mathcal{Q}_1$ and $\mathcal{Q}_2$. As a consequence, the mapping $(\alpha, \beta) \mapsto (\alpha, \beta)$ (which treats the element $(\alpha, \beta) \in \langle \rho_r, \ldots, \rho_s \rangle$ as an element of $\Gamma(\mathcal{Q}_1) \times \Gamma(\mathcal{Q}_2)$) is indeed an isomorphism $\langle \rho_r, \ldots, \rho_s \rangle \mapsto \Gamma(\mathcal{Q}_1, \mathcal{Q}_2)$. This proves part (a).

For the proof of part (b), it is convenient to extend the definition of $\Gamma = \Gamma(\mathcal{P}_1, \mathcal{P}_2)$ to allow the two cases $k = -1$ with $m \leqslant n$, and $k = m - n - 1$ with $m \geqslant n$. In the first case, the FAP condition on $\mathcal{P}_1$ is empty, $\mathcal{K} = \mathcal{P}_1$, and $N_k^- = N_{-1}^- := \{\varepsilon\}$; if also $m = n$, then $\mathcal{P}_1 = \mathcal{P}_2$, the FAP condition on $\mathcal{P}_2$ is also empty, and $N_m^+ = N_n^+ := \{\varepsilon\}$. In the second case, the FAP condition on $\mathcal{P}_2$ is empty, $\mathcal{K} = \mathcal{P}_2$, and $N_m^+ := \{\varepsilon\}$; if also $m = n$, then $\mathcal{P}_1 = \mathcal{P}_2$, the FAP condition on $\mathcal{P}_1$ is also empty, and $N_k^- = N_{-1}^- := \{\varepsilon\}$. In all other respects, the construction of Γ remains the same as before. In particular, in the first case, $\Gamma = \langle \rho_0, \ldots, \rho_{n-1} \rangle \cong \Gamma(\mathcal{P}_2)$, and in the second, $\Gamma = \langle \rho_0, \ldots, \rho_{m-1} \rangle \cong \Gamma(\mathcal{P}_1)$. To check this, we can argue as in the proof of Lemma 4F6, and note that in the first case the first component of an element $(\alpha, \beta) \in \Gamma$ is a homomorphic image of the second component, while in the second it is the other way around.

We can now prove the intersection property for $\Gamma = \Gamma(\mathcal{P}_1, \mathcal{P}_2)$, by induction on the number $k + n + 1$ of generators. If $k = -1$ with $m \leqslant n$, we have $\Gamma \cong \Gamma(\mathcal{P}_2)$, while if $k = m - n - 1$ with $m \geqslant n$, we have $\Gamma \cong \Gamma(\mathcal{P}_1)$; in each case, the intersection property holds trivially. Now, again let $0 \leqslant k \leqslant m - 2$ and $k \geqslant m - n$, and consider the two subgroups $\langle \rho_0, \ldots, \rho_{k+n-1} \rangle$ and $\langle \rho_1, \ldots, \rho_{k+n} \rangle$ of Γ with $k + n$ generators. By part (a), the first group is $\Gamma(\mathcal{P}_1, \mathcal{Q}_2)$ and the second is $\Gamma(\mathcal{Q}_1, \mathcal{P}_2)$, where $\mathcal{Q}_2$ is the facet of $\mathcal{P}_2$ and $\mathcal{Q}_1$ is the vertex-figure of $\mathcal{P}_1$. Since we have extended the definition of Γ, this is also true if $k = 0$ or $k = m - n$ (where $\Gamma(\mathcal{Q}_1, \mathcal{P}_2) \cong \Gamma(\mathcal{P}_2)$ or $\Gamma(\mathcal{P}_1, \mathcal{Q}_2) \cong \Gamma(\mathcal{P}_1)$, respectively). By the induction hypothesis, both subgroups have the intersection property. Therefore, to prove that Γ has the intersection property, it is enough to show that

$$\langle \rho_0, \ldots, \rho_{k+n-1} \rangle \cap \langle \rho_1, \ldots, \rho_{k+n} \rangle = \langle \rho_1, \ldots, \rho_{k+n-1} \rangle.$$

Now, by Lemma 4F6(a) applied to the pair $(\mathcal{P}_1, \mathcal{Q}_2)$, we have

$$\langle \rho_0, \ldots, \rho_{k+n-1} \rangle \cong (\{\varepsilon\} \times L_m^+) \rtimes \langle \rho_0, \ldots, \rho_{m-1} \rangle \cong L_m^+ \rtimes \Gamma(\mathcal{P}_1),$$

with L_m^+ as in (4F8) and $s = k + n - 1$. Similarly, from $(\mathcal{Q}_1, \mathcal{P}_2)$ we get

$$\langle \rho_1, \ldots, \rho_{k+n} \rangle \cong (\{\varepsilon\} \times N_m^+) \rtimes \langle \rho_1, \ldots, \rho_{m-1} \rangle \cong N_m^+ \rtimes \Gamma(\mathcal{Q}_1).$$

Then the intersection of the two subgroups is isomorphic to $(\{\varepsilon\} \times L_m^+) \rtimes \langle \rho_1, \ldots, \rho_{m-1} \rangle$, which by Lemma 4F6(a) is just $\Gamma(\mathcal{Q}_1, \mathcal{Q}_2)$. On the other hand, part (a) gives $\Gamma(\mathcal{Q}_1, \mathcal{Q}_2) \cong \langle \rho_1, \ldots, \rho_{k+n-1} \rangle$. Therefore, in Γ, the two subgroups intersect in $\langle \rho_1, \ldots, \rho_{k+n-1} \rangle$. This completes the proof. □

For our main result, we can translate the properties of Γ into the language of polytopes.

4F9 Theorem *Let $m, n \geqslant 2$, and let $0 \leqslant k \leqslant m - 2$ and $k \geqslant m - n$. Let $\mathcal{P}_1$ be a regular m-polytope, and let $\mathcal{P}_2$ be a regular n-polytope such that the co-k-faces $\mathcal{K}$ of $\mathcal{P}_1$ are isomorphic to the $(m - k - 1)$-faces of $\mathcal{P}_2$. Suppose that $\mathcal{P}_1$ has the FAP with respect to its co-k-faces, and that $\mathcal{P}_2$ has the FAP with respect to its $(m - k - 1)$-faces. Let $\Gamma(\mathcal{P}_1) = \langle \alpha_0, \ldots, \alpha_{m-1} \rangle$ and $\Gamma(\mathcal{P}_2) = \langle \beta_{k+1}, \ldots, \beta_{k+n} \rangle$ (where the indices of the distinguished generators of the latter have been shifted by $k + 1$). Further, let $\Gamma = \Gamma(\mathcal{P}_1, \mathcal{P}_2)$ be the group defined in (4F4), and let $\mathcal{P} := \mathcal{P}_1 \lozenge_{k+1} \mathcal{P}_2$ be the regular $(k + n + 1)$-polytope with group $\Gamma(\mathcal{P}) = \Gamma$. Then the following hold.*

(a) $\mathcal{P}$ is (k, m)-flat.

(b) The m-faces of $\mathcal{P}$ are isomorphic to $\mathcal{P}_1$, and the co-k-faces are isomorphic to $\mathcal{P}_2$. More generally, if $r \leqslant k,\ s \geqslant m$, and if $s - r \geqslant 4$, then the section F_s/F_r of $\mathcal{P}$ which is defined by the base r-face F_r and s-face F_s of $\mathcal{P}$ is isomorphic to the regular $(s - r - 1)$-polytope $\mathcal{Q} := \mathcal{Q}_1 \lozenge_{k-r} \mathcal{Q}_2$ which is constructed from the co-r-face $\mathcal{Q}_1$ of $\mathcal{P}_1$ and $(s - k - 1)$-face $\mathcal{Q}_2$ of $\mathcal{P}_2$ in the same way as $\mathcal{P}$ is from $\mathcal{P}_1$ and $\mathcal{P}_2$.

(c) $\mathcal{P}$ has the FAP with respect to both its co-k-faces and its m-faces. In particular,

$$\Gamma(\mathcal{P}) \cong N_m^+ \rtimes \Gamma(\mathcal{P}_1) \cong N_k^- \rtimes \Gamma(\mathcal{P}_2) \cong (N_k^- \times N_m^+) \rtimes \Gamma(\mathcal{K}),$$

where N_m^+ and N_k^- are the normal closures of $\{\beta_m, \ldots, \beta_{k+n}\}$ in $\Gamma(\mathcal{P}_2)$ and $\{\alpha_0, \ldots, \alpha_k\}$ in $\Gamma(\mathcal{P}_1)$, respectively.

Proof. By Lemma 4F7(b), $\Gamma = \langle \rho_0, \ldots, \rho_{k+n} \rangle$ is a string C-group, and so is the group of a polytope $\mathcal{P} = \mathcal{P}(\Gamma)$. For the proof of (a), we apply Proposition 4E4, and show that $\Gamma = \langle \rho_{k+1}, \ldots, \rho_{k+n} \rangle \langle \rho_0, \ldots, \rho_{m-1} \rangle$. But this follows from Lemma 4F6(a) if we note, as we did in its proof, that $\{\varepsilon\} \times N_m^+ \subseteq \langle \rho_{k+1}, \ldots, \rho_{k+n} \rangle$. For the proof of (b), we just need to appeal to Lemmas 4F6 and 4F7(a). In fact, $\langle \rho_0, \ldots, \rho_{m-1} \rangle \cong \Gamma(\mathcal{P}_1)$, $\langle \rho_{k+1}, \ldots, \rho_{k+n} \rangle \cong \Gamma(\mathcal{P}_2)$, and $\Gamma(F_s/F_r) \cong \langle \rho_{r+1}, \ldots, \rho_{s-1} \rangle \cong \Gamma(\mathcal{Q}_1, \mathcal{Q}_2)$. Note that $\Gamma(\mathcal{Q}_1, \mathcal{Q}_2)$ is defined because $s - r \geqslant 4$. Finally, part (c) just restates Lemma 4F6. □

We remark that Theorem 4F9 is a generalization of [368, Theorem 2], which covers the case $k = 0,\ m = n$, and gives ordinary flat (that is, $(0, n)$-flat) polytopes.

In practice, it is often convenient to construct the group Γ of the polytope $\mathcal{P}$ in Theorem 4F9 as a permutation group. Suppose that $\Gamma(\mathcal{P}_1) = \langle \alpha_0, \ldots, \alpha_{m-1} \rangle$ and

$\Gamma(\mathcal{P}_2) = \langle \beta_{k+1}, \ldots, \beta_{k+n} \rangle$ are faithfully represented as permutation groups on disjoint sets Z_1 and Z_2, respectively. Let $Z := Z_1 \cup Z_2$, and consider the faithful representation $\Gamma(\mathcal{P}_1) \times \Gamma(\mathcal{P}_2) \mapsto S_Z$, given by $(\alpha, \beta) \mapsto \alpha\beta$, where S_Z denotes the symmetric group on Z. Note that $\alpha\beta \in S_{Z_1} \times S_{Z_2}$. Then the restriction to the subgroup of $\Gamma(\mathcal{P}_1) \times \Gamma(\mathcal{P}_2)$ defined in (4F4) is also a faithful representation onto a subgroup of S_Z, which we again denote by Γ. If, as before, we label the generators of Γ by $\rho_0, \ldots, \rho_{k+n}$, then (4F5) translates into the following generators for Γ:

4F10
$$\rho_i := \begin{cases} \alpha_i & \text{if } 0 \leqslant i \leqslant k, \\ \alpha_i\beta_i & \text{if } k+1 \leqslant i \leqslant m-1, \\ \beta_i & \text{if } m \leqslant i \leqslant k+n. \end{cases}$$

Theorem 4F9 has the following interesting consequences for the amalgamation problem. The second corollary will be required in later chapters.

4F11 Corollary *Let $\mathcal{P}_1$ and $\mathcal{P}_2$ be two regular n-polytopes such that the vertex-figures of $\mathcal{P}_1$ are isomorphic to the facets of $\mathcal{P}_2$. If $\mathcal{P}_1$ has the FAP with respect to its vertex-figures and $\mathcal{P}_2$ has the FAP with respect to its facets, then $\langle \mathcal{P}_1, \mathcal{P}_2 \rangle \neq \emptyset$, and the class contains a flat $(n+1)$-polytope.*

Proof. Apply Theorem 4F9 with $m = n$ and $k = 0$. Then $\mathcal{P} \in \langle \mathcal{P}_1, \mathcal{P}_2 \rangle$. □

4F12 Proposition *Let $\mathcal{P}_1$ and $\mathcal{P}_2$ be two regular n-polytopes such that the vertex-figures of $\mathcal{P}_1$ are isomorphic to the facets of $\mathcal{P}_2$. Suppose that $\mathcal{P}_1$ is flat and has the FAP with respect to its vertex-figures, and that $\Gamma(\mathcal{P}_1)$ acts faithfully on the vertices of $\mathcal{P}_1$.*

(a) $\langle \mathcal{P}_1, \mathcal{P}_2 \rangle \neq \emptyset$ if and only if $\mathcal{P}_2$ has the FAP with respect to its facets.
(b) If $\langle \mathcal{P}_1, \mathcal{P}_2 \rangle \neq \emptyset$, then $\{\mathcal{P}_1, \mathcal{P}_2\}$ has the FAP with respect to its facets.

Proof. Let $\mathcal{P} \in \langle \mathcal{P}_1, \mathcal{P}_2 \rangle$, and let $\Gamma(\mathcal{P}) = \langle \rho_0, \ldots, \rho_n \rangle$. By Lemma 4E3, the $(n+1)$-polytope $\mathcal{P}$ is $(0, n-1)$-flat, and so each vertex of $\mathcal{P}$ is also a vertex of the base $(n-1)$-face F_{n-1} of $\mathcal{P}$. But the generator ρ_n acts trivially on (the faces of) F_{n-1}, and thus fixes each vertex of $\mathcal{P}$. It follows that each conjugate of ρ_n fixes each vertex of $\mathcal{P}$ (or, equivalently, each vertex of the base facet of $\mathcal{P}$), and hence so do all the elements of the normal closure N_n^+ of ρ_n in $\Gamma(\mathcal{P})$. Now, since $\Gamma(\mathcal{P}_1)$ acts trivially on the vertices of $\mathcal{P}_1$, this implies that $N_n^+ \cap \langle \rho_0, \ldots, \rho_{n-1} \rangle = \{\varepsilon\}$. It follows that $\mathcal{P}$ must have the FAP with respect to its facets. If $\mathcal{P} = \{\mathcal{P}_1, \mathcal{P}_2\}$, this proves part (b). For an arbitrary $\mathcal{P}$, Proposition 4E11(b) implies that the vertex-figures must also have the FAP with respect to their facets, and so must $\mathcal{P}_2$. This gives one direction of part (a). The other direction follows directly from Theorem 4F9 applied to $\mathcal{P}_1, \mathcal{P}_2$. This concludes the proof. □

5

Realizations

Regular polytopes can be discussed in a purely abstract setting. As we saw in Chapter 2, in that context, the theory of regular polytopes is equivalent to the theory of certain kinds of groups generated by involutions, namely, the string C-groups. However, much of the appeal of regular polytopes throughout their history has been their geometric symmetry. Even when the notion of symmetry was only hazily understood (at least from the present viewpoint of action of a group), it was very clearly appreciated that geometric regular polygons and polyhedra were (in some sense) as symmetric as they could possibly be.

It is this geometric picture of regular polytopes which we shall address in this chapter. More specifically, we shall define the idea of realizations of an abstract regular polytope, and discuss their various properties. In Section 5A, we shall consider the general theory of realizations, which is that part common to the whole subject. The realization theory for finite regular polytopes is fairly detailed, and we shall cover this in Section 5B. As yet, the theory of realizations of a regular apeirotope is a little sketchy; we shall describe what is known in Section 5C.

5A Realizations in General

There are many candidates for spaces in which regular polytopes might be realized geometrically, although the most natural are probably the spherical, euclidean and hyperbolic spaces. Of course, a spherical space can be thought of as embedded in a euclidean space, and both spherical and euclidean spaces can be embedded in hyperbolic spaces. However, while the other two kinds may be useful in certain specific contexts, it is in euclidean spaces that we obtain the richest structure of realizations. We shall therefore consider euclidean realizations exclusively.

A *realization* of an abstract polytope $\mathcal{P}$ is a mapping β of the vertex-set $\mathcal{P}_0$ of $\mathcal{P}$ into some euclidean space E. The image $V := \mathcal{P}_0\beta$ is called the *vertex-set* of the realization β.

It is not immediately clear that this definition gives us what we want – after all, it does not mention any faces of $\mathcal{P}$ of higher rank than 0, and so does not give the appearance of being particularly geometric. However, the required geometric structure of the realization is induced by β as follows. Writing 2^X for the family of subsets of the set X (that is, its *power set*), we then have

5A1 Theorem *Let β be a realization of an abstract n-polytope $\mathcal{P}$, and define $\beta_0 := \beta$ and $V_0 := V$. Then β recursively induces surjections $\beta_j : \mathcal{P}_j \to V_j$, for $j = 1, \ldots, n$, with $V_j \subseteq 2^{V_{j-1}}$ consisting of the elements*

$$F\beta_j := \{G\beta_{j-1} \mid G \in \mathcal{P}_{j-1} \text{ and } G \leqslant F\}$$

for $F \in \mathcal{P}_j$. Further, β_{-1} is given by $F_{-1}\beta_{-1} := \emptyset$.

Even though each β_j is determined by $\beta = \beta_0$, it is sometimes helpful to think of the realization as given by all the β_j. We say that β is *faithful* if each β_j is a bijection; otherwise, β is *degenerate*. If β_0 is bijective, then we say that β is *vertex-faithful*. If β is degenerate, it is often convenient to regard the sets V_j as multiply counted (ordered) sets. An extreme example of a degenerate realization is the *trivial* one, where V consists of a single point.

As we said in Section 2A, we allow little possibility of confusion if we identify a section F/F_{-1} of an abstract polytope $\mathcal{P}$ with the face F. If we regard the sets V_j of Theorem 5A1 as multiply counted, then we may make similar identifications in a realization of $\mathcal{P}$. In particular, if $\operatorname{rank}\mathcal{P} = n$, then $V_n = F_n\beta_n$ encapsulates (in this sense) all the structure of $\mathcal{P}$. We shall therefore use a notation such as $P := V_n$ to denote the realization of $\mathcal{P}$, and call P a (*geometric*) *polytope*. Furthermore, we shall often not distinguish between a face $F \in \mathcal{P}_j$ and the corresponding $F\beta_j$. If P is a polytope (thus understood), we shall then write $V(P) := V$ for its vertex-set.

There is usually no loss of generality in assuming that $E = \operatorname{aff} V$ is the affine hull of the vertex-set $V = V(P)$ of the polytope P (given by the realization β). We then call $\dim V := \dim \operatorname{aff} V$ the *dimension* of P, and denote it by $\dim P$.

The main reason for working in euclidean spaces, rather than in any others, is that we can then operate on realizations of an abstract polytope $\mathcal{P}$ in various ways. We first remark that the congruence class of a realization is usually of more interest to us than the actual realization itself; that is, we shall identify two realizations if the natural mapping between their vertex-sets induces an isometry between their ambient spaces. This simplification enables us, for example, to ignore the distinction between translates, and so we can always assume that a realization of a finite polytope has its centroid (as a multiply counted set) at the origin o.

The two main operations on realizations are the following. First, let $\beta : \mathcal{P}_0 \to L$ and $\gamma : \mathcal{P}_0 \to M$ be two realizations of the abstract polytope $\mathcal{P}$. Then the *blend* $\beta \# \gamma : \mathcal{P}_0 \to L \times M$ is defined by $F_0(\beta \# \gamma) := (F_0\beta, F_0\gamma)$ for $F_0 \in \mathcal{P}_0$. If, in addition, $L = M$ and $\lambda, \ \mu \in \mathbb{R}$, then the *linear combination* $\lambda\beta + \mu\gamma$ is defined by $F_0(\lambda\beta + \mu\gamma) := \lambda(F_0\beta) + \mu(F_0\gamma)$ for each $F_0 \in \mathcal{P}_0$. A particular case of this is the *scalar multiple* $\lambda\beta$ of a single realization.

Note also that the blend can be regarded as a special case of a linear combination, with the realizations β and γ as before thought of as acting on E itself in the obvious way. However, we prefer to keep the concepts of blend and linear combination separate, for reasons which will become clearer later.

As things stand at present, the structure of the space of all realizations of a given abstract polytope is somewhat inchoate. It is therefore appropriate to impose some restrictions before we discuss realizations any further.

We call a realization β of the abstract polytope $\mathcal{P}$ *symmetric* if each automorphism of $\mathcal{P}$ induces an isometric permutation of the vertex-set V of β. In this context, we should observe that an isometry of V always extends (uniquely) to an isometry of the ambient space $E = \text{aff}\, V$. The isometries of the vertex-set V induced by the automorphisms of $\mathcal{P}$ are then called the *symmetries* of the polytope P, and form its *symmetry group* $G(P)$. Clearly, $G(P)$ is an euclidean representation of the automorphism group $\Gamma(\mathcal{P})$; it will be faithful if the realization is faithful, but perhaps also in other cases. In the context of regular polytopes, we shall discuss $G(P)$ in more detail later in this section.

For finite polytopes, vertex-faithful symmetric realizations always exist (just map the vertices to those of a regular simplex of the appropriate dimension), but for apeirotopes even this question remains open. Whether a vertex-faithful realization is faithful is a purely combinatorial condition on the polytope $\mathcal{P}$, because it just comes down to whether, for each j, the j-faces of $\mathcal{P}$ are (uniquely) determined by the $(j-1)$-faces which they contain.

However, we do have

5A2 Theorem *A (faithful, symmetric) realization of a regular polytope $\mathcal{P}$ induces a (faithful, symmetric) realization of each section of $\mathcal{P}$.*

Proof. It is enough to see that we can realize the facet and vertex-figure of $\mathcal{P}$, because we may then use inductive arguments. The realization of the facet F_{n-1} explicitly occurs in the realization of $\mathcal{P}$ itself, as is evident from Theorem 5A1. The vertex-figure at a vertex $F_0 \in \mathcal{P}_0$ is also realized; its vertex-set is

5A3 $$V' := \{v' \in V \mid \{v, v'\} \in V_1\},$$

where $v := F_0\beta \in V$. Both these realizations will be faithful or symmetric whenever β itself is. □

We also have the following straightforward result.

5A4 Theorem *If $\mathcal{P}$ and $\mathcal{Q}$ are regular polytopes such that $\mathcal{P}$ covers $\mathcal{Q}$, then any realization of $\mathcal{Q}$ induces one of $\mathcal{P}$.*

From now on, we shall confine our attention to symmetric realizations of regular polytopes, and so, for brevity (and to conform with the earlier usage in [282, 288]), we shall drop the qualification "symmetric". We now investigate the geometric properties of the symmetry group $G(P)$ of a realization P of an abstract regular polytope $\mathcal{P}$.

Corresponding to each generating involution ρ_j of the automorphism group $\Gamma(\mathcal{P})$ of $\mathcal{P}$ is an isometry R_j of the euclidean space E, which is either also an involution, that is, a *reflexion*, or the identity, which we denote by the same symbol E as for the space. We identify a reflexion R_j with its *mirror*

5A5
$$\{x \in E \mid xR_j = x\}$$

of fixed points, which explains why we write the identity as E. We shall use this identification frequently, often without any comment. We first have

5A6 Lemma *If $R_j = E$ for some $j = 0, \ldots, n-2$, then $R_k = E$ for each $k > j$.*

Proof. To see this, suppose that $v \in V$ is the vertex of the realization P which corresponds to the vertex in the base flag Φ of $\mathcal{P}$. Then

$$\begin{aligned} V &= v\langle R_0, \ldots, R_{n-1}\rangle \\ &= v\langle R_0, \ldots, R_{j-1}\rangle\langle R_{j+1}, \ldots, R_{n-1}\rangle \\ &= v\langle R_0, \ldots, R_{j-1}\rangle, \end{aligned}$$

since $R_i R_k = R_k R_i$ and $vR_k = v$ for $i < j < k$. Thus R_k leaves $E = \operatorname{aff} V$ pointwise invariant for each $k > j$, so that $R_k = E$. □

If the polytope $\mathcal{P}$ has Schläfli type $\{p_1, \ldots, p_{n-1}\}$, we saw in Proposition 2B16 that its automorphism group $\Gamma(\mathcal{P})$ splits into a direct product if some $p_k = 2$. A similar argument to that of Lemma 5A6 yields

5A7 Theorem *If $p_k = 2$ for some $k = 1, \ldots, n-1$, then $\mathcal{P}$ covers its face F_k, and each realization of $\mathcal{P}$ is induced by a realization of F_k.*

As we remarked earlier, to each realization of $\mathcal{P}$ corresponds a euclidean representation of $\Gamma(\mathcal{P})$. We now consider the extent to which there is a converse of this.

Suppose that $G := \langle R_0, \ldots, R_{n-1}\rangle$ is an euclidean representation (that is, one by isometries) of the string C-group Γ, so that (in particular), for $j = 0, \ldots, n-1$, R_j is a reflexion in the euclidean space E, or possibly $R_j = E$. We now describe *Wythoff's construction*. The *Wythoff space* $W = W_G$ of G is defined to be

5A8
$$W := R_1 \cap \cdots \cap R_{n-1}.$$

If $W \neq \emptyset$, pick a point $v \in W$ (the *initial vertex*), and define $V := vG$; this is our vertex-set. Setting $F_0 := v$, we define recursively

5A9
$$F_j := F_{j-1}\langle R_0, \ldots, R_{j-1}\rangle$$

for $j = 1, \ldots, n-1$; this gives the base flag $\Phi := \{F_{-1}, F_0, \ldots, F_n\}$ (with $F_{-1} := \emptyset$) of a realization P of $\mathcal{P}(\Gamma)$, whose remaining flags are the Φg with $g \in G$.

5A10 Theorem *Every euclidean representation of the automorphism group $\Gamma(\mathcal{P})$ of an abstract regular polytope $\mathcal{P}$ with a non-empty Wythoff space gives rise to a realization of $\mathcal{P}$.*

Proof. This is just about self-evident. If $\Gamma = \langle \rho_0, \ldots, \rho_{n-1}\rangle$ in terms of its distinguished generators, so that the mapping $\rho_j \mapsto R_j$ induces the representation $\gamma \mapsto g$ of Γ, then mapping the vertex in the base flag of $\mathcal{P}(\Gamma)$ to v induces the mapping of the vertex-set $\mathcal{P}_0$ of $\mathcal{P}$ to V. □

However, whether or not the realization is non-trivial (let alone faithful) is not so easy to determine. We shall consider this problem in more detail later, but for the moment we merely mention a few possibilities.

For example, we may always suppose that $W \neq \emptyset$; otherwise we obtain no realization at all from the representation. Further, if G has a fixed point (in which case it is isomorphic to an orthogonal group), and v is this fixed point, then the realization is trivial.

Next, if $R_0 \cap W \neq \emptyset$, then the realization is necessarily bounded; in fact, it is trivial if W then consists of a single point. In these circumstances, if W is not a point-set, then an infinite realization is necessarily non-discrete. Conversely, for the realization to be bounded, all the mirrors R_j must contain a common point.

Third, suppose that $R_0 \cap W = \emptyset$. Assuming that the subgroup $G_0 := \langle R_1, \ldots, R_{n-1} \rangle$ is non-trivial, we then see that the realization is unbounded, and hence infinite. The initial vertex v is moved by R_0 to a point whose images under G_0 lie on a sphere centred at v, and these images are themselves centres of such spheres.

We have described the two basic ways of operating on realizations of regular polytopes. We may recall at this point that we only have these operations because we confine our attention to euclidean realizations, since these operations use the euclidean structure in an essential way.

The blend corresponds to the sum of two representations of a group.

5A11 Lemma *Let Γ be a C-group, let L, M be orthogonal complementary subspaces of an euclidean space E, and let $S_0, \ldots, S_{n-1} \subseteq L$ and $T_0, \ldots, T_{n-1} \subseteq M$ be the generating reflexions of representations of Γ in L and M. If $R_j := S_j \times T_j \subseteq L \times M = E$ for $j = 0, \ldots, n-1$, then $\langle R_0, \ldots, R_{n-1} \rangle$ is a representation of Γ.*

Proof. The proof is obvious, when we recall the convention of identifying a reflexion with its mirror. □

If we apply Lemma 5A11 to realizations P in L and Q in M of a regular polytope $\mathcal{P}$, there is a natural choice of an associated realization of $\mathcal{P}$ in E. If we write $G(P) := \langle S_0, \ldots, S_{n-1} \rangle$, $G(Q) := \langle T_0, \ldots, T_{n-1} \rangle$ and $G := \langle R_0, \ldots, R_{n-1} \rangle$, then the new Wythoff space is

5A12 $$W_G = W_{G(P)} \times W_{G(Q)}.$$

Indeed, if $u \in V(P) \cap W_{G(P)}$ and $v \in V(Q) \cap W_{G(Q)}$ are the initial vertices of P and Q, then $(u, v) \in W_G$ is a suitable initial vertex of the new regular polytope. The resulting polytope is denoted by $P \# Q$, and is called the *blend* of P and Q, since it is clear that the realization map is the blend of those of P and Q in the previously introduced sense.

At this point, it is worth remarking that there is an abstract version of the blending operation at the group theoretic level. This, however, operates on different string C-groups and so does not really correspond to what we are looking at here. It gives, in fact, what we call in Section 7A the "mix" of two polytopes. (The choice in [292, §4] of the same term "blend" for these distinct concepts is, in retrospect, unfortunate.)

The construction shows that we have $V(P \# Q) \subseteq V(P) \times V(Q)$, and if $\{F_0, \ldots, F_{n-1}\}$ and $\{G_0, \ldots, G_{n-1}\}$ are the base flags of P and Q, then the corresponding base flag of $P \# Q$ is $\{F_0 \# G_0, \ldots, F_{n-1} \# G_{n-1}\}$.

If a realization cannot be expressed as a blend in a non-trivial way (we shall discuss this further in the following), then we call it *pure*. We observe that a realization P will be pure precisely when $G(P)$ acts irreducibly (in the affine sense) on the ambient space aff $V(P)$ of the realization. Here, a group G of isometries of an euclidean space E acts *affinely reducibly* if it permutes two mutually perpendicular families of proper affine subspaces of E.

If G is a representation of the string C-group Γ, λ, μ are constants, and $u, v \in W_G$, then $\lambda u + \mu v \in (\lambda + \mu)W_G$, which is the Wythoff space of the representation of Γ with mirrors $(\lambda + \mu)R_j$ for $j = 0, \ldots, n-1$. In this context, it is natural to define $0R_j$ to be that translate of R_j which contains the origin o. If u, v are the initial vertices of the realizations P, Q of $\mathcal{P}(\Gamma)$, then $\lambda u + \mu v$ is the initial vertex of a realization which we denote by $\lambda P + \mu Q$, and call a *linear combination* of P and Q. If $\mu = 0$, we obtain the *scalar multiple* λP of P. Again, these operations correspond exactly to those introduced earlier in the general case.

We have already noted that the blend is a special case of a sum, because we can write the direct product as $S_j \times T_j = S_j + T_j$, regarded as subspaces of E. However, as we said before, we prefer to keep the notions of blend and sum separate. Observe, though, that $(\lambda P) \# (\mu P)$ is congruent to νP, where $\nu^2 = \lambda^2 + \mu^2$.

In discussing possible realizations of an abstract regular polytope $\mathcal{P}$, we have remarked earlier that we should not wish to distinguish between *congruent* realizations, where the natural bijection between their vertex-sets, regarded if necessary as ordered sets, is induced by an isometry between the ambient spaces. For the rest of this chapter, we shall use the same symbol $\mathcal{P}$ to denote the family of all classes under congruence of realizations of $\mathcal{P}$.

A pair of distinct vertices of $\mathcal{P}$ is called a *diagonal* of $\mathcal{P}$. Two diagonals are said to be *equivalent* if they are equivalent under $\Gamma(\mathcal{P})$. The diagonals of $\mathcal{P}$ thus fall into *diagonal classes*, consisting of equivalent diagonals. Since a polytope can have at most countably many vertices, and hence countably many diagonals, we may assume these diagonal classes ordered in some way; conventionally, we think of the first diagonal class as that of an edge of $\mathcal{P}$.

Let P be a (possibly degenerate) realization of $\mathcal{P}$, so that $P \in \mathcal{P}$ in our new notation. If the jth diagonal class of $\mathcal{P}$ is represented by the vertex-pair $\{u, v\}$ of P, write $\delta_j = \delta_j(P) := \|u - v\|^2$, the square of the length of the represented diagonal, and define the *diagonal vector* of P to be the (possibly infinite) vector $\Delta(P) := (\delta_1, \delta_2, \ldots)$.

The next result lies at the heart of realization theory.

5A13 Lemma *The diagonal vector $\Delta(P)$ determines the realization P up to congruence.*

Proof. This is straightforward. Pick out from the vertex-set V of P an affine basis $\{b_0, \ldots, b_d\}$ of the ambient space $E = \text{aff}\, V$; we then observe that the distances $\|x - b_j\|$ determine $x \in E$ completely. It follows at once that realizations with the same diagonal vector are congruent. □

If we identify $P \in \mathcal{P}$ with its diagonal vector $\Delta(P)$, then we easily see that

$$\Delta(P \# Q) = \Delta(P) + \Delta(Q),$$
$$\Delta(\lambda P) = \lambda^2 \Delta(P),$$

for $P, Q \in \mathcal{P}$ and $\lambda \in \mathbb{R}$. Recalling that a set C in some real linear space is a *convex cone* if $\lambda x + \mu y \in C$ whenever $x, y \in C$ and $\lambda, \mu \geqslant 0$, as an immediate consequence we have

5A14 Theorem *The family $\mathcal{P}$ has the structure of a convex cone.*

Moreover,

5A15 Theorem *The vertex-faithful realizations in $\mathcal{P}$ form a convex subset of $\mathcal{P}$.*

Proof. Indeed, the blend of two realizations of $\mathcal{P}$, at least one of which is vertex-faithful, will also be vertex-faithful. □

The pure realizations in $\mathcal{P}$ are obviously of some interest; they are easily characterized.

5A16 Theorem *The pure realizations of $\mathcal{P}$ correspond to points on extreme rays of the realization cone $\mathcal{P}$.*

Proof. This follows from the definition of an *extreme* ray C of a cone: if x, y are such that $x + y \in C$, then $x, y \in C$ also. □

At this point, we could describe in more detail the structure of the subcones of $\mathcal{P}$ corresponding to a given irreducible representation of its automorphism group. However, this is of most interest when $\mathcal{P}$ is finite (or at least has finitely many vertices), and so we shall discuss this structure in Section 5B. We shall then remark that almost the same result holds even when $\mathcal{P}$ is infinite.

5B The Finite Case

We now consider the realization cone of a finite regular polytope. We know much more about this case than we do about the realizations of regular apeirotopes (infinite polytopes), and, indeed, the description which follows is almost complete. Before we proceed, let us merely remark that, in fact, all we really appeal to here is that the regular polytopes have finitely many vertices, and so certain types of apeirotopes are also covered by these arguments.

Throughout the section, $\mathcal{P}$ will be a fixed finite abstract regular polytope. Recall first our convention that every realization P of $\mathcal{P}$ will be symmetric. Thus all the relevant representations of the automorphism group $\Gamma(\mathcal{P})$ are finite groups of isometries, in particular the symmetry group $G(P)$ of the realization. The *centre* $c(P)$ of P is defined to be the centroid of its vertices; hence $c(P) \in E := \operatorname{aff} P$, the ambient space of P. The following remark is often useful.

5B1 Lemma *The centre $c(P)$ of P is the unique point of E which is fixed by all of $G(P)$.*

Proof. Clearly $c(P)$ is fixed by $G(P)$. Now points in E are uniquely determined by their distances from the vertices of P. But $c(P)$ is equidistant from these vertices, and is thus the unique point of E with this property. □

It is usually convenient to take the centre $c(P)$ of a realization P of $\mathcal{P}$ to be the origin o of the ambient space E, so that its symmetry group $G(P)$ is a finite orthogonal group.

We shall denote by v the number of vertices of $\mathcal{P}$, and by r the number of its diagonal classes. An important special realization of $\mathcal{P}$ is the *simplex realization* T; the vertex-set of T is the set of v vertices of a regular $(v-1)$-simplex of side length 1. Our first result gives the dimension of the cone of realizations, which we have agreed also to denote by $\mathcal{P}$.

5B2 Theorem *The realization cone $\mathcal{P}$ is a closed cone of dimension r.*

Proof. Since the diagonal vectors $\Delta(P)$ of realizations $P \in \mathcal{P}$ have length r, it is clear that $\dim \mathcal{P} \leqslant r$.

For the opposite inequality, recall that a necessary and sufficient condition for there to exist a set of points $\{x_1, \ldots, x_k\}$ in some euclidean space whose mutual squared distances $\|x_i - x_j\|^2$ are given quantities $\delta_{ij} = \delta_{ji} \geqslant 0$ (with $\delta_{ii} = 0$) is that the *Cayley–Menger matrix*

$$D := \begin{bmatrix} 0 & 1 & \ldots & 1 \\ 1 & & & \\ \vdots & & \delta_{ij} & \\ 1 & & & \end{bmatrix}$$

have just 1 positive eigenvalue (and k non-positive eigenvalues); see [329] for further details. Moreover, D is non-singular if and only if $\{x_1, \ldots, x_k\}$ is affinely independent. It follows that, if $\Delta = (\delta_1, \ldots, \delta_r)$ is any vector sufficiently near $(1, \ldots, 1)$, then we can find a set of points in one-to-one correspondence with the vertices of $\mathcal{P}$, whose mutual squared distances are the appropriate δ_j, and this set gives rise to a realization of $\mathcal{P}$ with diagonal vector Δ. Hence $\dim \mathcal{P} \geqslant r$, and so we have equality.

Finally, if we consider realizations of $\mathcal{P}$ in some fixed bounded region (in a $(v-1)$-dimensional space), we can see that the limit of a convergent sequence of diagonal vectors is again the diagonal vector of some realization (possibly degenerate) in $\mathcal{P}$. That is, $\mathcal{P}$ is a closed cone, as claimed. □

We may observe that, in principle, the Cayley–Menger matrix condition could be used to describe all the possible diagonal vectors $\Delta(P)$ of realizations P of $\mathcal{P}$. In practice, however, the condition is far from easy to work with as it stands.

The cone $\mathcal{P}$ lies in the positive orthant of $\mathbb{E}^r$, because all its vectors have non-negative coordinates. We can therefore apply Carathéodory's theorem of convexity (compare [357, p. 3], which gives the affine version); this asserts that every point of $\mathcal{P}$ is a sum of at most r points on extreme rays of $\mathcal{P}$. In other words, because sum in $\mathcal{P}$ is equivalent to blending of realizations of $\mathcal{P}$, and since Theorem 5A16 tells us that pure realizations correspond to points on extreme rays of $\mathcal{P}$, we have

5B3 Theorem *Every realization of $\mathcal{P}$ is a blend of at most r pure realizations.*

The proof of Theorem 5A14 shows that the simplex realization T is an interior point of $\mathcal{P}$. We can now use the geometry of the cone $\mathcal{P}$ to prove

5B4 Theorem *Every pure realization P of $\mathcal{P}$ is similar to a component of T; that is, there are $\lambda > 0$ and $Q \in \mathcal{P}$, such that $T = (\lambda P) \# Q$.*

Proof. There is also a direct proof of Theorem 5B4. Consider the linear mapping Π which takes the vertices of T onto the corresponding vertices of P (bear in mind that all vertex-sets have centroid o). A little thought shows that the kernel of Π is invariant under $G(T)$, and hence that P is similar to an orthogonal projection of T. □

We now move on to discuss the realizations of $\mathcal{P}$ associated with a fixed irreducible representation G of $\Gamma(\mathcal{P})$. Suppose that the ambient space E on which G acts has dimension d (in other words, d is the degree of the representation). Further, let the Wythoff space W of G have dimension w. As we saw in Section 5A, we obtain a realization of $\mathcal{P}$ by applying Wythoff's construction to a point $p \in W$. The realization will be pure unless $p = o$, when it will be trivial. We denote by $\mathcal{P}_G$ the subfamily of $\mathcal{P}$ consisting of all possible blends of these realizations with symmetry group G.

An obvious remark is

5B5 Lemma *$\mathcal{P}_G$ is a closed subcone of $\mathcal{P}$.*

Our first observation is that the expression of a polytope in $\mathcal{P}_G$ as a blend need not be unique, even allowing for the fact that the blend is commutative.

5B6 Lemma *Let $P, Q \in \mathcal{P}$ have the same symmetry group G (acting on E), and let $\lambda^2 + \mu^2 = 1$. Then*

$$P \# Q = (\lambda P + \mu Q) \# (\mu P - \lambda Q).$$

Proof. In fact, in this lemma, G need not act irreducibly on E. Now G acts on $E \times E$ by

$$(x, y)g := (xg, yg),$$

with $x, y \in E$ and $g \in G$, and

$$L := \{(\mu x, -\lambda x) \mid x \in E\}$$

is clearly a subspace of $E \times E$ invariant under G. The orthogonal projection on L is (essentially) $(x, y) \mapsto \lambda x + \mu y$, from which we see that $\lambda P + \mu Q$ is also a component of $P \# Q$. The orthogonal complement of L is

$$M := \{(\lambda x, \mu x) \mid x \in E\},$$

and the complementary component of $P \# Q$ is $\mu P - \lambda Q$. That is,

$$P \# Q = (\lambda P + \mu Q) \# (\mu P - \lambda Q),$$

as claimed. □

Lemma 5B6 is our main tool for investigating the subcone $\mathcal{P}_G$. In order to exploit it, we introduce the following concept, which varies the idea of a row echelon matrix. We call a $k \times m$ real matrix $\Lambda := (\lambda_{ij})$ *canonical* if

(M1) each zero row of Λ lies beneath each non-zero row;
(M2) the leading (non-zero) entry $\lambda_{i,j(i)}$ in a non-zero row i (and in column $j(i)$) is positive;
(M3) if Λ has s non-zero rows, then $1 \leqslant j(1) < \cdots < j(s) \leqslant m$.

As an analogue to the well-known result about the reduction of a matrix to row echelon form, we have

5B7 Lemma *Let A be a real $k \times m$ matrix. Then there are an orthogonal $k \times k$ matrix U and a unique canonical $k \times m$ matrix Λ, such that $\Lambda = UA$.*

Proof. We sketch the proof. The core idea is the following. Suppose that we have reduced the first $t-1$ rows. If there is a further non-zero row in the matrix (which we continue to call A), pick one with a non-zero entry in the smallest column $j = j(t)$, and, if necessary, interchange it with row t and change signs to make α_{tj} positive (this multiplies A on the left by an orthogonal matrix). If some row i of A with $i > t$ has a non-zero entry α_{ij}, then replace the old rows a_t and a_i of A by the new rows

$$\begin{aligned} a_t' &:= \beta_t a_t + \beta_i a_i, \\ a_i' &:= \beta_i a_t - \beta_t a_i, \end{aligned}$$

where

$$\beta_t := \frac{\alpha_{tj}}{\sqrt{\alpha_{tj}^2 + \alpha_{ij}^2}},$$

and similarly for β_i. This again multiplies A by an orthogonal matrix, and has the effect of making the new (i, j)-entry zero. We repeat this on all appropriate rows i, and proceed in the same way until no more non-zero rows remain.

Finally, it should be obvious that the resulting canonical matrix is unique. □

Before we can apply our results to obtain counting formulae, we must discuss irreducible representations in more detail (see [173, 434] for general properties of group representations). We wish to find essentially unique representations of congruence classes in $\mathcal{P}$. We resume the previous conventions that G is an irreducible representation of the group Γ of $\mathcal{P}$ in the ambient space E and that W is the Wythoff space. Now it is clear that, if $p \in W$, then p and $-p$ always give congruent realizations; it is convenient to accept this ambiguity. However, the centralizer of G in the orthogonal group on E may have non-scalar elements and, in particular, may not be $\{\pm I\}$. This will happen if G is reducible as a complex representation of Γ, say $G = H + \bar{H}$, where $H, \bar{H}$ is a pair of complex conjugate representations of Γ. (Recall that an irreducible real orthogonal representation of a finite group either remains irreducible as a unitary complex representation, or decomposes as a sum of two irreducible unitary representations which are *complex conjugates* of each other. The latter means that the underlying complex space is the orthogonal sum of a pair of complex conjugate invariant subspaces on which the group elements can be described by pairs of complex conjugate matrices. From Schur's Lemma, if G is irreducible as a complex group, then the centralizer of G in the orthogonal group on E consists only of scalar elements.) If $G = H + \bar{H}$, then the representation H is centralized by the unitary mappings $z \mapsto e^{i\vartheta}z$, which induce (multiple) rotations through ϑ on E taking W into itself.

Indeed, G may have even more centralizers than this. Recall that the character χ of G is the (generally complex-valued) function on G which associates with each group element the trace of its matrix, taken relative to any basis of the underlying complex space. The *character norm* (χ, χ) of χ is defined by

5B8
$$(\chi, \chi) := |G|^{-1} \sum_{g \in G} \chi(g)\overline{\chi(g)}.$$

It is clear that (χ, χ) is real and positive; in fact, it turns out that (χ, χ) can only take the values

- 1, if G is irreducible as a unitary representation;
- 2, if $G = H + \bar{H}$, with H and $\bar{H}$ non-isomorphic complex irreducible representations (with non-real characters);
- 4, if $G = H + \bar{H} = 2H$, where H is a non-real representation (with a real-valued character) isomorphic to its conjugate.

In saying that two representations are *isomorphic*, we mean that they are conjugate (as groups) under some unitary mapping. We use this term here to distinguish it from *conjugate* representations, in which each unitary matrix is replaced by its complex conjugate. Then (χ, χ) gives the dimension of the subspace of E spanned by the orbit of a single point (other than o) under the action of the centralizer of G.

Thus, to give the uniqueness we seek, it is appropriate to pass to a subspace W^* of W, which contains only pairs $\pm p$ of points representing congruence classes, so that W^* is transverse to the action of the centralizer (see [291]). We shall call W^* an *essential Wythoff subspace* of the representation G. We then define $w^* := \dim W^*$, so that $w^* = w/(\chi, \chi)$.

Our main theorem follows at once.

5B9 Theorem *A polytope in the subcone $\mathcal{P}_G$ is a blend of at most w^* pure polytopes.*

Proof. In fact, we establish something rather stronger. Pick any basis $\{p_1, \ldots, p_{w^*}\}$ of an essential Wythoff subspace W^* of the representation G, and let $P_1, \ldots, P_{w^*}$ be the corresponding (pure) realizations of $\mathcal{P}$. The general pure polytope in $\mathcal{P}_G$ arises from a point $\alpha_1 p_1 + \cdots + \alpha_{w^*} p_{w^*}$ of W^*, and so is of the form $\alpha_1 P_1 + \cdots + \alpha_{w^*} P_{w^*}$, by definition of linear combinations of polytopes. The general polytope in $\mathcal{P}_G$ is then a blend of such polytopes, and so is of the form

$$P = \left(\sum_{j=1}^{w^*} \alpha_{i1} P_1 \right) \# \cdots \# \left(\sum_{j=1}^{w^*} \alpha_{ik} P_k \right).$$

Consider the $k \times w^*$ matrix $A := (\alpha_{ij})$. Bearing in mind that $-P = P$ in $\mathcal{P}$ (since they are congruent), and that # is commutative (as well as associative), because the basic operation in Lemma 5B7 (with $m = w^*$) corresponds to that of Lemma 5B6 we see that P is unchanged by multiplying A on the left by orthogonal matrices. If $\Lambda = (\lambda_{ij})$ is the unique canonical matrix corresponding to A, then we have

$$P = \left(\sum_{j=1}^{w^*} \lambda_{i1} P_1 \right) \# \cdots \# \left(\sum_{j=1}^{w^*} \lambda_{is} P_s \right),$$

which is a blend of $s \leqslant w^*$ pure polytopes in $\mathcal{P}_G$. □

We shall call this expression for P *canonical* in terms of the ordered set $(P_1, \ldots, P_{w^*})$ of pure realizations. The uniqueness of the canonical representation also shows

5B10 Theorem *A polytope in $\mathcal{P}_G$ has dimension at most w^*d, and w^*d is possible.*

Proof. In fact, the polytope P of Theorem 5B9 clearly has dimension sd; in particular, $P_1 \# \cdots \# P_{w^*}$ has dimension w^*d. □

The same circle of ideas leads to our next result.

5B11 Theorem *The cone $\mathcal{P}_G$ has dimension $\frac{1}{2}w^*(w^* + 1)$.*

Proof. Using the notation of Theorem 5B9, the diagonal vector of $\sum_{j=1}^{w^*} \alpha_j P_j$ is a quadratic polynomial in $\alpha_1, \ldots, \alpha_{w^*}$, and so is a linear combination (with coefficients the α_i^2 and $\alpha_i \alpha_j$) of at most $\frac{1}{2}w^*(w^* + 1)$ vectors, which depend only upon $P_1, \ldots, P_{w^*}$ (and not on $\alpha_1, \ldots, \alpha_{w^*}$). The same is therefore true of any blend of such polytopes; thus $\dim \mathcal{P}_G \leqslant \frac{1}{2}w^*(w^* + 1)$.

On the other hand, the diagonal map $P \mapsto \Delta(P)$ is one-to-one on the canonical representations of the $P \in \mathcal{P}_G$. In particular, a neighbourhood of the identity matrix in the set of real upper triangular $w^* \times w^*$ matrices is mapped into the cone $\mathcal{P}_G$ by a one-to-one quadratic polynomial, from which it follows that $\dim \mathcal{P}_G \geqslant \frac{1}{2}w^*(w^* + 1)$.

These two inequalities establish the theorem. □

We obtain a representative of each similarity class of pure realizations in $\mathcal{P}_G$ by starting from the unit sphere in an essential Wythoff subspace W^*. The diagonal mapping $P \mapsto \Delta(P)$ identifies antipodal points of this sphere, but no others (again, we appeal to the uniqueness of the canonical representation). We thus have

5B12 Theorem *For $w^* \geqslant 2$, the subcone $\mathcal{P}_G$ is the positive hull of a $(w^* - 1)$-dimensional real projective space embedded in an ellipsoid of dimension $\frac{1}{2}w^*(w^* + 1) - 2$.*

Observe that, when $w^* \geqslant 2$, a polytope $P \in \mathcal{P}_G$ will always be a blend of fewer than $\dim \mathcal{P}_G$ pure polytopes, so that the bound of Theorem 5B3 will not be achieved.

We are now in a position to describe the whole realization cone of a finite regular polytope $\mathcal{P}$. In particular, we shall obtain some useful numerical relationships for the cone. We first have

5B13 Theorem *The cone $\mathcal{P}$ is the direct sum of the subcones $\mathcal{P}_G$, where G ranges over the distinct irreducible orthogonal representations G of the automorphism group $\Gamma(\mathcal{P})$ of $\mathcal{P}$.*

Proof. Theorem 5B3 tells us that $\mathcal{P}$ is the sum of the cones $\mathcal{P}_G$. The sum must be direct, because the symmetry group $G(P)$ of a realization P of $\mathcal{P}$ has a unique expression as a sum of irreducible representations of $\Gamma(P)$. Finally, Theorem 5B10 shows how we collect together components of the realization belonging to the same irreducible representation. □

Let $\bar{W}$ denote the Wythoff space of the simplex realization T of $\mathcal{P}$, and let $\bar{w}$ be its dimension. If G is an irreducible orthogonal representation of $\Gamma(P)$, we denote by d_G its dimension (degree), by w_G the dimension of its Wythoff space W_G, and by w_G^* ($= w_G/(\chi, \chi)$) the dimension of a corresponding essential Wythoff subspace W_G^*. As before, we suppose that $\mathcal{P}$ has v vertices and r diagonal classes.

5B14 Theorem *The following relationships hold for $\mathcal{P}$, where in each case the sum ranges over the distinct irreducible orthogonal representations of $\Gamma(P)$:*

(a) $\sum_G w_G^* d_G = v - 1$;
(b) $\sum_G \frac{1}{2} w_G^*(w_G^* + 1) = r$;
(c) $\sum_G w_G^* w_G = \bar{w}$.

Proof. In view of Theorems 5B10 and 5B13, the sum on the left of (a) is the maximum dimension of a polytope in $\mathcal{P}$, namely, $\dim T = v - 1$, where T is the simplex realization of $\mathcal{P}$. The relation (b) just expresses Theorem 5B13, since $\dim \mathcal{P} = r$ by Theorem 5B2. For (c), we observe that all realizations of $\mathcal{P}$ which are strictly $(v-1)$-dimensional, and so correspond to points of the interior of the cone $\mathcal{P}$, can be chosen to have the same symmetry group as the simplex realization T, since the group is discrete, while the polytopes vary continuously. Thus $\bar{W}$ is the direct sum, over the different irreducible representations G of $\Gamma(\mathcal{P})$, of the Wythoff spaces of the components $P_1 \# \cdots \# P_{w_G^*}$, say, of T in $\mathcal{P}_G$, whose dimension is $w_G^* w_G$. This completes the proof. □

The important numbers v, r and $\bar{w}$ can be calculated in a different way. We return to the automorphism group $\Gamma := \Gamma(\mathcal{P})$. By Proposition 2B12, we may identify a vertex of $\mathcal{P}$ with a right coset $\Gamma_0 \sigma$ of the subgroup

5B15 $$\Gamma_0 := \langle \rho_1, \ldots, \rho_{n-1} \rangle$$

of Γ. Since there is a bijection between the vertices of $\mathcal{P}$ and the vertices of the regular $(v-1)$-simplex (associated with the simplex realization T), we shall therefore label the vertices of this simplex by cosets of Γ_0.

Using the transitivity of Γ, a typical diagonal of $\mathcal{P}$ can thus be thought of as a pair $\{\Gamma_0, \Gamma_0\sigma\}$ with $\sigma \notin \Gamma_0$. Then two such diagonals $\{\Gamma_0, \Gamma_0\sigma_1\}$ and $\{\Gamma_0, \Gamma_0\sigma_2\}$ are equivalent if and only if $\{\Gamma_0, \Gamma_0\sigma_1\}\tau = \{\Gamma_0, \Gamma_0\sigma_2\}$ for some $\tau \in \Gamma$. Considering the two possibilities $\Gamma_0\tau = \Gamma_0$ or $\Gamma_0\tau = \Gamma_0\sigma_2$, and eliminating τ, we see that the two diagonals are equivalent if and only if

5B16 $$\sigma_2 \in \Gamma_0\sigma_1\Gamma_0 \cup \Gamma_0\sigma_1^{-1}\Gamma_0.$$

Last, the Wythoff space $\bar{W}$ of T is the set of points invariant under Γ_0, acting on the vertices of T. This is spanned by the centroids of the orbits of the vertices of T under Γ_0, which correspond to the double cosets $\Gamma_0\sigma\Gamma_0$. Since these centroids are affinely independent, we see that we have proved

5B17 Theorem *Let $\mathcal{P}$ be a finite regular polytope. With the notation of Theorem 5B14:*

(a) $v - 1 = \text{card}\{\Gamma_0\sigma \mid \sigma \in \Gamma \setminus \Gamma_0\}$;
(b) $r = \text{card}\{\Gamma_0\sigma\Gamma_0 \cup \Gamma_0\sigma^{-1}\Gamma_0 \mid \sigma \in \Gamma \setminus \Gamma_0\}$;
(c) $\bar{w} = \text{card}\{\Gamma_0\sigma\Gamma_0 \mid \sigma \in \Gamma \setminus \Gamma_0\}$.

Parts (b) and (c) of Theorem 5B17 admit an interesting interpretation. Bearing in mind that the subcones $\mathcal{P}_G$ are not polyhedral if $w_G^* \geqslant 2$, and that Theorems 5B13, 5B14 and 5B17 take no account of those representations G of Γ for which $w_G = 0$, we conclude that we have

5B18 Theorem *Let $\mathcal{P}$ be a finite regular polytope. With the previous notation, $r \leqslant \bar{w}$. Moreover, the following conditions are equivalent:*

(a) $r = \bar{w}$;
(b) the realization cone $\mathcal{P}$ is polyhedral;
(c) $w_G^* \leqslant 1$ *for each irreducible orthogonal representation G of $\Gamma(\mathcal{P})$;*
(d) $\sigma^{-1} \in \Gamma_0\sigma\Gamma_0$ *for each $\sigma \in \Gamma(\mathcal{P})$.*

Theorem 5B18 prompts the introduction of some useful terminology. We have identified diagonal classes with unions of cosets $\Gamma_0\sigma\Gamma_0 \cup \Gamma_0\sigma^{-1}\Gamma_0$, while the Wythoff space is associated with simple double cosets $\Gamma_0\sigma\Gamma_0$. The latter obviously correspond to classes of *ordered* pairs of vertices, or *directed* diagonals. If a diagonal class of $\mathcal{P}$ is represented by the pair of vertices $\{u, v\} \subseteq \mathcal{P}_0$, then we call it *symmetric* if the ordered pairs (u, v) and (v, u) are equivalent under $\Gamma(\mathcal{P})$, and *asymmetric* otherwise. Of course, this definition does not just apply to finite regular polytopes, but it is in this context that we shall find it most useful.

In theory, one can read off much of the information about the realization cone $\mathcal{P}$ from the real character table of the group $\Gamma(\mathcal{P})$, in particular that about the dimensions of the various pure realizations. However, we can deduce certain information about dimensions of realizations in very general terms. Recall that P/v is the vertex-figure of P at its vertex v. We begin with a lemma.

5B19 Lemma *Let P be a non-trivial realization of a finite regular polytope, and let v be a vertex of P. Then $v \notin \text{aff}(P/v)$.*

Proof. Since P is non-trivial, we must have $v \neq c(P)$, the centre of P. Then P/v (identified with the set of its vertices) lies in the intersection of two spheres, one centred at v, and the other, which also contains v, centred at $c(P)$. Thus P/v lies on a hyperplane in the ambient space (parallel to the one which supports the second sphere at v). This hyperplane does not contain v, as we wished to show. □

5B20 Theorem *Let P be a faithful realization of a finite regular n-polytope, with group $G(P) = \langle R_0, \ldots, R_{n-1} \rangle$. Then*

(a) $\dim P \geqslant n$*;*
(b) $\dim R_j \geqslant j + 1$ *for* $j = 0, \ldots, n-2$*, and* $\dim R_{n-1} \geqslant n - 1$*;*
(c) if $\dim P = n$*, then* $\dim R_{n-2} = \dim R_{n-1} = n - 1$*.*

Proof. Note that we are not demanding that the realization be pure. Our proof proceeds by induction on n; the result is clearly trivial if $n \leqslant 1$, so let us suppose that $n \geqslant 2$. By Theorem 5A2, the realizations F of a facet and P/v of a vertex-figure of $\mathcal{P}$ (at a vertex $v \in P$) are also faithful; hence, in particular, $\dim(P/v) \geqslant n - 1$ by the inductive assumption. But $v \notin \operatorname{aff}(P/v)$ by Lemma 5B19, so that $\dim P \geqslant \dim \operatorname{aff}(\{v\} \cup P/v) > \dim(P/v) \geqslant n - 1$, which is part (a).

The inductive proof of part (b) is similar; again, we may assume that $n \geqslant 2$. Write $A := \operatorname{aff}(P/v)$; we have already observed that $v \notin A$. Let $S_j := A \cap R_{j+1}$ for $j = 0, \ldots, n-2$. Then $G(P/v) = \langle S_0, \ldots, S_{n-2} \rangle$. The obvious inductive assumption implies that, for $j = 1, \ldots, n-1$,

$$\dim R_j \geqslant \dim \operatorname{aff}(S_{j-1} \cup \{v\}) = \dim S_{j-1} + 1 \geqslant j + 1,$$

since $v \in R_j$ for such j. Finally, R_0 preserves the centre $c(P)$ of P, as well as the mid-point of the initial edge of P, which is distinct from $c(P)$ because P is not a segment. (The mid-points of the edges through v span an affine subspace parallel to A, and half-way between A and v.) Thus $\dim R_0 \geqslant 1$.

Part (c) is just a special case of part (b); note that each R_j is a proper subspace of $\operatorname{aff} P$. □

We remark that we have not used so much the finiteness of the abstract polytope in Theorem 5B20, as the boundednes of the realization.

We end this section with some examples to illustrate the theory we have developed.

We begin with polygons (see also [126, Chapter 1; 161, 198, §2]). For each $p \geqslant 3$, a regular p-gon has $\lfloor \frac{1}{2}p \rfloor$ diagonal classes, and, for each $k = 1, \ldots, \lfloor \frac{1}{2}p \rfloor$, there is a planar regular polygon $\{\frac{p}{k}\}$, which gives a degenerate realization if the greatest common divisor $(p, k) > 1$, reducing to a line-segment $\{\,\}$ when $k = \frac{1}{2}p$. It is easy to see directly that a general regular p-gon is a blend of these pure realizations. For each irreducible representation G, we have $w_G = 1$ in the notation we have used.

The regular n-simplex clearly has (up to similarity) a unique realization, since all its diagonals are edges.

If a regular polytope $\mathcal{P}$ has a central involutory automorphism which does not fix a vertex (and so fixes no vertex), this pairs up its v vertices, and so $\mathcal{P}$ has centrally symmetric realizations and, in particular, the cross-polytope realization X of dimension $\frac{1}{2}v$. We also obtain degenerate realizations from the covering of $\mathcal{P}/2$ by $\mathcal{P}$, which is obtained by identifying these paired vertices. All pure realizations of $\mathcal{P}$ will be components of X or of $X/2$ (the vertices of the latter are those of the simplex of dimension $\frac{1}{2}v - 1$), according as this central involution does not or does reduce to the identity. The regular cross-polytope itself, with its two pure realizations, is a particular example.

The regular n-cube C^n has n diagonal classes, and also has n pure realizations (up to similarity), which have dimensions $\binom{n}{k}$ for $k = 1, \ldots, n$. The first of these, for $k = 1$, is the ordinary n-cube C^n itself, while the last, for $k = n$, is the line-segment onto which C^n collapses when we identify vertices which are at an even distance apart along chains of edges. Moreover, those realizations for which k is even are also realizations of $C^n/2$, obtained by identifying diametrically opposite vertices of C^n.

We shall construct the realizations implicitly, by demonstrating that the simplex realization has appropriate invariant subspaces of dimensions $\binom{n}{k}$. In fact, it is convenient to take the vertices of the simplex as the 2^n standard basis vectors in euclidean space of that dimension. We label these vertices $e(\varepsilon_1, \ldots, \varepsilon_n)$, with $\varepsilon_i = \pm 1$ for $i = 1, \ldots, n$, corresponding to the vertices $(\varepsilon_1, \ldots, \varepsilon_n)$ of C^n. The invariant subspaces will be chosen to be spanned by vectors of the form

$$\sum_{\varepsilon_1, \ldots, \varepsilon_n} \pm e(\varepsilon_1, \ldots, \varepsilon_n);$$

indeed, the signs will be those of the row-entries of a $2^n \times 2^n$ Hadamard matrix, whose columns will be identified with the vertices of C^n. Since the rows of an Hadamard matrix are orthogonal, it will be plain that subspaces spanned by disjoints sets of such rows will be orthogonal. If we can find $n + 1$ invariant such subspaces, including that spanned by $(1, 1, \ldots, 1)$ (that is, all signs are "+"), whose dimensions sum to 2^n, then these must be those which yield the pure realizations of C^n (or the trivial one for $(1, 1, \ldots, 1)$).

We split the 2^n rows of the matrix of coefficients into $n + 1$ blocks $B_{n0}, B_{n1}, \ldots, B_{nn}$, with card $B_{nk} = \binom{n}{k}$. We begin with $B_{00} := \{(1)\}$, and define recursively

5B21
$$\begin{aligned} B'_{nk} &:= \{(v, -v) \mid v \in B_{n-1,k-1}\}, \\ B''_{nk} &:= \{(v, v) \mid v \in B_{n-1,k}\}, \end{aligned}$$

and $B_{nk} := B'_{nk} \cup B''_{nk}$. It is clear that card $B_{nk} = \binom{n}{k}$. The corresponding ordering of the vertices is the $e(1, \varepsilon_2, \ldots, \varepsilon_n)$ followed by the $e(-1, \varepsilon_2, \ldots, \varepsilon_n)$, with the sub-orderings in the last $n - 1$ coordinates being defined recursively.

With this ordering, it is convenient to take the generators of the symmetry group G of C^n in the form

5B22
$$\begin{aligned} R_j &: \xi_{j+1} \leftrightarrow \xi_{j+2} \quad \text{for } j = 0, \ldots, n-2, \\ R_{n-1} &: \xi_n \leftrightarrow -\xi_n. \end{aligned}$$

We shall show that the group G is transitive on the rows in each block B_{nk}. By the obvious inductive assumption, which is easy to check for small n, $\langle R_1, \ldots, R_{n-1} \rangle$ leaves each sub-block B'_{nk} and B''_{nk} invariant, and is transitive on the rows in them. For

R_0, we must write

5B23
$$B'_{nk} = \{(u, -u, -u, u) \mid u \in B_{n-2,k-2}\} \cup \{(u, u, -u, -u) \mid u \in B_{n-2,k-1}\},$$
$$B''_{nk} = \{(u, -u, u, -u) \mid u \in B_{n-2,k-1}\} \cup \{(u, u, u, u) \mid u \in B_{n-2,k}\}.$$

Considering the recursive ordering of the vertices of C^n shows that the effect of R_0 is to interchange the second and third blocks in each of the rows, so that

$$\begin{aligned} (u, -u, -u, u)R_0 &= (u, -u, -u, u) && \text{if } u \in B_{n-2,k-2}, \\ (u, u, -u, -u)R_0 &= (u, -u, u, -u) && \text{if } u \in B_{n-2,k-1}, \\ (u, u, u, u)R_0 &= (u, u, u, u) && \text{if } u \in B_{n-2,k}. \end{aligned}$$

This establishes our claims that G leaves each block B_{nk} invariant, and is transitive on the rows in each block.

Thus, the realization cone $\mathcal{P}$ of C^n must be as we have described it: $\mathcal{P}$ has n pure realizations, each with a 1-dimensional Wythoff space. We observe that the vertices of the realization corresponding to the block B_{nk} are a subset of those of the cube $C^{\binom{n}{k}}$; we read off the coordinates of the vertex corresponding to $(\varepsilon_1, \ldots, \varepsilon_n)$ as those in the column $e(\varepsilon_1, \ldots, \varepsilon_n)$ lying in B_{nk}. It is fairly straightforward to check our initial assertions, namely, that B_{n1} gives the vertices of the usual C^n, while B_{nn} gives the collapse of C^n onto the line-segment, and that the realizations for even k are those of the half-cube $C^n/2$.

The case $n = 3$ already illustrates much of what is happening; the half-cube $C^3/2$ has only the 3-dimensional realization $\{4, 3\}_3$, the Petrial of the regular tetrahedron. Note that the dual of $\{4, 3\}_3$ has no faithful realizations, since each pair of its vertices determines two edges.

We now treat the remaining regular convex polyhedra. The regular icosahedron $\{3, 5\}$ has 3 diagonal classes. Apart from the usual icosahedron $\{3, 5\}$ itself, there is another 3-dimensional pure realization, namely, the great icosahedron $\{3, \frac{5}{2}\}$. The final pure realization is induced by its covering of $\{3, 5\}/2 = \{3, 5\}_5$, the *hemi-icosahedron*, all of whose diagonals are edges; thus its vertices must be those of a 5-simplex. Each of these pure realizations must have a 1-dimensional Wythoff space, as we see from Theorem 5B14(a), which reads $3 + 3 + 5 = 12 - 1$. (We shall see this example in a more general context in Section 13C.)

The regular dodecahedron $\{5, 3\}$ has 5 diagonal classes. The two 3-dimensional realizations are the usual dodecahedron $\{5, 3\}$ itself and the great stellated dodecahedron $\{\frac{5}{2}, 3\}$. There is also a faithful 4-dimensional realization. Its 20 vertices are the mid-points of the 10 edges of the regular 4-simplex T^4, together with their 10 reflexions in the centre of T^4. The generating reflexions of the group of the realization may be written as

5B24
$$\begin{aligned} R_0 &= -(1\,3)(2\,4), \\ R_1 &= -(1\,2)(3\,5), \\ R_2 &= -(1\,2)(3\,4), \end{aligned}$$

where the "−" sign indicates that the permutation is to be compounded with the central reflexion. Observe, by the way, that this makes plain why the symmetry group of the dodecahedron is $A_5 \times C_2$; compare also [120, p. 50], where our generating reflexions permit the same interpretation as Coxeter's, although there he only considered the rotation subgroup.

There are two pure realizations of the *hemi-dodecahedron* $\{5, 3\}/2 = \{5, 3\}_5$, whose graph is the Petersen graph; these have dimensions 4 and 5. The 5-dimensional realization corresponds to that of the dual hemi-icosahedron $\{3, 5\}_5$. The vertices of the 4-dimensional realization are the mid-points of the edges of the 4-simplex T^4, and the generating reflexions are just R_0, R_1 and R_2 as in (5B24) with the "$-$" signs dropped.

Notice that the two 4-dimensional realizations of the dodecahedron stand alone, and illustrate the fact that a realization of a regular polytope need not lead to a realization of its dual. Of course, we should not expect that it would.

While the n-cube needs some work to find its realizations, with the other examples discussed so far, we need only look for solutions of the equations of Theorem 5B14 to find the numerical details of the possible realizations. For example, we already know two realizations in $\mathbb{E}^3$ of the isocahedron, and since we only have 3 diagonal classes (all symmetric), there can be only one remaining pure realization, which must have dimension 5. Then we find realizations of the dodecahedron of the same dimensions, and with 5 diagonal classes, and the need to find further realizations of $\{5, 3\}/2$ and of $\{5, 3\}$ itself (counting the components of the cross-polytope realization), we are forced to the two remaining realizations in $\mathbb{E}^4$.

The same ideas help us to analyse the realization cone of the 24-cell $\{3, 4, 3\}$. There are 4 diagonal classes, all of which are symmetric. We already have $\{3, 4, 3\}$ itself as a 4-dimensional pure realization. This naturally preserves the original central symmetry of $\{3, 4, 3\}$, and so is a component of the cross-polytope realization, which must then lead to a complementary pure realization of dimension $8 = 12 - 4$. We may identify this realization as follows. Define the 4×4 matrix A to be

5B25
$$A := \begin{bmatrix} 1 & 1 & 1 & 1 \\ 1 & 1 & -1 & -1 \\ 1 & -1 & 1 & -1 \\ -1 & 1 & 1 & -1 \end{bmatrix}.$$

Then the vertices of the original $\{3, 4, 3\}$ can be taken to be the rows of the 4×12 matrix

5B26
$$B := \begin{bmatrix} 2I \\ A \\ A^{\mathsf{T}} \end{bmatrix},$$

where I is the 4×4 identity matrix and A^{T} is the transpose of A, together with their negatives. The implied partition of the 24 vertices into three sets of eight is just that into the three inscribed 4-dimensional cross-polytopes. Moreover, we see that $\frac{1}{2}A$ is orthogonal, and since $A^2 = 2A^{\mathsf{T}}$, multiplication by $\frac{1}{2}A$ (on the right) permutes the partitions cyclically.

Now the columns of the matrix B form an orthogonal set of vectors of the same length. The complementary projection of the 24 vertices of the 12-dimensional cross-polytope is thus obtained by completing B to a 12×12 matrix whose columns have the same property. Actually, all that is really needed is that the last eight columns have the same length; we are not concerned to distinguish between similar realizations here. There are many ways of doing this; it turns out that one of the neater completions is by the 8×12 matrix

5B27
$$\begin{bmatrix} 4I & O \\ -A & \sqrt{3}A \\ -A^{\mathsf{T}} & -\sqrt{3}A^{\mathsf{T}} \end{bmatrix},$$

whose rows (together with their negatives) give the 24 vertices of the 8-dimensional realization. It is easy to verify that the 24 vertices are distinct, so that the realization is also faithful.

Further, we can observe that $\{3, 4, 3\}$ collapses onto the triangle $\{3\}$. In a similar way, since this collapse actually realizes $\{3, 4, 3\}/2$, there is a complementary realization of dimension $11 - 2 = 9$: here, of course, 11 is the dimension of the simplex realization of $\{3, 4, 3\}/2$. In actual fact, we shall describe this complementary realization in Section 14A in the context of the discussion of locally projective regular polytopes; all we shall say here is that the twelve vertices fall into three sets of four vertices of regular tetrahedra in mutually orthogonal subspaces.

The last two regular convex polytopes, namely, $\{3, 3, 5\}$ and $\{5, 3, 3\}$, have much more complicated realization cones, and we have not so far investigated them. The 600-cell $\{3, 3, 5\}$ may not be impossibly difficult; however, it has 8 diagonal classes, and 120 vertices. The two 4-dimensional realizations, $\{3, 3, 5\}$ and its isomorph $\{3, 3, \frac{5}{2}\}$, are components of the cross-polytope realization of dimension 60; moreover, the edges of $\{3, 3, 5\}/2$ and $\{3, 3, \frac{5}{2}\}/2$ do not coincide, so that a faithful realization of $\{3, 3, 5\}/2$ will actually yield another distinct one.

Hitherto, we have had $w_G \leqslant 1$ for each irreducible representation of the automorphism group $\Gamma(\mathcal{P})$ of $\mathcal{P}$; hence $w_G^* \leqslant 1$ also. We now turn to the Klein polyhedron $\{3, 7\}_8$. As with the icosahedron, we shall describe its realization cone in Section 13C in a more general context, and so we give it only the briefest mention here. The polyhedron has a degenerate 7-dimensional realization, with its 24 vertices coinciding in threes with those of the 7-simplex; here, $w_G = 1$. But it also has faithful 8-dimensional realizations, with $w_G = w_G^* = 2$. For $\{3, 7\}_8$, Theorem 5B14 takes the form

(a) $7 + 2 \cdot 8 = 24 - 1$;
(b) $1 + \frac{1}{2} \cdot 2 \cdot (2 + 1) = 4$;
(c) $1 + 2 \cdot 2 = 5$.

In particular, the realization cone $\mathcal{P}$ is not polyhedral.

Finally, let us give an example with $w_G^* \neq w_G$. The group of the polyhedron $\mathcal{P} = \{5, 5\}_5$ is $\mathrm{PSL}(2, 11)$ of order 660; thus $\mathcal{P}$ has 66 vertices. In fact, these can be identified with the 66 unordered pairs of 12 points on the corresponding projective line, and so with the 66 mid-points of the edges of the regular 11-simplex. Direct calculations show that $r = 7$ and $\bar{w} = 9$. Apart from the trivial one, PSL(2, 11) has the following real irreducible representations:

- just one which is complex reducible of the form $H + \bar{H}$, with $d = 10$, $w = 2$ and $w^* = 1$;
- two with $d = 10$, for one of which $w = 2$, while the other has $w = 0$;
- one with $d = 11$ (that described above), with $w = 1$;
- two with $d = 12$ for which also $w = 1$.

For $\{5, 5\}_5$, Theorem 5B14 takes the form

(a) $10 + 2 \cdot 10 + 11 + 12 + 12 = 65$,
(b) $1 + 3 + 1 + 1 + 1 = 7$,
(c) $1 \cdot 2 + 2 \cdot 2 + 1 + 1 + 1 = 9$.

Once again, of course, the realization cone $\mathcal{P}$ is not polyhedral.

So far, we have found no examples with $w^* = w/4$, but we have no reason to suppose that they do not exist.

For a discussion of the realizations of the toroidal polyhedra $\{4, 4\}_{(s,t)}$ and $\{3, 6\}_{(s,t)}$ we refer to [66–68, 325, 326].

If we relax the stringent requirement that each automorphism be realized by an isometry of the ambient space, then much more freedom is available to construct real models for polytopes; of course, in general, these will not be realizations in our sense. For example, polyhedral embeddings in ordinary 3-space of many well-known (abstract) regular polyhedra have been found. Examples include Coxeter's infinite series of polyhedra $\{4, n \,|\, 4^{\lfloor \frac{n}{2} - 1 \rfloor}\}$ and their duals (for $n \geqslant 3$), as well as Coxeter's finite polyhedra $\{4, 6 \,|\, 3\}$, $\{6, 4 \,|\, 3\}$, $\{4, 8 \,|\, 3\}$ and $\{8, 4 \,|\, 3\}$ (see Section 7B, and [105, 309–312, 376]). There are also polyhedral models for Klein's map $\{3, 7\}_8$ and Dyck's map $\{3, 8\}_6$ found by Schulte and Wills, and Bokowski and Brehm, respectively (see [34, 38, 46, 47, 375, 379]); a beautiful colour plate of the model for $\{3, 7\}_8$ can be found in [37, p. 132]. Some polyhedral models which admit self-intersections are described in [377, 378]. For an example of an abstract polyhedron with triangular faces which does not admit an embedding without self-intersections into ordinary space, see [36]. For further related work see also [64, 133, 201, 202, 380, 381, 455–457].

5C Apeirotopes

In order to focus more clearly on some of the salient features of the rest of the chapter, in which we shall treat realizations of regular apeirotopes (with infinitely many vertices), at this point it is a good idea to look at a simple example. The *apeirogon* is the infinite (regular) polytope $\{\infty\}$ of rank 2; we discuss some features of its realization cone.

Suppose that P is a realization of $\{\infty\}$. Its group G is generated by two reflexions R_0 and R_1. Then $T := R_0 R_1$ acts as a (combinatorial) translation, moving one vertex of P to the next, and the diagonal vector $\Delta(P)$ of P is clearly determined by T and the initial vertex of P.

We can express T, in a unique way, as a commuting product of a translation (by a vector t, say, which may be o) and an orthogonal transformation Θ acting on a hyperplane orthogonal to t (if $t \neq o$). Further, we can decompose Θ, again uniquely, into a product of rotations, by distinct angles $\vartheta_1, \ldots, \vartheta_k$, say, with $0 < \vartheta_j < \pi$ for $j = 1, \ldots, k$, and a reflexion, corresponding to an eigenvalue -1; some or all of these may be absent. Note that eigenvalues $\exp(\pm i\vartheta_j)$ for $j = 1, \ldots, k$ or -1 of Θ are not repeated, because of our assumption that the ambient space E is the affine hull of P.

This analysis demonstrates that the apeirogon P is a blend of pure polygons and apeirogons, with at least one of the latter if we confine our attention to the case when P has infinitely many vertices. We have a polygon corresponding to each rational angle ϑ_j (by *rational* we mean, of course, a rational multiple of π). To a non-zero translation vector t, and each irrational angle ϑ_j, corresponds an apeirogon, in the latter case bounded and non-discrete.

As one simple example, let R_0 and R_1 be two non-intersecting lines in $\mathbb{E}^3$, and let φ be the angle between them (or, rather, between translates of them which meet). Then a point on R_1, but not on the common perpendicular of both lines, is the initial vertex of a regular helical apeirogon, with twist-angle 2φ.

It is now obvious that the realization cone of $\{\infty\}$ has uncountably infinite (algebraic) dimension; this cone therefore cannot be closed. To see why, let $\vartheta_1, \vartheta_2, \ldots$ be distinct angles with $0 < \vartheta_j < \pi$, and let P_j be the apeirogon with vertices on the unit circle in $\mathbb{E}^2$ corresponding to the rotation through ϑ_j (thus if ϑ_j is rational, the apeirogon covers the appropriate polygon). Then the limit of the diagonal vector of

$$P_1 \# \frac{1}{2} P_2 \# \cdots \# \frac{1}{n} P_n$$

certainly exists, since each of its coordinates is a sum of non-negative terms which is bounded above by

$$\sum_{j=1}^{\infty} \left(\frac{2}{j}\right)^2 = \frac{2\pi^2}{3}.$$

However, this vector cannot be that of a realization. The reason is simple: the Cayley–Menger criterion for embedding a set of points in some euclidean space shows that the limit apeirogon, if it were to exist, would have an infinite affinely independent set of vertices. Notice also that, if we blend this example with the ordinary linear apeirogon, we have a sequence of discrete realizations whose diagonal vectors converge pointwise, but with limit which is not a diagonal vector of any realization.

The previous example shows that a genuinely infinite apeirotope may have bounded realizations. We now see how to recognize when a realization has a bounded component with respect to blending.

Let P be a realization of the regular apeirotope $\mathcal{P}$, with vertex-set $V := V(P)$, ambient space $E := \operatorname{aff} V$, and symmetry group $G := G(P)$. Suppose that $E \neq \operatorname{conv} V$, but that $\operatorname{conv} V$ is unbounded; here, $\operatorname{conv} X$ is as usual the convex hull of the subset X. We shall show that P is a blend in a non-trivial way, with one component bounded.

Write $K := \operatorname{conv} V$. Since E is the ambient space of P, we see that K is full-dimensional. Since all points of V are equivalent under G, it follows that $V \subseteq \operatorname{bd} K$, the boundary of K. For if not, then some $v \in V$ lies in the interior $\operatorname{int} K$ of K, say $B(v, \rho) \subseteq K$, where $B(x, \sigma)$ denotes the ball in E with centre x and radius $\sigma > 0$. Then the transitivity of G on V shows that $B(v', \rho) \subseteq K$ for every $v' \in V$, which would contradict $K = \operatorname{conv} V$.

Consider the recession cone $\operatorname{rec} K$ of K, which consists of the vectors $a \in E$ such that $x + a \in K$ for every $x \in \operatorname{int} K$. Now $\operatorname{rec} K$ is the orthogonal sum of a linear subspace L, say, and a pointed convex cone C, say; we must show that $\operatorname{rec} K = L$. Certainly, L must be invariant for G (as a set of directions), since G takes K into itself. Hence, if we project orthogonally along L, we shall obtain another realization of P, with corresponding recession cone C.

Finally, suppose, if possible, that $C \neq \{o\}$. Replacing P by its projection if necessary, we can take $\operatorname{rec} K = C$. Now C, again as a set of directions, is invariant under G. If we let g be the centre of gravity of the intersection $C \cap B$ of C with the unit ball of E, then $g \in \operatorname{int} C$ because C is pointed, and so is contained in some half-ball of B. Then g is clearly an invariant direction under G; thus the hyperplane H through o orthogonal to g is also invariant under G. Since C is pointed, it follows that $-g \notin C$, and we then see that H must support C (in $\{o\}$ alone). Then K has a support hyperplane H' parallel to H, which must meet V since $K = \operatorname{conv} V$. Using the transitivity of G on V, we thus have $V = (V \cap H')G \subset H'G = H'$. But since $C = \operatorname{rec} K \neq \{o\}$ by our assumption, K must also have vertices outside H'. This is the required contradiction, so that $C = \{o\}$ as claimed.

In summary, since our assumption on K implies that $\operatorname{rec} K \neq \{o\}$, we see that $\operatorname{rec} K = L$ is a subspace. The points of V are then equidistant from some translate of L, and projection along L yields a non-trivial bounded realization of $\mathcal{P}$. Thus we have

5C1 Theorem *Let P be an unbounded realization of a regular apeirotope $\mathcal{P}$ with vertex-set V. If* $\operatorname{conv} V \neq \operatorname{aff} V$, *then P is a non-trivial blend, one of whose components is bounded.*

While the discussion of the apeirogon shows that we cannot avoid considering non-discrete realizations, because these may occur as components under blending of discrete realizations, there is naturally greater interest in discrete realizations. We shall investigate these first. For many examples of discrete regular apeirotopes in euclidean spaces, see [120, 148, 150, 294–297, 304]; we shall also (in effect) give examples in Sections 7E and 7F.

We can suppose that our discrete realization P of a regular apeirotope $\mathcal{P}$ is infinite; otherwise, there is little to say about it. (Recall that finiteness in this context refers to the vertex-set V. When the realization is finite, the analysis of Section 5B can be followed with minor modifications.) There are two cases, according as the ambient space E is or is not the convex hull of P, or, rather, of its vertex-set V.

Suppose first that $\operatorname{conv} V \neq E$. As we saw in Theorem 5C1, we can translate P so that the points of V are equidistant from some linear subspace L of E. The orthogonal projection of V on L yields another realization of $\mathcal{P}$.

5C2 Lemma *The projection of V on L is a discrete set.*

Proof. The points of V are at a fixed distance $\rho > 0$ from L. Since the sphere is compact, any cluster point of the projection lifts to a cluster point of V itself. But V is discrete; hence no such cluster point can exist. □

Note that the projection along L need not be discrete (it is certainly bounded), as the examples of realizations of the apeirogon which we discussed at the beginning of the section demonstrate.

Before we discuss the other case, it is appropriate to prove a useful counterpart to Theorem 5B20.

5C3 Theorem *Let P be a discrete faithful realization of a regular n-apeirotope, with group $G(P) = \langle R_0, \ldots, R_{n-1} \rangle$. Then*

(a) $\dim P \geqslant n - 1$;
(b) $\dim R_j \geqslant j$ *for* $j = 0, \ldots, n-2$, *and* $\dim R_{n-1} \geqslant n - 2$;
(c) *if* $\dim P = n - 1$, *then* $\dim R_{n-1} = \dim R_n = n - 2$.

Proof. This is an easy consequence of Theorem 5B20, since the vertex-figure P/v of P at the initial vertex v is a faithfully realized finite regular n-polytope with centre v. □

Note, by the way, that equality forces P (or, rather, its vertex-set) to be infinite; otherwise we have a contradiction to Theorem 5B20. In fact, we can go further.

5C4 Theorem *Let P be a realization of a regular polytope $\mathcal{P}$, whose vertex-figure satisfies* $\dim(P/v) = \dim P$. *If P has vertex-set V and ambient space E, then* $\operatorname{conv} V = E$. *In particular, V is infinite and $\mathcal{P}$ is an apeirotope.*

Proof. Notice that we are not assuming discreteness or faithfulness here. The proof is straightforward. As we saw in the proof of Theorem 5C1, if $\operatorname{conv} V \neq E$, then $V \subseteq \operatorname{bd} \operatorname{conv} V$. However, since $\dim(P/w) = \dim E$ ($= \dim P$ by definition) for every vertex w, we have $w \in \operatorname{int} \operatorname{conv}(P/w) \subseteq \operatorname{int} \operatorname{conv} V$, and we have arrived at a contradiction. □

Resuming our discussion, we may now suppose that $\operatorname{conv} V = E$. Once more, of course, we are assuming that our realization P is discrete and faithful. The symmetry group G of P is transitive on V, and so, by a famous theorem of Bieberbach (see [25, 344, §7.4]), G contains a subgroup T of translations of full rank $\dim E$. This subgroup T is discrete, and so can be identified with a lattice in E. Note that G will then act as a group of automorphisms (that is, symmetries) of T.

We now come to a curiosity about discrete realizations. We call a realization P of a regular apeirotope $\mathcal{P}$ *translation-free* if its symmetry group G contains no translations. Of course, a bounded realization must be translation-free, but we also observed previously that we have unbounded translation-free realizations of the regular apeirogon. We shall prove the following somewhat surprising result.

5C5 Theorem *If a regular apeirotope $\mathcal{P}$ has a discrete realization, then it has a discrete translation-free realization.*

Proof. In view of the discussion just preceding, we may suppose that we have a discrete realization P of $\mathcal{P}$, whose vertex-set V satisfies $\operatorname{conv} V = \operatorname{aff} V = E$, the ambient space of the realization. So, the symmetry group G of P contains a lattice Λ of translations, of rank $\dim E$. If we take the initial vertex of P to be the origin o, then $\Lambda \subseteq V$, in the obvious sense. In fact, we see that the vertex-set V decomposes into finitely many translated copies of Λ.

Now G acts on Λ by conjugation; the effect is that of applying the rotation part of an element of G to the translation vector in Λ. Let Λ^* denote the reciprocal lattice of Λ; that is,

$$\Lambda^* := \{x^* \in E \mid \langle x, x^* \rangle \in \mathbb{Z} \text{ for all } x \in \Lambda\},$$

where $\langle \cdot, \cdot \rangle$ is the usual inner product on E. Then G similarly acts on Λ^*.

We now lift Λ^* in a symmetry-preserving way. Specifically, suppose that there are $2n$ nearest points $\pm p_1, \ldots, \pm p_n$ of Λ^* to the origin o, corresponding to the $2n$ facets of the Voronoï region of Λ^*. (Recall that the *Voronoï region* is the set of points of E no farther from o than from any other point of Λ^*.) Then these points are the images of the basis vectors $\pm e_1, \ldots, \pm e_n$ of the integer lattice $\mathbb{Z}^n$ under the linear map $\Pi : \mathbb{E}^n \to E$, with $e_i \Pi = p_i$ for $i = 1, \ldots, n$. The kernel of Π is given by the linear relations between the p_i, so that

$$\ker \Pi = \left\{ (\alpha_1, \ldots, \alpha_n) \in \mathbb{E}^n \,\middle|\, \sum_{i=1}^{n} \alpha_i p_i = o \right\}.$$

Since the vectors p_i clearly generate the lattice Λ^*, we see that $\ker \Pi$ has a basis consisting of rational vectors $(\alpha_1, \ldots, \alpha_n)$; hence $\ker \Pi$ is a rational subspace of $\mathbb{E}^n$, and so is spanned by points of $\mathbb{Z}^n$. We can now identify E with the orthogonal complement of $\ker \Pi$, and under this identification $\Lambda = E \cap \mathbb{Z}^n$.

Now let β be an irrational multiple of π, define $\psi : \mathbb{R} \to \mathbb{E}^3$ by $\xi\psi := (\xi, \cos(\beta\xi), \sin(\beta\xi))$, and let $\Psi : \mathbb{E}^n \to \mathbb{E}^{3n}$ be given by $(\xi_1, \ldots, \xi_n)\Psi := (\xi_1\psi, \ldots, \xi_n\psi)$, with the obvious meaning. Then Ψ respects the action of the group G; that is, there is an isomorphic group G' acting on E^3, such that $(xg)\Psi = (x\Psi)g'$ for $x \in E$ and corresponding $g \in G$ and $g' \in G'$. The crucial feature of Ψ, though, is that the lattice translations in $\mathbb{Z}^n$ are destroyed, and so the group G' acting on the new realization $V\Psi$ contains no translations. This proves the assertion of the theorem. □

In fact, this proof yields a little more.

5C6 Corollary *If a regular apeirotope has a discrete infinite realization, then it has uncountably many translation-free discrete realizations.*

The situation for unbounded non-discrete realizations of an apeirotope is quite different. We present a variety of examples to illustrate the kinds of thing that can happen; other examples can be found in [283].

As we remarked earlier, we cannot avoid non-discrete realizations. For instance, in the plane $\mathbb{E}^2$, the tessellation $\{8, \frac{8}{3}\}$ is a non-discrete realization of the hyperbolic honeycomb $\{8, 8\}$, as is its isomorph $\{\frac{8}{3}, 8\}$. If we blend these two realizations together, with an appropriate choice of scaling of each, the resulting set of vertices of the blended realization in $\mathbb{E}^4$ is $\mathbb{Z}^4$. Indeed, every non-discrete regular tessellation $\{p, q\}$ in $\mathbb{E}^2$ (with p and q rational, and at least one fractional) may be lifted to a discrete realization in an analogous way, although more than two components (isomorphic, but not similar) are generally required.

However, let β be an irrational multiple of π, and consider the reflexions in $\mathbb{E}^2$ defined by

5C7
$$\begin{aligned} xR_0 &:= (1 - \xi_1, \xi_2), \\ xR_1 &:= (\xi_1 \cos\beta + \xi_2 \sin\beta, \xi_1 \sin\beta - \xi_2 \cos\beta), \\ xR_2 &:= (\xi_1, -\xi_2), \end{aligned}$$

where $x = (\xi_1, \xi_2) \in \mathbb{E}^2$. Then $G := \langle R_0, R_1, R_2 \rangle$ is the symmetry group of a regular tessellation P of type $\{\infty, \infty\}$. However, note that P cannot be the universal $\{\infty, \infty\}$, because G contains translations (the product of any two distinct conjugates of R_0R_2 is one such); indeed, the translations in G form an abelian group of infinite rank, because the rotation of one translation by a multiple of β is another. Of course, P must be non-discrete, but in contrast to the previous examples, P cannot be lifted to a discrete realization, since its vertex-figure is infinite, and in any realization, the points of the vertex-figure at a vertex v are equidistant from v.

We remark that similar examples can be constructed in all dimensions greater than 2. If $d \neq 4$, the simplest way to see this is to observe that the regular d-cross-polytope $Q := \{3^{d-2}, 4\}$ does not tile $\mathbb{E}^d$ (even with multiplicity). So, if we fit copies of Q facet-to-facet, we obtain a regular apeirotope P of type $\{3^{d-2}, 4, \infty\}$, whose vertex-figure is infinite. Since the reflexion in the centre of each copy of Q is a symmetry, we see that the symmetry group of P contains a translation group of infinite rank. In $\mathbb{E}^4$, we may perform a similar construction, taking instead the 600-cell $Q := \{3, 3, 5\}$.

We now come to an example of a quite different nature in $\mathbb{E}^3$; this is a correction (and considerable modification) of one of [288, §12]. Let α be a number to be chosen later, which satisfies $0 < \alpha < 1/\sqrt{2}$. We define the following three half-turns:

5C8
R_0: about the line through $(1, 1, 0)$ in direction $(\alpha, -\alpha, \sqrt{1 - 2\alpha^2})$;
R_1: about the line through o in direction $(1, 0, 0)$;
R_2: about the line through o in direction $(1, 1, 0)$.

The axes of R_0 and R_1 do not meet, and the angle ϑ between them is given by $\cos(\vartheta) = \alpha$. Thus R_0R_1 is a rotatory-translation, with twist 2ϑ. Clearly, ϑ will be an irrational multiple of π for all but countably many α. The axis of R_2 meets those of R_0 and R_1 in angles $\pi/2$ and $\pi/4$, respectively, whence $(R_0R_2)^2 = (R_1R_2)^4 = E$.

It follows that $G := \langle R_0, R_1, R_2 \rangle$ is a homomorphic image of the Coxeter group $[\infty, 4]$. However, if R_0' is the half-turn about the line through o parallel to the axis of R_0, then the group $\langle R_0', R_1, R_2 \rangle$ can be made to be isomorphic to $[\infty, 4]$. The easiest way to see this is to consider the corresponding reflexion group $\langle -R_0', -R_1, -R_2 \rangle$. That is, we replace each half-turn R about a line by the reflexion $-R$ in the plane through o orthogonal to the line; we shall employ this useful trick again in Section 7E. Applying Wythoff's construction to $-R_0' \cap -R_1$ yields a spherical tessellation (usually non-discrete) by squares $\{4\}$. As α varies, we see that for only countably many of its values can fortuitous identifications of vertices of this tessellation occur; otherwise, it is isomorphic to $\{4, \infty\}$. (For such identifications to occur, α would have to satisfy some algebraic equation; it therefore suffices to take α transcendental.) Because the related group is a quotient of G, we see that $G \cong [\infty, 4]$.

If we apply Wythoff's construction with o as initial vertex, it is clear that the vertex-set V of the resulting regular apeirotope $P := \{\infty, 4\}$ satisfies conv $V = \mathbb{E}^3$. However, G can contain no translations, since $[\infty, 4]$ contains no commuting elements of infinite order (alternatively, a translation would have to correspond to a non-trivial relation in the quotient group $\langle R_0', R_1, R_2 \rangle$). We conclude that there is no analogue of Bieberbach's theorem for the non-discrete case.

Between these two extremes lies the case where the apeirotope has a translation group of finite positive rank. We now discuss this.

5C9 Lemma *Let P be a realization of a regular apeirotope $\mathcal{P}$ whose symmetry group G contains a normal translation subgroup Λ of finite positive rank. Then $\mathcal{P}$ has a realization whose group contains a translation subgroup isomorphic to Λ whose vectors span the ambient space.*

Proof. Indeed, the linear subspace L spanned by the translation vectors in Λ is invariant under G, and so orthogonal projection of P on L will yield the required realization. □

Henceforth, we shall work with realizations having the characteristic property of Lemma 5C9. Our last main result is

5C10 Theorem *Let $\mathcal{P}$ be a regular apeirotope, which has a realization P whose translation group is of finite positive rank, with its vectors spanning the ambient space. Then $\mathcal{P}$ has a discrete infinite realization.*

Proof. Informally, this means that we can lift P to a discrete realization; we can assume that P is not already discrete.

Each element of the symmetry group G of P can be written uniquely in the form $x \mapsto x\Phi + t$, where Φ is an orthogonal mapping, and t is a translation vector; the set of these associated orthogonal mappings Φ forms a group $\bar{G}$, which is a quotient of G, called the *special group* of G. Since $\bar{G}$ acts on the translation subgroup Λ of G, which is of finite rank with vectors spanning the ambient space E, we see that $\bar{G}$ must be finite.

Suppose that rank $\Lambda = m$; we can lift Λ into a discrete lattice in an m-dimensional space as follows (the proof bears a resemblance to part of that of Theorem 5C5). Choose a finite $\mathbb{Z}$-spanning subset $\{p_1, \ldots, p_n\}$ of Λ which is invariant under $\bar{G}$; then the image of the integer lattice $\mathbb{Z}^n$ (with usual basis $\{e_1, \ldots, e_n\}$) under the orthogonal projection in $\mathbb{E}^n$ with kernel

$$\left\{ \sum_{i=1}^{n} \alpha_i e_i \mid (\alpha_1, \ldots, \alpha_n) \in \mathbb{Z}^n \text{ and } \sum_{i=1}^{n} \alpha_i p_i = o \right\}$$

is a discrete lattice $\hat{\Lambda}$, on which acts a group isomorphic to $\bar{G}$ (our projection is designed to preserve these symmetries).

Our remaining task is to lift the whole of G. Write Π for the projection of the ambient spaces which takes $\hat{\Lambda}$ onto Λ. Taking the initial vertex of P to be the origin o of E, the symmetry group G_0 of the vertex-figure of P is a subgroup of $\bar{G}$, and this is already lifted (by Π^{-1}). So, we only have to lift the reflexion R_0, which takes o into an adjacent vertex v, say. The reflexion R_0' through o parallel to R_0 again lies in $\bar{G}$, and so this has also been lifted.

Our lifting of R_0 must satisfy three properties. First, it must be a translate of the lifting of R_0'. Second, it must lie in $R_0 \Pi^{-1}$. Third, it must meet the Wythoff space of the lifting of G_0. But all three conditions are easy to fulfil, and so we have lifted G in the required way. Applying Wythoff's construction now yields a discrete lifting of P, and therefore a discrete infinite realization of $\mathcal{P}$. □

As a consequence of these results, we have a generalization of Theorem 5C5.

5C11 Theorem *If a regular apeirotope $\mathcal{P}$ has an infinite realization, whose translation subgroup of its symmetry group has finite rank, then $\mathcal{P}$ has an infinite translation-free realization.*

6

Regular Polytopes on Space-Forms

In the traditional theory, the topological type of a polytope or tessellation is always determined by the geometry and topology of the ambient spherical, euclidean or hyperbolic space. In this chapter, we study the quotients of the regular tessellations in these spaces in the context of spherical, euclidean or hyperbolic space-forms.

After a short introduction to space-forms in Section 6A, we prove in Section 6B that the locally spherical abstract regular polytopes are precisely the (combinatorially) regular tessellations on space-forms. Then, in Section 6C, we briefly discuss the projective regular polytopes, which are the only regular tessellations on spherical space-forms which are not spheres.

In Sections 6D–6F, this is followed by a detailed investigation of the regular toroids of rank $n+1$ ($\geqslant 4$), which are the regular tessellations on topological n-tori; the regular toroids of rank 3 have already been discussed in Section 1D. In Section 6G, the tori are proved to be the only compact euclidean space-forms which admit regular tessellations. We saw in Section 1D, that there are toroidal polyhedra which are chiral. In contrast, in Section 6H, we shall show that there exist no chiral toroidal polytopes of rank at least 4.

Finally, Section 6J presents some results on regular tessellations on hyperbolic space-forms, which try to explain why so little is known about them.

6A Space-Forms

We begin with an informal introduction to space-forms. For more details the reader is referred to Ratcliffe [344, §§8.1, 8.2] or Wolf [467].

Let E be the spherical n-space $\mathbb{S}^n$, euclidean n-space $\mathbb{E}^n$ or hyperbolic n-space $\mathbb{H}^n$ ($n \geqslant 1$). As models for these spaces we can take: for $\mathbb{E}^n$, the n-dimensional real vector space $\mathbb{R}^n$ with the standard euclidean inner product; for $\mathbb{S}^n$, the euclidean unit n-sphere in $\mathbb{E}^{n+1}$; and for $\mathbb{H}^n$, the positive sheet of the hyperboloid $\xi_1^2 + \cdots + \xi_n^2 - \xi_{n+1}^2 = -1$ in $\mathbb{R}^{n+1}$, that is,

$$\mathbb{H}^n := \{(\xi_1, \ldots, \xi_{n+1}) \in \mathbb{R}^{n+1} \,|\, \xi_1^2 + \cdots + \xi_n^2 - \xi_{n+1}^2 = -1 \text{ and } \xi_{n+1} > 0\}.$$

We shall only consider these three spaces E, although some concepts extend to more general spaces. By $\mathcal{I}(E)$ we denote the group of all (spherical, euclidean or hyperbolic) isometries of E.

Let N be a group of homeomorphisms of E (or of any connected, locally pathwise connected topological space). In our applications, N will usually be a group of isometries of E. We say that N *acts freely* on E if each non-trivial element of N moves each point of E. Further, we say that N *acts properly discontinuously* on E if each point $x \in E$ has an open neighbourhood U in E which meets only finitely many of its transforms $U\varphi$ with $\varphi \in N$. The *orbit space*, or (*topological*) *quotient space*, E/N is the topological space whose points are the orbits of E under N and whose topology is the *quotient topology*, which is the strongest topology such that the *canonical projection* $\pi\colon E \to E/N$ is continuous (that is, a subset V of E/N is open in E/N if and only if $\pi^{-1}(V)$ is open in E).

If N acts freely and properly discontinuously on E, then π is a *topological covering map*, meaning that π is continuous and that every point $z \in E/N$ has an open neighbourhood V such that π maps each connected component of $\pi^{-1}(V)$ homeomorphically onto V. In fact, π is what is called a *normal covering projection*; that is, under the mapping $\pi_*\colon \pi_1(E) \to \pi_1(E/N)$ which is induced by π between the fundamental groups $\pi_1(E)$ of E and $\pi_1(E/N)$ of E/N, the group $\pi_1(E)$ of the original space is mapped onto a subgroup of $\pi_1(E/N)$ which is normal. The latter is trivial for the spaces we are considering, because $\mathbb{S}^n$ (for $n \geqslant 2$), $\mathbb{E}^n$ and $\mathbb{H}^n$ are simply connected. Further, N is just the group of *covering transformations* of π; this is the group of homeomorphisms $\varphi\colon E \to E$ such that $\varphi\pi = \pi$. It acts simply transitively on each fibre $\pi^{-1}(z)$ of π (because π is a normal covering projection). Also, if $E \neq \mathbb{S}^1$, then $\pi_1(E/N) \cong N$; again, this holds because E is simply connected [467, p. 38].

Recall that a group N of isometries of spherical, euclidean or hyperbolic space E is called *discrete* (or, more exactly, said to *act discretely* on E) if each of its orbits is a discrete subset of E. Then N is discrete if and only if it acts properly discontinuously on E. Thus these concepts apply to discrete groups of isometries which act freely on E. Note that a discrete group of isometries of $\mathbb{E}^n$ or $\mathbb{H}^n$ acts freely if and only if it is torsion-free (that is, it has no non-trivial element of finite order); in particular, a non-trivial such group is infinite. A discrete group of isometries of $\mathbb{S}^n$ is necessarily finite. These facts translate into corresponding statements for the finiteness or non-finiteness of the fundamental groups of space-forms. See also [23, 274, 413] for more about discrete groups acting on hyperbolic space.

Let N be a discrete group of isometries of $E (= \mathbb{S}^n, \mathbb{E}^n, \text{or } \mathbb{H}^n)$ which acts freely on E. Then the orbit space E/N is called an *n-dimensional spherical, euclidean or hyperbolic space-form*, respectively. These spaces have an outstanding history of study reaching back into the nineteenth century; for an excellent account see [467]. The Killing–Hopf theorem says that, up to isometry, they are precisely the complete connected riemannian manifolds of constant curvature; the euclidean space-forms are the complete connected flat riemannian manifolds [467, pp. 69, 97].

Two simple examples of space-forms are the projective or elliptic spaces and the topological tori. The *real projective* (or, perhaps, since we particularly have the metric in mind, *real elliptic*) n-space $\mathbb{P}^n$ is the quotient $\mathbb{S}^n/\{\pm I\}$ of the unit n-sphere $\mathbb{S}^n$ by $\{\pm I\}$ $(= N)$, where I denotes the identity mapping on $\mathbb{S}^n$. In this quotient, antipodal pairs of points on $\mathbb{S}^n$ are identified. In particular, $\mathbb{P}^n$ is an n-dimensional spherical space-form with fundamental group $\pi_1(\mathbb{P}^n) \cong C_2$ when $n \geqslant 2$.

The quotient $\mathbb{E}^n/\Lambda$ of euclidean n-space $\mathbb{E}^n$ by a lattice Λ $(= N)$ in $\mathbb{E}^n$ is called an *n-torus*, and is a compact euclidean space-form with $\pi_1(\mathbb{E}^n/\Lambda) \cong \mathbb{Z}^n$. Here, we identify the lattice Λ in $\mathbb{E}^n$, which is spanned by n linearly independent vectors, with the translation group which is generated by the translations by these vectors.

Each space-form E/N is equipped with a metric $d_{E/N}$ which is induced by the standard (spherical, euclidean or hyperbolic) metric d_E on E, and is defined for $x, y \in E$ by

$$d_{E/N}(xN, yN) := \inf\{d_E(x\varphi, y\psi) \mid \varphi, \psi \in N\}.$$

With respect to this metric, the group $\mathcal{I}(E/N)$ of all isometries (distance-preserving homeomorphisms) of E/N is isomorphic to the quotient group $N_{\mathcal{I}(E)}(N)/N$, where $N_{\mathcal{I}(E)}(N)$ denotes the normalizer of N in $\mathcal{I}(E)$ [344, p. 336]. In this context, note that each element $\sigma \in N_{\mathcal{I}(E)}(N)$ determines an isometry $\hat{\sigma} \in \mathcal{I}(E/N)$ which is defined by $(xN)\hat{\sigma} := (x\sigma)N$ for $x \in E$, and that each isometry of $\mathcal{I}(E/N)$ arises in this way. The isomorphism $N_{\mathcal{I}(E)}(N)/N \mapsto \mathcal{I}(E/N)$ is then given by $N\sigma \mapsto \hat{\sigma}$. (The reader will recall from Proposition 2D8 that there is a similar isomorphism which describes the automorphism group $\Gamma(\mathcal{P}/N)$ of a quotient polytope $\mathcal{P}/N$.)

We can further note that the covering projection $\pi\colon E \to E/N$ is locally an isometry, meaning that it maps a small neighbourhood of each point $x \in E$ isometrically onto a neighbourhood of $x\pi$ in E/N.

The isometry type of a space-form E/N is determined by the conjugacy class of N in $\mathcal{I}(E)$. In fact, two space-forms E/N and E/M are isometric (that is, there is an isometry which maps the first onto the second) if and only if the groups N and M are conjugate subgroups in $\mathcal{I}(E)$ [344, p. 335].

To discuss quotient relations between space-forms, let N and M be two discrete subgroups of $\mathcal{I}(E)$ which act freely on E. If N is a normal subgroup of M, then $M \subseteq N_{\mathcal{I}(E)}(N)$, and we may identify M/N with a subgroup of $\mathcal{I}(E/N)$ which acts freely and properly discontinuously on E/N. The corresponding quotient space $(E/N)/(M/N)$ is isometric to E/M, and an isometry is given by $(xN)M/N \mapsto xM$ for $x \in E$. In particular, we can identify these two spaces. Then the canonical projection $E/N \to (E/N)/(M/N)$ determines the covering map $E/N \to E/M$, with $xN \mapsto xM$ (for $x \in E$), between the space-forms given by N and M. In summary, we may think of E/M as a quotient of E/N under a suitable group of isometries.

For example, by Bieberbach's theorem, every n-dimensional crystallographic group N in $\mathbb{E}^n$ (that is, discrete subgroup of $\mathcal{I}(\mathbb{E}^n)$ with compact quotient $\mathbb{E}^n/N$) has a full (normal) subgroup T of translations such that the quotient group N/T is finite (see [344, §7.4]). Accordingly, every n-dimensional compact euclidean space-form $\mathbb{E}^n/N$ is finitely covered by an n-torus, namely, $\mathbb{E}^n/T$ (that is, the covering map $\mathbb{E}^n/T \mapsto \mathbb{E}^n/N$ has finite fibres).

A space-form E/N is orientable (as a manifold) if and only if each isometry in N is orientation preserving. Each non-orientable space-form E/N is doubly covered by the orientable space-form E/N', where N' is the subgroup of index 2 in N which consists of the orientation-preserving isometries.

For euclidean space-forms, affine equivalence provides a coarser classification than isometry [344, p. 340]. Two n-dimensional euclidean space-forms $\mathbb{E}^n/N$ and $\mathbb{E}^n/M$ are called *affinely equivalent* if there is a homeomorphism $\mathbb{E}^n/N \to \mathbb{E}^n/M$ which is induced by a non-singular affine transformation σ of $\mathbb{E}^n$; that is, $\sigma^{-1}N\sigma = M$, with conjugation in the group of all non-singular affine transformations of $\mathbb{E}^n$, and the homeomorphism is given by $xN \mapsto (x\sigma)M$ for $x \in \mathbb{E}^n$.

It is a consequence of Bieberbach's theorem that there are only finitely many affine equivalence classes of n-dimensional compact euclidean space-forms. For $n = 1, 2, 3, 4$, the exact numbers are 1, 2, 10, 74, respectively; for $n = 2$, the two classes are represented by the 2-torus and the Klein bottle. In particular, two n-dimensional compact space-forms $\mathbb{E}^n/N$ and $\mathbb{E}^n/M$ are affinely equivalent if and only if their fundamental groups N and M are isomorphic.

Two euclidean space-forms $\mathbb{E}^n/N$ and $\mathbb{E}^n/M$ are *similar* if N and M are conjugate in the group of all similarity transformations of $\mathbb{E}^n$; that is, in the above, we may take σ to be a similarity transformation. For example, two n-tori $\mathbb{E}^n/\Lambda_1$ and $\mathbb{E}^n/\Lambda_2$ are similar (respectively isometric) if and only if the two lattices Λ_1 and Λ_2 are similar (respectively congruent).

An important invariant for a space-form E/N is its volume. This can be defined in terms of a fundamental region for N in E. An open subset D of E is called a *fundamental region* for N in E if $E = \bigcup_{\varphi \in N} (\mathrm{cl}\, D)\varphi$, and the transforms $D\varphi$ (for $\varphi \in N$) of D are mutually disjoint; here, $\mathrm{cl}\, D$ denotes the closure of D in E. A fundamental region D is *proper* if its boundary $\mathrm{bd}\, D$ is a measurable set in E which has zero (spherical, euclidean or hyperbolic) volume. Every discrete group of isometries of E has a proper fundamental region in E, and any two such proper fundamental regions have the same volume in E [344, p. 235].

The *volume* $\mathrm{Vol}(E/N)$ of a space-form E/N is now defined to be the volume of any proper fundamental region for N in E. It is an important fact that any two isometric space-forms have the same volume [344, p. 337].

As a proper fundamental region for N we can take the (open) *Dirichlet domain*

$$D := \{x \in E \mid d_E(x, x_0) < d_E(x, x_0\varphi) \text{ for all } \varphi \in N \setminus \{\varepsilon\}\},$$

where x_0 is any point in E. Then its closure $\mathrm{cl}\, D$ is an n-dimensional (possibly non-compact) convex polyhedron in E bounded by hyperplanes which are perpendicular bisectors of line-segments $[x_0, x_0\varphi]$. Its boundary $\mathrm{bd}\, D$ carries a locally finite decomposition into (possibly non-compact) convex polyhedra of dimension less than n. The space-form E/N is then obtained from $\mathrm{cl}\, D$ by identifying points in $\mathrm{bd}\, D$ which are equivalent modulo N.

In conclusion, we introduce the concept of a tessellation on a real manifold without boundary. In our applications, this manifold will usually be a space-form. Here we shall only consider well-behaved tessellations which are locally finite and in which the tiles are topological cells. For our purposes, it will indeed be sufficient to discuss tessellations whose tiles are homeomorphic images of convex polytopes, and thus come equipped with a natural face structure. Our definition of tessellation will imply polytopality of the underlying face poset, and will exclude such examples as maps on surfaces which have only one 2-face.

Let X be any n-dimensional real manifold without boundary. A family $\mathcal{P}$ of proper subsets of X is called a (*locally finite*) *tessellation* in X if the following three conditions are satisfied. First, for each $F \in \mathcal{P}$ there exist a convex polytope F' and a homeomorphism $\gamma\colon F \to F'$ such that $G\gamma^{-1} \in \mathcal{P}$ for each face G of F'. The subsets in $\mathcal{P}$ are called the *faces* of $\mathcal{P}$, and the subsets $G\gamma^{-1}$ of F the *faces* of F. In particular, F is a j-*face* of $\mathcal{P}$ if F' is a j-polytope, and $G\gamma^{-1}$ is a j-*face* of F if G is a j-face of F'. The n-faces of $\mathcal{P}$ are also called the *tiles* or *facets* of $\mathcal{P}$. Second, if $F_1, F_2 \in \mathcal{P}$, then $F_1 \cap F_2$ is a union of faces of F_1 and F_2 (possibly the empty face $\emptyset$). Third, each point in X is contained in a tile of $\mathcal{P}$ and has a neighbourhood which meets only finitely many tiles (this last is the local finiteness condition).

We shall usually identify a tessellation $\mathcal{P}$ with the partially ordered set consisting of its faces ordered by inclusion. In this context, we shall often find it convenient to adjoin to $\mathcal{P}$ the underlying manifold X as an (improper) $(n+1)$-face. Then it is straightforward to check that $\mathcal{P}$ becomes an abstract polytope of rank $n+1$. In particular, the terminology for abstract polytopes can be applied to tessellations. Observe that the rank of the polytope is always one larger than the dimension of the manifold X. We say that $\mathcal{P}$ is a *face-to-face* tessellation if its face poset is a lattice; that is, if $F_1, F_2 \in \mathcal{P}$, then $F_1 \cap F_2$ is a common face of F_1 and F_2 (possibly $\emptyset$). A tessellation $\mathcal{P}$ on an n-dimensional manifold is called (*combinatorially*) *regular* if, as an abstract polytope, $\mathcal{P}$ is a regular $(n+1)$-polytope.

6B Locally Spherical Polytopes

In this section, we discuss the close connexion between locally spherical abstract polytopes and tessellations on n-dimensional space-forms. We do not investigate the relationship in full generality here; instead, we shall restrict ourselves to the most interesting case where the polytopes and tessellations are regular. Most of the material in this section is classical and can be traced in the literature. For recent accounts which also include more general results than we present here, see Kato [245] and Davis [145] for arbitrary n, and Edmonds, Ewing and Kulkarni [159] for $n = 2$ (see also [160]). However, their terminology and presentation are different from ours.

We begin with the concept of a globally or locally spherical polytope. Whenever appropriate, we shall assume here that the polytope is regular. Then our investigation proceeds in two steps. First, we describe how each locally spherical regular polytope can be viewed as a regular tessellation on a space-form (both in a topological and a geometric sense). This is summarized in Theorem 6B3. Then, in the second step, we consider the converse, and prove that every regular tessellation on a real manifold is a locally spherical regular polytope. We shall conclude the section with some general remarks.

In contrast to the traditional theory where a convex polytope and its local building blocks are topological spheres, it is a very subtle problem to define the global topological type of an abstract polytope $\mathcal{P}$. In general, this can only be done with considerable ambiguity.

On the other hand, there is always a standard way of finding at least one topological space for $\mathcal{P}$, by employing the order complex $\mathcal{C}(\mathcal{P})$. Recall from Section 2C that $\mathcal{C}(\mathcal{P})$ is the simplicial complex whose vertices are the proper faces of $\mathcal{P}$ and whose maximal simplices are the flags of $\mathcal{P}$ (with the improper faces removed). The automorphism group $\Gamma(\mathcal{P})$ acts faithfully as a group of homeomorphisms on the topological space $|\mathcal{C}(\mathcal{P})|$. In fact, any permutation of the flags which is induced by an automorphism of $\mathcal{P}$ clearly corresponds to a piecewise-linear homeomorphism of $|\mathcal{C}(\mathcal{P})|$.

This construction of a topological space for $\mathcal{P}$ is based on the correspondence between partially ordered sets and (vertex-labelled) simplicial complexes, which lies at the heart of many applications of topology to combinatorics and geometry (compare Björner [28–30] and Stanley [397, 398, Chapter 3]).

However, unless all facets and vertex-figures are spherical, the space $|\mathcal{C}(\mathcal{P})|$ only captures some of the interesting topological features which a polytope $\mathcal{P}$ can have, but leaves out many others. Here, we shall not attempt to give a detailed account of the various other ways in which a topological space can be associated with an abstract polytope. Instead, the reader is referred to Brehm, Kühnel and Schulte [49, §3]. At the end of this section we shall briefly comment on some of the difficulties which arise in this context. For our purposes, it is important to note that the following concept of a globally or locally spherical regular polytope will not be affected by any of these ambiguities. They are indeed ruled out by definition.

For notational reasons, in this section we prefer to denote the rank by $n + 1$. An abstract $(n + 1)$-polytope $\mathcal{P}$ is called *(globally) spherical* if it is isomorphic to the face-lattice of a convex $(n + 1)$-polytope $\mathcal{Q}$. By the results of Section 1B, if $\mathcal{P}$ is a regular polytope, then $\mathcal{Q}$ must be combinatorially regular and hence isomorphic to a regular convex polytope (spherical tessellation) $\mathcal{R}$ with $\Gamma(\mathcal{R}) \cong \Gamma(\mathcal{Q}) \cong \Gamma(\mathcal{P})$. Note, in particular, that each facet and vertex-figure, and more generally, section, of a spherical regular polytope is again a spherical regular polytope.

According to the last definition, the spherical regular $(n + 1)$-polytopes are precisely the finite universal $(n + 1)$-polytopes $\mathcal{P} = \{p_1, \ldots, p_n\}$ which have $p_i > 2$ for each i. Then $\Gamma(\mathcal{P}) = [p_1, \ldots, p_n]$, which is a finite irreducible Coxeter group. Occasionally we shall extend our terminology, and call *each* finite universal $(n+1)$-polytope $\mathcal{P}$ a spherical $(n + 1)$-polytope. This can be explained as follows.

The Coxeter group $\Gamma(\mathcal{P})$ is now the direct product of its irreducible components, which are determined by the occurrences of 2 in the Schläfli symbol. The order complex $\mathcal{C}(\mathcal{P})$ is isomorphic to the Coxeter complex for $\Gamma(\mathcal{P})$, and hence yields a triangulation of $\mathbb{S}^n$. Geometrically, its n-simplices can be obtained as the intersections of $\mathbb{S}^n$ with the $(n + 1)$-dimensional simplicial cones in the chamber complex for $\Gamma(\mathcal{P})$ (see Section 3A). We can now appeal to Theorem 3D7, which describes how the faces of $\mathcal{P}$ can be recovered from the chamber complex. When intersected with $\mathbb{S}^n$, this yields a "decomposition" of $\mathbb{S}^n$ which is "isomorphic" to $\mathcal{P}$. (However, in general, this will not be a tessellation as defined in Section 6A.) For example, the 3-polytope $\{p, 2\}$ gives a decomposition of $\mathbb{S}^2$ into two "regular p-gons"; this is a *dihedron* (see [120, p. 12]). Later in this section we shall further elaborate on these decompositions. In this context, it is interesting to note that each finite regular polytope $\mathcal{P}$, whose order complex is topologically a sphere, is indeed isomorphic to a finite universal polytope (Theorem 1B3 and [151, §§2, 4; 155]). We shall prove this in Theorem 6B2 under a slightly stronger assumption. In particular, if $\mathcal{P}$ is of type $\{p_1, \ldots, p_n\}$ with $p_i > 2$ for all i, then we have isomorphism with a convex regular $(n+1)$-polytope.

For notational convenience, unless specified otherwise, we shall use the notion of a spherical polytope in the more restrictive sense to mean isomorphism with a convex polytope.

Now, moving on to local topological type, we say that a polytope $\mathcal{P}$ is *locally spherical* if all its proper sections are spherical polytopes or, equivalently, if all its facets and vertex-figures are spherical. However, we do not require that $\mathcal{P}$ itself be spherical. Then, if $\mathcal{P}$ is regular, its facets and vertex-figures are isomorphic to convex regular polytopes. Easy examples of locally spherical regular polytopes are the globally spherical regular polytopes and the euclidean or hyperbolic regular tessellations. (Note that, if we allowed arbitrary finite universal polytopes as facets and vertex-figures of a locally spherical regular polytope, it would actually become spherical if its Schläfli symbol contained an entry 2.)

Another class of examples of locally spherical regular polytopes comprises the regular toroids, which we shall describe in detail in Sections 6D–6F. Here we give a brief introduction to them.

A (*globally*) *toroidal* $(n+1)$-polytope or, more briefly, an $(n+1)$-*toroid* is an abstract $(n+1)$-polytope $\mathcal{P}$ which is the quotient of a periodic tessellation $\mathcal{T}$ of euclidean n-space $\mathbb{E}^n$ (with convex polytopes as tiles) by a subgroup (lattice) Λ of its translational symmetries which is generated by n independent translations; the resulting toroid is written $\mathcal{P} = \mathcal{T}/\Lambda$. Topologically, $\mathcal{P}$ is then a tessellation on the n-torus $\mathbb{E}^n/\Lambda$, which is the orbit space of $\mathbb{E}^n$ modulo Λ (Section 6A). Note that the rank of the toroid is always one larger than the dimension of the torus. By Theorem 3D5, for a regular toroid $\mathcal{P}$, we may further assume that $\mathcal{T}$ is a regular tessellation of $\mathbb{E}^n$ and that Λ is a normal subgroup of the symmetry group $G(\mathcal{T})$ of $\mathcal{T}$. For rank 3, the regular toroids are regular maps on the 2-torus; see Section 1D or [131]. The facets and vertex-figures of each regular $(n+1)$-toroid $\mathcal{P}$ are necessarily spherical regular n-polytopes. They are indeed of the same kind as the facets and vertex-figures, respectively, of the corresponding regular euclidean tessellation $\mathcal{T}$. Now, if we consider $\mathcal{P} = \mathcal{T}/\Lambda$ as a tessellation on $\mathbb{E}^n/\Lambda$ and take the barycentric subdivision of the tiles (facets) in a consistent manner, we arrive at a triangulation of $\mathbb{E}^n/\Lambda$ which is isomorphic to the order complex $\mathcal{C}(\mathcal{P})$ of $\mathcal{P}$. It is now apparent that the corresponding space $|\mathcal{C}(\mathcal{P})|$ is just the n-torus $\mathbb{E}^n/\Lambda$ itself.

In summary, therefore, both the spherical and the toroidal regular polytopes are equipped with a natural topology which also coincides with that of the corresponding order complexes. As we shall see, this is more generally true for all locally spherical polytopes.

For later use we note the following two results. Recall from Section 2E the notion of a sparse subgroup.

6B1 Lemma *Let $\mathcal{P}$ be a locally spherical regular $(n+1)$-polytope of type $\{p_1, \ldots, p_n\}$, and let $\mathcal{T} := \{p_1, \ldots, p_n\}$, the universal $(n+1)$-polytope of its type. Then $\mathcal{P} = \mathcal{T}/N$, where N is a sparse normal subgroup of $\Gamma(\mathcal{T})(= G(\mathcal{T}) = [p_1, \ldots, p_n])$.*

Proof. Let $\Gamma(\mathcal{T}) = \langle \rho_0, \ldots, \rho_n \rangle$, and $\Gamma(\mathcal{P}) = \langle \sigma_0, \ldots, \sigma_n \rangle$, with the obvious meaning. Then the mappings $\rho_i \mapsto \sigma_i$ (for $i = 0, \ldots, n$) induce a homomorphism $\kappa\colon \Gamma(\mathcal{T}) \to \Gamma(\mathcal{P})$. Let $N := \ker(\kappa)$, so that $\Gamma(\mathcal{P}) = \Gamma(\mathcal{T})/N$. Then we know that $\Gamma(\mathcal{T})/N$ is a string C-group. Since the facets and vertex-figures of $\mathcal{T}$ and $\mathcal{P}$ are of the same kind, we must have

$$N \cap \langle \rho_0, \ldots, \rho_{n-1} \rangle = \{\varepsilon\} = N \cap \langle \rho_1, \ldots, \rho_n \rangle.$$

Then, by Lemma 2E22, N must be sparse. Hence also, by Proposition 2E19, $\mathcal{P} = \mathcal{T}/N$. □

6B2 Theorem *Let $n \geqslant 1$, and let $\mathcal{P}$ be a finite regular $(n+1)$-polytope of type $\{p_1, \ldots, p_n\}$. If the order complex of each section of $\mathcal{P}$ of rank at least 2 (including $\mathcal{P}$ itself) is topologically a sphere, then $\mathcal{P}$ is isomorphic to the universal $(n+1)$-polytope $\{p_1, \ldots, p_n\}$.*

Proof. The proof is by induction on n. The case $n = 1$ is trivial. Let $n \geqslant 2$. By the obvious inductive hypothesis, we can assume that the facets and vertex-figures of $\mathcal{P}$ are isomorphic to the finite universal n-polytopes $\{p_1, \ldots, p_{n-1}\}$ and $\{p_2, \ldots, p_n\}$, respectively, each with a spherical order complex. Let $\mathcal{T} := \{p_1, \ldots, p_n\}$. Then, by Proposition 2E19, $\mathcal{P} = \mathcal{T}/N$, where N is a normal subgroup of $\Gamma(\mathcal{T})$. In fact, the proof of Lemma 6B1 carries over to this situation and shows that N is sparse. In particular, the structure of the facets and vertex-figures is preserved under the quotient. It follows that N acts freely and properly discontinuously as a group of homeomorphisms on $|\mathcal{C}(\mathcal{T})|$, and determines a topological covering $\pi\colon |\mathcal{C}(\mathcal{T})| \to |\mathcal{C}(\mathcal{P})|$ between the spaces of the two order complexes (Section 6A). For the proof that N is properly discontinuous, note that, for each $x \in |\mathcal{C}(\mathcal{T})|$, we can find an open neighbourhood U of x which entirely contains a (closed) maximal simplex of $\mathcal{C}(\mathcal{T})$ that contains x, such that the transforms $U\varphi$ (for $\varphi \in N$) of U are mutually disjoint. Now, since $|\mathcal{C}(\mathcal{P})| = \mathbb{S}^n$ and $\mathbb{S}^n$ is simply connected, the covering must be trivial; hence $|\mathcal{C}(\mathcal{T})| = |\mathcal{C}(\mathcal{P})|$. Then N must be trivial and $\mathcal{P}$ must be isomorphic to $\mathcal{T}$. This completes the proof. □

We remark that the statement of Theorem 6B2 remains true if we replace the requirement that $\mathcal{P}$ be regular by the weaker requirement that $\mathcal{P}$ be equivelar (see [155, 279] and Theorem 1B9). Recall from Section 1B that an equivelar $(n+1)$-polytope $\mathcal{P}$ is one which has a Schläfli symbol $\{p_1, \ldots, p_n\}$; that is, for each flag $\Psi = \{G_{-1}, G_0, \ldots, G_{n+1}\}$ of $\mathcal{P}$ and for each j, the section G_{j+1}/G_{j-2} of $\mathcal{P}$ is a p_j-gon.

We now explain how a locally spherical regular polytope $\mathcal{P}$ determines a space-form. To this end, if $\mathcal{P}$ is such a polytope of rank $n+1$ and type $\{p_1, \ldots, p_n\}$, then its facets $\mathcal{P}_1$ and vertex-figures $\mathcal{P}_2$ are isomorphic to the convex regular n-polytopes

$\{p_1, \ldots, p_{n-1}\}$ and $\{p_2, \ldots, p_n\}$, respectively, and $\mathcal{P}$ itself is a quotient of the universal regular $(n+1)$-polytope $\mathcal{T} := \{p_1, \ldots, p_n\}$. In the present situation, $\mathcal{T}$ is just the (locally finite) regular tessellation with the same Schläfli symbol in $\mathbb{S}^n$, $\mathbb{E}^n$ or $\mathbb{H}^n$. By E we denote the underlying (natural) space of $\mathcal{T}$. Note that the local finiteness of $\mathcal{P}$ excludes the possibility that $\mathcal{T}$ is a regular hyperbolic honeycomb with infinite facets or vertex-figures.

We can now appeal to the concepts introduced in Section 6A. Write $\mathcal{P} = \mathcal{T}/N$, the quotient of $\mathcal{T}$ defined by a sparse normal subgroup N of $G(\mathcal{T})$ $(= \Gamma(\mathcal{T}))$. Then, as a group of isometries of E, the group N is discrete and acts freely. In fact, N does not impinge on the facets and vertex-figures of $\mathcal{T}$ because their structure is preserved under the quotient of $\mathcal{T}$ by N. In particular, given a point $x \in E$, we can find an open neighbourhood U of x which entirely contains a closed fundamental simplex for $G(\mathcal{T})$ that contains x, such that the transforms $U\varphi$ (for $\varphi \in N$) of U are mutually disjoint (Section 3A). It follows that E/N is an n-dimensional spherical, euclidean or hyperbolic space-form on which $\mathcal{P}$ lives as a regular tessellation.

In the spherical or hyperbolic case, the Schläfli symbol determines the original tessellation $\mathcal{T}$ up to congruence, and so the isometry type of the space-form E/N is not affected by the particular representative we picked for $\mathcal{T}$. However, in the euclidean case, $\mathcal{T}$ is only determined up to similarity. The isometry type, but not the similarity type, will thus depend on our choice of the representative.

Proceeding with the general discussion, just as in the case of the regular toroids on the n-torus, we can now construct a triangulation of E/N from the original polytope $\mathcal{P}$ by taking the barycentric subdivision of the tiles (facets) in a consistent manner. Equivalently, we can first take the barycentric subdivision of the tessellation $\mathcal{T}$ on E, which is invariant under $G(\mathcal{T})$, and then pass to its quotient modulo N. This triangulation of E/N is isomorphic to the order complex $\mathcal{C}(\mathcal{P})$, and so we can conclude again that $|\mathcal{C}(\mathcal{P})| = E/N$. In particular, we may note that $\mathcal{P}$ is a finite polytope if and only if E/N is a compact space-form.

We can further observe that the combinatorial $(n-1)$-covering $\mathcal{T} \searrow \mathcal{T}/N = \mathcal{P}$ of $(n+1)$-polytopes determines the topological covering projection $\pi\colon E \to E/N$ of the original space E onto the space-form E/N, and vice versa. Thus, for locally spherical regular polytopes, the two concepts of combinatorial covering of polytopes and topological covering of associated spaces are very closely related. There is no analogue of this for general regular polytopes.

In this situation, $\mathcal{P}$ is a tessellation on E/N which is regular in a strong geometric sense. Let $G(\mathcal{P})$ denote the (*geometric*) *symmetry group* of $\mathcal{P}$; by this we mean the group of all isometries of the space-form E/N which map the tessellation $\mathcal{P}$ onto itself. We claim that $G(\mathcal{P})$ acts simply flag-transitively on the tessellation $\mathcal{P}$ and that $G(\mathcal{P}) \cong \Gamma(\mathcal{P})$; that is, $\mathcal{P}$ is *geometrically regular*.

To prove this, recall from Section 6A that the isometry group $\mathcal{I}(E/N)$ of E/N is isomorphic to the factor group $N_{\mathcal{I}(E)}(N)/N$, where $N_{\mathcal{I}(E)}(N)$ is the normalizer of N in the isometry group $\mathcal{I}(E)$ of E. In particular, the isomorphism $N_{\mathcal{I}(E)}(N)/N \to \mathcal{I}(E/N)$ is given by $N\sigma \mapsto \hat{\sigma}$, where $\hat{\sigma}$ is defined by $(xN)\hat{\sigma} := (x\sigma)N$ for $x \in E$.

Now N is a normal subgroup of $G(\mathcal{T})$; hence $G(\mathcal{T})$ is contained in $N_{\mathcal{I}(E)}(N)$. It follows that the isomorphism maps the subgroup $G(\mathcal{T})/N$ of $N_{\mathcal{I}(E)}(N)/N$ onto the subgroup $G'(\mathcal{P}) := \{\hat{\sigma} \mid \sigma \in G(\mathcal{T})\}$ of $\mathcal{I}(E/N)$. By Proposition 2E18, $G(\mathcal{T})/N = \Gamma(\mathcal{T})/N = \Gamma(\mathcal{T}/N) = \Gamma(\mathcal{P})$. We can now conclude that $G'(\mathcal{P})$ is a subgroup of the symmetry group $G(\mathcal{P})$ of $\mathcal{P}$. In fact, each $\hat{\sigma}$ with $\sigma \in G(\mathcal{T})$ is now an isometry of E/N which preserves the tessellation $\mathcal{P} = \mathcal{T}/N$ on E/N. In particular, $G'(\mathcal{P})$ acts simply flag-transitively on $\mathcal{P}$, because it is isomorphic to $G(\mathcal{T})/N$ and hence also to $\Gamma(\mathcal{P})$.

It remains to be shown that the two groups $G(\mathcal{P})$ and $G'(\mathcal{P})$ actually coincide. To this end, if $\tau \in G(\mathcal{P})$, then $\tau = \hat{\sigma}$, for some $\sigma \in N_{\mathcal{I}(E)}(N)$, because τ is an isometry of E/N; then σ is uniquely determined modulo N. On the other hand, τ also induces an automorphism of $\mathcal{P}$, and hence, modulo N, the pull-back σ must be in $G(\mathcal{T})$. It follows that $\tau = \hat{\sigma} \in G(\mathcal{P})$, which completes the proof.

In summary, then, we see that we have proved the following

6B3 Theorem *Let $\mathcal{P}$ be a locally spherical regular $(n+1)$-polytope of type $\{p_1, \ldots, p_n\}$.*

(a) Combinatorially, $\mathcal{P}$ is a quotient $\mathcal{T}/N$ of the regular tessellation $\mathcal{T} = \{p_1, \ldots, p_n\}$ in spherical, euclidean or hyperbolic n-space E by a sparse normal subgroup N of $\Gamma(\mathcal{T})$ $(= G(\mathcal{T}))$, which, when considered as a group of isometries of E, is discrete and acts freely on E.

(b) Topologically, $\mathcal{P}$ can be viewed as a regular tessellation on the corresponding spherical, euclidean or hyperbolic space-form E/N, whose fundamental group is isomorphic to N. In particular, $|\mathcal{C}(\mathcal{P})| = E/N$, where $\mathcal{C}(\mathcal{P})$ is the order complex of $\mathcal{P}$, and $\mathcal{P}$ is a finite polytope if and only if E/N is a compact space-form.

(c) Geometrically, $\mathcal{P}$ can be viewed as a tessellation on the space-form E/N, which is (geometrically) regular in the strong sense that its symmetry group $G(\mathcal{P})$ is simply flag-transitive and isomorphic to the combinatorial automorphism group $\Gamma(\mathcal{P})$.

As we pointed out earlier, in the spherical or hyperbolic case, the space-form E/N constructed in the proof of Theorem 6B3 is uniquely determined up to isometry by $\mathcal{P}$. However, in the euclidean case, it is only determined up to similarity, and depends also on the choice of the representative for $\mathcal{T}$. For most considerations this will be irrelevant, so that we may pick a representative at our convenience.

The volume $\mathrm{Vol}(E/N)$ of the space-form E/N can be directly expressed in terms of $\mathcal{T}$ and $\mathcal{P}$. If F_n is a tile of the original tessellation $\mathcal{T}$ in E with volume $\mathrm{Vol}(F_n)$, and if f_n is the number of facets of $\mathcal{P}$, then

$$\mathrm{Vol}(E/N) = f_n \, \mathrm{Vol}(F_n).$$

In fact, for the action of N on E, we can clearly choose a proper fundamental region D whose closure is the union of f_n tiles of $\mathcal{T}$, one for each orbit under N on the set of tiles of $\mathcal{T}$ (Section 6A). Then $\mathrm{Vol}(E/N) = \mathrm{Vol}(D) = f_n \, \mathrm{Vol}(F_n)$, because each tile of $\mathcal{T}$ is congruent to F_n.

If $v_{\mathcal{T}}$ denotes the volume of the characteristic orthoscheme which is the fundamental simplex in E for the symmetry group $G(\mathcal{T})$, then

$$\mathrm{Vol}(F_n) = v_{\mathcal{T}} \cdot |[p_1, \ldots, p_{n-1}]|.$$

In fact, F_n decomposes into such orthoschemes, namely, one for each element of the symmetry group $[p_1, \ldots, p_{n-1}]$ of F_n.

If $E = \mathbb{E}^n$, then $v_{\mathcal{T}}$ depends on the representative we pick for $\mathcal{T}$ (and could take any positive value). However, if $E = \mathbb{S}^n$ or $\mathbb{H}^n$, then $\mathcal{T}$ is determined up to congruence and $v_{\mathcal{T}}$ takes a fixed positive value denoted by $v(p_1, \ldots, p_n)$. In this case, $\mathrm{Vol}(E/N)$ is determined solely by the Schläfli symbol and the number f_n of facets of $\mathcal{P}$. For $\mathbb{S}^n$, there is a close relation between $v(p_1, \ldots, p_n)$ and Schläfli's famous volume function for spherical orthoschemes (see [101, 246, 247, §1.3]; see also Section 3E, for elementary calculations of $v(p_1, \ldots, p_n)$). The following lemma is now an easy consequence of the fact that isometric space-forms have the same volume.

6B4 Lemma *If two locally spherical regular* $(n + 1)$*-polytopes of the same Schläfli type have distinct numbers of facets, then the two corresponding space-forms cannot be isometric (provided that in the euclidean case they have been derived from the same representative for* $\mathcal{T}$*).*

In essence, Theorem 6B3 says that a locally spherical regular polytope can be viewed as a regular tessellation on a space-form. We shall now prove that the converse is also true. In fact, we shall obtain a stronger result.

To this end, we begin with a tessellation $\mathcal{P}$ on a real manifold which is not *a priori* associated with a given regular tessellation $\mathcal{T}$ in a space E. In this context, recall from Section 6A that, in our definition of tessellation, we require that the tiles of $\mathcal{P}$ be homeomorphic images of convex polytopes, which fit together in such a way that the set of faces is an abstract polytope (again denoted by $\mathcal{P}$). Further, recall that a tessellation is (combinatorially) regular if, as a polytope, it is regular.

Now let $\mathcal{P}$ be a tessellation on a real n-manifold X without boundary. If $\mathcal{P}$ is regular, then its tiles, being homeomorphic images of convex n-polytopes, must also be combinatorially regular and hence, by Theorem 1B3, must be isomorphic to convex regular n-polytopes. Also, since a tessellation is by definition locally finite and X is an n-manifold, each vertex-figure of $\mathcal{P}$ must be a finite n-polytope whose order complex is topologically an $(n - 1)$-sphere. Now two cases can occur.

Let $\mathcal{P}$ be of type $\{p_1, \ldots, p_n\}$. If $\mathcal{P}$ has only two tiles, then $p_n = 2$ and X is the n-sphere which is decomposed into two copies of the polytope $\{p_1, \ldots, p_{n-1}\}$ sharing a common boundary; that is, $\mathcal{P}$ is a ditope on $\mathbb{S}^n$. This is the degenerate case. Recall that a *ditope* is an abstract polytope which has only two facets. For a decomposition of $\mathbb{S}^n$ with full geometric symmetry, the (boundary of the) convex regular n-polytope $\{p_1, \ldots, p_{n-1}\}$ must be placed at the equator of $\mathbb{S}^n$; then the upper and lower hemisphere can be considered as the two copies of $\{p_1, \ldots, p_{n-1}\}$ which tile $\mathbb{S}^n$. This is the decomposition of $\mathbb{S}^n$ for the finite universal polytope $\{p_1, \ldots, p_{n-1}, 2\}$ which we mentioned earlier.

In the non-degenerate case, $\mathcal{P}$ has more than two tiles. Then $p_n > 2$ and the vertex-figures are also isomorphic to convex regular n-polytopes, now with Schläfli symbol $\{p_2, \ldots, p_n\}$. This follows from Theorem 6B2, using the observation that the order complex of each proper section of $\mathcal{P}$ is topologically a sphere, and that 2 does not occur in the Schläfli symbol. Now we have both facets and vertex-figures of $\mathcal{P}$ isomorphic to convex regular polytopes. In particular, we have proved

6B5 Theorem *Let $\mathcal{P}$ be a regular tessellation on an n-dimensional real manifold X without boundary.*

(a) If $\mathcal{P}$ has only two tiles, then $X = \mathbb{S}^n$ and $\mathcal{P}$ is a (regular) ditope on $\mathbb{S}^n$.
(b) If $\mathcal{P}$ has more than two tiles, then it is a locally spherical regular $(n+1)$-polytope.

The following corollary is now a consequence of Theorems 6B3 and 6B5.

6B6 Corollary *Every regular tessellation on a real manifold without boundary is isomorphic to a geometrically regular tessellation on a space-form which is homeomorphic to the manifold. The isomorphism is induced by a homeomorphism between the two spaces.*

Proof. If $\mathcal{P}$ is a regular tessellation on a manifold X with more than two tiles, then Theorem 6B5 implies that it is a locally spherical polytope, and Theorem 6B3 yields the isomorphism with a geometrically regular tessellation on a space-form X'. However, the combinatorics of a tessellation determines the topology of its underlying space. In fact, a topological barycentric subdivision of a tessellation gives a triangulation which is isomorphic to the order complex, and so the space is just that of the order complex. In particular, the original manifold X is homeomorphic to X'. The homeomorphism which carries the order complexes of the tessellations on X and X' into each other also carries the tessellations themselves into each other. (Note that, if X itself is a space-form, it need not be isometric to X'.) This completes the proof. □

By Corollary 6B6, a regular tessellation on a space-form X determines X up to homeomorphism. It is a non-trivial fact that, in the interesting cases, the type of geometry is uniquely determined by the topology.

For spherical space-forms this is obvious; they are topologically distinguished from the euclidean and hyperbolic space-forms by the finiteness of their fundamental groups. In the euclidean or hyperbolic case, the basic spaces $\mathbb{E}^n$ and $\mathbb{H}^n$ are of course homeomorphic and so are not determined by the topology. On the other hand, no hyperbolic space-form of finite volume can be homeomorphic to an euclidean space-form. In particular, this includes compact space-forms. The Gromov invariant for manifolds can distinguish topologically between these two types of spaces (see [194, §0.2; 344, §11.4; 411, §6.1]); it takes a positive value for hyperbolic space-forms of finite volume, but is zero for all euclidean space-forms. For even dimensions n, the Euler characteristic χ achieves the same effect. In particular, $\chi \neq 0$ for each even-dimensional hyperbolic space-form of finite volume (see [246, p. 30] and (6J8)), but $\chi = 0$ for each euclidean space-form (of any dimension) which is not $\mathbb{E}^n$ itself (see [467, Corollary 3.3.5]). For example, a closed surface of Euler characteristic χ is a 2-dimensional spherical, euclidean or hyperbolic space-form if $\chi > 0$, $\chi = 0$ or $\chi < 0$, respectively. We shall further explain this in Section 6J. For now, we just note the following.

6B7 Lemma *Let $n \geqslant 2$, and let $\mathcal{P}$ be a regular tessellation of type $\{p_1, \ldots, p_n\}$ on an n-dimensional compact space-form X. Let $\mathcal{T} := \{p_1, \ldots, p_n\}$, the universal polytope of which $\mathcal{P}$ is a quotient. Then the geometry of X determines the geometry of the (natural) space E of $\mathcal{T}$; that is, $E = \mathbb{S}^n$, $\mathbb{E}^n$ or $\mathbb{H}^n$ according as X is a spherical, euclidean or hyperbolic space-form.*

Proof. If $p_n = 2$, then $\mathcal{P}$ has only two tiles. By Theorem 6B5, it follows that $X = E = \mathbb{S}^n$ and $\mathcal{P} = \mathcal{T}$. If $p_n > 2$, then, from Theorem 6B5 again, $\mathcal{P}$ is locally spherical. Also, by Theorem 6B3, $\mathcal{P} = \mathcal{T}/M$ for some M, and $\mathcal{P}$ can be regarded as a regular tessellation on the space-form E/M. Since X and E/M are homeomorphic and X is compact, the type of geometry is determined by the topology of the underlying space. It follows that $E = \mathbb{S}^n$, $\mathbb{E}^n$ or $\mathbb{H}^n$ according as X is spherical, euclidean or hyperbolic. □

Note that spherical space-forms are always compact. If $X = \mathbb{E}^n$ or $\mathbb{H}^n$, the statement analogous to that of Lemma 6B7 is not true. In fact, since $\mathbb{E}^n$ and $\mathbb{H}^n$ are topologically the same, each geometrically regular tessellation in $\mathbb{E}^n$ can be viewed as a (combinatorially) regular tessellation in $\mathbb{H}^n$, and vice versa.

Although the type of geometry is usually uniquely determined, there are in general two space-forms that need to be considered, namely, the original space-form E/N $(= X)$ (say) with tessellation $\mathcal{P}$, and the new space-form E/M which is defined by expressing $\mathcal{P}$ as a quotient $\mathcal{T}/M$. Topologically they are the same (and hence their fundamental groups N and M are isomorphic), but as space-forms they need not be isometric.

In the euclidean case, this is hardly surprising. In fact, as we remarked in Section 6A, the fundamental group determines an n-dimensional compact euclidean space-form only up to affine equivalence, but not isometry or even similarity. For example, all regular $(n + 1)$-toroids described in Sections 6D and 6E are tessellations in the same topological space, the n-torus. However, more than one similarity class of n-tori occurs, because the corresponding lattices in $\mathbb{E}^n$ are not mutually similar. As we shall see in Section 6G, there are no other compact euclidean space-forms which admit regular tessellations.

On the other hand, we shall also see in Section 6C that $\mathbb{S}^n$ and $\mathbb{P}^n$ are the only spherical space-forms which admit a regular tessellation. For these, homeomorphism does imply isometry.

Finally, for hyperbolic space-forms of finite volume and dimension $n \geqslant 3$, it is a consequence of Mostow rigidity in $\mathbb{H}^n$ that homeomorphic space-forms are also isometric (see [344, §11.6; 411, §5.7]). We shall elaborate on this in Section 6J. For $n = 2$, the corresponding statement is not true [344, Chapter 9].

We conclude this section with two more general remarks. The first concerns the decompositions of $\mathbb{S}^n$ related to finite universal polytopes, which we mentioned earlier in this section. The second deals with the general problem of associating topological spaces with an abstract polytope.

To begin with the first, in our definition of a tessellation on a space-form we require the tiles to be topological images of convex polytopes. Except for ditopes on $\mathbb{S}^n$, this rules out tessellations whose Schläfli symbol contains an entry 2. Nevertheless, many polytopes with a 2 in the Schläfli symbol can be realized topologically by "decompositions" of spaces which share many properties with tessellations. Here we shall revisit the finite universal $(n+1)$-polytopes $\mathcal{P} = \{p_1, \ldots, p_n\}$ whose Schläfli symbol contains a 2; they determine a decomposition of $\mathbb{S}^n$. Now $\Gamma(\mathcal{P}) = [p_1, \ldots, p_n]$ is a finite reducible Coxeter group.

We proceed by induction on the number of connected components of the underlying Coxeter diagram. If there are just two components and $p_k = 2$ (say), then $\{p_1, \ldots, p_{k-1}\}$ and $\{p_{k+1}, \ldots, p_n\}$ are isomorphic to convex regular polytopes, which can be realized topologically as tessellations on $\mathbb{S}^{k-1}$ or $\mathbb{S}^{n-k}$, respectively. On $\mathbb{S}^n$, which is the topological join $\mathbb{S}^{k-1} * \mathbb{S}^{n-k}$, we can then find the decomposition into "tiles" $\mathbb{S}^{k-1} * F$, where F is a tile of the tessellation $\{p_{k+1}, \ldots, p_n\}$ on $\mathbb{S}^{n-k}$, and $\mathbb{S}^{k-1} * F$ is the join of $\mathbb{S}^{k-1}$ and F in $\mathbb{S}^n$. These tiles $\mathbb{S}^{k-1} * F$ are topologically n-cells whose boundary is decomposed into "faces" of two kinds: the faces G of the tessellation $\{p_1, \ldots, p_{k-1}\}$ on $\mathbb{S}^{k-1}$, of rank less than k, and the faces $\mathbb{S}^{k-1} * G$, of rank $k + \operatorname{rank} G$, where G is a proper face of F. This face structure is a consequence of the fact that the universal polytope $\{p_1, \ldots, p_n\}$ with $p_k = 2$ is combinatorially (i, j)-flat for each i, j with $i < k \leqslant j$; this means that each of its i-faces is incident with each of its j-faces (Section 4F).

If $k = n$, we obtain a ditope on $\mathbb{S}^n$, or a dihedron if $n = 2$. If $k = 1$, we arrive at the decomposition by placing $\{p_2, \ldots, p_n\}$ at the equator of $\mathbb{S}^n$, and joining every proper face of it to both the north and south poles of $\mathbb{S}^n$. For $n = 2$, this consists of p_2 digons, and is known as the *hosohedron* (see [120, p. 12]).

If there are more than two connected components and k is the largest index such that $p_k = 2$, then we can construct the decomposition for $\{p_1, \ldots, p_{k-1}\}$ on $\mathbb{S}^{k-1}$ by the obvious induction, and then again obtain the decomposition for $\{p_1, \ldots, p_n\}$ on $\mathbb{S}^n$ in the same way. The faces of the tiles $\mathbb{S}^{k-1} * F$ are then determined as before. This completes the construction.

As we mentioned at the beginning of this section, there are often many ways in which a topological space can be associated with an abstract n-polytope $\mathcal{P}$. If $\mathcal{P}$ is locally spherical, then $\mathcal{P}$ can be recovered from its order complex $\mathcal{C}(\mathcal{P})$ in such a way that, in the space $|\mathcal{C}(\mathcal{P})|$, its facets and vertex-figures appear as topological $(n-1)$-cells. This makes $|\mathcal{C}(\mathcal{P})|$ the natural space for $\mathcal{P}$. However, if $\mathcal{P}$ is not locally spherical, the picture is more complicated.

For example, if $\mathcal{P}$ is an abstract regular 4-polytope whose facet is isomorphic to a toroidal map and whose vertex-figure is isomorphic to a Platonic solid, then $|\mathcal{C}(\mathcal{P})|$ is a 3-manifold on which the facet occurs as a cone over the 2-torus (not as a solid torus); the vertex-figure is still a 3-cell. However, in this situation, we may rather wish to find another topological model for $\mathcal{P}$ on a 3-manifold X for which the facets are solid tori.

If $\mathcal{P}$ has only two facets, then, in such a model, the two solid tori are glued together along their common boundary and give a "Heegaard splitting of genus 1" of the underlying manifold X. It is well known that X is determined by the way the solid tori are glued, and can indeed be any lens space [383, §62; 470]. An example is the standard decomposition of $\mathbb{S}^3$ into two solid tori. This already shows that for traditional Heegaard splittings of 3-manifolds X, where X is decomposed into only two handlebodies of the same genus, the topological structure of the two pieces does not determine X uniquely.

In the general situation, the desired model for $\mathcal{P}$ on X would give a "generalized Heegaard splitting" for X which is also combinatorially regular. Again, the topology of X would not be completely determined by the combinatorics of $\mathcal{P}$.

For a more systematic approach to topological models for polytopes on manifolds (possibly with singularities), we refer to Brehm et al. [49]. For some small polytopes, the admissible spaces X can be classified if certain natural assumptions on the models are made; the spaces include the 3-sphere $\mathbb{S}^3$, connected sums of handles $\mathbb{S}^1 \times \mathbb{S}^2$, euclidean and spherical space-forms, and other examples with non-trivial fundamental groups. Another interesting example is a decomposition of $\mathbb{S}^3$ into 20 solid tori, which was independently discovered in Coxeter and Shephard [132] and in Grünbaum [199, Table 3]; this is a topological model for the regular 4-polytope ${}_1\mathcal{T}^4_{(3,0)} := \{\{4,4\}_{(3,0)}, \{4,3\}\}$ (see Section 10B). For further related work see also Banchoff [19], Costa [94], Ferri, Gagliardi, and Grasselli [168], Kühnel [256], Montesinos [327], and Valverde [428].

From an abstract polytope $\mathcal{P}$, we can also obtain lower-dimensional "complexes" which are not themselves polytopes. For example, the 2-skeleton (set of all faces of rank at most 2) of $\mathcal{P}$ gives a polygonal complex on which $\Gamma(\mathcal{P})$ acts in a natural way. These complexes and their topologies are completely unexplored. For example, it is not even known which complexes can arise in this way from polytopes. For related work see also [17].

6C Projective Regular Polytopes

The regular tessellations $\mathcal{P}$ on n-dimensional spherical space-forms are obtained as quotients of regular tessellations $\mathcal{T} = \{p_1, \ldots, p_n\}$ on the unit n-sphere $\mathbb{S}^n$ (Lemma 6B7). Now $\mathcal{T}$ is finite, and isomorphic to a convex regular $(n+1)$-polytope in euclidean $(n+1)$-space $\mathbb{E}^{n+1}$. Clearly, there can be only finitely many such quotients $\mathcal{P}$. We shall see that, in each case, the corresponding space-form is the n-sphere $\mathbb{S}^n$ itself or the real projective n-space $\mathbb{P}^n$. In this context, it is interesting to note that, in even dimensions n, these two spaces are the only spherical space-forms which exist, and that, in odd dimensions, each spherical space-form is orientable [344, p. 338].

As in the previous sections, we denote the space-form by $\mathbb{S}^n/N$, and the quotient tessellation $\mathcal{P}$ on $\mathbb{S}^n/N$ by $\mathcal{T}/N$, where N is a normal subgroup of the symmetry group $G(\mathcal{T})$ $(= \Gamma(\mathcal{T}))$ of $\mathcal{T}$, which acts freely as a group of isometries on $\mathbb{S}^n$. If N is trivial, then the improper quotients $\mathbb{S}^n/N$ and $\mathcal{T}/N$ can be identified naturally with $\mathbb{S}^n$ and $\mathcal{T}$, respectively. On the other hand, the proper quotients can be determined by inspecting the normal subgroups N of $G(\mathcal{T})$, and verifying in each case that N acts on $\mathcal{T}$ in such a way that the structure of the facets and vertex-figures is preserved. In all but three cases this is straightforward.

In particular, the *projective* regular $(n+1)$-polytopes $\mathcal{P}$ occur when the original tessellation $\mathcal{T}$ on $\mathbb{S}^n$ is centrally symmetric (that is, invariant under the mapping $-I$, where I is the identity mapping on $\mathbb{S}^n$), and $N = \{\pm I\}$. Then $\mathcal{P}$ is a regular tessellation in projective n-space $\mathbb{P}^n = \mathbb{S}^n/\{\pm I\}$ (see [112]). The projective polytope corresponding to a centrally symmetric polytope $\mathcal{T} = \{p_1, \ldots, p_n\}$ is also called the *hemi*-$\{p_1, \ldots, p_n\}$.

A projective regular polytope can be conveniently described by a suffix to the Schläfli symbol of the corresponding $\mathcal{T}$, which is half the length (number of edges) of the Petrie polygon of $\mathcal{T}$. Recall that, inductively, a *Petrie polygon* of $\mathcal{T}$, or more generally, of an abstract $(n+1)$-polytope, is an edge-path such that any n consecutive edges, but no $n+1$, belong to a Petrie polygon of a tile (facet) ([120, p. 223], Section 3E). Here we begin the induction by declaring that the Petrie polygon of a polygon (2-polytope) is the polygon itself.

If h is the length of the Petrie polygon of $\mathcal{T}$, then h is even and the distinguished generators $\rho_0, \rho_1, \ldots, \rho_n$ for $G(\mathcal{T})$ satisfy the relation

$$(\rho_0\rho_1 \cdots \rho_n)^h = \varepsilon$$

[120, p. 225]. In fact, the element $(\rho_0\rho_1 \cdots \rho_n)^{h/2}$ is just the mapping $-I$ which, in the quotient, identifies antipodal faces of $\mathcal{T}$ in pairs. The element $\rho_0\rho_1 \cdots \rho_n$ is also known as the *Coxeter element* of the Coxeter group $G(\mathcal{T}) = [p_1, \ldots, p_n]$, and is an important element (see [111, 222, p. 74]). As just one example, the number of reflexions in $G(\mathcal{T})$ is $(n+1)h/2$ (see [120, §12.6; 399]); we remark that this number can also be expressed in terms of the degrees of the invariants of the group [396].

A presentation for the group $\Gamma(\mathcal{P})$ $(= G(\mathcal{P}))$ of the projective regular polytope $\mathcal{P} = \mathcal{T}/\{\pm I\}$ can then be obtained by adjoining to the standard relations for the Coxeter group $[p_1, \ldots, p_n]$ in terms of $\rho_0, \rho_1, \ldots, \rho_n$ the single extra relation

6C1 $$(\rho_0\rho_1 \cdots \rho_n)^{h/2} = \varepsilon.$$

In fact, $\Gamma(\mathcal{P}) = \Gamma(\mathcal{T}/\{\pm I\}) = \Gamma(\mathcal{T})/\{\pm I\}$, and $(\rho_0\rho_1 \cdots \rho_n)^{h/2} = -I$. We extend the notation of Section 1D, and write $\{p_1, \ldots, p_n\}_{h/2}$ as a convenient shorthand for $\mathcal{P}$. Table 6C1 lists all the symbols that actually occur. Note further that the suffix $h/2$ is just the length of the Petrie polygon of the projective polytope $\mathcal{P}$ itself; that is, in passing from $\mathcal{T}$ to $\mathcal{P}$, the length of the Petrie polygon is halved.

Table 6C1. *The Projective Regular $(n+1)$-Polytopes in Real Projective n-Space $\mathbb{P}^n$*

n	Symbol	Name
≥ 2	$\{3^{n-1}, 4\}_{n+1}$	Hemi-$(n+1)$-cross-polytope
	$\{4, 3^{n-1}\}_{n+1}$	Hemi-$(n+1)$-cube
2	$\{3, 5\}_5$	Hemi-icosahedron
	$\{5, 3\}_5$	Hemi-dodecahedron
3	$\{3, 4, 3\}_6$	Hemi-24-cell
	$\{3, 3, 5\}_{15}$	Hemi-600-cell
	$\{5, 3, 3\}_{15}$	Hemi-120-cell

The nature of the extra relation for $\Gamma(\mathcal{P})$ is also consistent with the observation that the fundamental group of $\mathbb{P}^n$ is generated by a loop which corresponds to a path on $\mathbb{S}^n$ connecting antipodal points, and is therefore determined by the mapping $-I$. This is a special case of the more general fact that, for a quotient tessellation $\mathcal{P} = \mathcal{T}/N$ on a space-form E/N, the fundamental group $\pi_1(E/N)$ determines the extra relations for its group $\Gamma(\mathcal{P})$ that have to be added to the standard relations given by $\Gamma(\mathcal{T})$. Indeed, this fundamental group is just N itself, and the group of $\mathcal{T}/N$ is $\Gamma(\mathcal{T})/N$. We shall meet this situation again in the context of the regular toroids where we also have just one extra relation.

6C2 Theorem *The only regular tessellations on n-dimensional spherical space-forms are the regular tessellations on the euclidean n-sphere $\mathbb{S}^n$ and the projective regular $(n+1)$-polytopes in real projective n-space $\mathbb{P}^n$.*

Proof. Of the regular tessellations $\mathcal{T}$ on $\mathbb{S}^n$ which occur in infinite series, only those that correspond to the $(n+1)$-cross-polytope or the $(n+1)$-cube can give proper quotients $\mathcal{P}$. Since the facets and vertex-figures must be of the same kind as the originals, this leaves only the dual pair of projective polytopes $\{3^{n-1}, 4\}_{n+1}$ and $\{4, 3^{n-1}\}_{n+1}$, where the suffix is again the length of the Petrie polygon of $\mathcal{P}$. In fact, if $\mathcal{T} = \{4, 3^{n-1}\}$, then a proper quotient $\mathcal{P} = \mathcal{T}/N$ must have at least 2^n vertices with simplicial vertex-figure and hence a group of order at least $2^n(n+1)!$. Since $G(\mathcal{T}) = [4, 3^{n-1}]$ has order $2^{n+1}(n+1)!$, the corresponding subgroup N must have order 2 and therefore, being normal in $G(\mathcal{T})$, must coincide with the centre $\{\pm I\}$ of $G(\mathcal{T})$. This gives the hemi-$(n+1)$-cube $\{4, 3^{n-1}\}_{n+1}$ a projective regular $(n+1)$-polytope in $\mathbb{P}^n$. Its dual is the hemi-$(n+1)$-cross-polytope $\{3^{n-1}, 4\}_{n+1}$ in $\mathbb{P}^n$.

For $n = 2$, it is easy to verify that there are no further possibilities besides the hemi-icosahedron or hemi-dodecahedron. Both are projective regular 3-polytopes. This is also in agreement with the fact that there are no other 2-dimensional spherical space-forms besides $\mathbb{S}^2$ and $\mathbb{P}^2$. In the remaining case $n = 3$, we clearly have the projective 4-polytopes derived from $\{p, q, r\} = \{3, 4, 3\}$, $\{3, 3, 5\}$ or $\{5, 3, 3\}$. We need to argue that there are no further possibilities for proper quotients.

For $\mathcal{T} = \{3, 4, 3\}$, a proper quotient $\mathcal{P} = \mathcal{T}/N$ with octahedral facets and cubical vertex-figures must clearly have at least 6 vertices and group order 288 $(= 6 \cdot 48)$. This implies that $|N| = 2$, 3 or 4. Now it is straightforward (if tedious) to show that necessarily $|N| = 2$ and therefore again $N = \{\pm I\}$. This gives the projective 4-polytope $\{3, 4, 3\}_6$ (with Petrie polygons of length 6).

For $\mathcal{T} = \{3, 3, 5\}$ or its dual $\{5, 3, 3\}$, it is rather tedious to prove a similar statement by hand; see [33] for a complete argument. (Of course, one could also instead appeal to computer generated lists of normal subgroups of [3, 3, 5]; see, for example, [211, Table 3].) It turns out that the projective polytopes are again the only possibilities. This concludes the proof of the theorem. □

We remark that there are also interesting tessellations on spherical space-forms which are not regular. These can be constructed as quotients of regular tessellations $\mathcal{T}$ on $\mathbb{S}^n$ by subgroups N of $G(\mathcal{T})$ which are not normal. For instance, in the group

$[3, 4, 3]$ of the 24-cell, we have $(\rho_0\rho_1\rho_2\rho_3)^{12} = I$, because $\{3, 4, 3\}$ has Petrie polygons of length 12. Here the element $(\rho_0\rho_1\rho_2\rho_3)^4$ generates a subgroup N of order 3 which is not normal. The quotient $\{3, 4, 3\}/N$ is now a 4-polytope which is not regular, but has 8 octahedral facets and 8 vertices with cubical vertex-figures. It is a tessellation on the 3-dimensional spherical space-form $\mathbb{S}^3/N$ with fundamental group C_3 ($\cong N$), which is a lens-space L(3, 1); see [467, p. 224].

From the 600-cell we also obtain quotients with interesting graph-theoretic properties. If $\mathcal{G}$ and $\mathcal{H}$ are two graphs, then $\mathcal{G}$ is said to be *locally* $\mathcal{H}$ if the induced subgraph on the neighbours of each vertex in $\mathcal{G}$ is isomorphic to $\mathcal{H}$. For example, the edge-graph of a regular polytope $\mathcal{P}$ with triangular faces yields a graph which is locally the edge-graph of the vertex-figure of $\mathcal{P}$. For certain finite graphs $\mathcal{H}$, including the edge-graph of the icosahedron, the finite graphs which are locally $\mathcal{H}$ have been completely enumerated (see [53]). In particular, there are just three *locally icosahedral* graphs, namely, the edge-graphs of the 600-cell $\{3, 3, 5\}$, the hemi-600-cell $\{3, 3, 5\}_{15}$, and a quotient of $\{3, 3, 5\}$ with 40 vertices [33, 71, 72]. For related work see also [70, 80, 230, 430, 440].

Highly symmetric graphs are, of course, of considerable interest in their own right. In the present context, the famous Petersen graph occurs as the edge-graph of the hemi-dodecahedron $\{5, 3\}_5$. We briefly mention here a generalization to a graph $G(p, d)$, defined as follows. It can be drawn in the plane (with crossings permitted) as a p-gon outside a concentric (p/d)-gon (which is allowed to split if the greatest common divisor $(p, d) > 1$), with each pair of corresponding vertices of the two polygons joined by a further edge. Further, it is insisted that the automorphism group of the graph be transitive on the vertices and on the edges. The Petersen graph itself is $G(5, 2)$; actually, $G(5, 2)$ has twice as many automorphisms as $\{5, 3\}_5$, since the latter is self-Petrie. It was shown in [172] that there are just seven such graphs $G(p, d)$, namely,

$$G(4, 1),\quad G(5, 2),\quad G(8, 3),\quad G(10, 2),\quad G(10, 3),\quad G(12, 5),\quad G(24, 5).$$

In particular, $G(4, 1)$ is the graph of the ordinary cube, and $G(10, 2)$ is the graph of the dodecahedron itself. It was proved in [286] that

6C3 Theorem *The graphs $G(p, d)$ are exactly the edge-graphs of the regular polyhedra of type $\{p, 3\}$ with $2p$ vertices.*

6D The Cubic Toroids

A *regular toroidal polytope* of rank $n + 1$ or, more briefly, a *regular* $(n + 1)$*-toroid*, is a quotient of a regular tessellation $\mathcal{T}$ of euclidean n-space $\mathbb{E}^n$ by a normal subgroup Λ of its translational symmetries; the resulting toroid is denoted $\mathcal{T}/\Lambda$. By Theorem 6B3, $\mathcal{T}/\Lambda$ can be viewed as a (geometrically) regular tessellation on the corresponding n-torus $\mathbb{E}^n/\Lambda$. Further, we know from Corollary 6B6 and Lemma 6B7 that each regular tessellation on an n-torus is indeed isomorphic to such a regular $(n + 1)$-toroid.

Since the Schläfli symbol determines a regular euclidean tessellation only up to similarity (rather than congruence), the isometry class of the torus $\mathbb{E}^n/\Lambda$ will also depend on the metrical representative which we pick for $\mathcal{T}$. However, the similarity type of $\mathbb{E}^n/\Lambda$ is uniquely determined by the Schläfli symbol (and Λ). For most considerations, this ambiguity will be irrelevant and can be ignored, so that we may choose $\mathcal{T}$ at our convenience.

The regular toroids of rank 3 were discussed in Section 1D. Here and in the next sections, we shall describe the regular toroids of higher rank. They fall into two families – there are three classes derived from the regular cubic tessellation for each rank, which we shall discuss in this section, and four isolated classes of rank 5 derived from the tessellation $\mathcal{T} = \{3, 3, 4, 3\}$ of $\mathbb{E}^4$ and its dual, which will be described in Section 6E.

There are various relationships among these toroids, involving both subgroups and quotient groups, and these will be considered in Section 6F.

In order to construct a regular toroid of rank $n + 1 \geqslant 4$, we must begin with a regular honeycomb of $\mathbb{E}^n$. Except when $n = 4$, the only such honeycomb is the tessellation $\{4, 3^{n-2}, 4\}$ of $\mathbb{E}^n$ by cubes; as before, q^k will be used to denote a string $q, \ldots, q$ of length k.

The vertex-set of the cubic tessellation $\{4, 3^{n-2}, 4\}$ $(= \mathcal{T})$ may be taken to be $\mathbb{Z}^n$, the set of points in $\mathbb{E}^n$ with integer cartesian coordinates; this set can also be regarded as its translation group. We obtain a toroid by factoring out the group $[4, 3^{n-2}, 4]$ by a subgroup Λ of its translations; we may think of Λ as a sublattice of $\mathbb{Z}^n$, for which reason we call it the *identification lattice*. Because we wish the resulting toroid to be regular, if we have the translation $\mathbf{s} \in \Lambda$, then we must also have all its conjugates under $[4, 3^{n-2}, 4]$, or, what amounts to the same thing, under the group $[3^{n-2}, 4]$ of its vertex-figure. This consists of all permutations of the coordinates of vectors with all changes of their signs.

Let s be the smallest positive integer among the coordinates of vectors in Λ. Permuting these coordinates if necessary, we can assume that $(s, s_2, \ldots, s_n) \in \Lambda$ for some $s_2, \ldots, s_n \in \mathbb{Z}$. Hence, $(-s, s_2, \ldots, s_n) \in \Lambda$, and so (by subtraction) $2se_1 \in \Lambda$, where as usual e_i is the ith standard basis vector of $\mathbb{E}^n$ for $i = 1, \ldots, n$. It then follows that $2se_i \in \Lambda$ for each $i = 1, \ldots, n$.

We can now add integer multiples of these $2se_i$ to a general vector in Λ, to produce one each of whose coordinates have absolute value at most s. We next permute these coordinates, and change their signs if necessary, to see that Λ is generated by all permutations of $(s^k, 0^{n-k})$ with all changes of sign, for some k with $1 \leqslant k \leqslant n$. Observe that k is unique, since we may subtract a vector with smaller k from one with larger k, so as further to reduce k.

However, there are still restrictions on k. If $3 \leqslant k \leqslant n - 1$, we may shift coordinates along to obtain $(0, s^k, 0^{n-k-1}) \in \Lambda$. After subtracting this from the original vector, and again permuting coordinates and changing their signs, we see that $(s, s, 0^{n-2}) \in \Lambda$. (If k is odd, it further follows that $(s, 0^{n-1}) \in \Lambda$.) In any case, we see that the only allowed values of k are $k = 1$, 2 or n. We shall write $\Lambda_{\mathbf{s}}$ for the translation group (or lattice) generated by $\mathbf{s} := (s^k, 0^{n-k})$ and its images under permutation and changes of sign of coordinates.

The lattices $\Lambda_{\mathbf{s}}$ are familiar objects. First observe that

$$\Lambda_{\mathbf{s}} = s\Lambda_{(1^k, 0^{n-k})},$$

when $\mathbf{s} = (s^k, 0^{n-k})$. Of course, $\Lambda_{(1,0^{n-1})}$ is just the cubic lattice $\mathbb{Z}^n$, the vertex set of $\{4, 3^{n-2}, 4\}$ itself. The lattice $\Lambda_{(1,1,0^{n-2})}$ is the root-lattice D_n (Section 3B); when $n = 3$, it is also known as the *face-centred cubic* lattice. It consists of all integral vectors whose coordinate sum is even, and has as basis $\{e_1 + e_2, e_2 - e_1, e_3 - e_2, \ldots, e_n - e_{n-1}\}$ (with determinant 2). When $n = 3$, the lattice $\Lambda_{(1,\ldots,1)}$ is known as the *body-centred cubic* lattice, and is a scaled copy of the lattice reciprocal to $\Lambda_{(1,1,0^{n-2})}$. Recall that the *reciprocal* (or *dual* or *polar*) Λ^* of a lattice Λ is defined by

$$\Lambda^* := \{y \in \mathbb{E}^n \mid \langle x, y\rangle \in \mathbb{Z} \text{ for all } x \in \Lambda\},$$

and thus consists of all vectors in $\mathbb{E}^n$ whose inner product with each vector in Λ is integral. Then

$$\Lambda_{(1,\ldots,1)} = 2\big(\Lambda_{(1,1,0^{n-2})}\big)^*.$$

A basis for $\Lambda_{(1,\ldots,1)}$ is $\{2e_1, \ldots, 2e_{n-1}, e_1 + \cdots + e_n\}$ (with determinant 2^{n-1}). The lattices $\Lambda_{(1,1,0^{n-2})}$ and $\Lambda_{(1,\ldots,1)}$ are sublattices of the (self-reciprocal) lattice $\Lambda_{(1,0^{n-1})}$ of index 2 and 2^{n-1}, respectively. For $n = 2$, there are only two lattices, and they are similar (equivalent under a similarity transformation of $\mathbb{E}^2$). Accordingly, any two quotients $\mathbb{E}^2/\Lambda_{\mathbf{s}}$ are similar 2-tori. If $n \geqslant 3$, no two lattices $\Lambda_{(1^k,0^{n-k})}$ are similar; hence no two quotients $\mathbb{E}^n/\Lambda_{\mathbf{s}}$ for different values of k are similar n-tori.

For more details about the lattices $\Lambda_{(1^k,0^{n-k})}$ (and their applications to sphere packings), see also Conway and Sloane [91, pp. 106–120].

The regular polytope which results from factoring by a lattice $\Lambda_{\mathbf{s}}$ is denoted by $\{4, 3^{n-2}, 4\}_{\mathbf{s}} := \{4, 3^{n-2}, 4\}/\Lambda_{\mathbf{s}}$, and the corresponding group is written $[4, 3^{n-2}, 4]_{\mathbf{s}} := [4, 3^{n-2}, 4]/\Lambda_{\mathbf{s}}$. The latter group is clearly a pre-C-group; in order that it satisfy the intersection property, we must actually have $s \geqslant 2$, but otherwise there are no further restrictions. (Since the geometry of these toroids is so simple, it is straightforward to verify the intersection property directly; otherwise, refer to the general discussion in Chapter 2.) Summarizing, we have

6D1 Theorem *For each $n \geqslant 3$, and $\mathbf{s} = (s^k, 0^{n-k})$ with $s \geqslant 2$ and $k = 1, 2$ or n, there is a regular toroid $\{4, 3^{n-2}, 4\}_{\mathbf{s}}$. Each such toroid is self-dual.*

We give the details of these polytopes in Table 6D1. The most important things we need subsequently are the numbers v of their vertices and f of their facets, and the orders g of their groups.

If we wish, we can express the group order for the case $\mathbf{s} = (s^k, 0^{n-k})$ as $g = 2^{n+k-1}s^n \cdot n!$, with similar common expressions for v and f, bearing in mind that the only allowable values for k are 1, 2 and n.

Table 6D1. *The Polytopes* $\{4, 3^{n-2}, 4\}_{\mathbf{s}}$

$\mathbf{s}$	v	f	g
$(s, 0^{n-1})$	s^n	s^n	$(2s)^n \cdot n!$
$(s, s, 0^{n-2})$	$2s^n$	$2s^n$	$2^{n+1}s^n \cdot n!$
(s^n)	$2^{n-1}s^n$	$2^{n-1}s^n$	$2^{2n-1}s^n \cdot n!$

We now describe the groups of these toroids. Let the group of $[4, 3^{n-2}, 4]$ be $\langle R_0, \ldots, R_n \rangle$, where the R_j are the reflexions with the following mirrors in $\mathbb{E}^n$. Throughout, $x = (\xi_1, \ldots, \xi_n)$ is a general vector of $\mathbb{E}^n$.

6D2
$$\begin{cases} R_0\colon \xi_1 = \frac{1}{2}, \\ R_j\colon \xi_j = \xi_{j+1} \text{ for } j = 1, \ldots, n-1, \\ R_n\colon \xi_n = 0. \end{cases}$$

Then, for each $k = 1, \ldots, n$,

$$x R_0 R_1 \cdots R_{k-1} = (\xi_2, \ldots, \xi_k, 1 - \xi_1, \xi_{k+1}, \ldots, \xi_n),$$
$$x R_k R_{k+1} \cdots R_{n-1} R_n R_{n-1} R_{n-2} \cdots R_k = (\xi_1, \ldots, \xi_{k-1}, -\xi_k, \xi_{k+1}, \ldots, \xi_n),$$

so that

6D3
$$x R_0 R_1 \cdots R_{n-1} R_n R_{n-1} \cdots R_k = (\xi_2, \ldots, \xi_k, \xi_1 - 1, \xi_{k+1}, \ldots, \xi_n).$$

Hence, $(R_0 R_1 \cdots R_{n-1} R_n R_{n-1} \cdots R_k)^k$ is the translation by $((-1)^k, 0^{n-k})$, and the definition of $\{4, 3^{n-2}, 4\}_{\mathbf{s}}$ (for the various $\mathbf{s}$) yields

6D4 Theorem *Let* $\mathbf{s} = (s^k, 0^{n-k})$*, with* $s \geqslant 2$ *and* $k = 1$*,* 2*, or* n*. Then the group* $[4, 3^{n-2}, 4]_{\mathbf{s}}$ *is the Coxeter group* $[4, 3^{n-2}, 4] = \langle \rho_0, \ldots, \rho_n \rangle$*, factored out by the single extra relation*

6D5
$$(\rho_0 \rho_1 \cdots \rho_{n-1} \rho_n \rho_{n-1} \rho_{n-2} \cdots \rho_k)^{sk} = \varepsilon.$$

In this context, the reader may like to prove the following result as an exercise. From the geometric point of view, it is fairly obvious, since σ_j (in the notation of the proposition) is a reflexion which interchanges the standard basis vectors e_1 and e_{j+1} of $\mathbb{E}^n$.

6D6 Proposition *Let* $[4, 3^{n-2}, 4] = \langle \rho_0, \ldots, \rho_n \rangle$*, and define* $\tau := \rho_0 \rho_1 \cdots \rho_n \rho_{n-1} \cdots \rho_1$ *and* $\sigma_j := \rho_1 \rho_2 \cdots \rho_j \rho_{j-1} \cdots \rho_1$ *for* $j = 1, \ldots, n-1$*. Then, for* $k = 1, \ldots, n$*,*

$$\tau \cdot \sigma_1 \tau \sigma_1 \cdot \sigma_2 \tau \sigma_2 \cdots \sigma_{k-1} \tau \sigma_{k-1} = (\rho_0 \rho_1 \cdots \rho_n \rho_{n-1} \cdots \rho_k)^k.$$

The cases $k = 1$ and 2 of Theorem 6D4 have interesting geometric interpretations, which are of importance for the future. Let us anticipate Section 7C, by introducing the concept of a cut of a regular n-polytope $\mathcal{P}$ (in that section, we shall discuss cuts in much more generality). Suppose that the centralizer Δ of the subgroup Δ' generated by some proper subfamily of the distinguished generators of the group $\Gamma(\mathcal{P})$ of $\mathcal{P}$ is itself a string C-group (with respect to a suitable choice of generators). A regular polytope $\mathcal{Q}$ (of lower rank) with $\Gamma(\mathcal{Q}) = \Delta$, whose vertices are invariant under Δ', is called a *cut* of $\mathcal{P}$. The nature of cuts will be illustrated here by some examples. For our purposes, the vertices of a cut $\mathcal{Q}$ will usually be vertices of $\mathcal{P}$ itself. With the notation of the theorem, let us define

6D7
$$\tau := \rho_2 \rho_3 \cdots \rho_{n-1} \rho_n \rho_{n-1} \rho_{n-2} \cdots \rho_2.$$

Then the defining relation of Theorem 6D4 can be written as

6D8
$$\begin{cases} (\rho_0\rho_1\tau\rho_1)^s = \varepsilon, & \text{if} \quad k = 1, \\ (\rho_0\rho_1\tau)^{2s} = \varepsilon, & \text{if} \quad k = 2. \end{cases}$$

The subgroup $\langle \rho_0, \rho_1, \tau \rangle$ of $[4, 3^{n-2}, 4]_{\mathbf{s}}$ is that of the cut of $\{4, 3^{n-2}, 4\}_{\mathbf{s}}$ invariant under $\rho_3, \ldots, \rho_n$, which is a toroid of type $\{4, 4\}$. More specifically, if $\tilde{\mathbf{s}} := (s^k, 0^{2-k})$ for each $\mathbf{s} = (s^k, 0^{n-k})$ with $k = 1$ or 2, then we have

6D9 Theorem *If $k = 1$ or 2, then the cut of $\{4, 3^{n-2}, 4\}_{\mathbf{s}}$ invariant under $\rho_3, \ldots, \rho_n$ is a toroid $\{4, 4\}_{\tilde{\mathbf{s}}}$.*

These cuts are universal, meaning that the relations on the corresponding cut of the universal polytope (in this case, $\{4, 4\}$) which determine it are just those which determine the whole polytope as a quotient of its universal polytope (here, $\{4, 3^{n-2}, 4\}$). We refer to Section 7C, for a general discussion of the universality of cuts.

The two "smallest" of these toroids will occur in a number of places later in the book. For that reason, we shall devote a little space here to describing them in more detail.

The Case $\mathbf{s} = (2, 0^{n-1})$

The polytope $\mathcal{P} := \{4, 3^{n-2}, 4\}_{(2,0^{n-1})}$ only has 2^n vertices, which is just the number of vertices of one of its facets. Therefore $\mathcal{P}$ must be flat, which we recall means that every one of its vertices belongs to every one of its facets.

The Case $\mathbf{s} = (2, 2, 0^{n-2})$

Now our polytope $\mathcal{P} := \{4, 3^{n-2}, 4\}_{(2,2,0^{n-2})}$ has 2^{n+1} vertices, and the same number of facets. In the description of the possible identification lattices Λ, we remarked that the vector $2se_1 \in \Lambda$; in our case, $s = 2$. We already have 2^{n+1} vertices, when we take those of the basic unit cube and of its translate by $2e_1$; these must therefore be all the vertices of $\mathcal{P}$. More generally, any cubical facet of $\mathcal{P}$ and its translates under e_i, $2e_i$ and $3e_i$ (factored out by $\Lambda_{\mathbf{s}}$, of course) form a *ring*. Moreover, since the basic translations which give $\Lambda_{\mathbf{s}}$ are by vectors of the form $\pm e_i \pm e_j$ with $1 \leqslant i < j \leqslant n$, we see that two rings which have one facet in common must actually share two, which are "opposite" facets in each ring.

In conclusion, we note that Theorem 6D4 is yet another illustration of the general fact, already mentioned in Section 6C, that, for a quotient tessellation $\mathcal{T}/N$ on a space-form E/N, the fundamental group $\pi_1(E/N)$ determines the extra relations for its group $\Gamma(\mathcal{T}/N)$ which have to be added to the standard relations given by $\Gamma(\mathcal{T})$. Here, $E = \mathbb{E}^n$, $\mathcal{T} = \{4, 3^{n-2}, 4\}$, $N = \Lambda$ and $\pi_1(E/N) \cong \mathbb{Z}^n$. The fundamental group in now generated by the loops which correspond to the generating translations of Λ, namely, $(R_0R_1 \cdots R_nR_{n-1} \cdots R_k)^k$ and its images under permutations and changes of sign of coordinates. The same phenomenon will occur again for the regular toroids derived from $\{3, 3, 4, 3\}$ or its dual.

6E The Other Toroids

There are just four other kinds of toroids, namely, those derived from the dual pairs of honeycombs $\{3, 3, 4, 3\}$ and $\{3, 4, 3, 3\}$ of $\mathbb{E}^4$. For most purposes, it is enough to consider the former. We may take the vertex-set of $\mathcal{T} = \{3, 3, 4, 3\}$ to be $\mathbb{Z}^4 \cup (\mathbb{Z}^4 + (\frac{1}{2}, \frac{1}{2}, \frac{1}{2}, \frac{1}{2}))$, the set of points of $\mathbb{E}^4$ whose cartesian coordinates are all integers or all halves of odd integers. (The alternative choice of coordinates, the points of $\mathbb{Z}^4$ the sum of whose coordinates is even, will occur in the following.) Much the same analysis as that in Section 6D applies, and, initially bearing in mind only the vertices of $\{3, 3, 4, 3\}$ in $\mathbb{Z}^4$, we conclude that the identification which yields the toroid is by the lattice $\Lambda_{\mathbf{s}}$ generated by a vector $(s^k, 0^{4-k})$, and its transforms under the group $[3, 4, 3]$ of the vertex-figure at o, for some integer $s \geqslant 2$ and some $k = 1$, 2 or 4; these transforms include the images under permutation and changes of sign of coordinates.

However, taking the full group of symmetries of $\{3, 3, 4, 3\}$ into account, we observe that (s, s, s, s) is equivalent to $(2s, 0, 0, 0)$, and so the last case has already been counted.

Using the same notation as that for the cubic toroids, and denoting the dual by the same suffix, we thus obtain

6E1 Theorem *For each* $\mathbf{s} = (s^k, 0^{4-k})$ *with* $s \geqslant 2$ *and* $k = 1$ *or 2, there are regular toroids* $\{3, 3, 4, 3\}_{\mathbf{s}}$ *and* $\{3, 4, 3, 3\}_{\mathbf{s}}$.

We list the details of these polytopes in Table 6E1. However, since the number of vertices of $\{3, 3, 4, 3\}_{\mathbf{s}}$ is the same as the number of facets of its dual $\{3, 4, 3, 3\}_{\mathbf{s}}$, and vice versa, we need only consider the former. Observe that a common expression for the group order in case $\mathbf{s} = (s^k, 0^{4-k})$ is $g = 1152k^2s^4$, with similar ways of writing down v and f. We have already noticed that the case $k = 4$ is "spurious"; however, it still gives the correct answer in these expressions.

Observe that $\Lambda_{\mathbf{s}} = s\Lambda_{(1^k, 0^{4-k})}$. Of course,

$$\Lambda_{(1,0,0,0)} = \mathbb{Z}^4 \cup \left(\mathbb{Z}^4 + \left(\tfrac{1}{2}, \tfrac{1}{2}, \tfrac{1}{2}, \tfrac{1}{2}\right)\right),$$

which is the vertex set of $\{3, 3, 4, 3\}$. This is (a self-reciprocal version of) the root lattice D_4 (Section 3B) which occurs here in a different form from that of the previous section (for $n = 4$); see Conway and Sloane [91, p. 118] for more details about D_4. A basis of $\Lambda_{(1,0,0,0)}$ is $\{e_1, e_2, e_3, \frac{1}{2}(e_1 + e_2 + e_3 + e_4)\}$, where $e_1, \ldots, e_4$ denotes, as before, the canonical basis of $\mathbb{E}^4$. (This shows again that $\Lambda_{(1,1,1,1)} = \Lambda_{(2,0,0,0)}$.) Further, $\Lambda_{(1,1,0,0)}$ is a sublattice of index 4 in $\Lambda_{(1,0,0,0)}$, and has $\{e_1 + e_2, e_2 - e_1, e_3 - e_2, e_4 - e_3\}$ as a basis. This is the copy of D_4 mentioned in the previous section. In particular, $\Lambda_{(1,0,0,0)} = \Lambda_{(1,1,0,0)}\alpha$, where α is the similarity transformation of $\mathbb{E}^4$ given

Table 6E1. *The Polytopes* $\{3, 3, 4, 3\}_{\mathbf{s}}$

$\mathbf{s}$	v	f	g
$(s, 0, 0, 0)$	s^4	$3s^4$	$1152s^4$
$(s, s, 0, 0)$	$4s^4$	$12s^4$	$4608s^4$

by the matrix

$$\frac{1}{2}\begin{bmatrix} 1 & 1 & 0 & 0 \\ 1 & -1 & 0 & 0 \\ 0 & 0 & 1 & 1 \\ 0 & 0 & 1 & -1 \end{bmatrix}.$$

It follows that, as space-forms, any two 4-tori $\mathbb{E}^4/\Lambda_{\mathbf{s}}$ are similar. Furthermore, for the parameter vector $\mathbf{s} = (s, s, 0, 0)$, the two lattices $\Lambda_{\mathbf{s}}$ for $\{4, 3, 3, 4\}$ and $\{3, 3, 4, 3\}$ coincide, and so their 4-tori are actually the same.

We now determine the groups $[3, 3, 4, 3]_{\mathbf{s}}$. We take $\{3, 3, 4, 3\}$ with the vertex set $\mathbb{Z}^4 \cup (\mathbb{Z}^4 + (\frac{1}{2}, \frac{1}{2}, \frac{1}{2}, \frac{1}{2}))$ as before, and write its group as $[3, 3, 4, 3] = \langle R_0, \ldots, R_4 \rangle$. Then, for $j = 0, \ldots, 4$, the reflexion R_j is in the following hyperplane:

6E2

$$\begin{aligned} R_0&: \xi_1 = \tfrac{1}{2}, \\ R_1&: \xi_1 = \xi_2 + \xi_3 + \xi_4, \\ R_2&: \xi_4 = 0, \\ R_3&: \xi_3 = \xi_4, \\ R_4&: \xi_2 = \xi_3. \end{aligned}$$

Because we wish to investigate certain subgroups in what follows, it is helpful to represent the defining relations of the group $[3, 3, 4, 3]_{\mathbf{s}} = \langle \rho_0, \ldots, \rho_4 \rangle$ in terms of three particular involutions. The first is ρ_0; the second and third are

6E3

$$\begin{aligned} \sigma &:= \rho_1\rho_2\rho_3\rho_2\rho_1, \\ \tau &:= \rho_4\rho_3\rho_2\rho_3\rho_4. \end{aligned}$$

Now R_0 and the reflexions $S := R_1R_2R_3R_2R_1$ and $T := R_4R_3R_2R_3R_4$ in $[3, 3, 4, 3]$ which correspond to σ and τ are given by

6E4

$$\begin{aligned} xR_0 &= (1 - \xi_1, \xi_2, \xi_3, \xi_4), \\ xS &= (\xi_2, \xi_1, \xi_3, \xi_4), \\ xT &= (\xi_1, -\xi_2, \xi_3, \xi_4). \end{aligned}$$

Hence we have

6E5

$$\begin{aligned} xR_0STS &= (\xi_1 - 1, \xi_2, \xi_3, \xi_4), \\ xR_0ST &= (\xi_2, \xi_1 - 1, \xi_3, \xi_4). \end{aligned}$$

Thus R_0STS is the translation by $(-1, 0, 0, 0)$, while $(R_0ST)^2$ is the translation by $(-1, -1, 0, 0)$. As a consequence, we have

6E6 Theorem *Let $\mathbf{s} = (s^k, 0^{4-k})$, with $s \geqslant 2$ and $k = 1$ or 2. Then the group $[3, 3, 4, 3]_{\mathbf{s}}$ of the polytope $\{3, 3, 4, 3\}_{\mathbf{s}}$ is the Coxeter group $[3, 3, 4, 3] = \langle \rho_0, \ldots,$*

$\rho_4\rangle$, *factored out by the relation*

6E7
$$(\rho_0\sigma\tau\sigma)^s = \varepsilon, \quad \textit{if} \quad k = 1,$$
$$(\rho_0\sigma\tau)^{2s} = \varepsilon, \quad \textit{if} \quad k = 2,$$

with σ, τ *given by (6E3).*

In parallel to Proposition 6D6, we make the following trivial observation: in the notation of Theorem 6E6,

$$(\rho_0\sigma\tau\sigma)\cdot\sigma(\rho_0\sigma\tau\sigma)\sigma = (\rho_0\sigma\tau)^2.$$

It is of geometric interest to describe the meaning of the relations (6E7); this will also have future relevance. Inspection of the reflexions R_0, S and T shows that they act effectively on the plane $\xi_3 = \xi_4 = 0$, which is that invariant under R_2 and R_3. The cut of $\{3, 3, 4, 3\}$ by this plane is a square tessellation $\{4, 4\}$, passing through diametral squares of facets $\{3, 3, 4\}$ of the honeycomb, whose symmetry group is just $\langle R_0, S, T\rangle$. Thus we have

6E8 Theorem *The quotient map which yields* $\{3, 3, 4, 3\}_{\mathbf{s}}$ *from* $\{3, 3, 4, 3\}$ *is induced by the quotient map of the cut* $\{4, 4\}$ *which yields* $\{4, 4\}_{\tilde{\mathbf{s}}}$.

The notation for $\tilde{\mathbf{s}}$ is that introduced before Theorem 6D9. Thus these cuts are universal.

Alternative defining relations are possible. For example, when $k = 1$, we can take

$$(\rho_0\rho_1(\rho_2\rho_3\rho_4)^3)^{2s} = \varepsilon,$$

while if $k = 2$, we can take

$$(\rho_0\rho_1\rho_2\rho_3\rho_4\rho_3\rho_2\rho_1\rho_2\rho_3\rho_2)^{2s} = \varepsilon.$$

There is a cut corresponding to the relation for $k = 1$ (it is of type $\{3, 6\}$), but we shall make no use of it.

6F Relationships Among Toroids

In this section, we shall discuss various quotient and subgroup relations among the groups of the toroids, and their geometric consequences. We begin with quotients, since these will subsequently prove more important.

A quotient relation between (for example) two groups of the form $[4, 3^{n-2}, 4]_{\mathbf{s}}$ arises from a corresponding subgroup relation, $\Lambda_1 \leqslant \Lambda_2$ (say), between two translation groups of the form $\Lambda_{\mathbf{s}}$. In turn, this subgroup relation implies a covering $\mathbb{E}^n/\Lambda_1 \searrow \mathbb{E}^n/\Lambda_2$ by the first torus of the second, in the manner described in Section 6A. This covering is determined by a group of isometries of $\mathbb{E}^n/\Lambda_1$ which is isomorphic to Λ_2/Λ_1, and acts freely and properly discontinuously.

Now we have

$$\Lambda_{(2s,0^{n-1})} \leqslant \left\{\begin{matrix}\Lambda_{(s^n)}\\ \Lambda_{(s^2,0^{n-2})}\end{matrix}\right\} \leqslant \Lambda_{(s,0^{n-1})},$$

for all $s \geqslant 2$. If n is even, there is also the relation

$$\Lambda_{(s^n)} \leqslant \Lambda_{(s^2,0^{n-2})}.$$

Moreover, if p is an odd prime, we obviously have $\Lambda_{p\mathbf{s}} \leqslant \Lambda_{\mathbf{s}}$ for every $\mathbf{s}$. It may be seen that every other subgroup relationship is a consequence of these. We deduce

6F1 Theorem *Let $n \geqslant 3$. For each $s \geqslant 2$, there are coverings*

$$\{4, 3^{n-2}, 4\}_{(2s,0^{n-1})} \searrow \left\{ \begin{array}{c} \{4, 3^{n-2}, 4\}_{(s^n)} \\ \{4, 3^{n-2}, 4\}_{(s^2,0^{n-2})} \end{array} \right\} \searrow \{4, 3^{n-2}, 4\}_{(s,0^{n-1})}.$$

In addition, if n is even, there is a covering

$$\{4, 3^{n-2}, 4\}_{(s^n)} \searrow \{4, 3^{n-2}, 4\}_{(s^2,0^{n-2})}.$$

Lastly, for each $\mathbf{s} = (s^k, 0^{n-k})$ (with $s \geqslant 2$ and $k = 1, 2$ or n) and every odd prime p, there is a covering

$$\{4, 3^{n-2}, 4\}_{p\mathbf{s}} \searrow \{4, 3^{n-2}, 4\}_{\mathbf{s}}.$$

Exactly similar considerations apply to the polytopes of type $\{3, 3, 4, 3\}$, and we obtain

6F2 Theorem *Let $s \geqslant 2$. Then there are coverings*

$$\{3, 3, 4, 3\}_{(2s,0,0,0)} \searrow \{3, 3, 4, 3\}_{(s,s,0,0)} \searrow \{3, 3, 4, 3\}_{(s,0,0,0)}.$$

Further, if p is an odd prime, there is a covering

$$\{3, 3, 4, 3\}_{p\mathbf{s}} \searrow \{3, 3, 4, 3\}_{\mathbf{s}},$$

with $\mathbf{s} = (s, 0, 0, 0)$ or $(s, s, 0, 0)$.

With subgroups, our interest is in being able to take the vertices of one polytope as a subset of those of another, in such a way that the group of the first polytope is a subgroup of that of the second; we briefly call this *inscribing* the first in the second, and represent the relationship by the subset sign $\subseteq$. The question is thus more complicated than that for quotients; we must have corresponding subgroup relationships for the euclidean honeycombs, and, additionally, compatibility between the identifying translation groups.

With $[4, 3^{n-2}, 4]$, the situation is still fairly straightforward. Since $n \geqslant 3$, the only way to inscribe one copy of $\{4, 3^{n-2}, 4\}$ in another (with, say, the origin o as a common vertex) is to take its vertices as $m\mathbb{Z}^n$ for some integer $m \geqslant 2$. A translation subgroup which will identify the first copy must be of the form $\mathbb{Z}_{m\mathbf{s}}$ for some $\mathbf{s}$ of the usual kind, so that we have

6F3 Theorem *Let $n \geqslant 3$. Then for each $\mathbf{s} = (s^k, 0^{n-k})$ with $s \geqslant 2$ and $k = 1, 2$ or n, and each integer $m \geqslant 2$,*

$$\{4, 3^{n-2}, 4\}_{\mathbf{s}} \subseteq \{4, 3^{n-2}, 4\}_{m\mathbf{s}}.$$

Particular care should be taken over the meaning of "inscribe". As an example of what the definition is designed to exclude, take the sublattice of $\mathbb{Z}^4$ generated by the four vectors $(1, 1, 1, 1)$, $(1, 1, -1, -1)$, $(1, -1, 1, -1)$ and $(1, -1, -1, 1)$. This sublattice is a congruent copy of $2\mathbb{Z}^4$; however, this does not lead to one copy of $[4, 3, 3, 4]$ being inscribed in another, since the group of the vertex-figure of the first is not a subgroup of that of the second.

However, we have a second copy of the larger lattice, obtained by changing 1 to -1 in the first coordinate of each of the four vectors. This results in $16 = 2 \cdot 8$ congruent copies of $2\mathbb{Z}^4$ inscribed in $\mathbb{Z}^4$, with the full symmetry of the latter. Replacing the original $\mathbb{Z}^4$ by the larger copy $s\mathbb{Z}^4$ (with $s \geqslant 2$) then leads to a compound of 16 copies of $\{4, 3, 3, 4\}_{(s,s,0,0)}$ inscribed in $\{4, 3, 3, 4\}_{(s,0,0,0)}$. Other such compounds may be constructed by varying the subsequent examples, but we shall make no further mention of them.

For $[3, 3, 4, 3]$, an additional possibility arises. Another choice of vertices for $\{3, 3, 4, 3\}$ is $h\mathbb{Z}^4$, the set of vectors in $\mathbb{Z}^4$ whose coordinates have even sum. We can clearly inscribe $h\mathbb{Z}^4$ in $\mathbb{Z}^4 \cup (\mathbb{Z}^4 + (\frac{1}{2}, \frac{1}{2}, \frac{1}{2}, \frac{1}{2}))$, and each symmetry of the smaller set is also one of the larger. If we repeat this, we find $2\mathbb{Z}^4 \cup (2\mathbb{Z}^4 + (1, 1, 1, 1))$ inscribed in $h\mathbb{Z}^4$. We then obtain

6F4 Theorem *Let $s \geqslant 2$. Then*

$$\{3, 3, 4, 3\}_{(s,0,0,0)} \subseteq \{3, 3, 4, 3\}_{(s,s,0,0)} \subseteq \{3, 3, 4, 3\}_{(2s,0,0,0)}.$$

Moreover, for each $\mathbf{s} = (s, 0, 0, 0)$ or $(s, s, 0, 0)$ and each integer $m \geqslant 2$,

$$\{3, 3, 4, 3\}_{\mathbf{s}} \subseteq \{3, 3, 4, 3\}_{m\mathbf{s}}.$$

Finally, we observe that we can inscribe both $\{4, 3, 3, 4\}$ and $\{3, 4, 3, 3\}$ in $\{3, 3, 4, 3\}$. For the former, $\mathbb{Z}^4 \subseteq \mathbb{Z}^4 \cup (\mathbb{Z}^4 + (\frac{1}{2}, \frac{1}{2}, \frac{1}{2}, \frac{1}{2}))$ is obvious, and a little thought establishes the subgroup property. However, the translation by $(s, 0, 0, 0)$ is conjugate to that by $\frac{1}{2}(s, s, s, s)$ in $[3, 3, 4, 3]$, and this is only permissible for $\{4, 3, 3, 4\}$ if s is even. For the latter, $(\mathbb{Z}^4 \cup (\mathbb{Z}^4 + (\frac{1}{2}, \frac{1}{2}, \frac{1}{2}, \frac{1}{2}))) \setminus h\mathbb{Z}^4$ is the set of vertices of $\{3, 4, 3, 3\}$, and a translation permissible for this must also be permissible for the $\{3, 3, 4, 3\}$ whose vertices form the complementary set $h\mathbb{Z}^4$. We conclude that we have

6F5 Theorem *For each $s \geqslant 2$, there are the following inscribed polytopes:*

$$\left.\begin{matrix}\{4, 3, 3, 4\}_{(s,s,0,0)} \\ \{3, 4, 3, 3\}_{(s,0,0,0)}\end{matrix}\right\} \subseteq \{3, 3, 4, 3\}_{(s,s,0,0)},$$

$$\left.\begin{matrix}\{4, 3, 3, 4\}_{(s,s,s,s)} \\ \{3, 4, 3, 3\}_{(s,s,0,0)}\end{matrix}\right\} \subseteq \{3, 3, 4, 3\}_{(2s,0,0,0)}.$$

It is worth noting that the mixing operations on the group $[3, 3, 4, 3]$ which yield these subgroups are

$$(\rho_0, \ldots, \rho_4) \mapsto (\rho_0, \rho_1, \rho_2\rho_3\rho_2, \rho_4, \rho_3) =: (\sigma_0, \ldots, \sigma_4)$$

for [3, 4, 3, 3], and

$$(\rho_0, \dots, \rho_4) \mapsto (\rho_0, \rho_1\rho_2\rho_3\rho_2\rho_1, \rho_4, \rho_3, \rho_2) =: (\tau_0, \dots, \tau_4)$$

for [4, 3, 3, 4].

6G Other Euclidean Space-Forms

In this section we complete the enumeration of the regular tessellations on compact euclidean space-forms. We follow [212, §3], and prove that the n-tori are the only topological types of n-dimensional compact euclidean space-forms which admit regular tessellations (namely, the regular toroids of rank $n+1$). For $n = 2$, this proves again the well-known fact that there are no regular maps on the Klein bottle (see [131, p. 116]).

To begin with, recall from Section 2E the notion of a sparse subgroup. The key is the following

6G1 Theorem *Let $\mathcal{T}$ be a regular tessellation in euclidean n-space $\mathbb{E}^n$, and let Λ be the translation subgroup of its symmetry group $G(\mathcal{T})$. Then a sparse subgroup N of $\Gamma(\mathcal{T})$ whose normalizer contains $\Gamma^+(\mathcal{T})$ is a subgroup of Λ.*

Proof. First observe that, if necessary, we may replace the given regular tessellation $\mathcal{T}$ in $\mathbb{E}^n$ by its dual. In fact, dual tessellations have the same translation subgroups and rotation subgroups, and the concept of sparseness is also invariant under duality. In particular, of a pair of dual tessellations, we choose $\mathcal{T}$ to be the one with a vertex-transitive translation subgroup Λ; we may then identify the vertex-set $\mathcal{T}_0$ with the translation vectors in Λ whenever it is convenient, and so write $\mathcal{T}_0 = \Lambda$. More precisely, $\mathcal{T} = \{4, 3^{n-2}, 4\}$ (for $n \geqslant 2$), $\{\infty\}$, $\{3, 6\}$ or $\{3, 3, 4, 3\}$, as appropriate.

Let $G(\mathcal{T}) = \langle \rho_0, \dots, \rho_n \rangle$, so that $\Gamma_0 := \langle \rho_1, \dots, \rho_n \rangle$ is the stabilizer of the base vertex o of $\mathcal{T}$. The corresponding rotation subgroups are thus $\Gamma^+(\mathcal{T}) = \langle \rho_0\rho_1, \rho_1\rho_2, \dots, \rho_{n-1}\rho_n \rangle$ and $\Gamma_0^+ = \langle \rho_1\rho_2, \dots, \rho_{n-1}\rho_n \rangle$. For $t \in \mathbb{E}^n$, let $\tau(t)$ denote the translation of $\mathbb{E}^n$ defined by $x\tau(t) := x + t$ for $x \in \mathbb{E}^n$.

If $\varphi \in G(\mathcal{T})$, then for $x \in \mathbb{E}^n$ we have $x\varphi = x\omega + t$, with $t \in \mathbb{E}^n$ and ω a linear mapping; in fact, $t = o\varphi$, which is thus a vertex of $\mathcal{T}$, and $\omega = \varphi\tau(-t)$. Since Λ is vertex-transitive, we also know that $\tau(t) \in \Lambda$, and therefore $\omega \in G(\mathcal{T})$. Then $\omega \in \Gamma_0$, because $o\omega = o$. This proves that $G(\mathcal{T})$ is the semi-direct product of Λ by Γ_0.

Recall from Lemma 2E22 that a subgroup N of $G(\mathcal{T})$ $(= \Gamma(\mathcal{T}))$ is sparse if and only if each orbit of N meets each proper section of the polytope $\mathcal{T}$ in at most one face. Suppose that N is a sparse normal subgroup. Let $\varphi \in N$, and let $x\varphi = x\omega + t$ (for $x \in \mathbb{E}^n$), as before. Then φ^{-1} is given by $x\varphi^{-1} = (x - t)\omega^{-1}$ for $x \in \mathbb{E}^n$. We claim that $\omega = \varepsilon$, the identity map on $\mathbb{E}^n$, and therefore that $\varphi = \tau(t) \in \Lambda$.

Assume, if possible, that $\omega \neq \varepsilon$. We define

$$\mathcal{D} := \{F \in \mathcal{T}_n \mid o \in F\},$$

the set of facets of $\mathcal{T}$ which contain the initial vertex o. Our strategy is to prove that, if $\Lambda' := \Lambda \cap N$ denotes the subgroup consisting of the translations in N, then $\mathcal{D} + \Lambda'$

covers the vertex-set $\mathcal{T}_0 = \Lambda$. We then show that $\mathcal{D}$ contains an image of o under an element $\varphi \in N \setminus \{\varepsilon\}$, which will contradict the assumption that N is sparse.

So, let v be any neighbouring vertex of o in $\mathcal{T}$. By our assumption on Λ, we know that $\tau(v) \in \Lambda \leqslant \Gamma^+(\mathcal{T})$. Set $\psi(v) := \tau(-v)\varphi^{-1}\tau(v)\varphi$. Writing this as $\psi(v) = \tau(v)^{-1}\varphi^{-1}\tau(v) \cdot \varphi$, which is a product of elements of N because N is normalized by $\Gamma^+(\mathcal{T})$, we see that $\psi(v) \in N$. For $x \in \mathbb{E}^n$, we have

$$\begin{aligned} x\psi(v) &= x\tau(-v)\varphi^{-1}\tau(v)\varphi = (x - v - t)\omega^{-1}\tau(v)\varphi \\ &= ((x - v - t)\omega^{-1} + v)\omega + t \\ &= x + (v\omega - v). \end{aligned}$$

Hence $\psi(v) = \tau(v\omega - v)$. Since $\omega \neq \varepsilon$, there is a neighbour v of o such that $v\omega \neq v$. We define $w := v\omega - v$; then $o\psi(v) = w$, and hence $w \in \mathcal{T}_0$.

Since $v\omega$ is also a neighbour of o and $\mathcal{T}$ is centrally symmetric, w is a neighbour of v. It follows that N contains a translation, namely, $\tau(w)$, which maps o onto some *second neighbour*, by which we mean a neighbour (different from o) of one of its neighbours.

Now N is normalized by $\Gamma^+(\mathcal{T})$, and so it is invariant under conjugation by the elements in the subgroup Γ_0^+. It follows that $\tau(w\sigma) = \sigma^{-1}\tau(w)\sigma \in N$ for all $\sigma \in \Gamma_0^+$, so that, by definition, $\tau(w\sigma) \in \Lambda'$ ($= \Lambda \cap N$, as defined earlier). We perform a similar analysis to that used in Sections 6D and 6E (see also [302]) in the construction of the regular toroids, working through the individual cases. As before, we define $\Lambda_a \leqslant \Lambda$ to be the subgroup generated by a vector $a \in \Lambda$ and its conjugates under $\Gamma(\mathcal{T})$ (or under Γ_0).

First let $\mathcal{T} = \{4, 3^{n-2}, 4\}$, with vertex-set $\mathcal{T}_0 = \Lambda = \mathbb{Z}^n$. The second neighbours of o are $\pm e_i \pm e_j$ (with $1 \leqslant i < j \leqslant n$) and $\pm 2e_i$ (with $i = 1, \ldots, n$), where $e_1, \ldots, e_n$ are the standard unit vectors in $\mathbb{Z}^n$. Since the generating vector w of $\Lambda_w \leqslant \Lambda'$ is such a second neighbour, then we can argue as before that $\Lambda' \geqslant \Lambda_{(1,1,0^{n-2})}$ or $\Lambda_{(2,0^{n-1})}$ (note that the case $n = 2$ is not exceptional, even though rotations only may be employed). In either case, $\mathcal{D} + \Lambda'$ covers $\mathcal{T}_0$.

Next, let $\mathcal{T} = \{3, 3, 4, 3\}$, with vertex set $\Lambda = \mathbb{Z}^4 \cup ((\frac{1}{2}, \frac{1}{2}, \frac{1}{2}, \frac{1}{2}) + \mathbb{Z}^4)$. The second neighbours of o now consist of the neighbours themselves, comprising the vectors $\pm e_i$ (with $i = 1, \ldots, 4$) and $(\pm\frac{1}{2}, \pm\frac{1}{2}, \pm\frac{1}{2}, \pm\frac{1}{2})$, the vectors obtained from $(\pm 1, \pm 1, 0, 0)$, $(\pm 1, \pm 1, \pm 1, 0)$ or $(\pm\frac{3}{2}, \pm\frac{1}{2}, \pm\frac{1}{2}, \pm\frac{1}{2})$ by permutations of the coordinates, and $2v$ for each neighbour v. Again arguing as before, we see that $\Lambda' \geqslant \Lambda_{(1,0,0,0)}$, $\Lambda_{(1,1,0,0)}$ or $\Lambda_{(2,0,0,0)}$, as the generating vector w is a neighbour, one of the vectors $(\pm 1, \pm 1, 0, 0)$, $(\pm 1, \pm 1, \pm 1, 0)$ or $(\pm\frac{3}{2}, \pm\frac{1}{2}, \pm\frac{1}{2}, \pm\frac{1}{2})$, or twice a neighbour. (For the second case, note the typical calculations $(0, 0, 1, 1) = (1, 1, 1, 0) - (1, 1, 0, -1)$ and $(0, 0, 1, 1) = (\frac{3}{2}, \frac{1}{2}, \frac{1}{2}, \frac{1}{2}) - (\frac{3}{2}, \frac{1}{2}, -\frac{1}{2}, -\frac{1}{2})$, which only use rotations. If we can employ the full symmetry, we actually obtain $\Lambda_a = \Lambda_{(1,0,0,0)}$ when $a = (\pm 1, \pm 1, \pm 1, 0)$ or $(\pm\frac{3}{2}, \pm\frac{1}{2}, \pm\frac{1}{2}, \pm\frac{1}{2})$.) As before, in each case $\mathcal{D} + \Lambda'$ covers $\mathcal{T}_0$.

For $\mathcal{T} = \{\infty\}$ with vertex set $\mathbb{Z}$, the second neighbours are $\pm 2e_1$, and trivially $\mathcal{D} + \Lambda'$ covers $\mathcal{T}_0$.

If $\mathcal{T} = \{3, 6\}$, then Λ is generated by two unit vectors v_1 and v_2 which are inclined at an angle $\pi/3$. The second neighbours then comprise the neighbours, the six images under rotation through multiples of $\pi/3$ of $v_1 + v_2$, and twice the neighbours. Then we have $\Lambda' \geqslant \Lambda_{(1,0)}, \Lambda_{(1,1)}$ or $\Lambda_{(2,0)}$, and once again $\mathcal{D} + \Lambda'$ covers $\mathcal{T}_0$. (Here, a suffix $a = (a_1, a_2)$ is shorthand for $a = a_1 v_1 + a_2 v_2$.)

We are now able to finish the proof. Consider the translation vector $t \in \Lambda$ of the element $\varphi \in N$ with which we began. Since $\mathcal{D} + \Lambda'$ covers $\mathcal{T}_0 = \Lambda$, there exists a translation $\tau(u) \in \Lambda' \leqslant N$ such that $t_0 := t\tau(u)$ $(= t + u)$ is a vertex in $\mathcal{D}$. Define $\varphi_0 := \varphi\tau(u) \in N$, so that $x\varphi_0 = x\omega + t_0$ for $x \in \mathbb{E}^n$. Now o and t_0 $(= o\varphi_0)$ are vertices of some common facet $F \in \mathcal{D}$. Since they belong to the same orbit of N, we must necessarily have $t_0 = o$; here we have used Lemma 2E22. But then $\omega = \varphi_0 \in N \setminus \{\varepsilon\}$. This is a contradiction, because $N \cap \Gamma_0 = N \cap \langle \rho_1, \ldots, \rho_n \rangle = \{\varepsilon\}$ when N is sparse. It therefore follows that $\omega = \varepsilon$, and hence N consists of translations alone, as was claimed. □

Using the results of Section 6B on regular tessellations on space-forms, we can now prove a straightforward consequence of Theorem 6G1.

6G2 Theorem *For $n \geqslant 1$, the n-tori are the only n-dimensional compact euclidean space-forms which admit regular tessellations.*

Proof. This is trivial for $n = 1$, because $\mathbb{S}^1$ is the only 1-dimensional compact euclidean space-form and each finite regular polygon $\{p\}$ determines a tessellation on it.

Now let $n \geqslant 2$. Let $\mathcal{P}$ be a regular tessellation on an n-dimensional compact euclidean space-form, and let $\mathcal{P}$ be of type $\{p_1, \ldots, p_n\}$. By Theorem 6B5, $\mathcal{P}$ must have more than two tiles, so that $p_n > 2$. Then $\mathcal{P}$ is locally spherical. Let $\mathcal{T} := \{p_1, \ldots, p_n\}$. By Theorem 6B3 and Lemma 6B7, $\mathcal{P} = \mathcal{T}/N$ and $\mathcal{T}$ is a euclidean tessellation. In particular, we may view $\mathcal{P}$ as a tessellation on the euclidean space-form $\mathbb{E}^n/N$. By Lemma 6B1, N must be a sparse normal subgroup of $G(\mathcal{T})$ $(= \Gamma(\mathcal{T}))$. Now Theorem 6G1 applies and shows that N can only consist of translational symmetries of $\mathcal{T}$. It follows that $\mathbb{E}^n/N$ is an n-torus. This completes the proof, because the original space-form is homeomorphic to $\mathbb{E}^n/N$. □

6H Chiral Toroids

In this section, we briefly digress from the general theme of the preceding discussion to consider chirality of polytopes. We recall from Section 2B, that a polytope $\mathcal{P}$ is called *chiral* if there are precisely two orbits of flags under its automorphism group $\Gamma(\mathcal{P})$, with adjacent flags lying in different orbits (see [372, §3]). Intuitively, this means that $\mathcal{P}$ has complete rotational symmetry, but not the full symmetry containing opposite automorphisms.

Here we shall show that there do not exist chiral toroids of rank greater than 3. As a consequence, there can be no chiral polytopes of rank greater than 4, whose facets or vertex-figures are such chiral toroids. This is in contrast to the existence of chiral locally toroidal 4-polytopes, for which see [83, 333–335, 373].

6H1 Theorem *There are no chiral toroids of rank greater than 3.*

Proof. We first treat the case of toroids of type $\{4, 3^{n-2}, 4\}$, with $n \geqslant 3$. Such a toroid is of the form $\mathcal{P} = \{4, 3^{n-2}, 4\}/\Lambda$, where $\Lambda \leqslant \mathbb{Z}^n$ is a lattice. Since $\mathcal{P}$ is chiral, Λ is invariant under the rotations of the original tessellation and, in particular, under the rotation group $[3^{n-2}, 4]^+$ of its vertex-figure. This consists of all even (odd) permutations of the coordinates of vectors in $\mathbb{E}^n$, together with an even (odd) number of changes of sign. So, in order to demonstrate the required non-existence, it suffices to show that the image of an arbitrary vector in Λ under some odd permutation of its coordinates is also in Λ.

Let $\mathbf{a} := (a_1, \ldots, a_n) \in \Lambda$. From $\mathbf{a}$, we deduce the existence of the following vectors in Λ; bear in mind that $n \geqslant 3$, and that any sign change in coordinates is permitted for a vector with at least one coordinate 0. We have

$$\begin{aligned}(a_1, \ldots, a_n) \in \Lambda &\Rightarrow (-a_2, a_1, a_3, \ldots, a_n) \in \Lambda \\ &\Rightarrow (a_1 + a_2, a_2 - a_1, 0, \ldots, 0) \in \Lambda \\ &\Rightarrow (a_1 + a_2, 0, a_1 - a_2, 0, \ldots, 0) \in \Lambda \\ &\Rightarrow (0, a_2 - a_1, a_2 - a_1, \ldots, 0) \in \Lambda \\ &\Rightarrow (a_2 - a_1, a_1 - a_2, 0, \ldots, 0) \in \Lambda \\ &\Rightarrow (a_2, a_1, a_3, \ldots, a_n) \in \Lambda,\end{aligned}$$

which was what we wished to show. (In the last step, we also used the original assumption that $\mathbf{a} \in \Lambda$.)

The remaining cases concern the toroids of type $\{3, 3, 4, 3\}$ and their duals; we need only treat the former. But the argument used previously shows that, if a given vector lies in the factoring lattice Λ, then so does its images under all permutations and changes of sign of coordinates. Hence Λ is invariant under the whole symmetry group $[3, 4, 3]$ of the vertex-figure of $\{3, 3, 4, 3\}$, so that the resulting toroid is regular. This completes the proof. □

We remark that Theorem 6H1 is the most important special case of a general non-existence result for chiral tessellations on compact euclidean space-forms of dimension at least 3. This is a generalization of Theorem 6G2. Thus, if $n \geqslant 3$, then there are no chiral tessellations on n-dimensional compact euclidean space-forms (see [212, §4]). On the other hand, in contrast to this, it is well known that the 2-torus admits infinite series of chiral maps of types $\{3, 6\}$, $\{6, 3\}$ and $\{4, 4\}$ (see [131, §8.3, 8.4]).

6J Hyperbolic Space-Forms

In the previous sections, we enumerated all regular tessellations on spherical or compact euclidean space-forms. We now discuss the regular tessellations on compact hyperbolic space-forms. We shall not attempt to give a comprehensive account here, but instead try to explain why so little is known about them. By Lemma 6B7, each such tessellation $\mathcal{P}$ is obtained as a quotient of a (locally finite) regular tessellation $\mathcal{T} = \{p_1, \ldots, p_n\}$ in hyperbolic n-space $\mathbb{H}^n$. Table 6J1 lists the Schläfli symbols $\{p_1, \ldots, p_n\}$ which can actually occur. In particular, $n = 2$, 3 or 4, because there are no locally finite regular tessellations in $\mathbb{H}^n$ with $n > 4$ [113, §5]. As further references about regular hyperbolic

Table 6J1. *(Locally Finite) Regular Tessellations in* $\mathbb{H}^n$

n	Symbols
2	$\{p, q\}$ with $\frac{1}{p} + \frac{1}{q} < \frac{1}{2}$
3	$\{3, 5, 3\}, \{4, 3, 5\}, \{5, 3, 4\}, \{5, 3, 5\}$
4	$\{3, 3, 3, 5\}, \{5, 3, 3, 3\}, \{4, 3, 3, 5\}, \{5, 3, 3, 4\}, \{5, 3, 3, 5\}$

tessellations, honeycombs and compounds we mention [115, 136, 178, 179, 181, 182, 235].

As before, we write $\mathcal{P} = \mathcal{T}/N$, where N is a normal subgroup of the symmetry group $G(\mathcal{T})$ $(= \Gamma(\mathcal{T}))$ of $\mathcal{T}$. By Lemma 6B1, N is sparse. Then $\mathcal{P}$ can be viewed as a regular tessellation on the corresponding n-dimensional hyperbolic space-form $\mathbb{H}^n/N$.

There is a host of interesting cases of regular tessellations on hyperbolic space-forms. An example is often constructed by specifying a presentation for its automorphism group, or, equivalently, by specifying elements which generate the corresponding group N as a normal subgroup of $G(\mathcal{T})$.

For instance, if $n = 2$ and $\mathcal{T} = \{p, q\}$, then we can find the regular map $\mathcal{P} = \{p, q\}_r$ by adding to the standard relations

$$\rho_0^2 = \rho_1^2 = \rho_2^2 = (\rho_0\rho_1)^p = (\rho_1\rho_2)^q = (\rho_0\rho_2)^2 = \varepsilon$$

for $\Gamma(\mathcal{T}) = [p, q] = \langle \rho_0, \rho_1, \rho_2 \rangle$, the single extra relation

$$(\rho_0\rho_1\rho_2)^r = \varepsilon;$$

see Section 7B or Coxeter and Moser [131, §8.6]. Then, as a normal subgroup, N is generated by $(\rho_0\rho_1\rho_2)^r$; that is, N is generated by $(\rho_0\rho_1\rho_2)^r$ and its conjugates in $G(\mathcal{T})$. The maps $\{p, q\}_r$ are finite for only a few values of p, q, r.

Similarly, there are also 3-dimensional polytopes $\{p, q, r\}_s$ and 4-dimensional polytopes $\{p, q, r, s\}_t$, which can be derived from the groups $[p, q, r]$ or $[p, q, r, s]$ by factoring out the extra relations $(\rho_0\rho_1\rho_2\rho_3)^s = \varepsilon$ or $(\rho_0\rho_1\rho_2\rho_3\rho_4)^t = \varepsilon$, respectively (see [83, 119, §12; 135, 429, 441, 442, 446]).

For $n = 2$, the polytopes $\mathcal{P} = \mathcal{T}/N$ are regular maps of type $\{p, q\}$ on surfaces $S := \mathbb{H}^2/N$ of negative Euler characteristic χ. Then $\mathcal{P}$ is finite if and only if S is compact. In this case, if f_0, f_1, f_2 are the numbers of vertices, edges and facets of $\mathcal{P}$, then $qf_0 = 2f_1 = pf_2$; hence

6J1
$$\chi = f_0 - f_1 + f_2 = \frac{|\Gamma(\mathcal{P})|}{2}\left(\frac{1}{p} + \frac{1}{q} - \frac{1}{2}\right).$$

Since $\{p, q\}$ is hyperbolic, this implies that

6J2
$$|\Gamma(\mathcal{P})| \leqslant 84|\chi|,$$

with equality occurring if and only if $\mathcal{P}$ is of type $\{3, 7\}$ or $\{7, 3\}$. In particular, this shows that there can only be finitely many regular maps on a given closed surface of non-zero Euler characteristic.

The study of maps on surfaces has a long history (see [43–45, 157, 158, 183, 214, 248, p. 22; 250, p. 461; 251, p. 260; 252, p. 109]). The regular maps on orientable surfaces are only classified for genus $g \leqslant 6$ (see [131, Chapter 8; 164, 174, 391, 410, p. 44]). On the other hand, each regular map $\mathcal{P}$ on a non-orientable surface of any Euler characteristic is doubly covered by a regular map $\mathcal{Q}$ of the same type $\{p, q\}$ on an orientable surface, and the corresponding covering $\mathcal{Q} \searrow \mathcal{P}$ is unique (see [459]). For some non-orientable surfaces, there are also enumeration results available (see [88, 189–192]). If S is orientable and of genus g, then (6J2) implies that

6J3 $$|\Gamma^{+}(\mathcal{P})| \leqslant 84(g-1),$$

with equality occurring as before, where $\Gamma^{+}(\mathcal{P})$ is the rotation subgroup of $\Gamma(\mathcal{P})$. Recall that $\Gamma^{+}(\mathcal{P})$ is the group of orientation preserving automorphisms of $\mathcal{P}$; it has index 2 in $\Gamma(\mathcal{P})$ and is generated by $\rho_0\rho_1$ and $\rho_1\rho_2$. The inequality (6J3) corresponds to Hurwitz's classical theorem, which says that the order of any group of orientation-preserving conformal automorphisms of a Riemann surface of genus g is bounded by $84(g-1)$ (see [223, 240, §5.11; 270, p. 103]). The groups $\Gamma^{+}(\mathcal{P})$ for which equality holds in (6J3) are also called *Hurwitz groups* (see [84–86]).

There is a wealth of further results available which relate to regular or chiral maps. Some deal with specific classes of maps (see [1, 87, 89, 109, 110, 142, 171, 175, 392, 395, 408, 409, 437], for example), others with general properties and construction principles (see [232, 233, 239, 241, 257, 268, 338, 458, 461, 463], for example). Some examples of small chiral maps are described in [176, 460]. The colouring of maps is discussed in [347]. Significant work has also been done on establishing connexions between the combinatorial theory of maps on compact orientable surfaces and the complex analytic theory of the Riemann surfaces underlying the maps (see [52, 239, 240, Chapters 4, 5; 250, 394] for example). For a general discussion of the interplay between maps, graphs and groups, we also refer to [26, 27, 93, 128, 177, 196, 435, 436].

The next case is $n = 3$. But here we cannot expect to gain much insight from a similar approach to that for maps, because the Euler characteristic of each compact 3-manifold is zero. (Interesting examples of regular tessellations of type $\{4, 3, 5\}$ on 3-dimensional hyperbolic space-forms can be obtained from the polytopes $2^{\mathcal{K}}$ and $2^{\mathcal{K},\mathcal{G}(s)}$, with $\mathcal{K} = \{3, 5\}$; see Section 8D.)

However, the method works again for $n = 4$. Then there are only five possible types $\{p, q, r, s\}$, namely, $\{5, 3, 3, s\}$ with $s = 3, 4$ or 5, and their duals. If $\mathcal{P}$ is finite and has f_i i-faces for $i = 0, \ldots, 4$, and if the corresponding 4-dimensional space-form $\mathbb{H}^4/N$ has Euler characteristic χ, then

6J4 $$\begin{aligned} \chi &= f_0 - f_1 + f_2 - f_3 + f_4 \\ &= |\Gamma(\mathcal{P})| \Big\{ \frac{1}{g_{q,r,s}} - \frac{1}{2g_{r,s}} + \frac{1}{4ps} - \frac{1}{2g_{p,q}} + \frac{1}{g_{p,q,r}} \Big\}, \end{aligned}$$

where we follow Coxeter [120, p. 130] in writing $g_{p,q,\ldots,s} := |[p, q, \ldots, s]|$. The denominator of the term in the parentheses which corresponds to f_i is just the order of the stabilizer of an i-face of $\mathcal{P}$. For the pair of duals $\{5, 3, 3, s\}$ and $\{s, 3, 3, 5\}$, this gives

6J5 $$\chi = c_s \mid \Gamma(\mathcal{P})|,$$

with

6J6 $$c_s = \begin{cases} \dfrac{1}{14400}, & \text{if } s = 3; \\ \dfrac{17}{28800}, & \text{if } s = 4; \\ \dfrac{13}{7200}, & \text{if } s = 5. \end{cases}$$

Now $\chi > 0$ (as it should be, by (6J8)). This proves again that there can only be finitely many regular tessellations $\mathcal{P}$ on a compact 4-dimensional space-form of a given Euler characteristic χ. (We obtain examples of regular tessellations of type $\{4, 3, 3, 5\}$ on 4-dimensional hyperbolic space-forms from the polytopes $2^{\mathcal{K}}$ and $2^{\mathcal{K},\mathcal{G}(s)}$, with $\mathcal{K} = \{3, 3, 5\}$; again, see Section 8D.)

We proceed with the following theorem about regular tessellations on euclidean or hyperbolic space-forms. For euclidean space-forms it is already subsumed in our enumeration of the regular toroids in the earlier sections. In the hyperbolic case, it indicates that in general we cannot expect to obtain any classification results very easily. Our proof is non-constructive.

6J7 Theorem *Let $n \geqslant 2$, and let $\{p_1, \ldots, p_n\}$ be the type of an euclidean or hyperbolic regular tessellation. Then there are infinitely many mutually non-isometric compact euclidean or hyperbolic space-forms, respectively, which admit a regular tessellation of type $\{p_1, \ldots, p_n\}$.*

Proof. We already know from Corollary 4C5 that there are infinitely many regular polytopes $\mathcal{P}$ which are finite and have facets $\{p_1, \ldots, p_{n-1}\}$ and vertex-figures $\{p_2, \ldots, p_n\}$. These polytopes are locally spherical polytopes of type $\{p_1, \ldots, p_n\}$. In particular, by Theorem 6B3, they can be viewed as regular tessellations on n-dimensional euclidean or hyperbolic space-forms, respectively.

We can now argue as follows. Since there are infinitely many such polytopes $\mathcal{P}$, there is also an infinite sequence of integers which occur as the number of facets of such polytopes. But then Lemma 6B4 applies, and consequently we must have infinitely many mutually non-isometric space-forms which admit regular tessellations of the required type. □

For hyperbolic space-forms of dimension 2 or 4, we can also use equations (6J1) and (6J5) to improve on Theorem 6J7, and obtain the stronger result that there are infinitely many mutually non-homeomorphic (rather than non-isometric) space-forms which admit regular tessellations. In fact, in the proof of Theorem 6J7, the infinite sequence of polytopes $\mathcal{P}$ gives an infinite sequence of corresponding Euler characteristics, and this, in turn, gives an infinite sequence of mutually non-homeomorphic space-forms. We conjecture that the stronger version is also true for 3-dimensional hyperbolic space-forms. The following considerations (which apply to all $n \geqslant 3$) indicate the difficulties. Note that the stronger version cannot hold in the euclidean case, because the tori are the only compact euclidean space-forms which admit regular tessellations.

Assume that two polytopes $\mathcal{P}_1$ and $\mathcal{P}_2$ from the infinite sequence in the proof of Theorem 6J7 belong to homeomorphic hyperbolic space-forms of dimension $n \geqslant 3$. For $i = 1, 2$, let $\mathcal{P}_i = \mathcal{T}/N_i$, where $\mathcal{T} = \{p_1, \ldots, p_n\}$ and N_i is a normal subgroup of $G(\mathcal{T})$. Since $\mathbb{H}^n/N_1$ and $\mathbb{H}^n/N_2$ are homeomorphic, their fundamental groups N_1 and N_2 are isomorphic. We can now appeal to the group-theoretic version of what is known as *Mostow rigidity* for hyperbolic spaces of dimension $n \geqslant 3$ [344, §11.6; 411, §5.7; 412]. This says that any two groups of hyperbolic isometries in $\mathbb{H}^n$ are abstractly isomorphic if and only if they are conjugate in the full isometry group $\mathcal{I}(\mathbb{H}^n)$ of $\mathbb{H}^n$; that is, abstract isomorphism of groups in $\mathbb{H}^n$ is equivalent to geometric isomorphism. One of the consequences is that the fundamental group of a hyperbolic space-form of dimension $n \geqslant 3$ determines it up to isometry.

Now, when applied to $\mathbb{H}^n/N_1$ and $\mathbb{H}^n/N_2$, this shows that N_1 and N_2 are conjugate in $\mathcal{I}(\mathbb{H}^n)$, and hence that $\mathbb{H}^n/N_1$ and $\mathbb{H}^n/N_2$ are isometric space-forms. The question is now whether or not N_1 and N_2 are also conjugate in the smaller group $G(\mathcal{T})$, of which they are subgroups. Then indeed, $N_1 = N_2$, because N_1 and N_2 are normal in $G(\mathcal{T})$, and therefore $\mathcal{P}_1 = \mathcal{P}_2$. In other words, if this line of argument applies to a sequence of polytopes $\mathcal{P}$, then the corresponding space-forms will be mutually non-homeomorphic.

In conclusion, we observe that it is difficult to compute the volume of hyperbolic space-forms in general. For even dimensions $n = 2k \geqslant 2$, the theorem of Gauss–Bonnet says that the volume of a hyperbolic space-form $\mathbb{H}^n/N$ of Euler characteristic $\chi(\mathbb{H}^n/N)$ is given by

6J8
$$\mathrm{Vol}(\mathbb{H}^n/N) = (-1)^k \tfrac{1}{2}\,\mathrm{Vol}(\mathbb{S}^n)\chi(\mathbb{H}^n/N).$$

In the situation of equations (6J1) and (6J4), this directly relates $\mathrm{Vol}(\mathbb{H}^n/N)$ to the order of $\Gamma(\mathcal{P})$. More exactly, if $n = 2$ and $\mathcal{P}$ is of hyperbolic type $\{p, q\}$, then the area of the closed surface is given by

$$\mathrm{Vol}(\mathbb{H}^2/N) = \pi\left(\frac{1}{2} - \frac{1}{p} - \frac{1}{q}\right)|\Gamma(\mathcal{P})|.$$

Similarly, if $n = 4$ and $\mathcal{P}$ is of hyperbolic type $\{5, 3, 3, s\}$ or $\{s, 3, 3, 5\}$, with $s = 3, 4$ or 5, then

$$\mathrm{Vol}(E/N) = \frac{4\pi^2 c_s}{3}|\Gamma(\mathcal{P})|,$$

with c_s as in (6J6). For a brief survey on hyperbolic volume, see also Kellerhals [246]; a more extensive account is given in Johnson et al. [236].

7

Mixing

In this chapter and the next, we shall discuss some general techniques for constructing new regular polytopes from old ones, or, rather, for constructing groups of regular polytopes from other groups, which will be those of regular polytopes or those closely related to them. These techniques fall into two broad categories. First, we may select certain elements of such a group, which themselves generate a string C-group. Second, we may augment a given group by adjoining outer automorphisms. The two chapters will be devoted to these topics in turn.

The general name for the first kind of technique will be *mixing*, while that for the second will be *twisting*. However, the two techniques are by no means exclusive, and we shall see that some regular polytopes can arise in both ways.

The discussion of this chapter will be split into six sections. In Section 7A, we consider the general idea of mixing operations. We then concentrate in Section 7B on the special case of polytopes of rank 3, where mixing sometimes permits a deeper investigation of the combinatorial structure of a regular polyhedron. We next explore in Section 7C the notion of a *cut* of a regular polytope, which yields, as the name might suggest, a regular polytope of lower rank embedded in a natural way in the original. Then, in Section 7D, we enumerate the classical regular polytopes of [120]. Finally, in Sections 7E and 7F, we illustrate the various techniques developed in the chapter by classifying all the regular polytopes (both finite and infinite) which admit discrete faithful realizations in $\mathbb{E}^3$.

7A General Mixing

The idea of a mixing operation is very general. Let $\varDelta$ be a group generated by involutions, say $\varDelta = \langle \sigma_0, \ldots, \sigma_{n-1} \rangle$; usually, but not necessarily, $\varDelta$ will be a C-group. A *mixing operation* then derives a new group $\varGamma$ from $\varDelta$, by taking as generators $\rho_0, \ldots, \rho_{m-1}$ for $\varGamma$ certain suitably chosen products of the σ_i, so that $\varGamma$ is a subgroup of $\varDelta$; it is denoted by

$$(\sigma_0, \ldots, \sigma_{n-1}) \mapsto (\rho_0, \ldots, \rho_{m-1}).$$

(We often loosely refer to mixing operations simply as *operations*.) We naturally wish Γ to be a string C-group, but unfortunately there are few general circumstances which will guarantee this, and it is often the case that the intersection property has to be addressed directly.

What we shall do in this section is illustrate mixing operations by means of a number of examples and discuss the notion of the mix of two regular polytopes (see [308]). The mixing technique is extremely versatile, and we shall see that many constructions for regular polytopes which have appeared in the literature can be subsumed under its heading. For instance, we may replace certain of the generators of a C-group by suitable conjugates; we shall explore such possibilities in Section 7B.

In our initial example, Δ is the group of a regular n-polytope $\mathcal{P}$. The operation

7A1 $$(\sigma_0, \ldots, \sigma_{n-1}) \mapsto (\sigma_0\sigma_2\sigma_4\cdots, \sigma_1\sigma_3\sigma_5\cdots) =: (\rho_0, \rho_1),$$

which results in the two products of alternate generators of Δ, yields a dihedral group, which is that of the *Petrie polygon* of $\mathcal{P}$. (The period of the Petrie polygon is usually taken to be that of the product $\sigma_0\sigma_1\cdots\sigma_{n-1}$ of the generators in their natural order, but it is a useful exercise for the reader to show that the products of the generators of a string C-group in any order are conjugate.)

For the second example, let $\Delta = \Gamma(\{4, 3^{n-2}\}) = [4, 3^{n-2}]$ be the group of the regular n-cube. The operation

7A2 $$(\sigma_0, \ldots, \sigma_{n-1}) \mapsto (\sigma_0, \sigma_1\sigma_3\sigma_5\cdots, \sigma_2\sigma_4\sigma_6\cdots) =: (\rho_0, \rho_1, \rho_2)$$

yields the regular polyhedron $\{4, n \mid 4^{\lfloor \frac{n}{2} \rfloor - 1}\}$, which was first described by Coxeter [105, p. 57]. The notation here will be defined in Section 7B. The polyhedron shares its 2^n vertices and $2^{n-1}n$ edges with the n-cube, and its $2^{n-2}n$ square faces occur among those of the cube; its vertex-figure is the Petrie polygon $\{n\}$ of the vertex-figure $\{3^{n-2}\}$ of the cube.

A very similar example arises from $[4, 3^{n-2}, 4] = \langle \sigma_0, \ldots, \sigma_n \rangle$, which is the group of the tiling of $\mathbb{E}^n$ by regular cubes (as usual in such cases, the rank will be $n+1$ instead of n). Let k be chosen with $0 < k < n$, and define

7A3 $$\begin{aligned} \rho_0 &:= \textstyle\prod_{j \geqslant 0} \sigma_{k-2j-1}, \\ \rho_1 &:= \textstyle\prod_{j} \sigma_{k+2j}, \\ \rho_2 &:= \textstyle\prod_{j \geqslant 0} \sigma_{k+2j+1}. \end{aligned}$$

In each case, we set $\sigma_i := \varepsilon$ unless $0 \leqslant i \leqslant n$; further, in the second product, j is any integer. Then the mixing operation

7A4 $$(\sigma_0, \ldots, \sigma_n) \mapsto (\rho_0, \rho_1, \rho_2)$$

yields a regular apeirotope of type $\{2(k+1), 2(n-k+1)\}$. The dual is obtained by replacing k by $n-k$, and taking the generators in the reverse order $\sigma_n, \ldots, \sigma_0$. For example, if $n = 3$ and $k = 1$, we have the Petrie–Coxeter apeirohedron $\{4, 6 \mid 4\}$ (see Section 7E).

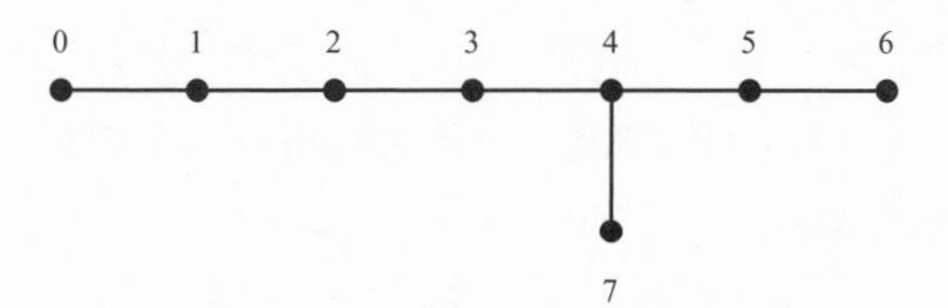

Figure 7A1. The Coxeter diagram of E_8.

Our next examples arise from the Coxeter group $\Delta = E_8 = [3^{4,2,1}]$ and certain of its subgroups. Here, we have Δ generated by $\sigma_0, \ldots, \sigma_7$, with relations $(\sigma_i\sigma_j)^{p_{ij}} = \varepsilon$, where $p_{ii} = 1$ for $i = 0, \ldots, 7$, $p_{ij} = 3$ for $j = i + 1$ with $i = 0, \ldots, 6$ and $\{i, j\} = \{4, 7\}$, and $p_{ij} = 2$ otherwise. (See Figure 7A1 for the Coxeter diagram of E_8, in which a label j indicates the generator σ_j.) The mixing operation is

7A5 $$(\sigma_0, \ldots, \sigma_7) \mapsto (\sigma_0\sigma_6, \sigma_1\sigma_5, \sigma_2\sigma_4, \sigma_3\sigma_7) =: (\rho_0, \rho_1, \rho_2, \rho_3).$$

It may be verified (see [320]) that $\langle \rho_0, \ldots, \rho_3 \rangle$ is isomorphic to the Coxeter group $[3, 3, 5]$; excluding $\rho_0 = \sigma_0\sigma_6$ exhibits how to obtain $[3, 5]$ from $B_6 = [3^{3,1,1}]$ by an analogous operation. The latter is worth comparing to the realizations of $\{3, 5\}$ discussed in Section 5B, since one blend of its two pure realizations has the vertices of the regular 6-dimensional cross-polytope.

We next consider a construction which is often useful, but which unfortunately does not always yield a polytope. Let $m \leqslant n$, and let $\Gamma = \langle \sigma_0, \ldots, \sigma_{n-1} \rangle$ and $\Delta = \langle \tau_0, \ldots, \tau_{m-1} \rangle$ be string C-groups. If we define

$$\rho_j := (\sigma_j, \tau_j) \in \Gamma \times \Delta,$$

for $j = 0, \ldots, n - 1$, with $\tau_j := \varepsilon$ for $j \geqslant m$, then the group

7A6 $$\Gamma \diamond \Delta := \langle \rho_0, \ldots, \rho_{n-1} \rangle$$

is called the *mix* of Γ and Δ. In Section 4F, we introduced a more general construction in the context of amalgamations; in that terminology, the mix is actually the 0-mix. Of course, we can express the mix as a mixing operation

$$(\sigma_0, \ldots, \sigma_{n-1}, \tau_0, \ldots, \tau_{m-1}) \mapsto (\sigma_0\tau_0, \ldots, \sigma_{n-1}\tau_{n-1}) =: (\rho_0, \ldots, \rho_{n-1}).$$

This certainly yields a group satisfying the relations $(\rho_i\rho_j)^2 = \varepsilon$ for $0 \leqslant i < j - 1 \leqslant n - 2$, but it will not generally be a C-group.

We first give an example which illustrates the latter point. Let $\mathcal{P} := \{3, 3, 3\}$ and $\mathcal{Q} := \{\{5, 3\}_5, \{3, 5\}_5\}$ (this self-dual polytope has 57 hemi-dodecahedral facets; see [121]), and define $\Gamma := \Gamma(\mathcal{P})$ and $\Delta := \Gamma(\mathcal{Q})$. Now we have

$$(\sigma_0\sigma_1\sigma_2)^4 = \varepsilon = (\tau_0\tau_1\tau_2)^5.$$

Within Γ, we have

$$\begin{aligned} \sigma_2\sigma_1\sigma_0 \cdot \sigma_2\sigma_1 \cdot \sigma_0\sigma_1\sigma_2 &= \sigma_2\sigma_1\sigma_2 \cdot \sigma_0\sigma_1\sigma_0\sigma_1 \cdot \sigma_2 \\ &= \sigma_2\sigma_1\sigma_2\sigma_1 \cdot \sigma_0\sigma_2 \\ &= \sigma_1\sigma_2\sigma_0\sigma_2 \\ &= \sigma_1\sigma_0. \end{aligned}$$

Similarly,

$$\sigma_2\sigma_1\sigma_0 \cdot \sigma_0\sigma_1 \cdot \sigma_0\sigma_1\sigma_2 = \sigma_2\sigma_0\sigma_1\sigma_2 = \sigma_0\sigma_1\sigma_2\sigma_1.$$

Thus, if $\alpha := \sigma_0\sigma_1\sigma_2 = (\sigma_0\sigma_1\sigma_2)^5$, then

$$\begin{aligned}(\sigma_2\sigma_1)^{\alpha^2}(\sigma_1\sigma_2)^{\alpha} &= (\sigma_1\sigma_0)^{\alpha} \cdot (\sigma_1\sigma_0)^{-1} \\ &= \sigma_1\sigma_2\sigma_1\sigma_0 \cdot \sigma_0\sigma_1 \\ &= \sigma_1\sigma_2.\end{aligned}$$

However, within Δ, the analogous calculations with $\beta := (\tau_0\tau_1\tau_2)^5 = \varepsilon$ yield

$$(\tau_2\tau_1)^{\beta^2}(\tau_1\tau_2)^{\beta} = \tau_2\tau_1 \cdot \tau_1\tau_2 = \varepsilon.$$

Hence, if $\gamma := (\rho_0\rho_1\rho_2)^5$, then

$$(\rho_2\rho_1)^{\gamma^2}(\rho_1\rho_2)^{\gamma} = (\sigma_1\sigma_2, \varepsilon),$$

so that $(\sigma_1\sigma_2, \varepsilon) \in \langle \rho_0, \rho_1, \rho_2 \rangle$. Analogous calculations (in effect, the same ones applied to the dual) show that $(\sigma_1\sigma_2, \varepsilon) \in \langle \rho_1, \rho_2, \rho_3 \rangle$. But it is clear that $(\sigma_1\sigma_2, \varepsilon) \notin \langle \rho_1, \rho_2 \rangle$; and hence we have the required violation of the intersection property in $\Gamma \diamondsuit \Delta$.

When $\Gamma := \Gamma(\mathcal{P})$ and $\Delta := \Gamma(\mathcal{Q})$ for some regular polytopes $\mathcal{P}$ and $\mathcal{Q}$, then we define

$$\mathcal{P} \diamondsuit \mathcal{Q} := \mathcal{P}(\Gamma \diamondsuit \Delta),$$

whether or not $\Gamma \diamondsuit \Delta$ is a C-group, and call it the *mix* of $\mathcal{P}$ and $\mathcal{Q}$. Here, $\mathcal{P}(\Gamma \diamondsuit \Delta)$ denotes the poset (indeed, pre-polytope) determined by $\Gamma \diamondsuit \Delta$ (see Section 2E). Clearly, $\mathcal{P} \diamondsuit \mathcal{Q}$ is polytopal if and only if $\Gamma \diamondsuit \Delta$ is a C-group. But, as we have just seen, the mix of two regular polytopes need not itself be a polytope. However, there are a couple of cases in which the mix is always polytopal.

7A7 Theorem *Let $\mathcal{P}$ be a regular n-polytope, and let $\mathcal{Q}$ be the m-face of $\mathcal{P}$. Then, at least in the cases $m = 1$ and $m = n - 1$, the mix $\mathcal{P} \diamondsuit \mathcal{Q}$ is polytopal.*

Proof. The proof is easy. Using the notation for the groups employed before, we just consider the surjective homomorphism π of $\Lambda := \Gamma(\mathcal{P}) \diamondsuit \Gamma(\mathcal{Q})$ onto $\Gamma(\mathcal{P})$ induced by the mappings

$$\rho_j = (\sigma_j, \tau_j) \mapsto \sigma_j$$

for $j = 0, \ldots, n - 1$, and appeal to the quotient criterion of Theorem 2E17. Note that π is the restriction to Λ of the natural projection of the direct product $\Gamma(\mathcal{P}) \times \Gamma(\mathcal{Q})$ onto its first factor $\Gamma(\mathcal{P})$. If $m = 1$, the homomorphism is one-to-one on the group of the vertex-figure, and if $m = n - 1$, then it is one-to-one on the group of the facet. By Theorem 2E17, this proves that the mix is polytopal. □

We can say more about the case $m = 1$, where $\mathcal{Q}$ is a segment (1-polytope).

7A8 Theorem *Let $\mathcal{P}$ be a regular n-polytope and $\mathcal{Q}$ a segment. Then $\Gamma(\mathcal{P} \diamondsuit \mathcal{Q}) \cong \Gamma(\mathcal{P})$ if all edge-circuits of $\mathcal{P}$ are even; otherwise $\Gamma(\mathcal{P} \diamondsuit \mathcal{Q}) \cong \Gamma(\mathcal{P}) \times C_2$.*

Proof. Using the notation of Theorem 7A7, we write each relation in $\Gamma(\mathcal{P})$ in the form

$$\sigma_0\alpha_1\sigma_0\alpha_2\cdots\sigma_0\alpha_k = \varepsilon,$$

with $\alpha_1, \ldots, \alpha_k \in \langle\sigma_1, \ldots, \sigma_{n-1}\rangle$, the group of the vertex-figure of $\mathcal{P}$. It is shown in Theorem 2F4 that this relation corresponds to an edge-circuit in $\mathcal{P}$ of length k. Since $\rho_i = (\sigma_i, \varepsilon)$ for $i \geqslant 1$, we have

$$\rho_0\beta_1\rho_0\beta_2\cdots\rho_0\beta_k = \left(\sigma_0\alpha_1\sigma_0\alpha_2\cdots\sigma_0\alpha_k,\, \tau_0^k\right) = \left(\varepsilon,\, \tau_0^k\right)$$

in $\Lambda = \Gamma(\mathcal{P} \diamondsuit \mathcal{Q})$, with $\beta_i := (\alpha_i, \varepsilon)$. Clearly, this is the identity in Λ if and only if k is even. It follows at once that $\Lambda \cong \Gamma(\mathcal{P})$ if and only if all edge-circuits of $\mathcal{P}$ are even; otherwise $(\varepsilon, \tau) \in \Lambda$, and $\Lambda \cong \Gamma(\mathcal{P}) \times C_2$. □

Before we proceed, let us give a couple of examples to illustrate Theorem 7A8; we shall appeal to these examples in Sections 10C and 11D.

First, consider the toroidal polyhedron $\mathcal{P} := \{4, 4\}_{\mathbf{s}}$, with $\mathbf{s} = (s, 0)$ or (s, s). The criterion of Theorem 7A8 says that $\mathcal{P} \diamondsuit \{\,\} \cong \mathcal{P}$ unless $\mathcal{P}$ has odd edge-circuits. The exceptions are thus those with $\mathbf{s} = (s, 0)$ and s odd. In this case, it is easy to see (just count the number of vertices) that

7A9 $$\{4, 4\}_{(s,0)} \diamondsuit \{\,\} = \{4, 4\}_{(s,s)} \qquad (s \text{ odd}).$$

Second, let $\mathcal{P} := \{3, 3\}$. Then $\mathcal{P} \diamondsuit \{\,\}$ is a toroidal polyhedron, namely,

7A10 $$\{3, 3\} \diamondsuit \{\,\} = \{6, 3\}_{(2,0)}.$$

To see this, observe that the mix is of type $\{6, 3\}$, and hence is a toroid; again, a count of its vertices identifies which one it is.

The result of Theorem 7A7 no longer holds in full generality when $1 < m < n - 1$. Nevertheless, it is still possible to obtain some fairly general results about polytopality of the mix in this case.

We first discuss the case when the mix is isomorphic to the original polytope. Recall from Section 4E that a regular n-polytope $\mathcal{P}$, with group $\Gamma(\mathcal{P}) = \langle\sigma_0, \ldots, \sigma_{n-1}\rangle$, has the FAP (flat amalgamation property) with respect to its m-faces if and only if

$$\Gamma(\mathcal{P}) \cong N_m^+ \rtimes \langle\sigma_0, \ldots, \sigma_{m-1}\rangle,$$

where $N_m^+ := \langle\varphi^{-1}\sigma_i\varphi \mid i \geqslant m,\ \varphi \in \Gamma(\mathcal{P})\rangle$. Then we have

7A11 Theorem *Let $\mathcal{P}$ be a regular n-polytope, and let $\mathcal{Q}$ be the m-face of $\mathcal{P}$. Then $\mathcal{P} \diamondsuit \mathcal{Q}$ is isomorphic to $\mathcal{P}$ if and only if $\mathcal{P}$ has the FAP with respect to its m-faces.*

Proof. Recall that two regular n-polytopes are isomorphic if and only if there is an isomorphism between the groups mapping the distinguished generators of the first group onto those of the second.

Consider again the surjective homomorphism $\pi\colon \Gamma(\mathcal{P}) \diamondsuit \Gamma(\mathcal{Q}) =: \Lambda \to \Gamma(\mathcal{P})$, which is the restriction to Λ of the natural projection of $\Gamma(\mathcal{P}) \times \Gamma(\mathcal{Q})$ onto $\Gamma(\mathcal{P})$. We now investigate when π is an isomorphism.

Let $\rho := \rho_{i_1}\rho_{i_2}\cdots\rho_{i_k} = (\varphi, \tau)$ be an element in Λ, where $\varphi := \sigma_{i_1}\sigma_{i_2}\cdots\sigma_{i_k}$ and $\tau := \tau_{i_1}\tau_{i_2}\cdots\tau_{i_k}$. Now consider φ modulo N_m^+. First note that a generator σ_j of $\Gamma(\mathcal{P})$ either belongs to $\langle\sigma_0, \ldots, \sigma_{m-1}\rangle$ if $j \leqslant m-1$ or to N_m^+ if $j \geqslant m$. Then we have

$$\varphi N_m^+ = \sigma_{i_1}\sigma_{i_2}\cdots\sigma_{i_k} N_m^+ = \textstyle\prod_l \sigma_{i_l} N_m^+,$$

where the product is taken over all l such that $i_l \leqslant m-1$. Since $\tau_j = \varepsilon$ if $j \geqslant m$, we also have $\tau = \prod_l \tau_{i_l}$, where the product is taken over the same indices l.

Now suppose that $\mathcal{P}$ has the FAP with respect to its m-faces, so that $\Gamma(\mathcal{P})$ is the semi-direct product of N_m^+ by $\langle\sigma_0, \ldots, \sigma_{m-1}\rangle$. If ρ is in the kernel of π, then $\varphi = \varepsilon$; hence $\prod_l \sigma_{i_l} = \varepsilon$, since the latter is the component of φ in $\langle\sigma_0, \ldots, \sigma_{m-1}\rangle$. But then also $\tau = \prod_l \tau_{i_l} = \varepsilon$, since this product represents the same element in $\Gamma(\mathcal{Q})$ as the former product. It follows that $\rho = \varepsilon$; hence π is an isomorphism.

Conversely, let π be an isomorphism. Then its kernel is trivial; that is, if $(\varepsilon, \tau) \in \Lambda$, then $\tau = \varepsilon$. To prove the FAP we show that the trivial element in $\Gamma(\mathcal{P})$ admits only the trivial decomposition as a product of an element in $\langle\sigma_0, \ldots, \sigma_{m-1}\rangle$ with an element in N_m^+. Let $\varepsilon = \varphi_1\varphi_2$, where $\varphi_1 \in \langle\sigma_0, \ldots, \sigma_{m-1}\rangle$ and $\varphi_2 \in N_m^+$. Write φ_1 as a product of generators σ_j with $j \leqslant m-1$, and φ_2 as a product of conjugates of generators σ_j with $j \geqslant m$ by elements in $\langle\sigma_0, \ldots, \sigma_{m-1}\rangle$ (see Lemma 4E7). Let $\varphi_1 = \sigma_{i_1}\sigma_{i_2}\cdots\sigma_{i_r}$ and $\varphi_2 = \sigma_{i_{r+1}}\sigma_{i_{r+2}}\cdots\sigma_{i_k}$ (say), and consider the element $\rho := \rho_{i_1}\rho_{i_2}\cdots\rho_{i_k} = (\varphi, \tau)$ in Λ. Then $\varphi = \varphi_1\varphi_2 = \varepsilon$; hence also $\tau_{i_1}\tau_{i_2}\cdots\tau_{i_k} = \tau = \varepsilon$, because π is an isomorphism. On the other hand, we also have $\tau_{i_{r+1}}\tau_{i_{r+2}}\cdots\tau_{i_k} = \varepsilon$; indeed, if we replace each generator σ_{i_s} in φ_2 by τ_{i_s}, and use the fact that $\tau_j = \varepsilon$ if $j \geqslant m$, then we arrive at the trivial element. But then also $\tau_{i_1}\cdots\tau_{i_r} = \varepsilon$; hence $\sigma_{i_1}\cdots\sigma_{i_r} = \varepsilon$, because both words in the generators represent the same element in $\Gamma(\mathcal{Q})$. It follows that $\varphi_1 = \varepsilon$; hence also $\varphi_2 = \varepsilon$. This proves the FAP. □

The sufficiency of the FAP in Theorem 7A11 can also be obtained from Theorem 4F9, applied with $\mathcal{P}_2 := \mathcal{P}$, $\mathcal{K} := \mathcal{Q}$, and $\mathcal{P}_1 := \left(2^{\mathcal{K}^*}\right)^*$, the dual of the power complex $2^{\mathcal{K}^*}$ of the dual $\mathcal{K}^*$ of $\mathcal{K}$. Then $\mathcal{P}$ is the vertex-figure of the polytope in Theorem 4F9, and inspection of the proof shows that this is just the mix $\mathcal{P} \diamondsuit \mathcal{Q}$. Hence, if the FAP holds, then the mix is isomorphic to the original polytope.

Next we discuss conditions which guarantee polytopality of the mix. We begin with the following

7A12 Lemma *Let $n \geqslant 3$ and $1 \leqslant m \leqslant n-1$. Let $\mathcal{P}$ be a regular n-polytope of type $\{p_1, \ldots, p_{n-1}\}$, and let $\mathcal{Q}$ be the m-face of $\mathcal{P}$. For $0 \leqslant i \leqslant m$ define $\Gamma_{i,m} := \langle\sigma_i, \ldots, \sigma_m\rangle$, where again $\Gamma(\mathcal{P}) = \langle\sigma_0, \ldots, \sigma_{n-1}\rangle$. If each group $\Gamma_{i,m}$ with $1 \leqslant i \leqslant m$ is generated by the conjugates of σ_m in $\Gamma_{i,m}$, then the mix $\mathcal{P} \diamondsuit \mathcal{Q}$ is polytopal. If the condition also holds for $i = 0$, then $\Gamma(\mathcal{P} \diamondsuit \mathcal{Q}) = \Gamma(\mathcal{P}) \times \Gamma(\mathcal{Q})$.*

Proof. We first prove the polytopality statement by induction on n, assuming that the condition on $\Gamma_{i,m}$ holds for $i = 1, \ldots, m$. Then, if $m = 1$ or $m = n-1$, we can directly appeal to Theorem 7A7. In particular, this settles the case $n = 3$.

Now let $n \geqslant 4$ and $1 < m < n-1$. Then we can make the obvious inductive assumption that the facet and vertex-figure of the mix $\mathcal{P} \diamondsuit \mathcal{Q}$ are polytopal. Indeed, the condition on the groups $\Gamma_{i,m}$ is hereditary and holds again for the facet and vertex-figure of $\mathcal{P}$, so that the inductive assumption for rank $n-1$ yields polytopality of the facet and vertex-figure of the mix.

We now appeal to Proposition 2E16 and verify a restricted intersection property for the group $\Lambda := \Gamma(\mathcal{P}) \diamondsuit \Gamma(\mathcal{Q})$ of the mix. Set $\Lambda_{n-1} := \langle \rho_0, \ldots, \rho_{n-2} \rangle$, $\Lambda_0 := \langle \rho_1, \ldots, \rho_{n-1} \rangle$, and $\Lambda_{0,n-1} := \langle \rho_1, \ldots, \rho_{n-2} \rangle$, where again $\rho_j := (\sigma_j, \tau_j)$, $\Gamma(\mathcal{Q}) = \langle \tau_0, \ldots, \tau_{m-1} \rangle$, and $\tau_j := \varepsilon$ for $j \geqslant m$. Then Λ is a C-group if $\Lambda_{n-1} \cap \Lambda_0 = \Lambda_{0,n-1}$.

Now, from the definition of the mix, it is clear that the two groups Λ_{n-1} and Λ_0 are contained in the direct products $\langle \sigma_0, \ldots, \sigma_{n-2} \rangle \times \langle \tau_0, \ldots, \tau_{m-1} \rangle$ and $\langle \sigma_1, \ldots, \sigma_{n-1} \rangle \times \langle \tau_1, \ldots, \tau_{m-1} \rangle$, respectively, and so $\Lambda_{n-1} \cap \Lambda_0$ is a subgroup of $\langle \sigma_1, \ldots, \sigma_{n-2} \rangle \times \langle \tau_1, \ldots, \tau_{m-1} \rangle$, by the intersection property of $\Gamma(\mathcal{P})$. We now use our hypothesis on $\Gamma_{i,m}$, with $i = 1$, to conclude that $\Gamma_{1,m} \times \langle \varepsilon \rangle$ is a subgroup of $\Lambda_{0,n-1}$. First, note that $\Lambda_{0,n-1}$ contains $\rho_m = (\sigma_m, \varepsilon)$ and all its conjugates by elements in $\langle \rho_1, \ldots, \rho_{m-1} \rangle$, and that each such conjugate is of the form (φ, ε) with $\varphi \in \Gamma_{1,m}$. But by assumption $\Gamma_{1,m}$ is generated by the conjugates of σ_m in $\Gamma_{1,m}$, and so we can achieve every $\varphi \in \Gamma_{1,m}$.

We now know that $\Lambda_{0,n-1}$ contains all the elements (σ_i, ε) for $i = 1, \ldots, n-2$, and hence also (ε, τ_i) for $i = 1, \ldots, m-1$, since the latter is the product with ρ_i. It follows that $\Lambda_{0,n-1} = \langle \sigma_1, \ldots, \sigma_{n-2} \rangle \times \langle \tau_1, \ldots, \tau_{m-1} \rangle$. But then $\Lambda_{n-1} \cap \Lambda_0$ is clearly a subgroup of $\Lambda_{0,n-1}$, because it is contained in the direct product. Since the opposite inclusion is trivial, this now proves that the mix is polytopal.

Now assume that the condition on $\Gamma_{i,m}$ also holds for $i = 0$. We can then replace $\Lambda_{0,n-1}$ by the group Λ_{n-1} of the facet and conclude similarly that it must contain $\Gamma_{0,m} \times \langle \varepsilon \rangle$ as a subgroup. But then Λ contains all the elements (σ_i, ε) for $i = 0, \ldots, n-1$, and hence also (ε, τ_i) for $i = 0, \ldots, m-1$. It follows that $\Lambda = \Gamma(\mathcal{P}) \times \Gamma(\mathcal{Q})$. □

7A13 Theorem *Let $n \geqslant 4$, let $\mathcal{P}$ be a regular n-polytope of type $\{p_1, \ldots, p_{n-1}\}$, and let $\mathcal{Q}$ be the m-face of $\mathcal{P}$. If $p_2, \ldots, p_m$ are odd, then the mix $\mathcal{P} \diamondsuit \mathcal{Q}$ is polytopal. If p_1 is also odd, then $\Gamma(\mathcal{P} \diamondsuit \mathcal{Q}) = \Gamma(\mathcal{P}) \times \Gamma(\mathcal{Q})$.*

Proof. We apply Lemma 7A12, and use the simple observation that the distinguished generators of a dihedral group D_p are conjugate in D_p if p is odd. Then, if $p_2, \ldots, p_m$ are odd, the two generators of $\langle \sigma_{j-1}, \sigma_j \rangle$ $(= D_{p_j})$ are conjugate in $\langle \sigma_{j-1}, \sigma_j \rangle$ for $j = 2, \ldots, m$; hence any two consecutive generators of $\Gamma_{i,m} = \langle \sigma_i, \ldots, \sigma_m \rangle$ are conjugate in $\Gamma_{i,m}$ for $1 \leqslant i \leqslant m$. But then each generator of $\Gamma_{i,m}$ is conjugate to σ_m, and so $\Gamma_{i,m}$ is generated by the conjugates of σ_m in $\Gamma_{i,m}$. By Lemma 7A12, this proves polytopality of the mix.

In addition, if p_1 is odd, then we can apply the same argument to the larger group $\Gamma_{0,m} := \langle \sigma_0, \ldots, \sigma_m \rangle$ to show that it is also generated by the conjugates of σ_m. Then another appeal to Lemma 7A12 yields $\Gamma(\mathcal{P} \diamondsuit \mathcal{Q}) = \Gamma(\mathcal{P}) \times \Gamma(\mathcal{Q})$. □

Before we give an example of a polytope whose mix with a face is not polytopal, we mention the following interesting special case. (This also admits a generalization to higher ranks m.)

7A14 Theorem *Let $\mathcal{P}$ be a regular n-polytope of type $\{p_1, \ldots, p_{n-1}\}$, and let p_2 be odd if $n \geqslant 4$. Let r be a multiple of p_1 (including $r = \infty$). Then the mix $\mathcal{P} \diamondsuit \{r\}$ is polytopal.*

Proof. We appeal to the quotient criterion. There is a natural covering

$$\mathcal{P} \diamondsuit \{r\} \searrow \mathcal{P} \diamondsuit \{p_1\}.$$

The latter is polytopal by Theorem 7A7 (with $n = 3$) or by Theorem 7A13 (with $m = 2$), and the covering mapping is one-to-one on the vertex-figure, which is the mix of the vertex-figure of $\mathcal{P}$ with the line segment { } (see Theorem 7A7). Hence $\mathcal{P} \diamondsuit \{r\}$ is polytopal. □

We now describe a regular 4-polytope $\mathcal{P}$ whose mix with a 2-face is not polytopal. Thus, the entry p_2 in its Schläfli symbol must necessarily be even. (Note that, when realized as a suitable section, this polytope will also give examples with $n > 4$ and $m > 2$.)

Indeed, if $\mathcal{P}$ is a regular 4-polytope with facets $\{3, 6\}_{(s,0)}$ and vertex-figures $\{6, 3\}_{(t,0)}$ with s, t odd, then the intersection property fails; thus the mix $\mathcal{P} \diamondsuit \{3\}$ is not polytopal. Examples of such polytopes $\mathcal{P}$ are obtained by Theorem 11H7. The defining relations for facets and vertex-figures now yield

$$(\rho_1\rho_2\rho_3\rho_2)^t = \left((\sigma_1\sigma_2\sigma_3\sigma_2)^t, \tau_1^t\right) = (\varepsilon, \tau_1)$$

and

$$(\rho_0\rho_1\rho_2\rho_1)^s = \left((\sigma_0\sigma_1\sigma_2\sigma_1)^s, \tau_0^s\right) = (\varepsilon, \tau_0);$$

hence also

$$\rho_0\rho_1(\rho_0\rho_1\rho_2\rho_1)^s\rho_1\rho_0 = (\varepsilon, \tau_0\tau_1\tau_0\tau_1\tau_0) = (\varepsilon, \tau_1).$$

But then $\langle\rho_0, \rho_1, \rho_2\rangle = \langle\sigma_0, \sigma_1, \sigma_2\rangle \times \langle\tau_0, \tau_1\rangle$ and $\langle\rho_1, \rho_2, \rho_3\rangle = \langle\sigma_1, \sigma_2, \sigma_3\rangle \times \langle\tau_1\rangle$, and these groups intersect in $\langle\sigma_1, \sigma_2\rangle \times \langle\tau_1\rangle$ $(= D_6 \times C_2)$. On the other hand, the latter group is twice as large as the dihedral subgroup $\langle\rho_1, \rho_2\rangle$ $(= D_6)$ of the mix. It follows that the mix cannot be polytopal.

Let us also observe that more general mixes of the kind discussed in Section 4F need not be polytopal either. A trivial example is provided by $\{3, 3\} \diamondsuit_1 \{\,\}$. Let τ_1 be an involution, set $\tau_j = \varepsilon$ for $j = 0$ and 2, and define $\rho_j := (\sigma_j, \tau_j)$ for $j = 0, 1, 2$. Then

$$(\varepsilon, \tau_1) = (\rho_0\rho_1)^3 = (\rho_1\rho_2)^3 \in \langle\rho_0, \rho_1\rangle \cap \langle\rho_1, \rho_2\rangle,$$

but clearly

$$(\varepsilon, \tau_1) \notin \langle\rho_1\rangle.$$

Hence we do not obtain a polytope.

It is appropriate to end this section with some remarks about the connexions between mixing and blending. The point of view in Chapter 5 was "internal"; that is, we looked at a realization of a regular polytope and asked whether it was a blend in a non-trivial way, or, in other words, whether its symmetry group was reducible. The subject of this chapter, however, is mixing, which may be regarded as an "external" construction. In a sense, blending and mixing are the geometric and combinatorial sides of the same coin; nevertheless, there are important distinctions between the two concepts. It should particularly be borne in mind that mixing is a combinatorial operation, which applies to different regular polytopes, whereas blending, on the other hand, is a geometric operation, which applies to different realizations of the same regular polytope.

As a simple example of this distinction, if $p \geqslant 3$ is an odd integer, then (combinatorially) $\{2p\} = \{p\} \diamond \{\,\}$. This mix, which is not isomorphic to either component, has pure faithful realizations as well as blended ones (see the description of the realization space of a polygon in Section 5B); we shall come across further instances in Section 7E.

We saw earlier that the mix of two polytopes (even of the same rank) need not be polytopal. However, the example we gave, namely,

$$\{3, 3, 3\} \diamond \{\{5, 3\}_5, \{3, 5\}_5\},$$

can be "realized" faithfully in $\mathbb{E}^{60}$, as follows. Take the simplex realizations of the two component polytopes in $\mathbb{E}^4$ and $\mathbb{E}^{56}$, respectively, with corresponding groups $\langle S_0, \ldots, S_3 \rangle$ and $\langle T_0, \ldots, T_3 \rangle$ (thus S_j corresponds to σ_j and T_j to τ_j in the notation used earlier). We then copy the construction of the blend; that is, for $j = 0, \ldots, 3$, we set

$$R_j := S_j \times T_j \subseteq \mathbb{E}^4 \times \mathbb{E}^{56} = \mathbb{E}^{60},$$

and apply Wythoff's construction. Since both $\{3, 3, 3\}$ and $\{\{5, 3\}_5, \{3, 5\}_3\}$ are covered by the (infinite) universal polytope $\{15, 3, 15\}$, this construction has yielded a realization of the latter which is not itself polytopal (that is, its symmetry group is not a C-group).

In Section 7E, we shall actually have the following situation. There will be blended realizations whose symmetry groups are C-groups, and the groups of the components will also be C-groups, which are often non-isomorphic homomorphic images of the original. The natural identifications induced by the projection mappings will then give coverings (again not usually isomorphisms) of polytopal components by the original realization. In this case, the realization is then also the mix of these components – certainly in the combinatorial sense that this holds of their symmetry groups. In such circumstances, it is, perhaps, pardonable to confuse mixing and blending, although we shall try to avoid doing so. Our attitude will be, as we said before, that blending is a geometric operation, while mixing is combinatorial.

In our applications, mixing operations will usually admit a geometric interpretation. However, mixing operations can also be viewed purely mechanically as a way of deriving regular polytopes from a group generated by involutions. It is then natural to ask for the enumeration of all (regular) polytopes which are associated with the given group. Similar problems have been studied in the more general context of diagram geometries, where computer programs (CAYLEY programs) have been developed to compile an atlas of diagram geometries for certain types of groups. The interest is usually in "small" simple or almost simple groups ([12, 14]), and only mild restrictions are placed on the geometries to be regarded as admissible. This effort is led by a team consisting mainly of Buekenhout, Cara, Dehon and Leemans, and has resulted in comprehensive lists with detailed information about such geometries (see [56, 57, 61–63, 65, 147, 216, 260–262]). These lists also include many new examples of regular polytopes. This computational approach is part of an ambitious program to understand the geometry of the finite simple groups (see also [13, 55, 90, 231, 253, 352, 419, 427]).

7B Operations on Regular Polyhedra

Mixing operations are particularly powerful when applied to a polyhedron (3-polytope) $\mathcal{Q}$, which we may also think of as a (finite or infinite) regular map. The underlying surface for such a map is, of course, the order complex $\mathcal{C}(\mathcal{Q})$ of $\mathcal{Q}$, or, more exactly, its underlying (topological) polyhedron $|\mathcal{C}(\mathcal{Q})|$. Recall that a triangle $T(\Psi)$ of $\mathcal{C}(\mathcal{Q})$ is associated with each flag Ψ of $\mathcal{Q}$, with each vertex of $T(\Psi)$ associated with a face of Ψ, and two triangles share an edge precisely when the corresponding flags are adjacent. Here, many of the operations have direct geometric interpretations; we have already informally introduced these in Section 1D.

The effect of a mixing operation on a polyhedron $\mathcal{Q}$ can often be pictured geometrically by applying Wythoff's construction (or, rather, its abstract analogue) in the underlying surface $\mathcal{C}(\mathcal{Q})$. However, observe that the new faces (that is, 2-faces) which are obtained, regarded as circuits of vertices and edges, will not usually bound discs in $\mathcal{C}(\mathcal{Q})$.

As usual, we shall take our regular polyhedron $\mathcal{Q}$ to have $\Gamma(\mathcal{Q}) = \Delta = \langle \sigma_0, \sigma_1, \sigma_2 \rangle$, and we suppose that $\mathcal{Q}$ is of Schläfli type $\{p, q\}$. Each operation μ will lead to a new group Γ, and a new polyhedron $\mathcal{P} := \mathcal{Q}^\mu$ with $\Gamma(\mathcal{P}) = \Gamma$.

Duality

Our first mixing operation is not commonly thought of as such. This is *duality*, denoted in this chapter by δ, and given by

7B1 $$\delta \colon (\sigma_0, \sigma_1, \sigma_2) \mapsto (\sigma_2, \sigma_1, \sigma_0) =: (\rho_0, \rho_1, \rho_2).$$

The dual of $\mathcal{Q}$ is thus denoted $\mathcal{Q}^\delta$ here, rather than $\mathcal{Q}^*$.

The Petrie Operation

Next, we have the *Petrie operation* π, defined by

7B2 $$\pi \colon (\sigma_0, \sigma_1, \sigma_2) \mapsto (\sigma_0\sigma_2, \sigma_1, \sigma_2) =: (\rho_0, \rho_1, \rho_2).$$

The resulting polyhedron $\mathcal{Q}^\pi$ is often called the *Petrie dual* or, more briefly, the *Petrial* of $\mathcal{Q}$. It has the same vertices and edges as $\mathcal{Q}$; however, its faces are the *Petrie polygons* of $\mathcal{Q}$, whose defining property is that two successive edges, but not three, are edges of a face of $\mathcal{Q}$. Thus the faces of $\mathcal{Q}^\pi$ are *zigzags*, leaving a face of $\mathcal{Q}$ after traversing two of its edges.

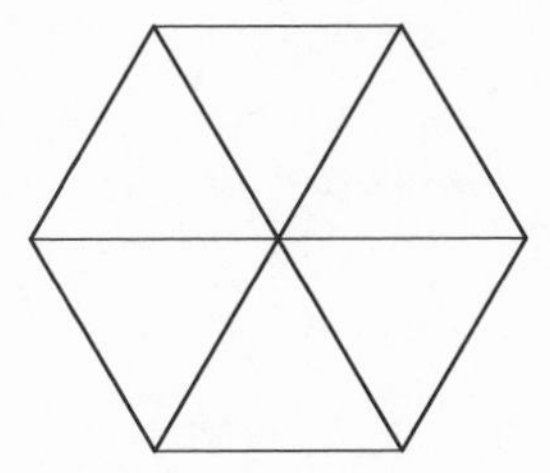

Figure 7B1. The regular polyhedron $\{3, 6\}_{(1,1)}$.

It is clear that the Petrie operation π is involutory, so that $\pi^{-1} = \pi$ and $(\mathcal{Q}^\pi)^\pi = \mathcal{Q}$. If $\mathcal{Q}^\pi$ is isomorphic to $\mathcal{Q}$, then we call $\mathcal{Q}$ *self-Petrie*; it should, however, be observed that a self-Petrie polyhedron and its Petrial do not coincide, since while they share the same vertices and edges, their faces are different. An example of a self-Petrie polyhedron is the *hemi-dodecahedron* $\{5, 3\}_5 = \{5, 3\}/2$, obtained from the dodecahedron $\{5, 3\}$ by identifying its faces of each dimension under the central involutory symmetry. We should recall from Section 1D that, in this context, a regular polyhedron of type $\{p, q\}$ is denoted $\{p, q\}_r$ if the length r of its Petrie polygons determines its combinatorial type; observe that $\{p, q\}_r^\pi = \{r, q\}_p$. We shall meet further examples later. The group of $\{p, q\}_r$ is denoted by $[p, q]_r$.

There are rare cases in which the Petrial of a polyhedron is not polytopal (that is, is not itself a polyhedron). One instance is the flat toroidal polyhedron $\{3, 6\}_{(1,1)}$. This has six triangles with a common vertex, which form a hexagon with opposite edges identified (see Figure 7B1), and so has 3 vertices, 9 edges and 6 faces. It is easy to check its polytopality, which is not vitiated by (for example) its edges falling into three triples with the same vertices. But its Petrial is not a polyhedron; between a face (illustrated by heavy lines in Figure 7B1) and one of its vertices are four edges rather than two.

What is happening here indicates the condition under which polytopality is retained.

7B3 Lemma *The Petrial $\mathcal{Q}^\pi$ of a regular polyhedron $\mathcal{Q}$ is polytopal if and only if a Petrie polygon of $\mathcal{Q}$ visits any given vertex at most once.*

Proof. Of course, this is the familiar interpretation of the intersection property, as was given in Section 2B. □

Curiously, the Petrial of the dual polyhedron $\{6, 3\}_{(1,1)}$ is polytopal, as may be verified by (for example) looking at Figure 7B2. (In this figure, the heavy lines indicate a typical identification to create the torus.)

In general, though, the Petrial of a regular polyhedron will also be a regular polyhedron (that is, it will also be polytopal). The polyhedra obtained from a given one by iterating the Petrie operation and duality then form a family of six; that is, we have

7B4 Lemma $(\pi\delta)^3 = \varepsilon$, *the identity operation on classes of polyhedra.*

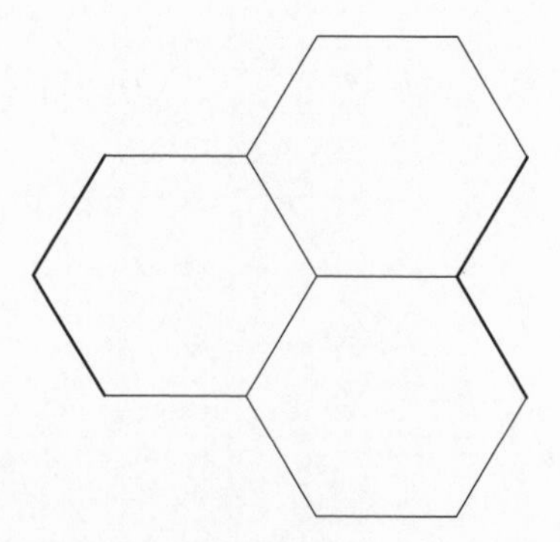

Figure 7B2. The regular map $\{6, 3\}_{(1,1)}$.

Proof. Indeed, considering the groups, we have

7B5
$$\begin{aligned}(\sigma_0, \sigma_1, \sigma_2) &\stackrel{\pi}{\longmapsto} (\sigma_0\sigma_2, \sigma_1, \sigma_2)\\ &\stackrel{\delta}{\longmapsto} (\sigma_2, \sigma_1, \sigma_0\sigma_2)\\ &\stackrel{\pi}{\longmapsto} (\sigma_0, \sigma_1, \sigma_0\sigma_2)\\ &\stackrel{\delta}{\longmapsto} (\sigma_0\sigma_2, \sigma_1, \sigma_0)\\ &\stackrel{\pi}{\longmapsto} (\sigma_2, \sigma_1, \sigma_0)\\ &\stackrel{\delta}{\longmapsto} (\sigma_0, \sigma_1, \sigma_2),\end{aligned}$$

as claimed. □

Facetting

We now have an operation which replaces σ_1 by some other reflexion (conjugate of σ_1 or σ_2) in $\langle\sigma_1, \sigma_2\rangle$. More specifically, the *kth facetting operation* φ_k is given by the operation:

7B6
$$\varphi_k\colon (\sigma_0, \sigma_1, \sigma_2) \mapsto (\sigma_0, \sigma_1(\sigma_2\sigma_1)^{k-1}, \sigma_2) =: (\rho_0, \rho_1, \rho_2).$$

We shall suppose that $2 \leqslant k < \frac{1}{2}q$, since φ_{q-k} has the same effect as φ_k up to isomorphism (actually, conjugation of the whole group by σ_2), and the case $k = \frac{1}{2}q$ can only yield a polyhedron $\{r, 2\}$ for some r. (In this last case, the number r may be of independent interest, but it is not in the present context. In some special circumstances, we shall also find it convenient to allow the case $k = 1$ as well, where φ_1 is just the identity mixing operation ε.) When the highest common factor $(k, q) = 1$, then $\rho_1\rho_2$ has the same period q as $\sigma_1\sigma_2$; indeed, the groups are the same, and φ_k is inverted by $\varphi_{k'}$, where $kk' \equiv \pm 1 \pmod q$. In fact, we have

7B7 Lemma *The facetting operations satisfy* $\varphi_k\varphi_m = \varphi_{km}$, *where the suffix is to be read as that number between* 0 *and* $\frac{1}{2}q$ *which is congruent to* $\pm km$ *modulo* q.

Proof. We apply φ_k and φ_m in succession to the group. Noting that only σ_1 changes, and writing $\rho_1 = \sigma_1(\sigma_2\sigma_1)^{k-1} = (\sigma_1\sigma_2)^k\sigma_2$, it becomes

7B8 $$\left((\sigma_1\sigma_2)^k\sigma_2^2\right)^m\sigma_2 = (\sigma_1\sigma_2)^{km}\sigma_2,$$

as required. □

This lemma covers all possible k and m. Generally speaking, we shall be rather less interested in the case $(k, q) > 1$, although it will occasionally be useful. In particular, we shall employ φ_2 with q even (actually, $q = 6$) in Section 7E; we also discuss the effect of φ_2 in Lemma 7B10, which deals with the theory we shall need.

Geometrically, φ_k has the following effect when $(k, q) = 1$. The new polyhedron $\mathcal{P} := \mathcal{Q}^{\varphi_k}$ has the same vertices and edges as $\mathcal{Q}$. However, a typical face of $\mathcal{P}$ is a *k-hole* of $\mathcal{Q}$, which is formed by the edge-path which leaves a vertex by the kth edge from which it entered, in the same sense (that is, keeping always to the left, say, in some local orientation of $\mathcal{C}(\mathcal{Q})$). The faces of $\mathcal{P}$ then comprise all the k-holes of $\mathcal{Q}$. Hence, if such a k-hole is an r-gon, so that r is the period of

$$\rho_0\rho_1 = \sigma_0 \cdot (\sigma_1\sigma_2)^{k-1}\sigma_1,$$

then $\mathcal{Q}^{\varphi_k}$ is of type $\{r, q\}$. If $\mathcal{Q}$ is infinite, then it is possible that $r = \infty$, even if p is finite.

Of course, we must not forget to verify the intersection property, but in this case it is much easier to do this "geometrically", thinking of $\mathcal{P} = \mathcal{Q}^{\varphi_k}$ as embedded in a surface. Naturally, this will not generally be the same as the original surface $|\mathcal{C}(\mathcal{Q})|$ underlying $\mathcal{Q}$, although in practice we shall be able to work with $|\mathcal{C}(\mathcal{Q})|$ instead of the new surface $|\mathcal{C}(\mathcal{P})|$, and employ Wythoff's construction, as we said earlier. Noting that $\Gamma(\mathcal{Q})$ is transitive on the k-holes of $\mathcal{Q}$, the actual condition is, as in Lemma 7B3, just the usual interpretation of the intersection property.

7B9 Lemma *Let $\mathcal{Q}$ be a regular polyhedron of type $\{p, q\}$, and let $2 \leqslant k < \frac{1}{2}q$ with $(k, q) = 1$. Then kth facetting operation φ_k applied to $\mathcal{Q}$ yields a regular polyhedron $\mathcal{P} = \mathcal{Q}^{\varphi_k}$ if and only if any vertex of $\mathcal{Q}$ is visited by a given k-hole at most once.*

When $(k, q) > 1$, the situation is similar, except that now, in general, a compound of several polyhedra of type $\{r, q/(k, q)\}$ will be formed. However, if $\mathcal{Q}^{\varphi_k}$ remains connected, then it will fail to be polytopal, since the vertex-figure at a vertex will no longer be connected. There is, however, one case of particular interest, when a polyhedron is usually obtained.

7B10 Lemma *Let $\mathcal{Q}$ be a regular polyhedron of type $\{3, 2s\}$ for some $s \geqslant 3$. If $\mathcal{P} := \mathcal{Q}^{\varphi_2}$ is a regular polyhedron, then it is of type $\{2s, s\}$. Moreover, $\Gamma(\mathcal{P})$ is then a subgroup of $\Gamma(\mathcal{Q})$ of index 1 or 3.*

Proof. We appeal to the geometry of the underlying surface $|\mathcal{C}(\mathcal{Q})|$ of the polyhedron $\mathcal{Q}$; we shall see that $|\mathcal{C}(\mathcal{P})| = |\mathcal{C}(\mathcal{Q})|$. Let $\Delta = \Gamma(\mathcal{Q}) = \langle\sigma_0, \sigma_1, \sigma_2\rangle$ as usual, and let Ψ be the base flag of $\mathcal{Q}$. If $T := T(\Psi)$ is the fundamental region of Δ in $|\mathcal{C}(\mathcal{Q})|$, then the corresponding fundamental region of $\Gamma = \Gamma(\mathcal{P})$ is, in general, $T \cup T\sigma_1 \cup T\sigma_0\sigma_1$.

Indeed, this is an example of a more general result obtained by simplex dissection; see Section 11G for similar methods applied to polytopes of rank 4. However, in certain special circumstances, when $|\mathcal{C}(\mathcal{Q})|$ has to be a torus, then it is possible that $\Gamma(\mathcal{P}) = \Gamma(\mathcal{Q})$; an example of this is $\mathcal{Q} = \{3, 6\}_{(t,0)}$, with $3 \not| t$. The conclusions of the lemma are then immediate. □

The Petrie operation and facetting are related as follows.

7B11 Lemma *The Petrie operation π and the facetting operation φ_k commute.*

Proof. This is easily verified algebraically. However, it is even more instructive to look at the geometry. Whether we apply π or φ_k first, the result will be (assuming that it exists) a polyhedron $\mathcal{P}$ whose vertices and edges are those of $\mathcal{Q}$, but whose typical face is a *k-zigzag*, given by an edge-path which (as for φ_k) leaves a vertex at the kth edge from the one by which it entered, but in the oppositely oriented sense at alternate vertices (see [461]). Thus the Petrie polygons themselves are 1-zigzags, which accounts for our nomenclature. □

At this point, it is appropriate to introduce some general notation; we have seen particular cases already. Recall from Section 1D that we denote by $\{p, q \mid h\}$ a regular polyhedron of type $\{p, q\}$, whose combinatorial type is determined by the fact that its (2-)holes are h-gons. Examples are the Petrie–Coxeter regular skew polyhedra $\{4, 6 \mid 4\}$, $\{6, 4 \mid 4\}$ and $\{6, 6 \mid 3\}$ in euclidean 3-space, as well as the finite regular skew polyhedra $\{4, 6 \mid 3\}$, $\{6, 4 \mid 3\}$, $\{4, 8 \mid 3\}$ and $\{8, 4 \mid 3\}$ in spherical 3-space (see [104, 105, 107, 235]). For a discussion of regular skew polyhedra in hyperbolic 3-space, see also [178, 180, 293, 447].

Analogously, if a regular polyhedron $\mathcal{P}$ of type $\{p, q\}$ is determined by the lengths h_j of its j-holes for certain j in the range $2 \leqslant j \leqslant k := \lfloor \frac{1}{2}q \rfloor$, then we denote it by

7B12 $$\mathcal{P} := \{p, q \mid h_2, \ldots, h_k\};$$

any unnecessary h_j (that is, one which is not needed for the specification) is replaced by "·", with those at the end of the sequence omitted. (Of course, the 1-holes of $\mathcal{P}$ are just its faces.) An example where all the h_j are required for the specification is provided by Coxeter's polyhedron $\{4, n \mid 4^{\lfloor \frac{n}{2} \rfloor - 1}\}$ (see [105, p. 57]).

Similarly, the notation $\{p, q\}_r$ for a regular polyhedron $\mathcal{P}$ of type $\{p, q\}$ determined by the length r of its Petrie polygons is generalized to

7B13 $$\mathcal{P} := \{p, q\}_{r_1, \ldots, r_k},$$

with $\mathcal{P}$ now determined by the lengths r_j of its j-zigzags for $j = 1, \ldots, k$, with k as before. The same conventions for unnecessary r_j apply.

These notations can be combined, to give regular polyhedra

$$\{p, q \mid h_2, \ldots, h_k\}_{r_1, \ldots, r_l}$$

of type $\{p, q\}$, determined by certain of its holes and zigzags. The notation is not symmetric between holes and zigzags; the 1-holes are, of course, just the faces $\{p\}$.

The corresponding defining relations for the groups of such regular polyhedra are easily obtained from the earlier discussion. Thus, $\mathcal{P}$ is forced to have j-holes of length h_j by imposing the relation

7B14 $$(\rho_0\rho_1(\rho_2\rho_1)^{j-1})^{h_j} = \varepsilon.$$

on the group $\Gamma(\mathcal{P}) = \langle \rho_0, \rho_1, \rho_2 \rangle$, while $\mathcal{P}$ is forced to have j-zigzags of length r_j by imposing the relation

7B15 $$(\rho_0(\rho_1\rho_2)^j)^{r_j} = \varepsilon.$$

Of course, it is a consequence of Lemma 7B11 that the Petrie operation interchanges j-holes and j-zigzags. (Note that in [106; 131, §8.6] the group of $\{p, q\}_r$ was denoted by $G^{p,q,r}$.)

Halving

The *halving operation* η applies only to a regular polyhedron $\mathcal{Q}$ of type $\{4, q\}$ for some $q \geqslant 3$, and turns it into a self-dual polyhedron $\mathcal{P} := \mathcal{Q}^\eta$ of type $\{q, q\}$. We define η by

7B16 $$\eta\colon (\sigma_0, \sigma_1, \sigma_2) \mapsto (\sigma_0\sigma_1\sigma_0, \sigma_2, \sigma_1) =: (\rho_0, \rho_1, \rho_2).$$

The intersection property is easily checked for $\Gamma := \langle \rho_0, \rho_1, \rho_2 \rangle$; it will become clear from the following discussion. When we think of $\Delta = \Gamma(\mathcal{Q})$ acting on the surface $|\mathcal{C}(\mathcal{Q})|$, the triangle $T = T(\Phi)$ associated with the base flag Φ of $\mathcal{Q}$ is a fundamental region for Δ, and σ_0, σ_1 and σ_2 act as reflexions in the sides of T. Now let $T' := T \cup T\sigma_0$. Then T' is the fundamental region for Γ, and Γ is similarly generated by the reflexions in the sides of T'.

If we now apply Wythoff's construction (or, rather, its abstract analogue, in the underlying surface $|\mathcal{C}(\mathcal{Q})|$), then we see that there are two possibilities.

First, suppose that the (edge-)graph of $\mathcal{Q}$ is bipartite, so that all the edge circuits of $\mathcal{Q}$ have even length. Then $\mathcal{P}$ will be a map on the same surface $|\mathcal{C}(\mathcal{Q})|$. It will have half as many vertices as $\mathcal{Q}$, namely, those in the same partition of the vertex-set $\mathcal{Q}_0$ of $\mathcal{Q}$ as the initial vertex in the base flag Φ. Further, Γ will have index 2 in Δ. As we asserted before, $\mathcal{P}$ will be self-dual, since $\sigma_0 \in \Delta$ acts as an automorphism of Γ, which interchanges ρ_0 and ρ_2 and leaves ρ_1 fixed. The vertices of the dual $\mathcal{P}^\delta$ will then be those in the other partition of $\mathcal{Q}_0$.

We may observe that Δ can be recovered from Γ by a twisting operation, of the kind we shall discuss in Chapter 8.

In the other case, the graph of $\mathcal{Q}$ is not bipartite. Unless $q = 4$ also, $\mathcal{P}$ will be a map on a different surface from $|\mathcal{C}(\mathcal{Q})|$. In actual fact, $|\mathcal{C}(\mathcal{P})|$ will be a double cover of $|\mathcal{C}(\mathcal{Q})|$ in every case; we must be careful to note that $\mathcal{Q}$ itself does not cover $\mathcal{P}$ in general. We shall now have $\Gamma = \Delta$, and $\mathcal{P}$ will have the same vertex-set $\mathcal{Q}_0$ as $\mathcal{Q}$. Finally, $\mathcal{P}$ will still be self-dual, although now the conjugating element σ_0 is in Γ.

In either case, we have

$$\rho_0\rho_1\rho_2\rho_1 = \sigma_0\sigma_1\sigma_0\sigma_2\sigma_1\sigma_2 = (\sigma_0\sigma_1\sigma_2)^2.$$

This shows that, if the original polyhedron $\mathcal{Q}$ has Petrie polygons of length h, then the new polyhedron $\mathcal{P}$ will have 2-holes of length h or $h/2$ according as h is odd or even. Note that the latter will be the case when the graph of $\mathcal{Q}$ is bipartite (but possibly in other cases also); then $[\Delta : \Gamma] = 2$.

In this spirit, in certain cases the combinatorial type of a polyhedron $\mathcal{P} = \mathcal{Q}^\eta$ is easily determined from that of $\mathcal{Q}$.

7B17 Theorem *Let $q \geqslant 4$ and $s \geqslant 2$. Then*

(a) $\{4, q \mid 2s\}^\eta = \{q, q\}_{2s}$*;*
(b) $(\{4, q\}_{2s})^\eta = \{q, q \mid s\}$.

Proof. The graphs of the two polyhedra $\{4, q \mid 2s\}$ and $\{4, q\}_{2s}$ are bipartite, since their defining circuits are faces and holes or zigzags, of lengths 4 and $2s$, respectively. The operation $\mathcal{Q} \mapsto \mathcal{Q}^\eta =: \mathcal{P}$ is given by

$$\eta\colon (\sigma_0, \sigma_1, \sigma_2) \mapsto (\sigma_0\sigma_1\sigma_0, \sigma_2, \sigma_1) =: (\rho_0, \rho_1, \rho_2).$$

In case (a), we therefore have

$$\begin{aligned}
\varepsilon &= (\sigma_0\sigma_1\sigma_2\sigma_1)^{2s} \\
&= (\sigma_0\sigma_1\sigma_2\sigma_1\sigma_0\sigma_1\sigma_2\sigma_1)^s \\
&= (\sigma_0\sigma_1\sigma_2 \cdot \sigma_0\sigma_1\sigma_0\sigma_1\sigma_0 \cdot \sigma_2\sigma_1)^s \\
&\sim (\sigma_1\sigma_2 \cdot \sigma_0\sigma_1\sigma_0 \cdot \sigma_1\sigma_2 \cdot \sigma_0\sigma_1\sigma_0)^s \\
&= (\sigma_1\sigma_2 \cdot \sigma_0\sigma_1\sigma_0)^{2s} \\
&= (\rho_2\rho_1\rho_0)^{2s} \\
&\sim (\rho_0\rho_1\rho_2)^{2s}.
\end{aligned}$$

Similarly, in case (b), we have

$$\begin{aligned}
\varepsilon &= (\sigma_0\sigma_1\sigma_2)^{2s} \\
&= (\sigma_0\sigma_1\sigma_2\sigma_0\sigma_1\sigma_2)^s \\
&= (\sigma_0\sigma_1\sigma_0 \cdot \sigma_2\sigma_1\sigma_2)^s \\
&= (\rho_0\rho_1\rho_2\rho_1)^s.
\end{aligned}$$

In each case, the defining relation of the original polyhedron $\mathcal{Q}$ is equivalent to the corresponding defining relation for the new polyhedron $\mathcal{P}$.

In support of this, we can also argue as follows. For part (a), let $\mathcal{Q} := \{4, q \mid 2s\}$ and $\mathcal{R} := \{q, q\}_{2s}$. By construction, it is immediate that $\mathcal{P} := \mathcal{Q}^\eta$ is a quotient of $\mathcal{R}$. On the other hand, $\mathcal{R}$ is self-dual, and its extended group $\bar{\Gamma}(\mathcal{R})$, consisting of all automorphisms and dualities of $\mathcal{R}$, can be identified as a quotient of $\mathcal{Q}$, which must therefore be $\mathcal{Q}$ itself. In fact, let $\Gamma(\mathcal{R}) = \langle \tau_0, \tau_1, \tau_2 \rangle$, and let ω be the polarity which fixes the base flag of $\mathcal{R}$, so that $\omega^2 = \varepsilon$ and $\omega\tau_j\omega = \tau_{2-j}$ for each j. If $\bar{\Gamma}(\mathcal{R}) = \Gamma(\mathcal{R}) \rtimes \langle \omega \rangle = \Gamma(\mathcal{R}) \rtimes C_2$, then $\bar{\Gamma}(\mathcal{R})$, with distinguished generators ω, τ_2, τ_1, is the group of a polyhedron $\mathcal{Q}'$ which is a quotient of $\mathcal{Q}$; note that $(\omega\tau_2\tau_1\tau_2)^2 = (\tau_0\tau_1\tau_2)^2$. It follows that $\mathcal{Q}' = \mathcal{Q}$ and $\mathcal{P} = \mathcal{R}$. Also, $\mathcal{Q}$ must be bipartite, because $[\Gamma(\mathcal{Q}) : \Gamma(\mathcal{P})] = [\bar{\Gamma}(\mathcal{R}) : \Gamma(\mathcal{R})] = 2$. The argument for part (b) is similar. □

Halving admits a generalization to an operation $\eta_{k,m}$ on a regular polyhedron of type $\{2n, q\}$ with $n = k + m$, which is defined by

7B18 $$\eta_{k,m}\colon (\sigma_0, \sigma_1, \sigma_2) \mapsto ((\sigma_0\sigma_1)^k\sigma_0, \sigma_2, (\sigma_1\sigma_0)^{m-1}\sigma_1) =: (\rho_0, \rho_1, \rho_2).$$

Thus $\eta_{1,1} = \eta$ in the case $n = 2$ which we have been considering. We may observe that

$$\delta\eta_{k,m}\delta = \eta_{m,k};$$

the duality is induced, as before, by conjugation under σ_0. However, even if the resulting polyhedron $\mathcal{P} := \mathcal{Q}^{\eta_{k,m}}$ exists at all (that is, if $\Gamma := \langle \rho_0, \rho_1, \rho_2 \rangle$ has the intersection property), it will appear that the structure of $\mathcal{P}$ is very complicated, unless k, m and q are carefully chosen.

Skewing

Our final operation is the *skewing* operation σ or, perhaps as it would better be named, *skew halving*. It applies to a regular polyhedron of $\mathcal{Q}$ type $\{p, 4\}$, and is defined by

7B19 $$\sigma\colon (\sigma_0, \sigma_1, \sigma_2) \mapsto (\sigma_1, \sigma_0\sigma_2, (\sigma_1\sigma_2)^2) =: (\rho_0, \rho_1, \rho_2).$$

It is remotely related to halving; in fact

$$\sigma = \pi\delta\eta\pi\delta,$$

since, because $(\sigma_1\sigma_2)^4 = \varepsilon$, we have

$$\begin{aligned} (\sigma_0, \sigma_1, \sigma_2) &\overset{\pi}{\longmapsto} (\sigma_0\sigma_2, \sigma_1, \sigma_2) \\ &\overset{\delta}{\longmapsto} (\sigma_2, \sigma_1, \sigma_0\sigma_2) \\ &\overset{\eta}{\longmapsto} (\sigma_2\sigma_1\sigma_2, \sigma_0\sigma_2, \sigma_1) \\ &\overset{\pi}{\longmapsto} ((\sigma_1\sigma_2)^2, \sigma_0\sigma_2, \sigma_1) \\ &\overset{\delta}{\longmapsto} (\sigma_1, \sigma_0\sigma_2, (\sigma_1\sigma_2)^2), \end{aligned}$$

as claimed. This also indicates that σ halves the order of the group just when η does, modulo the double of application of $\pi\delta$, that is, just when the graph of $\mathcal{Q}^{\pi\delta}$ is bipartite. If this is the case, then $\sigma_2 \notin \Gamma := \Gamma(\mathcal{Q}^\sigma) = \langle \rho_0, \rho_1, \rho_2 \rangle$, but acts on it as an automorphism. In any event, $\mathcal{P} := \mathcal{Q}^\sigma$ is self-Petrie; the isomorphism between $\mathcal{P}$ and $\mathcal{P}^\pi$ is given by conjugation of $\Gamma(\mathcal{P})$ by σ_2, since

$$\sigma_2\sigma_1\sigma_2 = \sigma_1 \cdot (\sigma_1\sigma_2)^2.$$

The type $\{s, t\}$ of $\mathcal{P} = \mathcal{Q}^\sigma$ is determined by the periods of

$$\rho_0\rho_1 = \sigma_1 \cdot \sigma_0\sigma_2 \sim \sigma_0\sigma_1\sigma_2$$

and

$$\rho_1\rho_2 = \sigma_0\sigma_2 \cdot (\sigma_1\sigma_2)^2 = \sigma_0\sigma_1\sigma_2\sigma_1.$$

Hence s is the length of the Petrie polygons of $\mathcal{Q}$, while t is the length of its 2-holes.

As in the case of halving, σ admits a generalization $\sigma_{k,m}$ to polyhedra of type $\{p, 2n\}$ with $n = k + m$. This is defined exactly analogously to $\eta_{k,m}$, bearing in mind the connexion between η and σ, namely,

7B20
$$\sigma_{k,m} := \pi\delta\eta_{k,m}\pi\delta.$$

In direct terms, this is the operation

$$\sigma_{k,m}: (\sigma_0, \sigma_1, \sigma_2) \mapsto ((\sigma_1\sigma_2)^{m-1}\sigma_1, \sigma_0\sigma_2, (\sigma_1\sigma_2)^n).$$

From either definition, we see that $\sigma_{1,1} = \sigma$ when $n = 2$. If $\sigma_{k,m}$ defines a new polyhedron $\mathcal{P} := \mathcal{Q}^{\sigma_{k,m}}$ of type $\{s, t\}$, then we may verify directly that s is the length of the m-zigzags of $\mathcal{Q}$, while t is the length of its n-holes.

Let us illustrate these techniques with a few examples. In particular, we shall see how they apply to the Platonic polyhedra. In this section, we shall treat them in a purely combinatorial way; in Section 7E, where we give further examples, we shall look at them more geometrically.

With the regular tetrahedron $\{3, 3\}$, only one of the operations yields anything new, namely, the Petrie operation π. The new polyhedron $\{3, 3\}^\pi$ is $\{4, 3\}_3$; it can also be obtained from the cube $\{4, 3\}$ by identifying its antipodal faces of each rank, so that an alternative notation for it is $\{4, 3\}/2$.

With the cube, again only the Petrie operation leads to a new polyhedron (we leave aside duality for the moment); we have $\{4, 3\}^\pi = \{6, 3\}_4 = \{6, 3\}_{(2,0)}$, a toroidal polyhedron. Similarly, for the octahedron, $\{3, 4\}^\pi = \{6, 4\}_3$. Duality completes the family; the cube and octahedron are dual, and the duals of the two Petrials, namely, $\{3, 6\}_4$ and $\{4, 6\}_3$, are each the Petrial of the other.

The operation φ_2 can be applied to the octahedron, but will only yield the degenerate polyhedron $\{4, 2\}$.

The icosahedron $\{3, 5\}$ and dual dodecahedron $\{5, 3\}$ give rise to a rich family. We begin with the icosahedron. Its Petrial is $\{3, 5\}^\pi = \{10, 5\}_3$. If we apply the facetting operation φ_2, we obtain $\{3, 5\}^{\varphi_2} = \{5, \frac{5}{2}\} \cong \{5, 5 \mid 3\}$, and applying both yields $\{3, 5\}^{\pi\varphi_2} \cong \{6, 5\}_{5,3}$; this polyhedron is actually $\{6, 5\}_{\cdot,3}$. Combinatorially, of course, $\{5, \frac{5}{2}\}$ is self-dual, and so the only new member in the family obtained from it by taking duals and Petrials is the dual $\{6, 5\}_{5,3}^\delta = \{5, 6\}_{5,*3}$. The notation for the second suffix has the following meaning; we prefix a $*$ to a number denoting a k-hole for $k \geqslant 3$, or a k-zigzag for $k \geqslant 2$, to mean that this is the size of the corresponding hole or zigzag of the dual polyhedron. Of course, (2-)holes and Petrie polygons (1-zigzags) are symmetric between a polyhedron and its dual, so the $*$ prefix is omitted. This polyhedron $\{5, 6\}_{5,*3}$ illustrates the situation we considered in Lemma 7B9; if we apply φ_2 to it, we do not obtain a polyhedron, but instead the dodecahedron $\{5, 3\}$ with antipodal vertices identified, but not antipodal edges or faces.

From the dodecahedron, we first obtain $\{5, 3\}^\pi = \{10, 3\}_5$. Its dual $\{3, 10\}_5$ is the Petrial of the dual $\{5, 10\}_3$ of $\{3, 5\}^\pi$. Now we can apply φ_3 to both these polyhedra with vertex-figures $\{10\}$; in each case the resulting polyhedron is isomorphic to the other. Of course, these polyhedra are also interchanged by the Petrie operation; it is interesting to see why.

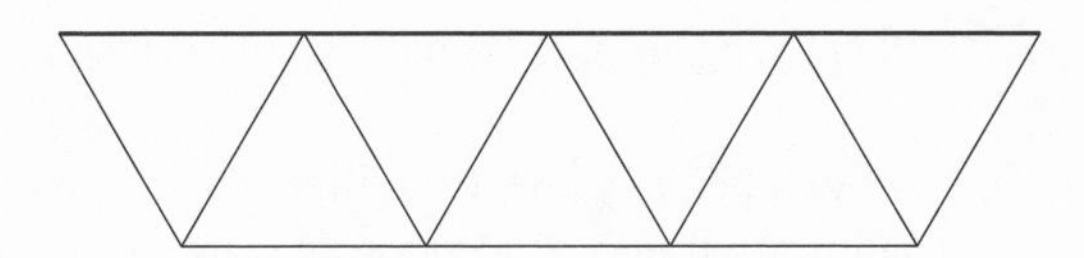

Figure 7B3. Petrie polygons and facetting.

7B21 Theorem *Let $\mathcal{Q}$ be a regular polyhedron of type $\{3, q\}$ for some q not divisible by 3, whose Petrie polygon has length h. Then $\mathcal{Q}^{\varphi_3}$ is a polyhedron of type $\{p, q\}$, where $p = h/2$ or h according as h is even or odd.*

Proof. This is easy to see. The edge-paths forming the faces of $\mathcal{Q}^{\varphi_3}$ are the 3-holes of $\mathcal{Q}$. These join alternate vertices of the Petrie polygons of $\mathcal{Q}$ (see Figure 7B3, where the face of $\mathcal{Q}^{\varphi_3}$ is indicated by heavy lines), whence the result. □

Operations on polyhedra or maps on surfaces have appeared in the works of many authors, either implicitly or explicitly (see, for example, [39, 131, §8.5, 8.6; 150, 198, 241, 268, 295, 304, 461]). From Lemma 7B4 we know that δ and π generate a group of operations on regular maps which is isomorphic to S_3. For arbitrary surface maps, there is again a group S_3 of invertible "operations" which is generated by the duality operation and a Petrie-type operation [268, 461]. Algebraically, these operations can be defined by requiring that they be induced by outer automorphisms of the Coxeter group $[\infty, \infty]$ acting on the flags [241]. There are also similar such "operations" for higher dimensional complexes, each (by definition) induced by an outer automorphism of the (universal string) Coxeter group $[\infty, \ldots, \infty]$ [232, 233, 421]. We should point out here that our use of the term "operation" for regular polyhedra or polytopes is more general than this; indeed, most operations we employ are not directly related to outer automorphisms of $[\infty, \ldots, \infty]$.

7C Cuts

The notion of a *cut* of a regular polytope is a special case of a mixing operation, and is potentially very useful. However, we shall see that there are basic problems in trying to employ the idea. What we mean will become clearer as we proceed; indeed, the initial definition is itself somewhat vague.

Let $\mathcal{P}$ be a regular n-polytope with group $\Gamma = \Gamma(\mathcal{P}) = \langle \sigma_0, \ldots, \sigma_{n-1} \rangle$, and let $S = S(I) := \{\sigma_i \mid i \in I\}$ be a subset consisting of $n - m$ of the distinguished generators of $\Gamma(\mathcal{P})$. In the underlying surface $|\mathcal{C}(\Gamma)|$ of Γ, the corresponding intersection $H_I := \bigcap\{H_i \mid i \in I\}$ of the mirrors H_i associated with the σ_i (thought of as the concrete mirrors of reflexions) is an $(m - 1)$-dimensional "subspace". We may also associate S with an $(m - 1)$-dimensional face $T(I)$ of the fundamental simplex $T(\Phi)$ corresponding to the base flag Φ of $\mathcal{P}$.

Let $\Gamma(I)$ be a subgroup of the centralizer of S in $\Gamma(\mathcal{P})$, which is itself a string C-group with respect to some suitably chosen set $\rho_0, \ldots, \rho_{m-1}$ of generators. Then the associated regular polytope $\mathcal{P}(I) := \mathcal{P}(\Gamma(I))$, or its dual, will be called a *cut* of $\mathcal{P}$. (We shall assume the distinguished generators to be taken in reverse order, if necessary, so that the cut is always $\mathcal{P}(I)$ itself; the context will usually determine the appropriate choice.)

In some examples, $\Gamma(I)$ will be the whole centralizer of $S(I)$, but generally it will be a proper subgroup. The informal idea of a cut of $\mathcal{P}$ is that it should be the part of $\mathcal{P}$ cut through by some subset of its generating involutions, but there will not always be such a neat geometric interpretation.

Let us begin by giving some examples. One which we frequently encounter is the following. Let $n \geqslant 2$, and consider the regular apeirotope $\mathcal{P}_n := \{4, 3^{n-2}, 4\}$ of rank $n+1$, the tiling of ordinary n-dimensional euclidean space $\mathbb{E}^n$ by cubes (as in Section 7A, the geometry accounts for our choice of the rank, as it will for our use of m rather than $m-1$ in the following). Let its group be $\langle \sigma_0, \ldots, \sigma_n \rangle$ in the usual way (that is, we make the natural correspondence between the generators). When $1 \leqslant m < n$, define the new reflexions ρ_j (for $j = 0, \ldots, m$) by

7C1
$$\begin{aligned} \rho_j &:= \sigma_j \text{ for } j = 0, \ldots, m-1, \\ \rho_m &:= \sigma_m \sigma_{m+1} \cdots \sigma_{n-1} \sigma_n \sigma_{n-1} \sigma_{n-2} \cdots \sigma_m. \end{aligned}$$

Then $\Gamma(I) := \langle \rho_0, \ldots, \rho_m \rangle$ is the group of a cut $\mathcal{P}_m = \{4, 3^{m-2}, 4\}$ of $\mathcal{P}_n$ (or $\mathcal{P}_1 := \{\infty\}$ in case $m = 1$). In fact, in geometric terms, the cut is the intersection of the tiling $\mathcal{P}_n$ by the m-dimensional space H_I containing one of its m-faces; the corresponding subset $I \subset N := \{0, \ldots, n\}$ is $I = \{m+1, \ldots, n\}$. Since $\rho_0, \ldots, \rho_{m-2}$ clearly commute with ρ_m, the only relation to be checked algebraically is that giving the period of $\rho_{m-1}\rho_m$. We appeal to the following useful remark.

7C2 Lemma *Let Γ be a group, and let $\rho, \sigma, \tau \in \Gamma$ be involutions such that $(\rho\sigma)^3 = (\rho\tau)^2 = \varepsilon$. Then*

$$\rho\sigma\tau\sigma \sim \sigma\tau.$$

Proof. We have

$$\begin{aligned} \rho\sigma\tau\sigma &\sim \sigma\rho\sigma\tau \\ &= \rho\sigma\rho\tau \\ &= \rho\sigma\tau\rho \\ &\sim \sigma\tau, \end{aligned}$$

as claimed. □

For our result, we apply Lemma 7C2 $n - m$ times to

$$\rho_{m-1}\rho_m = \sigma_{m-1}\sigma_m\sigma_{m+1} \cdots \sigma_n\sigma_{n-1} \cdots \sigma_m,$$

to obtain

7C3
$$\rho_{m-1}\rho_m \sim \sigma_{n-1}\sigma_n.$$

The space H_I of the cut is, as asserted by the notation, the intersection of the reflexion hyperplanes $H_{m+1}, \ldots, H_n$, and so is invariant under each α in the centralizer. In particular, $\rho_0, \ldots, \rho_m$ are in this centralizer, because their reflexion planes are perpendicular to those of $\sigma_{m+1}, \ldots, \sigma_n$. However, $\Gamma(I)$ is a proper subgroup of the centralizer, since the latter contains the central element $\alpha := (\sigma_{m+1} \cdots \sigma_n)^{n-m}$ of $\langle \sigma_{m+1}, \ldots, \sigma_n \rangle$ ($\cong [3^{n-m-3}, 4]$), which is not in $\Gamma(I)$. In fact, this element α fixes the base m-face of $\mathcal{P}_n$, but is not in the stabilizer $\langle \rho_0, \ldots, \rho_{m-1} \rangle$ of this face in $\Gamma(I)$.

There are two examples of cuts in the infinite group $\langle \sigma_0, \ldots, \sigma_4 \rangle = [3, 3, 4, 3]$, one of which will be very useful in our investigation of locally toroidal polytopes. This is

7C4 $$(\sigma_0, \ldots, \sigma_4) \mapsto (\sigma_0, \sigma_1\sigma_2\sigma_3\sigma_2\sigma_1, \sigma_4\sigma_3\sigma_2\sigma_3\sigma_4) =: (\rho_0, \rho_1, \rho_2).$$

Now we see that

$$\rho_0\rho_1 \sim \sigma_2\sigma_3,$$

by two applications of Lemma 7C2. Further

$$\begin{aligned} \rho_1\rho_2 &= \sigma_1\sigma_2\sigma_3\sigma_2\sigma_1\sigma_4\sigma_3\sigma_2\sigma_3\sigma_4 \\ &\sim \sigma_4\sigma_3\sigma_4\sigma_2\sigma_3\sigma_1\sigma_2\sigma_1\sigma_3\sigma_2 \\ &\sim \sigma_4\sigma_2\sigma_3\sigma_2\sigma_3\sigma_1\sigma_3\sigma_2\sigma_3\sigma_2 \\ &\sim \sigma_4\sigma_3\sigma_2\sigma_1\sigma_2\sigma_3 \\ &\sim \sigma_3\sigma_2\sigma_1\sigma_2 \\ &\sim \sigma_1\sigma_2\sigma_3\sigma_2 \\ &= \sigma_2\sigma_3, \end{aligned}$$

where we have applied Lemma 7C2 twice. Finally, ρ_0 and ρ_2 clearly commute. It follows that $\Gamma(I) := \langle \rho_0, \rho_1, \rho_2 \rangle \cong [4, 4]$, and we have $I = \{2, 3\}$. Actually, all that we have really shown is that the group is a quotient of [4, 4]. However, inspection of the geometry shows that the corresponding cut genuinely is {4, 4}; its vertices and edges lie among those of the original tessellation $\mathcal{P} = \{3, 3, 4, 3\}$, its faces are diametral squares of facets {3, 3, 4} of $\mathcal{P}$, and its vertex-figure is a diametral square of one of the cross-polytopes {3, 3, 4} inscribed in the vertex-figure {3, 4, 3} of $\mathcal{P}$. (We should recall that the 24 vertices of {3, 4, 3} split into three sets of eight vertices of such cross-polytopes.)

Similar calculations to those we have just employed show that ρ_0, ρ_1 and ρ_2 all commute with σ_2 and σ_3. The cut is thus determined by $I = \{2, 3\}$, as just stated.

The less useful cut of the two is the {3, 6}, whose vertices and edges are again among those of $\mathcal{P} = \{3, 3, 4, 3\}$, but whose vertex-figure this time is a diametral hexagon {6} of the vertex-figure {3, 4, 3}. (The neighbouring vertices to a given vertex of such a hexagon {6} are antipodal vertices in this vertex-figure {3, 4, 3}.) With $[3, 3, 4, 3] = \langle \sigma_0, \ldots, \sigma_4 \rangle$ as before, the cut is given by

$$(\sigma_0, \ldots, \sigma_4) \mapsto (\sigma_0, \sigma_1, (\sigma_2\sigma_3\sigma_4)^3) =: (\rho_0, \rho_1, \rho_2).$$

Thus ρ_2 is the central symmetry of a co-edge of $\mathcal{P}$. We may check that indeed $\langle \rho_0, \rho_1, \rho_2 \rangle \cong [3, 6]$, and that ρ_0, ρ_1, ρ_2 commute with σ_3 and σ_4, so that $I = \{3, 4\}$. However, since we do not use this particular cut for any specific purpose, we shall leave the details to the interested reader.

There are analogous examples derived from hyperbolic tilings, which we shall employ in what follows later. The first occurs in $\{3, 4, 3, 4\}$. If we write $[3, 4, 3, 4] = \langle \sigma_0, \sigma_1, \sigma_2, \sigma_3, \sigma_4 \rangle$, then we obtain an obvious cut through the operation

$$(\sigma_0, \sigma_1, \sigma_2, \sigma_3, \sigma_4) \mapsto (\sigma_0, \sigma_1, \sigma_2, \sigma_3\sigma_4\sigma_3) =: (\rho_0, \rho_1, \rho_2, \rho_3).$$

That is, we merely pass to the cut $\{4, 4\}$ in the vertex-figure $\{4, 3, 4\}$; this is clearly the cut determined by $I = \{4\}$.

More interesting are the cuts of the hyperbolic honeycombs of rank 6 (and so dimension 5). These are obtained in exactly the same way as the cuts of $\{3, 3, 4, 3\}$. Up to duality, there are three regular hyperbolic 5-honeycombs, namely, $\{3, 3, 3, 4, 3\}$, $\{3, 3, 4, 3, 3\}$ and $\{3, 4, 3, 3, 4\}$. Let us treat them in turn.

First, then, let $\langle \sigma_0, \ldots, \sigma_5 \rangle = [3, 3, 3, 4, 3]$. If we perform the analogous operation to that of (7C4) on the generators of the group of the vertex-figure, namely,

$$(\sigma_0, \sigma_1, \sigma_2, \sigma_3, \sigma_4, \sigma_5) \mapsto (\sigma_0, \sigma_1, \sigma_2\sigma_3\sigma_4\sigma_3\sigma_2, \sigma_5\sigma_4\sigma_3\sigma_4\sigma_5) =: (\rho_0, \rho_1, \rho_2, \rho_3),$$

then we obtain the cut whose group is $\Gamma(I) \cong [3, 4, 4]$, with $I = \{3, 4\}$.

We may also employ the second of our cuts of the vertex-figure, given by

$$(\sigma_0, \sigma_1, \sigma_2, \sigma_3, \sigma_4, \sigma_5) \mapsto (\sigma_0, \sigma_1, \sigma_2, (\sigma_3\sigma_4\sigma_5)^3) =: (\rho_0, \rho_1, \rho_2, \rho_3),$$

which gives a cut with group $\Gamma(I) \cong [3, 3, 6]$, where $I = \{4, 5\}$.

Second, with $\langle \sigma_0, \ldots, \sigma_5 \rangle = [3, 3, 4, 3, 3]$, we perform the operation given by (7C4) on the group of the facet instead, namely,

$$(\sigma_0, \sigma_1, \sigma_2, \sigma_3, \sigma_4, \sigma_5) \mapsto (\sigma_0, \sigma_1\sigma_2\sigma_3\sigma_2\sigma_1, \sigma_4\sigma_3\sigma_2\sigma_3\sigma_4, \sigma_5) =: (\rho_0, \rho_1, \rho_2, \rho_3);$$

then we obtain the cut whose group is $\Gamma(I) \cong [4, 4, 4]$, with $I = \{2, 3\}$. In this instance, there is no cut analogous to the second kind of $\{3, 3, 4, 3\}$. (It is tempting to think that there might be a cut of kind $\{3, 6, 3\}$ of $\{3, 3, 4, 3, 3\}$, but this appears not to be the case.)

With the last case $\langle \sigma_0, \ldots, \sigma_5 \rangle = [3, 4, 3, 3, 4]$, we perform the analogous dual operation on the group of the facet, yielding

$$(\sigma_0, \sigma_1, \sigma_2, \sigma_3, \sigma_4, \sigma_5) \mapsto (\sigma_0\sigma_1\sigma_2\sigma_1\sigma_0, \sigma_3\sigma_2\sigma_1\sigma_2\sigma_3, \sigma_4, \sigma_5) =: (\rho_0, \rho_1, \rho_2, \rho_3),$$

which leads to a cut with group $\Gamma(I) \cong [4, 4, 4]$ again, but now $I = \{1, 2\}$. Here, there is again a cut analogous to the second kind of $\{3, 3, 4, 3\}$, given by

$$(\sigma_0, \sigma_1, \sigma_2, \sigma_3, \sigma_4, \sigma_5) \mapsto ((\sigma_0\sigma_1\sigma_2)^3, \sigma_3, \sigma_4, \sigma_5) =: (\rho_0, \rho_1, \rho_2, \rho_3);$$

the group is now $\Gamma(I) \cong [6, 3, 4]$, with $I = \{0, 1\}$.

There is one important aspect of the theory of cuts which we have not so far mentioned. In our applications of cuts to problems of classifying locally toroidal regular polytopes, we should most like to know which cuts were *universal*, in the following sense.

Suppose that we wish to obtain a new regular polytope by imposing a set $\mathcal{R}$ of further relations on the group $\Gamma(\mathcal{P})$ of a given regular polytope $\mathcal{P}$. In many applications, $\mathcal{P}$ will be a universal polytope $\{p_1, \ldots, p_{n-1}\}$, for instance, a regular honeycomb in euclidean or hyperbolic space; then $\Gamma(\mathcal{P})$ will be a Coxeter group in the natural way. In some cases, it turns out that we are, in effect, actually imposing such relations $\mathcal{R}$ on the group $\Gamma(\mathcal{Q})$ of a suitable cut of the original polytope $\mathcal{P}$, the corresponding cut $\mathcal{Q} = \mathcal{P}(I)$ itself having a Coxeter group as its group. The question which must therefore be addressed is whether the imposition of these relations actually forces more relations on the quotient of $\Gamma(\mathcal{Q})$ induced by $\mathcal{R}$, by virtue of the embedding of $\Gamma(\mathcal{Q})$ as a subgroup of $\Gamma(\mathcal{P})$. (We shall see later that this is the likely situation.) If they do not, then we say that $\mathcal{Q}$ is a *universal cut* of $\mathcal{P}$ *relative to* $\mathcal{R}$.

The new relations $\mathcal{R}$ on $\Gamma(\mathcal{P})$ may indeed define a new polytope $\mathcal{P}'$. However, in many cases, the cut $\mathcal{Q}$ of $\mathcal{P}$ induces a cut $\mathcal{Q}'$ of $\mathcal{P}'$ whose group $\Gamma(\mathcal{Q}')$ satisfies the set $\mathcal{R}'$ of relations corresponding to $\mathcal{R}$, but with the addition of other independent relations. Thus, $\mathcal{Q}$ is a universal cut of $\mathcal{P}$ relative to $\mathcal{R}$ if and only if $\mathcal{R}'$ and the defining relations for $\Gamma(\mathcal{Q})$ give a complete presentation for $\Gamma(\mathcal{Q}')$.

Unfortunately, it turns out that very few cuts are actually universal. We shall shortly discuss the concept in general terms, but for the moment let us illustrate the phenomenon of non-universality by a simple example. (We shall encounter some more important examples in our investigation of locally toroidal regular polytopes in Chapter 12.) Let $n \geqslant 4$, and consider the group $[4, 3^{n-2}, 4] = \langle \sigma_0, \ldots, \sigma_n \rangle$ of the regular tessellation of $\mathbb{E}^n$ by cubes. There is a cut $\mathcal{Q} = \{4, 3^{n-3}, 4\}$ of $\mathcal{P} = \{4, 3^{n-2}, 4\}$, given by

$$(\sigma_0, \ldots, \sigma_n) \mapsto (\sigma_0, \ldots, \sigma_{n-2}, \sigma_{n-1}\sigma_n\sigma_{n-1}) =: (\rho_0, \ldots, \rho_{n-1}).$$

The relation which we impose on $\Gamma(\mathcal{Q})$ to yield $\{4, 3^{n-3}, 4\}_{(s^{n-1})}$ is

$$(\rho_0 \cdots \rho_{n-1})^{(n-1)s} = \varepsilon.$$

In terms of the generators of $\Gamma(\mathcal{P})$, this is

$$(\sigma_0 \cdots \sigma_{n-1}\sigma_n\sigma_{n-1})^{(n-1)s} = \varepsilon.$$

Now we saw in Section 6D that a toroidal quotient of $\mathcal{P}$ can only be of the form $\{4, 3^{n-2}, 4\}_{\mathbf{s}}$, with $\mathbf{s} = (s, 0^{n-1})$, $(s, s, 0^{n-2})$ or (s^n). Indeed, the analysis there (which relied quite heavily on the geometry) showed that the corresponding quotient $\mathcal{P}'$ of $\mathcal{P}$ must be $\{4, 3^{n-2}, 4\}_{\mathbf{s}}$, with $\mathbf{s} = (s, 0^{n-1})$ or $(s, s, 0^{n-2})$, as n is even or odd. In either case, the resulting relation on the cut is not the one originally imposed. In fact, the cut $\mathcal{Q}'$ of $\mathcal{P}'$ which is induced by $\mathcal{Q}$ is actually $\{4, 3^{n-2}, 4\}_{\mathbf{s}}$, with $\mathbf{s} = (s, 0^{n-2})$ or $(s, s, 0^{n-3})$. On the other hand, the new relation on $\Gamma(\mathcal{Q})$ is just the one which yields the toroid $\{4, 3^{n-2}, 4\}_{(s^{n-1})}$. It thus follows that $\mathcal{Q}$ is not universal relative to the relation which we imposed on $\Gamma(\mathcal{P})$.

What, then, is the general situation? Let us write, as before, $\Gamma = \Gamma(\mathcal{P})$ for the full group, and $\Delta := \Gamma(\mathcal{Q})$ for the group of the cut. We wish to impose on Γ a set of relations $\mathcal{R} \subseteq \Delta$. (Here we adopt the usual convention of identifying a relation $\omega = \varepsilon$, where ω is a word in the appropriate group, with the corresponding *relator* ω.) The resulting quotient group is thus $\Gamma / N_\Gamma(\mathcal{R})$, where $N_\Gamma(\mathcal{R})$ is the normal closure of $\mathcal{R}$ in Γ (that is, the smallest normal subgroup of Γ which contains $\mathcal{R}$). The induced quotient of the subgroup Δ is

$$\Delta/(\Delta \cap N_\Gamma(\mathcal{R})) \cong \Delta \cdot N_\Gamma(\mathcal{R})/N_\Gamma(\mathcal{R}).$$

It is then clear when the cut is universal.

7C5 Theorem *Let $\Gamma = \Gamma(\mathcal{P})$ be the group of a regular polytope $\mathcal{P}$, $\Delta = \Gamma(\mathcal{Q})$ the group of a cut $\mathcal{Q}$ of $\mathcal{P}$, and $\mathcal{R} \subseteq \Delta$ a set of relations such that $\Gamma/N_\Gamma(\mathcal{R})$ is a C-group. Then $\mathcal{Q}$ is universal relative to $\mathcal{R}$ if and only if $N_\Delta(\mathcal{R}) = \Delta \cap N_\Gamma(\mathcal{R})$.*

Proof. This is clear from the definition. Observe that $N_\Delta(\mathcal{R}) \leqslant \Delta \cap N_\Gamma(\mathcal{R})$ always holds, so that imposing the relations $\mathcal{R}$ on Γ will, in general, force more identifications on the cut $\mathcal{Q}$ than would have resulted from imposing $\mathcal{R}$ on Δ. □

7D The Classical Star-Polytopes

The polytopes of the heading of this section are those whose classification was central to the earlier theory of regular polytopes, and, in particular, to [120]. For a brief historical introduction, see also Section 1A.

We adopt an exclusively geometrical viewpoint here, confining our attention to faithfully realized finite regular n-polytopes P, with the additional property that, for each $j = 0, \ldots, n$ and each j-face F of P, we have $\dim F = j$. Recall that we distinguish between the combinatorial concept of rank of a polytope and the geometric concept of dimension of one of its realizations. We also refer to Theorem 5B20; since the dimension of a faithful realization cannot be less than the rank, such regular polytopes have an obvious minimality property. We shall call these polytopes *classical*, in view of their historical role in the subject. Of course, the motivation is provided by (among others) the regular convex polytopes, which certainly are classical in this technical sense as well.

Our first main result will illustrate the importance of the defining condition. However, we begin with a geometrical lemma concerning the *centre* $c(P)$ of a regular polytope P, that is, the centroid of its vertices.

7D1 Lemma *Let P be a classical regular polytope, and let F be a facet of P. Then $c(P) \notin \operatorname{aff} F$.*

Proof. We employ an induction argument on $n := \dim P$; the case $n = 1$ is trivial. If $n > 1$, let $G < F$ be a ridge (that is, $(n-2)$-face) of P; in such circumstances, we may always assume that faces under consideration belong to the base flag of P. By the inductive assumption, $c(F) \notin \operatorname{aff} G$, since F itself is classical. Were we to have $c(P) \in \operatorname{aff} F$, then $c(F) = c(P)$, by the uniqueness of the centre. But then the generating reflexion R_{n-1} of $G(P)$ would have to fix G and $c(F)$, and so would fix F itself, since $\operatorname{aff}(G \cup \{c(F)\}) = \operatorname{aff} F$. Since P is faithful, this is impossible. The contradiction thus yields the result. □

We are now in a position to prove

7D2 Theorem *The symmetry group $G(P)$ of a classical regular polytope P is generated by hyperplane reflexions.*

Proof. Again, we may use induction on $n := \dim P$, the case $n = 1$ being trivial. Let F and G be as in the proof of Lemma 7D1, and let $G(P) = \langle R_0, \ldots, R_{n-1} \rangle$ in the usual way. Within aff F, by the inductive assumption, (the mirrors of) $R_0, \ldots, R_{n-2}$ are hyperplanes; that is, $\dim(R_j \cap \operatorname{aff} F) = n - 2$ for each $j = 0, \ldots, n - 2$. Since $c(P) \notin \operatorname{aff} F$ by Lemma 7D1, we see that $R_j = \operatorname{aff}((R_j \cap \operatorname{aff} F) \cup \{c(P)\})$ is a hyperplane for $j = 0, \ldots, n - 2$. Finally, since R_{n-1} fixes G, $\dim G = n - 2$ and $c(P) \notin \operatorname{aff} G \subset \operatorname{aff} F$, we conclude that $\dim R_{n-1} \geqslant n - 1$. Using Theorem 5B20, or just remarking that $R_{n-1} \neq \mathbb{E}^n$, we see that the opposite inequality is trivial, and so we have proved the result. □

It is usually the case that an abstract regular polytope can only be described combinatorially by considerable modification of the symbol denoting its Schläfli type. For classical regular polytopes, on the other hand, a much more concise notation can be devised. Let P be a classical regular n-polytope, and let $G(P) = \langle R_0, \ldots, R_{n-1} \rangle$ be its symmetry group, generated by the hyperplane reflexions associated with a base flag of P. Taking the ambient space of P to be $\mathbb{E}^n$ and $c(P) = o$, for $j = 0, \ldots, n - 1$, we may write

$$R_j = \{x \in \mathbb{E}^n \mid \langle x, u_j \rangle = 0\},$$

with each u_j a unit vector. Further, we may change signs of the u_j, if necessary, to ensure that $\langle u_{j-1}, u_j \rangle < 0$ for $j = 1, \ldots, n - 1$ (we have $\langle u_j, u_k \rangle = 0$ if $j \leqslant k - 2$). Since $\langle R_{j-1}, R_j \rangle$ is a finite group, for each $j = 1, \ldots, n - 1$ there is a rational number $p_j > 2$ (written as a fraction in its lowest terms), such that $\langle u_{j-1}, u_j \rangle = -\cos(\pi/p_j)$. Then P is denoted by its *Schläfli symbol*

7D3 $$P := \{p_1, \ldots, p_{n-1}\}.$$

This new notation naturally accords with our informal earlier usage, such as $\{\frac{5}{2}\}$ for the regular pentagram, and $\{\frac{5}{2}, 5\}$ for the small stellated dodecahedron. We have shown (see Section 1B) that the classical regular polytopes whose Schläfli symbols contain only integer entries are precisely the convex regular polytopes (or, equivalently, the finite universal regular polytopes $\{p_1, \ldots, p_{n-1}\}$ of Section 3D). Those whose symbols contain at least one non-integer entry are therefore *starry*, or are *regular star-polytopes*, so called because their faces or co-faces are not those of the convex hull of the vertices.

There are important consequences of Theorem 7D2.

7D4 Corollary *A classical regular polytope $P = \{p_1, \ldots, p_{n-1}\}$ always has a dual, which is the classical regular polytope $P^* = \{p_{n-1}, \ldots, p_1\}$.*

Proof. We construct P^* geometrically. Just as P is obtained from its group $\langle R_0, \ldots, R_{n-1} \rangle$ by Wythoff's construction, starting with the initial vertex $v \in (R_1 \cap \cdots \cap R_{n-1}) \setminus \{o\}$, so we obtain P^* by Wythoff's construction, with initial vertex $v^* \in (R_{n-2} \cap \cdots \cap R_0) \setminus \{o\}$. In this context, observe that the Wythoff space is one-dimensional, because the generators R_j are hyperplane reflexions. Thus $G(P^*) = G(P)$, with the distinguished generators labelled in the reverse order. Evidently the Schläfli symbol of P^* is the reverse of that of P, as claimed. □

7D5 Corollary *For $n \geqslant 2$, the facet and vertex-figure of the regular n-polytope $P = \{p_1, \ldots, p_{n-1}\}$ are the regular $(n-1)$-polytopes $\{p_1, \ldots, p_{n-2}\}$ and $\{p_2, \ldots, p_{n-1}\}$, respectively.*

Proof. This is clear either directly, or by applying Wythoff's construction to the subgroup $\langle R_0, \ldots, R_{n-2} \rangle$ with the same initial vertex v of P, or to $\langle R_1, \ldots, R_{n-1} \rangle$ with initial vertex the other vertex w of the initial (base) edge E of P. □

The aim in this section is to classify completely all the classical regular polytopes; we shall demonstrate that there are no starry such polytopes in dimensions larger than 4. In fact, we shall establish a result first observed by Hess [215], and proved using a case by case enumeration by van Oss [337]. Several proofs are to be found in [120, Chapter 14], but here we follow the systematic approach in [280], which does not rely on any prior knowledge of the actual polytopes. For a discussion of the densities of the regular star-polytopes, the reader is referred to Coxeter [95, 97, 98, 120, §14.8]; see also Section 3E.

The observation made by Hess was

7D6 Theorem *Every regular star-polytope has the same vertices as some regular convex polytope.*

We shall prove this theorem in a stronger form, for which we need one further concept. Associated with the regular polytope $P = \{p_1, \ldots, p_{n-1}\}$ is its *Schläfli determinant* $\Delta(P)$, which is defined by

7D7
$$\Delta(P) := \begin{vmatrix} 1 & -\cos\frac{\pi}{p_1} & 0 & 0 & \cdots & 0 & 0 \\ -\cos\frac{\pi}{p_1} & 1 & -\cos\frac{\pi}{p_2} & 0 & \cdots & 0 & 0 \\ 0 & -\cos\frac{\pi}{p_2} & 1 & -\cos\frac{\pi}{p_3} & \cdots & 0 & 0 \\ \vdots & \vdots & \vdots & \vdots & \ddots & \vdots & \vdots \\ 0 & 0 & 0 & 0 & \cdots & 1 & -\cos\frac{\pi}{p_{n-1}} \\ 0 & 0 & 0 & 0 & \cdots & -\cos\frac{\pi}{p_{n-1}} & 1 \end{vmatrix}.$$

Thus the appropriate definition for the 1-polytope { } is

7D8
$$\Delta(\{\,\}) := 1.$$

If we write U for the $n \times n$ matrix whose rows are the unit vectors $u_0, \ldots, u_{n-1}$ associated with the mirrors $R_0, \ldots, R_{n-1}$, then we see that

7D9
$$\Delta(P) = \det(UU^{\mathsf{T}}) > 0,$$

since clearly $\{u_0, \ldots, u_{n-1}\}$ spans $\mathbb{E}^n$.

We now give a geometric interpretation of the Schläfli determinant (compare [120, §7.7]). Let P be a classical regular n-polytope in $\mathbb{E}^n$; as usual, we take $c(P) = o$. An edge E of P will subtend an angle $2\varphi(P)$, say, at o. We recall also that P/v denotes the vertex-figure of P (at some vertex v). Then we have

7D10 Lemma *If* $n > 1$, *then*

$$\Delta(P) = \Delta(P/v) \sin^2 \varphi(P).$$

Proof. Suppose that the vertices of P lie on a sphere of radius $\rho(P)$ about o, and that P has edge-length $\lambda(P)$; thus

$$\lambda(P) = 2\rho(P) \sin \varphi(P).$$

The vertices of the vertex-figure P/v are the other vertices of the edges of P through v, so that a typical edge E of P/v has length

$$\lambda(P/v) = 2\lambda(P) \cos \frac{\pi}{p_1},$$

the edge-length of the vertex-figure of the 2-face $\{p_1\}$ of P. Further, we clearly have

$$\rho(P/v) = \lambda(P) \cos \varphi(P).$$

Substituting $\lambda(P/v) = 2\rho(P/v) \sin \varphi(P/v)$, there follows at once

7D11 $$\cos \varphi(P) = \frac{\rho(P/v)}{\lambda(P)} = \frac{\lambda(P/v)}{\lambda(P)} \cdot \frac{\rho(P/v)}{\lambda(P/v)} = \frac{\cos(\pi/p_1)}{\sin \varphi(P/v)},$$

which we can rewrite as

7D12 $$\sin^2 \varphi(P) = 1 - \frac{\cos^2(\pi/p_1)}{\sin^2 \varphi(P/v)}.$$

We may now make the inductive assumption

$$\Delta(P/v) = \Delta(P/E) \sin^2 \varphi(P/v),$$

with E (as before) an edge of P. We observe that the lemma does hold if $n = 2$, as is easily checked. This then leads to

$$\begin{aligned} \Delta(P/v) \sin^2 \varphi(P) &= \Delta(P/v) \left[1 - \frac{\cos^2(\pi/p_1)}{\sin^2 \varphi(P/v)}\right] \\ &= \Delta(P/v) - \cos^2 \frac{\pi}{p_1} \Delta(P/E) \\ &= \Delta(P), \end{aligned}$$

by direct calculation from the definition (7D7). This concludes the proof. □

Proof. We shall now prove Theorem 7D6, in the following strong form. If P is a starry regular n-polytope, then

7D13
(a) P has the vertices of a unique regular convex polytope $\bar{P}$,
(b) moreover, $G(P) = G(\bar{P})$,
(c) and $\Delta(P) > \Delta(\bar{P})$.

The induction begins with the case $n = 2$. If $P = \{\frac{p}{d}\}$ is a regular star-polygon, with p and d integers satisfying $2 \leqslant d < \frac{1}{2}p$ and $(d, p) = 1$, then $\bar{P}$ is the convex regular p-gon $\{p\}$ with the same vertices. Clearly, $G(P) = G(\bar{P})$; further

$$\Delta(P) = \sin^2 \frac{d\pi}{p} > \sin^2 \frac{\pi}{p} = \Delta(\bar{P}).$$

Thus the conditions hold here.

Now suppose that $n \geqslant 3$, and that $P = \{p_1, \ldots, p_{n-1}\}$ is a regular star-polytope. Then at least one of $p_1, \ldots, p_{n-1}$ is not an integer; appealing to Corollary 7D4, we may dualize, if necessary, and suppose that it is one of $p_2, \ldots, p_{n-1}$. By Corollary 7D5, the vertex-figure $P/v = \{p_2, \ldots, p_{n-1}\}$ of P is a classical regular $(n-1)$-polytope, which is starry. Our inductive assumption then enables us to replace $P_0 := P/v$ by the unique regular convex polytope $\overline{P_0} = \{q_2, \ldots, q_{n-1}\}$, say, with the same vertices. Note that $q_2, \ldots, q_{n-1} \geqslant 3$ are integers.

Let $E = \{v, w\}$ be the initial edge of P. Since, also, $G(\overline{P_0}) = G(P_0) = G(P/v) = \langle R_1, \ldots, R_{n-1}\rangle$ (each considered as a reflexion group in $\mathbb{E}^n$), we can write $G(\overline{P_0}) = \langle S_1, \ldots, S_{n-1}\rangle$, where $S_1, \ldots, S_{n-1} \in G(P/v)$ are the distinguished generators (again hyperplane reflexions in $\mathbb{E}^n$) associated with a base flag of $\overline{P_0}$ which contains w. Define $S_0 := R_0$, where $G(P) = \langle R_0, \ldots, R_{n-1}\rangle$ with the usual convention. Since $v \in S_j$ for each $j = 1, \ldots, n-1$, and $w \in S_j$ for $j = 2, \ldots, n-1$, we see that $E \subseteq S_j$ for $j = 2, \ldots, n-1$. But $S_0 = R_0$ is the perpendicular bisector of E, so that $S_0 \perp S_j$ for $j = 2, \ldots, n-1$.

Our inductive assumption says that $G(\overline{P_0}) = G(P_0) = G(P/v)$; hence

$$\langle S_0, \ldots, S_{n-1}\rangle = \langle R_0, \ldots, R_{n-1}\rangle.$$

If we apply Wythoff's construction, with the same initial vertex v, we thus obtain a regular polytope

$$Q = \{q_1, \ldots, q_{n-1}\}$$

with vertex-figure $\overline{P_0}$. By construction, Q will have the same vertices and edges as P; observe that E is also the initial edge of Q. We call this process of obtaining Q from P *vertex-figure replacement*.

We calculate $\Delta(Q)$ and q_1 as follows. First, we have $\varphi(Q) = \varphi(P)$, since P and Q have the same vertices and edges. From Lemma 7D10, we have

$$\Delta(P) = \Delta(P_0) \sin^2 \varphi(P),$$
$$\Delta(Q) = \Delta(\overline{P_0}) \sin^2 \varphi(P),$$

so that

$$\frac{\Delta(Q)}{\Delta(P)} = \frac{\Delta(\overline{P_0})}{\Delta(P_0)}.$$

In view of the inductive assumption $\Delta(\overline{P_0}) < \Delta(P_0)$, we conclude that

$$\Delta(Q) < \Delta(P).$$

Second, (7D11) immediately implies that

$$\cos\frac{\pi}{p_1} = \cos\varphi(P)\,\sin\varphi(P/v).$$

Since $\varphi(P) = \varphi(Q)$, there follows

7D14
$$\frac{\cos(\pi/q_1)}{\cos(\pi/p_1)} = \frac{\sin\varphi(Q/v)}{\sin\varphi(P/v)},$$

whence q_1 can be calculated using

$$\sin^2\varphi(Q/v) = \frac{\Delta(Q/v)}{\Delta(Q/E)} = \frac{\Delta(\overline{P_0})}{\Delta(\overline{P_0}/w)};$$

recall that $E = \{v, w\}$.

If q_1 is also an integer, then we are finished. Otherwise, we dualize and replace the vertex-figure, repeating the process as often as necessary. Now $G(P)$ is finite, and so only finitely many choices are possible of distinguished systems of generating reflexions from among those of $G(P)$. Since $\Delta(Q) < \Delta(P)$, the Schläfli determinant strictly decreases at each stage. The only way the process can terminate is thus at a convex regular polytope $\bar{P}$, say.

Now $G(\bar{P}) = G(P)$. We recall the description in Section 1B of the group $G(\bar{P})$ of a convex regular n-polytope $\bar{P}$. This is a finite Coxeter group, generated by the reflexions $\bar{R}_j$ in hyperplanes $\bar{H}_j$ containing o, for $j = 0, \ldots, n-1$. Since $\bar{H}_0, \ldots, \bar{H}_{n-1}$ bound a simplicial cone C which is a fundamental region for the action of $G(\bar{P})$ on $\mathbb{E}^n$, any subgroup of $G(\bar{P})$ which is generated by hyperplane reflexions and fixes a line is conjugate to a subgroup of a distinguished subgroup $\langle \bar{R}_j \mid j \in J\rangle$, for some subset $J \subseteq N := \{0, \ldots, n-1\}$. In particular, the only subgroups of $G(\bar{P})$ which can be the symmetry groups of classical regular $(n-1)$-polytopes are (subgroups of) the groups of its facet or vertex-figure. Dualizing $\bar{P}$ if necessary, we may thus suppose that the subgroup $G(P/v)$ of $G(P) = G(\bar{P})$ is a subgroup of the group of the vertex-figure $\bar{P}$. It follows at once that $\bar{P}$ can be taken to have the same initial vertex v as P, so that the vertices of P and $\bar{P}$ coincide; this completes the inductive step of the argument. □

Theorem 7D6 shows how a regular star-polytope is related to some regular convex polytope. But the method of proof of Theorem 7D6 is, if anything, even more instructive, for it says that we may find all the regular star-polytopes, if we reverse the vertex-figure replacement process in all possible ways, starting from a suitable regular convex polytope. Thus we may replace any vertex-figure by another regular polytope with the same vertices.

Taking for granted the regular star-polygons, we first apply the procedure to 3-polytopes. In this case, (7D14) takes the form

$$\frac{\cos(\pi/q_1)}{\cos(\pi/p_1)} = \frac{\sin(\pi/q_2)}{\sin(\pi/p_2)}.$$

The only regular convex polyhedron to which the process is applicable is $Q = \{3, 5\}$, whose vertex-figure $\{5\}$ may be replaced by $\{\frac{5}{2}\}$. We then obtain the family

$$\begin{array}{llll}
\{5,3\} & \{3,5\} & & \\
 & \{5,\frac{5}{2}\} & \{\frac{5}{2},5\} & \\
 & & \{3,\frac{5}{2}\} & \{\frac{5}{2},3\}
\end{array}$$

Polyhedra in the same row are dual, while those in the same column are obtained by vertex-figure replacement. Considering the symmetry groups of the vertex-figures shows that, of the star-polyhedra, $\{\frac{5}{2},3\}$ has the same vertices as $\{5,3\}$, while the others have the same vertices as $\{3,5\}$.

For a regular 4-polytope $P = \{p_1, p_2, p_3\}$, an easy direct calculation from (7D11) (see also [120, 2.44]) yields

$$\cos\varphi(P/v) = \cos\frac{\pi}{p_2}\operatorname{cosec}\frac{\pi}{p_3}$$

for its vertex-figure P/v. Thus (7D14) takes the form

$$\frac{\cos(\pi/q_1)}{\cos(\pi/p_1)} = \frac{\sqrt{1-\cos^2(\pi/q_2)\operatorname{cosec}^2(\pi/q_3)}}{\sqrt{1-\cos^2(\pi/p_2)\operatorname{cosec}^2(\pi/p_3)}}.$$

The only possible starting point for applying the procedure is the 600-cell $Q = \{3,3,5\}$. With the same conventions as before, the resulting family is

$$\begin{array}{llllll}
\{5,3,3\} & \{3,3,5\} & & & & \\
 & \{3,5,\frac{5}{2}\} & \{\frac{5}{2},5,3\} & & & \\
 & \{5,\frac{5}{2},5\} & & & & \\
 & \{5,3,\frac{5}{2}\} & & \{\frac{5}{2},3,5\} & & \\
 & & & \{\frac{5}{2},5,\frac{5}{2}\} & & \\
 & & \{5,\frac{5}{2},3\} & \{3,\frac{5}{2},5\} & & \\
 & & & \{3,3,\frac{5}{2}\} & \{\frac{5}{2},3,3\} &
\end{array}$$

Considering the groups of the vertex-figures, we see that, among the star-polytopes, $\{\frac{5}{2},3,3\}$ will have the vertices of $\{5,3,3\}$, while the remainder will have the vertices of $\{3,3,5\}$. Note that, in the second and fourth columns, the first two polytopes share the same 2-faces (as well as vertices and edges), as do the last two.

Since the procedure is not applicable to any regular convex n-polytope with $n \geqslant 5$, we see that these two lists complete the enumeration of the classical regular polytopes.

Let us end the discussion of this section by describing the groups of the regular star-polytopes (see also [290]). We shall employ the mixing operations which we introduced in Section 7B, and variants of them. We begin with a result we have already quoted; it occurs first in [105, p. 52].

7D15 Theorem $\{5,\frac{5}{2}\} \cong \{5,5\,|\,3\}$.

Proof. Let $\Gamma(\{5,\frac{5}{2}\}) = \langle\rho_0,\rho_1,\rho_2\rangle$, in terms of its distinguished generators. Since $\{5,\frac{5}{2}\} = \{3,5\}^{\varphi_2}$, we see that the required group is obtained from the Coxeter group

$[3, 5] = \langle \sigma_0, \sigma_1, \sigma_2 \rangle$ by the mixing operation

$$(\sigma_0, \sigma_1, \sigma_2) \mapsto (\sigma_0, \sigma_1\sigma_2\sigma_1, \sigma_2) = (\rho_0, \rho_1, \rho_2).$$

The inverse of this operation is φ_3. This is equivalent to φ_2, but would not yield the original generators; in this context, the distinction is unimportant, but we shall wish to maintain it when we come to the starry 4-polytopes. Thus the inverse is

$$(\rho_0, \rho_1, \rho_2) \mapsto (\rho_0, \rho_1\rho_2\rho_1\rho_2\rho_1, \rho_2) = (\sigma_0, \sigma_1, \sigma_2).$$

Now the holes of any regular polyhedron $\mathcal{P}$ of type $\{3, q\}$ are q-gons, and so $\mathcal{P}^{\varphi_2}$ is of type $\{q, q\}$ if q is odd. Thus $\{5, \frac{5}{2}\}$ is indeed of type $\{5, 5\}$. Of the other defining relations for $[3, 5]$, there remains $(\sigma_0\sigma_1)^3 = \varepsilon$, which gives directly $(\rho_0\rho_1\rho_2\rho_1)^3 = \varepsilon$; this says that the holes are triangles. This is the conclusion required. □

Since $\{5, 5 \,|\, 3\}$ is self-dual, we see from the second half of the list of 3-polytopes that polyhedra obtained from each other by interchanging 5 and $\frac{5}{2}$ are isomorphic.

We now move on to the 4-polytopes. We shall soon see that the interchange of 5 and $\frac{5}{2}$ induces isomorphism of polytopes here as well. If we anticipate this result, and also confine our attention to one of each dual pair, we conclude that we need only consider the three polytopes $\{3, 5, \frac{5}{2}\}$, $\{5, \frac{5}{2}, 5\}$ and $\{5, 3, \frac{5}{2}\}$. For the last, we must introduce a generalization of the notation for holes. If the automorphism group $\Gamma(\mathcal{P})$ of the regular n-polytope $\mathcal{P}$ is obtained from the Coxeter group $[p_1, \ldots, p_{n-1}] = \langle \rho_0, \ldots, \rho_{n-1} \rangle$ by imposing the single extra relation

$$(\rho_0\rho_1 \cdots \rho_{n-1}\rho_{n-2} \cdots \rho_1)^r = \varepsilon,$$

then we write

$$\mathcal{P} = \{p_1, \ldots, p_{n-1} \,|\, r\}.$$

We may think of this extra relation as determining a *deep hole*. Just as with ordinary holes, a deep hole corresponds to a certain circuit of edges; its group is generated by the involutions ρ_0 and $\rho_1 \cdots \rho_{n-1}\rho_{n-2} \cdots \rho_1$; the former interchanges the two vertices of the initial edge in the base flag of $\mathcal{P}$, and the latter is a conjugate of ρ_{n-1} which fixes the initial vertex.

The relation for a deep hole is preserved under duality, so that

$$\mathcal{P}^* = \{p_{n-1}, \ldots, p_0 \,|\, r\}.$$

We remark that Lemma 7C2 and induction on k show that, for $k \geqslant 3$, a regular polytope of type $\{3^{k-2}, q\}$ has deep holes of type $\{q\}$.

The starry regular 4-polytopes are then given by

7D16 Theorem *The following isomorphisms hold:*

(a) $\{3, 5, \frac{5}{2}\} \cong \{\{3, 5\}, \{5, 5 \,|\, 3\}\}$;
(b) $\{5, \frac{5}{2}, 5\} \cong \{\{5, 5 \,|\, 3\}, \{5, 5 \,|\, 3\}\}$;
(c) $\{5, 3, \frac{5}{2}\} \cong \{5, 3, 5 \,|\, 3\}$.

Proof. Note that (a) and (b) say that the two polytopes are universal of their type, while (c) says that the type of $\{5, 3, \frac{5}{2}\}$ is determined by triangular deep holes. Of course, $\{5, 3, \frac{5}{2}\}$ cannot be universal of type $\{5, 3, 5\}$, since the latter is infinite. Another consequence of (c) is that $\{5, 3, \frac{5}{2}\}$ is isomorphic to its dual $\{\frac{5}{2}, 3, 5\}$, and this will justify the earlier remark about the interchange of 5 and $\frac{5}{2}$.

It is tedious rather than difficult to find the mixing operations which give the distinguished generators $\rho_0, \ldots, \rho_3$ of each new group in terms of the Coxeter group $[3, 3, 5] = \langle \sigma_0, \ldots, \sigma_3 \rangle$, and the inverse operations. We are guided by the process of vertex-figure replacement described earlier; the dissection theorems of [146] are also relevant in this context. It is helpful to remember that only the subgroup of the vertex-figure will actually change.

In the calculations that follow, we shall make frequent use of Lemma 7C2; we shall not note each occurrence of it.

(a) The first operation on $[3, 3, 5]$ is

$$(\sigma_0, \sigma_1, \sigma_2, \sigma_3) \mapsto (\sigma_0, \sigma_1, \sigma_2\sigma_3\sigma_2, \sigma_3) =: (\rho_0, \rho_1, \rho_2, \rho_3);$$

this employs the facetting operation φ_2 on the vertex-figure $\{3, 5\}$ of $\{3, 3, 5\}$. Most of the relations satisfied by the new generators are obvious (for example, they are all involutions), and so we concentrate on those which are not. We first observe that $\rho_2\rho_3 = (\sigma_2\sigma_3)^2$, reflecting the change from 5 to $\frac{5}{2}$ in the Schläfli symbol; thus $(\rho_2\rho_3)^5 = \varepsilon$. Further,

$$\begin{aligned} \rho_1\rho_2 &= \sigma_1\sigma_2\sigma_3\sigma_2 \\ &\sim \sigma_2\sigma_3, \end{aligned}$$

with $\sim$ as usual denoting conjugacy, since $(\sigma_1\sigma_2)^3 = \varepsilon$.

The inverse operation is

$$(\rho_0, \rho_1, \rho_2, \rho_3) \mapsto (\rho_0, \rho_1, \rho_2\rho_3\rho_2\rho_3\rho_2, \rho_3) = (\sigma_0, \sigma_1, \sigma_2, \sigma_3).$$

It follows that the automorphism group $\Gamma(\{3, 5, \frac{5}{2}\})$ of $\{3, 5, \frac{5}{2}\}$ is obtained from the Coxeter group $[3, 5, 5]$ by imposing the single extra relation arising from $(\sigma_1\sigma_2)^3 = \varepsilon$. Now

$$\begin{aligned} \sigma_1\sigma_2 &= \rho_1\rho_2\rho_3\rho_2\rho_3\rho_2 \\ &= \rho_1\rho_3\rho_2\rho_3\rho_2\rho_3 \\ &= \rho_3\rho_1\rho_2\rho_3\rho_2\rho_3 \\ &\sim \rho_1\rho_2\rho_3\rho_2. \end{aligned}$$

That is, $\{3, 5, \frac{5}{2}\} \cong \{\{3, 5\}, \{5, 5 \,|\, 3\}\}$, as claimed.

(b) The next operation on $[3, 3, 5]$ is

$$(\sigma_0, \sigma_1, \sigma_2, \sigma_3) \mapsto (\sigma_0, \sigma_1\sigma_2\sigma_3\sigma_2\sigma_1, \sigma_3, \sigma_2) =: (\rho_0, \rho_1, \rho_2, \rho_3).$$

We observe that

$$\begin{aligned} \rho_0\rho_1 &= \sigma_0\sigma_1\sigma_2\sigma_3\sigma_2\sigma_1 \\ &\sim \sigma_1\sigma_2\sigma_3\sigma_2 \\ &\sim \sigma_2\sigma_3, \end{aligned}$$

since $(\sigma_0\sigma_1)^3 = (\sigma_1\sigma_2)^3 = \varepsilon$. Next,

$$\begin{aligned}\rho_1\rho_2 &= \sigma_1\sigma_2\sigma_3\sigma_2\sigma_1\sigma_3\\ &= \sigma_1\sigma_2\sigma_3\sigma_2\sigma_3\sigma_1\\ &\sim (\sigma_2\sigma_3)^2.\end{aligned}$$

Finally

$$\begin{aligned}\rho_1\rho_3 &= \sigma_1\sigma_2\sigma_3\sigma_2\sigma_1\sigma_2\\ &= \sigma_1\sigma_2\sigma_3\sigma_1\sigma_2\sigma_1\\ &= \sigma_1\sigma_2\sigma_1\sigma_3\sigma_2\sigma_1\\ &= \sigma_2\sigma_1\sigma_2\sigma_3\sigma_2\sigma_1\\ &= \rho_3\rho_1.\end{aligned}$$

The inverse operation is

$$(\rho_0, \rho_1, \rho_2, \rho_3) \mapsto (\rho_0, \rho_3\rho_2\rho_1\rho_2\rho_1\rho_2\rho_3, \rho_3, \rho_2) = (\sigma_0, \sigma_1, \sigma_2, \sigma_3).$$

It thus follows that $\Gamma(\{5, \frac{5}{2}, 5\})$ is obtained from $[5, 5, 5]$ by imposing the two extra relations arising from $(\sigma_0\sigma_1)^3 = \varepsilon = (\sigma_1\sigma_2)^3$. Now

$$\begin{aligned}\sigma_0\sigma_1 &= \rho_0\rho_3\rho_2\rho_1\rho_2\rho_1\rho_2\rho_3\\ &= \rho_3\rho_2\rho_0\rho_1\rho_2\rho_1\rho_2\rho_3\\ &\sim \rho_0\rho_1\rho_2\rho_1,\end{aligned}$$

while

$$\begin{aligned}\sigma_1\sigma_2 &= \rho_3\rho_2\rho_1\rho_2\rho_1\rho_2\rho_3^2\\ &= \rho_3\rho_2\rho_1\rho_2\rho_1\rho_2\\ &= \rho_3\rho_1\rho_2\rho_1\rho_2\rho_1\\ &= \rho_1\rho_3\rho_2\rho_1\rho_2\rho_1\\ &\sim \rho_3\rho_2\rho_1\rho_2\\ &\sim \rho_1\rho_2\rho_3\rho_2.\end{aligned}$$

That is, $\{5, \frac{5}{2}, 5\} \cong \{\{5, 5 \,|\, 3\}, \{5, 5 \,|\, 3\}\}$, as claimed.

(c) The final operation on $[3, 3, 5]$ is

$$(\sigma_0, \sigma_1, \sigma_2, \sigma_3) \mapsto (\sigma_0, \sigma_1\sigma_2\sigma_3\sigma_2\sigma_1, \sigma_3\sigma_2\sigma_3, \sigma_2) =: (\rho_0, \rho_1, \rho_2, \rho_3),$$

which is the operation of (b), followed by the facetting operation φ_2 applied to the vertex-figure $\{\frac{5}{2}, 5\}$ of $\{5, \frac{5}{2}, 5\}$. We see that

$$\begin{aligned}\rho_0\rho_1 &= \sigma_0\sigma_1\sigma_2\sigma_3\sigma_2\sigma_1\\ &\sim \sigma_2\sigma_3,\end{aligned}$$

as in case (b). Next,

$$\begin{aligned}\rho_0\rho_2 &= \sigma_0\sigma_3\sigma_2\sigma_3\\ &= \sigma_3\sigma_2\sigma_3\sigma_0\\ &= \rho_2\rho_0.\end{aligned}$$

Then we have

$$\begin{aligned}\rho_1\rho_2 &= \sigma_1\sigma_2\sigma_3\sigma_2\sigma_1\sigma_3\sigma_2\sigma_3\\ &= \sigma_1\sigma_2\sigma_3\sigma_2\sigma_3\sigma_1\sigma_2\sigma_3\\ &\sim \sigma_1\sigma_2\sigma_3\sigma_1\sigma_2\sigma_3\sigma_2\sigma_3\\ &= \sigma_1\sigma_2\sigma_1\sigma_3\sigma_2\sigma_3\sigma_2\sigma_3\\ &= \sigma_2\sigma_1\sigma_2\sigma_3\sigma_2\sigma_3\sigma_2\sigma_3\\ &\sim \sigma_1\sigma_2\sigma_3\sigma_2\sigma_3\sigma_2\sigma_3\sigma_2\\ &= \sigma_1\sigma_3\sigma_2\sigma_3\\ &= \sigma_3\sigma_1\sigma_2\sigma_3\\ &\sim \sigma_1\sigma_2.\end{aligned}$$

We also have $\rho_1\rho_3 = \rho_3\rho_1$ as in case (b), and $\rho_2\rho_3 = (\sigma_3\sigma_2)^2 \sim (\sigma_2\sigma_3)^2$, so that $(\rho_2\rho_3)^5 = \varepsilon$.

The inverse operation is

$$(\rho_0, \rho_1, \rho_2, \rho_3) \mapsto (\rho_0, \rho_1\rho_2\rho_3\rho_2\rho_1\rho_2\rho_3\rho_2\rho_1, \rho_3, \rho_2\rho_3\rho_2\rho_3\rho_2) = (\sigma_0, \sigma_1, \sigma_2, \sigma_3).$$

It follows that $\Gamma(\{5, 3, \frac{5}{2}\})$ is obtained from $[5, 3, 5]$ by imposing the extra relation arising from $(\sigma_0\sigma_1)^3 = \varepsilon$. (We have already implicitly mentioned $(\sigma_1\sigma_2)^3 = \varepsilon$, since this comes from $(\rho_1\rho_2)^3 = \varepsilon$.) For this,

$$\begin{aligned}\sigma_0\sigma_1 &= \rho_0\rho_1\rho_2\rho_3\rho_2\rho_1\rho_2\rho_3\rho_2\rho_1\\ &= \rho_0\rho_1\rho_2\rho_3\rho_1\rho_2\rho_1\rho_3\rho_2\rho_1\\ &= \rho_0\rho_1\rho_2\rho_1\rho_3\rho_2\rho_3\rho_1\rho_2\rho_1\\ &= \rho_0\rho_2\rho_1\rho_2\rho_3\rho_2\rho_3\rho_2\rho_1\rho_2\\ &= \rho_2\rho_0\rho_1\rho_2\rho_3\rho_2\rho_3\rho_2\rho_1\rho_2\\ &\sim \rho_0\rho_1\rho_2\rho_3\rho_2\rho_3\rho_2\rho_1\\ &= \rho_0\rho_1\rho_3\rho_2\rho_3\rho_2\rho_3\rho_1\\ &= \rho_3\rho_0\rho_1\rho_2\rho_3\rho_2\rho_1\rho_3\\ &\sim \rho_0\rho_1\rho_2\rho_3\rho_2\rho_1.\end{aligned}$$

That is, $\{5, 3, \frac{5}{2}\} \cong \{5, 3, 5 \mid 3\}$, as claimed. □

7E Three-Dimensional Polyhedra

We now employ the techniques we have introduced to describe the possible 3-dimensional realizations of abstract regular polytopes, which are both discrete and faithful. These restrictions are the natural geometric ones; they will be assumed to hold henceforth, and will not be repeated. We then prove, by direct methods, that the enumeration is complete. In this section, we treat the polyhedra (finite and infinite), and in the next we deal with the apeirotopes of rank 4 (there can be no finite 4-polytopes in $\mathbb{E}^3$). Our treatment will follow [304].

We have already described the geometric regular polygons in Section 5B (and polytopes of lower rank are just points and segments). However, it is worth reminding ourselves about them. In $\mathbb{E}^3$, a finite regular polygon P can only be either planar, and thus a pure polygon $\{p\}$, or skew, being a blend. In the latter case, we tend to use terminology a little loosely. In writing a skew polygon as $P = \{p\} \# \{\,\}$, we mean that the natural identifications under the projections on the components yield $\{p\}$ and $\{\,\}$. Then P itself is a faithful realization of the abstract regular polygon $\mathcal{P} := \{\bar{p}\} \diamond \{\,\}$, where $\bar{p}$ is the numerator of p, regarded as a fraction. We further note that $\mathcal{P} \cong \{\bar{p}\}$ or $\{2\bar{p}\}$ as $\bar{p}$ is even or odd. When we speak geometrically, we shall use the language of blends; however, when the combinatorial aspects come to the fore, we shall then talk about mixes. We refer to the end of Section 7A for a general discussion about the relationship between blends and mixes.

In general, in a blend $\{p\} \# \{\,\}$, we allow $p > 2$ to be a fraction; in this section, however, p will always be an integer. The vertices of a skew polygon $\{p\} \# \{\,\}$ (with integral p) are among those of a p-gonal prism; they will form all the vertices if p is odd, in which case the blend is a $2p$-gon, and half of them if p is even; in each case, they will lie alternately on the two p-gonal faces of the prism. Similarly, an apeirogon (infinite regular polygon) is a linear one $\{\infty\}$, a (planar) skew one (zigzag apeirogon), which is the blend $\{\infty\} \# \{\,\}$ with a segment, or a *helix*, which is the blend of $\{\infty\}$ with a bounded regular polygon. Note that the bounded regular polygon in the last type need not itself be finite, although it will actually be so in this section and the next.

When we consider discrete regular polytopes in $\mathbb{E}^3$ of higher rank, we note that the three regular tessellations $\{3, 6\}$, $\{4, 4\}$ and $\{6, 3\}$ are planar. However, we can treat these cases with the others, all of which are genuinely 3-dimensional.

Finite Polyhedra

We begin the classification problem with the finite case. One reason why we are classifying these polytopes in this chapter is that we can then apply the various mixing operations we have described in the previous sections, particularly Section 7B. So we shall begin by listing the (finite) regular 3-polytopes in $\mathbb{E}^3$, with the relationships between pairs of them. These polytopes will have the same symmetry groups as the tetrahedron, octahedron or icosahedron. In our lists, we shall not repeat self-dual polytopes, for example. And we should emphasize again that we are really considering realizations of polytopes; thus a polytope may be realizable, while its dual is not (at least, not faithfully).

The three groupings are then:

Tetrahedral Symmetry

7E1 $$\{3, 3\} \stackrel{\pi}{\longleftrightarrow} \{4, 3\}_3$$

Octahedral Symmetry

7E2 $$\{6, 4\}_3 \stackrel{\pi}{\longleftrightarrow} \{3, 4\} \stackrel{\delta}{\longleftrightarrow} \{4, 3\} \stackrel{\pi}{\longleftrightarrow} \{6, 3\}_4$$

Icosahedral Symmetry

7E3
$$\begin{array}{ccccccc}
\{10, 5\} & \stackrel{\pi}{\longleftrightarrow} & \{3, 5\} & \stackrel{\delta}{\longleftrightarrow} & \{5, 3\} & \stackrel{\pi}{\longleftrightarrow} & \{10, 3\} \\
\updownarrow \varphi_2 & & \updownarrow \varphi_2 & & & & \\
\left\{6, \frac{5}{2}\right\} & \stackrel{\pi}{\longleftrightarrow} & \left\{5, \frac{5}{2}\right\} & \stackrel{\delta}{\longleftrightarrow} & \left\{\frac{5}{2}, 5\right\} & \stackrel{\pi}{\longleftrightarrow} & \{6, 5\} \\
 & & & & \updownarrow \varphi_2 & & \updownarrow \varphi_2 \\
\left\{\frac{10}{3}, 3\right\} & \stackrel{\pi}{\longleftrightarrow} & \left\{\frac{5}{2}, 3\right\} & \stackrel{\delta}{\longleftrightarrow} & \left\{3, \frac{5}{2}\right\} & \stackrel{\pi}{\longleftrightarrow} & \left\{\frac{10}{3}, \frac{5}{2}\right\}
\end{array}$$

We should say a few things about the last display. First, we have suppressed the exact description of most of the polyhedra. Instead, we have given symbols more akin to those used by Coxeter in [120, Chapters 2, 6]. In fact, the polyhedra occur in isomorphic pairs, given by symmetry of the display about its centre, or by interchanging 5 with $\frac{5}{2}$ and 10 with $\frac{10}{3}$. The remaining details are:

$$\begin{aligned}
\{10, 5\} &\cong \{10, 5\}_3, \\
\{10, 3\} &\cong \{10, 3\}_5, \\
\left\{\tfrac{5}{2}, 5\right\} &\cong \{5, 5 \mid 3\}, \\
\{6, 5\} &\cong \{6, 5\}_{5,3}.
\end{aligned}$$

For the last of these, recall from Section 7B that the subscripts denote the lengths of the 1- and 2-zigzags of the polyhedron.

The classification result here is

7E4 Theorem *The list of the* 18 *finite regular polyhedra in (7E1), (7E2) and (7E3) is complete.*

Proof. The enumeration is performed by a systematic investigation of the various possibilities. Let P be a regular polyhedron P in $\mathbb{E}^3$, and let $G(P) = \langle R_0, R_1, R_2 \rangle$ be its symmetry group. We may suppose that $G(P)$ is an orthogonal group, so that P has centre o. Further, we adopt our usual convention of identifying a reflexion with its mirror.

By Theorem 5B20, we have $\dim R_0 \geqslant 1$ and $\dim R_1 = \dim R_2 = 2$. That is, R_0 may be a line or a plane, and R_1 and R_2 must both be planes.

We now employ a trick that we have already used in Section 5C. If R_0 is a line, then we replace it by the orthogonal plane $S_0 := R_0^{\perp}$. As an orthogonal mapping, $S_0 = -R_0$, that is, the product of R_0 with the central reflexion. We note that S_0 still commutes with R_2, but does not with R_1. If R_0 is a plane, we set $S_0 := R_0$. Further, we set $S_j := R_j$ for $j = 1, 2$ in each case. Define $H := \langle S_0, S_1, S_2 \rangle$. Then H is an irreducible finite (plane) reflexion group in $\mathbb{E}^3$, which is the symmetry group of some classical regular polyhedron Q (see Section 7D).

We may clearly reverse the argument. If we take the group $H = \langle S_0, S_1, S_2 \rangle$ of a classical regular polyhedron Q, we may replace S_0 by $-S_0$, to obtain a new finite orthogonal group in $\mathbb{E}^3$ generated by involutions. This then adds another 9 regular polyhedra to the 9 classical ones; thus, we arrive at the 18 polyhedra listed. □

Let us make a remark about this pairing of polyhedra. When the group is [3, 5], a line which is the axis of a two-fold rotation is perpendicular to a reflexion plane; thus the two groups $G(P)$ and H are the same. But when the group of P is [3, 3] or [3, 4], something strange can happen: the two groups may be interchanged by the procedure of replacing S_0 by $-S_0$. However, it is clear that applying this procedure to Petrie duals will yield Petrie duals. The resulting pairings among the 18 polyhedra are

$$\begin{aligned} \{3,3\} &\longleftrightarrow \{6,3\}_4, \\ \{3,4\} &\longleftrightarrow \{6,4\}_3, \\ \{4,3\} &\longleftrightarrow \{4,3\}_3, \end{aligned}$$

7E5

$$\begin{aligned} \{5,3\} &\longleftrightarrow \left\{\tfrac{10}{3},3\right\}, \\ \{3,5\} &\longleftrightarrow \{6,5\}, \\ \left\{5,\tfrac{5}{2}\right\} &\longleftrightarrow \left\{\tfrac{10}{3},\tfrac{5}{2}\right\}, \\ \left\{\tfrac{5}{2},5\right\} &\longleftrightarrow \{10,5\}, \\ \left\{3,\tfrac{5}{2}\right\} &\longleftrightarrow \left\{6,\tfrac{5}{2}\right\}, \\ \left\{\tfrac{5}{2},3\right\} &\longleftrightarrow \{10,3\}. \end{aligned}$$

For the polyhedra with symmetry group [3, 5], we have adopted the same abbreviated notation as that in (7E3).

Finally, let us observe that two of the same pairings can be expressed as $\{P, P \diamond \{\,\}\}$, with a polyhedron paired with its mix with a segment (see the end of Section 7A). This emphasizes again the distinction between the combinatorial sense of mixing and the geometrical sense of blending. In fact, we have

$$\begin{aligned} \{6,3\}_4 &= \{3,3\} \diamond \{\,\}, \\ \{4,3\} &= \{4,3\}_3 \diamond \{\,\}. \end{aligned}$$

For the other seven pairs $\{P, Q\}$, we have the curious phenomenon that $P \diamond \{\,\} \cong Q \diamond \{\,\}$. In each case, the group $\Gamma(P) \cong \Gamma(Q)$ has a relation involving ρ_0 an odd number of times (a different one in the two cases), corresponding to an odd edge-circuit. Hence, mixing with a segment doubles the order of the group, giving a new group $\Gamma(P) \times C_2 \cong \Gamma(Q) \times C_2$. All the edge-paths now have even length; a path is doubled in length in going from P to $P \diamond \{\,\}$ precisely when the length in P was odd. But edge-paths in P and Q are also related in such a way; if a path in one is odd, then the corresponding path in the other is even. The claimed conclusion thus follows.

For higher-dimensional euclidean spaces, no complete classification of the regular polyhedra (faithful realizations of finite abstract 3-polytopes) is known. However, under the assumption of planarity of the 2-faces, a complete enumeration of such polyhedra in $\mathbb{E}^4$ has been achieved in Arocha, Bracho, and Montejano [11] and Bracho [42].

Apeirohedra

We now move on to the regular apeirohedra, or infinite polyhedra, in $\mathbb{E}^3$. We repeat our blanket assumptions of discreteness and faithfulness.

A key tool in settling the classification problems is a refinement of Bieberbach's theorem [25] (see also [344, §7.4]).

7E6 Lemma *An affinely irreducible discrete infinite group of isometries in $\mathbb{E}^2$ or $\mathbb{E}^3$ does not contain rotations of periods other than* 2, 3, 4 *or* 6.

Proof. Let G be an affinely irreducible discrete infinite group of the whole group $\mathcal{I}_n$ of isometries of $\mathbb{E}^n$. In saying that G is affinely irreducible, we recall that we mean that there is no proper *invariant* linear subspace L of $\mathbb{E}^n$, such that G permutes the affine subspaces of $\mathbb{E}^n$ which are translates of L. (Our use of the term "invariant" is thus a little irregular.) If such an invariant subspace L exists, then its orthogonal complement $L^\perp$ is also invariant in the same sense. This accords with the usual definition of irreducibility, if we associate with an affine mapping $x \mapsto xA + b$ (with A an $n \times n$ matrix and $b \in \mathbb{E}^n$) the linear mapping on $\mathbb{E}^{n+1}$ given by

$$(x, \eta) \mapsto (x, \eta) \begin{bmatrix} A & o^\mathsf{T} \\ b & 1 \end{bmatrix}.$$

In this context, of course, we do not count the invariant hyperplane of points $(x, 0)$, which arises from $\mathbb{E}^n$.

Bieberbach's theorem now tells us that such a group G contains a full subgroup T of the group $\mathcal{T}_n$ of translations of $\mathbb{E}^n$, and the quotient G/T is finite; in effect, T can be thought of as a lattice of rank n in $\mathbb{E}^n$. If $x \mapsto x\varphi + t$ is a general element of G, with $\varphi \in O_n$, the orthogonal group, and $t \in \mathbb{E}^n$ a translation vector (we may thus think of $t \in \mathcal{T}_n$), then the mappings φ clearly form a subgroup G_0 of $\mathcal{O}_n$, called the *special group* of G. Thus G_0 is the image of G under the homomorphism on $\mathcal{I}_n$, whose kernel is $\mathcal{T}_n$ (the image is, of course, $\mathcal{O}_n$). In other words,

$$G_0 = G\mathcal{T}_n/\mathcal{T}_n \cong G/(G \cap \mathcal{T}_n) = G/T.$$

There is no loss of generality in assuming that G_0 contains the central inversion $-I$; if it does not, then we adjoin it. Suppose now that $n = 2$ or 3, and that G_0 does contain a rotation with period $k \neq 2, 3, 4, 6$; then it contains such a rotation φ through an angle $2\pi/k$. We first consider the planar case $n = 2$. Since $-I \in G_0$, we may suppose that $k \geqslant 8$ is even. There is a minimal length δ among the translations of T. If $t \in T$ has $\|t\| = \delta$, then the distance between t and $t\varphi$ is $2\delta \sin(\pi/k) < \delta$, an obvious contradiction. In the case $n = 3$ of ordinary space, we argue similarly, except that we consider the minimal distance between parallel axes of k-fold symmetry. □

Planar Apeirohedra

We have already mentioned the regular planar tessellations. To these must be added their Petrials. We thus obtain the two families of planar apeirohedra:

7E7 $$\{4, 4\} \stackrel{\pi}{\longleftrightarrow} \{\infty, 4\}_4$$

and

7E8 $$\{\infty, 6\}_3 \stackrel{\pi}{\longleftrightarrow} \{3, 6\} \stackrel{\delta}{\longleftrightarrow} \{6, 3\} \stackrel{\pi}{\longleftrightarrow} \{\infty, 3\}_6.$$

The other operations we have described in Section 7B lead to no new polyhedra, even when they are applicable. In fact, we have $\{4, 4\}^\eta = \{4, 4\}$ (actually, another copy, whose edges are diagonals of the original squares), and $\{3, 6\}^{\varphi_2} = \{6, 3\}$ (a special case of Lemma 7B10).

7E9 Theorem *The list of six planar regular apeirohedra in (7E7) and (7E8) is complete.*

Proof. The argument is similar to that for the finite regular polyhedra, but a little easier. Let P be a planar regular apeirohedron, with group $G(P) = \langle R_0, R_1, R_2 \rangle$. The initial vertex v satisfies $v \in R_1 \cap R_2$, and R_1 and R_2 are non-commuting involutions in $\mathbb{E}^2$; hence, R_1 and R_2 must be intersecting lines. Lemma 7E6 then implies that the angle between R_1 and R_2 can only be $\pi/3$, $\pi/4$ or $\pi/6$. Next, R_0 may be a point or a line. In the latter case, $G(P)$ must just be one of [4, 4] or [3, 6], yielding the three ordinary planar tessellations. In the former case, since R_0 commutes with R_2, the point must lie in R_2, and we obtain the three Petrials of the planar tessellations (observe that $R_0 R_2$ will be the reflexion in a line perpendicular to R_2). □

Blended Apeirohedra

We now consider the genuinely 3-dimensional discrete regular apeirohedra in $\mathbb{E}^3$. We shall see that they fall into two families of 12 each. The first comprises those which are blends in a non-trivial way, while the second consists of the pure apeirohedra.

The non-pure 3-dimensional apeirohedra are derived from the six planar regular apeirohedra by mixing with either a segment { } or the linear apeirogon $\{\infty\}$. (This point of view builds up a blended apeirohedron from its components.) The apeirohedra in each of these two families form one-dimensional classes under similarity; the parameter is the ratio between the edge-length of the planar apeirohedron and either the length of the segment or the edge-length of the apeirogon. These apeirohedra are thus listed as follows. First, we have the blends with segments:

7E10

$$\{4,4\}\,\#\,\{\,\} \stackrel{\pi}{\longleftrightarrow} \{\infty,4\}_4\,\#\,\{\,\},$$
$$\{3,6\}\,\#\,\{\,\} \stackrel{\pi}{\longleftrightarrow} \{\infty,6\}_3\,\#\,\{\,\},$$
$$\{6,3\}\,\#\,\{\,\} \stackrel{\pi}{\longleftrightarrow} \{\infty,3\}_6\,\#\,\{\,\}.$$

Recall that, combinatorially, these blends are also mixes, whose planar and linear components are covered by the 3-dimensional apeirohedra. Then we have the blends with apeirogons (to which the same comments about mixes apply):

7E11

$$\{4,4\}\,\#\,\{\infty\} \stackrel{\pi}{\longleftrightarrow} \{\infty,4\}_4\,\#\,\{\infty\},$$
$$\{3,6\}\,\#\,\{\infty\} \stackrel{\pi}{\longleftrightarrow} \{\infty,6\}_3\,\#\,\{\infty\},$$
$$\{6,3\}\,\#\,\{\infty\} \stackrel{\pi}{\longleftrightarrow} \{\infty,3\}_6\,\#\,\{\infty\}.$$

If p is finite, the facets of a blend $\{p,q\}\,\#\,\{\,\}$ are skew polygons $\{p\}\,\#\,\{\,\}$, and their vertex-figures are flat polygons $\{q\}$, parallel to the plane of the tessellation $\{p,q\}$. Note that $\{3\}\,\#\,\{\,\}$ is actually a skew hexagon (see our earlier remark about polygons). For a blend $\{\infty,q\}_p\,\#\,\{\,\}$, the facets are (planar) zigzag apeirogons $\{\infty\}\,\#\,\{\,\}$, since we have

$$(\{\infty\}\,\#\,\{\,\})\,\#\,\{\,\} = \{\infty\}\,\#\,\{\,\},$$

if we ignore the implicit parameter giving the relative sizes of the components of the blend. The vertex-figures are again flat polygons $\{q\}$, since the Petrie operation preserves vertex-figures.

Similarly, the facets of a blend $\{p,q\}\,\#\,\{\infty\}$ with p finite are helical apeirogons $\{p\}\,\#\,\{\infty\}$ (spiralling around a cylinder with a p-gonal base), and the vertex-figures are skew polygons $\{q\}\,\#\,\{\,\}$. The facets of a blend $\{\infty,q\}_p\,\#\,\{\infty\}$ are now zigzag apeirogons, since

$$(\{\infty\}\,\#\,\{\,\})\,\#\,\{\infty\} = \{\infty\}\,\#\,\{\,\},$$

again ignoring any implicit parameters. The vertex-figures are still skew polygons $\{q\}\,\#\,\{\,\}$.

In preparation for the classification theorem for these apeirohedra, we need a preliminary result.

7E12 Lemma *Let P be a faithful discrete blended regular apeirohedron in $\mathbb{E}^3$. Then the components of the blend are discrete.*

Proof. It is not possible for a blended regular apeirohedron to be planar; hence, the apeirohedron P is genuinely 3-dimensional. Its components must thus have dimensions 2 and 1. Let its (geometric) symmetry group be $\langle R_0, R_1, R_2\rangle$, and write $R_j = S_j \times T_j$, with $S_j \subseteq L$ and $T_j \subseteq M$, in the standard notation for blends adopted in Section 5A (thus L is the plane and M the line in the decomposition). We may suppose that L and M contain the initial vertex o. We first concentrate on the 1-dimensional component. We must have $T_0 \neq M$, otherwise P is planar, contrary to assumption. Hence T_0 is a point, distinct from o. Further, T_1 and T_2 contain o, and T_2 commutes with T_0. This implies that $T_2 = M$.

If $T_1 = M$ also, then the component in M is a segment. It is at once clear that the projection on L must be discrete.

Alternatively, $T_1 = \{o\}$. Then the component in M is an apeirogon, and hence is discrete. From Bieberbach's theorem, the group G of P contains a full translation subgroup Λ; this can safely be identified with a subset of the vertex-set V of P. Since the projection of Λ on M is discrete, the kernel of the projection is spanned by rational, and hence linear, combinations of points of Λ; thus it is a sublattice which spans L.

Next, the group of the vertex-figure of P (at o) also respects the decomposition $\mathbb{E}^3 = L \times M$, and contains the element R_1R_2 of period at least 3. This acts on the kernel sublattice $L \cap \Lambda$, and so must have period 3, 4 or 6. More importantly, consider the mirrors S_0, S_1 and S_2. We must have S_1 and S_2 lines, since they are proper mirrors which contain o (we cannot have $S_2 = L$ since $R_2 \neq \mathbb{E}^3$, and so $S_1 \neq L$ as well). The angle between S_1 and S_2 is thus $\pi/3$, $\pi/4$ or $\pi/6$. If S_0 is a point rather than a line, perform the Petrie operation $S_0 \mapsto S_0S_2$ (induced by $R_0 \mapsto R_0R_2$); then S_0 can be taken to be a line also. It follows at once that R_0R_1 is a rotatory-translation with a rational twist (so all we really need is that S_0S_1 be of finite order); hence some power of it is a translation. It follows that M is spanned by a sublattice of Λ as well, and this is all that is required to ensure that the projection of P on L be discrete. □

7E13 Theorem *The list of 12 blended apeirohedra in (7E10) and (7E11) is complete.*

Proof. Again, for the first claim there is little to say here; we have done most of the work in Lemma 7E12. A regular apeirohedron P in $\mathbb{E}^3$ which is a non-trivial blend must have invariant subspaces of dimensions 2 and 1. Moreover, the 2-dimensional component must be one of the six planar apeirohedra listed previously. Lemma 7E6 eliminates any other possibilities; in particular, the projection of P on the corresponding plane must be discrete. The required classification is then immediate. □

Notice that each of the blends with an apeirogonal component can be regarded equally as the corresponding blend with an apeirogon. However, we only obtain a different apeirohedron by blending with a segment if the original has edge-paths of odd length, which is true for $\{3, 6\}$ and its Petrial alone.

Pure 3-Dimensional Apeirohedra

We finally come to the pure 3-dimensional apeirohedra. Let us begin by listing them, together with the various ways in which they are related. We shall see that, in a sense, they fall into a single family, derived from the regular honeycomb {4, 3, 4}:

7E14

$$
\begin{array}{ccccccc}
\{\infty,4\}_{6,4} & \stackrel{\pi}{\longleftrightarrow} & \{6,4\,|\,4\} & \stackrel{\delta}{\longleftrightarrow} & \{4,6\,|\,4\} & \stackrel{\pi}{\longleftrightarrow} & \{\infty,6\}_{4,4} \\
 & & \sigma\downarrow & & \downarrow\eta & & \\
 & & \{\infty,4\}_{\cdot,*3} & & \{6,6\}_4 & \stackrel{\varphi_2}{\longrightarrow} & \{\infty,3\}^{(a)} \\
 & & & & \pi\updownarrow & & \updownarrow\pi \\
 & & \{6,4\}_6 & \stackrel{\delta}{\longleftrightarrow} & \{4,6\}_6 & \stackrel{\varphi_2}{\longrightarrow} & \{\infty,3\}^{(b)} \\
 & & \sigma\delta\downarrow & & \downarrow\eta & & \\
 & & \{\infty,6\}_{6,3} & \stackrel{\pi}{\longleftrightarrow} & \{6,6\,|\,3\} & &
\end{array}
$$

The two apeirohedra of type $\{\infty, 3\}$ will be described later in terms of their groups. In addition to the relationships given in the diagram, the two apeirohedra of type {6, 6} are of course self-dual, since they are obtained from other apeirohedra by the operation η. Further, $\{6, 4\}_6$ is self-Petrie as its notation indicates, as is $\{\infty, 4\}_{\cdot,*3}$ since it is obtained by means of σ from another apeirohedron. Finally, from the definition $\sigma = \pi\delta\eta\pi\delta$, we also have

$$(\{\infty, 4\}_{6,4})^{\sigma} = \{6, 4\}_6.$$

There are no other new apeirohedra which can be derived from any of these. For example, further applications of φ_2 give

$$\{4, 6\,|\,4\}^{\varphi_2} = \{4, 3\},$$
$$\{6, 6\,|\,3\}^{\varphi_2} = \{3, 3\},$$

both finite polyhedra. Similarly, with the operation $\eta_{1,2}$ we obtain

$$\{6, 4\,|\,4\}^{\eta_{1,2}} = \{4, 4\},$$

a planar apeirohedron.

7E15 Theorem *The discrete pure 3-dimensional apeirohedra are just the 12 listed in (7E14).*

Proof. Let P be a pure 3-dimensional apeirohedron in $\mathbb{E}^3$, with symmetry group $G(P) = \langle R_0, R_1, R_2\rangle$. Thus R_0, R_1 and R_2 are involutory isometries of $\mathbb{E}^3$ such that R_0 and R_2 commute, while R_1 does not commute with R_0 or R_2. As usual, we identify a reflexion with its mirror.

We first show that each of R_0, R_1 and R_2 must be a line or a plane; in other words, generating reflexions in points are excluded. Without loss of generality, we may take the initial vertex of P to be o. We then have $o \in R_1 \cap R_2$, and so this latter intersection must be non-empty and strictly contained in both. Hence $\dim R_j \geqslant 1$ for $j = 1, 2$. For these j, we write $S_j = R_j$ or $-R_j$ as R_j is a plane or line (as before, $-R_j$ is the orthogonal complement of R_j), and H for the plane through o perpendicular to S_1 and S_2. Further, $o \notin R_0$. If $\dim R_0 = 0$, then it easily follows that $G(P)$ is reducible, since each R_j permutes the planes parallel to H, which is contrary to the assumption that P is pure. Indeed, we would have $R_0 \subset R_2$, since these reflexions commute. Thus $\dim R_0 \geqslant 1$ also.

We next exclude the possibility that $\dim R_j = 2$ for each $j = 1, 2$. In this case R_1 and R_2 would be planes through o, with some acute angle between them. Then R_0 is a line or plane, whose reflexion commutes with R_2, but not with R_1. If R_0 is a plane, then since $G(P)$ is irreducible it will follow that $R_0 \cap R_1 \cap R_2 \neq \emptyset$; hence $G(P)$ will be a discrete orthogonal group, and so finite. Similarly, if R_0 is a line, there are two possibilities. First, R_0 may lie in R_2, giving $R_0 \cap R_1 \cap R_2 \neq \emptyset$ since $G(P)$ is irreducible, and as before $G(P)$ is finite. Second, R_0 may be perpendicular to R_2; this makes the group $G(P)$ reducible, which is again not permitted.

There is now a further case to be excluded; we cannot have $\dim R_0 = 2$ and $\dim R_2 = 1$. If this were so, the line R_2 would have to be perpendicular to the plane R_0 (since $o \notin R_0$, the possibility $R_2 \subset R_0$ is forbidden). As in the previous case, the group $G(P)$ would then be reducible, which we do not allow.

In conclusion, then, the *dimension vector* $(\dim R_0, \dim R_1, \dim R_2)$ for the mirrors can take only four values, namely, $(2, 1, 2)$, $(1, 1, 2)$, $(1, 2, 1)$ and $(1, 1, 1)$.

We have already introduced the planar reflexions S_1 and S_2. We now define a third reflexion S_0, whose mirror is also a plane, as follows. We let R_0' be the translate of R_0 which contains the origin o, and then set $S_j := R_j'$ or $-R_j'$ as R_j is a plane or a line. In other words, we are employing the same trick which we used earlier. We write $G' := \langle R_0', R_1, R_2 \rangle$, which is the special group of $G(P)$ (see the proof of Lemma 7E6), and set $H := \langle S_0, S_1, S_2 \rangle$. Then H is a finite irreducible (plane) reflexion group, namely, one of $[3, 3]$, $[3, 4]$ or $[3, 5]$, and G' is either again one of these reflexion groups, or its rotation subgroup (this can happen only when $\dim R_j = 1$ for each j). Since $G(P)$ has to be discrete, Lemma 7E6 excludes 5-fold rotations; hence G' cannot be $[3, 5]$ or its rotation subgroup. In other words, H must be $[3, 3]$ or $[3, 4]$.

With the four possibilities for the dimension vector $(\dim R_0, \dim R_1, \dim R_2)$, and three for the group H (which can also be taken as $[4, 3]$, of course), we see that we have just twelve possibilities. These twelve all occur; we may reverse the method of the proof, observing that different positions of R_0 not containing o (but meeting R_2) lead to similar apeirohedra. □

We may now list these twelve apeirohedra, according to the different scheme given by the proof of Theorem 7E15.

	$\{3, 3\}$	$\{3, 4\}$	$\{4, 3\}$
$(2, 1, 2)$	$\{6, 6 \mid 3\}$	$\{6, 4 \mid 4\}$	$\{4, 6 \mid 4\}$
$(1, 1, 2)$	$\{\infty, 6\}_{4,4}$	$\{\infty, 4\}_{6,4}$	$\{\infty, 6\}_{6,3}$
$(1, 2, 1)$	$\{6, 6\}_4$	$\{6, 4\}_6$	$\{4, 6\}_6$
$(1, 1, 1)$	$\{\infty, 3\}^{(a)}$	$\{\infty, 4\}_{\cdot,*3}$	$\{\infty, 3\}^{(b)}$

In this table, the entries on the left are the dimension vectors $(\dim R_0, \dim R_1, \dim R_2)$.

The columns are indexed by the finite regular polyhedra to which the respective apeirohedra correspond.

It is appropriate to make one further comment here. The symmetry groups of the three apeirohedra associated with the dimension vector (1, 1, 1) are generated by rotations (half-turns) in $\mathbb{E}^3$; thus the whole groups contain only direct isometries. This implies that the three apeirohedra occur in enantiomorphic (mirror-image) pairs, with their facets consisting of either all left-hand helices or all right-hand helices. The other six apeirohedra with helical facets (three blended and three pure) contain both left- and right-handed helices, since there is a plane reflexion among the generators of each of their symmetry groups.

On an historical note, let us record that Petrie found $\{4, 6 \mid 4\}$ and $\{6, 4 \mid 4\}$ in 1926, and Coxeter immediately afterwards discovered $\{6, 6 \mid 3\}$ (see [105]). All the other 3-dimensional regular apeirohedra (including those which are blended), except for $\{\infty, 4\}_{\cdot,*3}$, were described by Grünbaum around 1975 (see [198]), and the final instance was discovered by Dress, who proved the completeness of the enumeration, around 1980 (see [148, 150]). Some apeirohedra and polyhedra had indeed been found earlier, although no classification had been attempted; for example, see [69, 339, 358, 359, 448–451, 454].

There are quite a few other classes of more or less symmetrical polyhedra and apeirohedra in ordinary space, whose symmetry groups have transitivity properties which are weaker than flag-transitivity (see [198, 273]). Interesting classes are defined by requiring the symmetry group to be transitive on the vertices, on the edges, or on the 2-faces, or even simultaneously on all three ("fully transitive"). However, except for the enumeration of the regular polyhedra and apeirohedra, and the papers by Farris [165, 166] on fully transitive polyhedra, no general discussion is available in the literature. On the other hand, many interesting figures are known; for examples, see [51, §4; 129, 130, 198, 201, 203, 264–266, 349, Chapter 6].

The Automorphism Groups

It remains for us to determine the groups of the regular apeirohedra. We may take those of the planar apeirohedra for granted. We must thus look at the 3-dimensional apeirohedra.

The Blended Case

We shall treat the blended apeirohedra somewhat briefly. As we have already emphasized, combinatorially speaking these apeirohedra are mixes, and the discussion of the groups will soon reflect that fact. Of the six blends with segments, four are easily dealt with, namely, those whose edge-circuits all have even length.

7E16
$$\begin{aligned}
\{4, 4\} \# \{\,\} &\cong \{4, 4\} \diamond \{\,\} \cong \{4, 4\},\\
\{\infty, 4\}_4 \# \{\,\} &\cong \{\infty, 4\}_4 \diamond \{\,\} \cong \{\infty, 4\}_4,\\
\{6, 3\} \# \{\,\} &\cong \{6, 3\} \diamond \{\,\} \cong \{6, 3\},\\
\{\infty, 3\}_6 \# \{\,\} &\cong \{\infty, 3\}_6 \diamond \{\,\} \cong \{\infty, 3\}_6.
\end{aligned}$$

There remain $\{3, 6\} \# \{\,\} \cong \{3, 6\} \diamond \{\,\}$ and its Petrial; it is clearly enough to consider the former. A face of $\{3, 6\} \diamond \{\,\}$ is a (geometrically skew) hexagon $\{3\} \# \{\,\} \cong \{3\} \diamond \{\,\} \cong \{6\}$; this forms one kind F of minimal edge-circuit. The other kind is a skew quadrilateral Q, constructed as follows. Go round two sides of a face, then switch to the adjacent face which does not contain either of the first two edges; going round two edges of this will return one to the starting point. However, there is another way of thinking about the relation. In the group $\langle \rho_0, \rho_1, \rho_2 \rangle$ of the apeirohedron, the "rotation" $(\rho_0\rho_1)^3$ leaves the component $\{3, 6\}$ of the mix invariant, but interchanges the two vertices of the segment component. Hence, if we conjugate this rotation with any element of the group, and take the product, we must get the identity. It is appropriate to choose the conjugating element to be ρ_2 (which takes the initial face to an adjacent one), and this leads to the relation

$$\begin{aligned}\varepsilon &= (\rho_0\rho_1)^3\rho_2(\rho_0\rho_1)^3\rho_2\\ &= (\rho_1\rho_0\rho_1\rho_0\rho_1\rho_2)^2,\end{aligned}$$

taking into account $(\rho_0\rho_1)^3 = (\rho_1\rho_0)^3$ and conjugacy. We then have the following, bearing in mind the $*$ notation, which refers to a (hole or zigzag) relation on the dual (see Section 7B).

7E17 Theorem *As an abstract regular polytope,*

$$\{3, 6\} \diamond \{\,\} \cong \{6, 6 \mid \cdot, *2\}.$$

Proof. We have to show that any edge-circuit C in the graph of $P := \{3, 6\} \diamond \{\,\}$ can be obtained by concatenating copies of the two kinds of basic circuits F and Q. We look at the projection C' of C on the plane of the component $\{3, 6\}$, noting that C alternates between two parallel planes. If C lies above a single triangle, it must be a copy of F (and C' traces the triangle twice). If C lies above only two triangles, then it is a copy of Q (of course, we always consider just the connected components). Thus we may assume that C' requires the edges of at least three triangles.

A vertex of C can be written $v^{\pm}$, labelled by the corresponding vertex $v \in C'$, and the copy of $\{3, 6\}$ in which it lies. Now C' has components of the following kinds. The first is a *loop*, over which C returns to its starting point but in the other component over $\{3, 6\}$; thus C goes from a vertex v^+ round to v^-, say – we shall call the pair $\{v^+, v^-\}$ the *hinge* of the loop. A loop necessarily contains an odd number of edges, and encloses an odd number of triangles. The second is a *string*, which is an edge-path doubly covered by C.

A loop which encloses two or more triangles can be reduced by concatenating with copies of Q. Begin at a boundary triangle, and observe that the loop also encloses an adjacent triangle, so that concatenating with an appropriate copy of Q whose projection encloses the two triangles reduces the total number of enclosed triangles by 2. On the other hand, the part of C over a loop which encloses a single triangle can be concatenated with a copy of Q sharing two of its edges; this has the effect of rotating the loop about the hinge. Repeating as necessary will rotate the triangle until it encounters another edge of C, when edges of C will cancel in pairs.

It is clear that such reductions eventually reduce C to a single vertex (observe that a circuit cannot consist entirely of a string), and this leads to the assertion of the theorem. □

The substitution $\rho_0 \mapsto \rho_0\rho_2$ gives the group of the Petrial $\{\infty, 6\}_3 \diamond \{\,\}$, but the resulting relation seems to be of little interest.

We now move on to the blends in which one component is (or, more strictly, covers) an apeirogon. Again, we shall treat only one of a pair of Petrials, choosing that one whose planar component has finite faces. This time, the distinction is between those whose planar components have even or odd polygons as vertex-figures.

Both an apeirohedron $\mathcal{P} := \{p, q\} \diamond \{\infty\}$ and its Petrial $\{\infty, q\}_p \diamond \{\infty\}$ have infinite faces, and so we must look for edge-circuits in their graphs which arise in different ways. The key observation is that, if $\langle \rho_0, \rho_1, \rho_2 \rangle$ is the (abstract) group of $\mathcal{P}$, then $(\rho_0\rho_1)^p$ is a translation parallel to the linear component, and so its product with any conjugate of its inverse (that by ρ_2 is usual) is the identity.

We begin with $\{4, 4\} \diamond \{\infty\}$. Take the conjugating element of the last paragraph to be ρ_2. We quickly deduce the relation

$$(\rho_1\rho_0\rho_1\rho_0\rho_1\rho_2)^2 = \varepsilon.$$

This corresponds to a circuit H of six edges. The helical facets $\{4\} \diamond \{\infty\}$ over adjacent square faces of $\{4, 4\}$ have opposite twists.

The circuit H goes along three edges of a helix, and then returns to its starting point via three edges of the helix over an adjacent square. We then have (and again recall the $*$ convention)

7E18 Theorem *As an abstract regular polyhedron,*

$$\{4, 4\} \diamond \{\infty\} \cong \{\infty, 4 \mid \cdot, *2\}.$$

Proof. In other words, we assert that a general edge-circuit in the graph of $\mathcal{P} := \{4, 4\} \diamond \{\infty\}$ can be obtained by concatenating copies of H. Let us call the *height* of a circuit C the number of (distinct) edges of its projection on the apeirogonal component of P. Thus, H itself has height 3.

We first reduce the height to at most 2. The two edges at a highest vertex v of an edge-circuit C descend one level by edges to opposite vertices of the vertex-figure $\{4\}$ of P. Concatenating C with a copy of H for which v is also the highest vertex, and cancelling other edges in pairs as necessary, replaces v by vertices on at most the next three lower levels. (This may split C into subcircuits, of course.) Repeating as often as necessary results in one or more circuits of height 2 or less.

Actually, a circuit cannot have height less than 2, because the projection of an edge-path whose height is 1 on the planar component lies in a straight line (of edges of $\{4, 4\}$). Moreover, if C does have height 2, then its projection encloses sets of 2×2 blocks of squares of $\{4, 4\}$. Concatenating two copies of H over such a block which share the central vertex gives a "basic" circuit of height 2 (this can be done in two ways), and a general such circuit is clearly obtained by concatenating these "basic" circuits. This ends the proof. □

The apeirohedron $\{3, 6\} \diamondsuit \{\infty\}$ succumbs to a similar argument. This time, the extra relation is

$$(\rho_1\rho_0\rho_1\rho_0\rho_1\rho_2)^2 = \varepsilon.$$

The basic circuit is now a quadrilateral Q which goes up round two sides of a helical face, and back down two sides of an adjacent face. There then results

7E19 Theorem *As an abstract regular polyhedron,*

$$\{3, 6\} \diamondsuit \{\infty\} \cong \{\infty, 6 \mid *2\}.$$

Proof. The argument is very similar to that in the previous case, and so we shall only sketch it. Defining *height* as before, we see that Q itself has height 2. Thus we may concatenate a circuit C with copies of Q to reduce its height to 1. A "basic" circuit of height 1 is a skew hexagon over the outside of six triangles of $\{3, 6\}$ containing a common vertex, and is obtained by concatenating three copies of Q (this can be done in two ways). A general circuit of height 1 is then obtained by concatenating such "basic" hexagonal circuits. □

Among the blended apeirohedra, we finally come to $\mathcal{P} := \{6, 3\} \diamondsuit \{\infty\}$. Now the vertex-figure is a skew hexagon $\{3\} \diamondsuit \{\,\}$, and we soon see that above each hexagon $\{6\}$ of the planar component $\{6, 3\}$ lie six distinct helical faces, three with each twist.

There are two basic kinds of relations. While we have ones analogous to those of the last two cases, it is more convenient to observe that a (2-)hole of P is a skew hexagon H lying over a hexagon of the planar component; this is one kind of basic circuit in the graph of P. The other is a quadrilateral circuit Q; go round two edges of a hole, and then return by two edges of another hole with the same pair of end vertices. The corresponding relation is explained as follows. Let P have (abstract) group $\langle \rho_0, \rho_1, \rho_2 \rangle$. Then $\sigma := (\rho_0\rho_1\rho_2\rho_1)^2$ takes us two steps around a hole, and is a rotation by $2\pi/3$ which leaves the linear component fixed. Similarly, $\tau := (\rho_1\rho_2)^3$ is just the reflexion in the base plane of the blend, and so leaves the planar component fixed. Thus σ and τ commute, and after some manipulation (involving the fact that ρ_1 and ρ_2 both commute with $(\rho_1\rho_2)^3$), we obtain the relation

$$(\rho_0\rho_1\rho_2\rho_1\rho_0(\rho_1\rho_2)^3)^2 = \varepsilon.$$

Our characterization in this case is

7E20 Theorem *The group of the apeirohedron* $\{6, 3\} \diamondsuit \{\infty\}$ *is the Coxeter group* $[\infty, 6] = \langle \rho_0, \rho_1, \rho_2 \rangle$, *with the additional relations*

$$(\rho_0\rho_1\rho_2\rho_1)^6 = (\rho_0\rho_1\rho_2\rho_1\rho_0(\rho_1\rho_2)^3)^2 = \varepsilon.$$

Proof. Again, we shall treat this very briefly. Concatenating a general circuit with copies of the quadrilateral Q reduces its height to at most 1. This resulting circuit is then the concatenation of hexagonal holes H, and the theorem follows. □

The Pure Case

For the pure 3-dimensional apeirohedra, it remains for us to verify that ten of the apeirohedra do have the automorphism groups that their notations signify, and to determine the groups of the remaining two.

For the moment, we shall leave aside $\{4, 6 \mid 4\}$ and its dual; we shall thus take their groups as given. In Corollary 7F7, we shall demonstrate that their groups are as indicated by the notation. We shall also obtain them by twisting arguments in later chapters. (The same assumption could apply to $\{6, 6 \mid 3\}$, but we shall actually obtain its group here, as it fits into our general scheme.) Since the Petrie operation interchanges k-holes and k-zigzags, we see that

$$\{4, 6 \mid 4\} \overset{\pi}{\longmapsto} \{\infty, 6\}_{4,4},$$
$$\{6, 4 \mid 4\} \overset{\pi}{\longmapsto} \{\infty, 4\}_{6,4},$$

as claimed. (Strictly speaking, perhaps we ought to replace "∞" here by "$\cdot$"!)

We next appeal to Theorem 7B17, to obtain

$$\{4, 6 \mid 4\} \overset{\eta}{\longmapsto} \{6, 6\}_4.$$

The Petrie operation and duality then yield $\{4, 6\}_6$ and $\{6, 4\}_6$. Another appeal to Theorem 7B17 results in

$$\{4, 6\}_6 \overset{\eta}{\longmapsto} \{6, 6 \mid 3\}.$$

From this last, just as before, we obtain

$$\{6, 6 \mid 3\} \overset{\pi}{\longmapsto} \{\infty, 6\}_{6,3}.$$

We have three apeirohedra remaining: one obtained by an application of σ ($= \pi\delta\eta\pi\delta$), and the other two by applications of φ_2. The first we can do directly:

$$\begin{aligned}\{6, 4 \mid 4\} &\overset{\pi}{\longmapsto} \{\infty, 4\}_{6,4}\\ &\overset{\delta}{\longmapsto} \{4, \infty\}_{6,*4}\\ &\overset{\eta}{\longmapsto} \{\infty, \infty \mid 3\}_4\\ &\overset{\pi}{\longmapsto} \{4, \infty\}_{\infty,3}\\ &\overset{\delta}{\longmapsto} \{\infty, 4\}_{\cdot,*3}.\end{aligned}$$

Recall that the $*$ prefix indicates (here) the 2-zigzag of the dual. We have replaced the ∞ in the suffix by "$\cdot$," to indicate that the corresponding value is unspecified. The only step left unexplained is the third, namely, the application of η to $\{4, \infty\}_{6,*4}$. The operation is

$$\eta\colon (\sigma_0, \sigma_1, \sigma_2) \mapsto (\sigma_0\sigma_1\sigma_0, \sigma_2, \sigma_1) =: (\rho_0, \rho_1, \rho_2).$$

The first suffix 6 is dealt with by Theorem 7B17; observe that the graph of $\{4, \infty\}_{6,*4}$

is indeed bipartite. For the second suffix $*4$, the relation gives the period of

$$\sigma_2(\sigma_1\sigma_0)^2 = \sigma_2\sigma_1\sigma_0\sigma_1\sigma_0 = \rho_1\rho_2\rho_0 \sim \rho_0\rho_1\rho_2,$$

namely, that of the Petrie polygon of the second apeirohedron.

The last two apeirohedra, those of type $\{\infty, 3\}$, must be characterized by direct methods. We shall work with $P^{(b)} := \{\infty, 3\}^{(b)}$ rather than $P^{(a)} := \{\infty, 3\}^{(a)}$, because its structure is a little easier to describe. The symmetry group of $\{4, 6\,|\,4\}$ is generated by the three involutions

$$\begin{aligned} S_0&: x \mapsto (1 - \xi_1, \xi_2, \xi_3),\\ S_1&: x \mapsto (\xi_2, \xi_1, -\xi_3),\\ S_2&: x \mapsto (\xi_1, \xi_3, \xi_2),\end{aligned}$$

in terms of $x = (\xi_1, \xi_2, \xi_3)$, and the initial vertex is o. These are all symmetries of the honeycomb $\{4, 3, 4\}$ of unit cubes in $\mathbb{E}^3$, whose vertices are the points with integer cartesian coordinates; hence all the groups which occur in the discussion are subgroups of its symmetry group $[4, 3, 4]$. Indeed, we shall see in Section 7F that, if $[4, 3, 4] = \langle T_0, \ldots, T_3\rangle$ in the natural way, then $S_0 = T_0$, $S_1 = T_1T_3$ and $S_2 = T_2$. As before, we have taken

$$\begin{aligned} T_0&: x \mapsto (1 - \xi_1, \xi_2, \xi_3),\\ T_1&: x \mapsto (\xi_2, \xi_1, \xi_3),\\ T_2&: x \mapsto (\xi_1, \xi_3, \xi_2),\\ T_3&: x \mapsto (\xi_1, \xi_2, -\xi_3).\end{aligned}$$

If we start from $\{4, 6\,|\,4\}$ and trace through the three mixing operations which lead to $\{\infty, 3\}^{(b)}$ (namely, $\eta\pi\varphi_2$), we find that the symmetry group of the latter has generators (reflexions in lines)

$$\begin{aligned} R_0 &= (S_0S_1)^2: x \mapsto (1 - \xi_1, 1 - \xi_2, \xi_3),\\ R_1 &= S_2S_1S_2: x \mapsto (\xi_3, -\xi_2, \xi_1),\\ R_2 &= S_1: x \mapsto (\xi_2, \xi_1, -\xi_3).\end{aligned}$$

We may picture $P^{(b)}$ in the following way. As we have already remarked, its facets are all helices with the same sense. The initial vertex is still o; hence all the vertices are points of $\mathbb{E}^3$ with integer cartesian coordinates. We easily see from the generators that, in fact, the sum of the coordinates of each vertex is even.

The initial edge of $P^{(b)}$ has vertices $o = (0, 0, 0)$ and $oR_0 = (1, 1, 0)$, which is a diagonal of a 2-face of $\{4, 3, 4\}$; hence all edges are such diagonals. Next, R_1R_0, which preserves the initial facet and takes o into $(1, 1, 0)$, is

$$R_1R_0: x \mapsto (1 - \xi_3, 1 + \xi_2, \xi_1),$$

which is a translation by $(0, 1, 0)$ together with a right-hand (or negative) twist of $\pi/2$ about the axis through $(\frac{1}{2}, 0, \frac{1}{2})$ in direction $(0, 1, 0)$. Hence the facets are helices of type $\{\infty\} \,\Diamond\, \{4\}$. Finally, R_1 takes $(1, 1, 0)$ into $(1, 1, 0)R_1 = (0, -1, 1)$, and R_2

takes $(0, -1, 1)$ into $(0, -1, 1)R_2 = (-1, 0, -1)$; indeed, $R_2 R_1 \colon x \mapsto (-\xi_3, -\xi_1, \xi_2)$ is a cyclic permutation of the signed basis vectors $e_1, -e_2, -e_3$).

It follows from this that $(R_1 R_0)^4$ is the translation by $(0, 4, 0)$. However, such translations in the directions of the three coordinate axes do not generate the whole translation group. Instead, we observe that

$$x R_2 R_1 R_0 = (1 + \xi_3, 1 + \xi_1, \xi_2),$$

so that $(R_2 R_1 R_0)^3$ is the translation by $(2, 2, 2)$. Since the image of $(2, 2, 2)$ under $R_1 R_2$ and its inverse are $(-2, 2, -2)$ and $(-2, -2, 2)$, we see that the translation subgroup is actually the lattice $\Lambda := \Lambda_{(2,2,2)}$ defined in Section 6D, which is generated by $(2, 2, 2)$ and its transforms under changes of signs of the coordinates. Incidentally, since $R_2 R_1 R_0$ is conjugate to the "translation" $R_0 R_2 \cdot R_1$ which takes one vertex of a facet of the Petrial $P^{(a)}$ of $P^{(b)}$ into the next vertex, we see that these facets are of type $\{\infty\} \diamond \{3\}$; this time, they are helices with a left-hand (positive) twist.

The axes of the helical facets of $P^{(b)}$ are parallel to the three coordinate axes; as we have seen, these three axes are permuted by $R_2 R_1 R_0$. To visualize the way in which the facets fit together, it is more convenient to concentrate on the vertical ones. The cubes in $\{4, 3, 4\}$ fall into vertical stacks or (infinite) towers. Just an eighth of these towers are associated with facets; they are all the images of one fixed tower under the translation lattice Λ. A typical facet winds upwards (or downwards) in a right-hand spiral around the tower, crossing its square faces diagonally; we may envisage it as a staircase. (In Figure 7E1, we are looking at the vertical towers from above. As we go around a

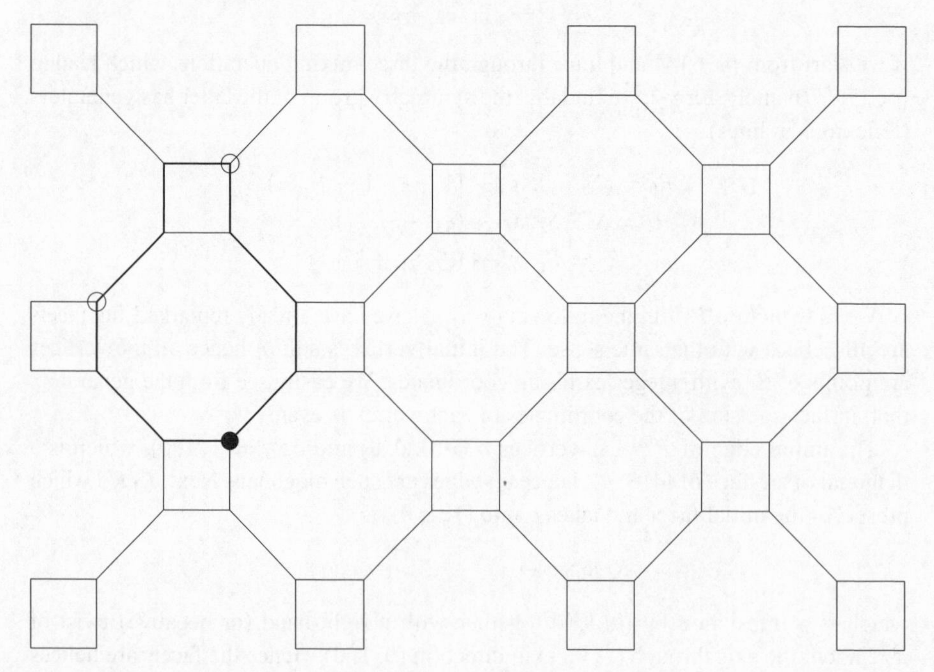

Figure 7E1. The apeirohedron $\{\infty, 3\}^{(b)}$.

tower in the clockwise direction, we rise by a floor each time we traverse an edge of a square.)

The origin o is a vertex of a vertical tower; we think of it as lying at ground level. Ascending four flights of stairs brings us to $(0, 0, 4)$ on the fourth floor, immediately above our starting point. At each floor is a single horizontal bridge, leading away from one tower to an "adjacent" tower, across the diagonal of a horizontal square of $\{4, 3, 4\}$. If we ascend one flight to the first floor, cross the bridge, descend one flight in the adjacent tower to the ground floor, and then cross the next bridge, we shall similarly have gone four edges along a facet of $P^{(b)}$ with a horizontal axis. Each bridge belongs to two such horizontal facets, of course, according as it was reached by an ascending or descending flight.

Theorem 2F4 shows that we can find a presentation of the automorphism group of $P^{(b)}$ by considering its edge-circuits (the vertex-figure is, of course, known). We regard circuits as undirected (closed) chains of edges; it is also convenient to allow them to contain more than one component (so, strictly speaking, we are talking about finite unions of simple circuits), and we ignore isolated vertices. A minimal or *basic* circuit is constructed as follows. Ascend four floors of a tower by the staircase, cross the bridge to the adjacent tower, descend four floors by its staircase, and then cross back over the bridge to the starting point. The other basic circuits are then the images of this one under the symmetry group $G^{(b)} := G(P^{(b)})$ of $P^{(b)}$. A typical basic circuit using horizontal towers is formed similarly, although the description superficially appears different. From a starting vertex, cross a bridge and ascend one floor of a staircase. Repeat this twice, then cross a final bridge and descend three floors. Of course, we may interchange "ascend" and "descend" in this description. Such a circuit uses four towers in a square formation; the circuit goes along the inside of three towers and around the outside of the fourth. (In Figure 7E1, such a horizontal basic circuit is indicated by heavy lines.)

If C and D are two edge-circuits, then we *concatenate* them by taking their symmetric difference $C \triangle D$. Observe that $(C \triangle D) \triangle D = C$, so that concatenating twice with a fixed circuit has no effect. The key result for categorizing $G^{(b)}$ is

7E21 Lemma *An arbitrary edge-circuit in* $\{\infty, 3\}^{(b)}$ *is a concatenation of basic circuits.*

Proof. It is clear that, at any stage, we may confine our attention to a single connected circuit C; if, after any concatenation, a circuit becomes disconnected, then we simply consider the resulting components.

We now reduce the circuit C to a vertex by means of two kinds of operation. First, if C uses two or more bridges between the same towers, we may concatenate with vertical basic circuits to eliminate these bridges in pairs. Thus we may assume that C contains no more than one bridge between any two towers.

We now look down on C from a vertical direction, as in Figure 7E1. Since the plane is simply connected, we may contract the projection of C to a single vertex. For this purpose, we can safely identify the vertices of C in any one tower, since there is now no more than one bridge between any two towers. A contraction over a diamond formed by four towers is achieved by concatenating with a horizontal basic circuit (like one of those indicated in Figure 7E1) which uses these four towers and shares one of the bridges of C. Of course, further reductions of the first kind will then also generally be needed, since horizontal bridges along the other three sides of the diamond may be introduced (and some may disappear). It is clear that systematic application of these two kinds of operation will eventually reduce C to a single vertex of $P^{(b)}$. Hence, if we reverse the successive concatenations, we shall recover the original circuit, as was claimed. □

A basic edge-circuit in $P^{(b)}$ corresponds to a relation in $G^{(b)}$ between its distinguished generators R_0, R_1 and R_2. Let us consider the following horizontal basic circuit. It starts from the initial vertex o and contains the first four successive edges of the initial facet F_2. This sequence of four edges is continued at each end by the two edges (corresponding to bridges) joining F_2 to the facet in an adjacent (horizontal) stack of cubes, and is completed by the four intermediate edges of that facet. The symmetry group of this basic circuit has two generators. The first is the conjugate $U_1 := (R_0R_1)^3R_0$ of R_1 by $(R_1R_0)^2$, which fixes F_2 and interchanges the two bridging edges. The second is the conjugate $U_2 := R_2R_1R_0R_1R_2$ of R_0 by R_1R_2, which interchanges F_2 and the "adjacent" facet, and fixes the two bridging edges. The relation which imposes this basic circuit is then $U_1U_2 = U_2U_1$, or $(U_1U_2)^2 = I_3$, the identity in $\mathbb{E}^3$. When expressed in terms of the generators R_0, R_1 and R_2, the relation $(U_1U_2)^2 = I_3$ involves R_0 ten times, in keeping with the fact that a basic circuit in $P^{(b)}$ has ten edges.

In order to state the main theorem, we provide alternative interpretations of the group relation given by this basic circuit. We split the circuit into the four edges in the initial horizontal stack of cubes of $\{4, 3, 4\}$ (those in the facet F_2), and two sets of three edges, formed by a bridge and two of the four edges in the adjacent stack. Each of the latter two sets forms three edges of a Petrie polygon of $P^{(b)}$. This partition of the edges is depicted for the basic circuit just considered in Figure 7E1, with the solid circle indicating the initial vertex o. It gives three translations (elements of the lattice Λ) in $G^{(b)}$; the first is $V_1 := (R_1R_0)^4$ along F_2, the second is $V_2 := R_1(R_0R_1R_2)^3R_1$ along a Petrie polygon from o, and the third is $V_3 := T_1V_2^{-1}T_1 = (R_0R_1)^4(R_2R_1R_0)^3(R_1R_0)^4 = U_1^{-1}(R_2R_1R_0)^3U_1$. (We leave the details of these calculations to the reader.) The consequent relation arises from

$$\begin{aligned}(U_1U_2)^2 &= (R_0R_1)^3R_0 \cdot R_2R_1R_0R_1R_2 \cdot (R_0R_1)^3R_0 \cdot R_2R_1R_0R_1R_2 \\ &= (R_0R_1)^4 \cdot R_2R_1R_2R_0R_1R_2 \cdot (R_0R_1)^3 \cdot R_2R_0R_1R_0R_2R_1R_2 \cdot R_1 \\ &= (R_0R_1)^4 \cdot (R_2R_1R_0)^3 \cdot R_1(R_0R_1R_2)^3R_1 \\ &= V_3 \cdot V_1^{-1} \cdot V_2,\end{aligned}$$

since $V_3 = V_1^{-1}(R_2R_1R_0)^3V_1$; we have used $R_0R_2 = R_2R_0$ and $R_1R_2R_1 = R_2R_1R_2$ several times. Rearranging, we conclude that the relation $(U_1U_2)^2 = I_3$ is equivalent to $V_1 = V_2V_3$. Indeed, further manipulations show that both relations are equivalent to the commutation of the translations

$$S := (R_0R_1)^4 = V_1^{-1}$$
$$T := (R_0R_1R_2)^3.$$

To see this, we rewrite

$$(U_1U_2)^2 = (R_0R_1)^4 \cdot (R_2R_1R_0)^3 \cdot (R_1R_0)^4 \cdot R_0R_1R_0R_2R_1R_2$$
$$= ST^{-1}S^{-1}T,$$

showing that $(U_1U_2)^2 = I_3$ is equivalent to $ST = TS$. It is this last relation which we shall employ.

7E22 Theorem *The automorphism groups of the two apeirohedra of type* $\{\infty, 3\}$ *in* $\mathbb{E}^3$ *are the Coxeter group* $[\infty, 3] = \langle \rho_0, \rho_1, \rho_2 \rangle$, *with the imposition of the single extra relation*

$$\sigma\tau = \tau\sigma,$$

with $\sigma := (\rho_0\rho_1)^3$ *and* $\tau := (\rho_0\rho_1\rho_2)^4$ *for* $\{\infty, 3\}^{(a)}$, *or* $\sigma := (\rho_0\rho_1)^4$ *and* $\tau := (\rho_0\rho_1\rho_2)^3$ *for* $\{\infty, 3\}^{(b)}$.

Proof. The given relation for $\{\infty, 3\}^{(b)}$ is equivalent to the last one given, since σ corresponds to S and τ corresponds to T. Any relation on the automorphism group $\Gamma^{(b)}$ of $P^{(b)}$ corresponds to an edge-circuit, and we have shown in Lemma 7E21 that these are composed of basic circuits, which are specified solely by the extra relation. Thus the automorphism group of $\{\infty, 3\}^{(b)}$ is as claimed.

The corresponding relation for $\{\infty, 3\}^{(a)}$ is obtained from that for $\{\infty, 3\}^{(b)}$ by means of the Petrie operation substitution of $\rho_0\rho_2$ for ρ_0. Indeed, in terms of the generators of $\Gamma^{(b)}$, and with σ and τ retaining (for the moment) their original definitions, we have

$$(\rho_0\rho_2\rho_1)^3 = \rho_2\tau\rho_2$$
$$(\rho_0\rho_2\rho_1\rho_2)^4 = \rho_2\sigma\rho_2.$$

Thus the commutation relations for the new σ and τ are just those for the old ones, conjugated by ρ_2. This then gives the automorphism group of $\{\infty, 3\}^{(a)}$. □

Let us make a further comment on this group. By definition, $\{\infty, 3\}^{(b)} = (\{4, 6\}_6)^{\varphi_2}$, by means of the operation

$$(\sigma_0, \sigma_1, \sigma_2) \mapsto (\sigma_0, \sigma_1\sigma_2\sigma_1, \sigma_2) =: (\rho_0, \rho_1, \rho_2)$$

on the automorphism group $[4, 6]_6 = \Gamma(\{4, 6\}_6) = \langle \sigma_0, \sigma_1, \sigma_2 \rangle$. If we substitute for ρ_0, ρ_1 and ρ_2 in this way, it is easy to check that (as we must) we do obtain a valid relation in $[4, 6]_6$.

We end with a remark on the groups of the pure apeirohedra with finite faces. We put them in a table, together with the finite regular polyhedra to which they correspond; as before, the entry in the first column of the table is the dimension vector $(\dim R_0, \dim R_1, \dim R_2)$.

(2, 2, 2)	$\{3, 3\}$	$\{3, 4\}$	$\{4, 3\}$
(1, 2, 2)	$\{6, 3\}_4$	$\{6, 4\}_3$	$\{4, 3\}_3$
(2, 1, 2)	$\{6, 6 \mid 3\}$	$\{6, 4 \mid 4\}$	$\{4, 6 \mid 4\}$
(1, 2, 1)	$\{6, 6\}_4$	$\{6, 4\}_6$	$\{4, 6\}_6$

A convex regular polyhedron of type $\{3, q\}$ (or $\{q, 3\}$) has holes $\{h\}$ with $h = q$, while its Petrie polygon is an r-gon with

$$r = \frac{2q + 2}{7 - q}$$

(this is derived from [120, 4.91], and is just one of many possible expressions; we write r here for Coxeter's h, for obvious reasons). For $p > 2$, we define p' by

$$\frac{1}{p} + \frac{1}{p'} = \frac{1}{2}.$$

The corresponding polyhedra in each column are then

$$\{p, q\}, \quad \{p', q\}_{r'}, \quad \{p', q' \mid h\}, \quad \{p', q'\}_r.$$

Of course, the fact that the holes or Petrie polygons of the derived polyhedra are those given follows from the relationship between the generating reflexions R_0, R_1 and R_2 of their groups, and the corresponding plane reflexions S_0, S_1 and S_2 which generate the group of the convex polyhedron.

7F Three-Dimensional 4-Apeirotopes

We may confine our attention to apeirotopes in our discussion of possible faithfully realized regular polytopes of rank at least 4 in $\mathbb{E}^3$, since Theorem 5B20 tells us that finite regular 4-polytopes admit no such realizations. Indeed, Theorem 5C3 tells us that, if P is a discrete faithfully realized regular n-apeirotope in $\mathbb{E}^m$, then $m \geqslant n - 1$. Moreover, when $m = n - 1$, then the theorem also gives lower bounds for the dimensions of the mirrors of the generating reflexions R_j. In our case ($n = 4$ and $m = 3$), we have $\dim R_j \geqslant j$ for $j = 0$ and 1, and $\dim R_2 = \dim R_3 = 2$.

Let P be a discrete faithfully realized regular 4-apeirotope in $\mathbb{E}^3$, with symmetry group $\langle R_0, R_1, R_2, R_3 \rangle$. Its facets are (finite or infinite) regular polyhedra. Furthermore, its ridges (2-faces) must be planar regular polygons, again finite or infinite; they cannot be 3-dimensional, because this would force a stabilizing element R_3 of the base ridge to be the identity, and no regular apeirohedron has linear apeirogons as facets (basically for the same reason in two-dimensional space).

It follows that R_3 must be the reflexion in the plane of the base ridge. Moreover, the facets cannot be planar, because then all the vertices would lie in the plane of this base ridge.

The vertex-figure P/v of P at its initial vertex v must be a finite regular polyhedron, and hence one of the eighteen in our list; see Theorem 7E4. However, Bieberbach's theorem and discreteness again exclude 5-fold rotational symmetries; see Lemma 7E6.

The vertex-figures must therefore belong to the crystallographic family

$$\{3,3\},\ \{4,3\}_3;\quad \{3,4\},\ \{6,4\}_3;\quad \{4,3\},\ \{6,3\}_4.$$

We have listed Petrie duals together.

The only polyhedra which can be vertex-figures of a regular 4-apeirotope P with finite planar ridges are $\{3,4\}$ and its Petrial $\{6,4\}_3$; the 2-faces are then squares $\{4\}$. (Note, incidentally, that blended regular apeirohedra cannot have finite planar facets.) To see this, we may use the same argument as that of [120, p. 69]. If P has 2-faces $\{p\}$ for some rational number p, its vertex-figure must be a finite regular polyhedron whose ratio of edge-length to circumradius is of the form $2\cos(\pi/p)$; the only instances are the octahedron $\{3,4\}$ and $\{6,4\}_3$, where $p = 4$. We thus obtain the two apeirohedra

7F1 $$\{4,3,4\} = \{\{4,3\},\{3,4\}\},\quad \{\{4,6\,|\,4\},\{6,4\}_3\}.$$

We shall justify the notation for the second apeirotope later; it is indeed the universal regular polytope of its kind.

In our listing of the regular apeirohedra in $\mathbb{E}^3$, we found that the only ones with planar (zigzag) apeirogons as facets are

$$\{\infty,6\}_3\,\#\,\{\,\},\qquad \{\infty,4\}_4\,\#\,\{\,\},\qquad \{\infty,3\}_6\,\#\,\{\,\};$$
$$\{\infty,6\}_3\,\#\,\{\infty\},\qquad \{\infty,4\}_4\,\#\,\{\infty\},\qquad \{\infty,3\}_6\,\#\,\{\infty\}.$$

In each case, the reflexion R_0 is that in a point (obtained as the product of two such reflexions, one for each component of the blend). It follows that the only other possibilities for discrete regular 4-apeirotopes are obtained by taking R_0 to be a point of $R_2 \cap R_3$, since R_0 must commute with R_2 and R_3. (Observe that the six possible vertex-figures do have $R_2 \cap R_3$ a line.) Since R_0 and R_1 must not commute, we have $R_0 \notin R_1$. In effect, this amounts to choosing R_0 to be a vertex w of the vertex-figure P/v at v, or, more strictly perhaps, half-way between v and w (this ensures that w is the image of v under R_0). Each possible choice will yield an apeirotope.

The resulting six apeirotopes are

7F2
$$\{\{\infty,3\}_6\,\#\,\{\,\},\{3,3\}\},\qquad \{\{\infty,4\}_4\,\#\,\{\infty\},\{4,3\}_3\};$$
$$\{\{\infty,3\}_6\,\#\,\{\,\},\{3,4\}\},\qquad \{\{\infty,6\}_3\,\#\,\{\infty\},\{6,4\}_3\};$$
$$\{\{\infty,4\}_4\,\#\,\{\,\},\{4,3\}\},\qquad \{\{\infty,6\}_3\,\#\,\{\infty\},\{6,3\}_4\}.$$

For the moment, the notation suppresses the exact definitions of these apeirotopes; we shall consider their groups later.

The identification of the apeirotopes in the list (7F2) is facilitated by the observation that the vertex-figures of an apeirohedron $\{\infty,q\}_p\,\#\,\{\,\}$ are planar polygons $\{q\}$, while those of $\{\infty,q\}_p\,\#\,\{\infty\}$ are skew polygons $\{q\}\,\#\,\{\,\}$. (The 2-faces of $\{4,3\}_3$ are skew polygons $\{4\}\,\#\,\{\,\}$, while those of $\{6,4\}_3$ and $\{6,3\}_4$ are skew polygons $\{6\}\,\#\,\{\,\}$.)

Summarizing this discussion, we see that we have proved

7F3 Theorem *The list of discrete faithful 3-dimensional 4-apeirotopes in (7F1) and (7F2) is complete.*

Before we go on to describe the groups of these eight apeirotopes, let us make some remarks. The two lists (7F1) and (7F2) group the apeirotopes in pairs; their vertex-figures are Petrie duals, and so their automorphism (or symmetry) groups are related by the involutory mixing operation

$$(\sigma_0, \sigma_1, \sigma_2, \sigma_3) \mapsto (\sigma_0, \sigma_1\sigma_3, \sigma_2, \sigma_3) =: (\rho_0, \rho_1, \rho_2, \rho_3).$$

This may also be seen to induce the appropriate changes of kind in the facets in (7F2), namely, that between blends with the segment { } and the apeirogon $\{\infty\}$. Geometrically, when the vertex-figure is a convex regular polyhedron $\{p, q\}$, with symmetry group $\langle R_1, R_2, R_3\rangle$ generated by plane reflexions, then R_0 is a point of the line $R_2 \cap R_3$, showing that a 2-face, with group $\langle R_0, R_1\rangle$, is a finite planar polygon. If we replace R_1 by $R_1R_3 = R_1 \cap R_3$, which is the reflexion in a line, we see that the new 2-face is a zigzag (planar) apeirogon. Further, if we have a presentation for the automorphism group of one of the pair, then we have it for the other, just by making the substitution of $\rho_1\rho_3$ for ρ_1 wherever the latter occurs.

Now we already know that $\{4, 3, 4\} = \{\{4, 3\}, \{3, 4\}\}$ is the universal regular polytope of its Schläfli type (see Proposition 4A9). When we replace its vertex-figure $\{3, 4\}$ by its Petrial $\{6, 4\}_3$, we obtain the following presentation for the automorphism group of $P := \{\{4, 6\,|\,4\}, \{6, 4\}_3\}$:

$$\begin{aligned}\rho_0^2 &= (\rho_1\rho_3)^2 = \rho_2^2 = \rho_3^2\\ &= (\rho_0\rho_1\rho_3)^4 = (\rho_0\rho_2)^2 = (\rho_0\rho_3)^2\\ &= (\rho_1\rho_3\rho_2)^3 = \left(\rho_1\rho_3^2\right)^2 = (\rho_2\rho_3)^4 = \varepsilon.\end{aligned}$$

Simplifying these relations and reordering them, we obtain

7F4
$$\begin{aligned}\rho_0^2 &= \rho_1^2 = \rho_2^2 = \rho_3^2\\ &= (\rho_0\rho_1)^4 = (\rho_0\rho_2)^2 = (\rho_0\rho_3)^2 = (\rho_1\rho_3)^2 = (\rho_2\rho_3)^4\\ &= (\rho_1\rho_2\rho_3)^3 = \varepsilon.\end{aligned}$$

The relations involving ρ_1, ρ_2 and ρ_3 certainly specify the group of $\{6, 4\}_3$, which must therefore be the vertex-figure.

The relation $(\rho_1\rho_2)^6 = \varepsilon$ is implied by the other relations, and it is only conventional to insert the number "6" in $\{\cdot, 4\}_3$, as its omission looks a little strange.

The relations involving ρ_0, ρ_1 and ρ_2, on the other hand, are clearly inadequate as they stand to specify the facet $\{4, 6\,|\,4\}$. It is curious, therefore, that the relations of (7F4) must serve to specify the group of P itself. In this context, we shall show in Corollary 7F7 that the mixing operation

$$(\sigma_0, \sigma_1, \sigma_2, \sigma_3) \mapsto (\sigma_0, \sigma_1\sigma_3, \sigma_2) =: (\rho_0, \rho_1, \rho_2)$$

applied to $[4, 3, 4]$ indeed yields the group of $\{4, 6\,|\,4\}$. Here, it is appropriate to demonstrate:

7F5 Theorem *An abstract regular 4-polytope of type* $\{4, 6, 4\}$ *with vertex-figure of type* $\{6, 4\}_3$ *is a quotient of* $\{\{4, 6\,|\,4\}, \{6, 4\}_3\}$.

Proof. In fact, we could really describe the type of the polytope as $\{4, \cdot, 4\}$. Under the given conditions, the group of such a polytope Q, say, satisfies the relations (7F4); we have observed that the vertex-figure then must be of type $\{6, 4\}_3$. We now reverse the Petrie operation on the vertex-figure, whereby we recover the relations for the Coxeter group [4, 3, 4].

It remains to show that we come to the same conclusion, even if we impose the extra relations which specify the facet $\{4, 6\,|\,4\}$. We may certainly set $(\rho_1\rho_2)^6 = \varepsilon$, if this is not already given to us. Under the (reverse) mixing operation,

$$\begin{aligned}\rho_0\rho_1\rho_2\rho_1 &= \sigma_0\sigma_1\sigma_3\sigma_2\sigma_1\sigma_3\\ &\sim \sigma_0\sigma_1\sigma_2\sigma_1\\ &= \sigma_0\sigma_2\sigma_1\sigma_2\\ &\sim \sigma_0\sigma_1,\end{aligned}$$

so that $(\rho_0\rho_1\rho_2\rho_1)^4 = \varepsilon$ is compatible with the presentation of [4, 3, 4]. □

There are two immediate consequences of this argument. The second fills in the last gap in the arguments of Section 7E about the groups of the regular apeirohedra derived from $\{4, 6\,|\,4\}$.

7F6 Corollary *The apeirotope of type* $\{\{4, 6\,|\,4\}, \{6, 4\}_3\}$ *in* $\mathbb{E}^3$ *is universal.*

7F7 Corollary *The mixing operation*

$$(\sigma_0, \sigma_1, \sigma_2, \sigma_3) \mapsto (\sigma_0, \sigma_1\sigma_3, \sigma_2) =: (\rho_1, \rho_1, \rho_2)$$

applied to [4, 3, 4] *yields the group of* $\{4, 6\,|\,4\}$.

For the other six apeirotopes, on the face of it the procedure appears very simple. To the group $\langle \rho_1, \rho_2, \rho_3 \rangle$ of the vertex-figure, we adjoin a new generator ρ_0 such that

$$\rho_0^2 = (\rho_0\rho_2)^2 = (\rho_0\rho_3)^2 = \varepsilon.$$

This is just the dual procedure to the construction of the free extension in Theorem 4D4. However, we do not actually obtain the free extension, because the translation subgroup of the apeirotope imposes extra relations on the group.

We need only consider one out of each pair whose vertex-figures are Petrie duals; the natural choice is to take that vertex-figure which is convex (Platonic).

There are two distinct kinds of argument employed here. We begin with the vertex-figure $\{3, 4\}$. Let the group of the apeirotope be $\langle \rho_0, \ldots, \rho_3 \rangle$. The half-turn about the base edge is $(\rho_2\rho_3)^2$; hence $\sigma_0 := \rho_0(\rho_2\rho_3)^2$ is the reflexion in the plane which bisects the base edge. Thus the operation

7F8 $$(\rho_0, \ldots, \rho_3) \mapsto (\rho_0(\rho_2\rho_3)^2, \rho_1, \rho_2, \rho_3) =: (\sigma_0, \ldots, \sigma_3)$$

yields a discrete group in $\mathbb{E}^3$ generated by plane reflexions, such that σ_0 commutes with σ_2 and σ_3, and $\langle \sigma_1, \sigma_2, \sigma_3 \rangle \cong [3, 4]$. It follows that $\langle \sigma_0, \ldots, \sigma_3 \rangle \cong [4, 3, 4]$ is the only possibility.

We therefore conclude

7F9 Theorem *The group of the 4-apeirotope of type* $\{\infty, 3, 4\}$ *in* $\mathbb{E}^3$ *is isomorphic to* $[4, 3, 4]$. *It is also the group* $\langle \rho_0, \ldots, \rho_3 \rangle$ *of* $[\infty, 3, 4]$, *with the single extra relation*

$$(\rho_0\rho_1(\rho_2\rho_3)^2)^4 = \varepsilon.$$

Proof. The remaining relations for $[\infty, 3, 4]$ need no comment. The extra relation arises from $(\sigma_0\sigma_1)^4 = \varepsilon$, since

$$\begin{aligned}\sigma_0\sigma_1 &= \rho_0(\rho_2\rho_3)^2\rho_1\\ &= (\rho_2\rho_3)^2\rho_0\rho_1\\ &\sim \rho_0\rho_1(\rho_2\rho_3)^2.\end{aligned}$$

This gives the result. □

While we do not need to comment on its Petrial, the group is nevertheless of interest. For this group, we replace ρ_1 by $\rho_1\rho_3$. Since $\rho_3(\rho_2\rho_3)^2 = \rho_2\rho_3\rho_2$, we see that the group of the 4-apeirotope of type $\{\infty, 6, 4\}$ in $\mathbb{E}^3$ is obtained by imposing on the Coxeter group $[\infty, 6, 4] = \langle \rho_0, \ldots, \rho_3 \rangle$ the extra relations

$$(\rho_0\rho_1\rho_2\rho_3\rho_2)^4 = (\rho_1\rho_2\rho_3)^4 = \varepsilon.$$

The other two cases turn out to be universal of their kind, that is, universal with the facets and vertex-figures which we know they have. We first have the apeirotope with vertex-figure $\{3, 3\}$, whose facets are isomorphic to $\{\infty, 3\}_6 = \{6, 3\}^\pi$. Thus we certainly have hexagonal edge-circuits in the graph of the apeirotope; we show that these suffice to generate all edge-circuits by concatenation.

In fact, we have in effect already solved this problem. Consider the mixing operation on the group $\langle \rho_0, \ldots, \rho_3 \rangle$ (say), given by

$$(\rho_0, \ldots, \rho_3) \mapsto (\rho_0, \rho_1\rho_3, \rho_2) =: (\sigma_0, \sigma_1, \sigma_2).$$

Then $\langle \sigma_0, \sigma_1, \sigma_2 \rangle$ is the group of a discrete regular apeirohedron in $\mathbb{E}^3$ of type $\{\infty, 4\}$, whose corresponding concrete mirrors S_0, S_1 and S_2 have dimensions $\dim S_j = j$ for each j. This identifies the apeirohedron unambiguously as

$$\{\infty, 4\}_4 \,\#\, \{\infty\} = (\{4, 4\} \,\#\, \{\infty\})^\pi.$$

Hence the further operation

$$(\sigma_0, \sigma_1, \sigma_2) \mapsto (\sigma_0\sigma_2, \sigma_1, \sigma_2) =: (\tau_0, \tau_1, \tau_2)$$

gives the group of $\{4, 4\} \,\#\, \{\infty\}$ itself. As we saw in Theorem 7E18, all circuits in the edge-graph of the last are concatenations of hexagonal basic circuits. (Note that the version of $\{4, 4\} \,\#\, \{\infty\}$ which we have here is special, since these basic hexagons generally only have symmetry group $C_2 \times C_2$.) Indeed, from Theorem 7E18, the relation corresponding to the hexagon is

$$((\tau_1\tau_0)^3\tau_1\tau_2)^2 = \varepsilon.$$

Performing the substitutions given by the mixing operations, we have

$$\begin{aligned}(\tau_1\tau_0)^3\tau_1\tau_2 &= ((\rho_1\rho_3\rho_0\rho_2)^3\rho_1\rho_3\rho_2)^2\\ &= \rho_1\rho_0\cdot\rho_3\rho_2\rho_3\cdot\rho_1\rho_0\rho_2\rho_1\rho_0\cdot\rho_3\rho_2\rho_3\cdot\rho_1\rho_2\\ &= \rho_1\rho_0\cdot\rho_2\rho_3\rho_2\cdot\rho_1\rho_0\rho_2\rho_1\rho_0\cdot\rho_2\rho_3\rho_2\cdot\rho_1\rho_2\\ &\sim \rho_0\rho_3\rho_2\rho_1\rho_0\rho_2\rho_1\rho_0\rho_2\rho_3\cdot\rho_2\rho_1\rho_2\rho_1\rho_2\\ &= \rho_0\rho_3\rho_2\rho_1\rho_0\rho_2\rho_1\rho_0\rho_2\rho_3\rho_1\\ &\sim (\rho_0\rho_2\rho_1)^3\sim(\rho_0\rho_1\rho_2)^3.\end{aligned}$$

We conclude that we have

7F10 Theorem *The group of the regular 4-apeirotope of type* $\{\infty, 3, 3\}$ *in* $\mathbb{E}^3$ *is the Coxeter group* $[\infty, 3, 3] = \langle\rho_0, \ldots, \rho_3\rangle$, *with the imposition of the single extra relation*

$$(\rho_0\rho_1\rho_2)^6 = \varepsilon.$$

In other words, the apeirotope is the universal polytope

$$\{\{\infty, 3\}_6, \{3, 3\}\}.$$

The Petrie operation on the vertex-figure of this apeirotope replaces ρ_1 by $\rho_1\rho_3$, and so we see that the apeirotope with vertex-figure $\{4, 3\}_3$ has group $[\infty, 4, 3] = \langle\rho_0, \ldots, \rho_3\rangle$, factored out by the additional relations

$$(\rho_1\rho_2\rho_3)^4 = (\rho_0\rho_1\rho_2\rho_3)^6 = \varepsilon.$$

Let us add a few words of description to the formal treatment we have just presented. The common edge-graph of the two apeirotopes is the famous *diamond net*; in the diamond crystal, the carbon atoms sit at the vertices, and the bonds between adjacent atoms are represented by the edges. Wells [451, pp. 117, 118] describes this graph as one of a number of *uniform nets*; however, his approach is far from systematic, and he does not emphasize the fact that the diamond net has such a high degree of symmetry.

We treat the apeirotope with vertex-figure $\{4, 3\}$ in a quite different way. We may take the vertices of this apeirotope to be the lattice $\Lambda_{(1,1,1)}$, generated by $(1, 1, 1)$ and its images under the symmetries of the cube (in effect here, all changes of sign of the coordinates). With initial vertex o, joined to $(1, 1, 1)$ by the initial edge, a typical basic circuit then goes to $(2, 0, 0)$, then to $(1, -1, 1)$, and finally back to o. This is indeed a Petrie polygon of a facet, and so corresponds to the relation $(\rho_0\rho_1\rho_2)^4 = \varepsilon$ imposed on the Coxeter group $[\infty, 4, 3]$.

The convex hull of such a circuit is a tetrahedron, with a pair of opposite edges of length 2, and the remaining lateral edges of length $\sqrt{3}$. It is easy to see that these tetrahedra tile $\mathbb{E}^3$; for example, the $6\cdot 4 = 24$ which contain o form a rhombic dodecahedron. There is then little difficulty in using this fact to show that any edge-circuit in the graph of the apeirotope is a concatenation of basic circuits. That is, we have

7F11 Theorem *The group of the 4-apeirotope in* $\mathbb{E}^3$ *with vertex-figure* $\{4, 3\}$ *is the Coxeter group* $[\infty, 4, 3] = \langle\rho_0, \ldots, \rho_3\rangle$, *with the additional relation* $(\rho_0\rho_1\rho_2)^4 = \varepsilon$.

The three apeirotopes we have just described fall near the beginning of three infinite families of apeirotopes. We introduce some (temporary) notation. We write

7F12
$$\begin{aligned} \mathcal{Q}_1^3 &:= \{\infty, 3\}_6, & \mathcal{Q}_1^4 &:= \{\infty, 3, 3\}, \\ \mathcal{Q}_2^3 &:= \{\infty, 4\}_4, & \mathcal{Q}_2^4 &:= \{\infty, 3, 4\}, \\ \mathcal{Q}_3^3 &:= \{\infty, 4\}_4, & \mathcal{Q}_3^4 &:= \{\infty, 4, 3\}. \end{aligned}$$

Of course, the notation used here is the abbreviated one. Note that $\mathcal{Q}_2^3 = \mathcal{Q}_3^3$; we may even begin all three series with $\mathcal{Q}_i^2 := \{\infty\}$ for each $i = 1, 2, 3$.

Now we may carry out exactly the same constructions in $\mathbb{E}^n$, to obtain regular $(n+1)$-apeirotopes with vertex-figures the three regular n-polytopes, namely, the simplex $\{3^{n-1}\}$, the cross-polytope $\{3^{n-2}, 4\}$, and the cube $\{4, 3^{n-1}\}$. The initial vertex lies at the centre of the vertex-figure, and the reflexion R_0 is the mid-point of the edge joining the initial vertex to a vertex of the vertex-figure. Let us denote these apeirotopes by $\mathcal{Q}_1^{n+1}$, $\mathcal{Q}_2^{n+1}$ and $\mathcal{Q}_3^{n+1}$, respectively. While we shall give few details here, it is possible to prove the following.

7F13 Theorem *The two apeirotopes $\mathcal{Q}_1^{n+1}$ and $\mathcal{Q}_3^{n+1}$ are universal, in that*

$$\mathcal{Q}_1^{n+1} \cong \{\mathcal{Q}_1^n, \{3^{n-1}\}\}, \quad \mathcal{Q}_3^{n+1} \cong \{\mathcal{Q}_3^n, \{4, 3^{n-2}\}\}.$$

The group of $\mathcal{Q}_2^{n+1}$ is isomorphic to $[4, 3^{n-2}, 4] = \langle \sigma_0, \ldots, \sigma_n \rangle$, *and is obtained from it by the involutory mixing operation*

$$(\sigma_0, \ldots, \sigma_n) \mapsto (\sigma_0(\sigma_2 \cdots \sigma_n)^{n-1}, \sigma_1, \ldots, \sigma_n) =: (\rho_0, \ldots, \rho_n).$$

Hence this group is the Coxeter group $[\infty, 3^{n-2}, 4] = \langle \rho_0, \ldots, \rho_n \rangle$, *with the imposition of the single extra relation*

$$(\rho_0 \rho_1 (\rho_2 \cdots \rho_n)^{n-1})^4 = \varepsilon.$$

Proof. We just comment on the last assertion, in noting that $(\sigma_2 \cdots \sigma_n)^{n-1}$ is the half-turn about the initial edge of $\{4, 3^{n-2}, 4\}$. The relation arises from $(\sigma_0 \sigma_1)^4 = \varepsilon$, of course. □

We now feel obliged to mention two similar examples of this kind, which do not fit into these series. First, we perform the same construction with vertex-figure the hexagon $\{6\}$ in $\mathbb{E}^2$. We then obtain $\{\infty, 6\}_3$.

If we begin instead with the 24-cell $\{3, 4, 3\}$ in $\mathbb{E}^4$, and then note that the reflexion in the mid-point of the initial edge of $\{3, 3, 4, 3\}$ is $\sigma_0(\sigma_2\sigma_3\sigma_4)^3$ (in terms of its group $[3, 3, 4, 3] = \langle \sigma_0, \ldots, \sigma_4 \rangle$), we conclude

7F14 Theorem *There is a discrete regular 5-apeirotope of type* $\{\infty, 3, 4, 3\}$ *in* $\mathbb{E}^4$, *whose group is isomorphic to* $[3, 3, 4, 3]$. *Abstractly, the group is obtained from* $[3, 3, 4, 3] = \langle \sigma_0, \ldots, \sigma_4 \rangle$ *by the involutory mixing operator*

$$(\sigma_0, \ldots, \sigma_4) \mapsto (\sigma_0(\sigma_2\sigma_3\sigma_4)^3, \sigma_1, \ldots, \sigma_4) =: (\rho_0, \ldots, \rho_4).$$

It is therefore the Coxeter group $[\infty, 3, 4, 3] = \langle \rho_0, \ldots, \rho_4 \rangle$, *with the imposition of the single extra relation* $(\rho_0\rho_1(\rho_2\rho_3\rho_4)^3)^3 = \varepsilon$.

We may further generalize these constructions. Let $\mathcal{P}$ be a finite regular n-polytope. If $\mathcal{P}$ admits any faithful realizations, then its simplex realization P is faithful. We now adjoin to the geometric group of P the reflexion in one of its vertices. This certainly leads to a discrete group, which is of a regular $(n+1)$-apeirotope with apeirogonal 2-faces and vertex-figure isomorphic to $\mathcal{P}$.

Indeed, all that we really need is some kind of "crystallographic" restriction on the symmetry group and initial vertex of the realization P. (We shall deliberately leave the meaning of this term vague, but the intention to obtain a discrete group is clear.) This freedom, though, indicates that the construction is far from being combinatorial; it depends crucially on the geometry of the realization. As a simple example, take $\mathcal{P}$ to be a hexagon. With the planar realization $\{6\}$, there results the Petrial $\{\infty, 6\}_3$ of the tessellation $\{3, 6\}$, as we remarked earlier. On the other hand, beginning with a blend $\{3\} \# \{\,\}$, we obtain the apeirohedron

$$\{\infty, 3\}_6 \diamond \{\infty\},$$

which we discussed earlier.

8

Twisting

We now discuss a general method for constructing regular polytopes $\mathcal{P}$ from certain groups W by what are called twisting operations. These operations provide an important tool in the classification of the locally toroidal regular polytopes in later chapters. In our applications, W will usually be a Coxeter group or a complex reflexion group with a diagram admitting certain symmetries. In this chapter, we shall mainly study twisting operations on Coxeter groups; we shall leave the case of complex reflexion groups to Chapter 11.

Our use of the term "twisting operation" is motivated by the analogy with the importance of diagram symmetries for the twisted simple groups (see [75]).

The present chapter is organized as follows. In Section 8A we explain the basic concept of twisting. Then in Section 8B we construct the regular polytopes $\mathcal{L}^{\mathcal{K},\mathcal{G}}$ and describe their basic properties. This construction is very general and leads to a remarkable class of polytopes. In Sections 8C and 8D we shall treat two particularly interesting special cases of this construction, namely, the polytopes $2^{\mathcal{K}}$ and $2^{\mathcal{K},\mathcal{G}(s)}$. Then in Section 8E we prove a universality property of $\mathcal{L}^{\mathcal{K},\mathcal{G}}$, and apply it to identify certain universal regular polytopes. Finally, in Section 8F we describe a method for finding polytopes with small faces as quotients of polytopes constructed in earlier sections.

8A Twisting Operations

The basic idea of a twisting operation is to extend a given group W by some of its group automorphisms to obtain a new group Γ which is then the group of a regular polytope. In this section, we explain this method in more detail. However, here and in the next sections, we shall not attempt to develop the twisting technique in full generality, but instead restrict ourselves to some particularly interesting cases.

Let W be a group generated by k involutions $\sigma_1, \ldots, \sigma_k$; we allow k to be infinite here. In most applications, W will actually be a C-group (see Section 2E), but we shall not impose this condition. Our operations will only be defined for those groups W which admit certain group automorphisms τ permuting the generators σ_i. If these automorphisms τ are themselves involutions, then we can augment W by adding them, and in suitable cases obtain a new group Γ with certain distinguished generators $\rho_0, \ldots, \rho_{n-1}$ which is a string C-group. Then, writing Λ for the group of automorphisms of W generated by the τ, we have

$$\Gamma = W \rtimes \Lambda,$$

a semi-direct product of W by Λ. In the examples, Λ may be of order 2 generated by a single involutory automorphism τ, or, in the other extreme case, may itself be the group of a regular polytope of rank at least 2 acting suitably on W.

Of particular interest are those groups W which are represented by diagrams. In this case, the automorphisms τ correspond to diagram symmetries. The most important examples are the Coxeter groups with finitely many generators σ_i, as well as certain complex reflexion groups preserving an hermitian form (see Chapter 9).

As to the admissible choice of the generators ρ_i for the new group Γ, there is the obvious restriction (2E2), namely, that

$$(\rho_i \rho_j)^2 = \varepsilon \qquad (0 \leqslant i < j - 1 \leqslant n - 2).$$

This cuts down many possibilities, but is still not sufficient for Γ to be the group of a regular polytope, unless Γ also satisfies the intersection property (2E1). However, if both (2E2) and (2E1) hold, then Γ is a string C-group and, by Theorem 2E11, is the group of the regular polytope $\mathcal{P}(\Gamma)$. In the applications, the intersection property for Γ will usually follow from the fact that both W and Λ already have this property with respect to their (distinguished) generators.

For a twisting operation κ on W, we introduce the following notation. We write

8A1 $$\kappa\colon (\sigma_1, \ldots, \sigma_k; \tau_1, \ldots, \tau_m) \mapsto (\rho_0, \ldots, \rho_{n-1}),$$

where $\sigma_1, \ldots, \sigma_k$ are the generators of W, $\tau_1, \ldots, \tau_m$ the generating automorphisms in Λ, and $\rho_0, \ldots, \rho_{n-1}$ the distinguished generators of $\Gamma = W \rtimes \Lambda$ constructed from the σ_i and the τ_j by κ. Here we usually consider Γ as an internal semi-direct product of W by Λ (so that Λ acts on W by conjugation).

The following simple lemma deals with an interesting special case where the semi-direct product Γ is also a direct product.

8A2 Lemma *In the previous situation, if Λ is generated by a single involutory automorphism τ which transforms W according to an inner automorphism, and if the centre of W only contains elements of odd (finite) order, then*

$$\Gamma \cong W \times C_2.$$

Proof. Let $\hat{\tau} \in W$ be such that

$$\hat{\tau}^{-1}\sigma_i\hat{\tau} = \tau^{-1}\sigma_i\tau \qquad (i = 1, \ldots, k).$$

Then $\alpha := \hat{\tau}\tau^{-1} \notin W$, $\alpha^2 \in W$, and α^2 belongs to the centre of W. Let r denote the order of α^2. Since r is odd, $\beta := \alpha^r \notin W$, $\beta^2 = \varepsilon$, and β commutes with each σ_i. Hence

$$\Gamma \cong W \times \langle\beta\rangle \cong W \times C_2. \qquad \square$$

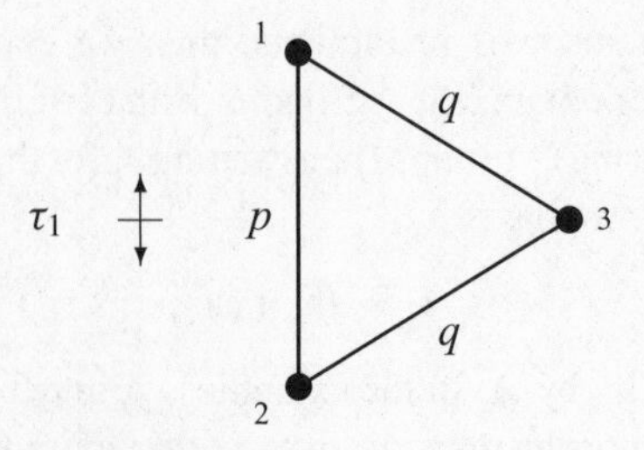

Figure 8A1. Twisting by τ_1.

In what follows, we shall assume that W is a Coxeter group with diagram $\mathcal{D}$ on the nodes $1, \ldots, k$ (where we allow $k = \infty$). In particular, we adopt the notation of Chapter 3, and write

$$W = W(\mathcal{D}) = \langle \sigma_1, \ldots, \sigma_k \rangle.$$

Then, by Theorem 3A2(b) and its analogue for groups with infinitely many generators, W is a C-group.

In describing or drawing diagrams $\mathcal{D}$ we adopt the following convention, which is in agreement with our notation in Chapter 3. Sometimes it is convenient to think of a pair of distinct nodes i, j of $\mathcal{D}$ which are not directly joined in $\mathcal{D}$ as being joined by an *improper* branch with label $m_{ij} = 2$. Accordingly, if a diagram $\mathcal{D}$ is drawn and some of its branches are labelled by 2, then these branches are improper branches and, strictly speaking, do not belong to $\mathcal{D}$. The diagram with k isolated nodes is called the *trivial diagram on k nodes*; here, each branch is improper,

$$W \cong C_2^k,$$

and each permutation of the nodes corresponds to an automorphism of W.

To give a simple example of twisting, let W be the triangle group with the diagram $\mathcal{D}$ in Figure 8A1. Here, and in similar figures, the symmetries of the diagram are indicated by arrows. Now, using τ_1 and twisting by

$$\kappa_1\colon (\sigma_1, \sigma_2, \sigma_3; \tau_1) \mapsto (\tau_1, \sigma_2, \sigma_3) =: (\rho_0, \rho_1, \rho_2)$$

gives the Coxeter group

$$\langle \rho_0, \rho_1, \rho_2 \rangle = W \rtimes C_2 = [2p, q],$$

and its corresponding regular 3-polytope $\{2p, q\}$.

On the other hand, if $p = q$, then the diagram of Figure 8A1 admits further symmetries, as shown in Figure 8A2. Now the operation

$$\kappa_2\colon (\sigma_1, \sigma_2, \sigma_3; \tau_1, \tau_2) \mapsto (\tau_1, \tau_2, \sigma_3) =: (\rho_0, \rho_1, \rho_2)$$

employs both symmetries τ_1 and τ_2, and leads to the Coxeter group

$$\langle \rho_0, \rho_1, \rho_2 \rangle = W \rtimes S_3 = [3, 2q],$$

and its corresponding regular 3-polytope $\{3, 2q\}$.

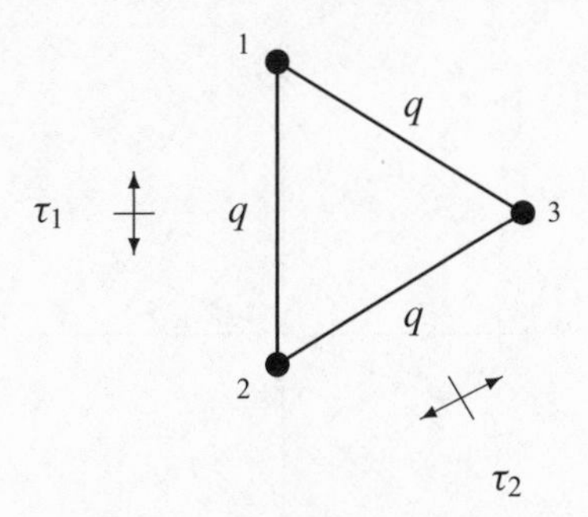

Figure 8A2. Twisting by τ_1 and τ_2.

8B The Polytopes $\mathcal{L}^{\mathcal{K},\mathcal{G}}$

The kind of twisting operation we shall discuss in this section is very general, and yields regular polytopes which are denoted $\mathcal{L}^{\mathcal{K},\mathcal{G}}$ (see [294, §3; 303, §4]). Here, $\mathcal{L}$ and $\mathcal{K}$ are regular polytopes of ranks m ($\geqslant 1$) and n ($\geqslant 1$), respectively, and $\mathcal{G}$ is a suitable diagram determined by $\mathcal{K}$ only. We shall frequently take $\mathcal{L}$ to be of rank 1 or $\mathcal{G}$ to be a trivial diagram; if $\mathcal{L}$ is of rank 1, we also write $2^{\mathcal{K},\mathcal{G}}$ for $\mathcal{L}^{\mathcal{K},\mathcal{G}}$, since $\mathcal{L}$ is determined by its 2 vertices (see Section 8C). In the general situation, $\mathcal{L}^{\mathcal{K},\mathcal{G}}$ is of rank $m+n$, with m-faces isomorphic to $\mathcal{L}$ and with co-$(m-1)$-faces isomorphic to $\mathcal{K}$. Its group is a semi-direct product of a Coxeter group $W = W(\mathcal{D})$ by the group $\Gamma(\mathcal{K})$, where $\mathcal{D}$ is a diagram which is defined by $\mathcal{L}$ and $\mathcal{K}$ and contains $\mathcal{G}$ as an induced subdiagram.

The most important property of $\mathcal{G}$ is that $\Gamma(\mathcal{K})$ acts suitably on $\mathcal{G}$ as a group of diagram automorphisms. This implies a certain restriction on $\mathcal{K}$, but still covers the most interesting cases. On the other hand, the structure of $\mathcal{G}$ and $\mathcal{D}$ essentially limits our construction to universal polytopes $\mathcal{L} = \{q_1, \ldots, q_{m-1}\}$, which is a restriction only if $m \geqslant 3$; if $m = 1$, then $\mathcal{L} = \{\,\}$, the unique 1-polytope.

In what follows, we shall need the notion of a $\mathcal{K}$-admissible diagram. Let $\mathcal{G}$ be the diagram of a group generated by involutions (typically a Coxeter group), and let $\mathcal{K}$ be a regular n-polytope with group

$$\Gamma(\mathcal{K}) = \langle \tau_0, \ldots, \tau_{n-1} \rangle.$$

Then $\mathcal{G}$ is called $\mathcal{K}$-*admissible* if $\mathcal{G}$ has more than one node, and $\Gamma(\mathcal{K})$ acts as a group of diagram automorphisms on $\mathcal{G}$ with the following properties. First, $\Gamma(\mathcal{K})$ acts transitively on the set $V(\mathcal{G})$ of nodes of $\mathcal{G}$. Second, the subgroup $\langle \tau_1, \ldots, \tau_{n-1} \rangle$ of $\Gamma(\mathcal{K})$ stabilizes a node F_0 (say) of $\mathcal{G}$, the *initial node* (it may stabilize more than one such node). Third, with respect to F_0, the action of $\Gamma(\mathcal{K})$ on $\mathcal{G}$ respects the intersection property for the generators $\tau_0, \ldots, \tau_{n-1}$ in that, if $V(\mathcal{G}, I)$ denotes the set of transforms of F_0 under $\langle \tau_i \mid i \in I \rangle$ for $I \subseteq \{0, \ldots, n-1\}$, then

8B1 $$V(\mathcal{G}, I) \cap V(\mathcal{G}, J) = V(\mathcal{G}, I \cap J) \quad \text{if } I, J \subseteq \{0, \ldots, n-1\}.$$

Note that we do not demand that $\Gamma(\mathcal{K})$ be faithfully represented as this group of diagram automorphisms. Observe that, if $j \leqslant n-1$ and $\mathcal{K}_j$ is the basic j-face of $\mathcal{K}$,

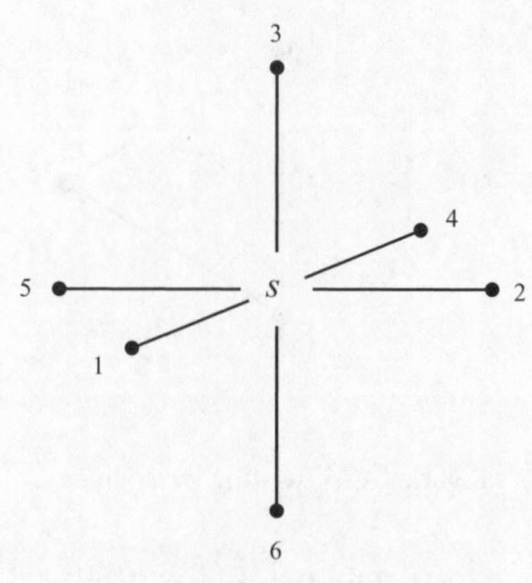

Figure 8B1

then the induced subdiagram of $\mathcal{G}$ with node set $V(\mathcal{G}, \{0, \ldots, j-1\})$ is a $\mathcal{K}_j$-admissible diagram with the same initial node. Note also that $F_0\tau_0 \neq F_0$, because otherwise $\Gamma(\mathcal{K})$ stabilizes F_0, which then would be the only node of $\mathcal{G}$.

We shall usually have

$$V(\mathcal{G}) = V(\mathcal{K}),$$

the vertex set of $\mathcal{K}$; then $\Gamma(\mathcal{K})$ acts in the natural way, and F_0 is the vertex in the base flag of $\mathcal{K}$. In this situation, (8B1) is always satisfied, and $\mathcal{K}$-admissibility of a diagram $\mathcal{G}$ just means that $\Gamma(\mathcal{K})$ acts on $\mathcal{G}$ as a group of diagram automorphisms. In particular, if $I \subseteq \{0, \ldots, n-1\}$ and j is the first index among $0, \ldots, n-1$ which is not in I, then

$$V(\mathcal{G}, I) = \{F \in V(\mathcal{K}) \mid F \leqslant F_j\},$$

where as usual $\{F_{-1}, F_0, \ldots, F_n\}$ denotes the base flag of $\mathcal{K}$.

For instance, consider the diagram $\mathcal{G}$ on the vertex set of the octahedron $\mathcal{K} = \{3, 4\}$ which connects antipodal vertices by a branch labelled s with $s \geqslant 3$, but has no other proper branches (or, equivalently, all other branches are improper); see Figure 8B1. Then $\Gamma(\mathcal{K})$ acts on $\mathcal{G}$ in a natural way as a group of diagram symmetries; equivalently, $\Gamma(\mathcal{K})$ acts on the corresponding Coxeter group $D_s \times D_s \times D_s$ as a group of automorphisms permuting the generators. If the vertices of $\mathcal{K}$ are $1, \ldots, 6$ with $i, i+3 \pmod 6$ adjacent, then we can take

$$\tau_0 := (1\ 2)(4\ 5), \quad \tau_1 := (2\ 3)(5\ 6), \quad \tau_2 := (3\ 6), \quad \text{and} \quad F_0 := 1.$$

Then F_0 is fixed by the group $\langle \tau_1, \tau_2 \rangle$ of the vertex-figure at F_0, and (8B1) is also satisfied. Hence $\mathcal{G}$ is $\mathcal{K}$-admissible. If we allow $s = 2$, then $\mathcal{G}$ is the *trivial* diagram with no (proper) branches. We shall revisit this example in Corollaries 8C6 and 8E18.

In the general situation where $\mathcal{K}$ is given as before and $\mathcal{G}$ is unspecified except for its node set $V(\mathcal{G}) = V(\mathcal{K})$, the number of possible choices for $\mathcal{G}$ depends essentially on the number of diagonal classes of $\mathcal{K}$. Recall that a pair of distinct vertices of $\mathcal{K}$ is called a *diagonal* of $\mathcal{K}$, and that two diagonals are said to be *equivalent* if they are equivalent under $\Gamma(\mathcal{K})$ (see Section 5A). The diagonals of $\mathcal{K}$ thus fall into *diagonal classes*, consisting of equivalent diagonals.

Now, given any diagonal class of $\mathcal{K}$, either none of its diagonals is represented by a (proper) branch in $\mathcal{G}$, or all diagonals are represented by branches with the same label. Clearly, if $\mathcal{G}$ is the trivial diagram on $V(\mathcal{K})$, then $\mathcal{G}$ is $\mathcal{K}$-admissible. By the edge-transitivity of $\Gamma(\mathcal{K})$, if any edge of $\mathcal{K}$ is a branch of $\mathcal{G}$, then all edges of $\mathcal{K}$ are branches of $\mathcal{G}$ with the same label $r \geqslant 3$; that is, $\mathcal{G}$ contains the r-labelled 1-skeleton of $\mathcal{K}$ as an induced subdiagram.

If $\mathcal{K}$ has only one diagonal class, then $\mathcal{K}$ is neighbourly, where we recall that a polytope is *neighbourly* if any two of its vertices are joined by an edge. (More strictly, this property is 2-neighbourliness, but we shall not need to consider its higher generalizations here; see Grünbaum [197, Chapter 7].) In this case, if $\mathcal{G}$ is not the trivial diagram, then $\mathcal{G}$ is the r-labelled 1-skeleton of $\mathcal{K}$. If $\mathcal{K}$ is neighbourly and regular, then any two vertices are joined by the same number of edges. But note that, in the diagram $\mathcal{G}$, only one of these edges is represented by a branch.

If $\mathcal{K}$ has at least two diagonal classes, then some classes may be represented by branches of $\mathcal{G}$, while others are not. This gives much freedom for the choice of $\mathcal{G}$, and thus for the corresponding (Coxeter) groups. Though these groups are generally infinite, there is at least one (non-trivial) important special case where they become finite. In fact, if $\mathcal{K}$ is *centrally symmetric*, in the sense that $\Gamma(\mathcal{K})$ has a *proper* central involution α, which does not fix any of its vertices, then α pairs up the vertices of $\mathcal{K}$, and these pairs of *antipodal* vertices form one diagonal class of $\mathcal{K}$; we shall elaborate on this in the next section. If this but no other diagonal class is represented by branches in $\mathcal{G}$, all with the same label, then $\mathcal{G}$ is a matching; thus the corresponding Coxeter group is finite if and only if $\mathcal{K}$ is finite, in which case it is a direct product of dihedral groups. The diagram of Figure 8B1 gives an example of this kind which is based on the octahedron $\mathcal{K} = \{3, 4\}$.

Now, to construct the regular polytope $\mathcal{L}^{\mathcal{K},\mathcal{G}}$, let $\mathcal{G}$ be any $\mathcal{K}$-admissible diagram with node set $V(\mathcal{G})$, and let

$$\mathcal{L} = \{q_1, \ldots, q_{m-1}\},$$

the universal regular m-polytope whose group is the Coxeter group

$$[q_1, \ldots, q_{m-1}] = \langle \sigma_0, \ldots, \sigma_{m-1} \rangle$$

with string diagram

8B2 $\quad$ 0 —q_1— 1 —q_2— 2 - - - $m{-}3$ —q_{m-2}— $m{-}2$ —q_{m-1}— $m{-}1$

We construct a new diagram $\mathcal{D}$ by adjoining this diagram to $\mathcal{G}$, as follows.

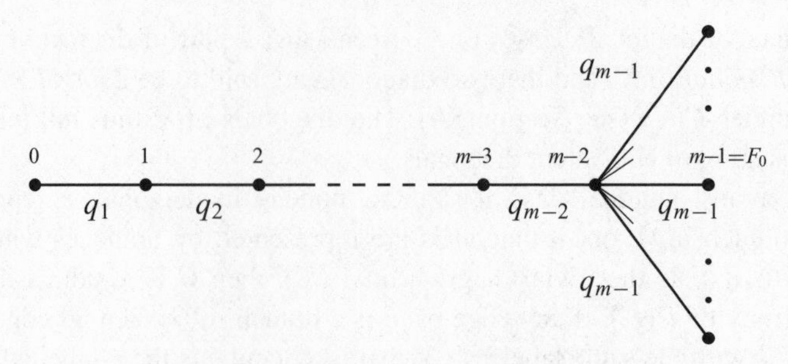

Figure 8B2

We identify the node $m-1$ of (8B2) with the initial node F_0 in $V(\mathcal{G})$, and take as the node set of $\mathcal{D}$ the disjoint union

8B3 $$V(\mathcal{D}) := V(\mathcal{G}) \cup \{0, \dots, m-2\}.$$

As branches and labels of $\mathcal{D}$ we take all the old branches and labels of both (8B2) and $\mathcal{G}$, and, in addition, for each node $F \neq F_0$ $(= m-1)$ in $V(\mathcal{G})$, one new branch with label q_{m-1} connecting F to the node $m-2$ of $\mathcal{D}$. Then $\mathcal{D}$ takes the form of the diagram in Figure 8B2, on which $\Gamma(\mathcal{K})$ acts as a group of diagram symmetries fixing the new nodes $0, 1, \dots, m-2$. But then $\Gamma(\mathcal{K})$ acts also on the corresponding Coxeter group

$$W := W(\mathcal{D}) = \langle \sigma_k \mid k \in V(\mathcal{D}) \rangle$$

as a group of automorphisms which permute the generators and fix $\sigma_0, \sigma_1, \dots, \sigma_{m-2}$. By construction, W contains $\Gamma(\mathcal{L}) = \langle \sigma_0, \dots, \sigma_{m-1} \rangle$ as a subgroup. Note that $\mathcal{D} = \mathcal{G}$ (with initial node F_0) if $\mathcal{L}$ is of rank 1.

We can now define the $(m+n)$-polytope $\mathcal{L}^{\mathcal{K},\mathcal{G}}$ by its group

$$\Gamma := W \rtimes \Gamma(\mathcal{K}),$$

the semi-direct product induced by the action of $\Gamma(\mathcal{K}) = \langle \tau_0, \dots, \tau_{n-1} \rangle$ on W, and the distinguished generators $\rho_0, \dots, \rho_{m+n-1}$ of Γ given by

8B4 $$\rho_i := \begin{cases} \sigma_i, & \text{for } i = 0, \dots, m-1, \\ \tau_{i-m}, & \text{for } i = m, \dots, m+n-1. \end{cases}$$

The transitivity of $\Gamma(\mathcal{K})$ on the nodes of $\mathcal{G}$ implies that these elements $\rho_0, \dots, \rho_{m+n-1}$ really generate Γ. In fact, for each $F \in V(\mathcal{G})$, there exists an element $\tau \in \Gamma(\mathcal{K})$ such that

$$\sigma_F = \sigma_{F_0 \tau} = \tau^{-1} \sigma_{F_0} \tau = \tau^{-1} \sigma_{m-1} \tau.$$

Then we have

8B5 Lemma *Γ is a string C-group with distinguished generators $\rho_0, \dots, \rho_{m+n-1}$.*

Proof. We need to check properties (2E1) and (2E2). Here, (2E2) takes the form

$$(\rho_i \rho_j)^2 = \varepsilon \qquad (0 \leqslant i < j-1 \leqslant m+n-2).$$

This is trivial if $i, j \leqslant m-1$ or $i, j \geqslant m$. If $i = m-1$, it follows from the fact that, by our choice of F_0, the element $\rho_j = \tau_{j-m}$ fixes F_0 in $\mathcal{G}$ and thus in $\mathcal{D}$. Similarly, if $i \leqslant m-2$ and $j \geqslant m$, then the element $\rho_j = \tau_{j-m}$ fixes the node i of $\mathcal{D}$; thus $\rho_j \rho_i \rho_j = \rho_i$.

To prove the intersection property (2E1), let $I \subseteq \{0, \ldots, m+n-1\}$. Write $I = I_1 \cup I_2$, with $I_1 \subseteq \{0, \ldots, m-1\}$ and $I_2 \subseteq \{m, \ldots, m+n-1\}$. Then the subgroup

$$\langle \rho_i \mid i \in I \rangle = \langle \sigma_i, \tau_{j-m} \mid i \in I_1 \text{ and } j \in I_2 \rangle$$

is given by

8B6
$$\begin{cases} \langle \sigma_i \mid i \in I_1 \rangle \, \langle \tau_{j-m} \mid j \in I_2 \rangle & \text{if } m-1 \notin I_1, \\ \langle \sigma_k \mid k \in V(\mathcal{G}, I_2 - m) \cup I_1 \rangle \, \langle \tau_{j-m} \mid j \in I_2 \rangle & \text{if } m-1 \in I_1. \end{cases}$$

Here, the first factorization is a direct product, while the second is a semi-direct product; in any case, the first factor is a subgroup of W and the second a subgroup of $\Gamma(\mathcal{K})$. Now, if $J \subseteq \{0, \ldots, m+n-1\}$ also, with $J = J_1 \cup J_2$ its corresponding decomposition, then $I \cap J = (I_1 \cap J_1) \cup (I_2 \cap J_2)$, and (2E1) now follows from the intersection property for W and $\Gamma(\mathcal{K})$, and our assumption (8B1). □

For later use, we note a special case of (8B6). If $m \leqslant j \leqslant m+n-1$, we have the semi-direct product

$$\langle \rho_0, \ldots, \rho_j \rangle = \langle \sigma_k \mid k \in V(\mathcal{G}, \{0, \ldots, j-m\}) \cup \{0, \ldots, m-1\} \rangle \, \langle \tau_0, \ldots, \tau_{j-m} \rangle.$$

8B7

Here, if $V(\mathcal{G}) = V(\mathcal{K})$, then

$$V(\mathcal{G}, \{0, \ldots, j-m\}) = \{F \in V(\mathcal{K}) \mid F \leqslant F_{j-m+1}\}.$$

Our next theorem states the basic properties of the regular polytope whose group is Γ. This polytope will be denoted by $\mathcal{L}^{\mathcal{K},\mathcal{G}}$. Strictly speaking, $\mathcal{L}^{\mathcal{K},\mathcal{G}}$ also depends on the action of $\Gamma(\mathcal{K})$ on the $\mathcal{K}$-admissible diagram $\mathcal{G}$ and on the choice of the initial node F_0, but we shall not make this part of our notation. Note that different choices of F_0 lead to inner automorphisms of the group. In our applications, $V(\mathcal{G}) = V(\mathcal{K})$ and F_0 is the vertex in the base flag of $\mathcal{K}$. Also, in this situation, we simply write $\mathcal{L}^{\mathcal{K}}$ for $\mathcal{L}^{\mathcal{K},\mathcal{G}}$ if $\mathcal{G}$ is the trivial diagram on $V(\mathcal{K})$.

8B8 Theorem *Let $\mathcal{L} = \{q_1, \ldots, q_{m-1}\}$ be a universal regular m-polytope, and let $\mathcal{K}$ be a regular n-polytope of type $\{p_1, \ldots, p_{n-1}\}$ with group $\Gamma(\mathcal{K}) = \langle \tau_0, \ldots, \tau_{n-1} \rangle$, where $m, n \geqslant 1$. Let $\mathcal{G}$ be a $\mathcal{K}$-admissible diagram with node set $V(\mathcal{G})$, initial node F_0 and subsets $V(\mathcal{G}, I)$. Further, let $\mathcal{D}$ be the diagram with node set $V(\mathcal{D})$ defined by $\mathcal{G}$ and $\mathcal{L}$ as in (8B3) and Figure 8B2, and let $W = W(\mathcal{D}) = \langle \sigma_k \mid k \in V(\mathcal{D}) \rangle$ be the corresponding Coxeter group. Then the regular $(m+n)$-polytope $\mathcal{L}^{\mathcal{K},\mathcal{G}}$ has the following properties.*

(a) The m-faces of $\mathcal{L}^{\mathcal{K},\mathcal{G}}$ are isomorphic to $\mathcal{L}$, and its co-$(m-1)$-faces are isomorphic to $\mathcal{K}$.

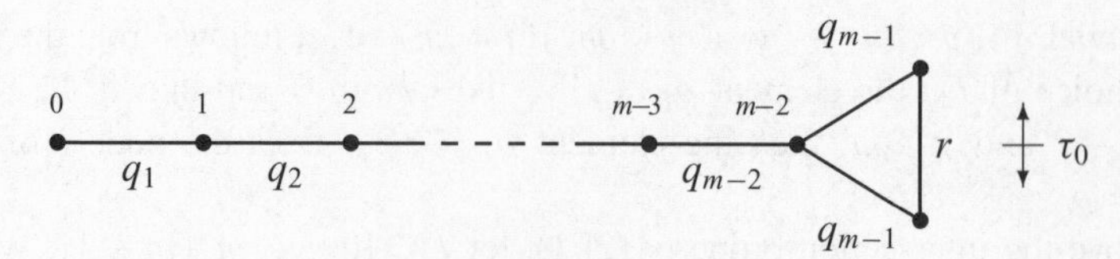

Figure 8B3

(b) *$\mathcal{L}^{\mathcal{K},\mathcal{G}}$ is of type $\{q_1, \ldots, q_{m-1}, 2r, p_1, \ldots, p_{n-1}\}$, where $r = 2$ if the two nodes of $\mathcal{G}$ in $V(\mathcal{G}, \{0\}) = \{F_0, F_0\tau_0\}$ are not connected by a branch of $\mathcal{G}$, or else $r \geqslant 3$ and r is the label of the connecting branch of these two nodes. If $\mathcal{K}$ is of rank 1, then $\mathcal{L}^{\mathcal{K},\mathcal{G}} = \{q_1, \ldots, q_{m-1}, 2r\}$ is universal.*

(c) *$\Gamma(\mathcal{L}^{\mathcal{K},\mathcal{G}}) = \Gamma = W \rtimes \Gamma(\mathcal{K})$, a semi-direct product, where the action of $\Gamma(\mathcal{K})$ on W is induced by that on $\mathcal{G}$. In particular, $\mathcal{L}^{\mathcal{K},\mathcal{G}}$ is finite if and only if W and $\mathcal{K}$ are finite.*

(d) *If $1 \leqslant i \leqslant n$ and if $\mathcal{K}_i$ denotes the i-face of $\mathcal{K}$ and $\mathcal{G}_i$ the induced subdiagram of $\mathcal{G}$ on the subset $V(\mathcal{G}, \{0, \ldots, i-1\})$ of $V(\mathcal{G})$, then the $(m+i)$-faces of $\mathcal{L}^{\mathcal{K},\mathcal{G}}$ are isomorphic to $\mathcal{L}^{\mathcal{K}_i,\mathcal{G}_i}$. Similarly, if $-1 \leqslant i \leqslant m-2$ and $\mathcal{L}_i = \{q_{i+2}, \ldots, q_{m-1}\}$ denotes the co-i-face of $\mathcal{L}$, then the co-faces of i-faces of $\mathcal{L}^{\mathcal{K},\mathcal{G}}$ are isomorphic to $\mathcal{L}_i^{\mathcal{K},\mathcal{G}}$.*

Proof. Lemma 8B5 says that Γ is a string C-group, so that Theorem 2E11 takes care of the various properties of the corresponding regular $(m+n)$-polytope $\mathcal{L}^{\mathcal{K},\mathcal{G}} = \mathcal{P}(\Gamma)$. In particular, part (a) follows from

$$\langle \rho_0, \ldots, \rho_{m-1} \rangle = \langle \sigma_0, \ldots, \sigma_{m-1} \rangle = \Gamma(\mathcal{L})$$

and

$$\langle \rho_m, \ldots, \rho_{m+n-1} \rangle = \langle \tau_0, \ldots, \tau_{n-1} \rangle = \Gamma(\mathcal{K}).$$

For the first part of (b), note that

$$(\rho_{m-1}\rho_m)^2 = (\sigma_{F_0}\tau_0)^2 = \sigma_{F_0}(\tau_0\sigma_{F_0}\tau_0) = \sigma_{F_0}\sigma_{F_0\tau_0},$$

which is of order r. If rank $\mathcal{K} = 1$, then $\mathcal{D}$ takes the form of the diagram in Figure 8B3, and the situation is similar to that of Figure 8A1; in particular, the second part of (b) can easily be verified.

Part (c) is trivially true by construction. The first part of (d) is a consequence of (8B7), with $j = m + i - 1$. Finally, the last part of (d) relies on the fact that the induced subdiagram of $\mathcal{D}$ on the node set $V(\mathcal{G}) \cup \{i+1, \ldots, m-2\}$ coincides with the diagram which is constructed from $\mathcal{L}_i$, $\mathcal{K}$ and $\mathcal{G}$ in the same way as $\mathcal{D}$ from $\mathcal{L}$, $\mathcal{K}$ and $\mathcal{G}$. This completes the proof. □

It is straightforward in principle, but often tedious in practice, to obtain a presentation for the group $\Gamma = W \rtimes \Gamma(\mathcal{K})$ of $\mathcal{L}^{\mathcal{K},\mathcal{G}}$ from that of W and $\Gamma(\mathcal{K})$ and the action of $\Gamma(\mathcal{K})$ on W. In some particularly interesting cases, this is done (implicitly) in Sections 8E and 8F, by identifying $\mathcal{L}^{\mathcal{K},\mathcal{G}}$ as the universal polytope in its class.

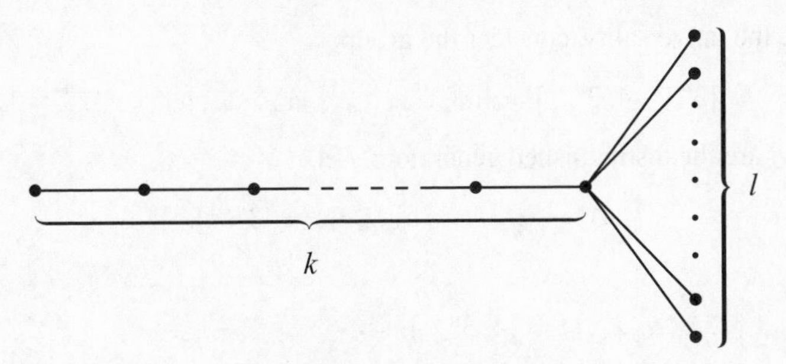

Figure 8B4. The diagram $\mathcal{D}_{k,l}$.

8B9 Remark Under the assumptions of Theorem 8B8, if $V(\mathcal{G}) = V(\mathcal{K})$ and $\Gamma(\mathcal{K})$ acts on $\mathcal{G}$ in the natural way, then the first part of (b) takes the form

(b′). $\mathcal{L}^{\mathcal{K},\mathcal{G}}$ is of type $\{q_1, \dots, q_{m-1}, 2r, p_1, \dots, p_{n-1}\}$, where $r = 2$ if the edges of $\mathcal{K}$ are not branches of $\mathcal{G}$, or else $r \geqslant 3$ and r is the label of those branches of $\mathcal{G}$ given by edges of $\mathcal{K}$.

Note that, in Theorem 8B8, we do not rule out the possibility that $\mathcal{K}$ is flat, where we recall that a polytope is *flat* if each vertex is incident with each facet. In this case, if $V(\mathcal{G}) = V(\mathcal{K})$ (with the natural action of $\Gamma(\mathcal{K})$), then $\mathcal{G}_{n-1} = \mathcal{G}$; moreover, $\mathcal{L}^{\mathcal{K},\mathcal{G}}$ is also flat (by Lemmas 4E3 and 4E1), and the facets of $\mathcal{L}^{\mathcal{K},\mathcal{G}}$ are in one-to-one correspondence with the facets of $\mathcal{K}$, where the facets of $\mathcal{K}$ correspond to the right cosets of $\langle \tau_0, \dots, \tau_{n-2} \rangle$ in $\Gamma(\mathcal{K})$, and those of $\mathcal{L}^{\mathcal{K},\mathcal{G}}$ correspond to the right cosets of

$$\langle \rho_0, \dots, \rho_{m+n-2} \rangle = W \rtimes \langle \tau_0, \dots, \tau_{n-2} \rangle$$

in Γ.

In most applications, $\mathcal{L}$ will be of rank 1; this case is further investigated in the next two sections. Here we illustrate our concepts with some interesting examples of polytopes $\mathcal{L}^{\mathcal{K},\mathcal{G}}$ with $\mathcal{L}$ and $\mathcal{K}$ of higher ranks.

For k, $l \geqslant 0$, let $\mathcal{D}_{k,l}$ denote the Coxeter diagram with $k + l$ nodes and $k + l - 1$ (unmarked) branches depicted in Figure 8B4. In particular, $\mathcal{D}_{0,l}$ is a trivial diagram and $\mathcal{D}_{k,0}$ a string diagram. Then we have

8B10 Corollary *Let $m, n \geqslant 1$. Let $\mathcal{D}_{m-1,n+1}$ be as in Figure 8B4, and let $W(\mathcal{D}_{m-1,n+1})$ be the corresponding Coxeter group. Then*

$$\{3^{m-1}\}^{\{3^{n-1}\}} = \{3^{m-1}, 4, 3^{n-1}\},$$

with group

$$[3^{m-1}, 4, 3^{n-1}] = W(\mathcal{D}_{m-1,n+1}) \rtimes S_{n+1}.$$

Proof. Apply Theorem 8B8 with $\mathcal{L} := \{3^{m-1}\}$, $\mathcal{K} := \{3^{n-1}\}$, and $\mathcal{G}$ the trivial diagram on $V(\mathcal{K})$; here, $\mathcal{L} = \{\,\}$ if $m = 1$. Then $\mathcal{D} = \mathcal{D}_{m-1,n+1}$, $\Gamma(\mathcal{K}) = S_{n+1}$, and $\mathcal{L}^{\mathcal{K}} = \mathcal{L}^{\mathcal{K},\mathcal{G}}$ is of the correct Schläfli type.

To prove the universality, consider the group

$$[3^{m-1}, 4, 3^{n-1}] = \langle \alpha_0, \ldots, \alpha_{m-1}, \alpha_m, \ldots, \alpha_{m+n-1} \rangle,$$

where the α_i are the distinguished generators. Let

$$\Lambda := \langle \alpha_m, \ldots, \alpha_{m+n-1} \rangle = [3^{n-1}].$$

Then

$$\begin{aligned} \langle \alpha_{m-1}, \ldots, \alpha_{m+n-1} \rangle &= [4, 3^{n-1}] \\ &= C_2^{n+1} \rtimes S_{n+1} = \langle \varphi^{-1}\alpha_{m-1}\varphi \mid \varphi \in \Lambda \rangle \rtimes \Lambda. \end{aligned}$$

Also, $\varphi^{-1}\alpha_i\varphi = \alpha_i$ if $i \leqslant m-2$ and $\varphi \in \Lambda$. Hence, if

$$\hat{W} := \langle \varphi^{-1}\alpha_i\varphi \mid i \leqslant m-1 \text{ and } \varphi \in \Lambda \rangle,$$

then we have the semi-direct product

$$[3^{m-1}, 4, 3^{n-1}] \cong \hat{W} \rtimes \Lambda,$$

with Λ acting on the $m+n$ generators of $\hat{W}$ by conjugation. But $\hat{W}$ is easily seen to satisfy the defining relations for $W(\mathcal{D}_{m-1,n+1})$, and thus is a quotient of this group. On the other hand, $\Gamma(\mathcal{L}^{\mathcal{K}})$ is a quotient of $[3^{m-1}, 4, 3^{n-1}]$, and so we must have $\hat{W} = W(\mathcal{D}_{m-1,n+1})$ and $\mathcal{L}^{\mathcal{K}} = \{3^{m-1}, 4, 3^{n-1}\}$, as claimed. □

The sequence of polytopes in Corollary 8B10 includes the following interesting special cases: the $(n+1)$-cube

$$\{4, 3^{n-1}\} = \{\,\}^{\{3^{n-1}\}} = 2^{\{3^{n-1}\}},$$

the $(m+1)$-cross-polytope

$$\{3^{m-1}, 4\} = \{3^{m-1}\}^{\{\,\}},$$

the 24-cell

$$\{3, 4, 3\} = \{3\}^{\{3\}},$$

the regular tessellations $\{3, 3, 4, 3\}$ and $\{3, 4, 3, 3\}$ in euclidean 4-space, and the regular honeycombs $\{3, 4, 3, 3, 3\}$, $\{3, 3, 4, 3, 3\}$ and $\{3, 3, 3, 4, 3\}$ in hyperbolic 5-space (see [113, §5; 120, Chapters 7, 8]).

There is another way in which the (star) diagrams $\mathcal{D}_{1,l}$ with $l+1$ nodes can be used to construct regular polytopes. For example, taking the diagram $\mathcal{D}_{1,5}$ in the form of Figure 8B5, then the twisting operation

$$(\sigma_0, \ldots, \sigma_5; \tau_0, \tau_1, \tau_2) \mapsto (\tau_0, \tau_1, \sigma_0, \sigma_5, \sigma_3, \tau_2) =: (\rho_0, \ldots, \rho_5)$$

on $W(\mathcal{D}_{1,5})$ defines the regular honeycomb $\{3, 4, 3, 3, 4\}$ in hyperbolic 5-space with group

8B11 $$\langle \rho_0, \ldots, \rho_5 \rangle = [3, 4, 3, 3, 4] = W(\mathcal{D}_{1,5}) \rtimes (S_3 \times C_2).$$

But $\mathcal{D}_{1,5} = \mathcal{D}_{2,4}$, and so we have the following

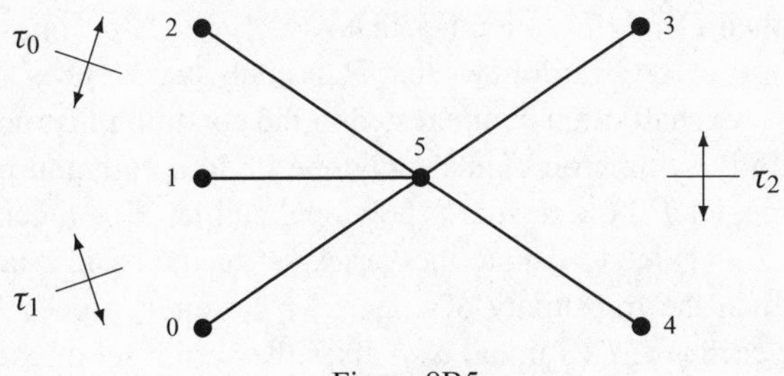

Figure 8B5

8B12 Corollary *The Coxeter group* $[3, 4, 3, 3, 3]$ *has subgroups* $[3, 3, 4, 3, 3]$ *and* $[3, 4, 3, 3, 4]$, *of indices 5 and 10, respectively.*

Proof. By Corollary 8B10 and by (8B11), the three groups mentioned are extensions of the same group $W(\mathcal{D}_{1,5}) = W(\mathcal{D}_{2,4})$, namely, by S_5, S_4 and $S_3 \times C_2$, respectively. But the latter two groups are subgroups of S_5 of indices 5 and 10, respectively. □

We remark that Corollary 8B12 can also be proved by simplex dissection of hyperbolic simplices.

We conclude by observing that the regular $(m + n)$-polytopes $\mathcal{L}^{\mathcal{K},\mathcal{G}}$ are almost always infinite if $m = \text{rank}\,\mathcal{L} > 1$. In fact, if $\mathcal{L} = \{q_1, \ldots, q_{m-1}\}$ and $q_1, \ldots, q_{m-1} \geqslant 3$, then Theorem 8B8(c) implies that there are only two possible ways to make $\mathcal{L}^{\mathcal{K},\mathcal{G}}$ finite, namely, $\mathcal{L}^{\mathcal{K},\mathcal{G}} = \{3^{m-1}, 4\}$ and $\mathcal{L}^{\mathcal{K},\mathcal{G}} = \{3, 4, 3\}$, with $\mathcal{G}$ trivial, and with rank $\mathcal{K} = 1$, $\mathcal{L} = \{3^{m-1}\}$, $r = 2$, and $\mathcal{K} = \mathcal{L} = \{3\}$, $r = 2$, respectively. Both examples were mentioned earlier.

8C The Polytopes $2^{\mathcal{K}}$ and $2^{\mathcal{K},\mathcal{G}(s)}$

We begin with a brief discussion of centrally symmetric regular polytopes. A regular polytope $\mathcal{P}$ is called *centrally symmetric* if its group $\Gamma(\mathcal{P})$ contains a *proper* central involution, which does not fix any of its vertices. Note that, if a central involution in $\Gamma(\mathcal{P})$ fixes one vertex, then it fixes every vertex, and thus acts on the vertex-set of $\mathcal{P}$ like the identity. For the definition of central symmetry it is therefore natural to leave such central involutions out of consideration. On the other hand, a proper central involution pairs up *antipodal* vertices of $\mathcal{P}$. Note that a central involution α in the group $\Gamma(\mathcal{P})$ of a regular polytope $\mathcal{P}$ whose faces are uniquely determined by their vertex-sets (as is the case, for example, when $\mathcal{P}$ is a lattice) must be proper, and so makes $\mathcal{P}$ centrally symmetric. In fact, if α is not proper, then it must fix every face of $\mathcal{P}$ because it fixes each of its vertices; hence α is the identity on $\mathcal{P}$.

Antipodal vertices of a centrally symmetric regular k-polytope $\mathcal{P}$ cannot be joined by an edge unless $\mathcal{P}$ has only two vertices, or, equivalently, $p_1 = 2$ in the Schläfli symbol $\{p_1, \ldots, p_{k-1}\}$ of $\mathcal{P}$. To see this, let $\Gamma(\mathcal{P}) = \langle \rho_0, \ldots, \rho_{k-1} \rangle$, with the ρ_i the distinguished generators with respect to the base flag $\{F_0, \ldots, F_{k-1}\}$ of $\mathcal{P}$, and let $\alpha \in \Gamma(\mathcal{P})$ be a proper central involution. If the vertices F_0 and $F_0\rho_0$ of the base edge

F_1 are antipodal, then $F_0\alpha = F_0\rho_0$ implies that $\rho_0 \in \langle \rho_1, \ldots, \rho_{k-1}\rangle\alpha$, and hence that $\Gamma(\mathcal{P}) \cong \langle \rho_1, \ldots, \rho_{k-1}\rangle \times C_2$. It follows that $\mathcal{P}$ has only two vertices, as claimed.

More generally, we shall often be interested in the condition that no two antipodal vertices of a centrally symmetric regular polytope lie in a common proper face. To see when this holds, let $\mathcal{P}$ be a regular k-polytope, and let α be a central involution of $\Gamma(\mathcal{P})$. For $i \leqslant k-1$, let V_i denote the vertex-set of the basic i-face of $\mathcal{P}$. Now, if $V_i\alpha \cap V_i \neq \emptyset$, then the transitivity of $\langle \tau_0, \ldots, \tau_{i-1}\rangle$ on V_i forces $V_i\alpha = V_i$; thus $(V_i\varphi)\alpha = V_i\varphi$ for each $\varphi \in \Gamma(\mathcal{P})$; that is, α fixes the vertex-set of each i-face of $\mathcal{P}$. Hence, if $i \leqslant k-1$ and there is a pair of antipodal vertices in an i-face of $\mathcal{P}$, then for each $j \geqslant i$ there is such a pair in V, and therefore α must fix the vertex-set of each j-face of $\mathcal{P}$. This property is usually strong enough to imply that no pair of antipodal vertices can lie in a common proper face of $\mathcal{P}$. The following lemma covers a fairly general case.

8C1 Lemma *Let $\mathcal{P}$ be a centrally symmetric regular polytope with the property that both its faces and those of its dual $\mathcal{P}^*$ are uniquely determined by their vertex-sets (as is the case, for example, if $\mathcal{P}$ is a lattice). Then a proper face of $\mathcal{P}$ cannot contain a pair of antipodal vertices.*

Proof. Our arguments show that, if a proper face of $\mathcal{P}$ contains a pair of antipodal vertices, then the vertex-set of each facet F of $\mathcal{P}$ is kept fixed by α; hence so is F itself, which is determined by its vertex-set. But, by our assumption on $\mathcal{P}^*$, every vertex of $\mathcal{P}$ is determined by the facets of $\mathcal{P}$ which contain it, and since each such facet is fixed, each vertex of $\mathcal{P}$ must then also be fixed by α. This is not possible, because α is a central involution which is proper. □

We shall now resume the discussion of Section 8B, and consider the regular polytopes $\mathcal{L}^{\mathcal{K},\mathcal{G}}$ with $\mathcal{L} = \{\,\}$, the unique polytope of rank 1. As before, let $\mathcal{K}$ be a regular n-polytope with group $\Gamma(\mathcal{K}) = \langle \tau_0, \ldots, \tau_{n-1}\rangle$, and let $\mathcal{G}$ be a $\mathcal{K}$-admissible diagram with node set $V(\mathcal{G})$. Again we have $V(\mathcal{G}) = V(\mathcal{K})$, the vertex-set of $\mathcal{K}$ with initial node the base vertex F_0 of $\mathcal{K}$. As before, since $\mathcal{L}$ is determined by its 2 vertices, we shall write $2^{\mathcal{K},\mathcal{G}}$ for $\mathcal{L}^{\mathcal{K},\mathcal{G}}$ and $2^{\mathcal{K}}$ for $\mathcal{L}^{\mathcal{K}}$ (if $\mathcal{G}$ is the trivial diagram on $V(\mathcal{K})$); that is,

$$2^{\mathcal{K},\mathcal{G}} := \{\,\}^{\mathcal{K},\mathcal{G}}, \qquad 2^{\mathcal{K}} := \{\,\}^{\mathcal{K}}.$$

Then the basic properties of $2^{\mathcal{K},\mathcal{G}}$ are as stated in Theorem 8B8. We shall discuss two choices of diagrams $\mathcal{G}$ which make the polytopes $2^{\mathcal{K},\mathcal{G}}$ finite. In particular, this leads to the regular polytopes $2^{\mathcal{K}}$ and $2^{\mathcal{K},\mathcal{G}(s)}$, which are the higher-rank analogues of Coxeter's regular maps

$$\{4, p\,|\,4^{[p/2]-1}\}$$

and

$$\{4, 2p\,|\,4^{p-2}, 2s\},$$

respectively [105, p. 57]; see Section 7B for the definition of these maps. The polytopes

$2^{\mathcal{K}}$ were first discovered by Danzer in the form in which they are discussed in Theorem 8D2 of the next section; see also [140, 367, §5].

Our first result concerns trivial diagrams and restates Theorem 8B8.

8C2 Theorem *Let $n \geqslant 1$, and let $\mathcal{K}$ be a regular n-polytope of type $\{p_1, \ldots, p_{n-1}\}$. For $1 \leqslant i \leqslant n$, let $\mathcal{K}_i$ denote the i-face of $\mathcal{K}$. Then the regular $(n+1)$-polytope $2^{\mathcal{K}}$ has the following properties.*

(a) $2^{\mathcal{K}}$ is of type $\{4, p_1, \ldots, p_{n-1}\}$. The vertex-figures are isomorphic to $\mathcal{K}$, and the $(i+1)$-faces to $2^{\mathcal{K}_i}$ for $i = 1, \ldots, n$.

(b) $\Gamma(2^{\mathcal{K}}) = C_2 \wr \Gamma(\mathcal{K})$, the wreath product of C_2 with $\Gamma(\mathcal{K})$, where $\Gamma(\mathcal{K})$ acts in the natural way on the vertex-set $V(\mathcal{K})$ of $\mathcal{K}$. In particular, $2^{\mathcal{K}}$ is finite if and only if $\mathcal{K}$ is finite, in which case

$$|\Gamma(2^{\mathcal{K}})| = 2^{|V(\mathcal{K})|}|\Gamma(\mathcal{K})|.$$

(c) If $\mathcal{K}$ has only finitely many vertices, then $2^{\mathcal{K}}$ is centrally symmetric.

Proof. Theorem 8B8 covers parts (a) and (b). Note that now $\mathcal{D} = \mathcal{G}$ and $W = C_2^{|V(\mathcal{K})|}$, so that the semi-direct product is actually a wreath product. For part (c), if $V(\mathcal{K})$ is finite, then

$$\alpha := \prod_{F \in V(\mathcal{K})} \sigma_F$$

is a central involution in $\Gamma(2^{\mathcal{K}})$. Moreover, α does not fix the base vertex of $2^{\mathcal{K}}$, because it is not contained in the stabilizer $\Gamma(\mathcal{K})$ $(= \langle \rho_1, \ldots, \rho_n \rangle)$ of this vertex. But then α must be proper, because a central involution fixes one vertex if and only if it fixes all vertices. □

8C3 Corollary *If $n \geqslant 1$, then $2^{\{3^{n-1}\}} = \{4, 3^{n-1}\}$, the $(n+1)$-cube with group $C_2 \wr S_{n+1}$, of order $2^{n+1}(n+1)!$.*

Proof. Choose $\mathcal{K} := \{3^{n-1}\}$. Then $2^{\mathcal{K}}$ is of type $\{4, 3^{n-1}\}$, and its group has the same order as $[4, 3^{n-1}]$. Now the corollary follows. Equivalently, we could use (8B10) with $m = 1$. □

The preceding corollary says that, in a sense, we may regard $2^{\mathcal{K}}$ as a generalization of the ordinary cube. We shall give further justification for this view in the next section.

The following corollary identifies the polytopes $2^{\mathcal{K}}$ with $\mathcal{K}$ of rank 2.

8C4 Corollary *If $2 \leqslant p < \infty$, then $2^{\{p\}} = \{4, p \,|\, 4^{\lfloor p/2 \rfloor - 1}\}$, with group $C_2 \wr D_p$, of order $2^{p+1}p$.*

Proof. It is straightforward to check that the group of $2^{\{p\}}$ satisfies the defining relations for $[4, p \,|\, 4^{\lfloor p/2 \rfloor - 1}]$, the group of the regular map $\{4, p \,|\, 4^{\lfloor p/2 \rfloor - 1}\}$. Since these groups have the same order, this proves the corollary. □

Our second choice of diagram assumes that $\mathcal{K}$ is centrally symmetric, with proper central involution α. As described in the preceding section, we can then construct a diagram by representing exactly the pairs of antipodal vertices by branches. If no pair of antipodal vertices is contained in a common i-face of $\mathcal{K}$, then the induced subdiagrams on the vertex-sets of i-faces are all trivial.

8C5 Theorem *Let $n \geqslant 1$, and let $\mathcal{K}$ be a centrally symmetric regular n-polytope of type $\{p_1, \ldots, p_{n-1}\}$ with $p_1 \geqslant 3$, and, for $1 \leqslant i \leqslant n$, let $\mathcal{K}_i$ denote its i-face. For $s \geqslant 3$, let $\mathcal{G}(s)$ denote the diagram on $V(\mathcal{K})$ which connects each pair of antipodal vertices of $\mathcal{K}$ by a branch labelled by s, but has no other (proper) branches. For $s = 2$, let $\mathcal{G}(s) = \mathcal{G}(2)$ be the trivial diagram on $V(\mathcal{K})$. Then the regular $(n+1)$-polytope $2^{\mathcal{K},\mathcal{G}(s)}$ has the following properties.*

(a) $2^{\mathcal{K},\mathcal{G}(s)}$ is of type $\{4, p_1, \ldots, p_{n-1}\}$ and has vertex-figures isomorphic to $\mathcal{K}$. If $i \leqslant n-1$ and no pair of antipodal vertices of $\mathcal{K}$ is contained in a common i-face of $\mathcal{K}$, then the $(i+1)$-faces of $2^{\mathcal{K},\mathcal{G}(s)}$ are isomorphic to $2^{\mathcal{K}_i}$. In particular, $2^{\mathcal{K},\mathcal{G}(2)} = 2^{\mathcal{K}}$.

(b) $\Gamma(2^{\mathcal{K},\mathcal{G}(s)}) = D_s^q \rtimes \Gamma(\mathcal{K})$, with $q := |V(\mathcal{K})|/2$, where the action of $\Gamma(\mathcal{K})$ on D_s^q $(= W)$ is induced by the action on $\mathcal{G}(s)$. In particular, $2^{\mathcal{K},\mathcal{G}(s)}$ is finite if and only if $\mathcal{K}$ is finite, in which case

$$|\Gamma(2^{\mathcal{K},\mathcal{G}(s)})| = (2s)^{|V(\mathcal{K})|/2}\,|\Gamma(\mathcal{K})|.$$

(c) If s is even and $\mathcal{K}$ has only finitely many vertices, then $2^{\mathcal{K},\mathcal{G}(s)}$ is also centrally symmetric.

Proof. If no pair of antipodal vertices of $\mathcal{K}$ is contained in a common i-face of $\mathcal{K}$, then the induced subdiagram $\mathcal{G}_i$ on the vertex-set of $\mathcal{K}_i$ is trivial. In particular, because $p_1 \geqslant 3$ this applies with $i = 1$, proving the statement about the Schläfli type. Also, by construction, $\mathcal{D} = \mathcal{G}(s)$ and $W \cong D_s^q$ (the direct product of q copies of D_s). Then parts (a) and (b) follow from Theorem 8B8. In particular, $2^{\mathcal{K},\mathcal{G}(2)} = 2^{\mathcal{K}}$, because $\mathcal{G}(2)$ is the trivial diagram.

If s is even, then each copy $\langle \sigma_F, \sigma_{F\alpha} \rangle$ (with $F \in V(\mathcal{K})$) of D_s has a central involution, namely, $(\sigma_F \sigma_{F\alpha})^{s/2}$. Hence, if $\mathcal{K}$ has only finitely many vertices, then the product β (say) of all these involutions is indeed a central involution of $\Gamma(2^{\mathcal{K},\mathcal{G}(s)})$. To check that β is proper, recall that the vertices of a regular polytope are in one-to-one correspondence with the right cosets in the full group of the group of the vertex-figure, which here is

$$\langle \rho_1, \ldots, \rho_n \rangle = \langle \tau_0, \ldots, \tau_{n-1} \rangle = \Gamma(\mathcal{K})$$

in $\Gamma(2^{\mathcal{K},\mathcal{G}(s)}) = \langle \rho_0, \ldots, \rho_n \rangle$. However, these cosets are represented by the elements of $W(\mathcal{D})$. Then (c) follows, and the proof of the theorem is complete. □

We note two interesting applications of Theorem 8C5. The first identifies certain polytopes $2^{\mathcal{K},\mathcal{G}(s)}$ as cubic toroids or as cubical tessellations, and the second deals with the case of rank 2.

8C6 Corollary *Let $n \geqslant 2$. If $2 \leqslant s < \infty$, then $2^{\{3^{n-2},4\},\mathcal{G}(s)} = \{4, 3^{n-2}, 4\}_{(2s,0^{n-1})}$, a cubical regular $(n+1)$-toroid, with group $D_s^n \rtimes [3^{n-2}, 4]$ of order $(4s)^n n!$. Moreover, $2^{\{3^{n-2},4\},\mathcal{G}(\infty)} = \{4, 3^{n-2}, 4\}$, the cubical regular tessellation in euclidean*

$(n+1)$-space, with group $[4, 3^{n-2}, 4] = D_{\infty}^{n} \rtimes [3^{n-2}, 4]$, where D_{∞} denotes the infinite dihedral group.

Proof. If $\mathcal{K} := \{3^{n-2}, 4\}$, then $2^{\mathcal{K},\mathcal{G}(s)}$ is of type $\{4, 3^{n-2}, 4\}$. Hence, if $s < \infty$, this is a cubical toroid with a group of order $(4s)^n n!$, which is easily identified as $\{4, 3^{n-2}, 4\}_{(2s,0^{n-1})}$ (see Section 6D). For $s = \infty$, we obtain the tessellation $\{4, 3^{n-2}, 4\}$, with its group occurring as $D_{\infty}^{n} \rtimes [3^{n-2}, 4]$. □

8C7 Corollary *Let $2 \leqslant p < \infty$ and $2 \leqslant s \leqslant \infty$. Then $2^{\{2p\},\mathcal{G}(s)} = \{4, 2p \,|\, 4^{p-2}, 2s\}$, with group $D_s^p \rtimes D_{2p}$, of order $(2s)^p \cdot 4p$ if $s < \infty$. If $p = 2$ and $s < \infty$, this is the torus map $\{4, 4\}_{(2s,0)}$, with group $(D_s \times D_s) \rtimes D_4$ of order $32s^2$.*

Proof. It is easy to see that the group $\Gamma(2^{\{2p\},\mathcal{G}(s)})$ satisfies all the defining relations for $[4, 2p \,|\, 4^{p-2}, 2s]$, the group of Coxeter's regular map $\{4, 2p \,|\, 4^{p-2}, 2s\}$, and is therefore a quotient of this group (see [105, p. 57] or Section 7B). Then a direct (but tedious) computation shows that the two groups are actually the same. Instead, if $s < \infty$, we could appeal to [105, p. 57] and note that the orders of the two (finite) groups are the same; hence, since one is a quotient of the other, they must coincide. (Observe that, for $s = \infty$, the relation represented by the last entry in the symbol for the map is void.) □

8D Realizations of $2^{\mathcal{K}}$ and $2^{\mathcal{K},\mathcal{G}(s)}$

We now discuss realizations in euclidean spaces of the regular polytopes $2^{\mathcal{K}}$ and $2^{\mathcal{K},\mathcal{G}(s)}$ defined in Section 8C. We use the same notation as before. For general results on realizations we refer to Chapter 5. Here we shall only describe realizations which can be constructed by Wythoff's construction from certain natural orthogonal representations of the groups. For $2^{\mathcal{K}}$, these realizations will give us a simple combinatorial description of the polytopes, and it was in this form that $2^{\mathcal{K}}$ was originally discovered by Danzer (see [140, 367, §5]).

In this section, all polytopes are assumed to be finite. Let $\mathcal{K}$ be a finite regular n-polytope with vertex-set

$$V(\mathcal{K}) = \{0, 1, \ldots, v-1\}.$$

We begin with the discussion of $2^{\mathcal{K}}$. By Theorem 8C2,

$$\Gamma(2^{\mathcal{K}}) = C_2 \wr \Gamma(\mathcal{K}),$$

where $\Gamma(\mathcal{K})$ acts naturally on $V(\mathcal{K})$. Thus $\Gamma(2^{\mathcal{K}}) = \langle \rho_0, \ldots, \rho_n \rangle$ can be embedded into the symmetry group

$$[4, 3^{v-2}] = C_2 \wr S_v = \langle \psi_0, \ldots, \psi_{v-1} \rangle$$

(with distinguished generators ψ_j) of the regular v-cube $\{4, 3^{v-2}\}$ in euclidean v-space $\mathbb{E}^v$, in such a way that $\rho_0 = \psi_0$ and $\langle \rho_1, \ldots, \rho_n \rangle \subseteq \langle \psi_1, \ldots, \psi_{v-1} \rangle$. Write

$$\hat{V} := \{0, 1\}^v$$

for the vertex-set of $\{4, 3^{v-2}\}$.

Now, to apply Wythoff's construction, we pick the vertex G_0 in the base flag of $\{4, 3^{v-2}\}$, say

$$G_0 := (0, \ldots, 0),$$

and define recursively

$$G_i := \{G_{i-1}\varphi \mid \varphi \in \langle \rho_0, \ldots, \rho_{i-1} \rangle\} \qquad (i = 1, \ldots, n+1).$$

This gives the base flag $\{G_0, \ldots, G_n\}$ for the realization $\mathcal{P}$ of $2^{\mathcal{K}}$, and all the other flags are the transforms of this flag by the elements of $\Gamma(2^{\mathcal{K}})$. In this way, the 2^v points in $\hat{V}$ become the vertices of $\mathcal{P}$; that is, $\mathcal{P}$ is a vertex-faithful realization of $2^{\mathcal{K}}$. Also, our method of realization commutes with passing to faces of $\mathcal{K}$, which means that, if $0 \leqslant i \leqslant n-1$ and $\mathcal{K}_i$ denotes the basic i-face of $\mathcal{K}$ with vertex-set $V(\mathcal{K}_i)$, then the face G_{i+1} of $\mathcal{P}$ is the realization of $2^{\mathcal{K}_i}$ on the subset $\hat{V}_i$ of $\hat{V}$ corresponding to $V(\mathcal{K}_i)$. In particular, the 2-faces of $\mathcal{P}$ belong (in an obvious sense) to the 2-skeleton of $\{4, 3^{n-2}\}$.

An equivalent construction of $\mathcal{P}$ can be found if the faces in $\mathcal{K}$ are uniquely determined by their vertex-sets; this is true if $\mathcal{K}$ is a lattice. We may then identify the faces in $\mathcal{K}$ with their vertex-sets, and we may also do so for $2^{\mathcal{K}}$ (and, in fact, for any polytope $2^{\mathcal{K},\mathcal{G}}$). The latter can be proved using Theorem 2B14. Now, making this identification, we take as j-faces of $\mathcal{P}$, for any $(j-1)$-face F of $\mathcal{K}$ and any $x := (x_0, \ldots, x_{v-1})$ in $\hat{V}$, the subsets $F(x)$ of $\hat{V}$ defined by

$$F(x) := \{(y_0, \ldots, y_{v-1}) \mid y_i = x_i \text{ if } i \text{ is not a vertex of } F\},$$

or, abusing notation, by the cartesian product

8D1
$$F(x) := \left(\bigotimes_{i \in F} \{0, 1\}\right) \times \left(\bigotimes_{i \notin F} \{x_i\}\right).$$

Then, if F, F' are faces of $\mathcal{K}$ and $x = (x_0, \ldots, x_{v-1})$, $x' = (x'_0, \ldots, x'_{v-1})$ are points in $\hat{V}$, we have $F(x) \subseteq F'(x)$ if and only if $F \leqslant F'$ in $\mathcal{K}$ and $x_i = x'_i$ for each i which is not a vertex of F'. The proof of the following theorem is now straightforward.

8D2 Theorem *Let $\mathcal{K}$ be a regular n-polytope whose faces are uniquely determined by their vertex-sets. Let $\mathcal{P}$ be the set of all faces $F(x)$ as in (8D1), with F a face of $\mathcal{K}$ and $x \in V$, and let $\mathcal{P}$ be ordered by inclusion. Then $\mathcal{P}$ is a regular $(n+1)$-polytope which is isomorphic to $2^{\mathcal{K}}$.*

An immediate consequence of Theorem 8D2 is that the polytope $\mathcal{P}$, and thus $2^{\mathcal{K}}$, is a lattice if the original polytope $\mathcal{K}$ is a lattice.

By Corollary 8C4, if Theorem 8D2 is applied with $\mathcal{K} = \{p\}$, we obtain a realization for Coxeter's regular map $\{4, p \mid 4^{[p/2]-1}\}$ in the 2-skeleton of the regular p-cube $\{4, 3^{p-2}\}$, whose edge-graph coincides with that of the cube. This is identical with the realization outlined in Coxeter [105, p. 57]. It is interesting to note that this map and its realization have been rediscovered several times in the literature.

For example, Ringel [346, §3] and Beineke and Harary [24] pointed out that the genus

$$2^{p-3}(p-4)+1$$

of $\{4, p\,|\,4^{[p/2]-1}\}$ is the smallest genus of any surface into which the edge-graph of the p-cube can be embedded without self-intersections. The map also occurred in Banchoff [18, Theorem A].

It is rather surprising that each map $\{4, p\,|\,4^{[p/2]-1}\}$ can be modelled by a polyhedron with convex faces and no self-intersections in ordinary 3-space (see [309, 311, §3]). This polyhedron has many remarkable topological properties, although it is not a realization of the map in the sense of Chapter 5. For example, if $p \geqslant 12$, its genus exceeds the number 2^p of its vertices, which is somewhat hard to visualize. Perhaps even more counterintuitive is the fact that the dual map $\{p, 4\,|\,4^{[p/2]-1}\}$ can similarly be modelled in $\mathbb{E}^3$; for $p \geqslant 12$, the resulting polyhedron has more holes than faces.

The construction of the polytopes $2^{\mathcal{K}}$ can also be extended to work for arbitrary (not necessarily regular) polytopes $\mathcal{K}$, and even more general complexes $\mathcal{K}$ such as simplicial complexes (see [140, 255, §3C; 367, §5]). The topological properties of the cubical complexes $2^{\mathcal{K}}$ which arise from simplicial complexes $\mathcal{K}$ have been studied in Kühnel and Schulz [258], Kühnel [255, §3C] and Brehm [48, 49, §2].

We next discuss realizations of the regular $(n+1)$-polytopes $2^{\mathcal{K},\mathcal{G}(s)}$. Here we shall restrict ourselves to centrally symmetric regular n-polytopes $\mathcal{K}$ which have the property that, in both $\mathcal{K}$ and $\mathcal{K}^*$, the faces are uniquely determined by their vertex-sets; this is true if $\mathcal{K}$ is a lattice. By Lemma 8C1 this implies that no two antipodal vertices can lie in a common proper face of $\mathcal{K}$. As before, we denote the proper central involution by α.

Let $\mathcal{K}$ be given. Let $v = 2l$ (say), and let the vertices in $V(\mathcal{K})$ be labelled in such a way that $\{k, k+l\}$ is an antipodal pair for each $k = 0, \ldots, l-1$. Then, if we take the vertices modulo $2l$, the central involution acts as $j \mapsto j + l$. In particular, each proper face of $\mathcal{K}$ shares at most one vertex with any pair $\{k, k+l\}$. For simplicity, we shall assume that, for $i \leqslant n$, the vertex-set of the basic i-face $\mathcal{K}_i$ of $\mathcal{K}$ is given by

$$V(\mathcal{K}_i) = \{0, 1, \ldots, v_i - 1\}, \quad v_i := |V(\mathcal{K}_i)|.$$

In particular,

$$V(\mathcal{K}_0) = \{0\} \quad \text{and} \quad V(\mathcal{K}_1) = \{0, 1\}.$$

Furthermore, let $e_0, \ldots, e_{v-1}$ denote any orthonormal basis of euclidean v-space $\mathbb{E}^v$.

Now, take the l mutually orthogonal planes

$$\mathbb{E}^2_k := \langle e_k, e_{k+l} \rangle \quad (k = 0, \ldots, l-1)$$

in $\mathbb{E}^v$, inside each $\mathbb{E}^2_k$ the regular $2s$-gon P_k with centre at the origin and one vertex at e_k, and form the rectangular product

$$\Pi_{l,s} := P_0 \times P_1 \times \cdots \times P_{l-1}$$

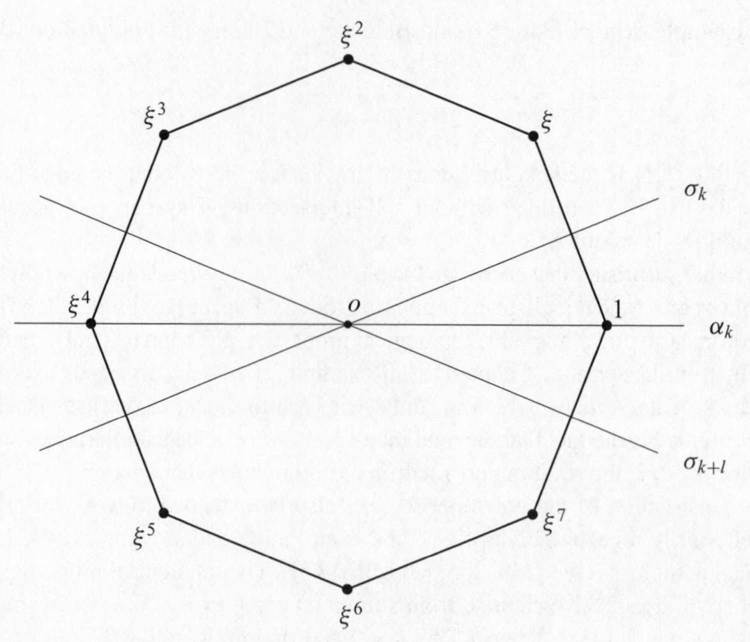

Figure 8D1. The polygon P_k for $s=4$; here, $\xi = e^{\frac{\pi i}{4}}$.

of these $2s$-gons in $\mathbb{E}^v$; see Figure 8D1. Note that we have taken $2s$-gons rather than s-gons. The vertex-set of $\Pi_{l,s}$ will now become the vertex set of our realization of $2^{\mathcal{K},\mathcal{G}(s)}$.

First we explain why $\Gamma(2^{\mathcal{K},\mathcal{G}(s)}) = \langle \rho_0, \ldots, \rho_n \rangle$ can be embedded in a natural way into the symmetry group $S(\Pi_{l,s})$ of $\Pi_{l,s}$, which is isomorphic to the wreath product $D_{2s} \wr S_l$.

Recall from Theorem 8C5 that

$$\Gamma\left(2^{\mathcal{K},\mathcal{G}(s)}\right) = W \rtimes \Gamma(\mathcal{K}) = \langle \sigma_i \mid i \in V(\mathcal{K}) \rangle \rtimes \langle \tau_0, \ldots, \tau_{n-1} \rangle.$$

Here, by our labelling of the vertices of $\mathcal{K}$, we have the direct product

$$W \cong \bigotimes_{k=0}^{l-1} \langle \sigma_k, \sigma_{k+l} \rangle \cong D_s^l.$$

In this product, each factor $\langle \sigma_k, \sigma_{k+l} \rangle$ can be realized as a subgroup of index 2 in the symmetry group D_{2s} of P_k by choosing the generators σ_k and σ_{k+l} as indicated in Figure 8D1; here we regard D_{2s} as being trivially extended from $\mathbb{E}_k^2$ to $\mathbb{E}^v$. In this way, W becomes a subgroup of $S(\Pi_{l,s})$.

On the other hand, for each τ in $\Gamma(\mathcal{K})$ we can find a unique symmetry $\hat{\tau}$ of $\Pi_{l,s}$ which operates on W (considered as a subgroup of $S(\Pi_{l,s})$) in the same way as τ acts on W (considered as an abstract group); that is,

$$\hat{\tau}^{-1} \sigma_i \hat{\tau} = \tau^{-1} \sigma_i \tau \qquad (i \in V(\mathcal{K})).$$

To see this, first observe that τ permutes the dihedral factors of W and maps each pair $\{\sigma_k, \sigma_{k+l}\}$ of generators onto another pair $\{\sigma_j, \sigma_{j+l}\}$. Hence, if $\tau^{-1}\sigma_k\tau = \sigma_j$ and thus $\tau^{-1}\sigma_{k+l}\tau = \sigma_{j+l}$, then the restriction $\hat{\tau}_k\colon \mathbb{E}_k^2 \to \mathbb{E}_j^2$ of $\hat{\tau}$ is the unique isometry $\beta_{kj}\colon \mathbb{E}_k^2 \to \mathbb{E}_j^2$ defined by $e_k\beta_{kj} = e_j$ and $e_{k+l}\beta_{kj} = e_{j+l}$; or, if $\tau^{-1}\sigma_k\tau = \sigma_{j+l}$ and thus $\tau^{-1}\sigma_{k+l}\tau = \sigma_j$, then $\hat{\tau}_k = \alpha_k\beta_{kj}$, where $\alpha_k\colon \mathbb{E}_k^2 \to \mathbb{E}_k^2$ is the reflexion of $\mathbb{E}_k^2$ defined by $e_k\alpha_k = e_k$ and $e_{k+l}\alpha_k = -e_{k+l}$. In particular, for the proper central involution α in $\Gamma(\mathcal{K})$ we have

$$\hat{\alpha} = \alpha_0 \perp \alpha_1 \perp \cdots \perp \alpha_{l-1},$$

which involves a reflexion in each plane $\mathbb{E}_k^2$.

Finally, if $\tau \in \Gamma(\mathcal{K})$ and $\hat{\tau}$ is the identity map on $\mathbb{E}^v$, then $\sigma_i = \tau^{-1}\sigma_i\tau = \sigma_{(i)\tau}$ for each i in $V(\mathcal{K})$; hence τ fixes each vertex of $\mathcal{K}$. By our assumptions on $\mathcal{K}$, this implies that $\tau = \varepsilon$. It follows that the representation $\tau \to \hat{\tau}$ is faithful, so that $\Gamma(\mathcal{K})$, and thus $\Gamma(2^{\mathcal{K},\mathcal{G}(s)})$, can indeed be embedded into $S(\Pi_{l,s})$. In particular, observe that the point $(1,\ldots,1,0,\ldots,0) =: (1^l, 0^l)$ of $\mathbb{E}^v$ is invariant under all elements $\hat{\tau}$ with $\tau \in \Gamma(\mathcal{K}) = \langle \rho_1, \ldots, \rho_{n-1}\rangle$.

Now, a realization $\mathcal{Q}$ of $2^{\mathcal{K},\mathcal{G}(s)}$ can be obtained by applying Wythoff's construction again with initial vertex $(1^l, 0^l)$. However, this time the situation is more difficult, due to the fact that $\Gamma(2^{\mathcal{K},\mathcal{G}(s)})$ will not contain the full symmetry group D_{2s} of each P_k as a subgroup (although the subgroups $\langle \sigma_k, \sigma_{k+l}, \alpha\rangle$ of $\Gamma(2^{\mathcal{K},\mathcal{G}(s)})$ act like the groups D_{2s} on $\mathbb{E}_k^2$). This means that we should not expect such an elegant description for $\mathcal{Q}$ as for the realization $\mathcal{P}$ of $2^{\mathcal{K}}$. However, we can still give a simple description for the base flag of $\mathcal{Q}$ and its transforms by the subgroup W of $\Gamma(2^{\mathcal{K},\mathcal{G}(s)})$.

For convenience, we use l-dimensional complex coordinates rather than v-dimensional real coordinates. The correspondence is now

$$a + ib \leftrightarrow (a, b) \qquad (a, b \in \mathbb{E}^l).$$

Define $\xi := e^{\pi i/s}$. Consider the subset

$$\hat{V} := \{(\xi^{j_0}, \ldots, \xi^{j_{l-1}}) \mid 0 \leqslant j_0, \ldots, j_{l-1} < 2s\}$$

of complex l-space $\mathbb{C}^l$; this is the vertex-set of the generalized l-cube

$$\gamma_l^{2s} = 2s\{4\}2\{3\}2\cdots 2\{3\}2,$$

a regular complex polytope (not an abstract polytope in our sense) in unitary space $\mathbb{C}^l$ (see [126, Chapter 12; 387]). In real $2l$-space, $\hat{V}$ corresponds to the vertex-set of $\Pi_{l,s}$.

To apply Wythoff's construction for $\mathcal{Q}$, pick the base vertex

$$G_0 := (1, \ldots, 1)$$

of $\hat{V}$, and define recursively

$$G_i := \{G_0\varphi \mid \varphi \in \langle \rho_0, \ldots, \rho_{i-1}\rangle\} \qquad (i = 1, \ldots, n);$$

define also

$$G_{n+1} := \hat{V}.$$

By our labelling of $V(\mathcal{K})$, this gives the cartesian products

$$G_i = \left(\bigotimes_{k=0}^{v_{i-1}-1} \{1, \xi\} \right) \times \left(\bigotimes_{k=v_{i-1}}^{l-1} \{1\} \right) \quad (i = 1, \ldots, n),$$

so that the transforms of G_i by W are the subsets

$$\left(\bigotimes_{k=0}^{v_{i-1}-1} \{\xi^{j_k}, \xi^{j_k+1}\} \right) \times \left(\bigotimes_{k=v_{i-1}}^{l-1} \{\xi^{j_k}\} \right)$$

of $\hat{V}$ with $0 \leqslant j_0, \ldots, j_{l-1} < 2s$. Now, for any given $\mathcal{K}$, it would be possible, in principle, also to find a description for the transforms of these sets by the elements τ of $\Gamma(\mathcal{K})$. In fact, if $\hat{\tau}$, $\hat{\tau}_k$ and α_k are as before (but taken in $\mathbb{C}^l$), and if $c = (c_0, \ldots, c_k, \ldots, c_{l-1})$ and $d = (d_0, \ldots, d_j, \ldots, d_{l-1})$ are in $\hat{V}$ with $c\hat{\tau} = d$, then for each $k = 0, \ldots, l-1$, we have $d_j = c_k \hat{\tau}_k = c_k$ if $\tau^{-1}\sigma_k\tau = \sigma_j$, or $d_j = c_k\hat{\tau}_k = c_k^{-1}$ if $\tau^{-1}\sigma_k\tau = \sigma_{j+l}$. Here, the latter is explained by $c_k\alpha_k = {c_k}^{-1}$.

All these considerations imply that $\hat{V}$ is the vertex-set of $\mathcal{Q}$. In fact, we even have

$$\hat{V} = \{G_0\sigma \mid \sigma \in W\},$$

since we can independently choose elements in the factors $\langle \sigma_k, \sigma_{k+l} \rangle$ so as to get any power ξ^j. Therefore the realization is again vertex-faithful, with $(2s)^l$ vertices. Further, when considered in real space, the 2-faces of $\mathcal{Q}$ belong (in an obvious sense) to the 2-skeleton of $\Pi_{l,s}$. The facets of $\mathcal{Q}$ are transforms of G_n, and our description immediately reveals the isomorphism with $2^{\mathcal{K}_{n-1}}$, in agreement with Theorem 8C5(a).

By Corollary 8C7, when applied with $\mathcal{K} = \{2p\}$, the construction gives a realization for the regular maps $\{4, 2p \,|\, 4^{p-2}, 2s\}$ in the 2-skeleton of the rectangular product $\Pi_{l,s}$ of l regular $2s$-gons. This realization is identical to that outlined in Coxeter [105, p. 57].

Another immediate consequence of our construction is that $\mathcal{Q}$, and hence $2^{\mathcal{K},\mathcal{G}(s)}$, is again a lattice if $\mathcal{K}$ is a lattice.

Further, if we choose $s = 2$, then we get (except for reordering and an appropriate change of coordinates) the same realization of $2^{\mathcal{K}} = 2^{\mathcal{K},\mathcal{G}(2)}$ as before. However, if $\mathcal{K}$ is centrally symmetric, it is sometimes useful to rather take $2^{\mathcal{K}}$ in its non-standard representation as $2^{\mathcal{K},\mathcal{G}(2)}$. This trick is employed in Section 8F.

Finally, if we replace ξ by any other primitive $2s$-th root of unity, then we can modify our construction so as to obtain other realizations of $2^{\mathcal{K},\mathcal{G}(s)}$ with the same vertex-set, which are, in a sense, "star-versions" of $\mathcal{Q}$.

8E A Universality Property of $\mathcal{L}^{\mathcal{K},\mathcal{G}}$

In this section, we prove a property of certain polytopes $2^{\mathcal{K},\mathcal{G}}$ and $\mathcal{L}^{\mathcal{K},\mathcal{G}}$, which allows us to identify them as universal polytopes in their classes. The crucial concept is that of an *extension* of a given diagram to a diagram on a larger set of nodes, which preserves the action of $\mathcal{K}$. It is striking that this method enables us to deal with many classes of locally toroidal regular polytopes and to find the finite universal polytopes in these classes (see Chapters 10–12). This also lays the foundations for Section 8F.

Let $\mathcal{K}$ be a regular n-polytope with group $\Gamma(\mathcal{K}) = \langle \tau_0, \ldots, \tau_{n-1} \rangle$ and (basic) facet $\mathcal{F} := \mathcal{K}_{n-1}$; then $\Gamma(\mathcal{F}) = \langle \tau_0, \ldots, \tau_{n-2} \rangle$. Let $\mathcal{H}$ be an $\mathcal{F}$-admissible diagram on the vertex set $V(\mathcal{F})$ of $\mathcal{F}$, with the base vertex F_0 of $\mathcal{K}$ as initial vertex. Then the regular n-polytope $2^{\mathcal{F},\mathcal{H}}$ is defined and has vertex-figures isomorphic to $\mathcal{F}$. Now, if $\mathcal{G}$ is any $\mathcal{K}$-admissible diagram on the vertex set $V(\mathcal{K})$ of $\mathcal{K}$ (with initial vertex F_0) such that $\mathcal{H}$ is the induced subdiagram of $\mathcal{G}$ on $V(\mathcal{F})$, then $2^{\mathcal{K},\mathcal{G}}$ is a regular $(n+1)$-polytope in the class $\langle 2^{\mathcal{F},\mathcal{H}}, \mathcal{K} \rangle$. However, given $\mathcal{H}$, such a diagram $\mathcal{G}$ need not exist in general. If $\mathcal{G}$ does exist, then we call $\mathcal{H}$ a *$\mathcal{K}$-extendable diagram* and $\mathcal{G}$ a *$\mathcal{K}$-extension* of $\mathcal{H}$.

For instance, let $\mathcal{K}$ be the 24-cell $\{3, 4, 3\}$ (with 24 octahedral facets), and let $\mathcal{H}$ be the diagram on the octahedron $\{3, 4\}$ which connects antipodal vertices by a branch labelled s (see Figure 8B1). Then we can take for $\mathcal{G}$ the diagram on the 24 vertices of $\mathcal{K}$ whose restriction to the vertex-set of each facet is $\mathcal{H}$. However, we could also add further branches in a suitable way.

To give an example of a diagram $\mathcal{H}$ which is not $\mathcal{K}$-extendable, choose for $\mathcal{K}$ the hemi-cube $\{4, 3\}_3$ (obtained by identifying antipodal vertices of the 3-cube), and for $\mathcal{H}$ the diagram on the four vertices of $\mathcal{F} := \{4\}$ which connects antipodal vertices of $\mathcal{F}$ by a branch with a label at least 3, but connects no other vertices of $\mathcal{F}$. Then $\mathcal{H}$ is not $\mathcal{K}$-extendable, because in $\mathcal{K}$ antipodal vertices of $\mathcal{F}$ are connected by edges of $\mathcal{K}$.

It is easy to see that a diagram $\mathcal{H}$ is $\mathcal{K}$-extendable if and only if it has the following property.

8E1 $\begin{cases} \text{If two diagonals (pairs of vertices) of } \mathcal{F} \text{ are equivalent under } \Gamma(\mathcal{K}), \text{ then} \\ \text{the corresponding branches in } \mathcal{H} \text{ have the same label (possibly 2).} \end{cases}$

Note that it is not required that the two diagonals be equivalent under $\Gamma(\mathcal{F})$. Given a $\mathcal{K}$-extendable diagram $\mathcal{H}$, there exists a *universal $\mathcal{K}$-extension* $\mathcal{D}$ of $\mathcal{H}$, in the sense that for any $\mathcal{K}$-extension $\mathcal{G}$ there is a homomorphism

$$W(\mathcal{D}) \to W(\mathcal{G})$$

between the corresponding Coxeter groups, which maps generators onto corresponding generators. This diagram $\mathcal{D}$ with node set $V(\mathcal{K})$ is obtained as follows.

First, take the branches and labels of $\mathcal{H}$ and all their transforms under elements of $\Gamma(\mathcal{K})$; by (8E1), this already gives a $\mathcal{K}$-extension. (This corresponds to the example for the 24-cell.) Second, complete the diagram $\mathcal{D}$ by adding (if possible) branches with label ∞, one for each diagonal of $\mathcal{K}$ which is not equivalent under $\Gamma(\mathcal{K})$ to a diagonal of $\mathcal{F}$. This gives a $\mathcal{K}$-extension of $\mathcal{H}$. Note that, in constructing the universal $\mathcal{K}$-extension, we have added as many branches with marks ∞ as possible, while preserving the $\mathcal{K}$-extension property. Note also that a $\mathcal{K}$-extendable diagram $\mathcal{H}$ coincides with all its $\mathcal{K}$-extensions if and only if $\mathcal{K}$ is flat.

The following theorem shows the significance of the universal $\mathcal{K}$-extension $\mathcal{D}$.

8E2 Theorem *Let $\mathcal{K}$ be a regular n-polytope with facet $\mathcal{F}$, and let $\mathcal{H}$ be an $\mathcal{F}$-admissible diagram on the vertex-set of $\mathcal{F}$ which is $\mathcal{K}$-extendable. Then the universal* *polytope* $\{2^{\mathcal{F},\mathcal{H}}, \mathcal{K}\}$ *exists, and*

$$\{2^{\mathcal{F},\mathcal{H}}, \mathcal{K}\} = 2^{\mathcal{K},\mathcal{D}},$$

where $\mathcal{D}$ *is the universal* $\mathcal{K}$*-extension of* $\mathcal{H}$.

Proof. By construction, $2^{\mathcal{K},\mathcal{D}}$ is in $\langle 2^{\mathcal{F},\mathcal{H}}, \mathcal{K}\rangle$, so that the universal polytope does indeed exist (see Theorem 4A2). It remains to be proved that $2^{\mathcal{K},\mathcal{D}}$ is itself universal.

Let $\mathcal{P} := \{2^{\mathcal{F},\mathcal{H}}, \mathcal{K}\}$ and $\Gamma(\mathcal{P}) =: \langle \alpha_0, \ldots, \alpha_n\rangle$. If N_0 denotes the normal closure of α_0 in $\Gamma(\mathcal{P})$, then

8E3 $$\Gamma(\mathcal{P}) = N_0 \cdot \langle \alpha_1, \ldots, \alpha_n\rangle \quad (= N_0 \cdot \Gamma(\mathcal{K}))$$

(see Lemma 4E7). For the polytope $2^{\mathcal{K},\mathcal{D}}$, the normal closure of $\rho_0 = \sigma_0$ in $\Gamma(2^{\mathcal{K},\mathcal{D}}) = \langle \rho_0, \ldots, \rho_n\rangle$ is the subgroup $W(\mathcal{D}) = \langle \sigma_i \mid i \in V(\mathcal{K})\rangle$; here 0 is the base vertex of $\mathcal{K}$. This shows that

8E4 $$\Gamma(2^{\mathcal{K},\mathcal{D}}) = W(\mathcal{D}) \cdot \langle \rho_1, \ldots, \rho_n\rangle \cong W(\mathcal{D}) \rtimes \Gamma(\mathcal{K}).$$

On the other hand, the mapping

$$\alpha_i \mapsto \rho_i \quad (i = 0, \ldots, n)$$

induces a surjective homomorphism

$$f \colon \Gamma(\mathcal{P}) \to \Gamma(2^{\mathcal{K},\mathcal{D}}),$$

which is one-to-one on $\langle \alpha_1, \ldots, \alpha_n\rangle$, and maps N_0 onto $W(\mathcal{D})$. It follows that the product in (8E3) is semi-direct.

Next, we observe that the action of $\Gamma(\mathcal{K})$ on N_0 is equivalent to that on $W(\mathcal{D})$. More precisely, if $\varphi, \psi \in \langle \alpha_1, \ldots, \alpha_n\rangle$, then

8E5 $$\begin{aligned} \varphi^{-1}\alpha_0\varphi = \psi^{-1}\alpha_0\psi \quad &\Longleftrightarrow \quad (\varphi f)^{-1}\sigma_0(\varphi f) = (\psi f)^{-1}\sigma_0(\psi f) \\ &(\Longleftrightarrow \quad (0)(\varphi f) = (0)(\psi f), \text{ as vertices of } \mathcal{K}). \end{aligned}$$

To prove the less obvious assertion, set $\tau := (\varphi\psi^{-1})f$. Then $\tau \in \langle \rho_1, \ldots, \rho_n\rangle$, and from the second equation, $\sigma_0 = \tau^{-1}\sigma_0\tau = \sigma_{(0)\tau}$; thus $(0)\tau = 0$. But then $(\varphi\psi^{-1})f = \tau \in \langle \rho_2, \ldots, \rho_n\rangle$, the stabilizer of the base vertex 0 in $\mathcal{K}$; hence $\psi^{-1}\varphi \in \langle \alpha_2, \ldots, \alpha_n\rangle$, because f is one-to-one on $\langle \alpha_1, \ldots, \alpha_n\rangle$. This implies the first equation, because $\psi^{-1}\varphi$ commutes with α_0.

By (8E5), the generators of N_0 are in one-to-one correspondence with the vertices of $\mathcal{K}$. For $i \in V(\mathcal{K})$, define $\beta_i := \varphi_i\alpha_0\varphi_i^{-1}$, where $\varphi_i \in \langle \alpha_1, \ldots, \alpha_n\rangle$ is such that $(0)(\varphi f) = i$. Then

$$N_0 = \langle \beta_i \mid i \in V(\mathcal{K})\rangle,$$

$\beta_0 = \alpha_0$, and $(\beta_i)f = \sigma_i$ for all $i \in V(\mathcal{K})$. To prove that f is also one-to-one on N_0, it suffices to check that N_0 belongs to the same diagram $\mathcal{D}$ as the Coxeter group $W(\mathcal{D})$, which is defined by $\mathcal{D}$. Two cases must be considered.

First, consider $\beta_i\beta_j$, with $\{i, j\}$ representing a diagonal class of $\mathcal{K}$ which is equivalent under $\Gamma(\mathcal{K})$ to a diagonal class of $\mathcal{F}$. Then we may assume that $i, j \in V(\mathcal{F})$. Since

$\mathcal{P}$ and $2^{\mathcal{K},\mathcal{D}}$ have the same facets $2^{\mathcal{F},\mathcal{H}}$, we know that f must be one-to-one on the subgroup

$$\langle \alpha_0, \ldots, \alpha_{n-1} \rangle = \langle \beta_k \mid k \in V(\mathcal{F}) \rangle \rtimes \langle \alpha_1, \ldots, \alpha_{n-1} \rangle$$

of $\Gamma(\mathcal{P})$, whose image under f is

$$\langle \rho_0, \ldots, \rho_{n-1} \rangle = \langle \sigma_k \mid k \in V(\mathcal{F}) \rangle \rtimes \langle \rho_1, \ldots, \rho_{n-1} \rangle.$$

It follows that f must also be one-to-one on $\langle \beta_k \mid k \in V(\mathcal{F}) \rangle$; hence $\beta_i \beta_j$ and $\sigma_i \sigma_j$ have the same order.

Second, if $\{i, j\}$ represents a diagonal class of $\mathcal{K}$ which is not equivalent under $\Gamma(\mathcal{K})$ to one of $\mathcal{F}$, then since $\mathcal{D}$ is universal, $\sigma_i \sigma_j$ already has infinite order, and so does $\beta_i \beta_j$.

It follows that f is one-to-one on $\Gamma(\mathcal{P})$; hence $\mathcal{P} \cong 2^{\mathcal{K},\mathcal{D}}$, as required. □

Call a polytope $\mathcal{P}$ *weakly neighbourly* if any two vertices of $\mathcal{P}$ lie in a common facet. Examples of such polytopes are the toroids $\{4, 4\}_{(3,0)}$ and $\{3, 3, 4, 3\}_{(2,0,0,0)}$; see Sections 6D and 6E. Other examples are the neighbourly polytopes, for which any two vertices are joined by an edge.

8E6 Corollary *Let $\mathcal{K}$ be a regular n-polytope with facet $\mathcal{F}$, and let $\mathcal{D}$ be the universal $\mathcal{K}$-extension of the trivial diagram on $\mathcal{F}$. Then $\{2^{\mathcal{F}}, \mathcal{K}\}$ exists and coincides with $2^{\mathcal{K},\mathcal{D}}$. In particular, $\{2^{\mathcal{F}}, \mathcal{K}\}$ is finite if and only if $\mathcal{K}$ is finite and weakly neighbourly; in this case, $\mathcal{D}$ is the trivial diagram on $\mathcal{K}$, and $\{2^{\mathcal{F}}, \mathcal{K}\} = 2^{\mathcal{K}}$, with group $C_2^{|V(\mathcal{K})|} \rtimes \Gamma(\mathcal{K})$.*

Proof. Apply Theorem 8E2 with $\mathcal{H}$ the trivial diagram on $\mathcal{F}$. Then $\mathcal{H}$ is $\mathcal{K}$-extendable by the trivial diagram on $\mathcal{K}$. From the definition of $\mathcal{D}$, the polytope $2^{\mathcal{K},\mathcal{D}}$ is finite if and only if $\mathcal{K}$ is finite, and each pair of vertices of $\mathcal{K}$ is equivalent under $\Gamma(\mathcal{K})$ to a pair of vertices of $\mathcal{F}$, that is, if and only if $\mathcal{K}$ is finite and weakly neighbourly; in this case, $W(\mathcal{D}) \cong C_2^{|V(\mathcal{K})|}$ because $\mathcal{D}$ is trivial. □

For instance, if $\mathcal{K} = \{4, 4\}_{(3,0)}$, then Corollary 8E6 gives

8E7 $$\big\{\{4, 4\}_{(4,0)}, \{4, 4\}_{(3,0)}\big\} = 2^{\{4,4\}_{(3,0)}},$$

with group $C_2^9 \rtimes [4, 4]_{(3,0)}$ of order 36 864. Here we used the fact that $2^{\{4\}} = \{4, 4\}_{(4,0)}$ (see Corollary 8C4). Similarly, if $\mathcal{K} = \{3, 3, 4, 3\}_{(2,0,0,0)}$, then by Corollary 8C6 (with $s = 2$) we obtain

8E8 $$\big\{\{4, 3, 3, 4\}_{(4,0,0,0)}, \{3, 3, 4, 3\}_{(2,0,0,0)}\big\} = 2^{\{3,3,4,3\}_{(2,0,0,0)}},$$

with group $C_2^{16} \rtimes [3, 3, 4, 3]_{(2,0,0,0)}$ of order $2^{20} \cdot 1152$.

Our next corollary restates Corollary 8E6 for regular polytopes $\mathcal{K}$ of rank 3. For $p \geqslant 3$, set $\mathcal{M}_p := \{4, p \,|\, 4^{\lfloor p/2 \rfloor - 1}\}$. Then we have

8E9 Corollary *Let $\mathcal{K}$ be a regular 3-polytope of type $\{p, q\}$. Then the regular 4-polytope $\{\mathcal{M}_p, \mathcal{K}\}$ exists, and is finite if and only if $\mathcal{K}$ is finite and weakly neighbourly. In this case, $\{\mathcal{M}_p, \mathcal{K}\} = 2^{\mathcal{K}}$, with group $C_2^{|V(\mathcal{K})|} \rtimes \Gamma(\mathcal{K})$.*

Proof. Use $\mathcal{M}_p = 2^{\{p\}}$ (see Corollary 8C4). □

For the next example, recall that a polytope is called *simplicial* if all its facets are isomorphic to simplices, and *cubical* if its facets are isomorphic to cubes.

8E10 Theorem *Let $\mathcal{K}$ be a simplicial regular n-polytope, and let $\mathcal{D}$ be the universal $\mathcal{K}$-extension of the trivial diagram on its facet $\mathcal{F}$. Then the cubical regular $(n+1)$-polytope $\{\{4, 3^{n-2}\}, \mathcal{K}\}$ exists, and coincides with $2^{\mathcal{K},\mathcal{D}}$. In particular, $\{\{4, 3^{n-2}\}, \mathcal{K}\}$ is finite (and then equal to $2^{\mathcal{K}}$, with group the semi-direct product $C_2^{|V(\mathcal{K})|} \rtimes \Gamma(\mathcal{K})$) if and only if $\mathcal{K}$ is finite and neighbourly.*

Proof. Apply Corollary 8E6 with $\mathcal{F} = \{3^{n-2}\}$, and use $2^{\mathcal{F}} = \{4, 3^{n-2}\}$. Further, observe that a simplicial polytope is neighbourly if and only if it is weakly neighbourly. □

Theorem 8E10 says that the only finite universal cubical regular polytopes are those in which the vertex-figure is finite and neighbourly. It would therefore be interesting to be able to characterize the finite neighbourly regular polytopes. This problem is related to the enumeration of 2-transitive permutation groups [73, Chapter 12; 144, §7.7]. In fact, it is easy to see that a regular polytope $\mathcal{K}$ is neighbourly if and only if $\Gamma(\mathcal{K})$ acts 2-transitively on the vertices of $\mathcal{K}$.

Before we move on, let us briefly mention a few results about neighbourly polyhedra and polytopes. A neighbourly polytope with v vertices obviously has the complete graph K_v as its edge-graph. But what about the converse? For which v is there a neighbourly regular polytope with v vertices different from the $(v-1)$-simplex? In particular, when is K_v the edge-graph of some regular polyhedron? There are more general variants of the latter question: one may merely ask for a triangulated 2-manifold with edge-graph K_v, and then further ask if it can be embedded geometrically in $\mathbb{E}^3$. A positive answer is known for $v = 7$ (the Császár torus [8, 35, 138], which is embeddable), and the 59 orientable examples with 12 vertices were classified in [6] (although the question of embeddability remains open). We can guess that the original question only rarely has a positive answer. One instance (with $v = 4$) is the Petrial $\{4, 3\}_3$ of the tetrahedron; two more examples (with $v = 6$) are the hemi-icosahedron $\{3, 5\}_5$ and its Petrial $\{5, 5\}_3$; another (with $v = 11$) is the 4-polytope $\{\{3, 5\}_5, \{5, 3\}_5\}$ with hemi-icosahedral facets. It is not too hard to prove that the only neighbourly regular polyhedra of some type $\{p, q\}$ with $p = 3$ or 4 are $\{3, 3\}$, $\{3, 5\}_5$ and $\{4, 3\}_3$ (see Lemma 11C6 or [249] for a proof when $p = 3$).

The following special case of Theorem 8E10 covers the classification of the locally toroidal regular 4-polytopes of type $\{4, 3, 6\}$ (see also Chapter 11).

8E11 Corollary *The regular 4-polytope $\{\{4, 3\}, \{3, 6\}_{(b,c)}\}$ exists for all (b, c) with $b = c \geqslant 1$ or $b \geqslant 2$, $c = 0$. It is finite if and only if $b = c = 1$ or $b = 2$, $c = 0$, with groups $C_2^3 \rtimes [3, 3]_{(1,1)}$ of order 288 and $C_2^4 \rtimes [3, 6]_{(2,0)}$ of order 768, respectively.*

Proof. The map $\mathcal{K} := \{3, 6\}_{(b,c)}$ is neighbourly if and only if $(b, c) = (1, 1)$ or $(2, 0)$. □

We proceed with further applications of Theorem 8E2. In all these cases, $\mathcal{K}$ will be a regular n-polytope with centrally symmetric facets $\mathcal{F}$ (see Section 8C). We write $\mathcal{H} = \mathcal{H}(\mathcal{F}, s)$ for the diagram on the vertex-set $V(\mathcal{F})$ of $\mathcal{F}$ which connects antipodal vertices of $\mathcal{F}$ by a branch marked s ($\geqslant 2$). In the cases we consider, $\mathcal{H}(\mathcal{F}, s)$ will be $\mathcal{K}$-extendable, and we denote its universal $\mathcal{K}$-extension by $\mathcal{D} := \mathcal{D}(\mathcal{K}, s)$. Note that, if $\mathcal{F} = \{3^{n-3}, 4\}$, then

8E12
$$2^{\mathcal{F},\mathcal{H}} = \{4, 3^{n-3}, 4\}_{(2s,0^{n-2})},$$

a toroid with group

$$[4, 3^{n-3}, 4]_{(2s,0^{n-2})} = D_s^{n-1} \rtimes [3^{n-3}, 4].$$

If $\mathcal{F} = \{3, 4\}$, then $\mathcal{H}(\mathcal{F}, s)$ is the diagram which connects antipodal vertices of the octahedron by a branch marked s, and $2^{\mathcal{F},\mathcal{H}}$ is the toroid $\{4, 3, 4\}_{(2s,0,0)}$ obtained from a $(2s \times 2s \times 2s)$ cubical grid by identifying opposite walls.

8E13 Corollary *Let $s \geqslant 2$, and let $\mathcal{K} := \{4, 4\}_{(b,c)}$ with $b \geqslant 2$ and $c = 0$ or b. Then*

$$\big\{\{4, 4\}_{(2s,0)}, \mathcal{K}\big\} = 2^{\mathcal{K},\mathcal{D}(\mathcal{K},s)},$$

which is finite if and only if $s \geqslant 2$ and $(b, c) = (2, 0)$ or $s = 2$ and $(b, c) = (3, 0)$. In the finite cases, the groups are $(D_s \times D_s) \rtimes [4, 4]_{(2,0)}$ and $C_2^9 \rtimes [4, 4]_{(3,0)}$, respectively.

Proof. Here, $\mathcal{F} = \{4\}$, and $\mathcal{H} = \mathcal{H}(\mathcal{F}, s)$ is $\mathcal{K}$-extendable for all (b, c). The facets are $2^{\mathcal{F},\mathcal{H}} = \{4, 4\}_{(2s,0)}$. The group of $2^{\mathcal{K},\mathcal{D}}$ is finite if and only if $W(\mathcal{D})$ is finite, that is, if and only if $s \geqslant 2$ and $(b, c) = (2, 0)$, or $s = 2$ and $(b, c) = (3, 0)$. In the finite cases, $W(\mathcal{D}) = D_s \times D_s$ or $W(\mathcal{D}) = C_2^9$, respectively, so that the groups are as described. □

For another construction of the polytopes in Corollary 8E13, the reader is referred to Chapter 10. The structure of the polytopes for $s = 1$ will be discussed in Section 8F.

8E14 Corollary *Let $s \geqslant 2$, and let $\mathcal{K} := \{3, 4, 3\}$. Then*

$$\big\{\{4, 3, 4\}_{(2s,0,0)}, \mathcal{K}\big\} = 2^{\mathcal{K},\mathcal{D}(\mathcal{K},s)},$$

which is infinite for all s.

Proof. Now $\mathcal{F} = \{3, 4\}$, and again $\mathcal{H} = \mathcal{H}(\mathcal{F}, s)$ is $\mathcal{K}$-extendable. The facets are 4-toroids $\{4, 3, 4\}_{(2s,0,0)}$, and the group $W(\mathcal{D})$ is infinite for all $s \geqslant 2$. □

8E15 Corollary *Let $s, t \geqslant 2$, $k = 1$ or 2, and let $\mathcal{K} := \{3, 3, 4, 3\}_{(t^k,0^{4-k})}$. Then*

$$\big\{\{4, 3, 3, 4\}_{(2s,0,0,0)}, \mathcal{K}\big\} = 2^{\mathcal{K},\mathcal{D}(\mathcal{K},s)},$$

which is an infinite polytope unless $(s, t, k) = (2, 2, 1)$. If $(s, t, k) = (2, 2, 1)$, then the group of the polytope is $C_2^{16} \rtimes [3, 3, 4, 3]_{(2,0,0,0)}$, of order 12079 59552.

Proof. We now have $\mathcal{F} = \{3, 3, 4\}$, the 4-cross-polytope. Note that antipodal vertices of facets of $\mathcal{K}$ are never joined by an edge, so that $\mathcal{H} = \mathcal{H}(\mathcal{F}, s)$ is always $\mathcal{K}$-extendable. Observe also that, if $(t, k) = (2, 1)$ or $(2, 2)$, then in $\mathcal{K}$ two vertices can be antipodal vertices of more than one facet (in fact, eight facets); hence, if we generate a $\mathcal{K}$-extension by applying all automorphisms of $\mathcal{K}$ to $\mathcal{H}$, we can only take one branch for each such pair of vertices. Now the facets are isomorphic to $\{4, 3, 3, 4\}_{(2s,0,0,0)}$. The group of the polytope is infinite unless $\mathcal{K}$ is weakly neighbourly and $s = 2$. This leaves the exceptional case $(s, t, k) = (2, 2, 1)$; here, $\mathcal{D}(\mathcal{K}, 2)$ pairs up the 16 vertices of $\mathcal{K}$ to give a group $C_2^{16} \rtimes [3, 3, 4, 3]_{(2,0,0,0)}$. (When $\mathcal{K} = \{3, 3, 4, 3\}_{(2,0,0,0)}$, the diagram $\mathcal{D}(\mathcal{K}, s)$ splits into four components, each given by a complete graph on 4 nodes with all branches marked s.) □

Our next result generalizes Theorem 8E2, and deals with the polytopes $\mathcal{L}^{\mathcal{K},\mathcal{G}}$ for universal polytopes $\mathcal{L}$ of rank at least 2. Here we do not need the concept of a $\mathcal{K}$-extension of a diagram.

8E16 Theorem *Let $\mathcal{K}$ be a regular n-polytope with facet $\mathcal{F}$, let $\mathcal{G}$ be a $\mathcal{K}$-admissible diagram on the vertex-set of $\mathcal{K}$, and let $\mathcal{H}$ denote the induced subdiagram of $\mathcal{G}$ on the vertex-set of $\mathcal{F}$. Let $m \geqslant 2$, and let $\mathcal{L} := \{q_1, \ldots, q_{m-1}\}$ and $\mathcal{L}_0 := \{q_2, \ldots, q_{m-1}\}$, the vertex-figure of $\mathcal{L}$. Then the universal regular $(n + m)$-polytope $\{\mathcal{L}^{\mathcal{F},\mathcal{H}}, \mathcal{L}_0^{\mathcal{K},\mathcal{G}}\}$ exists, and*

$$\{\mathcal{L}^{\mathcal{F},\mathcal{H}}, \mathcal{L}_0^{\mathcal{K},\mathcal{G}}\} = \mathcal{L}^{\mathcal{K},\mathcal{G}}.$$

Proof. The proof is similar to that of Theorem 8E2. Let $\mathcal{P} := \{\mathcal{L}^{\mathcal{F},\mathcal{H}}, \mathcal{L}_0^{\mathcal{K},\mathcal{G}}\}$. We choose N_0 to be the normal closure of $\langle \alpha_0, \ldots, \alpha_{m-1} \rangle$ in $\Gamma(\mathcal{P}) := \langle \alpha_0, \ldots, \alpha_{m+n-1} \rangle$. Now we can relate $\mathcal{P}$ to $\mathcal{L}^{\mathcal{K},\mathcal{G}}$, and conclude as before that

8E17 $$\Gamma(\mathcal{P}) \cong N_0 \rtimes \langle \alpha_m, \ldots, \alpha_{m+n-1} \rangle \cong N_0 \rtimes \Gamma(\mathcal{K}).$$

Then an analogue of (8E5) holds with the suffix 0 replaced by $m - 1$ (and the subgroup $\langle \alpha_1, \ldots, \alpha_n \rangle$ by $\langle \alpha_m, \ldots, \alpha_{m+n-1} \rangle$); the corresponding statement is also true for the suffixes $0, 1, \ldots, m - 2$, but here it is trivial because $\varphi, \psi \in \langle \alpha_m, \ldots, \alpha_{m+n-1} \rangle$ commute with each α_j with $j \leqslant m - 2$. It follows that $N_0 = \langle \beta_i \mid i \in V(\mathcal{D}) \rangle$, with $\mathcal{D}$ the diagram used to define $\mathcal{L}^{\mathcal{K},\mathcal{G}}$ (see Section 8B). Again it can be shown that N_0 belongs to the same diagram $\mathcal{D}$, because the polytopes have the same facets and the same vertex-figures. The details are left to the reader. □

8E18 Corollary *Let $\mathcal{K} := \{3, 4\}$, and let $\mathcal{G}(\mathcal{K}, s)$ be the diagram on the vertex-set of $\mathcal{K}$ connecting antipodal vertices by a branch labelled $s \geqslant 2$. Then*

$$\big\{\{3, 4, 3\}, \{4, 3, 4\}_{(2s,0,0)}\big\} = \{3\}^{\{3,4\},\mathcal{G}(\mathcal{K},s)},$$

which is infinite for all s.

Proof. In this case, the facets are $\{3\}^{\{3\}} = \{3, 4, 3\}$ and the vertex-figures are $2^{\mathcal{K},\mathcal{G}(\mathcal{K},s)} = \{4, 3, 4\}_{(2s,0,0)}$ (see Corollaries 8B10 and 8C6). □

Note that Corollary 8E18 gives another construction of the duals of the 5-polytopes occurring in Corollary 8E14. The diagram $\mathcal{D}$ which is involved in Corollary 8E18 will be used again in Section 12B.

We conclude this section with yet another application of our methods to an interesting extension problem for regular polytopes. Here we restrict ourselves to the most important case where all the entries in the Schläfli symbols are at least 3; the general case can easily be derived from this.

8E19 Theorem *Let $n \geqslant 2$, and let $\mathcal{K}$ be a regular n-polytope of type $\{p_1, \ldots, p_{n-1}\}$ with $p_1, \ldots, p_{n-1} \geqslant 3$. Let $r \geqslant 2$, $m \geqslant 1$ and $q_1, \ldots, q_{m-1} \geqslant 3$. Then there exists a regular $(m+n)$-polytope $\mathcal{P}$ with the following properties.*

(a) $\mathcal{P}$ is of type $\{q_1, \ldots, q_{m-1}, 2r, p_1, \ldots, p_{n-1}\}$, and its co-faces to $(m-1)$-faces are isomorphic to $\mathcal{K}$.

(b) $\mathcal{P}$ is "universal" among all regular $(m+n)$-polytopes which have the property of part (a); that is, any other such regular $(m+n)$-polytope is a quotient of $\mathcal{P}$.

(c) $\mathcal{P}$ is finite if and only if $m = 1$, $r = 2$ and $\mathcal{K}$ is finite and neighbourly, or $m = 2$, $p_1 = 3$, $r = 2$ and $\mathcal{K} = \{3\}$. In these cases, $\mathcal{P} = 2^{\mathcal{K}}$ or $\mathcal{P} = \{3, 4, 3\}$, with groups $C_2^{|V(\mathcal{K})|} \rtimes \Gamma(\mathcal{K})$ or $[3, 4, 3]$, respectively.

(d) $\mathcal{P} = \{q_1, \ldots, q_{m-1}\}^{\mathcal{K},\mathcal{G}}$, where $\mathcal{G}$ is the complete graph on the vertex set of $\mathcal{K}$, in which the branches representing edges of $\mathcal{K}$ are marked r, while all other branches are marked ∞. In particular, the m-faces of $\mathcal{P}$ are isomorphic to $\{q_1, \ldots, q_{m-1}\}$.

Proof. We use the same method of proof as for Theorems 8E2 and 8E16. Let $\mathcal{L} := \{q_1, \ldots, q_{m-1}\}$, and let $\mathcal{G}$ be as in (d). First note that $\mathcal{L}^{\mathcal{K},\mathcal{G}}$ (with group $\Gamma(\mathcal{L}^{\mathcal{K},\mathcal{G}}) = \langle \rho_0, \ldots, \rho_{m+n-1} \rangle$) is a polytope which satisfies (a). On the other hand, the existence of any regular polytope with property (a) implies the existence of a universal such polytope; a proof of this fact can be given by adapting the proof of Theorem 4A2 to the situation discussed here. Let $\mathcal{P}$ be this universal polytope, and let $\Gamma(\mathcal{P}) = \langle \alpha_0, \ldots, \alpha_{m+n-1} \rangle$. As in the proof of Theorem 8E16, if N_0 denotes the normal closure of $\langle \alpha_0, \ldots, \alpha_{m-1} \rangle$ in $\Gamma(\mathcal{P})$, then (8E17) holds and we again arrive at $N_0 = \langle \beta_i \mid i \in V(\mathcal{D}) \rangle$, with $\mathcal{D}$ the diagram used to define $\mathcal{L}^{\mathcal{K},\mathcal{G}}$. Here, $\beta_i = \alpha_i$ if $i \leqslant m-2$, or is the conjugate of α_{m-1} which corresponds to the vertex i of $V(\mathcal{K})$ otherwise. Again, we need to check that N_0 belongs to the same diagram $\mathcal{D}$ as the group $W(\mathcal{D}) = \langle \sigma_i \mid i \in V(\mathcal{D}) \rangle$, that is, that $\beta_i \beta_j$ and $\sigma_i \sigma_j$ have the same order for any two nodes i, j of $\mathcal{D}$.

If $i, j \leqslant m-1$, then $\beta_i \beta_j = \alpha_i \alpha_j$ and $\sigma_i \sigma_j = \rho_i \rho_j$ clearly have the same order, because $\mathcal{P}$ and $\mathcal{L}^{\mathcal{K},\mathcal{G}}$ have the same Schläfli symbols. Since $\Gamma(\mathcal{K}) = \langle \alpha_m, \ldots, \alpha_{m+n-1} \rangle$ acts on N_0 as it does on $W(\mathcal{D})$, the elements $\beta_i \beta_j$ and $\sigma_i \sigma_j$ are conjugate to $\beta_i \beta_{m-1}$ and $\sigma_i \sigma_{m-1}$, respectively, if $i \leqslant m-2$ and $j \in V(\mathcal{K})$; hence the orders are again the same.

Now let $i, j \in V(\mathcal{K})$. If i, j are joined by an edge of $\mathcal{K}$, then modulo $\Gamma(\mathcal{K})$ we may assume that $i = m-1 = F_0$, the base vertex of $\mathcal{K}$, and $j = \rho_m(F_0)$ $(= \alpha_m(F_0))$, the other vertex in the base edge of $\mathcal{K}$. But then

$$\sigma_i \sigma_j = \sigma_{F_0} \rho_m \sigma_{F_0} \rho_m = \left(\sigma_{F_0} \rho_m\right)^2 = (\rho_{m-1} \rho_m)^2,$$

and similarly

$$\beta_i \beta_j = \beta_{F_0} \alpha_m \beta_{F_0} \alpha_m = \left(\beta_{F_0} \alpha_m\right)^2 = (\alpha_{m-1} \alpha_m)^2.$$

It follows that both elements have order $2r$, because the corresponding entry in the Schläfli symbol is $2r$. Finally, if i, j are not joined by an edge of $\mathcal{K}$, then, by the construction of $\mathcal{G}$, the order of $\sigma_i\sigma_j$ is infinite, and so is the order of $\beta_i\beta_j$.

It follows that the homomorphism $\Gamma(\mathcal{P}) \to \Gamma(\mathcal{L}^{\mathcal{K},\mathcal{G}})$ is an isomorphism; thus $\mathcal{P}$ and $\mathcal{L}^{\mathcal{K},\mathcal{G}}$ are isomorphic polytopes. This proves parts (a), (b) and (d) of the theorem.

Furthermore, $\mathcal{L}^{\mathcal{K},\mathcal{G}}$ is finite if and only if $W(\mathcal{D})$ and $\Gamma(\mathcal{K})$ are finite. Let $\mathcal{K}$ be finite. If $m = 1$, then $W(\mathcal{D})$ is finite if and only if $\mathcal{G}$ has no branches marked $r \geqslant 3$ or ∞; that is, if and only if $\mathcal{K}$ is neighbourly and $r = 2$. In this case, $\mathcal{P} = 2^{\mathcal{K}}$, $W(\mathcal{D}) \cong C_2^{|V(\mathcal{K})|}$, and $\Gamma(\mathcal{P}) \cong C_2^{|V(\mathcal{K})|} \rtimes \Gamma(\mathcal{K})$. If $m \geqslant 2$, then $W(\mathcal{D})$ can only be finite if $r = 2$, and the node $m-2$ of $\mathcal{D}$ is contained in at most 3 branches (marked $q_{m-2} \geqslant 3$ or $q_{m-1} \geqslant 3$), that is, if and only if $r = 2$ and $\mathcal{K} = \mathcal{L} = \{3\}$ (and $m = n = 2$). But if $r = 2$ and $\mathcal{K} = \mathcal{L} = \{3\}$, then $\mathcal{P} = \{3, 4, 3\}$. This completes the proof. □

8F Polytopes with Small Faces

In enumeration problems we often run into the trouble of having to exclude the small polytopes of a certain class from our discussion. Typical examples are the (locally toroidal) regular 5-polytopes

$$\big\{\{4,3,4\}_{(2,2,0)}, \{3,4,3\}\big\}, \qquad \big\{\{4,3,4\}_{(2,2,2)}, \{3,4,3\}\big\},$$

whose facets are cubic 4-toroids $\{4,3,4\}_{(2,2,0)}$ (with 16 vertices) or $\{4,3,4\}_{(2,2,2)}$ (with 32 vertices), respectively, and whose vertex-figures are 24-cells $\{3,4,3\}$ (see Section 6D). In such cases, the Todd–Coxeter coset enumeration algorithm can be applied to find the order of the corresponding group from its presentation (see [131, Chapter 2; 134]); even so, the structure of the group has to be determined by other means.

In this section, we describe a method for finding polytopes with small faces as quotients of polytopes constructed in earlier sections (following [303, §7]). In particular, we identify the groups of certain universal regular polytopes which we shall encounter again in later chapters. This also includes the two examples just mentioned. For basic results about toroids we refer to Sections 6D and 6E.

The basic technique is to construct a polytope as a quotient $2^{\mathcal{K},\mathcal{D}}/N$ of a larger polytope $2^{\mathcal{K},\mathcal{D}}$, whose structure we know by Theorem 8E2 and its corollaries. Then the geometry of the vertex-figure is used to identify the new group explicitly. Our approach avoids prior knowledge of the order of the group; thus we need not appeal to the Todd–Coxeter algorithm, although it can always be used to check our results. For a general discussion of quotients of polytopes, see Section 2D.

We first describe the basic technique. Let $\mathcal{K}$ be a regular n-polytope whose facets $\mathcal{F}$ are centrally symmetric; this assumption on the structure of the facets is important. As before, we denote by $\mathcal{H}(\mathcal{F}, s)$ the diagram on the vertex set of $\mathcal{F}$ which connects antipodal vertices of $\mathcal{F}$ by a branch labelled s, and by $\mathcal{D}(\mathcal{K}, s)$ the universal $\mathcal{K}$-extension of $\mathcal{H}(\mathcal{F}, s)$ to the vertex set $V(\mathcal{K})$ of $\mathcal{K}$. We are particularly interested in the case $s = 2$, where $\mathcal{H}(\mathcal{F}, 2)$ is the trivial diagram; thus every restriction of $\mathcal{D}(\mathcal{K}, 2)$ to the vertex-set of a facet of $\mathcal{K}$ is also trivial. We can further note that, by the central symmetry of $\mathcal{F}$, the polytope $2^{\mathcal{F}}$ has a non-standard representation as $2^{\mathcal{F},\mathcal{H}(\mathcal{F},2)}$, which we shall use in our construction (see Theorem 8C5). Then, by Theorem 8E2,

$$\{2^{\mathcal{F}}, \mathcal{K}\} = \left\{2^{\mathcal{F},\mathcal{H}(\mathcal{F},2)}, \mathcal{K}\right\} = 2^{\mathcal{K},\mathcal{D}(\mathcal{K},2)}.$$

In these examples, we shall work with $\mathcal{K} = \{3, 4, 3\}$, $\mathcal{F} = \{3, 4\}$, and

$$\{2^{\mathcal{F}}, \mathcal{K}\} = \left\{\{4, 3, 4\}_{(4,0,0)}, \{3, 4, 3\}\right\}.$$

For the remainder of this section, we define

$$\mathcal{H} := \mathcal{H}(\mathcal{F}, 2), \qquad \mathcal{D} := \mathcal{D}(\mathcal{K}, 2);$$

further, we retain the notation of Section 8E. In particular, we have $\Gamma(\mathcal{K}) = \langle \tau_0, \ldots, \tau_{n-1} \rangle$, and $W(\mathcal{D}) = \langle \sigma_i \mid i \in V(\mathcal{K}) \rangle$, with $i = 0$ $(= F_0)$ the base vertex of $\mathcal{K}$. Then

$$\Gamma(2^{\mathcal{K},\mathcal{D}}) = \langle \rho_0, \ldots, \rho_n \rangle = W(\mathcal{D}) \rtimes \Gamma(\mathcal{K}).$$

We shall now construct quotients $2^{\mathcal{K},\mathcal{D}}/N$ of $2^{\mathcal{K},\mathcal{D}}$, where N is a subgroup of $W(\mathcal{D})$ which is normal in $\Gamma(2^{\mathcal{K},\mathcal{D}})$. We choose N in such a way that the facets $2^{\mathcal{F}}$ of $2^{\mathcal{K},\mathcal{D}}$ collapse to smaller facets (like $\{4, 3, 4\}_{(2,2,0)}$ or $\{4, 3, 4\}_{(2,2,2)}$ in our examples).

Many of these quotients, but not all, are actually also quotients of the polytopes $2^{\mathcal{K}}$. If $\hat{\mathcal{D}}$ is the trivial diagram on $\mathcal{K}$, and

$$W(\hat{\mathcal{D}}) = \langle \hat{\sigma}_i \mid i \in V(\mathcal{K}) \rangle$$

is the corresponding (abelian) Coxeter group, then

$$W(\hat{\mathcal{D}}) = C_2^v \qquad (v = |V(\mathcal{K})|),$$

and $\sigma_i \mapsto \hat{\sigma}_i$ $(i \in V(\mathcal{K}))$ defines a surjective homomorphism $f\colon W(\mathcal{D}) \to W(\hat{\mathcal{D}})$ mapping N onto a subgroup $\hat{N}$ of $W(\hat{\mathcal{D}})$ which is normal in $\Gamma(2^{\mathcal{K}})$. This gives us the quotient $2^{\mathcal{K}}/\hat{N}$. More informally, $\hat{N}$ will be defined with respect to $W(\hat{\mathcal{D}})$ in the same way as is N with respect to $W(\mathcal{D})$. Note that, if C_2^v is identified with the v-dimensional vector space $\mathrm{GF}(2)^v$ over the Galois field GF(2) with 2 elements, then $\hat{N}$ is a linear binary code (see also [369]).

The following considerations indicate how the subgroup N of $W(\mathcal{D})$ should be defined in order to induce the right collapse onto the facets. Assume that the vertex-set of $\mathcal{F}$ is

$$V(\mathcal{F}) = \{0, \ldots, 2m - 1\},$$

with $j, j + m$ antipodal vertices of $\mathcal{F}$ for each $j = 0, \ldots, m - 1$. If $\mathcal{F}$ is the $(n - 1)$-cross-polytope $\{3^{n-3}, 4\}$, then $m = n - 1$. Now recall from Section 8D that the polytope $2^{\mathcal{F}} = 2^{\mathcal{F},\mathcal{H}}$ admits a realization in euclidean $2m$-space $\mathbb{E}^{2m}$. In particular, if $e_0, \ldots, e_{2m-1}$ denotes an orthonormal basis of $\mathbb{E}^{2m}$, then the direct product

$$W(\mathcal{D}) = \bigotimes_{j=0}^{m-1} \langle \sigma_j, \sigma_{j+m} \rangle = (C_2 \times C_2)^m$$

is realized in such a way that $\langle \sigma_j, \sigma_{j+m} \rangle$ acts on the plane $\mathbb{E}^2_j = \langle e_j, e_{j+m} \rangle$ of

$$\mathbb{E}^{2m} = \bigoplus_{j=0}^{m-1} \mathbb{E}^2_j$$

by

$$e_j \sigma_j = e_{j+m},$$
$$e_j \sigma_{j+m} = -e_{j+m}.$$

(Recall that each σ_i is a linear involution and that $s = 2$.) Here, $\langle \sigma_j, \sigma_{j+m} \rangle$ occurs as a subgroup of index 2 in the symmetry group of the square P_j with vertices $\pm e_j, \pm e_{j+m}$. Topologically, one can think of the boundary ∂P_j of each P_j as a 1-sphere (subdivided by vertices), whose product is an m-torus $\mathbb{S}^1 \times \cdots \times \mathbb{S}^1$. Our identifications will now impinge on this m-torus to give a new m-torus.

Identification Vector $(2, 0^{n-2})$

First we study the smallest possible quotients $2^{\mathcal{K},\mathcal{D}}/N$, that is, we take the largest possible choice for N. In particular, we shall prove Theorem 8F3. In our examples, we shall have $\mathcal{F} = \{3^{n-3}, 4\}$ (with $n \geqslant 3$), giving quotients with toroidal facets $2^{\mathcal{F}} = \{4, 3^{n-3}, 4\}_{(4,0^{n-2})}$ (see Section 6D). If $\mathcal{F} = \{4\}$, then, in $\Gamma(2^{\mathcal{F}}) = \langle \rho_0, \rho_1, \rho_2 \rangle = \langle \sigma_0, \tau_0, \tau_1 \rangle$, we have

$$(\rho_0 \rho_1 \rho_2 \rho_1)^2 = \sigma_0 \sigma_2;$$

hence, by Theorem 6D4, to collapse the facets onto $\{4, 4\}_{(2,0)}$, we have to choose N in such a way that $\sigma_0 \sigma_2 \in N$. A similar remark applies more generally with respect to facets $\{4, 3^{n-3}, 4\}_{(2,0^{n-2})}$ with identification vector $(2, 0^{n-2})$, using

$$(\rho_0 \rho_1 \cdots \rho_{n-1} \rho_{n-2} \cdots \rho_1)^2 = \sigma_0 \sigma_{n-1} = \sigma_0 \sigma_m$$

(see Theorem 6D4 again). We therefore define

8F1 $N := \langle \varphi \sigma_i \sigma_j \varphi^{-1} \mid \varphi \in W(\mathcal{D}),$ and $\{i, j\}$ antipodal vertices of a facet of $\mathcal{K} \rangle$.

Thus N is the normal closure of $\sigma_0 \sigma_m$ in $\Gamma(2^{\mathcal{K},\mathcal{D}})$, and

$$\Gamma(2^{\mathcal{K},\mathcal{D}})/N \cong W(\mathcal{D})/N \rtimes \Gamma(\mathcal{K}).$$

Observe that, in $W(\mathcal{D})/N$, we have $\sigma_i N = \sigma_j N$ if i, j are antipodal vertices of a facet of $\mathcal{K}$.

To find the structure of $W(\mathcal{D})/N$, define the graph $\mathcal{G}_{\mathcal{K}}$ with vertex-set $V(\mathcal{K})$ as follows. In $\mathcal{G}_{\mathcal{K}}$, two vertices i, j of $V(\mathcal{K})$ are joined by an edge if and only if they are antipodal vertices of a facet of $\mathcal{K}$. Then

8F2 $$W(\mathcal{D})/N \cong C_2^{\omega(\mathcal{G}_{\mathcal{K}})},$$

with $\omega(\mathcal{G}_{\mathcal{K}})$ the number of connected components of $\mathcal{G}_{\mathcal{K}}$. For (8F2), note that, by the connectivity properties of $\mathcal{K}$, each connected component of $\mathcal{G}_{\mathcal{K}}$ in fact has a representative vertex in the base facet $\mathcal{F}$ of $\mathcal{K}$. But, if $i, j \in V(\mathcal{F})$, then σ_i and σ_j commute, so that $W(\mathcal{D})/N$ is abelian. Note that, if $\mathcal{F} = \{3^{n-3}, 4\}$ (as in our examples), then we can even find the representative vertex among the vertices $0, \dots, n-2$ of the base $(n-2)$-face of $\mathcal{F}$.

Further, $\Gamma(\mathcal{K})$ acts transitively on the components of $\mathcal{G}_{\mathcal{K}}$. But, if i, j are in the same component of $\mathcal{G}_{\mathcal{K}}$, and if $\tau \in \Gamma(\mathcal{K})$ with $\tau(i) = i$, then $i, \tau(j)$, and thus $i, j, \tau(j)$ are in the same component of $\mathcal{G}_{\mathcal{K}}$. If $\mathcal{F} = \{3^{n-3}, 4\}$, we can apply this with $i, j, \tau(j) \in \{0, \dots, n-2\}$, so that now either $\omega(\mathcal{G}_{\mathcal{K}}) = 1$ or $\omega(\mathcal{G}_{\mathcal{K}}) = n-1$.

8F3 Theorem *For $n \geqslant 3$, let $\mathcal{K}$ be a regular n-polytope with facets $\mathcal{F} = \{3^{n-3}, 4\}$, and let $\mathcal{G}_{\mathcal{K}}$ be the graph defined previously. Then the regular $(n+1)$-polytope $\{\{4, 3^{n-3}, 4\}_{(2,0^{n-2})}, \mathcal{K}\}$ exists if and only if $\omega(\mathcal{G}_{\mathcal{K}}) = n-1$. In this case the polytope is flat, has group $C_2^{n-1} \rtimes \Gamma(\mathcal{K})$, and is finite if and only if $\mathcal{K}$ is finite.*

Proof. We have $\omega(\mathcal{G}_{\mathcal{K}}) = 1$ or $n-1$. In either case, $\Gamma(2^{\mathcal{K},\mathcal{D}})/N$ is a string C-group. If $\omega(\mathcal{G}_{\mathcal{K}}) = 1$, then $W(\mathcal{D})/N \cong C_2$ and $\Gamma(2^{\mathcal{K},\mathcal{D}})/N$ is the group of the regular $(n+1)$-polytope $\{\{2, 3^{n-3}, 4\}, \mathcal{K}\}$. However, if $\omega(\mathcal{G}_{\mathcal{K}}) = n-1$, then the facets are flat toroids, and $\Gamma(2^{\mathcal{K},\mathcal{D}})/N$ is the group of $\{\{4, 3^{n-3}, 4\}_{(2,0^{n-2})}, \mathcal{K}\}$. (Recall that a polytope is flat if each vertex is incident with each facet.) The polytope itself is then flat, because it has flat facets (see Section 4E). □

8F4 Corollary *The regular 4-polytope $\{\{4, 4\}_{(2,0)}, \{4, 4\}_{(p,q)}\}$ exists for all p, q except for $q = 0$ with p odd. It is finite and flat, and its group is $C_2^2 \rtimes [4, 4]_{(p,q)}$, of order $32(p^2 + q^2)$.*

Proof. Apply Theorem 8F3 with $n = 3$, $\mathcal{K} = \{4, 4\}_{(p,q)}$ and $\mathcal{F} = \{4\}$. In particular, $\omega(\mathcal{G}_{\mathcal{K}}) = 1$ if p is odd and $q = 0$, and $\omega(\mathcal{G}_{\mathcal{K}}) = 2$ otherwise. □

8F5 Corollary *If $n \geqslant 3$, the regular $(n+1)$-polytope $\{\{4, 3^{n-3}, 4\}_{(2,0^{n-2})}, \{3^{n-3}, 4, 3\}\}$ exists and is flat. Its group is $C_2^{n-1} \rtimes [3^{n-3}, 4, 3]$.*

Proof. In this case, $\mathcal{K} = \{3^{n-3}, 4, 3\}$ and $\omega(\mathcal{G}_{\mathcal{K}}) = n-1$. □

8F6 Corollary *The regular 5-polytope $\{\{4, 3, 4\}_{(2,0,0)}, \{3, 4, 3\}\}$ exists and is finite and flat. Its group is $C_2^3 \rtimes [3, 4, 3]$, of order 9216.*

Proof. This result restates Corollary 8F5 for $\mathcal{K} = \{3, 4, 3\}$. □

Note that another way to construct the (dual of the) polytope in Corollary 8F6 is to let $s = 1$ in (12B4) and (12B5); then this implies that $\sigma_i = \sigma_{i+3}$ for $i = 1, 2, 3$.

The next corollary deals with 6-polytopes whose vertex-figures are regular 5-toroids of type $\{3, 3, 4, 3\}$ (see Section 6E).

8F7 Corollary *The regular 6-polytope $\{\{4, 3, 3, 4\}_{(2,0,0,0)}, \{3, 3, 4, 3\}_{(t^k,0^{4-k})}\}$ exists for all $t \geqslant 2$ and $k = 1, 2$, unless t is odd and $k = 1$. It is finite and flat, and its group is $C_2^4 \rtimes [3, 3, 4, 3]_{(t^k,0^{4-k})}$, of order $18432k^2t^4$.*

Proof. Now $\mathcal{K} = \{3, 3, 4, 3\}_{(t^k, 0^{4-k})}$, so that $\omega(\mathcal{G}_\mathcal{K}) = 1$ or 4. If $k = 1$ and t is odd, then $|V(\mathcal{K})| = t^4$ is odd and $V(\mathcal{K})$ cannot split into 4 components of the same size; hence $\omega(\mathcal{G}_\mathcal{K}) = 1$. In the remaining cases, we can prove that $\omega(\mathcal{G}_\mathcal{K}) = 4$. In fact, if t is even (and $k = 1$ or 2), then the covering

$$\mathcal{K} \searrow \{3, 3, 4, 3\}_{(2^k, 0^{4-k})} =: \mathcal{L}$$

induces an incidence-preserving mapping of $\mathcal{G}_\mathcal{K}$ onto $\mathcal{G}_\mathcal{L}$; thus, $\omega(\mathcal{G}_\mathcal{L}) \leqslant \omega(\mathcal{G}_\mathcal{K})$. But a direct calculation shows that $\omega(\mathcal{G}_\mathcal{L}) = 4$, so that $\omega(\mathcal{G}_\mathcal{K}) = 4$ also. If $k = 2$ and t is odd, we can use a covering from $\mathcal{K}$ onto the degenerate "polytope" $\mathcal{L} := \{3, 3, 4, 3\}_{(1,1,0,0)}$ instead. The 4 vertices of $\mathcal{L}$ are the vertices of a 3-face, and still represent the 4 connected components of $\mathcal{G}_\mathcal{L}$, so that $\omega(\mathcal{G}_\mathcal{K}) = 4$ again. □

Identification Vector $(2, 2, 0^{n-3})$

We next study quotients whose facets are $\{4, 3^{n-3}, 4\}_{(2,2,0^{n-3})}$, with identification vector $(2, 2, 0^{n-3})$. This is more complicated than the previous case. As before if $\mathcal{F} = \{4\}$, then in $\Gamma(2^\mathcal{F})$ we have

$$(\rho_0\rho_1\rho_2)^4 = \sigma_0\sigma_2\sigma_1\sigma_3$$

(the order of the latter terms is irrelevant); hence, by Theorem 6D4, we must have $\sigma_0\sigma_1\sigma_2\sigma_3 \in N$ to obtain $\{4, 4\}_{(2,2)}$. A similar remark applies to the $(n-1)$-cross-polytope $\mathcal{F} = \{3^{n-3}, 4\}$ and to the facets $\{4, 3^{n-3}, 4\}_{(2,2,0^{n-3})}$, using

$$(\rho_0\rho_1\rho_2 \cdots \rho_{n-1}\rho_{n-2} \cdots \rho_2)^4 = \sigma_0\sigma_{n-1}\sigma_1\sigma_n$$

(see Theorem 6D4 again). We therefore define

8F8
$$N := \langle \varphi\sigma_i\sigma_j\sigma_k\sigma_l\varphi^{-1} \mid \varphi \in W(\mathcal{D}), \text{ and } \{i, j\}, \{k, l\} \text{ pairs of antipodal vertices in a common facet of } \mathcal{K} \rangle.$$

If $\mathcal{F} = \{4\}$, this is the normal closure of $\sigma_0\sigma_1\sigma_2\sigma_3$ in $\Gamma(2^{\mathcal{K},\mathcal{D}})$, and the vertices i, j, k, l are in fact all the vertices of the corresponding facet of $\mathcal{K}$; this case would also fit into the following discussion about the identification vector (2^{n-1}). More generally, if $\mathcal{F} = \{3^{n-3}, 4\}$, then the vertices i, j, k, l are the vertices of a diametral square of a facet of $\mathcal{K}$. Then in $W(\mathcal{D})/N$, the product of any three generators from $\sigma_i N, \sigma_j N, \sigma_k N, \sigma_l N$ is the fourth. Again,

$$\Gamma(2^{\mathcal{K},\mathcal{D}})/N \cong W(\mathcal{D})/N \rtimes \Gamma(\mathcal{K}).$$

We now discuss several applications.

8F9 Theorem *The universal regular 4-polytopes* $\{\{4, 4\}_{(2,2)}, \{4, 4\}_{(p,q)}\}$ *exist for all* p, q. *The only finite instances occur for* $(p, q) = (2, 0), (3, 0), (2, 2)$ *and* $(3, 3)$, *with groups* $C_2^3 \rtimes [4, 4]_{(2,0)}$, $C_2^5 \rtimes [4, 4]_{(3,0)}$, $C_2^4 \rtimes [4, 4]_{(2,2)}$ *and* $C_2^6 \rtimes [4, 4]_{(3,3)}$, *of orders* 256, 2304, 1024 *and* 9216, *respectively.*

Proof. For the existence statement we refer to Section 10C, though a direct proof is not difficult here; in fact, in most cases we can appeal to Corollary 8F4 and the quotient criterion of Theorem 2E17. By construction, the polytopes $2^{\mathcal{K},\mathcal{D}}/N$ are indeed isomorphic to the universal polytopes. It is also known from Section 10C that finiteness can only occur for the four choices of (p, q). We shall now identify the groups in the finite cases. In particular, this leads to the corresponding entries in Table 10C1.

The case $\mathcal{K} = \{4, 4\}_{(p,0)}$ with $p = 2$ or 3 is special, because $\mathcal{D}$ is trivial and $W(\mathcal{D})$ and $W(\mathcal{D})/N$ are abelian. Now, in $W(\mathcal{D})/N$, the product of three generators on a facet of $\mathcal{K}$ is the fourth. It follows that $W(\mathcal{D})/N$ is generated by the 3 elements $\sigma_i N$ corresponding to 3 vertices on a facet if $p = 2$, and by the 5 elements corresponding to one vertex and its neighbours in $\mathcal{K}$ if $p = 3$. It is easy to check that fewer generators will not suffice, so that $W(\mathcal{D})/N \cong C_2^3$ or C_2^5, respectively. Note that the case $\{4, 4\}_{(2,0)}$ is also covered by Corollary 8F4.

Now let $\mathcal{K} = \{4, 4\}_{(p,p)}$. Recall that a *hole* (or, more exactly, 2-*hole*) of $\mathcal{K}$ is a path along edges of $\mathcal{K}$ which leaves at each vertex exactly 2 faces to the right (see Section 7B). Now through each vertex of $\mathcal{K}$ pass exactly two holes, each of length $2p$. After p steps, these holes meet again in another vertex which, together with the original vertex, dissects them into halves. It is now easy to see that we can generate $W(\mathcal{D})/N$ by taking such a pair of holes and choosing on each hole all the vertices from one half, including the two vertices where the holes meet. This gives a set of $2p$ generators. Further, if $\{i_1, i_2\}$ and $\{i_3, i_4\}$ are edges p steps apart on a hole of $\mathcal{K}$, then $\sigma_{i_1}\sigma_{i_2}N = \sigma_{i_3}\sigma_{i_4}N$; that is, the edges on a hole are identified in pairs, as indicated by parallel heavy edges (and the same labels) in Figure 8F1, which illustrates the case $p = 3$.

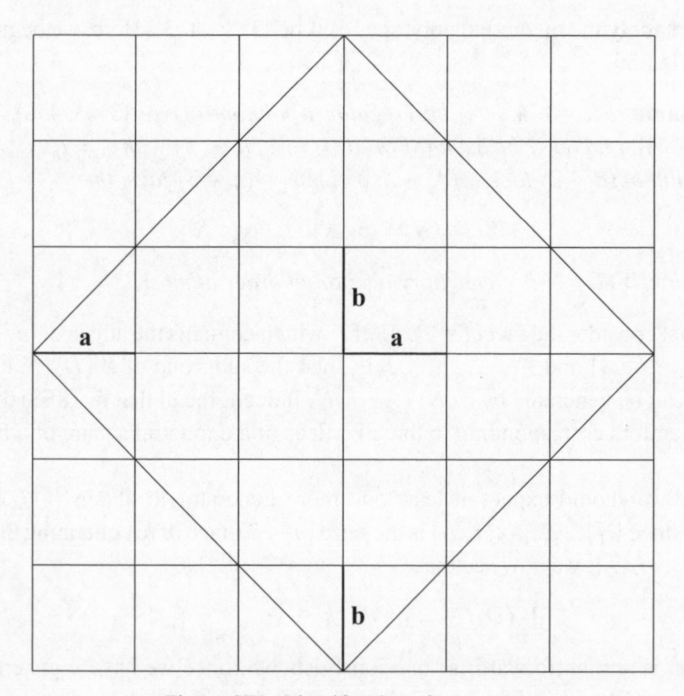

Figure 8F1. Identifications for $p = 3$.

For $p = 3$, we have generators $\sigma_i N$ for $i = 0, 1, 4, 5, 6, 7$. We shall prove that $W(\mathcal{D})/N$ is elementary abelian of order 2^6. Since $\Gamma(\mathcal{K})$ acts on $W(\mathcal{D})/N$, the following imply that $W(\mathcal{D})/N$ is elementary abelian:

$$\sigma_0\sigma_5 N = \sigma_0\sigma_1 N\sigma_1\sigma_4 N\sigma_4\sigma_5 N = \sigma_5\sigma_9 N\sigma_9\sigma_8 N\sigma_8\sigma_0 N = \sigma_5\sigma_0 N,$$

8F10 $$\sigma_0\sigma_4 N = \sigma_0\sigma_1 N\sigma_1\sigma_4 N = \sigma_5\sigma_9 N\sigma_9\sigma_8 N = \sigma_5\sigma_8 N = (\sigma_4\sigma_8\sigma_0)\sigma_8 N = \sigma_4\sigma_0 N,$$

$$\sigma_0(\sigma_{10}\sigma_{11}\sigma_1)N = \sigma_0\sigma_4 N = \sigma_4\sigma_0 N = (\sigma_{10}\sigma_{11}\sigma_1)\sigma_0 N = \sigma_{10}\sigma_0\sigma_{11}\sigma_1 N,$$

so that $\sigma_0\sigma_{10}N = \sigma_{10}\sigma_0 N$, or, equivalently, $\sigma_4\sigma_7 N = \sigma_7\sigma_4 N$. One can also check that no 5 generators suffice, so that indeed $W(\mathcal{D})/N \cong C_2^6$. Here it is helpful to observe that the covering $\mu\colon \mathcal{K} \to \{4, 4\}_{(3,0)}$ induces a surjective homomorphism

$$\hat{\mu}\colon W(\mathcal{D})/N \longrightarrow W\big(\mathcal{D}_{(3,0)}\big)/N_{(3,0)} \cong C_2^5,$$

with $\mathcal{D}_{(3,0)}$ and $N_{(3,0)}$ the corresponding diagram and normal subgroup, respectively. But $\ker(\hat{\mu})$ is non-trivial; in fact, if i, j are 3 steps apart on a hole of $\{4, 4\}_{(3,3)}$, then $(\sigma_i N)\hat{\mu} = (\sigma_j N)\hat{\mu}$; thus $\sigma_i\sigma_j N \in \ker(\hat{\mu})$. Let $\tilde{N}$ denote the normal subgroup of (8F1) defined with respect to $\mathcal{K}$, and let $\mathcal{G}_{\mathcal{K}}$ be the corresponding graph. Now if $\sigma_i\sigma_j N = N$, then $\sigma_i\sigma_j\tilde{N} = \tilde{N}$, contradicting the fact that i, j lie in different components of $\mathcal{G}_{\mathcal{K}}$. Hence $\sigma_i\sigma_j N$ is a non-trivial element of $\ker(\hat{\mu})$, proving that $W(\mathcal{D})/N \cong C_2^6$.

If $p = 2$, we have 4 generators, and an analogue of (8F10) shows that $W(\mathcal{D})/N$ is elementary abelian. Again one can check that 3 generators do not suffice. This completes the proof. □

We next apply the method to polytopes of type $\{3^{n-3}, 4, 3\}$. Here, we begin with the following lemma.

8F11 Lemma *For $n \geqslant 4$, let $\mathcal{K}$ be a regular n-polytope of type $\{3^{n-3}, 4, 3\}$ with facets $\mathcal{F} = \{3^{n-3}, 4\}$, and let N be as in (8F8). If $\{i_1, \ldots, i_{n-2}, j\}$ with $j = i_{n-1}, i_n, i_{n+1}$ are three simplicial $(n-2)$-faces of $\mathcal{K}$ with a common $(n-3)$-face, then*

$$W(\mathcal{D})/N = \big\langle \sigma_{i_1}N, \ldots, \sigma_{i_{n+1}}N \big\rangle.$$

In particular, $W(\mathcal{D})/N$ is elementary abelian of order at most 2^{n+1}.

Proof. First consider the facet $\{3^{n-3}, 4\}$ of $\mathcal{K}$ which contains the adjacent $(n-2)$-faces $\{i_1, \ldots, i_{n-2}, i_{n-1}\}$ and $\{i_1, \ldots, i_{n-2}, i_n\}$. Then the subgroup of $W(\mathcal{D})/N$ induced by this facet can be generated by $\sigma_{i_1}N, \ldots, \sigma_{i_n}N$. Indeed, recall that by (8F8) the product of the generators corresponding to three vertices of a diametral square of a facet is the fourth.

Clearly we should expect at least one more generator to obtain $W(\mathcal{D})/N$ itself. However, since $\{i_1, \ldots, i_{n-2}, i_{n+1}\}$ is the third $(n-2)$-face of $\mathcal{K}$ containing the $(n-3)$-face $\{i_1, \ldots, i_{n-2}\}$, we now see that

$$W(\mathcal{D})/N = \big\langle \sigma_{i_k}N \mid k = 1, \ldots, n+1 \big\rangle.$$

The reason is straightforward; as previously shown, once we have n generators on a facet corresponding to adjacent $(n-2)$-faces, we have them all, and so we can move from facet to facet of $\mathcal{K}$, starting with the three which contain $\{i_1, \ldots, i_{n-2}\}$. Since in $\mathcal{K}$ there are only three facets surrounding an $(n-3)$-face, this process covers all the facets of $\mathcal{K}$. But now each pair of $\sigma_{i_1}N, \ldots, \sigma_{i_{n+1}}N$ belongs to some common facet, and so they commute. It follows that $W(\mathcal{D})/N$ is elementary abelian, of order at most 2^{n+1}. □

By Lemma 8F11, if $\mathcal{K}$ is a regular n-polytope of type $\{3^{n-3}, 4, 3\}$ with cross-polytopes $\{3^{n-3}, 4\}$ as facets, then

$$\Gamma(2^{\mathcal{K},\mathcal{D}})/N \cong C_2^m \rtimes \Gamma(\mathcal{K})$$

with $m \leqslant n+1$. Using the natural identification of the automorphism group of C_2^m with the general linear group $\mathrm{GL}_m(2)$ over $\mathrm{GF}(2)$, this semi-direct product defines a representation

$$r\colon \Gamma(\mathcal{K}) \to \mathrm{GL}_m(2).$$

In particular, the generators $\tau_0, \ldots, \tau_{n-1}$ of $\Gamma(\mathcal{K})$ are represented by $m \times m$-matrices over $\mathrm{GF}(2)$. If $m = n+1$, then the generators $\sigma_{i_1}N, \ldots, \sigma_{i_{n+1}}N$ of $W(\mathcal{D})/N$ correspond to a basis of the $(n+1)$-dimensional vector space over $\mathrm{GF}(2)$, so that we can express $\tau_0, \ldots, \tau_{n-1}$ as matrices with respect to this basis. We shall use this approach later to prove existence and non-existence of certain polytopes.

8F12 Theorem *For $n \geqslant 3$, let $\mathcal{K}$ be a regular n-polytope of type $\{3^{n-3}, 4, 3\}$ with facets $\mathcal{F} = \{3^{n-3}, 4\}$, and let N be as in (8F8). If the $(n+1)$-polytope $\{\{4, 3^{n-3}, 4\}_{(2,0^{n-2})}, \mathcal{K}\}$ exists (with group $C_2^{n-1} \rtimes \Gamma(\mathcal{K})$), then $W(\mathcal{D})/N \cong C_2^m$ with $m = n+1$ or $n-1$. In particular, the $(n+1)$-polytope $\{\{4, 3^{n-3}, 4\}_{(2,2,0^{n-3})}, \mathcal{K}\}$ also exists if and only if $W(\mathcal{D})/N \cong C_2^{n+1}$, in which case its group is $C_2^{n+1} \rtimes \Gamma(\mathcal{K})$.*

Proof. Let $\tilde{N}$ be the normal subgroup defined as in (8F1), and let $\mathcal{G}_{\mathcal{K}}$ be the associated graph. By Theorem 8F3, $\omega(\mathcal{G}_{\mathcal{K}}) = n-1$ and $W(\mathcal{D})/\tilde{N} \cong C_2^{n-1}$. Clearly there is a homomorphism of $\Gamma(2^{\mathcal{K},\mathcal{D}})/N = \langle \rho_0, \ldots, \rho_n \rangle$ onto the group $\Gamma(2^{\mathcal{K},\mathcal{D}})/\tilde{N}$ of the polytope $\{\{4, 3^{n-3}, 4\}_{(2,0^{n-2})}, \mathcal{K}\}$. By the quotient criterion of Theorem 2E17, if the subgroup $\langle \rho_0, \ldots, \rho_{n-1} \rangle$ is isomorphic to $[4, 3^{n-3}, 4]_{(2,2,0^{n-3})}$, then $\{\{4, 3^{n-3}, 4\}_{(2,2,0^{n-3})}, \mathcal{K}\}$ exists, and $\Gamma(2^{\mathcal{K},\mathcal{D}})/N$ is its group. In this case, $N \neq \tilde{N}$, and the canonical projection

$$W(\mathcal{D})/N \to W(\mathcal{D})/\tilde{N}$$

is not an isomorphism. Conversely, if the polytope $\{\{4, 3^{n-3}, 4\}_{(2,2,0^{n-3})}, \mathcal{K}\}$ exists, then the subgroup $\langle \rho_0, \ldots, \rho_{n-1} \rangle$ of its group $\Gamma(2^{\mathcal{K},\mathcal{D}})/N$ is isomorphic to $[4, 3^{n-3}, 4]_{(2,2,0^{n-3})}$.

Let $\{i_1, \ldots, i_{n-2}\}$ be an $(n-3)$-face of $\mathcal{K}$, and let $\{i_1, \ldots, i_{n-2}, j\}$ with $j = i_{n-1}, i_n, i_{n+1}$ be the surrounding $(n-2)$-faces. By Lemma 8F11,

$$W(\mathcal{D})/N = \left\langle \sigma_{i_1}N, \ldots, \sigma_{i_{n+1}}N \right\rangle,$$

an elementary abelian group of order at most 2^{n+1}. The previous arguments show that its order is at least 2^{n-1}.

To begin with, we assume that

$$\langle \rho_0, \ldots, \rho_{n-1} \rangle \cong [4, 3^{n-3}, 4]_{(2,2,0^{n-3})}.$$

Then the subgroup of $W(\mathcal{D})/N$ induced by the base facet $\mathcal{F}$ of $\mathcal{K}$ has order 2^n, because it defines the base facet of $\{\{4, 3^{n-3}, 4\}_{(2,2,0^{n-3})}, \mathcal{K}\}$. We shall prove that $W(\mathcal{D})/N \cong C_2^{n+1}$.

First note that the stabilizer in $\Gamma(\mathcal{K})$ of the $(n-3)$-face $\{i_1, \ldots, i_{n-2}\}$ acts on the $n+1$ generators $\sigma_{i_1}N, \ldots, \sigma_{i_{n+1}}N$ like a group $S_{n-2} \times S_3$. If the order is not 2^{n+1}, then there is a non-trivial relation on the generators, and any such relation involves all the first $n-2$ generators or all the last three. In fact, if $i, j \in \{i_1, \ldots, i_{n-2}\}$ or $i, j \in \{i_{n-1}, i_n, i_{n+1}\}$, and if $\sigma_i N$ is involved but $\sigma_j N$ is not, then the transposition $(i\ j)$ in $S_{n-2} \times S_3$ maps the given relation onto a new relation in which i is replaced by j; taking the product gives the relation $\sigma_i \sigma_j N = N$, or $\sigma_i \sigma_j \in N$. But no relation can involve only the first n generators, so that this gives a contradiction if $i, j \in \{i_1, \ldots, i_{n-2}\}$. If $i, j \in \{i_{n-1}, i_n, i_{n+1}\}$, then σ_i, σ_j generate $\tilde{N}$ as a normal subgroup, so that $\sigma_i \sigma_j \in N$ implies that $N = \tilde{N}$, again a contradiction.

This leaves us with two possible relations. We prove that each leads to a contradiction modulo $\tilde{N}$. The first, $\sigma_{i_{n-1}} \sigma_{i_n} \sigma_{i_{n+1}} N = N$, cannot occur, because modulo $\tilde{N}$ the three generators are the same but are not trivial. The second, $\sigma_{i_1} \cdots \sigma_{i_{n+1}} N = N$, implies that

$$\sigma_{i_1} \cdots \sigma_{i_{n-1}} \tilde{N} = \sigma_{i_1} \cdots \sigma_{i_{n-1}} \sigma_{i_n} \sigma_{i_{n+1}} \tilde{N} = \tilde{N};$$

again this is a contradiction, because modulo $\tilde{N}$ there cannot be a relation involving only the first $n-1$ generators. It follows that $W(\mathcal{D})/N$ must indeed have order 2^{n+1}.

Finally, if the subgroup $\langle \rho_0, \ldots, \rho_{n-1} \rangle$ is a proper quotient of $[4, 3^{n-3}, 4]_{(2,2,0^{n-3})}$, then it must be isomorphic to $[4, 3^{n-3}, 4]_{(2,0^{n-2})}$. Now the subgroup $\langle \sigma_{i_1}N, \ldots, \sigma_{i_n}N \rangle$ representing a facet of $\mathcal{K}$ can already be generated by the generators $\sigma_{i_1}N, \ldots, \sigma_{i_{n-1}}N$ corresponding to the $(n-2)$-face $\{i_1, \ldots, i_{n-1}\}$ of this facet. It follows that this is so for all subgroups representing facets, and therefore

$$W(\mathcal{D})/N = \langle \sigma_{i_1}N, \ldots, \sigma_{i_{n-1}}N \rangle \cong C_2^{n-1}$$

(and $N = \tilde{N}$). This completes the proof. □

8F13 Corollary *For $n \geqslant 3$, the $(n+1)$-polytope* $\{\{4, 3^{n-3}, 4\}_{(2,2,0^{n-3})}, \{3^{n-3}, 4, 3\}\}$ *exists, and has group* $C_2^{n+1} \rtimes [3^{n-3}, 4, 3]$.

Proof. Apply Theorem 8F12 with $\mathcal{K} = \{3^{n-3}, 4, 3\}$, and use Corollary 8F5. We need to prove that $W(\mathcal{D})/N \cong C_2^{n+1}$.

Assume for now that $W(\mathcal{D})/N \cong C_2^{n+1}$, and consider the representation

$$r \colon \Gamma(\mathcal{K}) = [3^{n-3}, 4, 3] \longrightarrow \mathrm{GL}_{n+1}(2).$$

Let $i_1, \ldots, i_{n+1}$ be as in Lemma 8F11. Let $F_{n-1}^1, F_{n-1}^2, F_{n-1}^3$ be the facets of $\mathcal{K}$ surrounding the $(n-3)$-face $\{i_1, \ldots, i_{n-2}\}$, such that $\{i_1, \ldots, i_{n-2}, i_{n-2+j}\}$ and $\{i_1, \ldots, i_{n-2}, i_{n-1+j}\}$ are $(n-2)$-faces of F_{n-1}^j (with $j+1$ taken modulo 3). For $l = 1, \ldots, n-2$ and $j = 1, 2, 3$, let $i_{l,j}$ denote the vertex of F_{n-1}^j antipodal to i_l. The generators $\sigma_{i_1}N, \ldots, \sigma_{i_{n+1}}N$ of $W(\mathcal{D})/N$ correspond to a basis of $\mathrm{GF}(2)^{n+1}$, and the images under r of the generators $\tau_0, \ldots, \tau_{n-1}$ of $\Gamma(\mathcal{K})$ are uniquely determined by their effect on this basis. Thus, to find the matrices for $\tau_0 r, \ldots, \tau_{n-1} r$, it suffices to consider how $\tau_0, \ldots, \tau_{n-1}$ act on $\sigma_{i_1} N, \ldots, \sigma_{i_{n+1}} N$, with identification modulo N understood.

Using the geometry of $\mathcal{K}$, we can express $\tau_0, \ldots, \tau_{n-1}$ as permutations of $i_1, \ldots, i_{n+1}$ and of some of the $i_{l,j}$, as follows:

8F14 $$\tau_p = \begin{cases} (i_{p+1}\, i_{p+2}), & \text{if } 0 \leqslant p \leqslant n-1 \text{ and } p \neq n-3, \\ (i_{n-2}\, i_{n-1})(i_n\, i_{n-2,1})(i_{n+1}\, i_{n-2,3}), & \text{if } p = n-3. \end{cases}$$

The matrices of $\tau_p r$ with $p \neq n-3$ are now just the corresponding $(n+1) \times (n+1)$ permutation matrices. However, for τ_{n-3} we need the relations $\sigma_{i_{n-2,1}} N = \sigma_{i_{n-2}} \sigma_{i_{n-1}} \sigma_{i_n} N$ and $\sigma_{i_{n-2,3}} N = \sigma_{i_{n-2}} \sigma_{i_{n-1}} \sigma_{i_{n+1}} N$, to find the $(n+1) \times (n+1)$-matrix

$$\left[\begin{array}{c|c|c} I_{n-3} & & \\ \hline & K_2 & J_2 \\ \hline & & I_2 \end{array}\right]$$

of $\tau_{n-3} r$. Here, only non-zero entries are indicated; in particular, I_k is the $k \times k$ identity matrix, and

$$K_2 := \begin{bmatrix} 0 & 1 \\ 1 & 0 \end{bmatrix}, \qquad J_2 := \begin{bmatrix} 1 & 1 \\ 1 & 1 \end{bmatrix}.$$

Hence, if $W(\mathcal{D})/N \cong C_2^{n+1}$, then the representation r is given by these matrices for $\tau_0, \ldots, \tau_{n-1}$.

We can now proceed as follows. Since the given matrices satisfy all the defining relations for the Coxeter group $[3^{n-3}, 4, 3]$, they therefore define a representation

$$r' \colon \Gamma(\mathcal{K}) \to \mathrm{GL}_{n+1}(2),$$

and thus a semi-direct product $C_2^{n+1} \rtimes \Gamma(\mathcal{K})$. This semi-direct product satisfies all the defining relations for the group of $\{\{4, 3^{n-3}, 4\}_{(2,2,0^{n-3})}, \mathcal{K}\}$, and so must be isomorphic to this group. It follows that the polytope exists and has group $C_2^{n+1} \rtimes \Gamma(\mathcal{K})$, as required; further, $r' = r$. □

8F15 Corollary *The regular 5-polytope* $\{\{4, 3, 4\}_{(2,2,0)}, \{3, 4, 3\}\}$ *exists and is finite. Its group is* $C_2^5 \rtimes [3, 4, 3]$*, of order* 36864.

Proof. This result restates Corollary 8F13 for $\mathcal{K} = \{3, 4, 3\}$. □

8F16 Corollary *The regular 6-polytope* $\{\{4, 3, 3, 4\}_{(2,2,0,0)}, \{3, 3, 4, 3\}_{(t^k,0^{4-k})}\}$*, with* $t \geqslant 2$ *and* $k = 1$ *or* 2*, exists if and only if* t *is even. In this case, the polytope is finite, and its group is* $C_2^6 \rtimes [3, 3, 4, 3]_{(t^k,0^{4-k})}$ *of order* $73728k^2t^4$.

Proof. Now $\mathcal{K} = \{3, 3, 4, 3\}_{(t^k, 0^{4-k})}$ (see Section 6E) and $n = 5$. If t is even, then the covering

$$\mathcal{K} \searrow \{3, 3, 4, 3\}_{(2,0,0,0)}$$

induces a surjective homomorphism $W(\mathcal{D})/N \to W(\mathcal{D}')/N'$, with $\mathcal{D}'$ and N' the diagram and normal subgroup defined with respect to $\{3, 3, 4, 3\}_{(2,0,0,0)}$. The two polytopes

$$\big\{\{4, 3, 3, 4\}_{(2,0,0,0)}, \mathcal{K}\big\}, \qquad \big\{\{4, 3, 3, 4\}_{(2,0,0,0)}, \{3, 3, 4, 3\}_{(2,0,0,0)}\big\}$$

exist by Corollary 8F7; thus Theorem 8F12 applies. In particular, if $W(\mathcal{D}')/N' \cong C_2^6$, then $W(\mathcal{D})/N \cong C_2^6$, and the polytope $\{\{4, 3, 3, 4\}_{(2,2,0,0)}, \mathcal{K}\}$ also exists, with group $C_2^6 \rtimes \Gamma(\mathcal{K})$. But the case $\{3, 3, 4, 3\}_{(2,0,0,0)}$ can now be handled directly, using the same arguments as for $\{3, 3, 4, 3\}$ in the proof of Corollary 8F13. In particular, $W(\mathcal{D}')/N' \cong C_2^6$; to confirm this, the Todd–Coxeter algorithm predicts 2^6 vertices for the polytope. This settles the case where t is even.

Now let t be odd and $k = 2$. Again, the regular polytope $\{\{4, 3, 3, 4\}_{(2,0,0,0)}, \mathcal{K}\}$ exists by Corollary 8F7, so that Theorem 8F12 implies that $W(\mathcal{D})/N \cong C_2^m$ with $m = 4$ or 6. For the non-existence of $\{\{4, 3, 3, 4\}_{(2,2,0,0)}, \mathcal{K}\}$, it suffices to prove that $m \neq 6$.

Assume that $m = 6\ (= n + 1)$. We can now use the representation $r\colon \Gamma(\mathcal{K}) \to \mathrm{GL}_6(2)$, as in the proof of Corollary 8F13. In particular, (8F14) still holds (with $n = 5$), as does the representation of $\tau_0 r, \ldots, \tau_{n-1} r$ by matrices.

Once the matrices are found, we can proceed as follows. Consider the element

$$\alpha := (\tau_0 \cdot \tau_1\tau_2\tau_3\tau_2\tau_1 \cdot \tau_4\tau_3\tau_2\tau_3\tau_4)^2,$$

which is a "translation" of $\mathcal{K}$. By Theorem 6E6, this has order t; indeed, since $k = 2$, we know that $\alpha^t = \varepsilon$ is the only extra defining relation for $\Gamma(\mathcal{K})$. It is now straightforward to compute the matrix of αr using the matrices for $\tau_0 r, \ldots, \tau_{n-1} r$. However, we find that this matrix has period 2. This is a contradiction, because the period must also divide the period t of α. It follows that we cannot have $m = 6$. This settles the case where t is odd and $k = 2$.

Finally, let t be odd and $k = 1$. Now we cannot appeal to Corollary 8F7 and Theorem 8F12. Assume that

$$\big\{\{4, 3, 3, 4\}_{(2,2,0,0)}, \{3, 3, 4, 3\}_{(t,0,0,0)}\big\}$$

exists. Then, writing $\Gamma(2^{\mathcal{K},\mathcal{D}})/N =: \langle \rho_0, \ldots, \rho_5 \rangle$, we have $\langle \rho_0, \ldots, \rho_4 \rangle \cong [4, 3, 3, 4]_{(2,2,0,0)}$, so that the subgroup of $W(\mathcal{D})/N$ induced by a facet of $\mathcal{K}$ has order 2^5. Hence $W(\mathcal{D})/N \cong C_2^m$ with $m = 5$ or 6. Now the assumption $m = 6$ can be refuted as before using the same translation α. Note here that α still has order t, although $\alpha^t = \varepsilon$ is not the extra defining relation if $k = 1$.

If $m = 5$, we can argue as in the proof of Theorem 8F12 that any non-trivial relation on the generators $\sigma_{i_1} N, \ldots, \sigma_{i_6} N$ must involve all of the first three generators, or all of the last three. (Note that now $\omega(\mathcal{G}_{\mathcal{K}}) = 1$ and $W(\mathcal{D})/\tilde{N} \cong C_2$; hence $N \neq \tilde{N}$.) But any relation involving only three generators leads to a contradiction, because modulo $\tilde{N}$ the generators are the same but are not trivial. Hence we are left with $\sigma_{i_1} \cdots \sigma_{i_6} N = N$, or, equivalently, $\sigma_{i_6}N = \sigma_{i_1}\cdots\sigma_{i_5}N$. In particular,

$$W(\mathcal{D})/N = \langle \sigma_{i_1}N, \ldots, \sigma_{i_5}N \rangle.$$

We can now work with a representation $r\colon \Gamma(\mathcal{K}) \to \mathrm{GL}_5(2)$. Using the same notation F_4^j $(= F_{n-1}^j)$ and $i_{l,j}$ as before, and observing that $\tau_p r$ is now uniquely determined by the effect τ_p has on $\sigma_{i_1}N, \ldots, \sigma_{i_5}N$, we arrive at the following permutations: $\tau_0, \tau_1, \tau_3, \tau_4$ are as in (8F14) with $n = 5$, and $\tau_2 = (i_3\, i_4)(i_5\, i_{3,1})$. Now τ_0, τ_1, τ_3 are represented by the corresponding 5×5 permutation matrices. For τ_2 and τ_4, we can use the relations $\sigma_{i_{3,1}}N = \sigma_{i_3}\sigma_{i_4}\sigma_{i_5}N$ and $\sigma_{i_6}N = \sigma_{i_1}\cdots\sigma_{i_5}N$ to find the matrices

$$\begin{bmatrix} 1 & 0 & 0 & 0 & 0 \\ 0 & 1 & 0 & 0 & 0 \\ 0 & 0 & 0 & 1 & 1 \\ 0 & 0 & 1 & 0 & 1 \\ 0 & 0 & 0 & 0 & 1 \end{bmatrix}, \quad \begin{bmatrix} 1 & 0 & 0 & 0 & 1 \\ 0 & 1 & 0 & 0 & 1 \\ 0 & 0 & 1 & 0 & 1 \\ 0 & 0 & 0 & 1 & 1 \\ 0 & 0 & 0 & 0 & 1 \end{bmatrix},$$

respectively. If α is the same translation as before, then the matrix of αr again has period 2, which gives a contradiction. It follows that we cannot have $m = 5$. This completes the proof. □

For Corollary 8F16, note that faster non-existence proofs are available if t is an odd prime with $t \neq 3, 5, 7, 31$. Let $\mathcal{K} := \{3, 3, 4, 3\}_{(t,0,0,0)}$. By Lemma 8F11, we have a representation $r\colon \Gamma(\mathcal{K}) \to \mathrm{GL}_m(2)$ with $m \leqslant 6$. In $\mathcal{K}$, the vertex i_1 can be mapped to any of its neighbours by a "translation" β (say) of $\mathcal{K}$. If t is a prime, then βr is trivial or has order t; in the latter case, t divides the order of $\mathrm{GL}_m(2)$. But for $2 \leqslant m \leqslant 6$, the only odd primes t which can divide the order of $\mathrm{GL}_m(2)$ are 3, 5, 7 or 31, and these are excluded (see [12]). On the other hand, $i_2, \ldots, i_6$ are neighbours of i_1; hence, if $m \geqslant 2$ and β maps i_1 to a vertex from $i_2, \ldots, i_6$, then βr cannot be trivial. It follows that, for odd primes t with $t \neq 3, 5, 7, 31$, we must have $m = 1$; that is, $W(\mathcal{D})/N \cong C_2$ (and $N = \tilde{N}$).

For $\mathcal{K} = \{3, 3, 4, 3\}_{(t,t,0,0)}$, we can use instead a "translation" γ (say) of $\mathcal{K}$ which maps i_4 to i_5 or i_6. Then γr cannot be trivial if $m = 5$ or 6. Hence for odd primes t with $t \neq 3, 5, 7, 31$ we must have $m \leqslant 4$; thus $\{\{4, 3, 3, 4\}_{(2,2,0,0)}, \mathcal{K}\}$ cannot exist.

Identification Vector (2^{n-1})

Finally we discuss quotients whose facets are $\{4, 3^{n-3}, 4\}_{(2^{n-1})}$, with identification vector (2^{n-1}). This is the hardest of the three cases. In our examples, we have $n = 4$ or 5. If $\mathcal{F} = \{3, 4\}$, then in $\Gamma(2^{\mathcal{F}})$ we have

$$(\rho_0\rho_1\rho_2\rho_3)^6 = \sigma_0\sigma_1\cdots\sigma_5 = \sigma_0\sigma_3\cdots\sigma_1\sigma_4\cdots\sigma_2\sigma_5$$

(the order of these terms is immaterial). Hence, by Theorem 6D4, to obtain facets $\{4, 3, 4\}_{(2,2,2)}$ we must choose the normal subgroup N so that $\sigma_0\cdots\sigma_5 \in N$. A similar remark applies to facets $\{4, 3^{n-3}, 4\}_{(2^{n-1})}$, using

$$(\rho_0\rho_1\cdots\rho_{n-1})^{2n-2} = \sigma_0\sigma_{n-1}\sigma_1\sigma_n\cdots\sigma_{n-2}\sigma_{2n-3}$$

(again see Theorem 6D4). If F is a facet of $\mathcal{K}$, define σ_F in $W(\mathcal{D})$ by

$$\sigma_F := \prod_{i \in F} \sigma_i$$

(with the σ_i in any order). Then we define

8F17 $$N := \langle \varphi \sigma_F \varphi^{-1} \mid \varphi \in W(\mathcal{D}),\ F \text{ a facet of } \mathcal{K} \rangle,$$

the normal closure of σ_F in $\Gamma(2^{\mathcal{K},\mathcal{D}})$. As before, we have

$$\Gamma(2^{\mathcal{K},\mathcal{D}}/N) \cong W(\mathcal{D})/N \rtimes \Gamma(\mathcal{K}).$$

By construction, it is also clear that, if the universal polytope in question does exist, then it must be isomorphic to the quotient $2^{\mathcal{K},\mathcal{D}}/N$.

At present, we do not know the structure of the group of every regular 6-polytope

$$\{\{4, 3, 3, 4\}_{(2,2,2,2)}, \{3, 3, 4, 3\}_{(t^k, 0^{4-k})}\}$$

with $t \geqslant 2$ and $k = 1, 2$ (see Section 6E). For $(t, k) = (2, 1)$, the polytope is covered by the finite polytope

$$\{\{4, 3, 3, 4\}_{(4,0,0,0)}, \{3, 3, 4, 3\}_{(2,0,0,0)}\}$$

of Corollary 8E15, whose group is the semi-direct product $C_2^{16} \rtimes [3, 3, 4, 3]_{(2,0,0,0)}$. This fact is used in our next theorem.

8F18 Theorem *The regular* 6*-polytope* $\{\{4, 3, 3, 4\}_{(2,2,2,2)}, \{3, 3, 4, 3\}_{(2,0,0,0)}\}$ *exists and is finite. Its group is* $C_2^{10} \rtimes [3, 3, 4, 3]_{(2,0,0,0)}$, *of order* 188 74368.

Proof. We use the geometry of $\mathcal{K} = \{3, 3, 4, 3\}_{(2,0,0,0)}$ to show that $W(\mathcal{D})/N \cong C_2^{10}$. Our first observation is that the vertices of the vertex-figure $\{3, 4, 3\}$ of $\mathcal{K}$ coincide in antipodal pairs, so that $\mathcal{K}$ has the same vertices as those of the (locally projective) 5-polytope $\{\{3, 3, 4\}, \{3, 4, 3\}_6\}$ (see [287] or Section 14A). Recall that $\{3, 4, 3\}_6$ is the regular 4-polytope obtained from $\{3, 4, 3\}$ by identifying antipodal points; here the number 6 indicates the extra relation $(\rho_0\rho_1\rho_2\rho_3)^6 = \varepsilon$ for the group $[3, 4, 3]_6$ or, equivalently, the length of the Petrie polygon (see Section 6C). The 48 facets of $\mathcal{K}$ occur in pairs with the same vertices, so in what follows we are really referring to 24 pairs of facets.

For simplicity, we change our convention for the vertex labels in $\mathcal{F}$, so that we now label the 16 vertices of $\mathcal{K}$ by $0, \ldots, 15$, which we express as $a\,b$ in base 4 (with $a, b \in \{0, \ldots, 3\}$). Edges of $\mathcal{K}$ join two vertices with different labels a; vertices with the same label a are antipodal in some facet. If $e = \{b_0, b_1\} \subset \{0, \ldots, 3\}$ with $b_0 \neq b_1$, then we write $\bar{e} = \{0, \ldots, 3\} \setminus e$ for its complement. Then $a\,e$ stands for the pair $\{a\,b_0, a\,b_1\}$ of antipodal vertices, and so on. A (double) facet will have vertices $a\,e$ or $a\,\bar{e}$, where $a = 0, \ldots, 3$ and $e = \{0, 1\}, \{0, 2\}$ or $\{0, 3\}$; that is, the pairs e and $\bar{e}$ always go together. The 24 (double) facets then have:

$$e = \{0, 1\} \text{ or } \{0, 2\} \ - \text{ even number of } \bar{e}\text{ s};$$
$$e = \{0, 3\} \ - \text{ odd number of } \bar{e}\text{ s}.$$

The ten generators of $W(\mathcal{D})$ correspond to the vertices

$$00,\ 01,\ 03,\ 10,\ 11,\ 13,\ 20,\ 21,\ 23,\ 30.$$

The remainder are constructed by successively listing suitable facets:

$$\begin{aligned}
\{00,\ 01,\ 10,\ 11,\ 20,\ 21,\ 30,\ 31\} &\mapsto 31,\\
\{00,\ 03,\ 10,\ 13,\ 20,\ 23,\ 31,\ 32\} &\mapsto 32,\\
\{01,\ 03,\ 11,\ 13,\ 21,\ 23,\ 31,\ 33\} &\mapsto 33,\\
\{01,\ 02,\ 10,\ 13,\ 20,\ 23,\ 30,\ 33\} &\mapsto 02,\\
\{00,\ 03,\ 11,\ 12,\ 20,\ 23,\ 30,\ 33\} &\mapsto 12,\\
\{00,\ 03,\ 10,\ 13,\ 21,\ 22,\ 30,\ 33\} &\mapsto 22.
\end{aligned}$$

Bear in mind here that the product of any seven generators of $W(\mathcal{D})/N$ corresponding to vertices of a facet is the eighth. Of course, since $\mathcal{K}$ is weakly neighbourly, all generators commute. It follows that $W(\mathcal{D})/N$ has order at most 2^{10}, and a tedious check shows that no nine generators suffice. Finally, for the existence of the polytope we can appeal to Theorem 2E17 and Corollaries 8F7 or 8F16, using the fact that the subgroup of $W(\mathcal{D})/N$ induced by a facet of $\mathcal{K}$ has order 2^7. This completes the proof. □

We now further restrict our attention to polytopes of rank 5 with facets isomorphic to toroids $\{4, 3, 4\}_{(2,2,2)}$. We shall prove in Section 12B that there are only three universal regular 5-polytopes of type $\{4, 3, 4, 3\}$ (or dual type $\{3, 4, 3, 4\}$) which are finite. Together with Corollaries 8F6 and 8F15, the next theorem covers all of them. It is one of the most interesting examples obtained by the method of this section.

8F19 Theorem *The regular 5-polytope* $\{\{4, 3, 4\}_{(2,2,2)}, \{3, 4, 3\}\}$ *exists and is finite. Its group is* $(C_2^6 \rtimes C_2^5) \rtimes [3, 4, 3]$, *of order* 23 59296; *in the semi-direct product* $C_2^6 \rtimes C_2^5$, *the factor* C_2^6 *is its own centralizer.*

Proof. Now $\mathcal{K} = \{3, 4, 3\}$ and $\mathcal{F} = \{3, 4\}$. We show that $W(\mathcal{D})/N \cong C_2^6 \rtimes C_2^5$, a group of order 2^{11}.

First, we show that $W(\mathcal{D})/N$ is generated by 10 involutions. Let $0, \ldots, 23$ be the vertices of $\mathcal{K}$, in such a way that $l, l + 5$ $(l = 1, \ldots, 4)$ are the vertices of the cubical vertex-figure of $\mathcal{K}$ at 0 (this is a change from the previous notation). Define $I := \{1, \ldots, 9\}\backslash\{5\}$. By the definition of N, if k is the antipodal vertex to 0 in some facet, then $\sigma_k N \in \langle \sigma_l N \mid l \in I \cup \{0\}\rangle$. Let 5 be a vertex of the vertex-figure of the opposite vertex of $\mathcal{K}$ to 0. We claim that

$$W(\mathcal{D})/N = \langle \sigma_0 N, \ldots, \sigma_9 N\rangle.$$

Note first that, if i, j are adjacent vertices of this opposite vertex-figure, then $\{i, j, p, q, r, s\}$ is a facet for some $p, q \in I$ and r, s antipodal to 0 in facets of $\mathcal{K}$. It follows that

$$\sigma_i N \in \langle \sigma_l N \mid l \in I \cup \{0, j\}\rangle.$$

Now, if we migrate along the edges in the opposite vertex-figure from 5, we get all its vertices i, and hence the corresponding σ_i. Finally, we obtain the opposite vertex of

$\mathcal{K}$ to 0 from any of the facets which contain it, whose remaining vertices are already accounted for. Hence the assertion follows.

Next, we prove that the commutator subgroup has order at most 2, with $(\sigma_0\sigma_5)^2 N$ the only possible non-trivial commutator. Clearly, $\sigma_0 N$ commutes with $\sigma_j N$ for $j \in I$. Now 0 and 5 are opposite vertices in the vertex-figure of one of the vertices in I. Hence, by the action of $\Gamma(\mathcal{K})$ on $W(\mathcal{D})/N$, if $\sigma_0 N$ and $\sigma_5 N$ commute, then so do $\sigma_i N$ and $\sigma_j N$ for all $i, j \in I$; thus $W(\mathcal{D})/N$ is abelian. Therefore, if $W(\mathcal{D})/N$ is not abelian, then $(\sigma_0\sigma_5)^2 N \neq N$.

Let i, j be adjacent vertices of the opposite vertex-figure, and let p, q, r, s be as before. Since σ_0 commutes with $\sigma_p, \sigma_q, \sigma_r, \sigma_s$, it follows that $\sigma_0 N$ commutes with $\sigma_p\sigma_q\sigma_r\sigma_s N = \sigma_i\sigma_j N$, or $\sigma_0\sigma_i\sigma_j N = \sigma_i\sigma_j\sigma_0 N$. But

$$\sigma_0\sigma_i\sigma_j N = \sigma_i(\sigma_i\sigma_0)^2(\sigma_0\sigma_j)^2\sigma_j\sigma_0 N,$$

and equating this to $\sigma_i\sigma_j\sigma_0 N$ gives $(\sigma_i\sigma_0)^2(\sigma_0\sigma_j)^2 N = N$, or

$$(\sigma_0\sigma_i)^2 N = (\sigma_0\sigma_j)^2 N.$$

Starting with the vertex 5, and iterating such relations along edges $\{i, j\}$ of the opposite vertex-figure, gives $(\sigma_0\sigma_j)^2 N = (\sigma_0\sigma_5)^2 N$ for each of its vertices j. However, 0 and 5 are themselves antipodal vertices of the vertex-figure of one of the vertices in I. Thus, if we choose l to be antipodal to 5 in the opposite vertex-figure, then 0, 5, l are symmetrically related, so that

$$(\sigma_0\sigma_5)^2 N = (\sigma_0\sigma_l)^2 N = (\sigma_5\sigma_l)^2 N = (\sigma_5\sigma_0)^2 N,$$

or

$$(\sigma_0\sigma_5)^4 N = N.$$

Chasing arguments of this kind (using symmetry and connectivity properties of $\mathcal{K}$) show that $(\sigma_0\sigma_5)^2 N$ can indeed be the only non-trivial commutator between pairs of generators $\sigma_j N$; it then follows that the commutator subgroup is of order at most 2. In particular,

$$\kappa := (\sigma_0\sigma_5)^2 N$$

lies in the centre of $W(\mathcal{D})/N$. Note that the 10 generators thus commute in pairs, except for the five disjoint pairs $j, j+5$ ($j = 0, \ldots, 4$), for which $(\sigma_j\sigma_{j+5})^2 N = \kappa$.

We now prove that $W(\mathcal{D})/N$ is not abelian, and that

$$W(\mathcal{D})/N \cong C_2^6 \rtimes C_2^5.$$

First note that the vertex 5 in $\mathcal{K}$ is joined by an edge to exactly one vertex, 6 (say), of the vertex-figure at 0. Let 7, 8, 9 be the neighbouring vertices to 6 of this vertex-figure. If $j \in \{6, \ldots, 9\}$, then the vertices j and 5 lie in a common facet, so that $\sigma_j N$ and $\sigma_5 N$ commute. It follows that

$$B := \langle \sigma_5 N, \ldots, \sigma_9 N \rangle$$

is an abelian subgroup generated by (at most) 5 involutions.

Let $Z := W(\mathcal{D})/N$ and $\alpha_j := \sigma_j N$ for $j = 0, \ldots, 9$. Then, for all i, j, either $\alpha_j \alpha_i \alpha_j = \alpha_i$ or $\alpha_j \alpha_i \alpha_j = \kappa \alpha_i$. Now

$$A := \langle \alpha_0, \ldots, \alpha_4, \kappa \rangle$$

is a normal abelian subgroup of Z generated by (at most) 6 involutions, and the canonical projection $\pi\colon Z \to Z/A$ takes $B := \langle \alpha_5, \ldots, \alpha_9 \rangle$ onto Z/A. In particular, $|Z| \leqslant 2^{11}$.

Assume for the moment that we indeed have $|Z| = 2^{11}$. Then $A \cong C_2^6$ and $B \cong Z/A \cong C_2^5$, because the numbers of generators of A and B are at most 6 and 5, respectively. It follows that

$$Z = A \rtimes B \cong C_2^6 \rtimes C_2^5.$$

Consider now how B acts on A by conjugation. Here it is helpful to observe that we can identify A with a 6-dimensional vector space over GF(2), with basis vectors $e_j := \alpha_j$ for $j = 0, \ldots, 4$ and $e_5 := \kappa$ identified with those of the standard (row) basis, and represent conjugation with an element in B by a 6×6 matrix over GF(2). Then it is immediate that the generators α_j ($j = 5, \ldots, 9$) correspond to the matrices

8F20
$$\begin{bmatrix} I_5 & 0 \\ e_{j-5} & 1 \end{bmatrix},$$

with I_5 the 5×5-identity matrix. In particular, this implies that conjugation gives a faithful representation $B \mapsto \mathrm{Aut}(A)$ (where $\mathrm{Aut}(A)$ denotes the group of group automorphisms of A), or, equivalently, that A is its own centralizer in Z.

On the other hand, we can now complete the proof by observing that there is indeed a specific group $C_2^6 \rtimes C_2^5$, with the action of C_2^5 on C_2^6 defined by (8F20), which satisfies the given relations. Hence there can be no possibility of our group Z collapsing onto a group of smaller order than 2^{11}, since Z is just determined by these relations. The existence of the polytope now follows from Theorem 2E17 and Corollary 8F6 or 8F15, using the fact that the subgroup of $W(\mathcal{D})/N$ induced by a facet of $\mathcal{K}$ has order 2^5. □

We conclude the section with an application of our method to the projective 4-polytope $\{3, 4, 3\}_6$, giving us 5-polytopes with toroidal facets and projective vertex-figures (for the notation, see above or Section 6C).

8F21 Theorem *For* $k = 1, 2, 3$, *the 5-polytope* $\{\{4, 3, 4\}_{(2^k, 0^{3-k})}, \{3, 4, 3\}_6\}$ *exists and is finite. Its group is* $C_2^{m(k)} \rtimes [3, 4, 3]_6$, *with* $m(k) = 3, 5, 8$ *as* $k = 1, 2, 3$, *respectively.*

Proof. Let $\mathcal{K} := \{3, 4, 3\}_6$. Then Theorem 8F3 gives the polytope for $k = 1$, with group $C_2^3 \rtimes \Gamma(\mathcal{K})$; here we use the fact that $\omega(\mathcal{G}_{\mathcal{K}}) = 3$. If $k = 2$, Theorem 8F12 suggests that the group should be $C_2^5 \rtimes \Gamma(\mathcal{K})$. Indeed, as in the proof of Corollary 8F13, we can use the representation $\Gamma(\mathcal{K}) \mapsto \mathrm{GL}_5(2)$ defined by (8F14) (with $n = 4$) to identify $C_2^5 \rtimes \Gamma(\mathcal{K})$ as the group of the regular polytope $\{\{4, 3, 4\}_{(2,2,0)}, \mathcal{K}\}$.

Last, let $k = 3$. Since $\mathcal{K}$ is weakly neighbourly, $W(\mathcal{D})/N$ is an elementary abelian 2-group. Let $1, \ldots, 8$ be the vertices of the vertex-figure at 0, with opposite faces $\{1, \ldots, 4\}$, $\{5, \ldots, 8\}$. (Again, we have changed our earlier notation.) Let 9 be such that $\{0, \ldots, 4, 9\}$ is a facet of $\mathcal{K}$. Then $\{0, 5, \ldots, 9\}$ is also a facet of $\mathcal{K}$. Thus we can choose $\sigma_0 N, \ldots, \sigma_7 N$ as generators, obtaining $\sigma_9 N$ from the first facet and $\sigma_8 N$ from the second. We obtain the two remaining generators $\sigma_{10} N$ and $\sigma_{11} N$ as we found $\sigma_9 N$. Clearly, fewer generators will not serve, so that $W(\mathcal{D})/N \cong C_2^8$. Finally, since the subgroup of $W(\mathcal{D})/N$ induced by a facet of $\mathcal{K}$ has order 2^5, Theorem 2E17 now proves the existence of the polytope. □

9

Unitary Groups and Hermitian Forms

In the classical theory, the structure of a finite regular polytope or an infinite regular tessellation or honeycomb is governed by a real quadratic form. This form determines the geometry of the ambient space. The symmetry group of the polytope or tessellation is a group of isometries in this space, and is in fact the Coxeter group associated with the quadratic form as in Sections 3A and 3D.

As we shall see in Chapter 11, much of the correspondence between polytopes and forms remains true for an important class of abstract regular polytopes. Here the real quadratic form is replaced by a hermitian form on a finite-dimensional complex space. Moreover, a subgroup of finite index in the automorphism group $\Gamma = \Gamma(\mathcal{P})$ of such a polytope $\mathcal{P}$ is represented as a group of isometries with respect to this form. In particular, this subgroup (and thus $\mathcal{P}$ itself) is finite if and only if the hermitian form is positive definite, in which case it is a finite group generated by reflexions of period 2 in the ambient unitary space.

Unfortunately, we cannot generally be sure that the abstract and geometric groups are actually isomorphic; this may have to be shown on a case-by-case basis. However, an alternative approach will often settle our problem. It may happen that $\Gamma(\mathcal{P})$ has some subgroup (of such a kind as previously discussed), a quotient of which can be shown to be infinite; it then follows that $\mathcal{P}$ itself is infinite. This then restricts the discussion to a handful of possibly finite cases, which can be dealt with on an individual basis.

In this chapter, we provide the necessary background for the material developed in Chapter 11 (see [306, 307]). We shall not attempt to survey here the literature on complex reflexion groups. Instead we refer the reader to the original work of Shephard and Todd [390], who were the first to classify completely all the finite unitary groups generated by reflexions, or to the work of Coxeter [114, 116, 126] or Cohen [78]. Here we are mainly interested in unitary groups generated by reflexions of period 2.

In Section 9A we review some general results on finite unitary reflexion groups. Then in Section 9B we discuss representations of groups with involutory generators as complex reflexion groups which preserve a hermitian form. In Section 9C, we introduce some general considerations, in particular those concerning changes of generators in a reflexion group. In Section 9D, the techniques are applied to study certain "generalized triangle groups" and their corresponding hermitian forms. In certain cases, these groups are abstractly defined by a presentation which corresponds to a diagram based on a suitably labelled triangle. In Sections 9E and 9F, we explore variants of this method and obtain similar classification results for groups with more specific diagrams. Finally, in Section 9G, we discuss the general question, which asks whether there might be a useful theory of complex representations of groups generated by involutions, whose relators involve at most three generators (and if three, of the particular kind considered in Section 9D). We shall see that the approach is quite powerful for specific classes of groups, but that examples described in earlier sections suggest that the general answer is negative.

9A Unitary Reflexion Groups

In complex n-space $\mathbb{C}^n$, any finite group of linear transformations leaves invariant a positive definite hermitian form, and is therefore conjugate, in the general linear group $\mathrm{GL}_n(\mathbb{C})$, to a subgroup of the group $\mathrm{U}_n(\mathbb{C})$ of all unitary transformations in standard unitary n-space, which is $\mathbb{C}^n$ equipped with the standard positive definite hermitian form

9A1 $$\langle x, y\rangle := \sum_{i=1}^{n} \xi_i \bar{\eta}_i, \qquad \text{for } x = (\xi_1, \ldots, \xi_n),\ y = (\eta_1, \ldots, \eta_n) \in \mathbb{C}^n.$$

After a suitable change of basis, such a group can be thought of as a finite group of unitary transformations in this space. Therefore, for finite groups, the study of groups of linear transformations over $\mathbb{C}$ is equivalent to that of unitary groups over $\mathbb{C}$. However, it is sometimes more convenient to work with the corresponding groups of matrices over $\mathbb{C}$ (always acting on row vectors from the right).

In the real case, the finite groups of orthogonal transformations in euclidean n-space have been well studied. Among such groups, those generated by hyperplane reflexions are the most interesting, and are precisely the finite Coxeter groups classified by Theorem 3B2. In euclidean space, a hyperplane reflexion must necessarily have period 2.

In unitary n-space $\mathbb{C}^n$, a (*linear hyperplane*) *reflexion* R is a unitary transformation distinct from the identity, and always taken here to be of finite period, which leaves fixed each point of a hyperplane through the origin. Equivalently, R is a unitary transformation of finite period all of whose eigenvalues, save one, are equal to unity; the remaining eigenvalue ζ is a primitive rth root of unity if the order of R is r ($\geqslant 2$). Any such reflexion can be written in the form

9A2 $$xR = x + \frac{(\zeta - 1)\langle x, v\rangle}{\langle v, v\rangle}\, v \qquad (x \in \mathbb{C}^n),$$

where v is a normal vector to the reflecting hyperplane.

Each finite group G of unitary transformations in $\mathbb{C}^n$ is either *irreducible* or *completely reducible*; that is, either there is no non-trivial invariant subspace (other than $\{o\}$ or $\mathbb{C}^n$ itself), or $\mathbb{C}^n$ is the orthogonal sum of minimal non-trivial invariant subspaces.

Moreover, if G is reducible and generated by reflexions, then it is necessarily the direct product of its irreducible components, each of which is itself a finite unitary reflexion group in the corresponding invariant subspace. In fact, if H is a hyperplane with normal vector v and R is a unitary reflexion in H, then any invariant subspace either is contained in H, or is the orthogonal sum of the line $\text{lin}\{v\}$ spanned by v with another invariant subspace which is contained in H. Therefore, in a reducible unitary reflexion group G, the set of normal vectors of the generating reflexions splits into at least two mutually orthogonal subsets of vectors, each belonging to a set of generating reflexions for a direct factor of G. If these subsets are not further decomposable in this way, then each such direct factor acts irreducibly on the subspace spanned by the normal vectors in its subset, and acts trivially on the orthogonal complement of this subspace, which is just the intersection of the reflecting hyperplanes belonging to these normal vectors. Hence, for many questions about unitary groups, we can restrict ourselves to irreducible groups.

The finite unitary reflexion groups in $\mathbb{C}^n$ were completely enumerated by Shephard and Todd [390]. In particular, we have

9A3 Theorem *Each finite irreducible unitary reflexion group in $\mathbb{C}^n$ is generated by n or $n+1$ reflexions.*

This is different from the euclidean case, where the number of generating reflexions is always n, the dimension of the ambient space (see Chapter 3). For $n = 1$, there is only one kind of unitary group – a finite cyclic group generated by a single reflexion. For $n \geqslant 2$, the enumeration can be accomplished through the classification of certain complex collineation groups (see [390, p. 275]). Any finite unitary reflexion group G determines a collineation group G' on complex projective $(n-1)$-space $\mathbb{CP}^{n-1}$, which is generated by the homologies corresponding to the generating reflexions of G. (Recall that a homology is a collineation – of finite period, so far as we are concerned – which leaves fixed all points of a hyperplane in $\mathbb{CP}^{n-1}$, and a point not lying on that hyperplane. A homology of period 2 is called harmonic.) Abstractly, G' is the quotient of G by its centre, which is the cyclic subgroup consisting of those elements of G which are represented by scalar matrices. Conversely, it can be proved that each finite collineation group generated by homologies is associated with only finitely many finite unitary reflexion groups. (In fact, a homology acting on $\mathbb{CP}^{n-1}$ lifts to a unique reflexion on $\mathbb{C}^n$, except when $n = 2$.)

A group G of unitary transformations on $\mathbb{C}^n$ is called *imprimitive* if $\mathbb{C}^n$ is the direct sum

$$\mathbb{C}^n = E_1 \oplus \cdots \oplus E_k$$

of non-trivial proper linear subspaces $E_1, \ldots, E_k$, such that the family $\{E_1, \ldots, E_k\}$ is invariant under G; then $\{E_1, \ldots, E_k\}$ is called a *system of imprimitivity* for G. If G is not imprimitive, then G is called *primitive* and is then necessarily irreducible. For an irreducible imprimitive reflexion group G, we necessarily have $\dim(E_i) = 1$ for each $i = 1, \ldots, k$. To see why this is so, suppose that $\dim(E_i) > 1$ for some i. Since G is irreducible, there is a reflexion R in G such that $E_i R = E_j$ for some $j \neq i$. But then

$$\dim(E_j \cap E_i) = \dim(E_i R \cap E_i) = \dim(H \cap E_i) > 0,$$

where H is the mirror of the reflexion R, which contradicts $E_j \cap E_i = \{o\}$. Further, for an irreducible imprimitive reflexion group G, we can prove that $k = n$ and that $E_1, \ldots, E_n$ are spanned by the vectors in an orthonormal basis of $\mathbb{C}^n$. Therefore, up to conjugacy by a unitary matrix, we can assume that $E_i = \text{lin}\{e_i\}$ (for $i = 1, \ldots, n$), where $\{e_1, \ldots, e_n\}$ is the standard basis of $\mathbb{C}^n$.

Adopting the notation of Shephard and Todd [390, p. 277], the unitary group $G(m, p, n)$ is defined as follows. Let $m \geqslant 1, n \geqslant 2$, and let p be a divisor of m, so that $m = pq$ (say). Let ϑ be a primitive mth root of unity. Then $G(m, p, n)$ is the group of all monomial transformations in $\mathbb{C}^n$ of the form

9A4 $$\eta_i = \vartheta^{\nu_i} \xi_{i\sigma} \quad (i = 1, \ldots, n), \qquad \text{with } \sigma \in S_n \text{ and } \sum_{i=1}^{n} \nu_i \equiv 0 \,(\text{mod } p),$$

where $(\eta_1, \ldots, \eta_n)$ denotes the image of $(\xi_1, \ldots, \xi_n)$ in $\mathbb{C}^n$, and S_n is the symmetric group. This is a unitary reflexion group of order $qm^{n-1}n!$, which contains both 2-fold reflexions and r-fold reflexions with $r \mid q$ whenever $q > 1$. The set of 2-fold reflexions given by

9A5 $$\eta_j = \vartheta^{\nu} \xi_k, \qquad \eta_k = \vartheta^{-\nu} \xi_j, \qquad \eta_i = \xi_i \ \ (i \neq j, k)$$

generates the group $G(m, m, n)$, which is therefore a normal subgroup of $G(m, p, n)$ of order $m^{n-1}n!$. The other reflexions in $G(m, p, n)$ (if any) are all of the form

9A6 $$\eta_j = \vartheta^{\nu m/r} \xi_j, \qquad \eta_i = \xi_i \text{ for } i \neq j,$$

for some $j = 1, \ldots, n$, where $(\nu, m/r) = 1$, and are r-fold with $r \mid q$ (for $q > 1$).

If $m = 1$, then $G(m, p, n)$ leaves the hyperplane $\sum_{i=1}^{n} \xi_i = 0$ invariant and hence is reducible; in fact, $G(1, 1, n) \cong S_n$. We therefore suppose that $m > 1$. The group $G(m, m, 2)$ is the dihedral group of order $2m$ and is reducible if $m = 2$ (leaving invariant the line $\xi_1 - \xi_2 = 0$). All other groups $G(m, p, n)$ are irreducible. In particular, we have

9A7 Theorem *If $n \geqslant 2$, then, up to conjugacy within the group of all unitary transformations, the only finite irreducible unitary reflexion groups in $\mathbb{C}^n$ which are imprimitive are the groups $G(m, p, n)$ with $m \geqslant 2$, $p \mid m$, and $(m, p, n) \neq (2, 2, 2)$.*

Note that $G(m, p, n)$ is the symmetry group of Shephard's [388, §6] fractional complex polytope $\frac{1}{p}\gamma_n^m$. If $p = 1$, this is the generalized complex n-cube γ_n^m with m^n vertices, which is

$$m\{4\}2\{3\}2 \cdots 2\{3\}2$$

in the notation to be introduced later. In general, $\frac{1}{p}\gamma_n^m$ is not an abstract polytope in our sense.

There is a far richer variety of finite unitary groups generated by reflexions in the complex plane $\mathbb{C}^2$ than in higher dimensions. The reader is referred to Coxeter [126,

Chapter 10] for a detailed discussion of these groups. The collineation groups on the complex projective line $\mathbb{CP}^1$ which correspond to the primitive reflexion groups in $\mathbb{C}^2$ are now the tetrahedral, octahedral and icosahedral (rotation) groups; these are indeed the only primitive collineation groups on $\mathbb{CP}^1$ (see [390, p. 279]). As we have remarked, it is only in this case that a homology can lift to more than one reflexion.

Most irreducible unitary reflexion groups require only n generating reflexions [390, Table vii]. In the plane there are 7 exceptions to this with 3 generators; these groups are not symmetry groups of regular complex polygons. If $n \geqslant 3$, except in one case, each irreducible group requiring $n+1$ generators is an imprimitive group $G(m, p, n)$ with $p \neq 1$ or m. The single exception is a group of order 46080 in $\mathbb{C}^4$ with 5 generators (see [116, p. 134]).

For the irreducible n-generator groups G, the reflecting hyperplanes can be chosen as the coordinate hyperplanes for a suitable coordinate system. These groups fall into two families which are, to some extent, overlapping (see [116, p. 129; 126, Chapters 12, 13]).

In the first family, called by Orlik and Terao [336] *Shephard groups*, the generating reflexions of G occur in a natural order $R_0, \ldots, R_{n-1}$ (say), such that non-adjacent generators commute (but adjacent generators do not). In particular, if we denote the identity transformation by I, then G has a presentation of the form

9A8
$$\begin{cases} R_i^{r_i} = I & \text{for } i = 0, \ldots, n-1, \\ (R_i R_j)^{p_{ij}/2} = (R_j R_i)^{p_{ij}/2} & \text{for } 0 \leqslant i < j \leqslant n-1. \end{cases}$$

The convention here is that an expression $(R_i R_j)^{p_{ij}/2}$ consists of p_{ij} terms in all, beginning with R_i; for example, $(R_0 R_1)^{5/2} = R_0 R_1 R_0 R_1 R_0$. Hence the commutation property implies that $p_{ij} = 2$ unless $i = j-1$. Such a group can conveniently be represented by a string diagram. As for Coxeter groups, each generator is symbolized by a node, but now the node i is marked with the period r_i of the corresponding generator R_i. Similarly, the two nodes i and j are joined by a branch marked p_{ij}; as usual, the branch is omitted when $p_{ij} = 2$, in which case the corresponding generators commute. By convention, a node is not marked when the generator is involutory, and the mark of a branch is omitted when it takes the most prevalent value 3. We follow our previous convention of writing $p_i := p_{i-1,i}$. Then the diagram has the form

9A9

r_0 r_1 r_2 r_{n-3} r_{n-2} r_{n-1}

•———•———• – – – •———•———•

p_1 p_2 p_{n-2} p_{n-1}

and is connected, because $p_i > 2$ for each i. This group is also denoted

9A10
$$r_0[p_1]r_1[p_2]r_2 \cdots r_{n-2}[p_{n-1}]r_{n-1},$$

and is the symmetry group of the regular complex n-polytope

$$r_0\{p_1\}r_1\{p_2\}r_2 \cdots r_{n-2}\{p_{n-1}\}r_{n-1},$$

which is, in general, not an abstract polytope (again in our sense; see [126, Chapters 12, 13; 387]).

Table 9A1. *The Finite Non-Real Irreducible Groups* $r_0[p_1]r_1[p_2]r_2 \ldots r_{n-2}[p_{n-1}]r_{n-1}$

Dimension n	Group $p[4]2[3]2 \cdots 2[3]2$	Order $p^n n!$	Vertices p^n	Facets pn
2	3[3]3	24	8	8
	3[6]2	48	24	16
	3[4]3	72	24	24
	4[3]4	96	24	24
	3[8]2	144	72	48
	4[6]2	192	96	48
	4[4]3	288	96	72
	3[5]3	360	120	120
	5[3]5	600	120	120
	3[10]2	720	360	240
	5[6]2	1 200	600	240
	5[4]3	1 800	600	360
3	3[3]3[3]3	648	27	27
	3[3]3[4]2	1 296	72	54
4	3[3]3[3]3[3]3	155 520	240	240

A unitary group is called *real* if all its matrices can be made real orthogonal by a suitable choice of coordinate system. Up to conjugacy, the finite unitary reflexion groups which are real are precisely the finite Coxeter groups. In particular, the group $r_0[p_1]r_1 \cdots r_{n-2}[p_{n-1}]r_{n-1}$ is real if and only if $r_i = 2$ for each i, in which case it is the symmetry group of the regular convex n-polytope $\{p_1, \ldots, p_{n-1}\}$ in euclidean space.

9A11 Theorem *Up to reversing the order of the generators, the only finite non-real irreducible groups* $r_1[p_1]r_2[p_2]r_3 \ldots r_{n-1}[p_{n-1}]r_n$ *are precisely those listed in Table 9A1.*

In this table, the entries "Vertices" and "Facets" refer to the numbers of vertices and facets of the related complex polytope $r_0\{p_1\}r_1\{p_2\}r_2 \cdots r_{n-2}\{p_{n-1}\}r_{n-1}$.

In the second family, all the generating reflexions are involutory. This is the type of group we are mainly interested in; this family includes all the finite irreducible Coxeter groups.

The groups in this family can also be described by a connected diagram, but now with unmarked nodes representing involutory generators. Each pair $\{R_i, R_j\}$ of distinct non-commuting generators is joined by a branch marked with the period p_{ij} of their product. Again, the mark is omitted when $p_{ij} = 3$. It was proved in Shephard [388] and Shephard and Todd [390] that, for any irreducible group, the generators R_i may be so chosen that the underlying graph either is a tree, in which case the group is real, or contains just one triangular circuit with at least two unmarked branches (that is, branches with label 3). However, in general there are also other ways of generating the same group – this creates difficulties in attempts at a direct classification of the groups.

If the graph contains one triangular circuit with at least two unmarked branches, then the graphical symbol for the group G is completed by writing a mark s inside

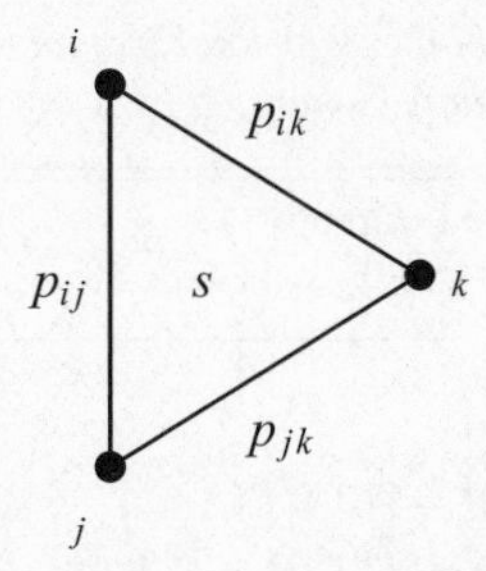

Figure 9A1. The general triangular diagram.

the triangle; see Figure 9A1. If i, j and k are the nodes in this circuit, then s is just the period of $R_i R_k R_j R_k$ in G. A presentation for G is now obtained by adding to the standard relations

9A12 $$(R_i R_j)^{p_{ij}} = I \qquad (1 \leqslant i < j \leqslant n)$$

of the underlying Coxeter diagram, the one extra relation

9A13 $$(R_i R_k R_j R_k)^s = I$$

for this triangular circuit. In later sections, we shall further investigate groups in which similar such relations are imposed on each triple of generators. Here we just mention that, since the triangle in Figure 9A1 has at least two unmarked branches (that is, at least two of the marks p_{ij}, p_{jk}, p_{ik} are 3), we can regard the relation (9A13) as assigning the period s to any one of the six products

$$R_i R_j R_k R_j, \quad R_i R_k R_j R_k, \quad R_j R_i R_k R_i, \quad R_j R_k R_i R_k, \quad R_k R_i R_j R_i, \quad R_k R_j R_i R_j.$$

In fact, these products are conjugate in pairs, since we always have (for example)

$$R_i R_j R_k R_j \sim R_k R_j R_i R_j,$$

exhibiting the symmetry between i and k, and additionally, if $p_{jk} = 3$, then

$$R_i R_j R_k R_j = R_i R_k R_j R_k.$$

It turns out that each finite non-real irreducible unitary group in $\mathbb{C}^n$ generated by n involutory reflexions is an instance of a group $[k\,l\,m^p]^q$ with $n := k + l + m$, whose diagram is shown in Figure 9A2. However, the actual values of k, l, m, p, q

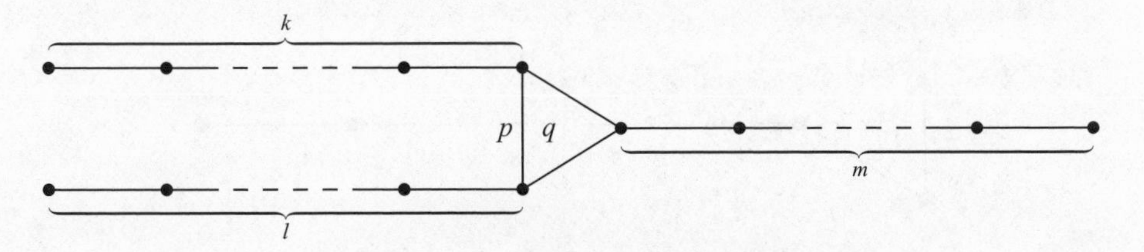

Figure 9A2. The group $[k\,l\,m^p]^q$.

Table 9A2. *The Finite Non-Real Irreducible Unitary Reflexion Groups Generated by n Involutions*

Dimension $n(\geq 3)$	Group $[1\,1\,(n-2)^p]^3 = [1\,1\ldots 1]^p$	Order $p^{n-1}n!$	Centre (p,n)
3	$[1\,1\,1^4]^4$	336	2
	$[1\,1\,1^5]^4 = [1\,1\,1^4]^5$	2160	6
4	$[2\,1\,1^4]^3 = [2\,1\,1]^4 = [1\,1\,2]^4$	$64\cdot 5!$	4
5	$[2\,1\,2]^3$	$72\cdot 6!$	2
6	$[2\,1\,3]^3$	$108\cdot 9!$	6

are restricted to only a few choices; see Theorem 9A15. In the diagram and the group name, the mark p is omitted if $p=3$, so that $[k\,l\,m^3]^q = [k\,l\,m]^q$ involves the three numbers k, l and m symmetrically. Clearly,

$$[k\,l\,m^p]^q \cong [l\,k\,m^p]^q.$$

Further, if $m=1$, then p and q are interchangeable; that is,

9A14 $$[k\,l\,1^p]^q \cong [k\,l\,1^q]^p.$$

In fact, if $m=1$ and the nodes in the circuit are denoted as in Figure 9A1, then we obtain an isomorphism by substituting for R_j its conjugate $R_kR_jR_k = R_jR_kR_j$, leaving all the other generators alone. (A more general change of generators will be discussed in Section 9C.)

The following theorem can be proved by methods similar to those discussed in the next two sections.

9A15 Theorem *The only finite non-real irreducible unitary reflexion groups in $\mathbb{C}^n$ which are generated by n involutory reflections are those listed in Table 9A2.*

Figure 9A3 shows the original diagram for the groups $[1\,1\,(n-2)^p]^3$. These groups are just the (imprimitive) groups $G(p,p,n)$. In particular, writing $\varphi := e^{2\pi i/p}$, we can take as generators

9A16
$$R_1\colon (\xi_1,\ldots,\xi_n) \mapsto (\varphi\xi_2, \bar{\varphi}\xi_1, \xi_3,\ldots,\xi_n),$$
$$R_i\colon (\xi_1,\ldots,\xi_n) \mapsto (\xi_1,\ldots,\xi_{i-2},\xi_i,\xi_{i-1},\xi_{i+1},\ldots,\xi_n) \quad (i=2,\ldots,n).$$

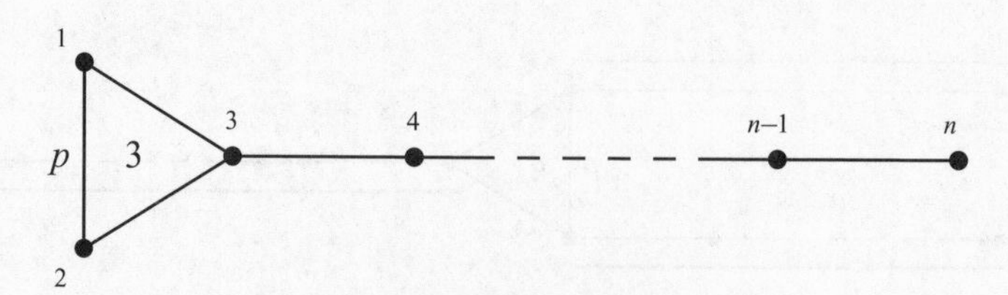

Figure 9A3. The group $[1\,1\,(n-2)^p]^3$.

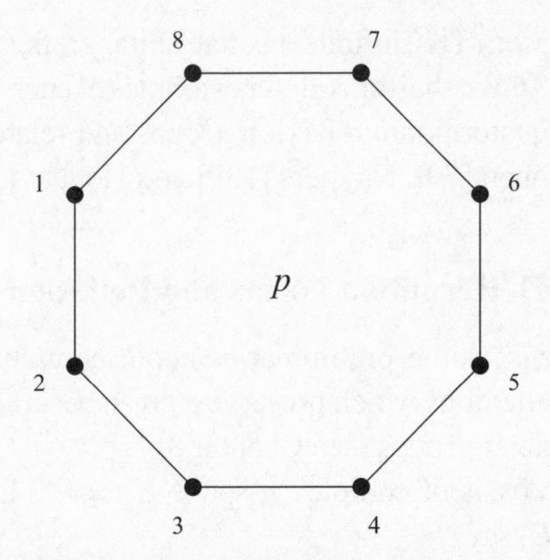

Figure 9A4. The octagonal diagram $[1\,1\,1\,1\,1\,1\,1\,1]^p$ for $[1\,1\,6^p]^3$.

The group $[k\,l\,m^p]^q$ has an alternative diagram representation (based on other generators). In the special case $k = l = 1$ (and $m = n - 2$) and $q = 3$, this consists of an n-gon with unmarked branches, which has a mark p inside the circuit to indicate any one of n equivalent relations such as

9A17 $$(R_1 R_2 R_3 \cdots R_{n-1} R_n R_{n-1} R_{n-2} \cdots R_2)^p = I$$

[114, p. 251]; Figure 9A4 illustrates the case $n = 8$ (that is, $m = 6$). In fact, if the numbering of the old generators S_i (say) of $[1\,1\,(n-2)^p]^3$ is as in Figure 9A3, then we can take as new generators

9A18 $$R_j := \begin{cases} S_n S_{n-1} \cdots S_4 S_3 S_1 S_3 S_4 \cdots S_{n-1} S_n, & \text{if } j = 1, \\ S_j, & \text{if } j = 2, \ldots, n, \end{cases}$$

and arrive at a presentation for $[1\,1\,(n-2)^p]^3$ given by the standard relations for the underlying n-gonal Coxeter diagram and the extra relation (9A17). Thus we can use the alternative symbol $[1\,1\,\ldots\,1]^p$ for $[1\,1\,(n-2)^p]^3$, with n digits 1. If $p = \infty$, this gives a real (euclidean) Coxeter group. (Again, we shall discuss this in more generality in Section 9C.)

To conclude this section, we comment on the intersection property for unitary groups. Many important properties of finite or infinite Coxeter groups $\Gamma = \langle \sigma_1, \ldots, \sigma_k \rangle$ depend in an essential way on the intersection property

$$\langle \sigma_i \mid i \in I \rangle \cap \langle \sigma_i \mid i \in J \rangle = \langle \sigma_i \mid i \in I \cap J \rangle,$$

for the distinguished subgroups $\langle \sigma_i \mid i \in I \rangle$ of Γ (see Theorem 3A2). For the finite unitary reflexion groups, no general proof of the corresponding property seems to be known, nor has this fact been pointed out explicitly anywhere in the literature. For most groups, the intersection property can be proved directly (if rather tediously) using explicit coordinate representations for the generators (taken from Shephard [388, 389], Shephard and Todd [390], and Coxeter [114]). However, it would be desirable to have a general proof.

A complete classification of the infinite discrete unitary reflexion groups was recently announced in Popov [343]. We shall encounter examples of such groups in later sections. For a discussion of the quaternionic reflexion groups and related regular quaternionic polytopes we refer to Cohen [79], Cuypers [139] and Hoggar [220].

9B Hermitian Forms and Reflexions

In this section, we discuss some preliminaries about complex groups generated by involutory hyperplane reflexions which preserve a given hermitian form. The approach is similar to that for Coxeter groups (see Chapter 3).

Let $\{e_1, \ldots, e_n\}$ be a basis of complex n-space $E := \mathbb{C}^n$. Let $\langle \cdot, \cdot \rangle$ be a hermitian (sesquilinear) form on E, say

9B1
$$\langle x, y \rangle = \sum_{i,j=1}^{n} \alpha_{ij} \xi_i \overline{\eta_j},$$

where $x = \sum_{i=1}^{n} \xi_i e_i$ and $y = \sum_{i=1}^{n} \eta_i e_i$, with $\alpha_{ij} = \overline{\alpha_{ji}}$ for each i and j. We assume that

9B2
$$\alpha_{ii} = 1 \quad \text{for } i = 1, \ldots, n,$$

so that each e_i is a unit vector.

We consider the group generated by the n involutory reflexions in the hyperplanes orthogonal to the base vectors. For S_i, the ith involutory reflexion, the reflecting hyperplane is the orthogonal complement $e_i^{\perp}$ of the ith base vector e_i. In particular,

9B3
$$x S_i = x - 2\langle x, e_i \rangle e_i \quad \text{for } x \in E,$$

so that S_i leaves invariant all the coordinates ξ_j of x except ξ_i, and changes ξ_i to

9B4
$$\xi_i - 2 \sum_{j=1}^{n} \alpha_{ji} \xi_j.$$

Moreover, S_i preserves $\langle \cdot, \cdot \rangle$. We should emphasize that the term "reflexion" is meant here with respect to the geometry on the ambient space, which is defined by the corresponding hermitian form.

The resulting reflexion group

$$G := \langle S_1, \ldots, S_n \rangle$$

is a subgroup of $\mathrm{GL}(E)$ which preserves the given hermitian form. In particular, G leaves pointwise fixed the radical $E^{\perp}$ of $\langle \cdot, \cdot \rangle$, which is the intersection of the reflecting hyperplanes $e_i^{\perp}$ of the generators S_i. It follows that G is reducible (but in general not completely reducible) if $\langle \cdot, \cdot \rangle$ is degenerate.

We call the system $\{S_1, \ldots, S_n\}$ of generators of G *reducible* if there exist non-empty sets I and J such that $I \cup J = \{1, \ldots, n\}$, $I \cap J = \emptyset$ and $\alpha_{ij} = 0$ if $i \in I$, $j \in J$. The system is *irreducible* if it is not reducible.

The proof of the following lemma is similar to that for Coxeter groups [222, p. 131].

9B5 Lemma *If the hermitian form $\langle \cdot, \cdot \rangle$ is non-degenerate, then the system of generators $\{S_1, \ldots, S_n\}$ is irreducible if and only if $G = \langle S_1, \ldots, S_n \rangle$ acts irreducibly on E.*

Proof. First, let the system of generators be irreducible. We show that each proper invariant subspace E' for G is contained in the radical $E^\perp$ of $\langle \cdot, \cdot \rangle$, or, equivalently, in each reflecting hyperplane $e_i^\perp$. For each i, the subspace E' is invariant under S_i; thus $E' \subseteq e_i^\perp$, or $E' = E_i \oplus \text{lin}\{e_i\}$ with $E_i \subseteq e_i^\perp$. Assume that, for some i, we have $e_i \in E'$. Let j be such that $\alpha_{ij} \neq 0$. Then

$$e_i S_j = e_i - 2\alpha_{ij} e_j \in E',$$

and therefore $e_j \in E'$. It follows that $e_i \in E'$ implies that $e_j \in E'$ whenever $\alpha_{ij} \neq 0$. Using the irreducibility of $\{S_1, \ldots, S_n\}$, we can now conclude that $e_k \in E'$ for each k, contradicting the assumption that E' was a proper subspace. It follows that $e_i \notin E'$ for each i; hence $E' \subseteq e_i^\perp$.

On the other hand, if the system of generators is reducible, we have non-empty sets I and J such that $I \cup J = \{1, \ldots, n\}$, $I \cap J = \emptyset$, and $\alpha_{ij} = 0$ if $i \in I$, $j \in J$. Then E is the orthogonal sum of the two non-degenerate subspaces $\text{lin}\{e_i \mid i \in I\}$ and $\text{lin}\{e_j \mid j \in J\}$, each of which is invariant under each generating reflexion. (Recall that $\text{lin}\, V$ denotes the linear hull of a subset V of E.) Hence G acts reducibly on E. □

Later, Lemma 9B5 will often be applied in connexion with the following proposition, which is of interest in its own right.

9B6 Proposition *Let h be a non-degenerate indefinite hermitian form on $\mathbb{C}^n$. If G is a subgroup of the isometry group of h which acts irreducibly on $\mathbb{C}^n$, then G is an infinite group.*

Proof. Indeed, were G to be finite, it would leave invariant a positive definite hermitian form g, and thus any linear combination of g and h. However, g and h can be diagonalized simultaneously, so that, for some $\lambda \in \mathbb{C}$, the form $f := g - \lambda h$ is degenerate but not zero. But then G will leave invariant the radical of f, which contradicts the irreducibility. □

Since we are ultimately interested in finite groups, we can already make here the following observation, which we usually apply in the situation when the hermitian form is known to be positive definite on an $(n-1)$-subspace of E. Let $A := (\alpha_{ij})$ denote the matrix of the form, and let $\Delta := \det A$.

9B7 Lemma *If $\Delta < 0$, then $G = \langle S_1, \ldots, S_n \rangle$ cannot be finite.*

Proof. If $\Delta < 0$, the form is non-degenerate and indefinite. Suppose that G is finite. By Proposition 9B6 and Lemma 9B5, it must then act reducibly on E, and its system of generators $\{S_1, \ldots, S_n\}$ must be reducible. Hence, there are non-empty subsets I, J with $I \cup J = \{1, \ldots, n\}$ and $I \cap J = \emptyset$, such that $\langle e_i, e_j \rangle = a_{ij} = 0$ if $i \in I$, $j \in J$. Let G_I and G_J denote the subgroups of G generated by the reflexions S_k with $k \in I$ or $k \in J$, respectively, and let A_I and A_J be the corresponding matrices obtained by restricting A to rows and columns in I or J. Then $\Delta = \det A = \det A_I \det A_J < 0$; hence $\det A_I < 0$ and $\det A_J > 0$, or vice versa. Let $\det A_I < 0$ (say). We can now work with the finite group G_I, on the subspace $\mathrm{lin}\{v_k \mid k \in I\}$ of dimension $|I|$, and apply the same argument as before. Continuing in this way, we obtain, at the final stage, a subset I consisting of a single element i (say). But then $\langle e_i, e_i \rangle = \det A_I < 0$, which contradicts $\langle e_i, e_i \rangle = 1$. Hence, G must be infinite. □

Notice that Lemma 9B7 says nothing about the case $\Delta = 0$. Depending on the type of the (degenerate) form, the group can be finite or infinite. The conditions we later impose on the form will imply that the groups are always infinite in this case.

In our applications in Chapter 11, the reflexion group G is usually the image of an abstract group Γ under a linear representation

$$r\colon \Gamma \to \mathrm{GL}(E).$$

In particular, Γ will be generated by n involutions $\sigma_1, \ldots, \sigma_n$, and $S_i = \sigma_i r$ for $i = 1, \ldots, n$. In a typical situation, Γ is given by a presentation whose relators employ two or three generators (we shall allow more generators later on, but the present case is the most interesting one), thereby imposing conditions on the *basic* 2-generator and 3-generator subgroups $\langle \sigma_i, \sigma_j \rangle$ and $\langle \sigma_i, \sigma_j, \sigma_k \rangle$ of Γ, respectively. In this context, we need to know when products of generators of G, such as $S_i S_j$ or $S_i S_j S_k S_j$, have finite period.

The condition for $S_i S_j$ to have finite period is that

$$|\alpha_{ij}| = |\langle e_i, e_j \rangle| = \cos \frac{\pi}{p_{ij}}$$

for some rational number $p_{ij} \geqslant 2$. (By Lemma 9B7 we must have $|\alpha_{ij}| \leqslant 1$ in this case, with $|\alpha_{ij}| = 1$ yielding infinite period.) Indeed, we can scale e_i or e_j to obtain $\langle e_i, e_j \rangle = -\cos \frac{\pi}{p_{ij}}$, and thus recognize $\langle S_i, S_j \rangle$ as a real (dihedral) group; we may then appeal to the well-known condition for real groups. When p_{ij} is expressed in its lowest terms, the period of $S_i S_j$ is given by the numerator of p_{ij}. In particular, S_i and S_j commute if and only if $p_{ij} = 2$. Note that $S_i S_j$ has infinite period if and only if $|\alpha_{ij}| = |\langle e_i, e_j \rangle| = 1$ (now scaling yields $\langle e_i, e_j \rangle = -1$).

Notice that the reflexion group G is real if all the normal vectors e_i can be rescaled simultaneously to yield real inner products $\langle e_i, e_j \rangle$. (This explains the fact, mentioned in Section 9A, that a unitary group is real if its diagram is a tree.)

We shall see in the next section (see Lemma 9C16) that the order of the products $S_i S_j S_k S_j$ is determined by the "cyclic" products of normal vectors,

$$\alpha_{ij}\alpha_{jk}\alpha_{ki} = \langle e_i, e_j \rangle \langle e_j, e_k \rangle \langle e_k, e_i \rangle,$$

and, of course, by the inner products themselves.

In Section 9D, we shall discuss generalized triangle groups and their representations as reflexion groups. These are the prototypes of groups which will occur as 3-generator subgroups in groups with n generators, both in the abstract and in the geometric context. In a geometric group G, the 3-generator subgroups fit together in a coherent way, each acting as a 3-dimensional reflexion group, such that the corresponding hermitian forms on 3 variables are just restrictions of the original form. Therefore, for an abstract group Γ to admit a representation r as before, we must impose conditions on the 3-generator subgroups which are consistent with the occurrence of 3-generator reflexion subgroups within an n-generator reflexion group. To this end, we must take, for each 3-generator subgroup of Γ, a representation (if it exists) as a 3-generator reflexion group (with a corresponding hermitian form), and then piece together all these 3-dimensional representations to obtain an n-dimensional representation of Γ as an n-generator reflexion group, such that the resulting hermitian form in n variables comprises all the forms in 3 variables. In general, however, this will not yield a faithful representation (indeed, the 3-dimensional representations may not even be faithful). On the other hand, any such representation will often yield enough information to decide whether Γ and G are finite or infinite.

The technique is most powerful when the 3-dimensional representations are faithful, which often is the case when the corresponding 3-generator subgroups are finite unitary groups. Indeed, we will frequently make this assumption. In this situation, a positive definite hermitian form will usually determine a finite group Γ, and vice versa. Thus we can enumerate the finite groups by calculating the determinant Δ. For a positive definite form, the ambient space then becomes a unitary space on which G acts as a unitary reflexion group.

In this context, whether the group is finite or not, we shall often require the intersection property, at least for 3-generator subgroups. For geometric groups, this will follow directly from the geometry; for example, to prove that

$$\langle S_i, S_j, S_k \rangle \cap \langle S_i, S_j, S_l \rangle = \langle S_i, S_j \rangle,$$

with i, j, k, l distinct, observe that an element in the intersection must fix both coordinates ξ_k and ξ_l, and must therefore belong to the subgroup $\langle S_i, S_j \rangle$ of the unitary group $\langle S_i, S_j, S_k \rangle$. For abstract groups, we then only need to appeal to the quotient criterion of Theorem 2E17; indeed, if the restricted 3-dimensional representation on $\langle \sigma_i, \sigma_j, \sigma_k \rangle$ or $\langle \sigma_i, \sigma_j, \sigma_l \rangle$ is faithful, then it is trivial that we also have

$$\langle \sigma_i, \sigma_j, \sigma_k \rangle \cap \langle \sigma_i, \sigma_j, \sigma_l \rangle = \langle \sigma_i, \sigma_j \rangle$$

if i, j, k, l are distinct.

In practice, when we work with a positive definite form, we will generally assume that this is the standard hermitian form on $\mathbb{C}^n$, usually expressed in terms of the basis of $\mathbb{C}^n$ given by the unit normals of the mirrors of the generating reflexions.

We shall also discuss the case when the hermitian form is positive semi-definite and has a 1-dimensional radical. Then the group acts as an n-generator reflexion group on a unitary $(n-1)$-space (see Lemma 9B19), and so it is often more convenient to work directly in standard unitary $(n-1)$-space, with the n generating reflexions determined by n linearly *dependent* normal vectors. In this context, the following comment clarifies the terminology we shall use. Among the hermitian forms on $\mathbb{C}^n$ which are positive definite on an $(n-1)$-subspace, the case $\Delta = 0$ (obtained for a positive semi-definite form) can be viewed as the *transitional* case between the cases $\Delta > 0$ (for positive definite forms) and $\Delta < 0$ (for forms of hyperbolic type). However, when the corresponding groups are acting directly as n-generator reflexion groups on standard unitary n-space, then this case occurs precisely when the normal vectors for the n reflexion planes become linearly dependent, so that the action is indeed on an $(n-1)$-space. Hence, while transitional in the first setting, $\Delta = 0$ is now the *degenerate* case (of linearly dependent normals). We shall further elaborate on this in Section 9C.

In the remainder of this section, we briefly discuss contragredient representations, and explain how they can be utilized to treat the case $\Delta = 0$.

Let $r\colon \Gamma \to \mathrm{GL}(E)$ be a linear representation as before, where again $\Gamma = \langle \sigma_1, \ldots, \sigma_n \rangle$ and $S_i = \sigma_i r$ for $i = 1, \ldots, n$. In this situation, it is sometimes useful to pass from r to its *contragredient* (or *dual*) representation r^* on the dual vector space E^* of E, which consists of the linear functionals on E. If $w \in \Gamma$, then we write wr^* for the dual map of $w^{-1}r$. In other words, $r^*\colon \Gamma \to \mathrm{GL}(E^*)$ is given by

9B8 $$y^*(wr^*) = (w^{-1}r)y^* \qquad \text{for } y^* \in E^*.$$

(Recall here that y^* is regarded as a linear mapping, which thus acts on the right of its argument.)

Write

$$S_i^* := \sigma_i r^* \quad (i = 1, \ldots, n), \qquad G^* := \Gamma r^* = \langle S_1^*, \ldots, S_n^* \rangle.$$

Then, using the dual pairing $(\cdot, \cdot)$ of E and E^* (defined by $(x, y^*) := xy^*$), and remembering that S_i is an involution, we find that

9B9 $$y^* S_i^* = S_i y^* = y^* - 2(e_i, y^*)\langle \cdot, e_i \rangle \qquad (y^* \in E^*),$$

where $\langle \cdot, e_i \rangle$ denotes the linear form on E defined by $x \mapsto \langle x, e_i \rangle$. It follows that S_i^* is the "affine reflexion" in the hyperplane $(e_i, y^*) = 0$ of E^* which maps $\langle \cdot, e_i \rangle$ to $-\langle \cdot, e_i \rangle$. Note that the reflecting hyperplanes for $S_1^*, \ldots, S_n^*$ are always in general position in E^*, unlike those for the original reflexions $S_1, \ldots, S_n$ in E, which intersect in $E^\perp$.

Let $\{e_1^*, \ldots, e_n^*\}$ be the basis of E^* dual to the original basis $\{e_1, \ldots, e_n\}$ of E. Then

9B10 $$\langle \cdot, e_i \rangle = \sum_{j=1}^{n} \alpha_{ji} e_j^* \qquad \text{for } i = 1, \ldots, n.$$

If we write a linear functional y^* as $y^* = \sum_{j=1}^{n} \eta_j^* e_j^*$, with coordinates $\eta_1^*, \ldots, \eta_n^*$ with respect to $\{e_1^*, \ldots, e_n^*\}$, then (9B9) implies that S_i^* changes each coordinate η_j^* to

9B11 $$\eta_j^* - 2\alpha_{ji}\eta_i^*.$$

In particular, $\eta_i^* = 0$ is the reflecting hyperplane of S_i^* in E^*.

If the underlying hermitian form $\langle \cdot, \cdot \rangle$ on E is non-degenerate, then the two representations r and r^* are semi-linearly equivalent. In fact, the map $L\colon E \to E^*$, defined by

9B12 $$xL := \langle \cdot, x \rangle$$

is a semi-linear bijection (compare (3A12) – note that $(\lambda x)L = \bar{\lambda}(xL)$ for $\lambda \in \mathbb{C}$), and r and r^* are related by

$$wr^* = L^{-1}(wr)L \quad (w \in \Gamma).$$

In particular, $S_i^* = L^{-1}S_iL$ for each i.

We can also find a hermitian form on the dual space E^*, which is invariant under the reflexions S_i^*. Indeed, suppose that the form $\langle \cdot, \cdot \rangle$ on E is non-degenerate, so that the matrix $A := (\alpha_{ij})$ is invertible. Let A_{ji} denote the cofactor of α_{ij} in the determinant $\det A$, so that $(-1)^{i+j}A_{ji}$ is the minor obtained when the ith row and jth column are deleted from A. The matrix $\operatorname{adj} A := (A_{ij})$ is called the *adjoint matrix* of A. Then

9B13
$$A \cdot \operatorname{adj} A = \operatorname{adj} A \cdot A = \det A \cdot I,$$

with I the identity matrix. Thus $\operatorname{adj} A$ is also invertible; indeed, $\operatorname{adj} A = \det A \cdot A^{-1}$.

Using (9B11) and (9B13), it is straightforward to verify that each reflexion S_i^* leaves invariant the (non-degenerate) *adjoint hermitian form*

9B14
$$\langle x^*, y^* \rangle_{\text{adj}} := \sum_{i,j=1}^{n} A_{ji}\xi_i^*\overline{\eta_j^*}$$

on E^*, where $x^* = \sum_{i=1}^n \xi_i^* e_i^*$ and $y^* = \sum_{i=1}^n \eta_i^* e_i^*$; it is indeed an orthogonal reflexion with respect to this form. (Note that the coefficient in (9B14) must be A_{ji}, not A_{ij}.) It follows that the adjoint form is invariant under the whole group

$$G^* := \langle S_1^*, \ldots, S_n^* \rangle,$$

which is the image of Γ in $\mathrm{GL}(E^*)$ under r^*.

Further, since $\langle \cdot, \cdot \rangle$ is non-degenerate on E, the set of linear functionals $\{y_1^*, \ldots, y_n^*\}$, with

$$y_i^* := e_iL = \langle \cdot, e_i \rangle \qquad (i = 1, \ldots, n),$$

is also a basis of E^*, which is related to $\{e_1^*, \ldots, e_n^*\}$ by

9B15
$$\begin{cases} y_i^* = \displaystyle\sum_{j=1}^{n} \alpha_{ji} e_j^*, & \text{for } i = 1, \ldots, n, \\ e_j^* = (\det A)^{-1} \displaystyle\sum_{i=1}^{n} A_{ij} y_i^*, & \text{for } j = 1, \ldots, n. \end{cases}$$

Each of these formulae can be verified by evaluating both sides on the base vectors $e_1, \ldots, e_n$ of E and using (9B13).

Notice that the semi-linear mapping L of (9B12) acts almost like an anti-isometry between (hermitian) metrical vector spaces; more precisely, we have

$$\langle xL, yL \rangle_{\text{adj}} = \det A \, \langle y, x \rangle$$

for all x, y in E, with $\det A$ a real scalar. Indeed, we can even say more. If f (say) is the hermitian form on E^* which actually makes L an anti-isometry between E and E^*, then necessarily,

$$f(y_i^*, y_j^*) = f(e_i L, e_j L) = \langle e_j, e_i \rangle = \alpha_{ji} \qquad (i, j = 1, \ldots, n),$$

and therefore

$$\begin{aligned} f(e_k^*, e_l^*) &= (\det A)^{-2} \sum_{i,j=1}^{n} A_{ik} \bar{A}_{jl} f(y_i^*, y_j^*) \\ &= (\det A)^{-2} \sum_{i=1}^{n} A_{ik} \left(\sum_{j=1}^{n} A_{lj} \alpha_{ji} \right) \\ &= (\det A)^{-1} \sum_{i=1}^{n} A_{ik} \delta_{li} \\ &= (\det A)^{-1} A_{lk} \; = \; (\det A)^{-1} \langle e_k^*, e_l^* \rangle_{\text{adj}} \end{aligned}$$

for each $k, l = 1, \ldots, n$. It follows that, necessarily, $\langle \cdot, \cdot \rangle_{\text{adj}} = \det A\, f$. In particular, up to a real scalar and complex conjugation, the adjoint form is equivalent to the original (non-degenerate) hermitian form $\langle \cdot, \cdot \rangle$ and is therefore of the same kind, definite or indefinite.

In conclusion, the semi-linear equivalence between the two representations r on E and r^* on E^*, which is furnished by (9B12), is therefore as geometric as one can expect.

We also need to consider the important special case where the original form $\langle \cdot, \cdot \rangle$ is positive semi-definite, and has a 1-dimensional radical $E^\perp$. This case is also familiar from the theory of Coxeter groups (Section 3B).

Let v be a non-zero vector in $E^\perp$. Then

9B16 $$T := \{y^* \in E^* \mid (v, y^*) = 0\}$$

is a linear hyperplane in E^*. The map L of (9B12) has kernel $E^\perp$ and image T, so that we can equip T with the positive definite hermitian form h which makes the induced semi-linear map from $E/E^\perp$ onto T an anti-isometry (now we ignore the real scalar); that is, if $x^* = \langle \cdot, x \rangle$, $y^* = \langle \cdot, y \rangle \in T$, then $h(x^*, y^*) := \langle y, x \rangle$. On the other hand, T acts by translation on the affine hyperplane

9B17 $$H := \{y^* \in E^* \mid (v, y^*) = 1\}$$

of E^*. Using the metric on T, we can now introduce a unitary metric on its translate H in the standard way; that is, the distance between y_1^* and y_2^* in H is the length (with respect to h) of $y_1^* - y_2^*$ in T.

Now, by (9B9), if $1 \leqslant i \leqslant n$ and $y^* \in E^*$, then

$$(v, y^* S_i^*) = (v, y^*) - 2\,(e_i, y^*)\langle v, e_i \rangle = (v, y^*);$$

hence T and H are both invariant under each S_i^*. It follows that, by restriction, r^* induces representations of Γ on the $(n-1)$-subspaces T and H; we shall use the same notation for these representations. On T, the representation r^* is given by

9B18 $$\langle \cdot, x \rangle (w r^*) = \langle \cdot, x(wr) \rangle \qquad (w \in \Gamma,\ x \in E);$$

this can be verified by evaluating both sides on elements of E. In particular, it implies that each wr^* preserves the form h on T. On the other hand, if $w \in \Gamma$ and $y_1^*,\ y_2^* \in H$, then $y_1^* - y_2^* \in T$; thus

$$\begin{aligned} h(y_1^*(wr^*) - y_2^*(wr^*), y_1^*(wr^*) - y_2^*(wr^*)) &= h((y_1^* - y_2^*)(wr^*), (y_1^* - y_2^*)(wr^*)) \\ &= h(y_1^* - y_2^*, y_1^* - y_2^*). \end{aligned}$$

It follows that on H, each wr^* is an "affine unitary" transformation, that is, a linear unitary transformation followed by a translation by some vector in T.

We are mainly interested in the representation on H, where the group $G^* = \langle S_1^*, \ldots, S_n^* \rangle$ is again a reflexion group, now generated by n reflexions in affine hyperplanes which do not intersect in a common point. In particular, the reflecting hyperplanes in H for the n generators S_i^* are the intersections of H with the original reflecting hyperplanes in E^*. Each of these intersections is non-empty, because T itself is not the reflecting hyperplane in E^* of any generator S_i^*; in fact, $\langle e_i, e_i \rangle = 1$ for each i, and hence v cannot be a multiple of e_i. These n hyperplanes in H are in general position, just like the bounding walls of an $(n-1)$-simplex.

Further, we can see that G^* must be an infinite unitary group on H. Suppose to the contrary that G^* is finite on H. Since every finite group of affine transformations does have a fixed point, there must exist a vector $y^* \in H$ which is invariant under each S_i^* acting on E^*. Therefore y^* must be contained in the reflecting hyperplane of each S_i^*, so that $y^* = 0$. But this is a contradiction to $y^* \in H$.

We summarize our results in the following lemma.

9B19 Lemma *In this situation, if the hermitian form $\langle \cdot, \cdot \rangle$ on E is positive semi-definite with 1-dimensional radical $E^\perp = \mathrm{lin}\{v\}$, then the restriction of $r^*\colon \Gamma \to$* $\mathrm{GL}(E^*)$ *to the affine $(n-1)$-space H of E^* induces an action of Γ on H as an infinite unitary group generated by n reflexions of period 2 in affine hyperplanes of H which are in general position.*

9C General Considerations

We now change our viewpoint a little from the previous sections, in that we take a more directly geometric approach. Since we ultimately wish to consider finite groups generated by involutory hyperplane reflexions in $\mathbb{C}^n$, we take a group G of the form $G = \langle S_1, \ldots, S_n \rangle$, with

$$xS_j = x - 2(\langle x, v_j \rangle - \beta_j)v_j$$

for $j = 1, \ldots, n$. In this context, $\langle \cdot, \cdot \rangle$ is an hermitian sesquilinear form, not necessarily non-singular, β_j is a scalar, and $\langle v_j, v_j \rangle = 1$ for each j. It follows that S_j is an involutory reflexion, which preserves the hyperplane (mirror) $\{x \mid \langle x, v_j \rangle = \beta_j\}$, which we identify with S_j itself. We allow the possibility that the unit vectors v_j are linearly dependent, to account for the infinite discrete groups in unitary space (see Lemma 9B19).

We are particularly interested in the case when G is *locally unitary* (*locally finite*), by which we mean that each subgroup generated by $n-1$ of the reflexions S_j is actually a finite reflexion group, acting on an $(n-1)$-space on which the form is positive definite. We adopt the convenient abbreviation *luggir* for locally unitary group generated by involutory (hyperplane) reflexions. Note that local finiteness is a property defined relative to a distinguished set of generators.

Let $\{e_1, \cdots, e_n\}$ be a basis of $\mathbb{C}^n$, and suppose that $v_i = \sum_{j=1}^n \beta_{ij} e_j$ for $i = 1, \ldots, n$. Then the Gram matrix is given by

$$(\langle v_i, v_j \rangle) = B \cdot M \cdot B^*,$$

where $B = (\beta_{ij})$ is the transformation matrix, $B^* := \bar{B}^\mathsf{T}$ is its conjugate transpose, and $M = (\langle e_i, e_j \rangle)$. Its determinant

$$\Delta := \begin{vmatrix} \langle v_1, v_1 \rangle & \langle v_1, v_2 \rangle & \cdots & \langle v_1, v_n \rangle \\ \langle v_2, v_1 \rangle & \langle v_2, v_2 \rangle & \cdots & \langle v_2, v_n \rangle \\ \vdots & \vdots & \ddots & \vdots \\ \langle v_n, v_1 \rangle & \langle v_n, v_2 \rangle & \cdots & \langle v_n, v_n \rangle \end{vmatrix}$$

is independent of the choice of unit vectors v_j for $S_1, \ldots, S_n$, and is called the *Schläfli determinant*. If the form is positive definite and $\{e_1, \ldots, e_n\}$ is an orthonormal basis, then the Gram matrix is just $B \cdot B^*$; hence $\Delta = |\det B|^2 > 0$. This is the case in which we are most interested, because it holds when the group G is finite, although the condition is not sufficient (see Lemma 9B7). Similarly, if the form is of hyperbolic type, and again $\{e_1, \ldots, e_n\}$ is an orthonormal basis, then $\Delta = -|\det B|^2 < 0$. The group is necessarily infinite in this case (again see Lemma 9B7).

If $\Delta \neq 0$ (and hence M is non-singular), then the vectors $v_1, \ldots, v_n$ are linearly independent and the reflexion planes for $S_1, \ldots, S_n$ all pass through a common point, which we can take to be the origin; that is, we may assume that $\beta_j = 0$ for each j.

For locally unitary groups, we also need to consider the case when the form on $\mathbb{C}^n$ is positive definite and $v_1, \ldots, v_n$ are linearly *dependent*. Then it is natural to regard G as acting on $\mathbb{C}^{n-1}$ instead. We shall exclude the possibility that the mirror planes all pass through a common point; otherwise G could only be discrete if it was finite, and so we are back at finite groups. But if the mirrors do not have a point in common, then any $n-1$ of the vectors v_j must necessarily be independent and span $\mathbb{C}^{n-1}$, and at least one β_j must be non-zero; indeed, we need take only one β_j non-zero. Notice that this is just the case described in Lemma 9B19, and so must necessarily yield an infinite group; in the context of the previous section, this case occurred with linear (rather than affine) reflexions preserving a positive semi-definite form with 1-dimensional radical on $\mathbb{C}^n$. In the present context of affine reflexions, the specific value of the non-zero β_j is, strictly speaking, irrelevant; indeed, different values of β_j yield groups which are conjugate in the group of all similarity transformations on $\mathbb{C}^{n-1}$.

The case $n = 1$ being trivial, let us look first at the case $n = 2$. We take $\{v_1, v_2\}$ to be linearly independent, in which event the condition for the product $S_1 S_2$ to have finite period is that

$$|\langle v_1, v_2 \rangle| = \cos \frac{\pi}{p}$$

for some rational number $p \geqslant 2$ (see Section 9B). (Of course, we must have $|\langle v_1, v_2 \rangle| < 1$ in order that $\Delta > 0$, the latter being a necessary condition for finiteness.)

Now suppose that $n \geqslant 3$. Much of our discussion will depend on being able to choose new generators for the group G, or being able to pass to certain subgroups. Under the assumption that G is finite, we shall arrive at a standardized way of choosing (possibly new) generating reflexions for G. However, much of what we say here will apply in a more general context.

We shall employ two kinds of change of generators. The first is mainly applied in the context of finite groups; we shall therefore suppose that $\beta_j = 0$ for each j. Let $j \neq k$, and replace the generator S_j by the new reflexion

9C1
$$S'_j := (S_j S_k)^{m-1} S_j,$$

while leaving all the other $n - 1$ generators unchanged. Thus $S'_j S_k = (S_j S_k)^m$. If $S_j S_k$ has finite period $q > 2$ (say), and $(m, q) = 1$, then $S_j \in \langle S'_j, S_k \rangle$, and the operation is invertible. Indeed, if $mm' \equiv 1 \pmod{q}$, then

$$(S'_j S_k)^{m'} = (S_j S_k)^{mm'} = S_j S_k,$$

so that

$$S_j = (S'_j S_k)^{m'-1} S'_j,$$

and it follows that the replacement leads to new generators for G. If $(m, q) > 1$, then generally we shall pass to a proper subgroup of G; nevertheless, this will often yield useful information.

Now let us look at the effect of the replacement (9C1) on the Schläfli determinant Δ, if Δ is positive. It is easier to calculate the effect by looking at a related determinant. If V is the $n \times n$ matrix over $\mathbb{C}$, whose rows are the coordinates of the vectors v_j with respect to some orthonormal basis of $\mathbb{C}^n$, then $\Delta = |\det V|^2$ (so, in effect, we have replaced B by V). We shall therefore work with $\det V$ instead. Now we may scale the vector v_j so that $\langle v_j, v_k \rangle = -\cos(\pi/p)$ (this is the usual convention when the vectors are real). It is then a routine matter to check that a unit normal v'_j to the new reflexion mirror S'_j is

$$v'_j = \frac{\sin(m\pi/p)}{\sin(\pi/p)} v_j + \frac{\sin((m-1)\pi/p)}{\sin(\pi/p)} v_k,$$

giving the new matrix V', say, from which easily follows

$$\det V' = \frac{\sin(m\pi/p)}{\sin(\pi/p)} \det V,$$

or

9C2
$$\Delta' = \frac{\sin^2(m\pi/p)}{\sin^2(\pi/p)} \Delta.$$

The second, closely related, kind of change of generator is even more important and applies to all groups. Again with $j \neq k$, we replace S_j by $S'_j := S_k S_j S_k$ and leave the other $n-1$ generators unchanged. This is obviously involutory, in that $(S'_j)' = S_j$, and so leads to a new set of generators of G. This operation is denoted by

9C3 $$(S_1, \ldots, S_n) \mapsto (S_1, \ldots, S'_j, \ldots, S_n),$$

and is called a *basic operation*. The unit normal v'_j to the new reflexion mirror S'_j is now given by

$$v'_j = v_j S_k = v_j - 2\langle v_j, v_k \rangle v_k,$$

and so the Schläfli determinant Δ is unaffected by a basic operation (actually, $\det V' = \det V$).

Two sets of generators of G which can be obtained from each other by sequences of basic operations are called *basically equivalent*. If two sets of generators are basically equivalent, then their Schläfli determinants are necessarily the same. Note that local finiteness need not be preserved by basic equivalence; we shall give a simple example in Section 9D.

We now associate with a set $\mathcal{S}$ of generators S_j of G (or normals v_j) a *diagram* $\mathcal{D}$. It has n *nodes* (vertices) labelled $1, \ldots, n$, with j corresponding to S_j. As we noted earlier, for each $j \neq k$, there is a rational number $p_{jk} = p_{kj} \geqslant 2$ such that

$$|\langle v_j, v_k \rangle| = \cos \frac{\pi}{p_{jk}}.$$

The *branch* (edge) joining j and k is then labelled p_{jk}; we follow the standard conventions in excising a branch which would be labelled 2 (the corresponding generators commute), and omitting the label 3 on branches because of its frequency. Observe, however, that we explicitly permit fractional marks, so that our diagrams are more akin to those of the regular star-polytopes in $\mathbb{E}^3$ or $\mathbb{E}^4$ than to those of Coxeter groups. Notice that the diagram for $\mathcal{S}$ does not depend on the particular choice of unit normals for the mirror planes of the reflexions in $\mathcal{S}$. For a diagram $\mathcal{D}$, we write $\Delta = \Delta(\mathcal{D})$ for the corresponding Schläfli determinant.

In Section 9D, we shall label triangles in the diagram as well (at least in certain cases), but for the moment we find it more convenient not to do so.

We shall also refer to diagrams corresponding to basically equivalent sets of generators as *basically equivalent*. We now come to the core result of this section.

9C4 Theorem *Let G be a finite group acting on $\mathbb{C}^n$, which preserves a positive definite hermitian form, and is generated by n involutory hyperplane reflexions with linearly independent normals. Then a set $\mathcal{S}$ of generators of G can be chosen so that the branches of any diagram basically equivalent to that of $\mathcal{S}$ bear only integer marks.*

Proof. Since the group is finite, it has only finitely many distinct generating sets; hence, the Schläfli determinant Δ takes only finitely many (positive) values. We choose the generators so that Δ is minimal, but suppose that some basically equivalent generating set $\mathcal{S}$ has a diagram with a fractional mark $p > 2$. Let this mark be $p = p_{jk}$ on the branch $\{j, k\}$, so that $p = q/a$, with q and a integers such that $(q, a) = 1$ and $1 < a < \frac{1}{2}q$. Let m be such that $ma \equiv 1 \pmod q$. Then the operation which replaces S_j by $S'_j := (S_j S_k)^{m-1} S_j$ (and leaves the other generators unchanged) gives a new set of generators of G; by (9C2), the new Schläfli determinant is

$$\Delta' = \frac{\sin^2(\pi/q)}{\sin^2(a\pi/q)}\,\Delta < \Delta,$$

the obvious contradiction which we are seeking. In other words, the marks on branches of each basically equivalent diagram must, after all, be integers. □

When $n \geqslant 3$, apart from infinite families which are easy to treat (namely, those of the imprimitive groups $G(p, p, n)$), there are actually only three finite irreducible groups which admit generating sets with different Schläfli determinants; two are the real groups [3, 5] and [3, 3, 5], and the third is the triangle group $[1\ 1\ 1^4]^5$ which we shall describe in Section 9D. For all of these, it happens that generating sets with the same Schläfli determinant are basically equivalent. However, a general proof of this fact is, at the moment, lacking.

Similar arguments actually apply to the remaining cases, when $\Delta \leqslant 0$. When $\Delta < 0$, with care, almost the same argument as that of Theorem 9C4 can be used; now we make $|\Delta|$ as small as possible. For $\Delta = 0$, a different approach yields rather more; we shall discuss this case shortly.

Nothing we have done so far has needed to address reducibility. However, if the hermitian form is non-degenerate and G is reducible, then any diagram of G must have proper components, and conversely. Each component of the diagram corresponds to a reflexion group acting effectively on a lower dimensional linear subspace (the orthogonal complement of the intersection of its generating mirrors); hence we may employ any appropriate argument using induction on dimension. In effect, we have already made the following remark in Lemma 9B5, which particularly applies to finite groups.

9C5 Lemma *Let the hermitian form on $\mathbb{C}^n$ be non-degenerate, and let $G = \langle S_1, \ldots, S_n \rangle$ be a group generated by involutory reflexions S_j in hyperplanes through o whose normals v_j span $\mathbb{C}^n$ (that is, $\Delta \neq 0$). Then G is irreducible if and only if the diagram of G associated with $S_1, \ldots, S_n$ is connected.*

In fact, for finite groups, we can say slightly more.

9C6 Corollary *Let $G = \langle S_1, \ldots, S_n \rangle$ be an irreducible finite group generated by involutory reflexions S_j in hyperplanes through o whose normals v_j span $\mathbb{C}^n$, and let $x \in \mathbb{C}^n$. Then the centroid of xG is o.*

Proof. This centroid is a fixed point of G, and hence is the sole fixed point o. □

Next, we look at infinite discrete groups in unitary space. In our context, this case occurs when the form on $\mathbb{C}^n$ is positive definite and the n normals span the subspace $\mathbb{C}^{n-1}$, such that any $n - 1$ of them are linearly independent. For notational convenience, we often prefer to let these groups act on $\mathbb{C}^n$, so that they now have $n + 1$ generators. We first have an analogue of Bieberbach's Theorem in a special case; rather than appealing to general results about (arbitrary) discrete *euclidean* groups (see [344, §5.4]), we shall give a direct proof. A more general result can also be found in [343].

9C7 Theorem *Let $G = \langle S_0, \ldots, S_n \rangle$ be an infinite discrete group, which preserves a positive definite hermitian form on $\mathbb{C}^n$, and is generated by involutory reflexions S_j in hyperplanes whose normals v_j span $\mathbb{C}^n$, with any n of them linearly independent, and with a corresponding connected diagram. Then the set T of vectors corresponding to translations in G is discrete and spans $\mathbb{C}^n$ linearly. Moreover, regarded as a set of real vectors in $\mathbb{R}^{2n}$, either T spans $\mathbb{R}^{2n}$ linearly, or T spans an n-dimensional subspace, and G is then conjugate to a real group.*

Proof. For convenience, we may assume that we have the standard positive definite hermitian form on $\mathbb{C}^n$.

First note that G is affinely irreducible on $\mathbb{C}^n$, because its original diagram is connected; indeed, any subgroup corresponding to a connected subdiagram on n generators is (linearly) irreducible. (Recall that G is affinely irreducible on $\mathbb{C}^n$ if there is no non-trivial linear subspace K of $\mathbb{C}^n$, such that G permutes the translates of K.)

The discreteness implies at once that G is locally finite; recall that local finiteness means that every subgroup $G_j := \langle S_0, \ldots, S_{j-1}, S_{j+1}, \ldots, S_n \rangle$ is finite. Clearly, we may relabel if necessary, and assume that the subgroup $G_0 := \langle S_1, \ldots, S_n \rangle$ is itself irreducible, and has the origin o as its single fixed point. Then $V := oG$ is discrete and spans $\mathbb{C}^n$ affinely; indeed, the images of oS_0 under G_0 span $\mathbb{C}^n$ linearly.

Now regard G as a group of real euclidean isometries acting on $\mathbb{R}^{2n}$. Let D be the Voronoĭ region of o, namely the set of points of $\mathbb{R}^{2n}$ no farther from o than from any other point of V. Then D is invariant under G_0. There are two possibilities.

First, D is bounded. Then G is a discrete group acting effectively on $\mathbb{R}^{2n}$, to which the usual Bieberbach Theorem applies (see [344, §7.4] or Section 6A). Hence there is a translation subgroup T of G of full rank $2n$.

Otherwise, D is unbounded. Consider any (non-zero) direction of recession x of D, meaning that $\lambda x \in D$ for each $\lambda \geqslant 0$. Lemma 9C5 shows that the images of x under G_0 span $\mathbb{C}^n$ linearly, and so span a real G_0-invariant subspace L of $\mathbb{R}^{2n}$ of dimension at least n; indeed, they must span L positively (because o is the centroid of xG_0 by Corollary 9C6), and so $L \subseteq D$. Now consider any normal vector u to a hyperplane which determines a facet of D. The same argument shows that the images of u under G_0 must span positively a real G_0-invariant subspace M of $\mathbb{R}^{2n}$ of dimension at least n. Since L and M are orthogonal (because u is perpendicular to L), we conclude that they are complementary real subspaces of $\mathbb{R}^{2n}$, each of dimension n. Further, D is then the product with L of a bounded region D_M in M. Moreover, $V \subseteq M$; indeed, if $y \in V$, and H is the perpendicular bisector of o and y in $\mathbb{R}^{2n}$, then L must be parallel to H; hence $y \in M$ because y is orthogonal to L. Bearing in mind that G acts transitively on the Voronoĭ regions of elements in V, it now follows that the Voronoĭ region of each $y \in V$ must be the product with L of a bounded region in M, the latter being the image of D_M under any element in G which maps o to y. In particular, G leaves M invariant and permutes the translates of L. The action of G on M is faithful, with bounded fundamental region; indeed, if an element of G acts like the identity on M, then it must fix V pointwise, and must therefore be the identity on $\mathbb{C}^n$. We may now appeal to Bieberbach's theorem again, this time in M, to conclude that G has a subgroup of translations of rank n, whose direction vectors span M linearly; since the complexification of M is $\mathbb{C}^n$, they also span $\mathbb{C}^n$ linearly. Finally, for the conjugacy of G to a real group, just observe that the generating reflexions S_j permute V, and so suitably chosen normal vectors to their mirrors are real scalar multiples of vectors (again thought of as in $\mathbb{R}^{2n}$) joining pairs of points of V, and so are vectors in M. We thus identify M with $\mathbb{R}^n$ (as the set of real linear combinations of a basis of M), and we are done. □

In fact, a rather more general result than Theorem 9C7 holds; we have stated it only for reflexion groups, because this is what we need here.

Recall that the *special group* $\bar{G}$ of an infinite unitary group G acting on $\mathbb{C}^n$ consists of those unitary mappings φ such that $x \mapsto x\varphi + t$ is in G for some $t \in \mathbb{C}^n$. We may observe the following consequence of Theorem 9C7.

9C8 Corollary *Let $G = \langle S_0, \ldots, S_n \rangle$ be an infinite discrete group, which preserves a positive definite hermitian form on $\mathbb{C}^n$, and is generated by involutory reflexions S_j in hyperplanes whose normals v_j span $\mathbb{C}^n$, with any n of them linearly independent, and with a corresponding connected diagram. Then the special group $\bar{G}$ of G is finite.*

Proof. Whichever case occurs in Lemma 9C7, the subgroup T is a (discrete) translation group of full rank acting on an appropriate real subspace of $\mathbb{R}^{2n}$, and $\bar{G}$ acts faithfully on T as a group of automorphisms. Thus $\bar{G}$ is finite. □

Notice that $\bar{G}$ is a finite group generated by hyperplane reflexions, namely the translates of $S_0, \ldots, S_n$ (as usual identified with their mirrors) which contain o.

We now have the "crystallographic" criterion for infinite unitary reflexion groups.

9C9 Theorem *Let $G = \langle S_0, \ldots, S_n \rangle$ be an infinite discrete group, which preserves a positive definite hermitian form on $\mathbb{C}^n$, and is generated by involutory reflexions S_j in hyperplanes whose normals v_j span $\mathbb{C}^n$, with any n of them linearly independent, and with a corresponding connected diagram. Then the marks on the branches of any diagram of G can only be* 2, 3, 4, 6 *or* ∞.

Proof. It is clear that the diagram of every set of $n+1$ generators of G must be connected; indeed, a non-connected diagram would yield a pair of complementary orthogonal subspaces whose translates are permuted by G.

Now, if the (finite) period of any product of two reflexions in the group is p, then by taking a suitable power of this product, we may suppose that G contains a complex rotation through an angle $2\pi/p$. Theorem 9C7 then implies that there is a translation $x \mapsto x + b$ in G whose vector $b \in T$ (in the notation of that theorem) is not parallel to the axis of the rotation. In particular, we may choose coordinates so that we have a rotation

$$\varphi\colon (\xi_1, \xi_2, \ldots, \xi_n) \mapsto (\omega\xi_1, \omega^{-1}\xi_2, \xi_3, \ldots, \xi_n),$$

with $\omega := \exp(2i\pi/p)$. Then $b = (\beta_1, \ldots, \beta_n)$ is such that one (at least) of β_1 and β_2 is non-zero. Choose b so that $|\beta_1|^2 + |\beta_2|^2 > 0$ is minimal (here we appeal to the discreteness of T). For $p \geqslant 7$, we have $b\varphi \in T$, so that $b\varphi - b \in T$ also. But

$$|(\omega - 1)\beta_1|^2 + |(\omega^{-1} - 1)\beta_2|^2 = 4\sin^2\frac{\pi}{p}\,(|\beta_1|^2 + |\beta_2|^2) < |\beta_1|^2 + |\beta_2|^2,$$

a contradiction to the choice of b. For $p = 5$, the argument is only a little different; the same contradiction is obtained from the translation by $b\varphi + b\varphi^{-1} \in T$, for which

$$|(\omega + \omega^{-1})\beta_1|^2 + |(\omega^{-1} + \omega)\beta_2|^2 = \tau^{-2}(|\beta_1|^2 + |\beta_2|^2) < |\beta_1|^2 + |\beta_2|^2,$$

with (as usual) $\tau := \frac{1}{2}(1 + \sqrt{5})$. □

Note that the arguments of the proof actually show that G cannot contain a p-fold complex rotation (with axis of codimension 2) when $p \neq 2, 3, 4, 6$ or ∞.

Observe that Theorem 9C9 contains the appropriate infinite version of Theorem 9C4: in this case, *no* diagram of the group can bear fractional marks on its branches. Notice also that irreducibility is needed in order to exclude (for example) a group which is a direct product of an infinite discrete group acting on a proper subspace with a finite group.

It will not be a main purpose of this chapter to classify the finite subgroups of $\mathrm{U}_n(\mathbb{C})$ generated by n involutory hyperplane reflexions, although we shall actually discuss them all. However, we shall see how to do this when $n = 3$, and we shall look at certain aspects of the question when $n \geqslant 4$. In general, we encounter two main problems. The first is that, as we have seen, a reflexion group can be generated in many different ways, and it is not easy to check whether two such groups are, in fact, isomorphic. Another stumbling block is that, if we are not careful, we run up against notational complexity.

Ultimately, we wish to phrase all the properties of our groups, and in particular of their diagrams, in terms of the periods of the products of certain pairs of reflexions. These (integer) periods will correspond to marks on diagrams basically equivalent to the original diagram.

As before, we take $\{v_1, \ldots, v_n\}$ to be the set of unit normals to the mirrors of the generating reflexions for G, which again preserve the hermitian form $\langle\cdot,\cdot\rangle$. For the moment, we shall not assume that G is finite. At the back of our mind will be the fact that, in the real case and with the natural generators, we shall have

$$\langle v_j, v_k\rangle = -\cos\frac{\pi}{p_{jk}};$$

the minus sign here motivates some of the definitions in the following.

Let $C = (j(1), \ldots, j(m))$ be a *cycle* (or *m-cycle*) of distinct numbers in $\{1, \ldots, n\}$, so that C is an equivalence class of sequences under the action of the cyclic group, and we do not regard as distinct from C the same cycle beginning at a different point. Thus we naturally take indices in a cycle modulo its *length* m. We then define

9C10 $$\gamma(C) := \prod_{i=1}^{m} \langle v_{j(i)}, v_{j(i+1)} \rangle,$$

9C11 $$\omega(C) := \begin{cases} 1, & \text{if } m = 1, \\ -\gamma(C), & \text{if } m = 2, \\ 2(-1)^{m-1}\Re\gamma(C), & \text{if } m \geqslant 3, \end{cases}$$

9C12 $$\vartheta(C) := \arg(-1)^m \gamma(C),$$

as long as $\gamma(C) \neq 0$, where $\Re z$ and $\arg z$ denote the real part and argument of a complex number z, with $-\pi < \arg z \leqslant \pi$ (if $z \neq 0$). Note that $\gamma(C) = |\langle v_{j(1)}, v_{j(2)} \rangle|^2$ if $m = 2$.

Clearly, each of $\gamma(C)$, $\omega(C)$ and $\vartheta(C)$ is invariant under cyclic permutation of the numbers in C. On the other hand, if we reverse C to obtain $\bar{C}$ (say), then we have

$$\gamma(\bar{C}) = \overline{\gamma(C)}, \quad \omega(\bar{C}) = \omega(C), \quad \vartheta(\bar{C}) = -\vartheta(C).$$

In particular, $\omega(C)$ is invariant under the action of the dihedral group D_m on the indices; if C represents a circuit $\mathcal{C}$ in the diagram of G, we may then unambiguously define $\omega(\mathcal{C}) := \omega(C)$. Indeed, if $m \geqslant 3$ and $p_{j(i),j(i+1)} > 2$ for each i, then we have

9C13 $$\omega(C) = -2\left(\prod_{i=1}^{m} \cos \frac{\pi}{p_{j(i),j(i+1)}}\right) \cos \vartheta(C).$$

(The convenience of incorporating the minus sign in the definition (9C11) of $\omega(C)$ will become clear in Theorem 9C14.) The angle $\vartheta(C)$ changes sign on reversing the cycle. However, for many purposes, it is the absolute value of the angle which is more important, so that we often allow reversal of order in a cycle as well. If C represents a circuit $\mathcal{C}$ in the diagram of G, then the absolute value of $\vartheta(C)$ (and, when convenient, $\vartheta(C)$ itself) is called the *turn* of $\mathcal{C}$ and is denoted by $\vartheta(\mathcal{C})$.

Let $j(1), \ldots, j(m)$ be distinct numbers. If its period is finite, the product

$$T_{j(1),\ldots,j(m)} := S_{j(1)} S_{j(2)} \cdots S_{j(m)} S_{j(m-1)} \cdots S_{j(2)}$$

is a complex rotation through an angle $2\pi/q$ (we ignore orientations here), where $q \geqslant 2$ is a rational number; in this case we define $p_{j(1),\ldots,j(m)} := q$. Observe, in this context, that the change of generators $S_{j(m)} \mapsto S_{j(2)} \cdots S_{j(m)} S_{j(m-1)} \cdots S_{j(2)}$ corresponds to a sequence of basic operations on the generators. For a finite group G, it then follows from Theorem 9C4 that the original generators may be chosen so that q is actually a positive integer. However, it is convenient in several contexts to allow q ($\geqslant 2$) merely to be rational (even if the group is finite). In any case, q is such that the two eigenvalues different from 1 of the rotation $T_{j(1),\ldots,j(m)}$ are $e^{\pm 2\pi i/q}$. Note that $p_{j(1),\ldots,j(m)}$ is only invariant under reversal of the indices, but generally not under cyclic permutation (however, there are important exceptions to this).

Our first observation is that the Schläfli determinant $\Delta = \Delta(\mathcal{D})$ can be calculated directly from the circuits in the diagram $\mathcal{D}$ and their turns. The fact that we want to have a sum (rather than an alternating sum) in the formula explains the inclusion of the minus sign in (9C11). The formula expresses Δ as a sum of products, with each product corresponding to a (*complete*) *circuit matching* $\mathcal{M}$ of $\mathcal{D}$, meaning a collection of node-disjoint circuits of $\mathcal{D}$ such that each node of $\mathcal{D}$ occurs in exactly one circuit; we allow here a circuit to consist of a single node (and no branch), or two nodes joined by a branch traversed in both directions. We denote by $\mathcal{M}(\mathcal{D})$ the family of all complete circuit matchings $\mathcal{M}$ of $\mathcal{D}$.

9C14 Theorem *Let $\mathcal{D}$ be a diagram of a group G acting on $\mathbb{C}^n$, which preserves a hermitian form, and is generated by n involutory hyperplane reflexions. Then*

9C15
$$\Delta(\mathcal{D}) = \sum_{\mathcal{M}\in\mathcal{M}(\mathcal{D})} \prod_{\mathcal{C}\in\mathcal{M}} \omega(\mathcal{C}).$$

Proof. We see this by a direct calculation of the determinant using the alternating sum formula

$$\Delta = \sum_{\rho\in S_n} \operatorname{sign}\rho \, \langle v_1, v_{1\rho}\rangle \cdots \langle v_n, v_{n\rho}\rangle,$$

where $\operatorname{sign}\rho$ denotes the sign of the permutation ρ in the symmetric group S_n. Then ρ gives rise to a non-zero term in the determinant if and only if each cycle in the cycle expression for ρ corresponds to a circuit in $\mathcal{D}$; indeed, recall that $\langle v_i, v_j\rangle = 0$ if the nodes i, j are not joined by a branch. For a non-zero term, each fixed point of ρ corresponds to a 1-circuit $\mathcal{C}$ in $\mathcal{D}$ which contributes $\omega(\mathcal{C}) = 1$ to the product, and each 2-cycle $C = (i\ j)$ of ρ gives a 2-circuit $\mathcal{C}$ which contributes a real factor $\omega(\mathcal{C}) = \omega(C) = -|\langle v_i, v_j\rangle|^2$, where the minus sign corresponds to the contribution of the 2-cycle to $\operatorname{sign}\rho$. Now we just group together with ρ those permutations which are obtained from ρ by reversing the order of any of its cycles of length at least 3; if ρ has k such cycles, then there are 2^k such permutations. (Of course, there are no such reversals for 1- and 2-cycles.)

More precisely, let $\rho = \sigma_1 \cdots \sigma_{k+l}$ be the cycle decomposition of ρ, where σ_i is a cycle of length at least 3 if $i \leqslant k$, and of length 2 if $i > k$. Let C_i denote the cycle corresponding to σ_i, and let C_i^{-1} $(:= \bar{C}_i)$ denote the reverse cycle corresponding to σ_i^{-1}. Then the contribution to Δ arising from the 2^k permutations is given by

$$s = \left(\prod_{i=k+1}^{k+l} \omega(C_i)\right) \sum_{\varepsilon_1,\ldots,\varepsilon_k=\pm 1} \operatorname{sign}\sigma_1 \cdots \operatorname{sign}\sigma_k \, \gamma\left(C_1^{\varepsilon_1}\right)\cdots\gamma\left(C_k^{\varepsilon_k}\right).$$

But $\gamma(C_i^{-1}) = \overline{\gamma(C_i)}$ for each i, and so we may group together pairs of summands which differ only in their terms for $i = k$ (say), and then pull out the factor

$$\operatorname{sign}\sigma_k \left(\gamma(C_k) + \gamma\left(C_k^{-1}\right)\right) = 2(-1)^{m_k-1}\Re(\gamma(C_k)) = \omega(C_k),$$

where m_k is the length of the cycle σ_k. This leads to an expression for s similar to the original one, except that k has been replaced by $k-1$ and l by $l+1$. We now apply the same argument to the terms with $i = k-1$, and so on. At the final stage, we obtain the expression

$$s = \prod_{i=1}^{k+l} \omega(C_i) = \prod_{\mathcal{C} \in \mathcal{M}} \omega(\mathcal{C}),$$

where $\mathcal{M}$ is the circuit matching determined by $C_1, \ldots, C_{k+l}$ and the fixed points of ρ.

Conversely, every circuit matching $\mathcal{M}$ of $\mathcal{D}$ arises in this way from a permutation ρ by grouping together with ρ those permutations which are obtained from ρ by reversing the order of any of its cycles of length at least 3; the 1-circuits and 2-circuits in $\mathcal{M}$ then correspond to the fixed points and 2-cycles of ρ. This proves the desired formula for the determinant. □

We next calculate certain of the numbers $p_{j(1),\ldots,j(m)}$. We call a circuit $\mathcal{C}$ in a diagram $\mathcal{D}$ *diagonal-free* if each branch of $\mathcal{D}$, which connects two nodes of $\mathcal{C}$, is a branch of $\mathcal{C}$.

9C16 Lemma *As before, let G be a group on $\mathbb{C}^n$, which preserves a hermitian form and is generated by n involutory hyperplane reflexions. Let $C = (j(1), \ldots, j(m))$ be a cycle which induces a diagonal-free circuit $\mathcal{C}$ in the diagram $\mathcal{D}$ of G. If the product $S_{j(1)} S_{j(2)} \cdots S_{j(m)} S_{j(m-1)} \cdots S_{j(2)}$ is a genuine rotation (this is true if it has finite period), then its rotation angle $2\pi/q$ depends only on the numbers $p_{j(i),j(i+1)}$ and the turn $\vartheta(\mathcal{C})$, the absolute value of $\vartheta(C)$.*

Proof. Let us simplify the notation by writing $j(i) = i$ for each $i = 1, \ldots, m$. Keeping to the notation used before, we need to calculate

$$\cos \frac{\pi}{q} = |\langle v_1, v_m S_{m-1} \cdots S_2 \rangle|.$$

(Clearly, the product $T_{1,\ldots,m} := S_1 S_2 \cdots S_m S_{m-1} \cdots S_2$ is a genuine rotation if and only if this inner product has absolute value less than 1; also, q must be rational if the period is finite.) Since $\mathcal{C}$ is diagonal-free, we can rescale the vectors v_j to make all but one inner product $\langle v_i, v_j \rangle$ real. Thus, there is no loss of generality in supposing that

$$\langle v_i, v_{i+1} \rangle = -\cos \frac{\pi}{p_{i,i+1}}$$

for $i = 1, \ldots, m-1$; in other words, we *load* the whole turn onto the last branch $\{1, m\}$, giving

$$\langle v_1, v_m \rangle = -e^{i\vartheta(C)} \cos \frac{\pi}{p_{1m}}$$

(then the argument of $(-1)^m \gamma(C)$ is indeed $\vartheta(C)$, as required). Because the circuit has no diagonals, it is straightforward to prove by induction that

$$v' := v_m S_{m-1} \cdots S_2 = \sum_{j=0}^{m-2} 2^j \left(\prod_{i=1}^{j} \cos \frac{\pi}{p_{m-i,m-i+1}} \right) v_{m-j}.$$

Since $\langle v_1, v_k\rangle = 0$ unless $k = 2$ or m, it follows that

$$\begin{aligned}
\cos^2\frac{\pi}{q} &= |\langle v_1, v'\rangle|^2 \\
&= \left|\langle v_1, v_m\rangle + 2^{m-2}\left(\prod_{k=2}^{m-1}\cos\frac{\pi}{p_{k,k+1}}\right)\langle v_1, v_2\rangle\right|^2 \\
&= \left|-e^{i\vartheta(C)}\cos\frac{\pi}{p_{1m}} - 2^{m-2}\prod_{k=1}^{m-1}\cos\frac{\pi}{p_{k,k+1}}\right|^2
\end{aligned}$$

9C17 $$= \cos^2\frac{\pi}{p_{1m}} + 2^{m-1}\left(\prod_{k=1}^{m}\cos\frac{\pi}{p_{k,k+1}}\right)\cos\vartheta(C) + 2^{2m-4}\left(\prod_{k=1}^{m-1}\cos^2\frac{\pi}{p_{k,k+1}}\right);$$

in the second term on the right, suffixes are as usual taken modulo m, so that $p_{m,m+1} = p_{1m}$. This is the result promised. □

For a circuit which has diagonals, the formula corresponding to (9C17) is obviously more complicated, since there will be terms involving turns of circuits which trace such diagonals and certain branches of the given circuit. For diagonal-free circuits, (9C17) can be written alternatively as

$$2^{m-1}\left(\prod_{k=1}^{m}\cos\frac{\pi}{p_{k,k+1}}\right)\cos\vartheta(C) = \cos^2\frac{\pi}{q} - \cos^2\frac{\pi}{p_{1m}} - 2^{2m-4}\left(\prod_{k=1}^{m-1}\cos^2\frac{\pi}{p_{k,k+1}}\right),$$

giving the turn in terms of the rotation angle, rather than the other way round.

Before we move on, let us note an important case of (9C17). If $p_{k,k+1} = 3$ for all but one k, with the remaining one p, say, then

9C18 $$\cos^2\frac{\pi}{q} = \cos^2\frac{\pi}{p} + \cos\frac{\pi}{p}\cos\vartheta(C) + \frac{1}{4},$$

irrespective of which branch carries the mark p. We leave it as an exercise for the reader to show that, in this case, all the corresponding products

$$T_{i,i+1,\dots,i+m-1} := S_i S_{i+1}\cdots S_{i+m-2}S_{i+m-1}S_{i+m-2}\cdots S_{i+2}S_{i+1},$$

with indices taken modulo m, are actually conjugate in the group (see Section 9A for the case $m = 3$); hence all the numbers $p_{i,i+1,\dots,i+m-1}$ are actually the same. In the even more special case with all $p_{k,k+1} = 3$, the equation (9C18) takes the form

$$\cos^2\frac{\pi}{q} = \frac{1}{4} + \frac{1}{2}\cos\vartheta + \frac{1}{4} = \cos^2\frac{\vartheta(C)}{2},$$

so that we have

$$\vartheta(C) = \pm\frac{2\pi}{q}.$$

If $q = 2$, then the circuit itself represents a real group; indeed, we have $\vartheta(C) = \pi$, and so all inner products $\langle v_i, v_j \rangle$ can be made real.

Notice that, in general, the expression in (9C17) is not invariant under cyclic permutation of the elements in C, so that the rotation angles of the products $T_{i,i+1,\ldots,i+m-1}$, and hence the numbers $p_{i,i+1,\ldots,i+m-1}$, will not all be the same.

More generally, while the diagonal-free circuits do not determine the periods of the other rotations (meaning here the products of two arbitrary reflexions), they do strongly restrict what values they can take. We see this as follows. We can concatenate two circuits $\mathcal{C}'$ and $\mathcal{C}''$ in the obvious way: if $\mathcal{C}'$ and $\mathcal{C}''$ share a single contiguous set of branches, then their *concatenation* is $\mathcal{C} := \mathcal{C}' \triangle \mathcal{C}'' = (\mathcal{C}' \cup \mathcal{C}'') \setminus (\mathcal{C}' \cap \mathcal{C}'')$, where we delete the common branches and any resulting isolated nodes. We then have

9C19 Lemma *Let the circuit $\mathcal{C} = \mathcal{C}' \triangle \mathcal{C}''$ be the concatenation of two circuits in a diagram. Let C' and C'' be cycles which represent $\mathcal{C}'$ and $\mathcal{C}''$, such that C' and C'' traverse the branches common to $\mathcal{C}'$ and $\mathcal{C}''$ in opposite directions. Then*

$$\vartheta(C) = \vartheta(C') + \vartheta(C'') \pmod{2\pi}.$$

Proof. Let

$$C' = (i(1), \ldots, i(m-1), k(0), \ldots, k(l)), \quad C'' = (j(1), \ldots, j(n-1), k(l), \ldots, k(0)),$$

so that

$$C = (i(1), \ldots, i(m-1), k(0), j(1), \ldots, j(n-1), k(l))$$

is a cycle of length $m + n$ which represents $\mathcal{C}$. In $\gamma(C')$ and $\gamma(C'')$, the common branches of $\mathcal{C}'$ and $\mathcal{C}''$ yield complex conjugate terms, so that $\gamma(C')\gamma(C'') = a\gamma(C)$ with $a > 0$; hence also

$$(-1)^{m+l}\gamma(C') \cdot (-1)^{n+l}\gamma(C'') = (-1)^{m+n} a\gamma(C).$$

It follows that $\vartheta(C) = \vartheta(C') + \vartheta(C'')$. □

It is clear that circuits may be concatenations of other circuits in different ways, which will lead to syzygies among their turns. We shall discuss this further in Section 9E.

While we could always appeal to (9C15), it is also convenient to set up a recursive calculation for certain Schläfli determinants; the result is routine.

9C20 Lemma *Let G be a group on $\mathbb{C}^n$ with diagram $\mathcal{D}$, which preserves a hermitian form, and is generated by n involutory hyperplane reflexions. Let the node 1 of $\mathcal{D}$ belong to a single branch $\{1, 2\}$ marked t. If Δ is the Schläfli determinant of $\mathcal{D}$, and if Δ_1 and Δ_{12} are those of the subdiagrams obtained by deleting node 1 or nodes 1 and 2, respectively, then*

$$\Delta = \Delta_1 - \cos^2 \frac{\pi}{t} \Delta_{12}.$$

A branch of the kind considered in Lemma 9C20 is usually part of a tail. A useful reduction shows that, for the most part, we need only consider tails in diagrams which are unmarked.

9C21 Lemma *Let G be a group on $\mathbb{C}^n$ with diagram $\mathcal{D}$, which preserves a hermitian form, and is generated by n involutory hyperplane reflexions. Let 1 be the end-node of a tail of $\mathcal{D}$ containing at least two branches $\{1, 2\}$ and $\{2, 3\}$. If $p_{23} = 3$, then G has a subgroup H, whose diagram is obtained from $\mathcal{D}$ by deleting node 1 (and branch $\{1, 2\}$), and transferring the mark $p := p_{12}$ from branch $\{1, 2\}$ to branch $\{2, 3\}$ (while leaving all other branches and marks unchanged).*

Proof. This is straightforward; we have seen the same argument earlier. Let $G = \langle S_1, S_2, S_3, \ldots, S_n\rangle$ (with the standard correspondence of generators S_j to nodes j), and consider the operation

$$(S_1, S_2, S_3, \ldots, S_n) \mapsto (S_2 S_1 S_2, S_3, \ldots, S_n) =: (R_2, R_3, \ldots, R_n).$$

Then

$$R_2 R_3 = S_2 S_1 S_2 S_3 \sim S_1 S_2$$

and

$$R_2 R_i = S_2 S_1 S_2 S_i \sim S_2 S_1 S_i S_2 \sim S_1 S_i \quad (i \geqslant 4),$$

and the claim follows. □

We now comment on the groups $[k\, l\, m^p]^q$ introduced in Section 9A. A suitable operation (change of generators) replaces the triangle and tail of $m - 1$ nodes and branches in the diagram for $[k\, l\, m^p]^q$ with a circuit of $m + 2$ nodes and branches. The operation is as follows. Label the nodes of the triangle belonging to the other two tails 1 and 2, and the nodes of the tail of $m - 1$ branches 3 to $m + 2$, as in Figure 9C1 (the labels on the two tails to the left are unimportant). With $S_1, S_2, \ldots, S_{m+2}$ the corresponding generators, the operation is

$$\begin{aligned}(\ldots, S_1, S_2, \ldots, S_{m+2}) \mapsto (\ldots, S_{m+2} S_{m+1} \cdots S_3 S_1 S_3 \cdots S_{m+1} S_{m+2}, S_2, \ldots, S_{m+2})\\ =: \quad (\ldots, R_1, R_2, \ldots, R_{m+2}).\end{aligned}$$

This is clearly obtained by a sequence of basic operations, and so leaves the Schläfli determinant unchanged. The nodes in the circuit are labelled (in order) $1, 2, \ldots, m + 2$; all branches are unmarked save $\{1, 2\}$ (from whose nodes leave the two tails) which is

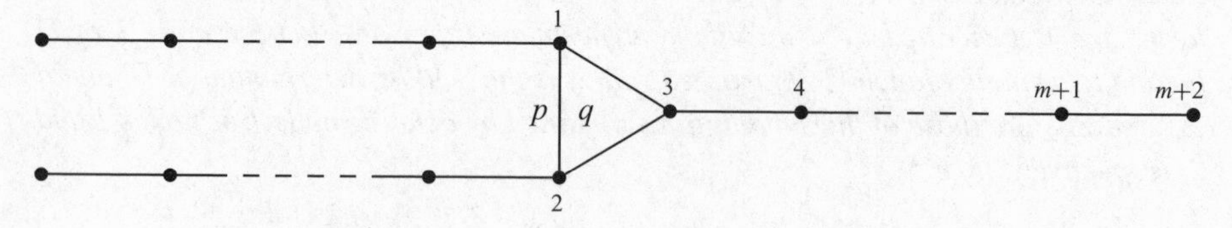

Figure 9C1. The standard diagram for $[k\, l\, m^p]^q$.

marked q, while the circuit is marked p. Indeed, we have

$$\begin{aligned} R_1R_2 &= S_{m+2}\cdots S_4S_3S_1S_3S_2S_4\cdots S_{m+2} \sim S_3S_1S_3S_2, \\ R_1R_{m+2} &= S_{m+2}S_{m+1}\cdots S_4S_1S_3S_1S_4\cdots S_{m+1} \\ &\sim S_{m+2}S_{m+1}\cdots S_4S_3S_4\cdots S_{m+1} \\ &\sim\cdots\sim S_{m+2}S_{m+1}S_mS_{m+1}, \end{aligned}$$

and, for $i = 3, \ldots, m+1$,

$$\begin{aligned} R_1R_i &= S_{m+2}\cdots S_3S_1S_3\cdots S_{m+2}S_i \\ &\sim S_{i+1}S_i\cdots S_3S_1S_3\cdots S_iS_{i+1}S_i \\ &\sim S_{i-1}\cdots S_3S_1S_3\cdots S_{i-1}S_{i+1} \\ &= S_{i-1}\cdots S_3S_1S_{i+1}S_3\cdots S_{i-1} \\ &\sim S_1S_{i+1}. \end{aligned}$$

Moreover, bearing in mind that all the numbers $p_{i,i+1,\ldots,i+m+1}$ for the $(m+2)$-circuit are the same, we can now calculate the interior mark of the circuit from

$$R_2R_3R_4\cdots R_{m+2}R_1R_{m+2}\cdots R_4R_3 = S_2S_1,$$

giving the mark $p\ (= p_{2,3,\ldots,m+2,1})$.

So, for example, the new diagram for the group $[3\,2\,4^5]^7$ (which is actually infinite, and not even locally finite!) would be as in Figure 9C2.

It is worth noting two special cases of this operation. First, $q = 2$ implies that $p = 3$ (because then $S_1S_2 \sim S_3S_1$), and so we see from the new diagram that

$$[k\,l\,m^3]^2 = [k\,l\,m]^2 \cong [3^{k+l+m-1}]$$

is always finite. Second, if $p = 2$ (so that the corresponding branch is missing) then $q = 3$ (because $S_1S_2S_3S_2 \sim S_1S_3$), and so we obtain an alternative diagram for $[k\,l\,m^2]^3 = [3^{k,l,m-1}]$, with a single circuit marked 2 (and no marked branches).

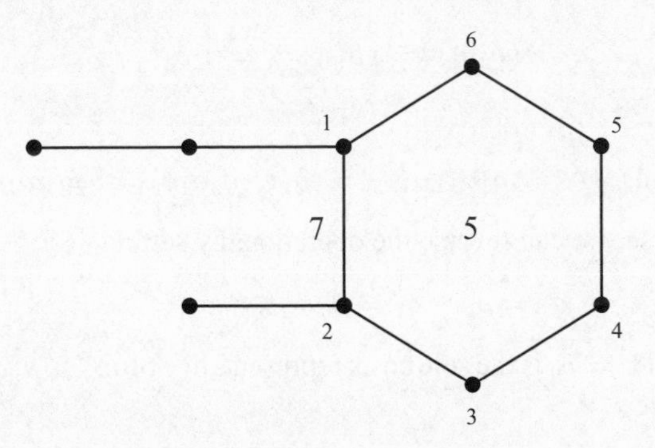

Figure 9C2. The new diagram for $[3\,2\,4^5]^7$.

9D Generalized Triangle Groups

For $2 \leqslant p, q, r, s \leqslant \infty$, the *generalized triangle group* $\Gamma^3(p,s;q,r)$ is the abstract group with presentation

9D1 $$\sigma_1^2 = \sigma_2^2 = \sigma_3^2 = (\sigma_1\sigma_2)^p = (\sigma_1\sigma_3)^q = (\sigma_2\sigma_3)^r = (\sigma_1\sigma_3\sigma_2\sigma_3)^s = \varepsilon.$$

Thus $\Gamma^3(p,s;q,r)$ is a quotient of the (usually infinite) group obtained by dropping the last relation $(\sigma_1\sigma_3\sigma_2\sigma_3)^s = \varepsilon$. Note that the interchange of generators $\sigma_1 \leftrightarrow \sigma_2$ and the basic (involutory) change of generators $\sigma_2 \leftrightarrow \sigma_3\sigma_2\sigma_3$ show that

$$\Gamma^3(p,s;r,q) \cong \Gamma^3(p,s;q,r) \cong \Gamma^3(s,p;q,r);$$

more precisely, $\sigma_1 \leftrightarrow \sigma_2$ corresponds to $q \leftrightarrow r$, and $\sigma_2 \leftrightarrow \sigma_3\sigma_2\sigma_3$ to $p \leftrightarrow s$.

In general, the definition of $\Gamma^3(p,s;q,r)$ says nothing about the periods of the other two products of involutions analogous to $\sigma_1\sigma_3\sigma_2\sigma_3$, namely, $\sigma_2\sigma_1\sigma_3\sigma_1$ and $\sigma_3\sigma_2\sigma_1\sigma_2$. However, if, say, $r = 3$, then

$$\sigma_1\sigma_3\sigma_2\sigma_3 = \sigma_1\sigma_2\sigma_3\sigma_2 \sim \sigma_3\sigma_2\sigma_1\sigma_2,$$

showing that the latter also has period s. If p or q is 3 as well, then all three (essentially) distinct such products have the same period (compare the discussion in Section 9A). In this context, it is usual to take $q = r = 3$ and denote the group by

$$[1\ 1\ 1^p]^s := \Gamma^3(p,s;3,3).$$

(Strictly speaking, this should denote the geometric group $G^3(p,s;3,3)$ which we shall describe later. In [114], l and m are used instead of p and s.)

Before we go on, let us also note the special case $q = r$. Here, there is an obvious involutory automorphism τ of the group, such that

$$\tau\sigma_1\tau = \sigma_2, \quad \tau\sigma_3\tau = \sigma_3.$$

Then the operation

9D2 $$(\sigma_1,\sigma_2,\sigma_3;\tau) \mapsto (\tau,\sigma_1,\sigma_3) =: (\rho_0,\rho_1,\rho_2)$$

yields the group $[2p,q]_{2s}$ of the regular polyhedron $\{2p,q\}_{2s}$ (see Section 7B). To see this, first,

$$(\rho_0\rho_1)^2 = \tau\sigma_1\tau\cdot\sigma_1 = \sigma_2\sigma_1,$$

and second,

$$(\rho_0\rho_1\rho_2)^2 = \tau\sigma_1\sigma_3\tau\sigma_1\sigma_3 = \tau\sigma_1\tau\cdot\sigma_3\sigma_1\sigma_3 = \sigma_2\sigma_3\sigma_1\sigma_3.$$

In turn, of course, we can reverse the operation, by setting

$$\sigma_1 = \rho_1, \quad \sigma_2 = \rho_0\rho_1\rho_0, \quad \sigma_3 = \rho_2.$$

It is easy to see that $\tau = \rho_0$ is the sole coset representative of $\Gamma^3(p,s;q,q)$ in $[2p,q]_{2s}$, and we thus deduce

9D3 Theorem *If* $2 \leqslant p,q,s \leqslant \infty$, *then* $[2p,q]_{2s} \cong \Gamma^3(p,s;q,q) \rtimes C_2$.

We now consider finite irreducible subgroups of $U_3(\mathbb{C})$ generated by involutory plane reflexions, with a view to seeing which groups $\Gamma^3(p,s;q,r)$ have natural linear represention in $U_3(\mathbb{C})$. We shall adopt a more geometric approach than that of Coxeter [114, §3], in that we always work within $U_3(\mathbb{C})$ itself, rather than constructing a general linear representation which preserves a hermitian form, and only afterwards asking whether or not the group is finite. In other words, the context is that of Section 9C.

So, let $G = \langle S_1, S_2, S_3 \rangle$ be a reflexion group in $\mathbb{C}^3$ which preserves a hermitian form and is generated by the involutory hyperplane reflexions S_j. Usually, we take the standard positive definite hermitian form on $\mathbb{C}^3$, so that G is a subgroup of $U_3(\mathbb{C})$, but it is convenient to allow other choices. Employing the conventions of Section 9C, we simplify the notation by writing

$$p := p_{12}, \quad q := p_{13}, \quad r := p_{23};$$

recall that $p, q, r \geqslant 2$ (and are finite rational numbers), but for the moment, we shall not insist that p, q, r be integers. Actually, since the real cases are known, we may suppose that p, q, $r > 2$, although the discussion will include all possible cases. Thus, if

$$xS_j = x - 2\langle x, v_j \rangle v_j,$$

with v_j a unit vector in $\mathbb{C}^3$, then

$$|\langle v_1, v_2 \rangle| = \cos\frac{\pi}{p}, \quad |\langle v_1, v_3 \rangle| = \cos\frac{\pi}{q}, \quad |\langle v_2, v_3 \rangle| = \cos\frac{\pi}{r}.$$

The diagram $\mathcal{D}$ of G is a triangle with marks p, q, r on the branches $\{1, 2\}$, $\{1, 3\}$, $\{2, 3\}$, respectively (we first ignore any interior marks on the triangle). As before, we shall denote the turn in the triangle by ϑ; then $\vartheta = \arg(-\gamma)$, with $\gamma = \langle v_1, v_2 \rangle \langle v_2, v_3 \rangle \langle v_3, v_1 \rangle$ (that is, we take the actual turn, not its absolute value).

As we saw in Lemma 9B7, the group G can only be finite if the Schläfli determinant Δ corresponding to $\{S_1, S_2, S_3\}$ is non-negative. If $\Delta > 0$, we then have a positive definite form, which we may take to be the standard form. For a finite group, the case $\Delta = 0$ can only occur in one of two ways, which we exclude from now on; indeed, either the form is degenerate (and then trivially $\Delta = 0$ for any set of normal vectors), or the form is positive definite, the normal vectors $\{v_1, v_2, v_3\}$ are linearly dependent, and $\{S_1, S_2, S_3\}$ generate a (finite) group (bear in mind that the reflexion planes all pass through the origin).

Later on, we shall also investigate infinite discrete groups in the unitary plane which are generated by 3 involutory reflexions. For these groups, we again take the standard positive definite form on $\mathbb{C}^3$, but with linearly dependent normal vectors $\{v_1, v_2, v_3\}$, spanning the plane on which the group is acting; moreover, local finiteness will require that any pair of these vectors be linearly independent.

Proceeding with the general case, by (9C15) the Schläfli determinant corresponding to $\{S_1, S_2, S_3\}$ is

9D4 $$\Delta = 1 - \cos^2\frac{\pi}{p} - \cos^2\frac{\pi}{q} - \cos^2\frac{\pi}{r} - 2\cos\frac{\pi}{p}\cos\frac{\pi}{q}\cos\frac{\pi}{r}\cos\vartheta.$$

Indeed, the diagram $\mathcal{D}$ has three types of circuit matchings: the trivial matching (consisting of three single nodes) contributes the 1, the three matchings with one 2-circuit contribute the three square terms, and finally the 3-circuit itself yields the last term (see (9C13)). If the hermitian form is positive definite, then necessarily $\Delta \geqslant 0$, with $\Delta = 0$ if and only if $\{v_1, v_2, v_3\}$ are linearly dependent.

If the period of $S_1S_3S_2S_3$ in G is finite (which is true if G itself is finite), then $S_1S_3S_2S_3$ is a complex rotation through an angle $2\pi/s$, with $s := p_{132}$ rational, which only depends on p, q, r and ϑ, and is given by

9D5 $$\cos^2\frac{\pi}{s} = \cos^2\frac{\pi}{p} + 4\cos\frac{\pi}{p}\cos\frac{\pi}{q}\cos\frac{\pi}{r}\cos\vartheta + 4\cos^2\frac{\pi}{q}\cos^2\frac{\pi}{r}$$

(see Lemma 9C16 and (9C17)). Of course, $2\pi/s$ is also the angle for the conjugate rotation $S_2S_3S_1S_3$.

Eliminating ϑ between the equations (9D4) and (9D5) yields

$$\begin{aligned} 2\Delta &= 2 - 2\cos^2\frac{\pi}{p} - 2\cos^2\frac{\pi}{q} - 2\cos^2\frac{\pi}{r} + \cos^2\frac{\pi}{p} + 4\cos^2\frac{\pi}{q}\cos^2\frac{\pi}{r} - \cos^2\frac{\pi}{s} \\ &= 2 - 2\cos^2\frac{\pi}{q} - 2\cos^2\frac{\pi}{r} + 4\cos^2\frac{\pi}{q}\cos^2\frac{\pi}{r} - \cos^2\frac{\pi}{p} - \cos^2\frac{\pi}{s}. \end{aligned}$$

Multiplying by 2 again, and using $2\cos^2\psi - 1 = \cos 2\psi$, results in the fundamental equation

9D6 $$4\Delta = 2\cos\frac{2\pi}{q}\cos\frac{2\pi}{r} - \cos\frac{2\pi}{p} - \cos\frac{2\pi}{s}.$$

This only depends on p, q, r and s (but not explicitly on ϑ), so that we can write $\Delta = \Delta(p, s; q, r)$. Of course, it is also symmetric between p and s, in view of the basic change of generators

$$(S_1, S_2, S_3) \mapsto (S_1, S_3S_2S_3, S_3).$$

Similarly, if the periods are finite, we also have rotation angles $2\pi/t$ for the conjugate pair $S_1S_2S_3S_2$ and $S_3S_2S_1S_2$, and $2\pi/t'$ for the pair $S_2S_1S_3S_1$ and $S_3S_1S_2S_1$; as before, $t := p_{123}$ and $t' := p_{213}$ are rational. Let us briefly note how this number t (and similarly, t') can be calculated from the data we already have. Interchanging p with q and t with s in the fundamental equation (9D6) (that is, interchanging the indices 2 and 3 of the generators) gives

$$4\Delta = 2\cos\frac{2\pi}{p}\cos\frac{2\pi}{r} - \cos\frac{2\pi}{q} - \cos\frac{2\pi}{t};$$

hence, subtracting (9D6) from this, we find that

9D7 $$\cos\frac{2\pi}{t} - \cos\frac{2\pi}{s} = \left(2\cos\frac{2\pi}{r} + 1\right)\left(\cos\frac{2\pi}{p} - \cos\frac{2\pi}{q}\right).$$

Observe that, as we expect, if $r = 3$ then $t = s$ (because $S_1S_3S_2S_3 = S_1S_2S_3S_2$), irrespective of the values of p and q.

At this point, we introduce the notation $G^3(p, s; q, r)$ for the (geometric) group G; this displays the symmetry between q and r, and (to a lesser extent – see the following) between p and s. There is a corresponding diagram $\mathcal{T}^3(p, s; q, r)$, namely,

9D8 p (s) q r $=$ $p\,(s)$ q r

where the latter alternative form emphasizes that $s\ (= p_{132})$ is *attached* to p, meaning that, in the product with rotation angle $2\pi/s$, the generator represented by the node opposite to the branch marked p is the only one which occurs twice. The latter form of the diagram enables us to attach (if we wish) the corresponding label $t\ (= p_{123})$ to q, and similarly $t'\ (= p_{213})$ to r. The former form now has an interior mark on the triangle, placed in parentheses to indicate that it is attached to p (and is generally not the same as t or t').

The symmetries between the parameters can be made more explicit. Indeed, we have

$$G^3(p, s; r, q) = G^3(p, s; q, r) = G^3(s, p; q, r),$$

meaning that a group expressed in one form can also be expressed in the two other forms; here, the interchanges of marks $p \leftrightarrow s$ and $q \leftrightarrow r$ on the diagram are furnished by the basic equivalence $S_2 \leftrightarrow S_3S_2S_3$ and the interchange of generators $S_1 \leftrightarrow S_2$, respectively.

Suppose next that $p = \bar{p}/a$, say, as a fraction in its lowest terms, with $\bar{q}$, $\bar{r}$ and $\bar{s}$ defined similarly. We then clearly have

9D9 Lemma *With the convention just introduced, the geometric group $G^3(p, s; q, r)$ is a quotient of the abstract group $\Gamma^3(\bar{p}, \bar{s}; \bar{q}, \bar{r})$.*

As we saw for the abstract group and in (9D7), the case where at least two of the marks p, q and r are 3 is rather special. The interior mark s is then the same, whichever pair of the other marks we care to call p and q; it is usual now to set $q = r = 3$. We then simplify the notation for the diagram, and write instead

9D10

There is another case where we may unambiguously write the mark s inside the triangle (and again omit the parentheses). This is when $p = q = r$; the geometry of the reflexion group clearly imposes the symmetry, with (9D7) implying $t = s$, and so on.

Now consider the case when G is finite and each of p, q, r, s (and t, t') is an integer; we have seen in Theorem 9C4 that we can assume this in order to classify the finite groups (recall that each of the interchanges $p \leftrightarrow s$, $q \leftrightarrow t$ and $r \leftrightarrow t'$ leads to a basically equivalent diagram). Since we know the real groups, we may suppose that $p, q, r > 2$ (although the excluded cases may turn up in other guises).

First, we claim that we can obtain a diagram with a branch marked at most 4. To this end, we apply (repeatedly if need be) the following argument until we arrive at a diagram with such a branch. If $r \geqslant 5$ (say), then $p < r$ if $p \leqslant s$, and $q < r$ if $q \leqslant t$. Note that we may obviously suppose that $p \leqslant s$ or $q \leqslant t$, so this will indeed apply. The argument itself can be verified as follows. Let $p \leqslant s$ and $p \geqslant r \geqslant 5$ (say). From (9D6) we then obtain

$$0 < 4\Delta \leqslant 2\cos\frac{2\pi}{r}\left(\cos\frac{2\pi}{q} - 1\right) < 0,$$

whatever q may be, a contradiction. In other words, if $r \geqslant 5$, then $p < r$ and, by symmetry, $q < r$ also, if we assume that $q \leqslant t$. Thus we arrive at a diagram with marks p, q less than r. We are done if one of them is less than 5; otherwise, relabel and apply the same argument as before.

We may thus suppose that we have a branch marked at most 4. As before, we take $p \leqslant s$. If, say, $r = 4$, then

$$0 < 4\Delta = -\cos\frac{2\pi}{p} - \cos\frac{2\pi}{s}$$

(obtained again from (9D6)) implies the familiar condition

$$\frac{1}{p} + \frac{1}{s} > \frac{1}{2},$$

corresponding to the Platonic polyhedra. Indeed, the operation

9D11 $$(S_1, S_2, S_3) \mapsto (S_1, S_2, S_3S_2S_3)$$

yields a subgroup $[p, s]$ (with standard generators $S_2, S_1, S_3S_2S_3$), whose new Schläfli determinant is $\Delta' = 2\Delta$; the latter is consistent with (9D6), with p, q, r, s replaced by $p, s, 2, p$, respectively. Similarly, we also obtain a subgroup $[q, t]$ (with standard generators), again with Schläfli determinant 2Δ. We now appeal to the fact that the (standard) Schläfli determinants $\Delta(l, m)$ (say) of the finite Coxeter groups $[l, m]$ are distinct:

$$\Delta(3,3) = \tfrac{1}{2},$$
$$\Delta(3,4) = \tfrac{1}{4},$$
$$\Delta(3,5) = \tfrac{1}{4}\tau^{-2},$$

with $\tau = \frac{1}{2}(1+\sqrt{5})$ as in Section 9C. This implies at once that the pairs $\{p, s\}$ and $\{q, t\}$ must be the same, and there is thus no loss of generality in taking $p = q = 3$ (we have decided to exclude marks 2 on branches of the original diagram). Then necessarily $s = t = 3, 4$ or 5.

Last, consider the case $r = 3$, with $p \leqslant s$ as before. From (9D6) we now have

$$0 < 4\Delta = -\cos\frac{2\pi}{q} - \cos\frac{2\pi}{p} - \cos\frac{2\pi}{s},$$

from which it follows at once that at least one other mark p or q is also 3. Supposing it to be q, we finally arrive at

$$0 < 4\Delta = \frac{1}{2} - \cos\frac{2\pi}{p} - \cos\frac{2\pi}{s},$$

whose solutions (with $p \leqslant s$) are $(p, s) = (3, m)$ for any $m \geqslant 2$, (4, 4) and (4, 5). Of course, those groups with $p = 4$ have already been counted. Groups of this kind occur frequently; following Coxeter [114, §3], we introduce the alternative notation

$$[1\ 1\ 1^p]^s := G^3(p, s; 3, 3).$$

This is consistent with the earlier notation, because the geometric group $G^3(p, s; 3, 3)$ and abstract group $\Gamma^3(p, s; 3, 3)$ are indeed isomorphic (see [114, §3]).

Summarizing this discussion, and restoring the real groups to the list, we see that we have proved

9D12 Theorem *The finite irreducible reflexion groups in unitary* 3*-space* $\mathbb{C}^3$ *generated by 3 planar reflexions are* $[p, 3]$ *with* $p = 3, 4, 5$, *and* $[1\ 1\ 1^p]^s$ *with* $\{p, s\} = \{3, m\}$ *for any* $m \geqslant 2$, $\{4, 4\}$ *or* $\{4, 5\}$. *In each case, the geometric group is isomorphic to the corresponding abstract group.*

This confirms the entries for $n = 3$ in Table 9A2; we observed in Section 9C that

$$[1\ 1\ 1]^2 = [1\ 1\ 1^3]^2 \cong [3, 3].$$

We shall see that there is indeed only one geometric group for each case of Theorem 9D12. In other words, any two groups with the same parameters are conjugate in the group of all unitary transformations of $\mathbb{C}^3$ (see Lemma 9D15).

For future reference, in Table 9D1 we list the (absolute value of the) turns ϑ for the triangle groups $[1\ 1\ 1^p]^s$ (recall that the angle itself is between $-\pi$ and π). The turns can be calculated from (9C18), with q replaced by s (and $m = 3$). In Table 9D2, we also give generators S_1, S_2, S_3 for these groups in standard cartesian coordinates (listed is the image of (ξ, η, ζ) under S_i); here, $\omega = \frac{1}{2}(-1 + i\sqrt{3})$ is a cube root of unity.

It is particularly important to note that, while the quotient mapping

$$\Gamma^3(p, s; q, r) \mapsto G^3(p, s; q, r)$$

Table 9D1. *The Turns for the Groups* $[1\ 1\ 1^p]^s$

p	s	ϑ
p	3	$\pi - \pi/p$
3	s	$2\pi/s$
4	4	$\arccos(-1/2\sqrt{2})$
4	5	$\arccos(-\tau^{-2}/2\sqrt{2})$
5	4	$2\pi/3$

Table 9D2. *Generators for the Finite Groups* $[1\,1\,1^p]^s$

(p,s)	Generators
$(3,s)$	$S_1 : (\xi, \zeta, \eta)$
	$S_2 : (\zeta, \eta, \xi)$
	$S_3 : (c\eta, \bar{c}\xi, \zeta), \quad c := e^{2\pi i/s}$
$(p,3)$	$S_1 : (c\eta, \bar{c}\xi, \zeta), \quad c := e^{2\pi i/p}$
	$S_2 : (\eta, \xi, \zeta)$
	$S_3 : (\xi, \zeta, \eta)$
$(4,4)$	$S_1 : (-\xi, \eta, \zeta)$
	$S_2 : (\eta, \xi, \zeta)$
	$S_3 : \frac{1}{2}(\xi + c\eta - \zeta, \bar{c}(\xi + \zeta), -\xi + c\eta + \zeta), \quad c := (-1 + i\sqrt{7})/2$
$(4,5)$	$S_1 : \frac{1}{2}(\xi + \omega\tau\eta + \omega^2\tau^{-1}\zeta, \omega^2\tau\xi - \tau^{-1}\eta - \omega\zeta, \omega\tau^{-1}\xi - \omega^2\eta + \tau\zeta)$
	$S_2 : \frac{1}{2}(-\tau^{-1}\xi - \tau\eta + \zeta, -\tau\xi + \eta + \tau^{-1}\zeta, \xi + \tau^{-1}\eta + \tau\zeta)$
	$S_3 : \frac{1}{2}(\xi - \omega\tau\eta - \omega^2\tau^{-1}\zeta, -\omega^2\tau\xi - \tau^{-1}\eta - \omega\zeta, -\omega\tau^{-1}\xi - \omega^2\eta + \tau\zeta)$
$(5,4)$	$S_1 : \frac{1}{2}(\xi + \omega\tau\eta + \omega^2\tau^{-1}\zeta, \omega^2\tau\xi - \tau^{-1}\eta - \omega\zeta, \omega\tau^{-1}\xi - \omega^2\eta + \tau\zeta)$
	$S_2 : (-\omega^2\eta, -\omega\xi, \zeta)$
	$S_3 : \frac{1}{2}(\xi - \omega\tau\eta - \omega^2\tau^{-1}\zeta, -\omega^2\tau\xi - \tau^{-1}\eta - \omega\zeta, -\omega\tau^{-1}\xi - \omega^2\eta + \tau\zeta)$

(with integers p, q, r, s) is known to be an isomorphism for the finite groups with $q = r = 3$, it need not be so in general. Let us give an example. First, $G^3(5, 3; 5, 5)$ is finite (there is no contradiction to the enumeration we carried out previously). By (9D6), its Schläfli determinant is given by

$$4\Delta(5,3;5,5) = 2\cos\frac{2\pi}{5}\cos\frac{2\pi}{5} - \cos\frac{2\pi}{5} - \cos\frac{2\pi}{3} = \tfrac{1}{2}(\tau^{-2} - \tau^{-1} + 1) = \tau^{-2};$$

indeed, $G^3(5, 3; 5, 5)$ is an alternative way of writing the group $[3, 5]$ of the icosahedron (with generators the reflexions in the three planes spanned by the edges of a face of the great icosahedron $\{3, \frac{5}{2}\}$ and its centre). We may write the diagram in the symmetric form

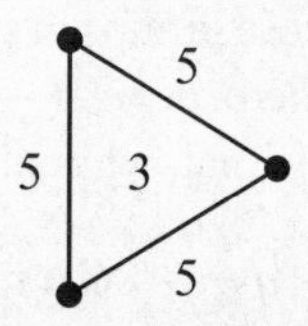

As before, the symmetry is imposed by the geometry, and enables us to omit the parentheses around the interior mark. To verify that the group is indeed $[3, 5]$, note that, from (9D7), the value of t (with the same notation as before) for the group in the equivalent form $G^3(3, 5; 5, 5)$ is given by

$$\begin{aligned}\cos\frac{2\pi}{t} &= \cos\frac{2\pi}{5} + \left(2\cos\frac{2\pi}{5} + 1\right)\left(\cos\frac{2\pi}{3} - \cos\frac{2\pi}{5}\right)\\ &= \tfrac{1}{2}\tau^{-1} + (\tau^{-1} + 1)\left(-\tfrac{1}{2} - \tfrac{1}{2}\tau^{-1}\right)\\ &= -1,\end{aligned}$$

or $t = 2$ as required. (Following the implied basic operations leads to the standard diagram for [3, 5].)

However, as we saw in Theorem 9D3, the abstract group $\Gamma^3(5, 3; 5, 5)$ is a subgroup of index 2 in $[10, 5]_6$. The latter is not listed in Table 8 of [131], and so is almost certainly infinite; in any event, it does not have order 240, as would be required for the isomorphism of $\Gamma^3(5, 3; 5, 5)$ with $G^3(5, 3; 5, 5)$ to hold.

We now discuss the degenerate case $\Delta = 0$, with a corresponding discrete unitary group in $\mathbb{C}^3$ generated by reflexions of the form

$$xS_j = x - 2(\langle x, v_j \rangle - \beta_j)v_j \quad (j = 1, 2, 3).$$

As before, $\langle \cdot, \cdot \rangle$ is the standard positive definite hermitian form on $\mathbb{C}^3$. When $\Delta = 0$, the normal vectors $\{v_1, v_2, v_3\}$ all lie in a plane, and since we are only interested in locally finite groups, we can further assume that any two of them are linearly independent. We can also take all but one β_j equal to zero. Then it is clear that $G := \langle S_1, S_2, S_3 \rangle$ acts as a reflexion group on the unitary plane spanned by $\{v_1, v_2, v_3\}$, which we identify with $\mathbb{C}^2$.

Now Theorem 9C9 tells us that the marks on any diagram of G can only be the crystallographic numbers $2,\ 3,\ 4,\ 6$ or ∞. These diagrams only permit basic operations for changing generators; indeed, operations like those in (9C1) are excluded, because if $p \in \{2, 3, 4, 6\}$, then the only m for which $(p, m) = 1$ satisfy $m \equiv \pm 1 \pmod p$. (To the contrary, notice that for $p = 5$ or $p \geqslant 7$, there always is an $m \not\equiv \pm 1 \pmod p$ such that $(p, m) = 1$, which provides a striking contrast with the crystallographic numbers.)

At this point, it is worth remarking that local finiteness is not necessarily preserved by basic operations. As examples, we have

$$[4, 4] = \Gamma^3(2, \infty; 4, 4), \qquad [3, 6] = \Gamma^3(2, \infty; 3, 6).$$

More subtle examples occur with more generators.

We now enumerate the discrete unitary groups G in $\mathbb{C}^2$ which are *strongly locally finite*, meaning that G is locally finite with respect to not only the original set of generators $\{S_1, S_2, S_3\}$, but also the three sets of generators derived from $\{S_1, S_2, S_3\}$ by the three basic operations $S_2 \leftrightarrow S_3S_2S_3$, $S_3 \leftrightarrow S_2S_3S_2$ or $S_3 \leftrightarrow S_1S_3S_1$, which interchange $p \leftrightarrow s, q \leftrightarrow t$ or $r \leftrightarrow t'$, respectively. In other words, G is strongly locally finite if and only if all six numbers we introduced earlier, namely, p, q, r, s, t and t', are finite. We shall assume this from now on.

When $\Delta = 0$, the formula (9D6) is still valid and shows that

9D13
$$2\cos\frac{2\pi}{q}\cos\frac{2\pi}{r} - \cos\frac{2\pi}{p} - \cos\frac{2\pi}{s} = 0.$$

If a mark 2 occurs among p, q, r or s, then the group is real. If this mark is p, q or r, then we have a linear diagram; in this case, we first make the non-zero β_j real so that the translation part of the generator is real, and then rescale the normal vectors for the other two generators so that the inner products of all three vectors are real. If (only) $s = 2$, we first obtain a linear diagram by applying the basic change of generators $S_2 \leftrightarrow S_3S_2S_3$, and then proceed as before. (In changing the generators, we may also have to change the origin to ensure that our conditions on the v_j's and β_j's continue to hold.) Thus the group is real if a mark 2 occurs, and since the infinite discrete euclidean groups are known, we may exclude this case from now on.

We may now take $p, q, r, s \in \{3, 4, 6\}$. To begin with, if (say) $q = 4$, then we have

$$\cos\frac{2\pi}{p} + \cos\frac{2\pi}{s} = 0;$$

hence $1/p + 1/s = 1/2$, so that

$$\{p, s\} = \{6, 3\} \text{ or } \{4, 4\}.$$

Let us treat the latter subcase first. We now have two marks 4 on the branches, so we relabel, and call these q and r (this leads to a permutation of the labels s, t, t'). The same argument shows that (the new) $\{p, s\} = \{6, 3\}$ or $\{4, 4\}$. We give explicit generators for the two possible groups, and check discreteness.

In fact, we can treat the two groups together. Let $p,\ s$ be such that $1/p + 1/s = 1/2$. The normals $v_1,\ v_2$ and v_3 can be taken to be

$$v_1 := \frac{1}{\sqrt{2}}(-1, 1), \quad v_2 := \frac{1}{\sqrt{2}}(-1, \varphi), \quad v_3 := (1, 0),$$

where $\varphi := e^{2i\pi/p}$, with the corresponding constants β_j given by $\beta_1 = \beta_2 := 0$, $\beta_3 := \frac{1}{2}$. Thus, with $x = (\xi, \eta)$, we have

$$\begin{aligned} xS_1 &= (\eta, \xi), \\ xS_2 &= (\bar{\varphi}\eta, \varphi\xi), \\ xS_3 &= (1 - \xi, \eta). \end{aligned}$$

For discreteness, we need $p = 3,\ 4$ or 6, but we know that we can subsume the first case under the last. We observe that we have a translation $T := (S_1S_2S_3)(S_2S_1S_3)$, which is such that

$$xT = x + (1 - \varphi, 0).$$

When $p = 4$, it is not too hard to check that the set of images of $o = (0, 0)$ under the group $G := \langle S_1, S_2, S_3\rangle$ is precisely $\mathbb{D}^2$, with $\mathbb{D} := \mathbb{Z}[i]$ the Gaussian integers. It follows that G is discrete (indeed, if one orbit of a group is infinite and discrete, then all orbits are discrete). Further, a little calculation shows that the translation subgroup of G, generated by T and its conjugates, is naturally to be identified with $(1 + i)\mathbb{D}^2$.

For $p = 6$, with only a little more effort than that in the previous case, we may check that the set of the images of $o = (0, 0)$ under the group $G := \langle S_1, S_2, S_3\rangle$ is precisely $\mathbb{G}^2$, with $\mathbb{G} := \mathbb{Z}[\omega]$ the Eisenstein integers. It follows that G is discrete. Further, $\mathbb{G}^2$ is also naturally to be identified with the translation subgroup of G (again generated by T and its conjugates).

With q and r both different from 4 (and so equal to 3 or 6), we now have

$$\cos\frac{2\pi}{p} + \cos\frac{2\pi}{s} = \pm\frac{1}{2},$$

according as $q = r$ or $q \neq r$. For $\{q, r\} = \{3, 3\}$ or $\{6, 6\}$ the solution is $\{p, s\} = \{4, 6\}$, while that for $\{q, r\} = \{3, 6\}$ is $\{p, s\} = \{3, 4\}$. Tracing through diagrams by basic operations actually shows that these are all different ways of representing the same group. In fact, these give all five essentially different ways of generating the group; note that the diagrams corresponding to $(p, s; q, r) = (6, 4; 3, 3)$ and $(3, 4; 3, 6)$ coincide after suitable relabelling.

If we take the diagram in the form $(p, s; q, r) = (6, 4; 3, 3)$, the normals v_1, v_2 and v_3 can be taken to be

$$v_1 := \tfrac{1}{2}(-1, \sqrt{2} + i), \quad v_2 := \tfrac{1}{2}(-1, \sqrt{2} - i), \quad v_3 := (1, 0),$$

with the corresponding constants β_j given by $\beta_1 = \beta_2 := 0$, $\beta_3 := 1$. Thus,

$$\begin{aligned} xS_1 &= \tfrac{1}{2}(\xi + (\sqrt{2} - i)\eta, (\sqrt{2} + i)\xi - \eta), \\ xS_2 &= \tfrac{1}{2}(\xi + (\sqrt{2} + i)\eta, (\sqrt{2} - i)\xi - \eta), \\ xS_3 &= (2 - \xi, \eta). \end{aligned}$$

It is useful to note that complex conjugation interchanges v_1 and v_2, and hence acts on the group $G := \langle S_1, S_2, S_3 \rangle$ as an outer automorphism which fixes S_3 and interchanges S_1 and S_2. With this choice, we eventually find that the image-set of $o = (0, 0)$ under G is

$$V := \{(\alpha + i\beta\sqrt{2}, \gamma\sqrt{2} + i\delta) \mid \alpha, \beta, \gamma, \delta \in \mathbb{Z} \text{ and } \alpha \equiv \delta \equiv \beta + \gamma \ (\mathrm{mod}\ 2)\}.$$

Since $-I = (S_1 S_2)^3 \in G$, it follows that V must be a lattice (that is, a discrete group under addition). Further, $T := (S_2 S_1 S_2 S_3)(S_1 S_2 S_1 S_3)$ is the translation

$$xT = x + (2, 2i),$$

and this and its conjugates generate the translation subgroup of G, which can be identified with a subgroup of index 4 in V.

If we identify V with real vectors in $\mathbb{E}^4$, so that

$$(\alpha + i\beta\sqrt{2}, \gamma\sqrt{2} + i\delta) \mapsto (\alpha, \beta\sqrt{2}, \gamma\sqrt{2}, \delta),$$

and perform the rotation

$$(\xi_1, \xi_2, \xi_3, \xi_4) \mapsto \left(\xi_1, \frac{1}{\sqrt{2}}(\xi_2 + \xi_3), \frac{1}{\sqrt{2}}(\xi_2 - \xi_3), \xi_4\right),$$

then we obtain the set

$$\{(\alpha, \beta + \gamma, \beta - \gamma, \delta) \mid \alpha, \beta, \gamma, \delta \in \mathbb{Z} \text{ and } \alpha \equiv \delta \equiv \beta + \gamma \ (\mathrm{mod}\ 2)\}.$$

This just consists of the integer vectors with all coordinates either even or odd; in other words, this is the vertex-set of the regular honeycomb $\{3, 3, 4, 3\}$.

Summarizing this discussion, we now have

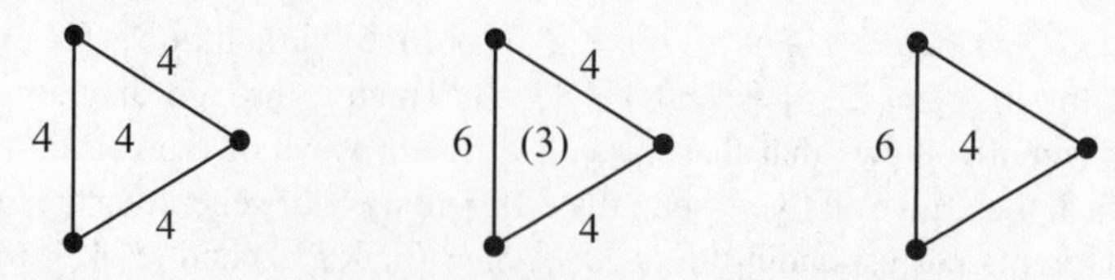

Figure 9D1. The infinite discrete unitary groups in $\mathbb{C}^2$, which are generated by 3 involutory reflexions in $\mathbb{C}^2$, and are irreducible, non-real, and strongly locally finite.

9D14 Theorem *The infinite discrete unitary groups in $\mathbb{C}^2$, which are generated by* 3 *involutory reflexions, and are irreducible, non-real, and strongly locally finite, are the three groups with the diagrams of Figure 9D1.*

In the first and third group, the numbers s, t and t' all coincide, and no parentheses around the interior mark are needed; for the second group, we have $s = 3$ and $t = t' = 4$, and so parentheses are required in this case. For these diagrams, we do not know if the corresponding geometric groups $G^3(p, s; q, r)$ and abstract groups $\Gamma^3(p, s; q, r)$ are isomorphic (but we suspect not).

We should observe that the first and second of these infinite groups have not previously appeared in the literature, although they were announced at a conference several years ago (and found by McMullen many years earlier). It might initially be thought that the second and third groups (in which 3, 4 and 6 all occur as marks) could be the same. However, to see that they are indeed different, simply note that there are 12 translation vectors of minimal length for the second, and 24 for the third.

It remains to show that these infinite groups, as well as the finite groups listed in Theorem 9D12, are indeed uniquely determined by their diagrams. In the finite case this would mean that any two geometric groups with the same diagram are conjugate in the group of all unitary transformations of $\mathbb{C}^3$. In the discrete case we must of course allow conjugacy in the group of all similarity transformations of $\mathbb{C}^2$ (to account for different values of the non-zero β_j). Both cases are covered by the following

9D15 Lemma *Let p, q, r, s be rational numbers with $p, q, r > 2$ and $s \geqslant 2$, and let $q = r$. Let $G = \langle S_1, S_2, S_3 \rangle$ and $H = \langle T_1, T_2, T_3 \rangle$ be geometric groups with a diagram as in (9D8), where G and H act on either unitary 3-space $\mathbb{C}^3$ if $\Delta > 0$, or the unitary plane $\mathbb{C}^2$ if $\Delta = 0$. Then G and H are conjugate in either the group of all unitary transformations of $\mathbb{C}^3$ if $\Delta > 0$, or the group of all similarity transformations of $\mathbb{C}^2$ if $\Delta = 0$.*

Proof. Let

$$\begin{aligned} xS_j &= x - 2(\langle x, v_j \rangle - \beta_j)v_j, \\ xT_j &= x - 2(\langle x, u_j \rangle - \gamma_j)u_j, \end{aligned}$$

with unit vectors v_j, u_j (in $\mathbb{C}^3$ or $\mathbb{C}^2$) and scalars β_j, γ_j. Again, if $\Delta > 0$, we take $\beta_j = \gamma_j = 0$ for all j; if $\Delta = 0$, we take $\beta_3, \gamma_3 \neq 0$, and $\beta_j = \gamma_j = 0$ for $j = 1, 2$. We now load the turns ϑ_G and ϑ_H (say) for the triangle onto the branch $\{1, 2\}$. In other words, we take

$$\langle v_1, v_2\rangle = -\cos\frac{\pi}{p}e^{i\vartheta_G}, \quad \langle v_1, v_3\rangle = -\cos\frac{\pi}{q}, \quad \langle v_2, v_3\rangle = -\cos\frac{\pi}{r},$$

$$\langle u_1, u_2\rangle = -\cos\frac{\pi}{p}e^{i\vartheta_H}, \quad \langle u_1, u_3\rangle = -\cos\frac{\pi}{q}, \quad \langle u_2, u_3\rangle = -\cos\frac{\pi}{r},$$

where $\vartheta_G = \arg(-\gamma_G)$ and $\vartheta_H = \arg(-\gamma_H)$, with $\gamma_G := \langle v_1, v_2\rangle\langle v_2, v_3\rangle\langle v_3, v_1\rangle$ and $\gamma_H := \langle u_1, u_2\rangle\langle u_2, u_3\rangle\langle u_3, u_1\rangle$, respectively. We now appeal to (9C17), which relates the rotation angles $2\pi/s$ of $S_1S_3S_2S_3$ and $T_1T_3T_2T_3$ to ϑ_G and ϑ_H, respectively. Indeed, the cosine of the turn is completely determined by p, q, r, s because $p, q, r > 2$. It follows that $\cos\vartheta_G = \cos\vartheta_H$; hence $\vartheta_H = \pm\vartheta_G$.

We first treat the case $\Delta > 0$ (recall that Δ depends only on p, q, r, s, so that the Schläfli determinants for G and H are necessarily the same). Now $\{v_1, v_2, v_3\}$ and $\{u_1, u_2, u_3\}$ are bases of $\mathbb{C}^3$. If $\vartheta_G = \vartheta_H$, we have $\langle u_i, u_j\rangle = \langle v_i, v_j\rangle$ for all i, j. But then $v_j \mapsto u_j$ $(j = 1, 2, 3)$ induces a linear isometry R of $\mathbb{C}^3$ such that $R^{-1}S_jR = T_j$ for all j; hence $R^{-1}GR = H$. If $\vartheta_H = -\vartheta_G$, then we relabel the generators of H and set $T_1' := T_2$, $T_2' := T_1$ and $T_3' := T_3$, with new normal vectors $u_1' := u_2$, $u_2' := u_1$ and $u_3' := u_3$. Since $q = r$, we again have $\langle u_i', u_j'\rangle = \langle v_i, v_j\rangle$ for all i, j. Then $v_j \mapsto u_j'$ $(j = 1, 2, 3)$ defines an isometry R such that $R^{-1}S_1R = T_2$, $R^{-1}S_2R = T_1$ and $R^{-1}S_3R = T_3$, and so again $R^{-1}GR = H$.

If $\Delta = 0$, then the groups act on $\mathbb{C}^2$. If $v_3 = \lambda v_1 + \mu v_2$ (say), then the two linear equations obtained from $\langle v_1, v_3\rangle$ and $\langle v_2, v_3\rangle$ show that λ and μ are determined by p, q, r and ϑ_G. Hence, if $\vartheta_G = \vartheta_H$, then we also have $u_3 = \lambda u_1 + \mu u_2$, with the same λ, μ as for v_3. Set $\nu := \gamma_3/\beta_3$. Then $v_j \mapsto \nu u_j$ $(j = 1, 2, 3)$ defines a similarity transformation R which maps the reflexion plane of S_i onto the plane of T_i for each i; therefore, $R^{-1}S_iR = T_i$ for each i, and hence again $R^{-1}GR = H$. Finally, if $\vartheta_H = -\vartheta_G$, we first relabel the generators of H as before and obtain a new set of vectors $\{u_1', u_2', u_3'\}$ such that $\langle u_i', u_j'\rangle = \langle v_i, v_j\rangle$ for all i, j. Now $v_j \to \nu u_j'$ $(j = 1, 2, 3)$, with the same μ as before, yields a similarity transformation R with $R^{-1}S_1R = T_2$, $R^{-1}S_2R = T_1$ and $R^{-1}S_3R = T_3$, and so again $R^{-1}GR = H$, as required. □

We conclude this section with remarks about the case $\Delta < 0$, which has basically been ignored so far. However, this case comes up naturally when we want to construct a representation of an arbitrary abstract group $\Gamma^3(p, s; q, r)$ as a reflexion group $G^3(p, s; q, r)$, with integers $p, q, r, s \geqslant 2$. We later need the existence of such groups for more general parameters than those of the finite or infinite discrete groups. As before, this involves solving the corresponding equations for the unit vectors $\{v_1, v_2, v_3\}$, namely,

9D16 $$\langle v_1, v_2\rangle = -\cos\frac{\pi}{p}e^{i\vartheta}, \quad \langle v_1, v_3\rangle = -\cos\frac{\pi}{q}, \quad \langle v_2, v_3\rangle = -\cos\frac{\pi}{r},$$

where the turn ϑ (loaded on the branch $\{1, 2\}$) is related to p, q, r, s by an equation as in (9C17). (We are ignoring here other possibilities arising from choosing non-standard generators in the dihedral subgroups $\langle S_i, S_j\rangle$ and $\langle S_i, S_jS_kS_j\rangle$ of the geometric group.)

We concentrate on the case $q = r = 3$, which is the most interesting for us. Now we can take (9C18) rather than (9C17), and obtain

$$\cos\frac{\pi}{p}\cos\vartheta = \cos^2\frac{\pi}{s} - \cos^2\frac{\pi}{p} - \frac{1}{4},$$

giving the turn ϑ in terms of p, s. For the finite groups, this yields the turns listed in Table 9D1. We also have an analogue of Lemma 9D15 in this case.

We now reverse our approach (see also [114, §2,3]). Rather than finding the unit vectors which satisfy the equations for a given hermitian form (in the above, the standard positive definite form), we now pick any basis $\{v_1, v_2, v_3\}$ of $\mathbb{C}^3$ (we use the same notation as before), and define a hermitian form by specifying its Gram matrix on this basis as in (9D16). If the resulting form is positive definite, we are back in the previous case. If it is positive semi-definite, then its radical must necessarily be 1-dimensional; now Lemma 9B19 shows that we can view the group as acting on $\mathbb{C}^2$, equipped with a positive definite form; hence we are in the degenerate case $\Delta = 0$. Finally, if the form on $\mathbb{C}^3$ is indefinite, then the group must necessarily be infinite (see Lemma 9B7). In either case, the corresponding geometric group is denoted by $[1\ 1\ 1^p]^s$. The finite groups are just those of Theorem 9D12. There is only one non-real infinite discrete unitary group (with $q = r = 3$), obtained with $(p, s) = (4, 6)$ or $(6, 4)$ (see Theorem 9D14). All other groups $[1\ 1\ 1^p]^s$ preserve an indefinite hermitian form and are infinite. In particular, the abstract group $\Gamma^3(p, s; 3, 3)$ is infinite whenever the geometric group $G^3(p, s; 3, 3) = [1\ 1\ 1^p]^s$ is infinite. We already know that the two groups are isomorphic when they are finite.

In summary, we have proved

9D17 Theorem *Geometric groups $G^3(p, s; 3, 3)$ with a diagram as in (9D10) exist for all $p, s \geqslant 3$ and for $\{p, s\} = \{3, 2\}$. Both $\Gamma^3(p, s; 3, 3)$ and $G^3(p, s; 3, 3)$ are infinite unless $\{p, s\} = \{3, m\}$ for $m \geqslant 2$, $\{4, 4\}$ or $\{4, 5\}$.*

9E Tetrahedral Diagrams

As we have already intimated, when $n \geqslant 4$ our methods do not easily lead to a solution of the analogous problem of classifying the finite unitary groups in $\mathbb{C}^n$ generated by n involutory hyperplane reflexions. To a certain extent, inductive techniques will work. For instance, each subgroup generated by $n - 1$ of the reflexions can be transformed (in practice, if perhaps not in theory) into a standard form by basic operations. This then severely restricts the forms which the remaining subgroups of this kind can take. However, there is no way of ensuring that *all* these subgroups can be nicely presented simultaneously. Nevertheless, in certain cases, these methods do lead to significant insights.

One small problem which we encounter is the following. If we decrease the (integer) mark on any branch of the diagram of a finite Coxeter group, then we obtain the diagram of another finite Coxeter group. Actually, a bit more is true. If the group is locally finite (that is, any subgroup generated by $n - 1$ of the generating reflexions is finite), but is infinite and acts discretely on $\mathbb{E}^m$ for some m, then decreasing any mark again yields the diagram of a finite group. Unfortunately, for our unitary groups, the analogous results are false. For example, the discrete infinite group $[1\,1\,1^4]^6$ can be represented by a diagram with all branches marked 6, and the triangle marked 4, namely, the following:

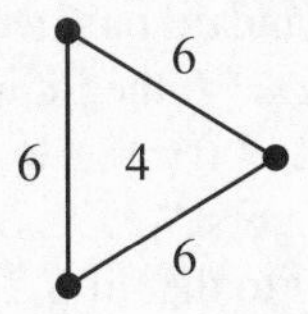

However, if we lower the triangle mark to 3, while the new Schläfli determinant is now positive (namely, $1/8$), we do not obtain a finite group. Indeed, the calculation for the corresponding turn ϑ from (9D5) gives

$$\begin{aligned}\cos\vartheta &= \left(\cos^2\frac{\pi}{3} - \cos^2\frac{\pi}{6} - 4\cos^2\frac{\pi}{6}\cos^2\frac{\pi}{6}\right)\Big/ 4\cos\frac{\pi}{6}\cos\frac{\pi}{6}\cos\frac{\pi}{6} \\ &= -\frac{11}{6\sqrt{3}} < -1,\end{aligned}$$

which is nonsensical! So, while as a rule of thumb similar considerations to those for Coxeter groups do apply, care must be taken to check each instance.

Our main purpose in this section is to investigate certain groups $\Gamma(\mathcal{D})$ defined by tetrahedral diagrams $\mathcal{D}$. These groups are of importance for the enumeration of corresponding locally toroidal regular polytopes (see Chapter 11); the relationship is that such a group $\Gamma(\mathcal{D})$ can be twisted to yield the appropriate C-group. (It should be emphasized that the twists may be "abstract", and so not necessarily realizable geometrically by means of, say, complex conjugation; however, see also Section 11D.) With these applications in mind, by and large we shall not attempt to classify groups which do not admit suitable twists.

However, we begin the discussion with those diagrams consisting of a triangle with a tail (these are degenerate tetrahedral diagrams). Many of these do permit a twist, and are important to the investigations of Chapters 10 and 11. Nevertheless, since we know that all the finite reflexion groups have diagrams of this kind, they are clearly of great importance, and so naturally get referred to frequently.

The general triangle with tail is as in Figure 9E1, with integers $p, q, r, s, t \geqslant 2$. As usual, the (involutory) generator S_j of the corresponding geometric group $G := G^4(p, s; q, r; t)$ is associated with the node labelled j. The group now acts on $\mathbb{C}^4$, with

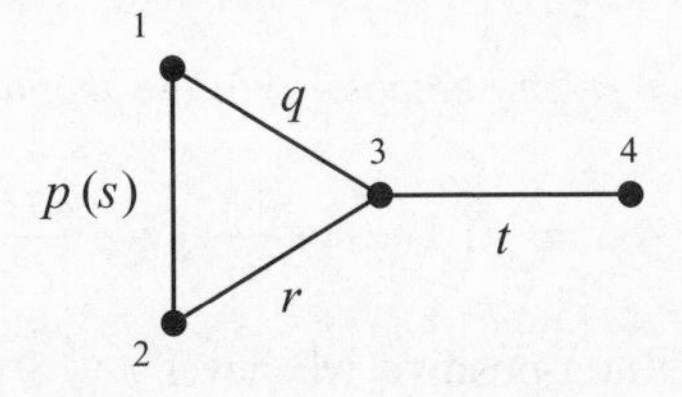

Figure 9E1. The diagram $\mathcal{T}_4(p, s; q, r; t)$.

generators given by

$$xS_j = x - 2(\langle x, v_j\rangle - \beta_j)v_j,$$

where, as before, $\langle\cdot,\cdot\rangle$ is a hermitian form, v_j a unit vector and β_j a scalar. In our applications, we usually take the standard positive definite form on $\mathbb{C}^4$, so that $\Delta > 0$ gives the finite groups in $\mathbb{C}^4$, and $\Delta = 0$ the infinite discrete groups in $\mathbb{C}^3$. Again we may take $\beta_j = 0$ for either all j if $\Delta > 0$ (or $\Delta \neq 0$), or all save one j if $\Delta = 0$. Recall that, for a finite group G, the generators $S_1, \ldots, S_4$ can be chosen so that the branches of any diagram basically equivalent to that of $S_1, \ldots, S_4$ bear only integer marks (see Theorem 9C4). For infinite discrete groups G, the branches on any diagram of G can only be 2, 3, 4, 6 or ∞ (see Theorem 9C9).

Now, after a little simplification, from (9D6) and Lemma 9C20, or directly from (9C15), we find for the Schläfli determinant $\Delta = \Delta(\mathcal{T}_4(p, s; q, r; t))$ the expression

9E1 $$4\Delta = 2\cos\frac{2\pi}{q}\cos\frac{2\pi}{r} - \cos\frac{2\pi}{s} - \left(1 - \cos\frac{2\pi}{p}\right)\cos\frac{2\pi}{t} - 1,$$

which only depends on p, q, r, s, t.

In fact, we can make some initial observations about strong local finiteness. (As in the previous section, we call G *strongly locally finite* if G is locally finite with respect to not only the original generators $\{S_1, \ldots, S_4\}$ but also those obtained by applying a single basic change of generators $S_j \mapsto S_kS_jS_k$ to $\{S_1, \ldots, S_4\}$.) If the branch $\{3, 4\}$ carries a mark $t > 3$, then $\{1, 3\}$ and $\{2, 3\}$ cannot, nor can such a mark be brought on to $\{1, 3\}$ by the basic operation $S_1 \mapsto S_2S_1S_2$ (the new mark would be q if $r = 2$, or s if $q = r = 3$). Thus any such non-linear diagram (with $t > 3$) can only be of the form

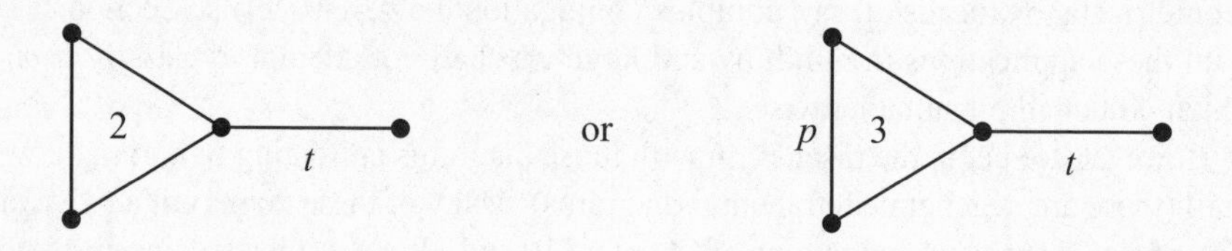

The first is the diagram for $[3, 3, t]$, as the previous basic operation shows, and so yields a finite group only for $t = 3, 4, 5$ (we have restored $t = 3$ to this list); indeed, since formula (9E1) can be simplified to

$$8\Delta = 1 - 3\cos\frac{2\pi}{t},$$

the condition $\Delta > 0$ gives $t < 2\pi/\arccos\frac{1}{3}$. For the second, (9E1) gives

$$4\Delta = -\left(1 - \cos\frac{2\pi}{p}\right)\cos\frac{2\pi}{t}.$$

Clearly, for any p, this is non-positive whenever $t > 3$ (but positive when $t = 3$). When $t = 4$, so that $\Delta = 0$, the crystallographic restriction of Theorem 9C9 allows only $p = 2, 3, 4$ or 6, the first case being real; one can show that the corresponding group genuinely is discrete (see [114, §8]).

When $t = 3$, the original diagram is

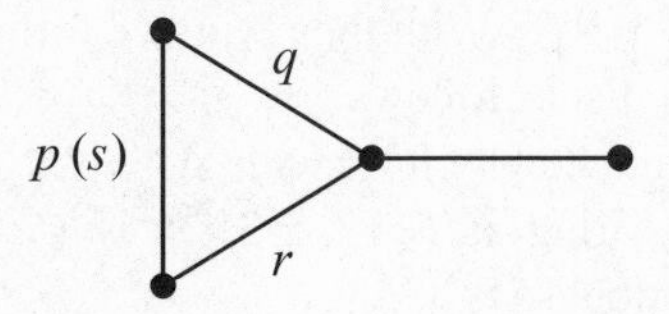

9E2

and the Schläfli determinant is given by

9E3 $$8\Delta = 4\cos\frac{2\pi}{q}\cos\frac{2\pi}{r} - \cos\frac{2\pi}{p} - 2\cos\frac{2\pi}{s} - 1.$$

In practice, it is often easier to use Lemma 9C20 directly, and write

$$\Delta = \Delta_4 - \frac{1}{4}\sin^2\frac{\pi}{p},$$

where Δ_4 is the Schläfli determinant of the triangle group. This has the advantage that the basically equivalent diagrams obtained by the change of generators $S_j \mapsto S_k S_j S_k$, with $\{j, k\} = \{1, 2\}$, are treated together.

In effect, we already noted that $G^4(p, 3; 3, 3; 3) = [1\ 1\ 2^p]^3$ is always finite; the Schläfli determinant is

$$\Delta = \frac{1}{8}\left(1 - \cos\frac{2\pi}{p}\right) = \frac{1}{4}\sin^2\frac{\pi}{p} > 0.$$

The case $p = 2$ is the real (Coxeter) group D_4. Further, with $p = q = r = 3$, we have

$$8\Delta = \frac{1}{2} - 2\cos\frac{2\pi}{s};$$

if $\Delta > 0$ (or even if $\Delta \geqslant 0$), we must have $s \leqslant 4$. This yields the finite groups $[1\ 1\ 2]^s$ with $s = 2, 3, 4$.

For the remaining finite groups with $t = 3$, we have $p \leqslant 5$ and $s \leqslant 5$, and the triangle is obtainable from [3, 4], [3, 5], $[1\ 1\ 1^4]^4$ or $[1\ 1\ 1^4]^5$ by basic changes of generators. Indeed, since the triangle group is finite and necessarily $q, r \leqslant 5$, we first eliminate choices for p, q, r, s by appealing to arguments similar to those used in the proof of Theorem 9D12, and then employ the recursive formula for the determinant to reduce the list further. Excluding the real groups, which we can handle directly, the details for the first step are as follows.

If, say, $r = 4$, then $\Delta_4 > 0$ implies that $\{p, s\} = \{q, t_q\} = \{3, 3\}$, $\{3, 4\}$ or $\{3, 5\}$, where t_q is the mark on the triangle attached to q (we cannot have $q = 2$ or $t_q = 2$ for a non-real group).

If, say, $r = 3$, then $\Delta_4 > 0$ yields a condition symmetric in p, q, s, which implies that at least one of p, q, s is also 3 and that the pair of the other two is $\{3, m\}$ $(m \geqslant 3)$, $\{4, 4\}$ or $\{4, 5\}$ (again, p, q or s cannot be 2 for a non-real group).

Finally, if $q = r = 5$, then $\Delta_4 > 0$ implies that $p < 5$ if $p \leqslant s$ (and $q < 5$ if $q < t_q$); now relabelling leads us back to the two previous cases.

For the remaining possibilities, a basic change of generators will take the triangle into a finite group $[1\ 1\ 1^l]^m$ (with the standard diagram), and so Δ_4 must be the (standard) Schläfli determinant of $[1\ 1\ 1^l]^m$. With the recursive formula for the determinant, we can now further reduce the list as follows.

For $p = 5$, we can eliminate $[1\ 1\ 1^5]^4$ with $\Delta_4 = \frac{1}{8}\tau^{-2}$ as a possibility. However, the real group $[5, 3]$, appearing as $G^3(5, 5; 3, 2)$, $G^3(5, 2; 5, 3)$ or $G^3(5, 3; 5, 5)$, is permitted; the resulting group is $[5, 3, 3]$.

With $p = 4$, we can also eliminate $[1\ 1\ 1^4]^s$ with $s = 4$ or 5; for the former $\Delta = 0$, and we obtain a discrete infinite group (for example, in the form $[1\ 1\ 2^4]^4$). But the real group $[4, 3]$, appearing as $G^3(4, 2; 4, 3)$, is permitted, yielding the group $[4, 3, 3]$.

Finally, with $p = 3$, the only possibilities for the triangle which have not been mentioned so far are $[1\ 1\ 1]^4$ and $[3, 4]$, which can appear additionally as $G^3(3, 3; 3, 4)$ and $G^3(3, 3; 4, 4)$, respectively, resulting in the groups $[1\ 1\ 2^4]^3$ and $[3, 4, 3]$. This now exhausts the list of possible groups.

Let us summarize this discussion. We write $\Gamma^4(p, s; q, r; t)$ for the abstract group determined by the diagram relations; this has generators $\sigma_1, \ldots, \sigma_4$ (say), and is defined by the standard Coxeter-type relations (for the marked branches), and the single extra relation $(\sigma_1\sigma_3\sigma_2\sigma_3)^s = \varepsilon$ (represented by the interior mark of the triangle). We now have

9E4 Theorem *A finite irreducible reflexion group $G^4(p, s; q, r; t)$ in unitary 4-space $\mathbb{C}^4$ whose diagram $T_4(p, s; q, r; t)$ is a triangle with a tail is one of the following (some groups are listed more than once):*

(a) $G^4(3, 2; 3, 3; t) \cong [3, 3, t]$ *for* $t = 3, 4, 5$;
(b) $G^4(p, 3; 3, 3; 3) = [1\ 1\ 2^p]^3$ *for* $p \geqslant 2$;
(c) $G^4(3, s; 3, 3; 3) = [1\ 1\ 2^3]^s = [1\ 1\ 2]^s$ *for* $s = 2, 3, 4$, *and* $G^4(3, 3; 3, 4; 3) = [1\ 1\ 2]^4$;
(d) $[3, 3, 4]$ *and* $[3, 3, 5]$, *but not as in the first part, or* $[3, 4, 3]$.

In the first three parts, the groups are isomorphic to the abstract groups determined by the diagram relations.

In Chapter 11, we also require the existence of certain reflexion groups $G^4(p, s; q, r; t)$ which are infinite. We concentrate again on the case $q = r = 3$, which is the most interesting for us. Similar arguments to those for the triangle group $G^3(p, s; 3, 3)$ apply in this case and prove that a group $G^4(p, s; 3, 3; t)$ exists whenever $G^3(p, s; 3, 3)$ exists and $t \geqslant 2$. In other words, we pick a basis $\{v_1, \ldots, v_4\}$ of $\mathbb{C}^4$ and specify the Gram matrix of an hermitian form on this basis as dictated by the diagram. For $G^4(p, s; 3, 3; t)$, the defining equations comprise those of (9D16) (with $q = r = 3$) and, in addition,

$$\langle v_1, v_4\rangle = \langle v_2, v_4\rangle = 0, \quad \langle v_3, v_4\rangle = -\cos\frac{\pi}{t}.$$

If the triangle represents a finite group (this is the only case we really need), then we choose the turn ϑ in (9D16) as in Table 9D1. Then three cases can occur. If the resulting form on $\mathbb{C}^4$ is positive definite, we are back in the previous enumeration. If it is positive semi-definite, the radical must be 1-dimensional, and the group can be viewed as acting on $\mathbb{C}^3$, with a positive definite form (see Lemma 9B19); this corresponds to the degenerate case $\Delta = 0$. Finally, if the form on $\mathbb{C}^4$ is indefinite, then the group must necessarily be infinite (see Lemma 9B7).

The following theorem summarizes the results for those diagrams which occur in the enumeration of locally toroidal regular polytopes.

9E5 Theorem *Let* $(p, s) = (p, 3), (3, s), (4, 4), (4, 5)$ *or* $(5, 4)$*, and let* $t = 3, 4$ *or* 5*. Then there exists an infinite geometric group* $G^4(p, s; 3, 3; t)$*, and so the abstract group* $\Gamma^4(p, s; 3, 3; t)$ *is also infinite, unless* p, s *and* t *are as in the first three parts of Theorem 9E4.*

We need one further comment, which applies in the context of a general hermitian form. The geometric groups, corresponding to the diagrams, are C-groups, as we can see from the geometry (see Section 9B). The corresponding abstract group $\Gamma^4(p, s; q, r; t)$ is then also a C-group; this follows from a variant of the quotient criterion Theorem 2E17, since the natural homomorphism which identifies the generators σ_i of the abstract group with the generating reflexions S_i of the geometric group is one-to-one on at least one 3-generator subgroup. This establishes

9E6 Theorem *For the diagram* $\mathcal{T}_4(p, s; q, r; t)$*, if the geometric group* $G^4(p, s; q, r; t)$ *exists, then both* $G^4(p, s; q, r; t)$ *and* $\Gamma^4(p, s; q, r; t)$ *are C-groups.*

We now come to general tetrahedral diagrams, which play an important role in Chapter 11. When $n = 4$, we have the following obvious remark; here and elsewhere, we employ the notation and terminology of Section 9C. In particular, if $C := (j(1), \ldots, j(m))$ is a cycle in a diagram, we set $\vartheta_{j(1)\ldots j(m)} := \vartheta(C)$.

9E7 Lemma *If all the branch marks on a tetrahedral diagram with nodes* $1, 2, 3, 4$ *are at least* 3*, then the turns of the triangles satisfy*

9E8
$$\vartheta_{234} - \vartheta_{134} + \vartheta_{124} - \vartheta_{123} \equiv 0 \,(\text{mod } 2\pi).$$

Proof. The 4-cycle $C := (1, \ldots, 4)$ in the diagram can be viewed in two ways as a concatenation of 3-cycles, each determined by a diagonal of C. From Lemma 9C19 we then have

$$\vartheta_{123} + \vartheta_{134} \equiv \vartheta_{1234} \equiv \vartheta_{234} + \vartheta_{124} \,(\text{mod } 2\pi);$$

hence the result follows at once. □

We can now put our finger on one of the main problems. When we specify a triangle group by means of a diagram, say on the nodes 1, 2 and 3, we may give the marks p_{12}, p_{13}, p_{23} and (let us suppose) p_{132}. If all these marks are at least 3, then this only yields the group and its generators geometrically up to a choice of sign of the turn ϑ_{132} (see the proof of Lemma 9D15). It follows that, when we attempt to construct a

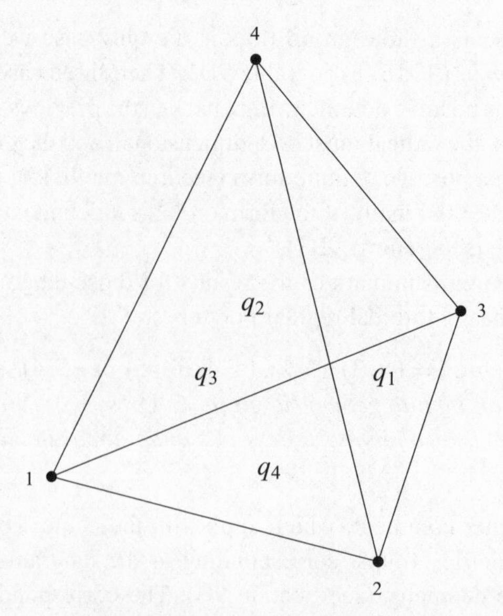

Figure 9E2. The diagram $\mathcal{T}_4(q_1, \ldots, q_4)$.

tetrahedral diagram out of compatible triangular diagrams, then this ambiguity of sign may possibly yield different groups. (In fact, we shall give a specific example of this phenomenon later.)

Consider first the tetrahedral diagram $\mathcal{D} = \mathcal{T}_4(q_1, \ldots, q_4)$ of Figure 9E2, whose branch marks are all 3 (and are therefore omitted), and whose triangle marks (which are now unambiguous) are integers q_1, q_2, q_3, q_4 ($\geqslant 2$), with q_i the mark on the face opposite to the node i. Let $\Gamma(\mathcal{D}) = \langle \sigma_1, \ldots, \sigma_4 \rangle$ be the abstract group corresponding to $\mathcal{D}$, which is the Coxeter group determined by the unmarked tetrahedron, factored by the extra relations

$$(\sigma_i \sigma_j \sigma_k \sigma_j)^{q_m} = \varepsilon, \quad \text{with } \{i, j, k, m\} = \{1, 2, 3, 4\}.$$

Any attempt at constructing a (locally unitary) representation

$$r\colon \Gamma(\mathcal{D}) \to G = \langle S_1, \ldots, S_4 \rangle \subseteq \mathrm{GL}_4(\mathbb{C}),$$

with G preserving an hermitian form, must bear in mind (9E8) when it applies, which it certainly does here. As we saw in Section 9D, the turn on the triangle opposite node i is $\pm 2\pi/q_i$; we conclude that the q_i must satisfy the relation

9E9
$$\frac{1}{q_1} \pm \frac{1}{q_2} \pm \frac{1}{q_3} \pm \frac{1}{q_4} \equiv 0 \,(\mathrm{mod}\ 1),$$

for suitable choices of signs.

Now there are just two (essentially) distinct cases where the diagram may permit a twist in order to yield a string C-group. First, the q_i are equal in pairs, say $q_1 = q_2 = s$ and $q_3 = q_4 = q$ (the notation is chosen to conform with a more general diagram in the following). It is trivial to choose signs so as to satisfy (9E9). Indeed, unless $s = q = 2$

or 4, or (say) $s = 3$ and $q = 6$, the only solution (up to permutation if $s = q$) is

$$\pm\left(\frac{1}{s} - \frac{1}{s}\right) \pm \left(\frac{1}{q} - \frac{1}{q}\right) = 0.$$

(Note that, if $s = 2$, then we could change a sign in the first term and sum to ± 1; however, this yields nothing new.) The case $s = q = 2$ conceals the group $[3, 3, 3]$ of the 4-simplex, as we shall see. For $s = q = 4$, we also have

9E10
$$\frac{1}{4} + \frac{1}{4} + \frac{1}{4} + \frac{1}{4} = 1,$$

and for $s = 3$ and $q = 6$ we have

9E11
$$\frac{1}{3} + \frac{1}{3} + \frac{1}{6} + \frac{1}{6} = 1;$$

we shall comment further on these later.

In the general case, given a geometric group G, we can load the turns $\pm 2\pi/s$ on the branch $\{3, 4\}$, and the turns $\pm 2\pi/q$ into $\{1, 2\}$. Indeed, by rescaling the normal vectors (if need be), we can ensure that $\langle v_i, v_j \rangle = -\frac{1}{2} (= -\cos\frac{\pi}{3})$ for all $\{i, j\}$, except $\{1, 2\}$ and $\{3, 4\}$; the latter determine the signs of the turns. (There are four possibilities for choices of sign each leading to a group G. Any two such groups are conjugate in the group of unitary transformations of $\mathbb{C}^4$, by arguments similar to those used in the proof of Lemma 9D15.) We evaluate the Schläfli determinant Δ, using (9C15). The only extra information to be noted is that the three circuits of four branches have turns 0 and $\pm(2\pi/q \pm 2\pi/s)$. Thus we have

$$\Delta = 1 - 6 \cdot \frac{1}{4} + 3 \cdot \frac{1}{16} - 2 \cdot \frac{1}{4} \cos\frac{2\pi}{q} - 2 \cdot \frac{1}{4} \cos\frac{2\pi}{s}$$
$$- \frac{1}{8} - \frac{1}{8} \cos\left(\frac{2\pi}{q} + \frac{2\pi}{s}\right) - \frac{1}{8} \cos\left(\frac{2\pi}{q} - \frac{2\pi}{s}\right),$$

from which we easily deduce that

9E12
$$16\Delta = 9 - 4\left(\cos\frac{2\pi}{s} + 2\right)\left(\cos\frac{2\pi}{q} + 2\right).$$

For $\Delta > 0$ (that is, for the finite groups), we see easily that the only solutions of (9E12) are $\{s, q\} = \{2, 2\}$, $\{2, 3\}$ and $\{2, 4\}$. Here, we have actually just found an alternative diagram for $[1\ 1\ 2]^q$, as we can see using the basic change of generators $S_4 \mapsto S_3 S_4 S_3$. Further, of course, the case $q = 2$ as well gives $[3, 3, 3]$; the generator S_i can be represented by the transposition $(i\ 5)$, for $i = 1, \ldots, 4$. Observe that $s = q = 3$ gives $\Delta = 0$, a specific example of the same more general case which we shall meet later.

The other case is, say, $q_1 = p, q_2 = q_3 = q_4 = q$. Excluding the case $p = q$ which was covered previously (but we shall need to bear it in mind for applications), we see that the only choice of signs to satisfy (9E9) is effectively

$$\frac{1}{p} - \frac{1}{q} - \frac{1}{q} - \frac{1}{q} = 0,$$

which results in $q = 3p$. (Changing the signs of the fractions $1/q$ and summing to 1 actually yields nothing new, because of the restriction that p and q be integers, with the exception of the excluded case $p = q = 4$. Note that the case $(p, q) = (2, 6)$ can occur here in two ways, because the terms can sum to 0 or 1.) By rescaling the normals of G (if need be), we can now load the turns $2\pi/q$ on the branches $\{2, 3\}$, $\{3, 4\}$ and $\{4, 2\}$ (we write the last branch this way to hint at the corresponding orientation). We again use (9C15) to calculate the Schläfli determinant. The three circuits of four branches each have turn $\pm 4\pi/3p$, and so Δ is now given by

$$\Delta = 1 - 6 \cdot \frac{1}{4} + 3 \cdot \frac{1}{16} - 3 \cdot \frac{1}{4} \cos \frac{2\pi}{3p} - \frac{1}{4} \cos \frac{2\pi}{p} - 3 \cdot \frac{1}{8} \cos \frac{4\pi}{3p},$$

which after simplification yields

9E13
$$16\Delta = 1 - 12 \cos^2 \frac{2\pi}{3p} - 16 \cos^3 \frac{2\pi}{3p}.$$

Bearing in mind that $p \geqslant 2$ is an integer, we easily see that (9E13) has no solutions with $\Delta > 0$ (or even $\Delta = 0$).

If we wish to consider diagrams $\mathcal{T}_4(q_1, q_2, q_3, q_4)$ which have less symmetry, then many more possibilities present themselves. We have no interest in fully investigating them here; suffice it to remark that there is evidence that none of them corresponds to a finite group. For instance, a comparatively crude estimate shows that the Schläfli determinant $\Delta < 0$ if all $q_i \geqslant 5$ (whatever signs are carried by the turns), so such a group must necessarily be infinite. If we then set (say) $q_1 = 2$, 3 or 4 in turn, and exclude the trivial cases where $q_i = q_j$ occur with opposite signs, then a little work shows that this effectively restricts the set $\{q_1, q_2, q_3, q_4\}$ to a finite (if rather long) list. It would be tedious to go into more detail, particularly since we do not need such diagrams. However, to illustrate the general principle, take $q_1 = 2, q_2 = 3$, and $q_3 = r$, $q_4 = s$ with $r \leqslant s$. The two equations

$$\frac{1}{2} + \frac{1}{3} + \frac{1}{r} \pm \frac{1}{s} = 1,$$

which imply that $1/r \pm 1/s = 1/6$, have integer solutions

$$(r, s) = (7, 42), (8, 24), (9, 18), (10, 15), (12, 12),$$

or

$$(r, s) = (2, 3), (3, 6), (4, 12), (5, 30),$$

respectively. In any event,

$$\Delta = -\frac{1}{4}\left(\cos \frac{2\pi}{r} + \cos \frac{2\pi}{s}\right),$$

which is negative unless $r \leqslant 3$ (and is zero when $r = 3$). The only case with $\Delta > 0$ is $r = 2$, which we have already met.

We can summarize this stage of the discussion as follows.

9E14 Theorem *Geometric groups in $\mathbb{C}^4$ exist for all diagrams $\mathcal{T}_4(s, s, q, q)$ with $s, q \geqslant 2$, and for all diagrams $\mathcal{T}_4(p, 3p, 3p, 3p)$ with $p \geqslant 2$. The only groups which are finite (indeed, finite irrespective of compatible choices of sign of the turns) are those with $s = 2$ and $q = 2, 3, 4$ (up to interchange of s and q). The corresponding finite group is $[1\ 1\ 2]^q = [1\ 1\ 2^3]^q$, and is isomorphic to the abstract group defined by the diagram relations. Furthermore, whether finite or not, the geometric and abstract groups with diagrams $\mathcal{T}_4(s, s, q, q)$ and $\mathcal{T}_4(p, 3p, 3p, 3p)$ are C-groups.*

Proof. The parts of the theorem which have not been mentioned hitherto are the first and the last two. For the last two, note that the basic change of generators $S_4 \leftrightarrow S_3S_4S_3$ transforms the diagram $\mathcal{T}_4(2, 2, q, q)$ into that of $[1\ 1\ 2]^q$. The abstract group defined by $\mathcal{T}_4(2, 2, q, q)$ permits the same basic change of generators, which again transforms the diagram into that for $[1\ 1\ 2]^q$; since we know from Coxeter [114, §4; 116, §4] that the abstract and geometric groups are isomorphic in these cases, this completes the proof. Both abstract and geometric groups are C-groups, for the same reason as that in Theorem 9E6.

For the first part, we employ the same technique as that for triangles with tails (see Theorem 9E5). Having chosen a basis $\{v_1, \ldots, v_4\}$ of $\mathbb{C}^4$, each triangle in the diagram determines a set of three equations for $\langle v_i, v_j \rangle$ as in (9D16) (here applied with $p = q = r = 3$). In the equations for $\mathcal{T}_4(s, s, q, q)$, we load the turns $\vartheta = 2\pi/q$ or $2\pi/s$ on the branches $\{1, 2\}$ or $\{3, 4\}$, respectively; for $\mathcal{T}_4(p, 3p, 3p, 3p)$, the turns $2\pi/3p$ are loaded on $\{2, 3\}$, $\{3, 4\}$ and $\{4, 2\}$. In each case the six equations specify the Gram matrix of a hermitian form on $\mathbb{C}^4$. Now we are in the same situation as before. In particular, the positive definite forms give the finite groups we have already enumerated. The positive semi-definite forms yield infinite groups in $\mathbb{C}^3$, namely, those obtained previously for $\Delta = 0$; the only possibility is $\mathcal{T}_4(3, 3, 3, 3)$, giving a discrete group. Finally, if the form is indefinite, then the group is infinite. □

Before we move on, let us also consider the anomalous cases of (9E10) and (9E11). The first, in fact, can be treated in the same way as the case $q_1 = p, q_2 = q_3 = q_4 = q$, but now with $p = \frac{4}{3}$ and $q = 4$ (note that $\frac{3}{4} \equiv -\frac{1}{4} \pmod 1$); in particular, the assignment of turns to branches is the same. The corresponding Schläfli determinant is just that of (9E13) with $p = \frac{4}{3}$, and so $\Delta = 1/16$. Though this is far from obvious, what we have here is another concealed form of the group $[1\ 1\ 2^3]^4 = [1\ 1\ 2]^4$, represented as a group with diagram $\mathcal{T}_4(4, 4, 4, 4)$. We now have two geometric groups with diagram $\mathcal{T}_4(4, 4, 4, 4)$; the new group is finite, and the other (obtained from Theorem 9E14 with $s = q = 4$) is infinite.

For the second anomalous case, we rescale the normals (if need be) such that the contributions to turns are loaded on three branches, namely, $\pi/3$ on $\{1, 3\}$ and $\{2, 4\}$ (again, this indicates the orientations), and π on $\{3, 4\}$. Then two of the circuits of length 4 have turns π, while the third has turn $2\pi/3$. From (9C15), we obtain $\Delta = 0$, so that even with a non-standard choice of turns we cannot get a finite group.

The *abstract* group with diagram $\mathcal{T}_4(p, q, q, q)$ is important in applications to regular polytopes of type $\{3, 6, 3\}$. The corresponding geometric diagram $\mathcal{T}_4(p, q, q, q)$,

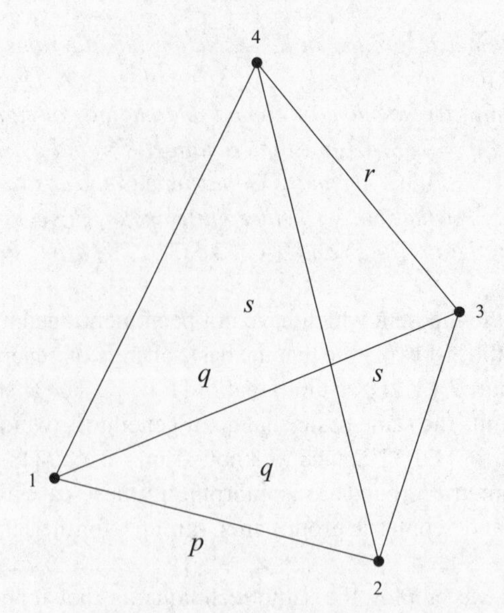

Figure 9E3. The diagram $\mathcal{S}_4(p, q; r, s)$.

and hence locally unitary group, can only exist if $q = p$ or $3p$. This is unfortunate, because it renders the techniques we shall describe in Section 11E of limited utility. However, since we know that $\mathcal{T}_4(p, p, p, p)$ yields an infinite group whenever $p \geqslant 3$, and that $\mathcal{T}_4(p, 3p, 3p, 3p)$ yields an infinite group whenever $p \geqslant 2$, we can appeal to the obvious quotient relations to assert

9E15 Theorem *The abstract group which satisfies the relations induced by the (abstract) diagram* $\mathcal{T}_4(q_1, \ldots, q_4)$ *is infinite in at least the following cases:*

(a) $p \mid q_1, \ldots, q_4$ *for some* $p \geqslant 3$;
(b) $p \mid q_1$ *and* $3p \mid q_2, q_3, q_4$ *for some* $p \geqslant 2$.

We now discuss groups G with the more general diagram $\mathcal{S}_4(p, q; r, s)$ illustrated in Figure 9E3. Here all edge marks are 3, except for those on the edges $\{1, 2\}$ and $\{3, 4\}$, which are p and r, respectively. For obvious reasons, we shall only consider those marks p, q, r, s which correspond to finite 3-generator groups $[1\ 1\ 1^p]^q$ and $[1\ 1\ 1^r]^s$. Thus, by Theorem 9D12,

$$(p, q) = (p, 3),\ (3, q),\ (4, 4),\ (4, 5) \text{ or } (5, 4),$$

with $p, q \geqslant 2$, and similarly for (r, s). We allow $p = 2$ or $r = 2$, in which case we regard the corresponding edge as missing. Note also that, by our diagram conventions, $\mathcal{S}_4(3, q; 3, s) = \mathcal{T}_4(s, s, q, q)$.

The calculation for the corresponding Schläfli determinant Δ is very similar to that of (9E12), if rather more elaborate. We write ϑ for the turn of the subdiagram $[1\ 1\ 1^p]^q$ and φ for that of $[1\ 1\ 1^r]^s$; we can appeal to Table 9D1 for the actual values of ϑ and φ. As before, given G we can ensure that the turns are loaded on the branches $\{1, 2\}$ and

$\{3, 4\}$ (again the orientation does not matter). The turns for the two 4-circuits which contain the edges $\{1, 2\}$ and $\{3, 4\}$ are $\vartheta \pm \varphi$, while that for the third is 0. Thus (9C15) gives

$$\begin{aligned}
\Delta = 1 - 4 \cdot \frac{1}{4} - \cos^2 \frac{\pi}{p} - \cos^2 \frac{\pi}{r} + 2 \cdot \frac{1}{16} \\
+ \cos^2 \frac{\pi}{p} \cos^2 \frac{\pi}{r} + 2 \cdot \left(-2 \cdot \frac{1}{4} \cos \frac{\pi}{p} \cos \vartheta \right) + 2 \cdot \left(-2 \cdot \frac{1}{4} \cos \frac{\pi}{r} \cos \varphi \right) \\
- \frac{1}{2} \cos \frac{\pi}{p} \cos \frac{\pi}{r} \cos(\vartheta + \varphi) - \frac{1}{2} \cos \frac{\pi}{p} \cos \frac{\pi}{r} \cos(\vartheta - \varphi) - \frac{1}{8}.
\end{aligned}$$

Using $\cos(\vartheta + \varphi) + \cos(\vartheta - \varphi) = 2 \cos \vartheta \ \cos \varphi$, and substituting

$$\cos \frac{\pi}{p} \cos \vartheta = \cos^2 \frac{\pi}{q} - \cos^2 \frac{\pi}{p} - \frac{1}{4}$$

from (9C18) (and similarly for φ), we see that this expression simplifies to

9E16 $$\Delta = \sin^2 \frac{\pi}{p} \sin^2 \frac{\pi}{r} \left[1 - \left(1 + \frac{3 - 4 \sin^2(\pi/q)}{4 \sin^2(\pi/p)} \right) \left(1 + \frac{3 - 4 \sin^2(\pi/s)}{4 \sin^2(\pi/r)} \right) \right].$$

An alternative, but somewhat less useful, way of expressing this determinant is

$$\begin{aligned}
16\Delta = &- 5 + 2 \cos \frac{2\pi}{p} + 2 \cos \frac{2\pi}{r} - 6 \cos \frac{2\pi}{q} - 6 \cos \frac{2\pi}{s} \\
&+ 4 \cos \frac{2\pi}{p} \cos \frac{2\pi}{s} + 4 \cos \frac{2\pi}{q} \cos \frac{2\pi}{r} - 4 \cos \frac{2\pi}{q} \cos \frac{2\pi}{s}.
\end{aligned}$$

Now, if one of r or s is 2, then the other must be 3, with a similar restriction on p and q. Let us consider this case first. With $r = 3$ and $s = 2$, the criterion $\Delta > 0$ reduces to

$$2 \sin^2 \frac{\pi}{p} + 4 \sin^2 \frac{\pi}{q} > 3.$$

A case by case check (recall that $q = 2$ implies that $p = 3$) shows that the admissible values are

$$(p, q) = (3, 2), \quad (p, 3) \text{ for any } p \geqslant 2, \quad (3, 4).$$

Indeed, the basic change of generators $S_4 \mapsto S_3 S_4 S_3$ yields a diagram consisting of a triangle with a tail, namely, that of the group $[1\ 1\ 2^p]^q$. Note additionally that (again as it must) $(p, q) = (4, 4)$ gives $\Delta = 0$; the corresponding infinite group $[1\ 1\ 2^4]^4$ is known to be discrete (see [114]).

With $s = 3$, then, independently of p and r, the criterion $\Delta > 0$ reduces at once to $\sin^2 \frac{\pi}{q} > \frac{3}{4}$, or $q < 3$; that is, $q = 2$. As we have just pointed out, this then implies that $p = 3$, and so we are reduced to the previous case.

We may now suppose that $p, q, r, s \geqslant 3$. With $q = s = 3$, (9E16) implies that $\Delta = 0$, no matter what p or r may be. In fact, the corresponding group is infinite and can be viewed as acting on $\mathbb{C}^3$ (equipped with the standard positive definite form). In particular, we can choose specific generating reflexions S_i with unit normals v_i given by

$$v_1 = (e^{2i\pi/p}/\sqrt{2}, -1/\sqrt{2}, 0),$$
$$v_2 = (1/\sqrt{2}, -1/\sqrt{2}, 0),$$
$$v_3 = (0, 1/\sqrt{2}, -e^{2i\pi/r}/\sqrt{2}),$$
$$v_4 = (0, 1/\sqrt{2}, -1/\sqrt{2}),$$

ensuring that the reflexion hyperplanes do not contain a common point. However, it is generally non-discrete; appealing to Theorem 9C9 shows that the only permissible values of p and r for discreteness are 2, 3, 4 or 6, and, in fact, 4 and 3 or 6 cannot occur together (if they do, then the special group contains a rotation of period 12, and so the translation subgroup cannot be discrete). The possible pairs $\{p, r\}$ are thus

$$\{p, r\} = \{2, 2\}, \quad \{2, 3\}, \quad \{2, 4\}, \quad \{2, 6\}, \quad \{3, 3\}, \quad \{3, 6\}, \quad \{4, 4\}, \quad \{6, 6\}.$$

In fact, the group is generated by the conjugates of the involutory reflexions R_1 and R_2 in the complex group $p[4]2[3]2[4]r$ (see (9A10) for the notation, and (9A8) for the group presentation); it is just for these values of $\{p, r\}$ that this infinite group is known to be discrete (see [276, §7.2] or [126, §§12.7, 13.2]).

Finally, with $q \geqslant 3$ and $s \geqslant 3$, we see that $\Delta < 0$ if $q > 3$ or $s > 3$, again irrespective of the values of p and r. Nevertheless, the geometric groups indeed exist in these cases, by arguments similar to those for the diagrams $\mathcal{T}_4(s, s, q, q)$. In the defining equations for the hermitian form, we must again load the turns for $[1\ 1\ 1^p]^q$ and $[1\ 1\ 1^r]^s$ on the branches $\{1, 2\}$ or $\{3, 4\}$, respectively. Then the resulting form is indefinite if $q > 3$ or $s > 3$, yielding an infinite geometric group.

We can summarize this discussion as follows: once again, the fact that all groups are C-groups follows for the same reason as that of Theorem 9E6.

9E17 Theorem *For the diagram of Figure 9E3, geometric groups G exist (at least) whenever* $(p, q) = (p, 3)$, $(3, q)$, $(4, 4)$, $(4, 5)$ *or* $(5, 4)$, *with* $p, q \geqslant 2$, *and* $(r, s) = (r, 3)$, $(3, s)$, $(4, 4)$, $(4, 5)$ *or* $(5, 4)$, *with* $r, s \geqslant 2$. *Apart from the groups listed in Theorem 9E14, the only finite examples are those with* $(p, q; r, s) = (p, 3; 3, 2)$ *for* $p \geqslant 2$ *(up to interchange of* (p, q) *and* (r, s)*); then* $G \cong [1\ 1\ 2^p]^3$. *Whether finite or not, the geometric and abstract groups are C-groups.*

We shall not consider more general tetrahedral diagrams in which all edges are present (that is, carry a mark other than 2). As we have said, the classification problem using our techniques is complicated. Indeed, for tetrahedral diagrams, listing the different kinds of diagram corresponding to choices of generators of the group is a problem akin to that of listing the Goursat tetrahedra, the reflexions in whose faces generate a finite subgroup of the orthogonal group of euclidean 4-space (see [120, §14.8; 235, Table H]). Indeed, it is little different in the real case – it is merely that our approach does not distinguish between the up to eight spherical tetrahedra bounded by the same planes.

Again because they do not admit twists which relate them to groups which occur later, we shall not look at those tetrahedral diagrams with a single missing edge; the case

$p = 2$ of the diagram in Figure 9E3 covers all we need subsequently. However, there is one further class of tetrahedral diagrams which does deserve further investigation, because of the light it sheds on the general classification problem, namely, that consisting of the single circuits (with integer marks).

For a single 4-circuit $\mathcal{D}$ to give a locally finite group $G = G(\mathcal{D})$, it must be of the form

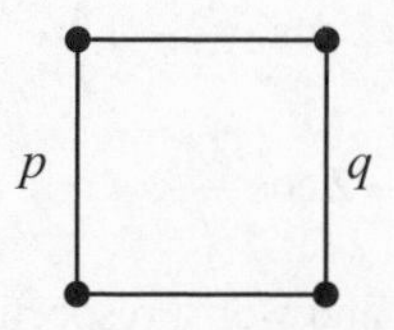

(no two marks greater than 3 can occur on adjacent branches). Moreover, to avoid the real cases and obviously infinite subgroups, we must have $3 \leqslant p, q \leqslant 5$. (Actually, for completeness of the discussion, it is convenient to allow $p, q = 2$ here.) Now (9C17) gives two different numbers for the period p_{1234} of the product $S_1S_2S_3S_4S_3S_2$ of the generators $S_1, \ldots, S_4$ taken in cyclic order around the circuit, depending upon the starting point, except when one of p or q is 3. In this latter case, when (say) $q = 3$, the group is actually $[1\ 1\ 2^s]^p$, with $s := p_{1234}$ the mark on the circuit, and we have already enumerated the finite groups of this kind in Theorem 9E4. Indeed, if $q = 3$, the change of generators described at the end of Section 9C replaces the diagram for $[1\ 1\ 2^s]^p$ by the given diagram $\mathcal{D}$ (with interior mark s).

More generally, the periods are different; we are now confined to the cases $p, q = 4$ or 5, although the discussion still covers all possibilities. If the period is s, say, when the branch $\{1, 4\}$ carries the mark p (or q), then, in terms of the turn ϑ of the 4-circuit, we have

$$\cos^2 \frac{\pi}{s} = \cos^2 \frac{\pi}{p} + \cos^2 \frac{\pi}{q} + 2 \cos \frac{\pi}{p} \cos \frac{\pi}{q} \cos \vartheta,$$

while if it is t when $\{1, 2\}$ is marked p (or q), then

$$\cos^2 \frac{\pi}{t} = \frac{1}{4} + 4 \cos^2 \frac{\pi}{p} \cos^2 \frac{\pi}{q} + 2 \cos \frac{\pi}{p} \cos \frac{\pi}{q} \cos \vartheta.$$

Of course, as we have remarked, if (say) $q = 3$, then $s = t$. In any case, we see that

$$\cos^2 \frac{\pi}{t} - \cos^2 \frac{\pi}{s} = 4 \cos^2 \frac{\pi}{p} \cos^2 \frac{\pi}{q} - \cos^2 \frac{\pi}{p} - \cos^2 \frac{\pi}{q} + \frac{1}{4},$$

or

9E18
$$\cos \frac{2\pi}{t} - \cos \frac{2\pi}{s} = 2 \left(\cos \frac{2\pi}{p} + \frac{1}{2} \right) \left(\cos \frac{2\pi}{q} + \frac{1}{2} \right).$$

On the other hand, a straightforward calculation using (9C15) shows that the Schläfli determinant Δ is given by

$$16\Delta = 9 - 16 \cos^2 \frac{\pi}{p} - 16 \cos^2 \frac{\pi}{q} + 16 \cos^2 \frac{\pi}{p} \cos^2 \frac{\pi}{q} - 8 \cos \frac{\pi}{p} \cos \frac{\pi}{q} \cos \vartheta.$$

Eliminating the term involving the turn then gives the two parallel formulae

9E19

$$16\Delta = \begin{cases} 9 - 12\cos^2\dfrac{\pi}{p} - 12\cos^2\dfrac{\pi}{q} + 16\cos^2\dfrac{\pi}{p}\cos^2\dfrac{\pi}{q} - 4\cos^2\dfrac{\pi}{s}, \\ 10 - 16\cos^2\dfrac{\pi}{p} - 16\cos^2\dfrac{\pi}{q} + 32\cos^2\dfrac{\pi}{p}\cos^2\dfrac{\pi}{q} - 4\cos^2\dfrac{\pi}{t}, \end{cases}$$

$$= \begin{cases} \left(2\cos\dfrac{2\pi}{p} - 1\right)\left(2\cos\dfrac{2\pi}{q} - 1\right) - 2\left(\cos\dfrac{2\pi}{s} + 1\right), \\ 8\cos\dfrac{2\pi}{p}\cos\dfrac{2\pi}{q} - 2\cos\dfrac{2\pi}{t}, \end{cases}$$

which express Δ in terms of p, q and s or t, respectively.

We first treat the case $p = q = 5$. We quickly see that, for $\Delta > 0$, we must have $s = 2$. A little work (whose details are not worth including – but consider the effect of the implied basic operations) then shows that we have found [3, 3, 5] in a different guise.

Finally, we have the case $q = 4$, say. This implies that

$$16\Delta = \begin{cases} -2\cos\dfrac{2\pi}{p} - 2\cos\dfrac{2\pi}{s} - 1, \\ -2\cos\dfrac{2\pi}{t}, \end{cases}$$

the latter irrespective of the value of p. We could carry out the analysis from this point (indeed $\Delta > 0$ implies that $t = 2$ or 3, and $\{p, s\} = \{2, 3\}, \{2, 4\}, \{2, 5\}$ or $\{3, 3\}$), but it is easier to appeal to Theorem 9E4. Indeed, the reflexions S_1, $S_2S_3S_2$, S_4, S_3 (in this order) generate a subgroup $G^4(4, s; p, 3; 3)$ of G, which is finite if G is finite. But we know from Theorem 9E4 (and its proof) that the only finite groups of this kind are $G^4(4, 3; 3, 3; 3) \cong [1\ 1\ 2^4]^3$ and $G^4(4, 2; 4, 3; 3) \cong [4, 3, 3]$. Inspection of these cases shows that the original first group is $[1\ 1\ 2]^4 = [1\ 1\ 2^3]^4$ (now $p = 3$ implies that $t = s$), while the second is isomorphic to [3, 4, 3] (apply the change of generators $S_4 \mapsto S_2S_3S_4S_3S_2$).

If $q = 4$ and $t = 4$, then we have the degenerate case $\Delta = 0$. The expression of Δ in terms of p, s then gives

$$\cos\frac{2\pi}{p} + \cos\frac{2\pi}{s} + \frac{1}{2} = 0,$$

which we can write in the form

$$\cos\frac{2\pi}{p} + \cos\frac{2\pi}{s} + \cos\frac{\pi}{3} = 0.$$

Now we know that all the solutions of *Gordan's equation*

$$\cos x\pi + \cos y\pi + \cos z\pi = 0 \quad (0 \leqslant x, y, z \leqslant 1)$$

in rational numbers are permutations of $(x, \frac{1}{2}, 1 - x)$ with $0 \leqslant x \leqslant \frac{1}{2}$, $(x, \frac{2}{3} - x, \frac{2}{3} + x)$ with $0 \leqslant x \leqslant \frac{1}{3}$, $(\frac{1}{5}, \frac{3}{5}, \frac{2}{3})$ or $(\frac{1}{3}, \frac{2}{5}, \frac{4}{5})$ (see [120, p. 274]). The second kind will not contribute any solutions to our equation, except possibly when $(p, s) = (2, 6)$ or $(6, 2)$; however, bearing in mind that $p = 2$ would give the *finite* group $[3, 4, 3]$, we can then rule out these possibilities entirely by observing that the group relation for $s = 2$ (that is, $(S_1S_2S_3S_4S_3S_2)^2 = I$) would already force $p = 2$ or 4 (that is, $(S_1S_4)^4 = I$). Finally, eliminating all but the *integer* values of p and s, we find the two solutions

$$(p, s) = (3, 4) \text{ or } (4, 3).$$

The case $(p, s) = (3, 4)$ (which implies that $t = s$, and thus $t = 4$ anyway) is familiar; this is just the discrete group $[1\ 1\ 2^4]^4$ in another guise. Strangely enough, this is also true for $(p, s) = (4, 3)$, though the isomorphism is less obvious.

9F Circuit Diagrams with Tails

We next come to the finite (irreducible) groups G generated by n involutory reflexions in $\mathbb{C}^n$ with $n \geqslant 5$. If the group is not (essentially) real, then each of its diagrams must contain at least one circuit. In practice, we know that, for a finite group, we may always choose a diagram which has a single triangular circuit, to which are appended a number of tails. (We shall be more precise about our terminology a little later.) Since the degenerate case, that of a discrete infinite (locally finite) group acting on $\mathbb{C}^{n-1}$, is also of interest, we shall consider rather more general diagrams here. However, we shall largely confine our attention to manageable diagrams, namely, those with a single circuit, which may now consist of more than three nodes and branches; we shall then allow tails to be added to these circuits.

As an aside, and to point out the problems which a classification of the finite groups based on the ideas of Section 9C alone would face, let us merely note that one possible diagram for the Coxeter group E_8 of order $192 \cdot 10!$ is

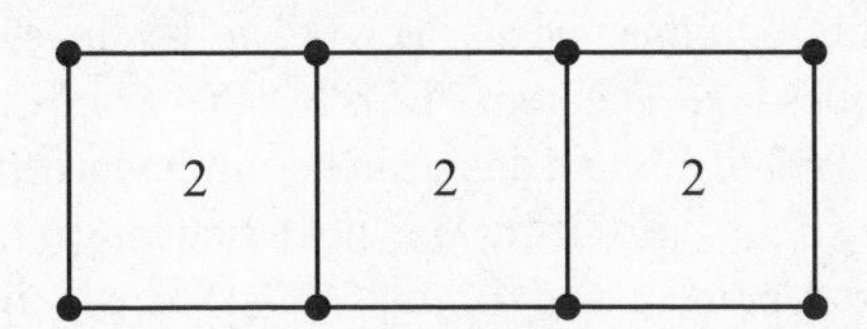

We shall obviously want to exclude from the outset those diagrams which contain proper (induced) subdiagrams corresponding to infinite groups. Here, the list of those Coxeter groups which are not finite (that is, the complement of those which are) is invaluable. In other words, we bear in mind a list of excluded proper subdiagrams. Of course, we shall also exclude branches marked 2 (that is, whose nodes are not joined), because either the diagram is disconnected (and so is an internal direct product, which enables us to appeal to induction), or it reduces to a tree (in which case the group is real, and so has been classified already).

By a *tail* in a diagram with a single circuit, we shall mean a (generally maximal) path of branches and nodes which is disjoint from the circuit except for its initial node, which lies in the circuit. In this context, we allow two tails to share some initial nodes or branches. Therefore, if we delete any node of the circuit, and the branches which contain it, then we obtain a diagram whose connected components are trees. We then mentally refer to the exclusion list.

We have already said in Section 9A that, for $n \geqslant 3$, a finite irreducible subgroup of $U_n(\mathbb{C})$ generated by n involutory reflexions is a group $[k\,l\,m^p]^q$, with diagram shown in Figure 9A2. These groups will be involved in everything we do in the following, and so it is appropriate to consider them first.

Starting from the initial value

$$8\Delta([1\,1\,1^p]^q) = 5 - 4\cos^2(\pi/p) - 4\cos^2(\pi/q),$$

which can be obtained from (9D6), the recursive Lemma 9C20 easily leads to

9F1 Theorem *For $k,\ l,\ m \geqslant 1$, the Schläfli determinant of the group $[k\,l\,m^p]^q$ with the diagram of Figure 9A2 is given by*

$$2^n\Delta([k\,l\,m^p]^q) = k + l + m + 1 - kl\left(4\cos^2\frac{\pi}{p} + 4m\cos^2\frac{\pi}{q} - 1\right),$$

with $n = k + l + m$.

It is a routine matter now to check that the non-real examples with positive Schläfli determinant (and whose subdiagrams have the same property) are precisely those in Table 9A2. We see that $q = 2$ (so that $p = 3$) always gives a positive value (we shall return to this later); more significantly, if $k, l, m \geqslant 2$ and $p, q \geqslant 3$, then some subdiagram has non-positive Schläfli determinant, and so corresponds to an infinite group.

Before we move on, a general comment addressing the last statement is appropriate. As we have seen repeatedly, the Schläfli determinant and changes of generators are the main tools to determine in a family of diagrams those which correspond to finite groups. In general, however, we must establish the existence of geometric groups for *all* diagrams in a family by other means. For $[k\,l\,m^p]^q$, the same technique as that for $\mathcal{T}_4(p, s; q, r; t)$ applies (see Theorem 9E5). In other words, with $n := k + l + m$, we pick a basis $\{v_1, \ldots, v_n\}$ of $\mathbb{C}^n$, and define an hermitian form by specifying its Gram matrix; that is, we take $\langle v_i, v_j\rangle = 0$ if $\{i, j\}$ is not a branch, $\langle v_i, v_j\rangle = -\cos\frac{\pi}{3} = -\frac{1}{2}$ if $\{i, j\}$ is an unmarked branch, and $\langle v_i, v_j\rangle = -e^{i\vartheta}\cos\frac{\pi}{p}$ if $\{i, j\}$ is the branch marked p, with ϑ the turn of $[1\ 1\ 1^p]^s$ (see Table 9D1). Then the group generated by the reflexions with normals v_i is indeed represented by the corresponding diagram. Again, if the form is positive definite, we are back in the original case. Otherwise, the group must be infinite. We thus have a group for every diagram in the family.

The initial cases of $[k\,l\,m^p]^q$ with $k = l = 1$, corresponding to a single tail, are of particular interest. With $m = n - 2$, let us slightly generalize by marking the last branch of the tail r (instead of 3) in order to give the diagram $\mathcal{D}_n(p, q, r)$ of Figure 9F1 (again, a geometric group exists for each such diagram). In effect, we have already treated the cases $n = 3$ or 4, but we still bear them in mind. Before we proceed, however, it is worth explaining why, among general triangles with single tails, we can confine our attention to this apparently very restricted class of diagrams.

Figure 9F1. The diagram $\mathcal{D}_n(p, q, r)$.

First, for a non-real finite group, we cannot have any mark greater than 3 on a branch $\{j-1, j\}$ with $j \geqslant 4$ of the tail. For $n = 4$ (and $j = 4$) we proved this in Theorem 9E4. If $n \geqslant 5$, we appeal to Lemma 9C21 and induction on the length of the tail, and note that we have already excluded a mark 4 (or larger) on branch $\{3, 4\}$. (If the inductive application of the lemma produces a real group for $n = 4$, then the original group must also have been real; see again Theorem 9E4.) Moreover, for a locally finite group, if the last branch of the tail is marked $r \geqslant 4$ (while the others are unmarked), then we cannot mark either $\{1, 3\}$ or $\{2, 3\}$; for a finite group, we can even go further in this case, because the basic change of generators $S_1 \mapsto S_2 S_1 S_2$ brings the mark q onto $\{1, 3\}$ (when $\{1, 3\}$ and $\{2, 3\}$ are unmarked), and so $q = 3$ is the only possible mark on the triangle. As we saw in Section 9E for the finite groups with $n = 4$, if all branches of the tail are unmarked, then we can indeed have marks on $\{1, 3\}$ or $\{2, 3\}$; however, then the group either is real or also admits a representation by a diagram with no marks on $\{1, 3\}$ or $\{2, 3\}$.

Before we perform the actual calculations, we recall once again that the degenerate case $\Delta = 0$ is of interest, especially when the associated infinite group acts discretely on $\mathbb{C}^{n-1}$. We also recollect that already for the finite groups with $n = 4$ we may have any $p \geqslant 3$ (or even $p = 2$ in appropriate circumstances), but that $q \leqslant 4$ must be imposed on us.

We define $\Delta_n(p, q, r) := \Delta(\mathcal{D}_n(p, q, r))$, bearing in mind our inductive assumption that, for the group to be finite, we must have $r = 3$. Repeating the calculations of Lemma 9C20, we find

$$\Delta_n(p, q, r) = \Delta_{n-1}(p, q, 3) - \cos^2 \frac{\pi}{r} \, \Delta_{n-2}(p, q, 3).$$

Our starting point is $n = 5$, but the formula also applies with $n = 4$, as long as we make the interpretations

$$\Delta_2(p, q, 3) = \sin^2 \frac{\pi}{p}, \quad \Delta_3(p, q, 3) = \Delta([1\,1\,1^p]^q) = \frac{1}{8}\left(5 - 4\cos^2 \frac{\pi}{p} - 4\cos^2 \frac{\pi}{q}\right).$$

Our initial observations are as follows.

- Starting with $\Delta_3(p, 3, 3) = \Delta([1\,1\,1^p]^3) = \frac{1}{2} \sin^2 \frac{\pi}{p}$, we conclude that $\Delta_n(p, 3, 3) = 2^{2-n} \sin^2 \frac{\pi}{p}$.
- Thence, if $n \geqslant 4$, we have $\Delta_n(p, 3, 4) = 0$, and $\Delta_n(p, 3, r) < 0$ for $r > 4$. For $\mathcal{D}_n(p, 3, 4)$, the crystallographic restriction implies that $p = 2, 3, 4$ or 6 for the group to be discrete.

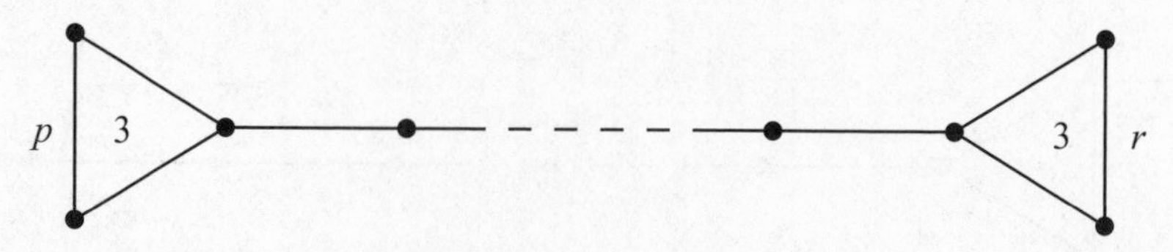

Figure 9F2. The diagram $\mathcal{E}_n(p,r)$.

- Since $\Delta_3(4,4,3)=\frac{1}{8}$, we see that $\Delta_4(4,4,3)=0$; the corresponding group is $[1\,1\,2^4]^4$, and is discrete (we noted this already in Section 9E).
- Since $\Delta_3(3,4,3)=\frac{1}{4}$, we see that $\Delta_4(3,4,3)=\frac{1}{4}-\frac{1}{4}\cdot\frac{3}{4}=\frac{1}{16}$, and then $\Delta_5(3,4,3)=\frac{1}{16}-\frac{1}{4}\cdot\frac{1}{4}=0$. The group corresponding to $\mathcal{D}_5(3,4,3)$ is $[3\,1\,1^4]^3$ and is discrete (see [114, p. 258; 388, p. 383]).

In view of the foregoing list, there are thus severe restrictions to adding two or more tails to a triangular diagram. Let us also note some infinite families with $\Delta=0$ whose diagrams contain two triangles; for these, $n\geqslant 5$. Their diagrams $\mathcal{E}_n(p,r)$ are shown in Figure 9F2 (again, there is a geometric group for each such diagram). For discreteness, we must have $p,r\in\{2,3,4,6\}$, with the further restriction that $\{p,r\}\neq\{3,4\}$ or $\{4,6\}$; these are the same qualifications as those for the groups with diagrams $\mathcal{S}_4(p,3;r,3)$, which extend the families down to $n=4$, and hold for the same reason. Indeed, the group is generated by the conjugates of the involutory reflexions $R_1,\ldots,R_{n-2}$ in $p[4]2[3]2\cdots2[3]2[4]r$ (see (9A10) for the notation for this group, and (9A8) for its presentation); again, reference to [276, §7.2] or [126, §§12.7, 13.2] shows that it is only for these values of $\{p,r\}$ that we obtain discrete groups. Notice that, if $r=2$, such groups occur as subgroups of index 2 in the groups with diagram $\mathcal{D}_n(p,3,4)$ shown in Figure 9F3. If $n=5$, this also gives an example of two tails emanating from the same node of a triangle.

For triangles with one or more tails, if we are concerned with finiteness, we are thus finally confined to the groups $[k\,l\,m^p]^q$, with diagrams as in Figure 9A2; we have $k+l+m=n\geqslant 4$.

Next, we consider single circuits. We already discussed circuits on 4 nodes at the end of the previous section, and so we may assume that we have 5 or more nodes. We also know that the finite group $[1\,1\,(n-2)^p]^3$ can be represented by a circuit on n nodes whose branches are unmarked (see Section 9A).

As before, let G be a group acting on $\mathbb{C}^n$, which preserves a hermitian form, and is generated by n involutory hyperplane reflexions.

Figure 9F3. The diagram $\mathcal{D}_n(p,3,4)$.

9F2 Lemma *If the diagram of G is a circuit on $n \geqslant 5$ nodes, then G is infinite unless each branch is unmarked.*

Proof. Clearly we cannot have two or more marked branches, because otherwise the group would contain an infinite real subgroup represented by a suitable subdiagram. So, let one branch have a possible mark $p > 3$. Now we have seen in Section 9C that the circuit can be marked unambiguously q, say, with p and q related as in (9C18). As we explained at the end of Section 9C, a change of generators then yields

$$G \cong [1\,1\,(n-2)^q]^p.$$

But, as we saw in Theorem 9F1, if $n \geqslant 5$ then $[1\,1\,(n-2)^q]^p$ is infinite unless $p = 3$, a contradiction to our assumption. □

We can now exclude many possible diagrams as corresponding to infinite groups. Bear in mind here that a marked branch carries a label greater than 3, and that Lemma 9F2 already excludes marks on the branches of a circuit with 5 or more nodes.

9F3 Lemma *Let the diagram $\mathcal{G}$ of G contain a circuit with m nodes and at least one tail. Then G is infinite if*

(a) $m \geqslant 4$ and a tail of $\mathcal{G}$ has at least one marked branch,
(b) $m \geqslant 7$ and $\mathcal{G}$ has two tails,
(c) $m \geqslant 4$ and two tails of $\mathcal{G}$ begin at the same node of the circuit,
(d) $m \geqslant 4$ and any two branches of $\mathcal{G}$ are marked.

Proof. Lemma 9C21 shows that, by passing to a suitable subgroup if necessary, we may assume that any marked tail has a mark adjacent to the circuit. Other than that, the proof just consists of a routine check to identify (Coxeter) subdiagrams associated with known infinite groups. □

Before we treat the case of a circuit with at most two tails and all branches unmarked, we briefly comment on extensions of 4-circuits. For a finite group, a marked branch of the circuit cannot be adjacent to the (necessarily unmarked) tail, and the circuit itself must represent a finite group $[1\,1\,2^q]^p$, with p the mark on the (single) marked branch and q the interior mark of the circuit. Since $[1\,1\,2^q]^p$ is infinite if $p, q \geqslant 4$, the only circuit with a marked branch which we need consider is

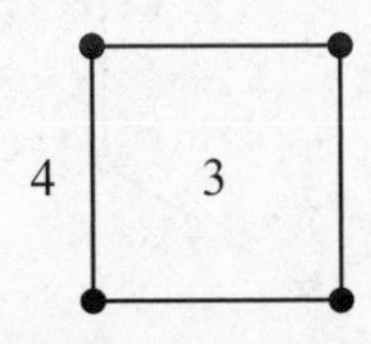

It turns out that the only extensions are infinite. In particular, with zero Schläfli

determinant, we have

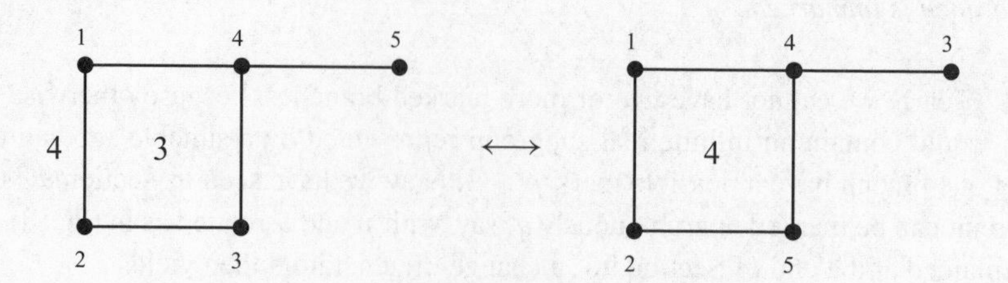

The operation which takes the first diagram into the second is

$$(S_1, \ldots, S_5) \mapsto (S_1, S_5S_4S_3S_2S_3S_4S_5, S_3, S_4, S_5) =: (R_1, \ldots, R_5).$$

We recognize the second diagram as an alternative form of that for $[2\,1\,2^4]^3$, which is infinite, and acts discretely on $\mathbb{C}^4$ (see [114, p. 258; 388, p. 382]).

Finally, then, we come to circuits with two tails and all branches unmarked. We denote by $\mathcal{J}_n(h, k; l, m; q)$ the following diagram on $n := h + k + l + m$ nodes. It has a circuit with $l + m$ nodes and branches. Two tails with h and k extra nodes and branches leave from nodes of the circuit separated by l (or m) branches. All branches are unmarked, but the circuit is labelled q. Note that h and k are interchangeable, as are l and m. So, for example, $\mathcal{J}_{11}(2, 3; 2, 4; 5)$ is the diagram of Figure 9F4. We allow $h = 0$ or $k = 0$ (when there is only one tail), in which case we may take l and m to be any non-negative integers whose sum is equal to the number of nodes in the circuit. Note that the tails leave from adjacent nodes of the circuit if $l = 1$ (or $m = 1$). As before, there is a geometric group for each diagram $\mathcal{J}_n(h, k; l, m; q)$ (here we can load the turn of the circuit on one of its branches).

A recursive calculation for the Schläfli determinant Δ of $\mathcal{J}_n(h, k; l, m; q)$ yields

$$\begin{aligned} 2^n \Delta &= 4(h+1)(k+1)\sin^2\frac{\pi}{q} - (2hk + h + k)(l + m) + hklm \\ &= hk(2 - l)(2 - m) + (h + k)(4 - l - m) + 4 - 4(h + 1)(k + 1)\cos^2\frac{\pi}{q}, \end{aligned}$$

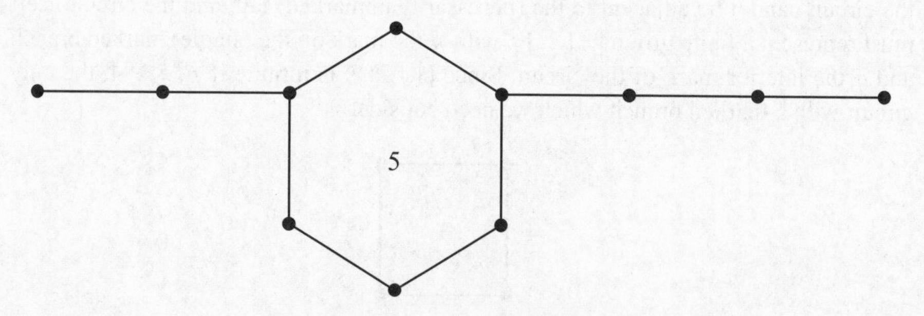

Figure 9F4. The diagram $\mathcal{J}_{11}(2, 3; 2, 4; 5)$.

with $n = h + k + l + m$, which is valid whenever $q \geqslant 2$, $h, k \geqslant 0$ and $l, m \geqslant 1$ with $l + m \geqslant 3$. (We shall find both forms for Δ useful.) Observe, though, that we can only get a corresponding group $G = G(\mathcal{J}_n(h, k; l, m; q))$ which we have not previously considered if $h, k \geqslant 1$ and $l, m \geqslant 2$, since (for instance)

$$G(\mathcal{J}_n(0, k; l, m; q)) \cong [(k+1)\ 1\ (l+m-2)^q]^3,$$

and

$$G(\mathcal{J}_n(h, k; 1, m; q)) \cong [(h+1)\ (k+1)\ (m-1)^q]^3.$$

The calculation for Δ is based on Lemma 9C20, using the initial values for $\mathcal{J}_n(0, k; l, m; q)$ of Theorem 9F1, and

$$2^{k+l+m-1}\Delta([3^{k,l-1,m-1}]) = k + l + m - k(l-1)(m-1).$$

For finiteness of the corresponding group, we must have $\Delta > 0$; however, bear in mind that the groups corresponding to each of the subdiagrams must be finite as well. As usual, the degenerate case $\Delta = 0$ is also of interest.

The first obvious remark is that Δ is monotonic decreasing in q, whatever h, k, l or m may be. Our classification of those groups $G(\mathcal{J}_n(h, k; l, m; q))$ which are finite will fix q first. Next, the second form for Δ suggests that the case $l = m = 2$ may be important; this is indeed so. Since we are taking $h, k \geqslant 1$, another obvious case for study is $h = k = 1$, because these groups must be finite in order that those with larger h or k are. Notice as well that, if $h, k \geqslant 1$ and (say) $l \geqslant 4$, then there is a subdiagram corresponding to an infinite group; thus we have $l, m \leqslant 3$ in such a case (see also the second part of Lemma 9F3).

When $q = 2$, if we obtain a finite group, then it is a real reflexion group, and hence one of the groups $[3^{r,s,t}]$, with $r + s + t = n - 1$. Since

$$2^n\Delta([3^{r,s,t}]) = n + 1 - rst,$$

it is straightforward to identify which group it actually is. Indeed, this even carries over to the degenerate case $\Delta = 0$, because these groups (with $\{r, s, t\} = \{2, 2, 2\}, \{3, 3, 1\}$ and $\{5, 2, 1\}$) are now specified by the number n of generating reflexions. We shall therefore leave this case to the interested reader.

It is appropriate, however, to make some additional remarks about this case. Here alone it is possible to add more than two tails to a circuit, to obtain the diagram for a finite group. So, for example, a further alternative diagram for $E_8 = [3^{4,2,1}]$ is

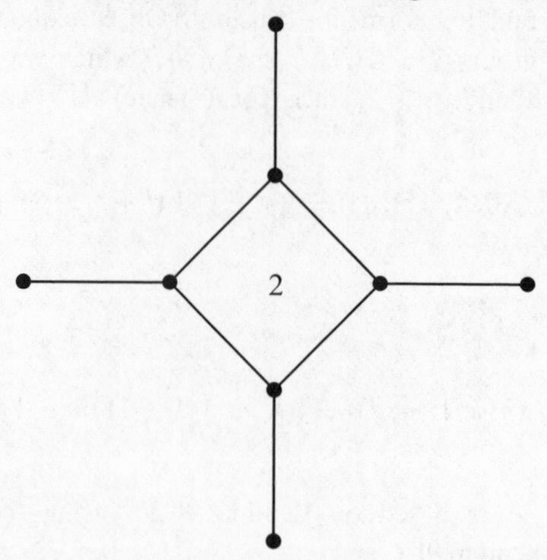

(Its connected subdiagrams on 7 nodes give $E_7 = [3^{3,2,1}]$, while those on 6 nodes give $E_6 = [3^{2,2,1}]$ or $B_6 = [3^{3,1,1}]$.)

We thus come down to the following case. With fixed $q \geqslant 3$, $l, m = 2$ or 3 and $h = k = 1$, we have

$$2^n \Delta = (4 - l)(4 - m) - 16 \cos^2 \frac{\pi}{q}.$$

It is clear that there are no solutions to $\Delta > 0$ at all, and even the degenerate case $\Delta = 0$ has only the single solution $q = 3$ and $l = m = 2$. The diagram $\mathcal{J}_6(1, 1; 2, 2; 3)$ of Figure 9F5 corresponds to an infinite unitary group which acts discretely on $\mathbb{C}^5$. This group stands alone among the discrete infinite groups we have studied, in that it has no diagram formed by adding tails to a triangle; perhaps this accounts for it having been overlooked for so long (it was found by McMullen more than thirty years ago, but only communicated to other people within the past few years). In view of this, and to check the discreteness, it may be appropriate to give a few more details.

With our usual conventions, and $\omega = \frac{1}{2}(-1 + i\sqrt{3})$, we take the generating reflexions (corresponding to the labels on the diagram in Figure 9F5) to be

$$S_j : x \mapsto x - 2(\langle x, v_j \rangle - \beta_j) v_j,$$

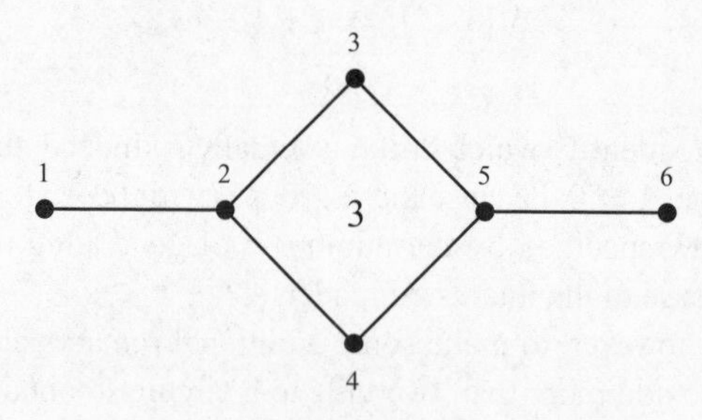

Figure 9F5. The diagram $\mathcal{J}_6(1, 1; 2, 2; 3)$.

where

$$\begin{aligned}
v_1 &:= -\tfrac{1}{2}(1, 1, \omega^2, \omega, 0),\\
v_2 &:= (1, 0, 0, 0, 0),\\
v_3 &:= -\tfrac{1}{2}(\omega, 1, 0, 1, \omega^2),\\
v_4 &:= -\tfrac{1}{2}(\omega^2, 1, 1, 0, \omega),\\
v_5 &:= (0, 0, 0, 0, 1),\\
v_6 &:= -\tfrac{1}{2}(0, 1, \omega, \omega^2, 1),
\end{aligned}$$

and $\beta_1 = 1$, $\beta_j = 0$ for $j = 2, 3, 4, 5, 6$. At this stage, it is helpful to note that

$$x \mapsto (\bar{\xi}_1, \bar{\xi}_2, \bar{\xi}_4, \bar{\xi}_3, \bar{\xi}_5)$$

(with $x = (\xi_1, \xi_2, \xi_3, \xi_4, \xi_5)$) acts as an outer automorphism of the group $G := \langle S_1, \ldots, S_6 \rangle$, interchanging S_3 and S_4, and fixing S_1, S_2, S_5 and S_6. With $\mathbb{G} = \mathbb{Z}[\omega]$ as before, the image-set $V := oG$ of o under G is

$$V = \{x \in \mathbb{G}^5 \mid \langle x, w_j \rangle \in 2\mathbb{G} \text{ for } j = 1, 2, 3\},$$

where

$$\begin{aligned}
w_1 &:= (1, \omega, -\omega, 0, -\omega^2),\\
w_2 &:= (1, \omega^2, 0, -\omega^2, -\omega),\\
w_3 &:= (1, 0, -1, -1, 1).
\end{aligned}$$

(As usual, we only need to verify that the images of a single point form a discrete set.) The translates by the vectors in V form the translation subgroup of G, since the reflexion

$$S_1' \colon x \mapsto x - 2\langle x, v_1 \rangle v_1$$

belongs to $\langle S_2, \ldots, S_6 \rangle$.

Finally, let us remark that this group G is the subgroup of the known infinite discrete reflexion group $[4\ 2\ 1]^3$ in $\mathbb{C}^6$ (see [114, p. 258; and 388, p. 381]), induced as a cut by one of the latter's generating mirrors.

9G Abstract Groups and Diagrams

The dihedral groups are the basic 2-generator subgroups of Coxeter groups; these have the property that presentations for the groups involve at most pairs of involutory generators. The obvious next step, motivated by what we have done in the previous sections of this chapter, is to consider groups generated by involutions, where now relators can employ two or three generators. It might initially be thought that the generalized triangle groups of Section 9D provide suitable paradigms for such 3-generator subgroups, giving a class of groups designated by diagrams which are suitably labelled simplicial 2-complexes.

However, the next natural question asks whether the obvious representations of such groups as complex matrices are faithful. We have seen, in Section 9D, that even in the "real" case, the natural homomorphism $\Gamma^3(5,3;5,5) \to G^3(5,3;5,5)$ is not an isomorphism. Moreover, in Section 9E, we saw that the 3-generator subgroups need not determine the geometry of the whole group; that is, it can happen that more than one geometric group is associated with the same diagram.

We therefore come back to what we said in the introduction to the chapter. The central criterion seems to be that, if we can find an infinite quotient of a subgroup (of finite index) of a group in which we are interested, then the original group itself must be infinite. Only when the corresponding quotient must necessarily be finite do we then look further into the problem, to determine if the corresponding unitary representation is faithful, and thus to establish that the original group must now itself be finite.

The core of the problem is that a complex representation imposes very strong conditions on the abstract group. First, a single triangle relator (of the form $S_i S_k S_j S_k$) determines the corresponding turn, and hence any other (non-Coxeter-type) relators for that triangle. (We may make exactly the same assumption here as we saw we could make in Section 9C, namely, that all the marks on diagrams basically equivalent to one implied by the abstract group relations are integers. While there may be representations of an abstract group corresponding to diagrams with fractional marks, to determine its finiteness we need not consider them.) Second, the turns on individual triangles are not independent – they must satisfy the compatibility relations (9E8) for any 4 nodes of the diagram. Third, as we have seen for circuit diagrams (notably the case of $\mathcal{J}_6(1,1;2,2;3)$), it may not be the case that triangle relators are actually most appropriate – we may need those on longer circuits. The cumulative effect of these objections is that groups generated by involutions with relators involving three as well as two generators are somewhat rarely faithfully represented as unitary groups.

All that notwithstanding, in the context of regular polytopes, their groups are usually abstract, even though we are often interested in their realizations (or, more generally, models). For that reason, it is appropriate to consider abstract groups with presentations corresponding to those we have discussed in this chapter, and their representations as complex linear groups.

We begin by discussing abstract diagrams; let us emphasize that we have been following Coxeter [114, 116] in associating diagrams with *geometric* groups. The most general abstract situation is the following. We set $\mathcal{N} := \{1, \ldots, n\}$, the set of *nodes*; each $i \in \mathcal{N}$ will be associated with an involutory generator σ_i of a group $\Gamma := \langle \sigma_1, \ldots, \sigma_n \rangle$. Let $\mathcal{S}$ be the set of finite sequences of elements of $\mathcal{N}$, with successive elements distinct (this eliminates some obvious trivialities below). A *marking* (or *labelling*) on $\mathcal{N}$ is a mapping $p\colon \mathcal{S} \to \{2, 3, \ldots, \infty\}$, such that $p_i = 2$ (for a 1-element sequence) and $p_{i(1),\ldots,i(k)} = p_{i(k),\ldots,i(1)}$ for sequences in $\mathcal{S}$ in reverse order; the pair $\mathcal{D} := (\mathcal{N}, p)$ is called an *abstract diagram*. (We write the argument of p as a suffix, to accord with the notation of Section 9C; the next definition mimics what we saw for geometric groups in that section.) The group $\Gamma = \Gamma(\mathcal{N}, p)$ is then defined by

9G1 $$\Gamma := \langle \sigma_1, \ldots, \sigma_n \mid (\sigma_{i(1)}\sigma_{i(2)} \cdots \sigma_{i(k)}\sigma_{i(k-1)} \cdots \sigma_{i(2)})^{p_{i(1),\ldots,i(k)}} = \varepsilon \text{ for each } (i(1), \ldots, i(k)) \in \mathcal{S}\rangle.$$

In theory, what we are doing here is specifying the periods of all products of pairs of conjugates of the generators. Thus a mark "∞" does not necessarily mean that the corresponding group element has infinite period, merely that it is unspecified. In applications, only a small number of sequences in $\mathcal{S}$ will receive a finite mark, including the 2-element sequences which specify Coxeter-type relations.

We could, in fact, further generalize this definition, by allowing $p_i \geqslant 2$, which would permit generators which are not involutions; however, the way we have specified p would then make less sense (compare the discussion in Section 9A).

Since each defining relation in (9G1) involves an even number of generators, the subgroup Γ^+ of Γ consisting of the *even* elements (products of an even number of generators σ_i) has index 2; hence $\Gamma = \Gamma^+ \rtimes C_2$. As generators of Γ^+ we can take the products $\sigma_i\sigma_j$ (with $i, j \in \mathcal{N}$), but usually a small subset of these will suffice; then (9G1) translates into a corresponding presentation for Γ^+. Note that, in general, these presentations are not equivalent to those studied in [74, 382, 425, 426].

It is immediately obvious that we may impose certain compatibility conditions on the marking p. For instance, suppose that $p_{jk} = 3$; then $\sigma_i\sigma_j\sigma_k\sigma_j = \sigma_i\sigma_k\sigma_j\sigma_k$ for each i, so that the actual period of $\sigma_i\sigma_j\sigma_k\sigma_j$ is a common divisor of p_{ijk} and p_{ikj}.

With certain of these groups $\Gamma(\mathcal{N}, p)$, we can associate a diagram $\mathcal{D}$ in a more concrete sense. First, we join each pair of distinct nodes $i, j \in \mathcal{N}$ by a *branch* marked p_{ij} $(= p_{ji})$; we employ our standard conventions, in omitting a branch labelled 2, and omitting a mark 3 on any remaining branch. If $i(1), \ldots, i(k)$ are the nodes in cyclic order of a diagonal-free circuit $\mathcal{C}$ in $\mathcal{D}$, and at most one branch in $\mathcal{C}$ is labelled, then the periods of the elements $\sigma_{i(1)}\sigma_{i(2)} \cdots \sigma_{i(k)}\sigma_{i(k-1)} \cdots \sigma_{i(2)}$ are independent of the starting point $i(1)$ (because all these elements are conjugate), and so we may take the corresponding marks $p_{i(1),\ldots,i(k)}$ to be the same. In such a case, we may unambiguously give $\mathcal{C}$ the mark $p_{i(1),\ldots,i(k)}$. Thus $\mathcal{D}$ is a diagram with marked or unmarked branches, and a mark for every diagonal-free circuit in which at most one branch is marked.

There are several questions one might wish to ask about such a group $\Gamma := \Gamma(\mathcal{N}, p)$. As far as we are concerned, two are important. First, what conditions guarantee that Γ does not collapse, particularly to a group with generators corresponding to a proper subset of $\mathcal{N}$? As a related problem we also mention that of pre-assigning the structure of the *basic* 3-generator subgroups $\langle \sigma_i, \sigma_j, \sigma_k \rangle$ for Γ. Second, when is Γ finite, and if so, is it naturally isomorphic to a reflexion group? In certain circumstances, we can answer these questions.

9G2 Theorem *Let $n \geqslant 3$, and let $2 \leqslant s \leqslant \infty$. Let p be defined by*

$$p_{jk} := 3 \quad (1 \leqslant j < k \leqslant n), \qquad p_{jkl} = s \quad (1 \leqslant j < k < l \leqslant n).$$

Then each basic 3-generator subgroup of $\Gamma = \Gamma(\mathcal{N}, p)$ is isomorphic to the unitary *group* $[1\,1\,1]^s$. *Moreover,* Γ *is finite in just two cases:*

(a) $n = 3$ *and* $s < \infty$, *when* $\Gamma \cong [1\,1\,1]^s$;
(b) $n \geqslant 3$ *and* $s = 2$, *when* $\Gamma \cong S_{n+1}$, *the symmetric group on* $n + 1$ *elements.*

Proof. We construct a representation of Γ as a reflexion group $G = \langle S_1, \ldots, S_n \rangle$. It is easiest here to specify the Gram matrix for the corresponding hermitian form; this has entries α_{jk} $(:= \langle v_j, v_k \rangle)$ given by

9G3
$$2\alpha_{jk} := \begin{cases} 2, & \text{if } j = k, \\ -e^{2i\pi/s}, & \text{if } j < k, \\ -e^{-2i\pi/s}, & \text{if } j > k. \end{cases}$$

Of course, $S_j S_k$ has period 3 whenever $j \neq k$, which is consistent with $p_{jk} = 3$. Further, if $j < k < l$, then the turn in the cycle (j, k, l) is

$$\frac{2\pi}{s} + \frac{2\pi}{s} - \frac{2\pi}{s} = \frac{2\pi}{s},$$

so that $S_j S_k S_l S_k$ has period s, which again is consistent with $p_{jkl} = s$. Then it follows that $\langle \sigma_j, \sigma_k, \sigma_l \rangle \cong \langle S_j, S_k, S_l \rangle \cong [1\,1\,1]^s$, as required.

The rest of the proof follows at once. If $n = 3$, then for finiteness of Γ we clearly have $s < \infty$. If $n \geqslant 4$, then Theorem 9E17 says that each 4-generator subgroup is infinite if $s \geqslant 3$; that is, if Γ is finite, then $s = 2$. If $s = 2$, we change the generators of Γ to $\tau_n := \sigma_n$ and $\tau_i := \sigma_{i+1}\sigma_i\sigma_{i+1}$ for $i < n$, and obtain the presentation $\tau_i^2 = (\tau_i\tau_{i+1})^3 = (\tau_i\tau_j)^2 = \varepsilon$, with $|i - j| \geqslant 2$, which shows that Γ is a quotient of S_{n+1}. On the other hand, the generating transpositions $(i\ n+1)$, with $i = 1, \ldots, n$, for S_{n+1} satisfy the relations for the generators σ_i of Γ, so that indeed $\Gamma \cong S_{n+1}$. □

In a somewhat similar way, we can also deal with the following more general kind of group, which we will need later. If $\mathcal{E}$ and $\mathcal{F}$ are subsets of pairs (2-subsets) and triples (3-subsets) in the node set $\mathcal{N} = \{1, \ldots, n\}$, we say that $\mathcal{E}$ is *compatible* with $\mathcal{F}$ if each 2-element subset of every triple in $\mathcal{F}$ belongs to $\mathcal{E}$. In our applications, $\mathcal{E}$ and $\mathcal{F}$ will consist of edge- or face-sets of a simplicial 2-complex on n vertices, for instance, a regular map with triangular faces on some surface.

9G4 Theorem *Let* $\mathcal{E}$ *and* $\mathcal{F}$ *be subsets of pairs and triples in* $\mathcal{N} = \{1, \ldots, n\}$, *such that* $\mathcal{E}$ *is compatible with* $\mathcal{F}$. *Let* $2 \leqslant s \leqslant \infty$, *and define* p *by*

$$p_{jk} = 3 \quad (\{j, k\} \in \mathcal{E}); \qquad p_{jkl} = s \quad (\{j, k, l\} \in \mathcal{F}).$$

Then the basic 3-generator subgroups of $\Gamma := \Gamma(\mathcal{N}, p)$ *which correspond to triples in* $\mathcal{F}$ *are isomorphic to the unitary group* $[1\,1\,1]^s$. *If* Γ *is finite, then* $\mathcal{E}$ *consists of all pairs in* $\mathcal{N}$, *and one of the following holds:*

(a) $n = 3$ *and* $s < \infty$;
(b) $n \geqslant 3$ *and* $s = 2$.

In case (b), if $\mathcal{F}$ *consists of all triples in* $\mathcal{N}$, *then* $\Gamma \cong S_{n+1}$.

Proof. We modify the Gram matrix whose entries are those of (9G3), by setting $\alpha_{jk} = -1$ if $j \neq k$ with $\{j, k\} \notin \mathcal{E}$ (otherwise α_{jk} is as before). Thus $S_j S_k$ has infinite period for such pairs $\{j, k\}$, which is consistent with $p_{jk} = \infty$, and so Γ is infinite unless $\mathcal{E}$ contains all pairs in $\mathcal{N}$. Moreover, the group of Theorem 9G2, with all branches marked 3 and all triangles marked s, is clearly a quotient of Γ, and so is finite if Γ is finite. The current assertions now follow at once from Theorem 9G2. □

Notice that the only cases which Theorem 9G4 leaves open are those where $\mathcal{E}$ consists of all pairs in $\mathcal{N}$, some triples in $\mathcal{N}$ are marked 2, while others are unmarked. In our application of Theorem 9G4 in Section 11C, when $\mathcal{F}$ will correspond to the set of faces of a regular map with vertex-set $\mathcal{N}$, we shall see that only one case needs further attention.

10

Locally Toroidal 4-Polytopes: I

An important problem in classical geometry is the complete description of all regular polytopes and tessellations in spherical, euclidean or hyperbolic space. When asked within the theory of abstract regular polytopes, such an enumeration problem must necessarily take a different form, because an abstract polytope is not *a priori* embedded into the geometry of an ambient space. One appropriate substitute now calls for the classification of abstract regular polytopes by their "local" or "global" topological type.

The traditional theory of regular polytopes and tessellations is concerned with, and solves, the case where the topology is spherical. This is well known. In Chapter 6, we already moved on to other topological types and enumerated the regular toroids, which are the regular polytopes whose global topology is toroidal. Now, in this and the next two chapters, we investigate the locally toroidal regular polytopes. As we explain in Section 10A, such polytopes can only exist in ranks 4, 5 or 6.

In this and the next chapter, we treat the polytopes of rank 4, and obtain a nearly complete classification of the universal regular polytopes which are locally toroidal and finite. Then, in Chapter 12, we enumerate all such polytopes of rank 5, and produce a list of such polytopes of rank 6 which we strongly conjecture to be complete.

This classification problem for locally toroidal regular polytopes has triggered much of the theory of abstract polytopes; it was originally posed by Grünbaum [199, p. 196].

10A Grünbaum's Problem

Historically, abstract polytope theory grew out of the attempt to find a convenient framework for the study of polytope-like structures which are topologically more complicated than convex polytopes [141, 199]. As an important step in this direction, Grünbaum [199, p. 196] posed the challenging problem, as yet unsolved, of completely classifying the regular polytopes which are locally toroidal.

The study of topological properties of polytopes must necessarily begin with the fundamental question of how a natural (global) topology can be defined for a given polytope. As we explained in Section 6B, this problem is subtle and cannot be uniquely solved in general [49, §3]. On the other hand, there are many interesting classes of polytopes which admit a natural global topology. Important examples are the regular tessellations on space-forms, which we completely enumerated for spherical or compact euclidean space-forms in Chapter 6.

In this section, for reasons which will soon become apparent, we prefer to denote the rank by $n + 1$. There are several ways in which a polytope can be described locally. In this and the next two chapters, we use the following basic definition. Given a topological type X, we say that an $(n + 1)$-polytope $\mathcal{P}$ is *locally of topological type* X if its facets and vertex-figures which are not spherical are of topological type X. This is a strong condition on the sections of $\mathcal{P}$ of rank n. More general terminology may only require the minimal sections of $\mathcal{P}$ which are not spherical to be of the required topological type; this will be the context of Section 14A. However, in the present setting, our terminology always refers to the sections of rank n (that is, the facets and vertex-figures).

The traditional theory of regular polytopes and tessellations deals with the locally spherical case (see Section 6B). Each tessellation is a locally spherical polytope; that is, its local building blocks (tiles, or facets, and vertex-figures) are topologically spheres. Indeed, up to isomorphism, the classical regular tessellations $\{p_1, \ldots, p_n\}$ in spherical, euclidean or hyperbolic n-space are the only abstract regular $(n + 1)$-polytopes which are locally spherical and universal. Among these, only the spherical regular tessellations (convex regular polytopes) are finite and also globally spherical. A generic locally spherical regular polytope is then a quotient of such a tessellation, and can be viewed as a regular tessellation on a space-form.

A major effort is now to describe and classify abstract regular polytopes with more complicated local topologies like that of arbitrary spherical, euclidean or hyperbolic space-forms. In this generality the enumeration problem is wide open, yet significant progress has been made in the case where the space-form is compact euclidean. We saw in Section 6G that, among the compact euclidean space-forms, the tori are the only ones which can admit regular tessellations; these are the regular toroids. Therefore, if a regular polytope $\mathcal{P}$ is locally a compact euclidean space-form, then it must actually be *locally toroidal*; that is, its facets and vertex-figures are globally spherical or globally toroidal, with at least one kind toroidal. Thus the facets and vertex-figures of a locally toroidal regular polytope $\mathcal{P}$ must be isomorphic to convex regular polytopes or regular toroids (see Chapter 6).

In this and the next two chapters, we shall describe the present state of the classification of locally toroidal regular polytopes. The term "classification" is used here in the same sense as in Section 4B; that is, it refers to the enumeration of *all universal regular* $(n+1)$*-polytopes* $\{\mathcal{P}_1, \mathcal{P}_2\}$ *which are locally toroidal and finite*. Also, for each universal polytope which is locally toroidal, we shall determine the structure of its group if it is known.

The following simple lemma explains why locally toroidal regular polytopes can only exist in ranks 4, 5 and 6. In fact, in higher ranks there are no hyperbolic honeycombs to derive them from.

10A1 Lemma *Each locally toroidal regular* $(n + 1)$*-polytope* $\mathcal{P}$ *is a quotient of a regular honeycomb in hyperbolic* n*-space* $\mathbb{H}^n$ *listed in Table 10A1, or of its dual. In particular, the rank of* $\mathcal{P}$ *is* 4, 5 *or* 6.

Table 10A1. *The Non-Locally Finite Regular Honeycombs in* $\mathbb{H}^n$

n	Symbol
3	$\{4, 4, r\}, \quad r = 3, 4$
	$\{6, 3, p\}, \quad p = 3, 4, 5, 6$
	$\{3, 6, 3\}$
4	$\{3, 4, 3, 4\}$
5	$\{3, 3, 3, 4, 3\}, \{3, 3, 4, 3, 3\}, \{3, 4, 3, 3, 4\}$

Note: only one from each dual pair is listed.

Proof. The facets $\mathcal{P}_1$ and vertex-figures $\mathcal{P}_2$ of $\mathcal{P}$ are spherical or toroidal, with at least one kind toroidal. If $\mathcal{P}$ is of type $\{p_1, \dots, p_n\}$, then the Schläfli symbols $\{p_1, \dots, p_{n-1}\}$ and $\{p_2, \dots, p_n\}$ are of spherical or euclidean type, with at least one of the latter. Up to duality, this leaves only the Schläfli symbols $\{p_1, \dots, p_n\}$ occurring in Table 10A1 (see Sections 3B and 3C). The corresponding universal regular $(n+1)$-polytope $\{p_1, \dots, p_n\}$ is then a hyperbolic honeycomb in $\mathbb{H}^n$ (Theorem 3D6). Its facets or vertex-figures, or both, are isomorphic to regular euclidean tessellations. □

In hyperbolic 3-space $\mathbb{H}^3$, there are fifteen regular honeycombs (Coxeter [113, §3]). Of these, only eleven can yield locally toroidal polytopes. The honeycombs $\{3, 4, 4\}$, $\{3, 3, 6\}$, $\{4, 3, 6\}$, and $\{5, 3, 6\}$ have spherical facets and have all their vertices on the absolute of $\mathbb{H}^3$. Their duals $\{4, 4, 3\}$, $\{6, 3, 3\}$, $\{6, 3, 4\}$ and $\{6, 3, 5\}$ have spherical vertex-figures and have all their facets inscribed in horospheres of $\mathbb{H}^3$ instead of infinite spheres (see [118, §11.3]). Finally, the self-dual honeycombs $\{4, 4, 4\}$, $\{6, 3, 6\}$ and $\{3, 6, 3\}$ have all their vertices on the absolute and all their facets inscribed in the absolute. All these eleven types occur as Schläfli symbols of locally toroidal regular polytopes of rank 4. The remaining four honeycombs $\{3, 5, 3\}$, $\{4, 3, 5\}$, $\{5, 3, 4\}$ and $\{5, 3, 5\}$ are locally spherical and are (locally finite) tessellations of $\mathbb{H}^3$. Their quotient polytopes are necessarily locally spherical or locally projective.

In $\mathbb{H}^4$, there are seven regular honeycombs. Of these, only $\{3, 4, 3, 4\}$ and its dual $\{4, 3, 4, 3\}$ are not locally spherical and can occur as the type of a locally toroidal regular 5-polytope. The first has 24-cells as facets, and its vertices are all on the absolute of $\mathbb{H}^4$. The second has 24-cells as vertex-figures, and its facets are cubic tessellations inscribed in horospheres of $\mathbb{H}^4$.

In $\mathbb{H}^5$, there are five regular honeycombs, namely,

$$\{3, 3, 3, 4, 3\}, \quad \{3, 4, 3, 3, 3\}, \quad \{3, 3, 4, 3, 3\}, \quad \{3, 4, 3, 3, 4\}, \quad \{4, 3, 3, 4, 3\}.$$

The first honeycomb has spherical facets (which are cross-polytopes), and its vertices are all on the absolute of $\mathbb{H}^5$. The second, the dual of the first, has spherical vertex-figures (which are cubes), and all its facets are inscribed in horospheres of $\mathbb{H}^5$. For the remaining three honeycombs, all the vertices are at infinity, and all the facets are euclidean tessellations inscribed in the absolute. These are the only types which lead to locally toroidal regular 6-polytopes.

In $\mathbb{H}^n$ with $n \geqslant 6$, there are no regular honeycombs.

In this and the next chapter we shall treat the locally toroidal polytopes of rank 4; in particular, we shall follow [294, 296, 297, 299, 300]. According to Lemma 10A1, this involves analysis of the Schläfli types $\{4, 4, r\}$ with $r = 3, 4$, $\{6, 3, p\}$ with $p = 3, 4, 5, 6$, and $\{3, 6, 3\}$, and their duals. In particular, we shall obtain a complete classification for all types except $\{4, 4, 4\}$ and $\{3, 6, 3\}$. For $\{4, 4, 4\}$ the enumeration is almost complete, and for $\{3, 6, 3\}$ partial results are known.

The picture is particularly satisfactory for the types $\{6, 3, p\}$ and $\{3, 6, 3\}$. The structure of the universal polytope is then often governed by a complex hermitian form, and we can appeal to the results of Chapter 9. In particular, any such polytope is finite if and only if the corresponding hermitian form is positive definite. This generalizes the well-known classical situation, where the structure of a regular tessellation is determined by a real quadratic form which defines the geometry of the ambient space; this correspondence sets up a beautiful link between geometry and algebra (see Chapter 11 and Coxeter [120, Chapters 10, 11]).

Then, in Chapter 12, we deal with the ranks 5 and 6; we shall follow [302, 303]. For rank 5, only one Schläfli type occurs, namely, $\{3, 4, 3, 4\}$ (and its dual); our enumeration is complete in this case. Finally, for rank 6, the types are $\{3, 3, 3, 4, 3\}$, $\{3, 3, 4, 3, 3\}$ and $\{3, 4, 3, 3, 4\}$, and their duals. Here, rather less is known. However, we shall describe a list of finite polytopes which we strongly conjecture to be complete.

On the group level, the classification of the locally toroidal regular polytopes $\{\mathcal{P}_1, \mathcal{P}_2\}$ translates into an enumeration problem for certain groups which are defined in terms of generators and relations. By Lemma 10A1, these groups are quotients of certain hyperbolic Coxeter groups and are obtained by adding to the standard Coxeter relations either one or two new relations. Each toroidal polytope $\mathcal{P}_1$ or $\mathcal{P}_2$ contributes exactly one such extra relation, of the kind described in Theorems 6D4 or 6E6 (or suitably dualized). This extra relation imposes the toroidal structure on the facets or vertex-figures. On the other hand, a spherical facet or vertex-figure does not contribute any extra relation.

Since the appearance of Grünbaum's original paper [199], there has appeared a significant number of publications which deal with locally toroidal regular or chiral polytopes. To those references already cited we can add [366, 368], which were the first papers to describe general construction techniques of such polytopes, and [83, 132, 370, 443–445]. In particular, representations of hyperbolic reflexion groups as finite projective linear groups have been employed to obtain many finite examples of locally toroidal polytopes (see [237, 322–324, 333–335, 373, 445]). We shall meet examples in Section 13D.

10B The Type $\{4, 4, 3\}$

In this section, we discuss the locally toroidal regular 4-polytopes of type $\{4, 4, 3\}$, as well as their duals, of type $\{3, 4, 4\}$. In particular, we shall enumerate the finite universal regular polytopes

$$_1\mathcal{T}_{\mathbf{s}}^4 := \{\{4, 4\}_{\mathbf{s}}, \{4, 3\}\},$$

where $\mathbf{s} := (s^k, 0^{2-k})$ with $2 \leqslant s < \infty$ and $k = 1$ or 2. Here, the superscript indicates the rank, and the subscript 1 on the left refers to $\{4, 4, 3\}$ as the *first* type of Schläfli symbol which we investigate. (In [199, p. 196], these polytopes were denoted by $\mathcal{L}_{s,0}$ or $\mathcal{L}_{s,s}$, respectively.) Our basic construction technique for these polytopes is twisting; see Chapter 8 for the notation employed.

In fact, each polytope can be directly obtained from a twisting operation on a Coxeter group.

Recall from Theorem 6D4 (see also [131, §8.3]) that the group of the regular torus map $\{4, 4\}_{(s^k, 0^{2-k})}$ is the Coxeter group $[4, 4] = \langle \rho_0, \rho_1, \rho_2 \rangle$, factored out by the single extra relation

10B1
$$\begin{aligned} (\rho_0\rho_1\rho_2\rho_1)^s &= \varepsilon \quad \text{if } k = 1, \\ (\rho_0\rho_1\rho_2)^{2s} &= \varepsilon \quad \text{if } k = 2. \end{aligned}$$

The extra relation corresponds to the alternative expression of this map as $\{4, 4 \mid s\}$ or $\{4, 4\}_{2s}$, respectively (see Section 7B). We shall denote this group by $[4, 4]_{\mathbf{s}}$, where $\mathbf{s} = (s^k, 0^{2-k})$.

The search for the universal polytope ${}_1\mathcal{T}^4_{\mathbf{s}}$ involves the analysis of ${}_1\Gamma^4_{\mathbf{s}}$, which denotes the Coxeter group $[4, 4, 3] = \langle \rho_0, \ldots, \rho_3 \rangle$ factored out by the corresponding extra relation (10B1). This relation involves only the first three generators ρ_0, ρ_1, ρ_2, and collapses the facet $\{4, 4\}$ of the honeycomb $\{4, 4, 3\}$ to its finite quotient $\{4, 4\}_{\mathbf{s}}$. In particular, if the polytope ${}_1\mathcal{T}^4_{\mathbf{s}}$ exists, then

$$\Gamma\big({}_1\mathcal{T}^4_{\mathbf{s}}\big) = {}_1\Gamma^4_{\mathbf{s}}.$$

This follows from Proposition 4A8, because each of these groups can easily be identified as the free product of $[4, 4]_{\mathbf{s}} = \Gamma(\{4, 4\}_{\mathbf{s}})$ and $[4, 3] = \Gamma(\{4, 3\})$ with amalgamation along their subgroups $[4] = \Gamma(\{4\})$, factored out by the commutation relation corresponding to $(\rho_0\rho_3)^2 = \varepsilon$.

There are several problems which need to be addressed. First, is ${}_1\Gamma^4_{\mathbf{s}}$ indeed a C-group, and, if so, are its subgroups $\langle \rho_0, \rho_1, \rho_2 \rangle$ and $\langle \rho_1, \rho_2, \rho_3 \rangle$ indeed isomorphic to $[4, 4]_{\mathbf{s}}$ and $[4, 3]$, respectively, rather than to proper quotients of these groups? Both questions are crucial for the existence of the related polytope ${}_1\mathcal{T}^4_{\mathbf{s}}$. In fact, by Proposition 4A8, this polytope exists if and only if these questions have a positive answer. Moreover, there is also the task of explicitly constructing the groups ${}_1\Gamma^4_{\mathbf{s}}$, and deciding which of them are finite.

As we shall see, these groups often have a natural structure as a twisted group, and so there is a natural twisting operation to begin with. (We could provide further motivation for these twisting operations as we will do in Chapter 11 for the regular polytopes of types $\{6, 3, p\}$, but here we prefer just to describe them directly.)

The Polytopes ${}_1\mathcal{T}^4_{(s,0)} = \{\{4, 4\}_{(s,0)}, \{4, 3\}\}$

We begin with the case $k = 1$, where the twisting operation is quite simple. For the construction of ${}_1\Gamma^4_{(s,0)}$, consider the Coxeter group $W = \langle \sigma_0, \ldots, \sigma_4 \rangle$ with the diagram of Figure 10B1, with $s \geqslant 2$. (Here, and in the following, if a mark $s = 2$ occurs in a

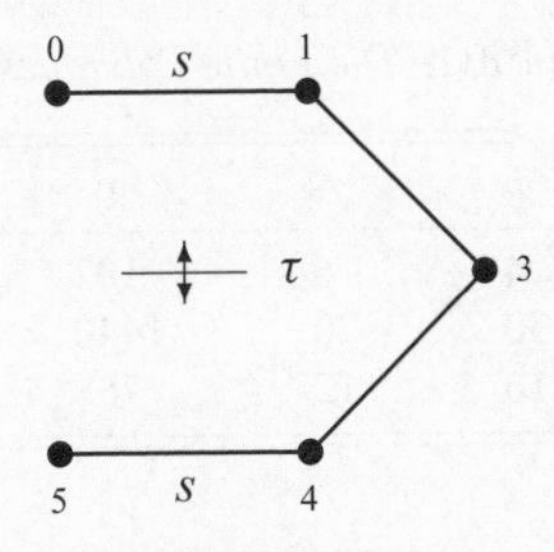

Figure 10B1

diagram, then we regard the corresponding branch as missing; that is, the two generators commute.) The symmetry of this diagram determines a group automorphism τ of W, and so we can apply the twisting operation

10B2 $$(\sigma_0, \ldots, \sigma_4; \tau) \mapsto (\sigma_0, \tau, \sigma_3, \sigma_2) =: (\rho_0, \rho_1, \rho_2, \rho_3).$$

Then the new group $\Gamma := \langle \rho_0, \ldots, \rho_3 \rangle$ is a semi-direct product of W by C_2 (generated by τ); in fact, we have

$$\sigma_1 = \tau\sigma_3\tau = \rho_1\rho_2\rho_1, \quad \sigma_4 = \tau\sigma_0\tau = \rho_1\rho_0\rho_1,$$

and so each generator of W belongs to Γ.

The generators $\rho_0, \ldots, \rho_3$ of Γ satisfy all the defining relations for ${}_1\Gamma^4_{(s,0)}$, that is, the standard relations for [4, 4, 3] as well as the extra relation (10B1) with $k = 1$. This extra relation follows from

$$\rho_0\rho_1\rho_2\rho_1 = \sigma_0\tau\sigma_3\tau = \sigma_0\sigma_1.$$

On the other hand, using the standard presentation for W, it is straightforward to check that these relations suffice for a presentation of Γ. Therefore

$${}_1\Gamma^4_{(s,0)} = \Gamma.$$

The construction of the polytope ${}_1\mathcal{T}^4_{(s,0)}$ (or, more exactly, of its group) is complete if we can verify the intersection property for Γ, and check that its subgroups $\langle \rho_0, \rho_1, \rho_2 \rangle$ and $\langle \rho_1, \rho_2, \rho_3 \rangle$ are of the right kind. To begin with, the intersection property for Γ with respect to its generators $\rho_0, \ldots, \rho_3$ follows easily from that of W with respect to $\sigma_0, \ldots, \sigma_4$. For example, we have

$$\begin{aligned} \langle \rho_0, \rho_1, \rho_2 \rangle \cap \langle \rho_1, \rho_2, \rho_3 \rangle &= \langle \sigma_0, \sigma_1, \sigma_3, \sigma_4, \tau \rangle \cap \langle \sigma_1, \sigma_2, \sigma_3, \tau \rangle \\ &= \langle \sigma_1, \sigma_3, \tau \rangle \\ &= \langle \rho_1, \rho_2 \rangle, \end{aligned}$$

as required. Finally, the subgroups $\langle \rho_0, \rho_1, \rho_2 \rangle$ and $\langle \rho_1, \rho_2, \rho_3 \rangle$ are isomorphic to $(D_s \times D_s) \rtimes C_2$ and $[3, 3] \rtimes C_2$, respectively, and are thus the groups of $\{4, 4\}_{(s,0)}$ and $\{4, 3\}$.

10B3 Theorem *The universal regular 4-polytope* ${}_1\mathcal{T}^4_{(s,0)} := \{\{4, 4\}_{(s,0)}, \{4, 3\}\}$ *exists for all* $s \geqslant 2$. *Its group is* $W \rtimes C_2$, *where* W *is the Coxeter group with the diagram of*

Table 10B1. *The Finite Polytopes* ${}_1\mathcal{T}^4_{\mathbf{s}}$

$\mathbf{s}$	v	f	g	Group
(2, 0)	4	6	192	$D_4 \rtimes S_4$
(3, 0)	30	20	1440	$S_6 \times C_2$
(2, 2)	16	12	768	$C_2 \wr D_6$

Figure 10B1. In particular, this polytope is finite if and only if $s = 2$ or 3, and then its group is $D_4 \times S_4$ or $S_6 \times C_2$, respectively.

Proof. It remains to identify the finite groups. Clearly, ${}_1\mathcal{T}^4_{(s,0)}$ is finite if and only if W is finite. This only leaves $s = 2$ or 3, with $W = C_2 \times C_2 \times S_4$ or $W = S_6$, respectively.

In the second case we can appeal to Lemma 8A2. Assume that S_6 acts on $1, \ldots, 6$ in the standard way, with $\sigma_i = (i+1\; i+2)$ for $i = 0, \ldots, 4$. Let $\hat{\tau} := (1\;6)(2\;5)(3\;4)$. Then we have $\hat{\tau}\sigma_i\hat{\tau} = \tau\sigma_i\tau$ for each i. Since the centre of S_6 is trivial, we must have $\Gamma = S_6 \times C_2$, with the last factor generated by $\tau\hat{\tau}$.

In the first case we can work with permutations on $0, \ldots, 7$, where the generators of W are given by $\sigma_0 = (0\;1)$, $\sigma_4 = (6\;7)$, and $\sigma_i = (i+1\; i+2)$ for $i = 1, 2, 3$. Then W has three orbits, namely, $\{0, 1\}$, $\{6, 7\}$, and $\{2, 3, 4, 5\}$; on the last, W acts as a copy of S_4. The symmetry of the diagram in Figure 10B1 can now be realized by the permutation $\tau := (0\;7)(1\;6)(2\;5)(3\;4)$, which is not an element of W. Then Γ is generated by W and τ, and it has two orbits, namely, $\{0, 1, 6, 7\}$ and $\{2, 3, 4, 5\}$. Now, since $(2\;5)(3\;4) \in W$, we also have $(0\;7)(1\;6) \in \Gamma$. But $(0\;1)$, $(6\;7)$, and $(0\;7)(1\;6)$ generate a subgroup of Γ isomorphic to D_4. In particular, the restriction of a permutation in Γ to $\{0, 1, 6, 7\}$ is an element of this subgroup. It follows that $\Gamma = D_4 \times S_4$, with the first factor acting on 0, 1, 6, 7 and the second on 2, 3, 4, 5. This completes the proof. □

In Table 10B1, we list the finite polytopes ${}_1\mathcal{T}^4_{\mathbf{s}}$, and give the numbers v of their vertices and f of their facets, and the orders g of their groups. For the first two rows we can appeal to Theorem 10B3, and for the last row to Theorem 10B5.

If $s = 2t$ (say) is even, then we can also obtain ${}_1\mathcal{T}^4_{(s,0)}$ from the general construction of the polytopes $\mathcal{L}^{\mathcal{K},\mathcal{G}}$ discussed in Chapter 8. More exactly, if $t \geqslant 2$ and $\mathcal{G}(t)$ is the diagram on the vertex-set of a square $\{4\}$ which connects antipodal vertices by a branch labelled t (but has no other branches), then Theorem 8E16 yields the infinite polytope

$$ {}_1\mathcal{T}^4_{(2t,0)} = \left(\{3\}^{\{4\},\mathcal{G}(t)}\right)^*, $$

the dual of $\{3\}^{\{4\},\mathcal{G}(t)}$. See also Corollary 8E18 (and Section 12B) for a similar construction based on the octahedron $\{3, 4\}$.

We can also employ Theorem 8E2, and obtain

$$ {}_1\mathcal{T}^4_{(2t,0)} = \left\{2^{\{4\},\mathcal{G}(t)}, \{4, 3\}\right\} = 2^{\{4,3\},\mathcal{D}(t)}, $$

where $\mathcal{G}(t)$ is as above and $\mathcal{D}(t)$ is the universal $\{4, 3\}$-extension of $\mathcal{G}(t)$ (that is, $\mathcal{D}(t)$

is the diagram on the vertices of a cube $\{4, 3\}$, which connects antipodal vertices of the square faces, or of the cube itself, by a branch with label t or ∞, respectively).

Moreover, if $t = 1$, we can also appeal to the methods of Section 8F, and identify ${}_1\mathcal{T}^4_{(2,0)}$ as a finite quotient of ${}_1\mathcal{T}^4_{(4,0)}$. In particular, Theorem 8F3 applies and gives (the dual of) ${}_1\mathcal{T}^4_{(2,0)}$, now with its group occurring in the form $(C_2 \times C_2) \rtimes [4, 3]$.

In topological terms, the finiteness of the polytopes ${}_1\mathcal{T}^4_{(2,0)}$ and ${}_1\mathcal{T}^4_{(3,0)}$ can also be explained as follows; see the discussion at the end of Section 6B. If we begin constructing a topological model for ${}_1\mathcal{T}^4_{(s,0)}$ by sticking together 3 solid tori $\{4, 4\}_{(s,0)}$ around an edge (as we must, because 3 is the last entry in the Schläfli symbol), and by successively extending this arrangement with further solid tori, then we observe that only for $s \leqslant 3$ will this arrangement close up locally. For example, in constructing ${}_1\mathcal{T}^4_{(3,0)}$, we find that there is precisely one torus $\{4, 4\}_{(3,0)}$ which surrounds the original 3 tori like a belt. This does not occur if $s \geqslant 4$. In general, the resulting topological space is a real 3-manifold which depends on the actual way the facets of ${}_1\mathcal{T}^4_{(s,0)}$ are glued together.

For example, the possible 3-manifolds which admit a topological model for ${}_1\mathcal{T}^4_{(2,0)}$ include connected sums of handles $\mathbb{S}^1 \times \mathbb{S}^2$ as well as certain spherical or euclidean space-forms (see [49, §4.1]). For ${}_1\mathcal{T}^4_{(3,0)}$, a topological model on $\mathbb{S}^3$ was described in [132] and [199, p. 196]; this is given by a decomposition of $\mathbb{S}^3$ into 20 solid tori. See [49] for a systematic approach to topological models for abstract polytopes.

The regular polytope ${}_1\mathcal{T}^4_{(3,0)}$ can also be employed to give a simple reason why the related universal polytope $\{\{4, 4\}_{(3,0)}, \{4, 3\}_3\}$ cannot exist; that is, identifying opposite vertices in the cubical vertex-figure of ${}_1\mathcal{T}^4_{(3,0)}$, to obtain the (projective) hemi-cube $\{4, 3\}_3$ as vertex-figure, does not lead to a polytope.

In fact, if this polytope were to exist, then its group would be obtained from ${}_1\Gamma^4_{(s,0)} = \langle \rho_0, \ldots, \rho_3 \rangle = S_6 \times C_2$ by imposing the extra relation

$$(\rho_1 \rho_2 \rho_3)^3 = \varepsilon,$$

which collapses $\{4, 3\}$ to $\{4, 3\}_3$ (see Section 6C). Since the alternating group A_6 is the only non-trivial normal subgroup of S_6, this limits the choices for corresponding normal subgroups N of ${}_1\Gamma^4_{(s,0)}$ to the second (central) factor in $S_6 \times C_2$. As in the proof of Theorem 10B3, we write $\hat{\tau}$ for the element of S_6 which defines an inner automorphism that acts on the generators σ_i of S_6, in the same way as the given automorphism ρ_1 $(= \tau)$. In ${}_1\Gamma^4_{(s,0)}$, we must then have $N = \langle \rho_1 \hat{\tau} \rangle$. But $\rho_1 \hat{\tau} \notin \langle \rho_1, \rho_2, \rho_3 \rangle$, so that taking the quotient by N does not lead to a collapse of the vertex-figure $\{4, 3\}$ of ${}_1\mathcal{T}^4_{(3,0)}$. In fact, ${}_1\Gamma^4_{(3,0)}/N$ is just the group of another regular polytope which also belongs to the class $\langle \{4, 4\}_{(3,0)}, \{4, 3\} \rangle$; in a sense, this is an "elliptic" quotient of ${}_1\mathcal{T}^4_{(3,0)}$ which could be denoted by ${}_1\mathcal{T}^4_{(3,0)}/2$.

The Polytopes ${}_1\mathcal{T}^4_{(s,s)} = \{\{4, 4\}_{(s,s)}, \{4, 3\}\}$

We now move on to the construction of the polytopes ${}_1\mathcal{T}^4_{(s,s)}$; this is the case $k = 2$ in our earlier notation. Now we shall work with the Coxeter group $W = \langle \sigma_0, \ldots, \sigma_5 \rangle$ of Figure 10B2 and employ three of its group automorphisms, which generate a dihedral

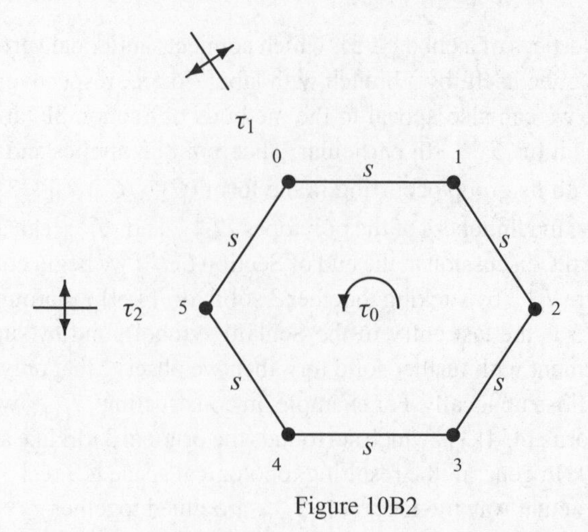

Figure 10B2

group D_6. These automorphisms are denoted by τ_0, which is induced by the half-turn of the diagram, and τ_1 and τ_2, which are again induced by reflexions as indicated in the figure. From the operation

10B4 $$(\sigma_0, \ldots, \sigma_5; \tau_0, \tau_1, \tau_2) \mapsto (\tau_0, \sigma_2, \tau_1, \tau_2) =: (\rho_0, \ldots, \rho_3),$$

we can now obtain a new group $\Gamma := \langle \rho_0, \ldots, \rho_3 \rangle$, which is a semi-direct product of W by D_6. Then it is easy to verify that Γ is a C-group. In particular, we have

$$\langle \rho_0, \rho_1, \rho_2 \rangle = \langle \sigma_1, \sigma_2, \sigma_4, \sigma_5 \rangle \rtimes \langle \tau_0, \tau_1 \rangle \cong (D_s \times D_s) \rtimes (C_2 \times C_2)$$

and

$$\langle \rho_1, \rho_2, \rho_3 \rangle = \langle \sigma_0, \sigma_2, \sigma_4 \rangle \rtimes \langle \tau_1, \tau_2 \rangle \cong [4, 3] \quad \left(\cong C_2^3 \rtimes S_3\right).$$

Hence, for the polytope with group Γ, the facets must be toroidal maps $\{4, 4\}_{(s,s)}$ and the vertex-figures must be cubes $\{4, 3\}$. Again, the standard presentation for W in terms of the σ_i translates into a presentation for Γ in terms of the ρ_i, which now consists of the standard Coxeter relations for $[4, 4, 3]$ and the one extra relation (10B1) with $k = 2$. Note that this extra relation follows from $\sigma_1\sigma_2 = (\rho_0\rho_1\rho_2)^2$. Hence

$$\Gamma = {}_1\Gamma^4_{(s,s)},$$

and the corresponding polytope is just ${}_1\mathcal{T}^4_{(s,s)}$.

Again we need to identify the finite polytopes and their groups. But clearly, ${}_1\mathcal{T}^4_{(s,s)}$ is finite if and only if W is finite, that is, if and only if $s = 2$. If indeed $s = 2$, then the diagram has no branches and $W = C_2^6$; the semi-direct product Γ is then the wreath product $C_2 \wr D_6$. This leads to the entries in the last row of Table 10B1.

We summarize our results as follows.

10B5 Theorem *The universal regular 4-polytope* ${}_1\mathcal{T}^4_{(s,s)} := \{\{4, 4\}_{(s,s)}, \{4, 3\}\}$ *exists for all* $s \geqslant 2$. *Its group is* $W \rtimes D_6$, *where* W *is the Coxeter group with the diagram of*

Figure 10B2. In particular, this polytope is finite if and only if $s = 2$, and then its group is $C_2 \wr D_6$.

We remark that the same operation (10B4) can also be applied to the finite unitary reflexion group $W := [1\ 1\ 4^q]^3 = \langle \sigma_0, \ldots, \sigma_5 \rangle$ (with $q \geqslant 2$) in complex 6-space (see Section 9A). This group can be represented by a hexagonal diagram with unmarked branches (that is, $s = 3$), which has a mark q inside to indicate the single extra relation

$$(\sigma_0\sigma_1 \cdots \sigma_4\sigma_5\sigma_4 \cdots \sigma_2)^q = \varepsilon.$$

This leads to an infinite family of finite regular 4-polytopes $\mathcal{P}$ in $\langle \{4, 4\}_{(3,3)}, \{4, 3\} \rangle$, which are all covered by ${}_1\mathcal{T}^4_{(3,3)}$. In particular,

$$\Gamma(\mathcal{P}) \cong [1\ 1\ 4^q]^3 \rtimes D_6,$$

of order $8640q^5$.

10C The Type $\{4, 4, 4\}$

In this section, we treat the locally toroidal regular 4-polytopes of type $\{4, 4, 4\}$. With the exception of a small number of parameter values, we shall enumerate the finite universal regular polytopes

$${}_2\mathcal{T}^4_{\mathbf{s},\mathbf{t}} := \{\{4, 4\}_{\mathbf{s}}, \{4, 4\}_{\mathbf{t}}\},$$

where $\mathbf{s} = (s^k, 0^{2-k})$ and $\mathbf{t} = (t^l, 0^{2-l})$, with $2 \leqslant s, t < \infty$ and $k, l = 1$ or 2. Again, the superscript indicates the rank, and the subscript 2 on the left refers to $\{4, 4, 4\}$ as the *second* type of Schläfli symbol which we investigate. The basic construction technique is again twisting; see Chapter 8 for the notation we are using.

We proceed as in the previous section, and consider the group ${}_2\Gamma^4_{\mathbf{s},\mathbf{t}}$, which now denotes the Coxeter group $[4, 4, 4] = \langle \rho_0, \ldots, \rho_3 \rangle$, factored out by two extra relations as in (10B1). The first relation involves only the first three generators, and depends on k and s; its effect is to collapse the facet $\{4, 4\}$ of the honeycomb $\{4, 4, 4\}$ to its finite quotient $\{4, 4\}_{\mathbf{s}}$. Similarly, the second relation involves only the last three generators, and is determined by l and t; this now turns the vertex-figure $\{4, 4\}$ of $\{4, 4, 4\}$ into $\{4, 4\}_{\mathbf{t}}$. Thus the group in question now depends on two parameter vectors $\mathbf{s}$ and $\mathbf{t}$, unlike in the previous section where it was determined by only one.

In particular, by Proposition 4A8, the related polytope ${}_2\mathcal{T}^4_{\mathbf{s},\mathbf{t}}$ exists if and only if ${}_2\Gamma^4_{\mathbf{s},\mathbf{t}}$ is a C-group whose subgroups $\langle \rho_0, \rho_1, \rho_2 \rangle$ and $\langle \rho_1, \rho_2, \rho_3 \rangle$ are isomorphic to $[4, 4]_{\mathbf{s}}$ and $[4, 4]_{\mathbf{t}}$, respectively. If indeed ${}_2\mathcal{T}^4_{\mathbf{s},\mathbf{t}}$ exists, then

$$\Gamma\left({}_2\mathcal{T}^4_{\mathbf{s},\mathbf{t}}\right) = {}_2\Gamma^4_{\mathbf{s},\mathbf{t}}.$$

There are now three cases which need to be considered, namely $(k, l) = (1, 1), (1, 2)$, or $(2, 2)$. By duality, this then also covers the case $(k, l) = (2, 1)$.

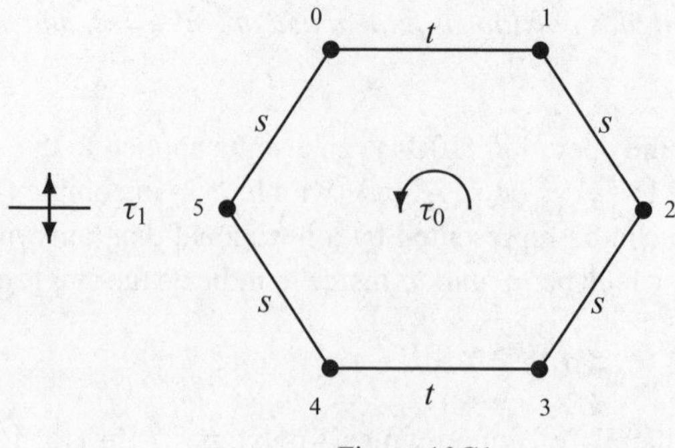

Figure 10C1

The Polytopes $_2\mathcal{T}^4_{(s,0),(t,t)} = \{\{4,4\}_{(s,0)}, \{4,4\}_{(t,t)}\}$

We begin with the construction of the polytopes $_2\mathcal{T}^4_{(s,0),(t,t)}$ with $s, t \geqslant 2$; this is the case $(k, l) = (1, 2)$. (We start here because this is the case which we can solve completely.) If s is even, we could directly appeal to Corollaries 8F4 and 8E13, and enumerate the finite polytopes. However, here we prefer to describe a method which applies to all s and t simultaneously.

Again we shall work with a Coxeter group $W = \langle \sigma_0, \ldots, \sigma_5 \rangle$ with a hexagonal diagram, but now this is taken in the form of Figure 10C1. Then the polytope is constructed by the operation

10C1 $$(\sigma_0, \ldots, \sigma_5; \tau_0, \tau_1) \mapsto (\sigma_3, \tau_0, \sigma_1, \tau_1) =: (\rho_0, \ldots, \rho_3),$$

where τ_0 and τ_1 are the group automorphisms of W corresponding to the half-turn and the reflexion in the horizontal axis of the diagram. The new group $\Gamma := \langle \rho_0, \ldots, \rho_3 \rangle$ is then a semi-direct product of W by $C_2 \times C_2$ $(= \langle \tau_0, \tau_1 \rangle)$. This is indeed a C-group, because Γ has the intersection property for the generators ρ_i, which follows from the corresponding property for the generators σ_i of W. In particular, we have

$$\langle \rho_0, \rho_1, \rho_2 \rangle = \langle \sigma_0, \sigma_1, \sigma_3, \sigma_4 \rangle \rtimes \langle \tau_0 \rangle \cong (D_s \times D_s) \rtimes C_2,$$

and

$$\langle \rho_1, \rho_2, \rho_3 \rangle = \langle \sigma_1, \sigma_2, \sigma_4, \sigma_5 \rangle \rtimes \langle \tau_0, \tau_1 \rangle \cong (D_t \times D_t) \rtimes (C_2 \times C_2).$$

Hence, for the polytope with group Γ, the facets and vertex-figures are toroidal maps $\{4,4\}_{(s,0)}$ and $\{4,4\}_{(t,t)}$, respectively.

Moreover, the standard presentation for W translates into a presentation for Γ, which consists of the standard relations for $[4, 4, 4]$, and the two extra relations

$$(\rho_0\rho_1\rho_2\rho_1)^s = \varepsilon, \quad (\rho_1\rho_2\rho_3)^{2t} = \varepsilon.$$

These extra relations follow directly from the equations

$$\rho_0\rho_1\rho_2\rho_1 = \sigma_3\tau_0\sigma_1\tau_0 = \sigma_3\sigma_4,$$
$$(\rho_1\rho_2\rho_3)^2 = \tau_0\sigma_1\tau_1\tau_0\sigma_1\tau_1 = \tau_0\sigma_1\tau_0\tau_1\sigma_1\tau_1 = \sigma_4\sigma_5.$$

Table 10C1. *The Finite Polytopes* ${}_2\mathcal{T}^4_{\mathbf{s},\mathbf{t}}$

s	t	v	f	g	Group
(2, 0)	$(t,t), t \geq 2$	4	$2t^2$	$64t^2$	$(D_t \times D_t \times C_2 \times C_2) \rtimes (C_2 \times C_2)$
(2, 0)	$(2m, 0), m \geq 1$	4	$4m^2$	$128m^2$	$(C_2 \times C_2) \rtimes [4,4]_{(2,0)}$ if $m = 1$ $(D_m \times D_m) \rtimes [4,4]_{(2,0)}$ if $m \geq 2$
(3, 0)	(3, 0)	20	20	1440	$S_6 \times C_2$
(3, 0)	(4, 0)	288	512	36864	$C_2 \wr [4,4]_{(3,0)}$
(3, 0)	(2, 2)	36	32	2304	$(S_4 \times S_4) \rtimes (C_2 \times C_2)$
(2, 2)	(2, 2)	16	16	1024	$C_2^4 \rtimes [4,4]_{(2,2)}$
(2, 2)	(3, 3)	64	144	9216	$C_2^6 \rtimes [4,4]_{(3,3)}$
(3, 0)	(5, 0)	19584	54400	3916800	see (10C8)

Since these two relations are just those which determine the group ${}_2\Gamma^4_{\mathbf{s},\mathbf{t}}$, we must have

$$\Gamma = {}_2\Gamma^4_{\mathbf{s},\mathbf{t}}.$$

In particular, ${}_2\mathcal{T}^4_{(s,0),(t,t)}$ does indeed exist, and is just the polytope with group Γ.

It is now straightforward to identify the finite polytopes and their groups. Clearly ${}_2\mathcal{T}^4_{(s,0),(t,t)}$ is finite if and only if W is finite. But the latter occurs only if $s = 2$ or $(s,t) = (3,2)$, and then $W = D_t \times D_t \times C_2 \times C_2$ or $W = S_4 \times S_4$, respectively.

Summarizing this discussion, we obtain the following.

10C2 Theorem *The universal regular 4-polytope* ${}_2\mathcal{T}^4_{(s,0),(t,t)} := \{\{4,4\}_{(s,0)}, \{4,4\}_{(t,t)}\}$ *exists for all* $s, t \geqslant 2$. *Its group is* $W \rtimes (C_2 \times C_2)$, *where* W *is the Coxeter group with the diagram of Figure 10C1. In particular, this polytope is finite if and only if* $s = 2$ *or* $(s,t) = (3,2)$, *and then its group is* $(D_t \times D_t \times C_2 \times C_2) \rtimes (C_2 \times C_2)$ *or* $(S_4 \times S_4) \rtimes (C_2 \times C_2)$, *respectively.*

In Table 10C1, we list the finite polytopes ${}_2\mathcal{T}^4_{\mathbf{s},\mathbf{t}}$, including an "exceptional" case which we shall investigate later in this section. We also include the numbers v of their vertices and f of their facets, and the orders g of their groups. The table is based on Theorems 10C2, 10C5, and 10C12.

If $t = s$, the Coxeter diagrams in Figures 10B2 and 10C1 are the same, and so are the corresponding Coxeter groups W. This gives us an obvious inclusion of ${}_2\Gamma^4_{(s,0),(s,s)}$ in ${}_1\Gamma^4_{(s,s)}$. More exactly, we have

10C3 Proposition ${}_2\Gamma^4_{(s,0),(s,s)}$ *is a subgroup of* ${}_1\Gamma^4_{(s,s)}$ *of index* 3.

At the end of the previous section, we employed a marked hexagonal diagram for the unitary reflexion groups $[1\ 1\ 4^q]^3$ to construct an infinite sequence of finite regular 4-polytopes in the class $\langle\{4,4\}_{(3,0)}, \{4,3\}\rangle$. When the diagram of Figure 10C1, with $t = s = 3$, is marked in a similar way, then the operation (10C1) again yields an infinite sequence of finite polytopes, which now are in the class $\langle\{4,4\}_{(3,0)}, \{4,4\}_{(3,3)}\rangle$, and have groups $[1\ 1\ 4^q]^3 \rtimes (C_2 \times C_2)$, of order $2880q^5$ (with $q \geqslant 2$). The subgroup relation of Proposition 10C3 carries over to this situation.

We have already mentioned that, for even s, the polytopes ${}_2\mathcal{T}^4_{(s,0),(t,t)}$ can also be obtained from the construction of Corollaries 8F4 and 8E13. If $s \geqslant 4$, the key idea here is to express ${}_2\mathcal{T}^4_{(s,0),(t,t)}$ as a certain polytope $2^{\mathcal{K},\mathcal{G}}$, which we can then prove to be infinite. On the other hand, if $s = 2$, then Corollary 8F4 shows that ${}_2\mathcal{T}^4_{(2,0),(t,t)}$ is a finite polytope, whose group can be expressed as

$$ {}_2\Gamma^4_{(2,0),(t,t)} = (C_2 \times C_2) \rtimes [4,4]_{(t,t)}. $$

The Polytopes ${}_2\mathcal{T}^4_{(s,s),(t,t)} = \big\{\{4,4\}_{(s,s)}, \{4,4\}_{(t,t)}\big\}$

For this class, the enumeration must proceed in a completely different manner, because general results about the group structure are not yet available. Nevertheless, we can describe all the finite polytopes and their groups. We begin with the case where both s and t are even. Then we have

10C4 Lemma ${}_2\mathcal{T}^4_{(s,s),(t,t)}$ *exists for all even s and t.*

Proof. Let $\mathcal{P}_1 := \{4,4\}_{(s,s)}$ and $\mathcal{P}_2 := \{4,4\}_{(t,t)}$, so that ${}_2\mathcal{T}^4_{(s,s),(t,t)} = \{\mathcal{P}_1, \mathcal{P}_2\}$. Then we know that the universal polytope $\{\mathcal{P}_1, \mathcal{P}_2\}$ exists if and only if the corresponding class $\langle \mathcal{P}_1, \mathcal{P}_2 \rangle$ is non-empty. Hence, to prove the lemma it suffices to construct a polytope in this class. This can be accomplished by the methods of Sections 4E and 4F, which yield a flat polytope.

Recall from Section 4E that a regular n-polytope $\mathcal{P}$ with group $\Gamma(\mathcal{P}) = \langle \rho_0, \ldots, \rho_{n-1} \rangle$ has the flat amalgamation property (FAP) with respect to its vertex-figures $\mathcal{K}$ if and only if

$$ \Gamma(\mathcal{P}) = N_0 \rtimes \langle \rho_1, \ldots, \rho_{n-1} \rangle \quad (= N_0 \rtimes \Gamma(\mathcal{K})), $$

where N_0 is the normal subgroup generated by the conjugates of ρ_0 in $\Gamma(\mathcal{P})$. Then Lemma 4E10 says that $\mathcal{P}$ has the FAP with respect to its vertex-figures if and only if the group $\Gamma(\mathcal{P})$, factored out by the additional relation $\rho_0 = \varepsilon$, is just the group $\Gamma(\mathcal{K})$ of the vertex-figure. Further, recall from Section 4E that we had also defined the FAP with respect to facets, such that a regular polytope $\mathcal{P}$ has the FAP with respect to its facets if and only if its dual has the FAP with respect to its vertex-figures.

All we need here is that the 3-polytopes $\mathcal{P}_1$ and $\mathcal{P}_2$ have the FAP, namely, $\mathcal{P}_1$ with respect to its vertex-figures and $\mathcal{P}_2$ with respect to its facets. To verify this, note that $\Gamma(\mathcal{P}_1)$ has a presentation consisting of the standard presentation for the Coxeter group $[4,4]$ and the extra relation $(\rho_0\rho_1\rho_2)^{2s} = \varepsilon$. Hence, if we now add $\rho_0 = \varepsilon$ as a new relation, then the latter becomes $(\rho_1\rho_2)^{2s} = \varepsilon$. Because we are assuming that s is even, this is a valid relation for the dihedral group $[4]$ $(= \Gamma(\mathcal{K}))$, which proves the FAP for $\mathcal{P}_1$. Since $\mathcal{P}_2$ is self-dual and t is also even, we can derive the FAP for $\mathcal{P}_2$ in the same way.

Finally, since $\mathcal{P}_1$ has the FAP with respect to its vertex-figures and $\mathcal{P}_2$ with respect to its facets, we can now directly appeal to Corollary 4F11 and obtain a flat polytope in $\langle \mathcal{P}_1, \mathcal{P}_2 \rangle$. This completes the proof. □

Now, for all (even or odd) s and t, we can relate ${}_2\mathcal{T}^4_{(s,s),(t,t)}$ to the polytope ${}_2\mathcal{T}^4_{(s,0),(t,t)}$, whose structure we determined earlier in this section. In fact, since the map $\{4,4\}_{(s,0)}$ is a quotient of $\{4,4\}_{(s,s)}$, we know that ${}_2\mathcal{T}^4_{(s,0),(t,t)}$ must also be a quotient of ${}_2\mathcal{T}^4_{(s,s),(t,t)}$. In particular, the group ${}_2\Gamma^4_{(s,0),(t,t)}$ of ${}_2\mathcal{T}^4_{(s,0),(t,t)}$ is the factor group derived from the group ${}_2\Gamma^4_{(s,s),(t,t)}$ of ${}_2\mathcal{T}^4_{(s,s),(t,t)}$ by imposing the extra relation

$$(\rho_0\rho_1\rho_2\rho_1)^s = \varepsilon$$

which defines $\{4,4\}_{(s,0)}$. This already proves that ${}_2\mathcal{T}^4_{(s,s),(t,t)}$ is an infinite polytope whenever ${}_2\mathcal{T}^4_{(s,0),(t,t)}$ is infinite.

In particular, if both s and t are even, then Theorem 10C2 implies that ${}_2\mathcal{T}^4_{(s,s),(t,t)}$ is an infinite polytope provided that $s \geqslant 4$ or $t \geqslant 4$. Hence, finiteness can only occur if $s = t = 2$. On the other hand, by Theorem 8F9 we already know that ${}_2\mathcal{T}^4_{(2,2),(2,2)}$ is finite, with group

$${}_2\Gamma^4_{(2,2),(2,2)} = C_2^4 \rtimes [4,4]_{(2,2)},$$

of order 1024.

Next, let at least one parameter, s (say), be odd. First, we again establish the existence of ${}_2\mathcal{T}^4_{(s,s),(t,t)}$. But this is straightforward. We appeal to Theorem 7A8, or, rather, to one of the examples which follows it. Since (7A9) tells us that

$$\{4,4\}_{(s,0)} \diamond \{\,\} = \{4,4\}_{(s,s)}$$

in case s is odd, it follows at once that

$${}_2\mathcal{T}^4_{(s,0),(t,t)} \diamond \{\,\}$$

is a polytope of type ${}_2\mathcal{T}^4_{(s,s),(t,t)}$. In particular, this establishes the existence of the universal polytope ${}_2\mathcal{T}^4_{(s,s),(t,t)}$ in this class.

But now we can argue as before. For odd s, the polytope ${}_2\mathcal{T}^4_{(s,s),(t,t)}$ must be infinite unless $(s,t) = (3,2)$, because ${}_2\mathcal{T}^4_{(s,0),(t,t)}$ is infinite. On the other hand, by Theorem 8F9 we already know that ${}_2\mathcal{T}^4_{(3,3),(2,2)}$ is finite, with group

$${}_2\Gamma^4_{(3,3),(2,2)} = C_2^6 \rtimes [4,4]_{(3,3)},$$

of order 9216, which is indeed twice the order of ${}_2\Gamma^4_{(3,0),(2,2)}$.

We summarize our results in the following theorem (see also Table 10C1).

10C5 Theorem *The universal regular* 4*-polytope* ${}_2\mathcal{T}^4_{(s,s),(t,t)} := \{\{4,4\}_{(s,s)}, \{4,4\}_{(t,t)}\}$ *exists for all* $s, t \geqslant 2$. *Up to duality, the only finite polytopes are* ${}_2\mathcal{T}^4_{(2,2),(2,2)}$ *and* ${}_2\mathcal{T}^4_{(3,3),(2,2)}$, *with groups* $C_2^4 \rtimes [4,4]_{(2,2)}$ *and* $C_2^6 \rtimes [4,4]_{(3,3)}$, *of orders* 1024 *and* 9216, *respectively.*

The Polytopes ${}_2\mathcal{T}^4_{(s,0),(t,0)} = \{\{4,4\}_{(s,0)}, \{4,4\}_{(t,0)}\}$

The enumeration of the finite regular polytopes ${}_2\mathcal{T}^4_{(s,0),(t,0)}$ appears to be more difficult, and so is not yet complete.

The case where s or t is even has already been settled by Corollaries 8F4 and 8E13. Suppose that s is even. Then Corollary 8F4 expresses ${}_2\mathcal{T}^4_{(s,0),(t,0)}$ with $s \geqslant 4$ as a certain polytope $2^{\mathcal{K},\mathcal{G}}$, and its group as a semi-direct product of a certain Coxeter group by the group $[4,4]_{(t,0)}$ of the vertex-figure. In particular, finiteness occurs only if $t = 2$ or $(s,t) = (4,3)$, and then the groups are given by

$${}_2\Gamma^4_{(s,0),(2,0)} = (D_s \times D_s) \rtimes [4,4]_{(2,0)} \qquad (s \text{ even}, s \geqslant 4)$$

and

$${}_2\Gamma^4_{(4,0),(3,0)} = C_2^9 \rtimes [4,4]_{(3,0)} = C_2 \wr [4,4]_{(3,0)},$$

of orders $128s^2$ and 36864, respectively. Recall here from (8E7) that

$${}_2\mathcal{T}^4_{(4,0),(3,0)} = 2^{\{4,4\}_{(3,0)}},$$

which explains the statement about the second group.

On the other hand, if $s = 2$, then we can apply Corollary 8F4 (or the arguments in the following). In particular, ${}_2\mathcal{T}^4_{(2,0),(t,0)}$ can only exist if t is even, and then it is finite and its group is given by

$${}_2\Gamma^4_{(2,0),(t,0)} = C_2^2 \rtimes [4,4]_{(t,0)},$$

of order $32t^2$. Moreover, if we view the latter as the group of the dual polytope and appeal to Corollary 8F4, then we also obtain

$${}_2\Gamma^4_{(2,0),(t,0)} = (D_m \times D_m) \rtimes [4,4]_{(2,0)} \qquad (t = 2m).$$

We give another argument, independent of Corollary 8F4, which shows that the polytope ${}_2\mathcal{T}^4_{(2,0),(t,0)}$ cannot exist if t odd. Suppose to the contrary that ${}_2\mathcal{T}^4_{(2,0),(t,0)}$ exists. Then ${}_2\mathcal{T}^4_{(2,0),(t,0)}$ must be a quotient of ${}_2\mathcal{T}^4_{(2,0),(t,t)}$; the latter polytope also exists, is finite, and has ${}_2\Gamma^4_{(2,0),(t,t)}$ as its group (see Theorem 10C2). Let ${}_2\Gamma^4_{(2,0),(t,t)} = \langle \rho_0, \ldots, \rho_3 \rangle$, where $\rho_0, \ldots, \rho_3$ denote the distinguished generators defined in (10C1). When we take the quotient, the vertex-figure $\{4,4\}_{(t,t)}$ collapses to $\{4,4\}_{(t,0)}$, and so the corresponding normal subgroup N of ${}_2\Gamma^4_{(2,0),(t,t)}$ must contain the element

$$\psi := (\rho_1\rho_2\rho_3\rho_2)^t.$$

Let $t = 2m+1$. Then, in the notations of Figure 10C1 and (10C1),

$$\psi = (\tau_0\sigma_1\tau_1\sigma_1)^{2m+1} = (\sigma_4\sigma_5)^m\sigma_4(\sigma_2\sigma_1)^m\sigma_2\tau_0\tau_1.$$

Since $s = 2$, we also have

$$\sigma_3\psi\sigma_3 = (\sigma_4\sigma_5)^m\sigma_4(\sigma_2\sigma_1)^m\sigma_2(\sigma_3\tau_0\tau_1\sigma_3) = \psi\sigma_0\sigma_3 \in N,$$

so that $\sigma_0\sigma_3 \in N$. But

$$\sigma_0\sigma_3 = (\tau_0\sigma_3)^2 = (\rho_1\rho_0)^2 \in \langle \rho_0, \rho_1, \rho_2 \rangle,$$

so that taking the quotient by N also leads to a collapse of the facets of ${}_2\mathcal{T}^4_{(2,0),(t,t)}$. This

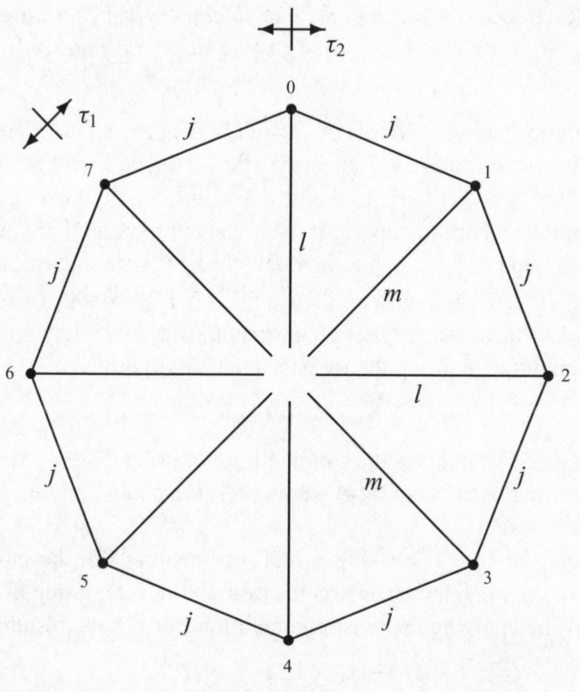

Figure 10C2

contradiction disproves our original assumption; hence the polytope ${}_2\mathcal{T}^4_{(2,0),(t,0)}$ cannot exist if t is odd. (Note that $\langle \psi, \sigma_0\sigma_3 \rangle$ itself is a normal subgroup of ${}_2\Gamma^4_{(2,0),(t,t)}$.)

If both s and t are even, then we can use a simple twisting operation to construct an infinite family of regular 4-polytopes $\mathcal{P}_j$ (for $j \geqslant 2$) in the class $\langle \{4, 4\}_{(s,0)}, \{4, 4\}_{(t,0)} \rangle$. Let $s = 2l$ and $t = 2m$ (say), and let $j \geqslant 2$. Consider the Coxeter group $W = \langle \sigma_0, \ldots, \sigma_7 \rangle$ of Figure 10C2, whose diagram consists of an octagon with edges marked j, as well as its four main diagonals which are alternately marked l or m; the centre of the octagon is not a node of the diagram. Again, τ_1 and τ_2 are the group automorphisms of W which are induced by the indicated reflexions. The polytopes are now constructed from the operation

10C6 $$(\sigma_0, \ldots, \sigma_7; \tau_1, \tau_2) \mapsto (\sigma_0, \tau_1, \tau_2, \sigma_3) =: (\rho_0, \ldots, \rho_3).$$

Indeed, it is straightforward to check that $\langle \rho_0, \ldots, \rho_3 \rangle = W \rtimes D_4$ is a C-group, and that the corresponding polytope $\mathcal{P}_j$ (say) has the desired properties. Note that $\mathcal{P}_j$ is finite only if $j = 2$ (in this case the branches with mark j are missing), and then $\Gamma(\mathcal{P}_j) = (D_l^2 \times D_m^2) \rtimes D_4$.

The remaining case where both s and t are odd seems to be somewhat more difficult. We shall see later how the case $s = t$ is related to the construction of the universal polytopes ${}_1\mathcal{T}^4_{(s,0)}$ described in Section 10B. However, if s and t are odd and distinct, we know very little about the polytopes ${}_2\mathcal{T}^4_{(s,0),(t,0)}$. In fact, all our present knowledge is compatible with the following

10C7 Conjecture *Let $s, t \geqslant 3$ be odd and distinct. Then the universal regular 4-polytope* ${}_2\mathcal{T}^4_{(s,0),(t,0)} := \{\{4,4\}_{(s,0)}, \{4,4\}_{(t,0)}\}$ *exists, but is finite only if* $(s,t) = (3,5)$ *or* $(5,3)$.

In our discussion, we shall frequently exclude the case where s and t are odd and distinct, namely, that covered by Conjecture 10C7. We shall refer to this as the *exceptional* case for the type $\{4,4,4\}$. In the next section, we shall explain the conjecture and provide supporting arguments for it using the cut method. For now we only mention that the polytope ${}_2\mathcal{T}^4_{(3,0),(5,0)}$ and its dual ${}_2\mathcal{T}^4_{(5,0),(3,0)}$ are indeed finite, with group ${}_2\Gamma^4_{(3,0),(5,0)}$ $(= {}_2\Gamma^4_{(5,0),(3,0)})$, of order $2^{10} \cdot 3^2 \cdot 5^2 \cdot 17 = 39\,16800$. The group order was determined by G. Havas using coset enumeration. Moreover, Leytem [263, 267] employed the computational group theory package GAP to establish

10C8 $${}_2\Gamma^4_{(3,0),(5,0)} = \mathrm{Sp}_4(4) \times C_2 \times C_2,$$

where $\mathrm{Sp}_4(4)$ $(= \mathrm{S}_4(4))$ denotes the simple group of order $2^8 \cdot 3^2 \cdot 5^2 \cdot 17$, which consists of all 4×4 matrices over GF(4) which preserve a non-singular symplectic form [90, p. 44; 360].

Next we treat the case $s = t$ with s odd. In Section 10B, the universal polytope ${}_1\mathcal{T}^4_{(s,0)} = \{\{4,4\}_{(s,0)}, \{4,3\}\}$ was constructed from the Coxeter group $W = \langle \sigma_0, \ldots, \sigma_4 \rangle$ of Figure 10B1, by applying the twisting operation (10B2); the resulting group was

$$\Gamma = \langle \rho_0, \ldots, \rho_3 \rangle = {}_1\Gamma^4_{(s,0)}.$$

Now consider the new operation

10C9 $$(\rho_0, \ldots, \rho_3) \mapsto (\rho_3, \rho_2\rho_1\rho_2, \rho_0, \rho_1) =: (\alpha_0, \ldots, \alpha_3)$$

on Γ, which employs the halving operation for the vertex-figure of the dual of ${}_1\mathcal{T}^4_{(s,0)}$ (see Section 7B). Since s is odd, we can use (10B2) to recover the ρ_i's from the α_j's; in particular,

$$\rho_2 = \sigma_3 = \sigma_3(\sigma_1\sigma_0)^s = (\sigma_3\sigma_1\sigma_0)^s = (\sigma_3\tau\sigma_3\tau\sigma_0)^s = (\alpha_1\alpha_3\alpha_2)^s = (\alpha_3\alpha_1\alpha_2)^s.$$

Hence $\alpha_0, \ldots, \alpha_3$ are indeed generators for Γ, and $\langle \alpha_1, \alpha_2, \alpha_3 \rangle$ is the group of the torus map $\{4,4\}_{(s,0)}$.

If we compute a presentation for Γ in terms of its new generators $\alpha_0, \ldots, \alpha_3$ from a presentation in terms of its old generators $\rho_0, \ldots, \rho_3$, we find that Γ is defined by the standard Coxeter relations for the group $[4,4,4]$ and the following three extra relations:

10C10 $$(\alpha_0\alpha_1\alpha_2\alpha_1)^s = (\alpha_1\alpha_2\alpha_3\alpha_2)^s = \varepsilon,$$

10C11 $$(\alpha_0(\alpha_3\alpha_1\alpha_2)^s)^3 = \varepsilon.$$

Here we have used the equations

$$\alpha_1\alpha_0 = \rho_2\rho_1\rho_2\rho_3 = \rho_2(\rho_1\rho_2\rho_3\rho_2)\rho_2$$

and

$$\begin{aligned}\alpha_2\alpha_1\alpha_0\alpha_1 &= \rho_0\rho_2\rho_1(\rho_2\rho_3\rho_2)\rho_1\rho_2 \\ &= (\rho_0\rho_2\rho_1\rho_3)(\rho_2\rho_3\rho_1\rho_2) = \rho_2\rho_3(\rho_0\rho_1\rho_2\rho_1)\rho_3\rho_2,\end{aligned}$$

as well as

$$\alpha_0(\alpha_3\alpha_1\alpha_2)^s = \rho_3\rho_2.$$

The proof of the intersection property for Γ and its generators $\alpha_0, \ldots, \alpha_3$ is straightforward but very tedious, and so we omit it here; it appeals to the same property for the Coxeter group W of Figure 10B1. Then this shows that Γ is a C-group with distinguished generators $\alpha_0, \ldots, \alpha_3$.

Now we can prove that the universal polytope ${}_2\mathcal{T}^4_{(s,0),(s,0)}$ exists. In fact, the regular 4-polytope $\mathcal{Q}$ with group Γ belongs to the class $\langle\{4,4\}_{(s,0)}, \{4,4\}_{(s,0)}\rangle$, which therefore is non-empty and so contains a universal member, which also covers $\mathcal{Q}$. This is just the desired polytope ${}_2\mathcal{T}^4_{(s,0),(s,0)}$. In particular, if $s > 3$, this must be an infinite polytope, because $\mathcal{Q}$ is also infinite. If $s = 3$, then we can prove that ${}_2\mathcal{T}^4_{(3,0),(3,0)} = \mathcal{Q}$ and

$${}_2\Gamma^4_{(3,0),(3,0)} = S_6 \times C_2 = \Gamma.$$

First observe that in this case the original polytope ${}_1\mathcal{T}^4_{(3,0)}$ is finite and has $\Gamma = S_6 \times C_2$ as its group. But then the new polytope $\mathcal{Q}$ must also be finite and have the same group (but with new distinguished generators). On the other hand, an application of the Todd–Coxeter coset enumeration algorithm shows that ${}_2\Gamma^4_{(3,0),(3,0)}$ is a group of order 1440 (see [83, Table 1]). Therefore, since Γ is a quotient of ${}_2\Gamma^4_{(3,0),(3,0)}$, we must indeed have ${}_2\Gamma^4_{(3,0),(3,0)} = \Gamma$ and ${}_2\mathcal{T}^4_{(3,0),(3,0)} = \mathcal{Q}$. In particular, the relation (10C11) can be omitted from the presentation if $s = 3$.

If $s > 3$, then $\mathcal{Q}$ seems to be a proper quotient of the universal polytope ${}_2\mathcal{T}^4_{(s,0),(s,0)}$. This is indicated by the existence of certain 3-cycles in the edge graph of $\mathcal{Q}$, which are not likely to exist in the universal polytope. To describe these 3-cycles, set $\beta := (\alpha_3\alpha_1\alpha_2)^s$, and let F_0 and F_1 be the base vertex and base edge of $\mathcal{Q}$, respectively. Then (10C11) shows that $\beta\alpha_0\beta = \alpha_0\beta\alpha_0$ stabilizes the vertex $F_0\alpha_0$ of F_1 and interchanges the two vertices F_0 and $F_0\alpha_0\beta$ of $F_1\beta$; moreover, α_0 stabilizes $F_0\alpha_0\beta$ and interchanges the two vertices F_0 and $F_0\alpha_0$ of F_1. Hence F_0, $F_0\alpha_0$ and $F_0\alpha_0\beta$ are just the vertices in a 3-cycle, whose edges are given by F_1, $F_1\beta$ and $F_1\beta\alpha_0$.

To conclude this section, we summarize our results in the following theorem (see also Table 10C1).

10C12 Theorem *Let $s, t \geqslant 2$, and let $s = t$ if both s and t are odd.*

(a) The universal regular 4-polytope ${}_2\mathcal{T}^4_{(s,0),(t,0)} := \{\{4,4\}_{(s,0)}, \{4,4\}_{(t,0)}\}$ exists, except when $s = 2$ and t is odd, or $t = 2$ and s is odd.

(b) Up to duality, the only finite polytopes are ${}_2\mathcal{T}^4_{(2,0),(t,0)}$ with t even, ${}_2\mathcal{T}^4_{(3,0),(3,0)}$, and ${}_2\mathcal{T}^4_{(3,0),(4,0)}$. The corresponding finite groups are $(D_m \times D_m) \rtimes [4,4]_{(2,0)}$, with $t = 2m$, $S_6 \times C_2$, and $C_2^9 \rtimes [4,4]_{(3,0)}$, of orders $32t^2$, 1440, and 36864, respectively.

Recall, though, that we also know that ${}_2\mathcal{T}^4_{(3,0),(5,0)}$ and its dual are finite.

10D Cuts for the Types $\{4, 4, r\}$

In Sections 10B and 10C, we enumerated the finite locally toroidal regular 4-polytopes of type

$$_1\mathcal{T}^4_{\mathbf{s}} = \{\{4, 4\}_{\mathbf{s}}, \{4, 3\}\},$$

and (probably completely) those of type

$$_2\mathcal{T}^4_{\mathbf{s},\mathbf{t}} = \{\{4, 4\}_{\mathbf{s}}, \{4, 4\}_{\mathbf{t}}\}.$$

In this section, we describe a powerful geometric tool, the *cut method*, which sheds some light on why certain parameter vectors **s**, **t** give finite polytopes, while others give infinite ones. In general, this method reduces enumeration problems for polytopes of a given rank to similar such problems for polytopes of lower rank. At present, this method is only developed to the extent that it explains many of our results rather than proving them. We do not know under what conditions the cut method can actually be turned into a proof technique. In particular, the cut method will not in general lead to existence results for polytopes.

Recall from Section 7C the concept of a cut of a regular polytope $\mathcal{P}$. Informally, a cut $\mathcal{M}$ of $\mathcal{P}$ is a regular polytope of lower rank, whose group $\Gamma(\mathcal{M})$ is a subgroup of $\Gamma(\mathcal{P})$ which is contained in the centralizer of a subset of the distinguished generators. It is often the case that the vertices of $\mathcal{M}$ lie among those of $\mathcal{P}$ or its dual $\mathcal{P}^*$.

We shall investigate cuts of regular 4-polytopes of type {4, 4, 3} and {4, 4, 4}, beginning with the latter. For $_1\mathcal{T}^4_{\mathbf{s}}$ or $_2\mathcal{T}^4_{\mathbf{s},\mathbf{t}}$, these cuts are regular maps which in many cases are isomorphic to spherical, euclidean, or hyperbolic tessellations. In a sense which we shall further explain, these maps cut right through the polytope.

The Type {4, 4, 4}

Let $\mathcal{P}$ be a regular 4-polytope of type $\{p, q, r\}$, and let $\Gamma(\mathcal{P}) = \langle \rho_0, \ldots, \rho_3 \rangle$, where $\rho_0, \ldots, \rho_3$ are the distinguished generators. Assume that the facet of $\mathcal{P}$ has 2-holes of length l, and that its vertex-figure has Petrie polygons of length m (see Sections 1D and 7B). Then the operation

10D1 $$(\rho_0, \ldots, \rho_3) \mapsto (\rho_0, \rho_1\rho_2\rho_1, \rho_3\rho_2\rho_3) =: (\beta_0, \beta_1, \beta_2)$$

gives a regular map $\mathcal{M} = \mathcal{M}(\mathcal{P})$ with group $\Gamma(\mathcal{M}) = \langle \beta_0, \beta_1, \beta_2 \rangle$, which is of type $\{l, m\}$ or $\{l, \frac{m}{2}\}$ according as m is odd or even. In fact, we have

$$\beta_0\beta_1 = \rho_0\rho_1\rho_2\rho_1,$$

of order l, and

$$\beta_1\beta_2 = \rho_1\rho_2\rho_1\rho_3\rho_2\rho_3 = (\rho_1\rho_2\rho_3)^2.$$

of order m or $\frac{m}{2}$, respectively.

Let $\{F_0, \ldots, F_3\}$ be the base flag of $\mathcal{P}$, where F_i denotes an i-face. Then $\mathcal{M}$ can be constructed from $\mathcal{P}$ by (an abstract version of) Wythoff's construction, now applied to the order complex $\mathcal{C}(\mathcal{P})$ of $\mathcal{P}$, with F_0 as initial vertex (see Sections 2C and 5A). In particular, the base flag $\{G_0, G_1, G_2\}$ of $\mathcal{M}$ is then given by $G_0 := F_0$, $G_1 := F_1$, and

$$G_2 := \{F_0(\rho_0\rho_1\rho_2\rho_1)^j \mid j = 0, \ldots, l-1\},$$

which is a 2-hole of the facet F_3 of $\mathcal{P}$. As usual, the other faces of the map are then the transforms of the G_i's under the elements of $\Gamma(\mathcal{M})$. In particular, the vertices

$$F_0\beta_0(\beta_2\beta_1)^j = F_0\rho_0(\rho_3\rho_2\rho_1)^{2j},$$

with $j = 1, \ldots, m$ or $j = 1, \ldots, \frac{m}{2}$ according as m is odd or even, are the neighbours of F_0 in $\mathcal{M}$; that is, as we go around F_0 to find its neighbours in $\mathcal{M}$, we pick every other vertex from a Petrie polygon of the vertex-figure of $\mathcal{P}$ at F_0, and so eventually all the vertices of the Petrie polygon if m is odd. Finally, if we span the 2-faces of $\mathcal{M}$ by topological discs, then we arrive at a surface which (in a sense) cuts right through the original polytope $\mathcal{P}$.

If $\mathcal{P}$ is of type $\{p, 4, 4\}$, then ρ_2 commutes with β_0, β_1, and β_2, and so $\mathcal{M}$ is invariant under ρ_2. This is consistent with our definition of a cut in Section 7C, which required the new group to be contained in the centralizer of a subset of the distinguished generators (which here is just $\{\rho_2\}$ itself). Hence, in some sense, we can think of $\mathcal{M}$ as lying on the reflexion wall of ρ_2.

Now we can investigate the polytopes ${}_2\mathcal{T}^4_{\mathbf{s},\mathbf{t}}$, and determine their cuts. Here we use the fact that the torus maps $\{4, 4\}_{(s,0)}$ and $\{4, 4\}_{(s,s)}$ have Petrie polygons of length $2s$ and that their 2-holes have length s or $2s$, respectively.

First we consider $\mathcal{P} := {}_2\mathcal{T}^4_{(s,0),(t,t)}$. Then $l = s$ and $m = 2t$, so that the resulting map $\mathcal{M}(\mathcal{P})$ is of type $\{s, t\}$. Now we can employ the construction of these polytopes in Section 10C to show that $\mathcal{M}(\mathcal{P})$ is isomorphic to the tessellation $\{s, t\}$. In fact, in the notation of Figure 10C1 and (10C1), we have

$$(\beta_0, \beta_1, \beta_2) = (\sigma_3, \tau_0\sigma_1\tau_0, \tau_1\sigma_1\tau_1) = (\sigma_3, \sigma_4, \sigma_5).$$

But these are just the generators of a Coxeter subgroup $[s, t]$ of the original Coxeter group; hence $\mathcal{M}(\mathcal{P}) = \{s, t\}$.

In a similar way, we can find the map $\mathcal{M}(\mathcal{P}^*)$ which is related to the dual $\mathcal{P}^* = {}_2\mathcal{T}^4_{(t,t),(s,0)}$ of $\mathcal{P}$. Here the generators of its group are given by

$$(\beta_0, \beta_1, \beta_2) = (\tau_1, \sigma_1\tau_0\sigma_1, \sigma_3\tau_0\sigma_3) = (\tau_1, \tau_0\sigma_4\sigma_1, \tau_0\sigma_0\sigma_3),$$

and the map itself is isomorphic to the tessellation $\{2t, s\}$. The latter can be proved as follows. Consider the subgroup

$$\Lambda := \langle \tau_0\sigma_1\sigma_4, \tau_0\sigma_2\sigma_5, \tau_0\sigma_0\sigma_3 \rangle$$

of $\Gamma(\mathcal{M}(\mathcal{P}^*)) = \langle \beta_0, \beta_1, \beta_2 \rangle$, which is described by the diagram in Figure 10D1. This is a subgroup of index 2, on which τ_1 acts as a group automorphism as indicated. Indeed, Λ is just the Coxeter group with this diagram, because the same is true for the related subgroup $\langle \sigma_1\sigma_4, \sigma_2\sigma_5, \sigma_0\sigma_3 \rangle$. Now, if we adjoin τ_1 to Λ, then on the one hand we recover $\Gamma(\mathcal{M}(\mathcal{P}^*))$, and on the other we obtain a Coxeter group $[2t, s]$. Hence $\Gamma(\mathcal{M}(\mathcal{P}^*)) = [2t, s]$, as required.

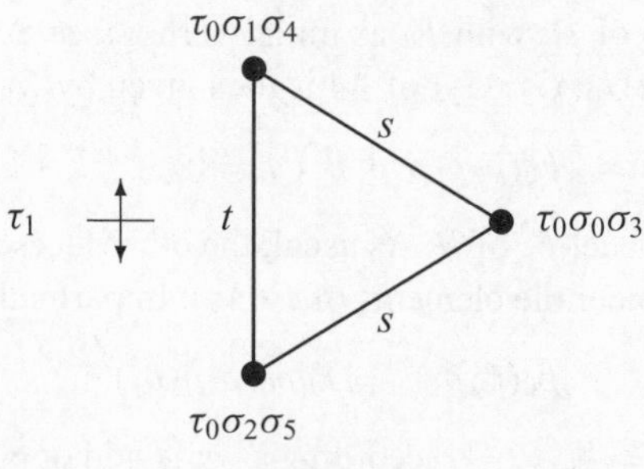

Figure 10D1

In this context, it is perhaps worth remarking that (when it is finite) the order of ${}_2\Gamma^4_{(t,t),(s,0)}$ is given by

$$|{}_2\Gamma^4_{(t,t),(s,0)}| = 16\left(\frac{1}{2t}+\frac{1}{s}-\frac{1}{2}\right)^{-2} = |[2t,s]|^2.$$

This is easily verified from the construction of the group (see Theorem 10C2 and Figure 10C1), but we have no general explanation for the formula.

The polytopes $\mathcal{P} := {}_2\mathcal{T}^4_{(s,s),(t,t)}$ can be dealt with by a simple argument involving coverings. Indeed, since $\mathcal{P}$ covers the polytope $\mathcal{K} := {}_2\mathcal{T}^4_{(s,s),(t,0)}$, we also have a covering map of $\mathcal{M}(\mathcal{P})$ onto $\mathcal{M}(\mathcal{K})$. More precisely, there is a homomorphism of $\Gamma(\mathcal{P})$ onto $\Gamma(\mathcal{K})$, which maps generators to generators and also carries the subgroup $\Gamma(\mathcal{M}(\mathcal{P}))$ of $\Gamma(\mathcal{P})$ onto the subgroup $\Gamma(\mathcal{M}(\mathcal{K}))$ of $\Gamma(\mathcal{K})$.

But $\mathcal{M}(\mathcal{P})$ is of type $\{2s,t\}$ and covers $\mathcal{M}(\mathcal{K})$, which is already known to be a tessellation $\{2s,t\}$. Hence this covering must be trivial, and $\mathcal{M}(\mathcal{P}) = \{2s,t\}$.

For the polytopes $\mathcal{P} := {}_2\mathcal{T}^4_{(s,0),(t,0)}$, with $s = 2k$ even and t arbitrary, we can again work with coverings. Since the torus map $\{4,4\}_{(k,k)}$ has 2-holes of length $2k$, there is a covering map of $\mathcal{P}$ onto $\mathcal{K} := {}_2\mathcal{T}^4_{(k,k),(t,0)}$, as well as a corresponding homomorphism of $\Gamma(\mathcal{P})$ onto $\Gamma(\mathcal{K})$. The latter also determines a homomorphism of $\Gamma(\mathcal{M}(\mathcal{P}))$ onto $\Gamma(\mathcal{M}(\mathcal{K}))$, and in turn a corresponding covering map of $\mathcal{M}(\mathcal{P})$ onto $\mathcal{M}(\mathcal{K})$. But now we can argue as before. Indeed, since $\mathcal{M}(\mathcal{P})$ is of type $\{2k,t\} = \{s,t\}$ and covers $\mathcal{M}(\mathcal{K})$, which now is a tessellation $\{s,t\}$, the cut $\mathcal{M}(\mathcal{P})$ must also be a tessellation $\{s,t\}$.

Summarizing, for those polytopes $\mathcal{P} = {}_2\mathcal{T}^4_{\mathbf{s},\mathbf{t}}$ which we have considered, the cuts are given by

10D2 $$\mathcal{M}(\mathcal{P}) = \begin{cases} \{2s,t\} & \text{if } \mathcal{P} = {}_2\mathcal{T}^4_{(s,s),(t,0)} \text{ or } {}_2\mathcal{T}^4_{(s,s),(t,t)}, \\ \{s,t\} & \text{if } \mathcal{P} = {}_2\mathcal{T}^4_{(s,0),(t,0)} \ (s \text{ even}), \text{ or } {}_2\mathcal{T}^4_{(s,0),(t,t)}. \end{cases}$$

In particular, note that in general the cut of the dual polytope $\mathcal{P}^*$ is not the dual of the cut of $\mathcal{P}$.

Our next theorem describes a simple criterion for the finiteness of such a polytope $\mathcal{P}$, which is, in effect, just a finiteness criterion for its cut and that of its dual.

10D3 Theorem *Let $s, t \geqslant 2$. The following is a necessary and sufficient condition for the universal polytope $\mathcal{P}$ to be finite:*

(a) $\frac{1}{2s} + \frac{1}{t} > \frac{1}{2}$, *if* $\mathcal{P} = {}_2\mathcal{T}^4_{(s,s),(t,0)}$;

(b) $\frac{1}{2s} + \frac{1}{t} > \frac{1}{2}$ *and* $\frac{1}{s} + \frac{1}{2t} > \frac{1}{2}$, *if* $\mathcal{P} = {}_2\mathcal{T}^4_{(s,s),(t,t)}$;

(c) $\frac{1}{s} + \frac{1}{t} > \frac{1}{2}$, *if* $\mathcal{P} = {}_2\mathcal{T}^4_{(s,0),(t,0)}$ *with s even.*

Proof. Indeed, if $\mathcal{P}$ is a finite polytope, then both $\mathcal{M}(\mathcal{P})$ and $\mathcal{M}(\mathcal{P}^*)$ must be (finite) spherical tessellations; this proves the necessity of the theorem. On the other hand, for the sufficiency we can appeal to the classification of the polytopes $\mathcal{P}$ described in the previous section, which yields finiteness under the stated conditions. Note that the condition of part (a) is not symmetric in s and t; in fact, here the condition on $\mathcal{M}(\mathcal{P})$ is sufficient for a finiteness condition on $\mathcal{P}$. On the other hand, part (b) requires conditions for both $\mathcal{M}(\mathcal{P})$ and $\mathcal{M}(\mathcal{P}^*)$. □

The following example shows that the condition of Theorem 10D3(a) cannot be replaced by the weaker condition $\frac{1}{s} + \frac{1}{t} > \frac{1}{2}$, which expresses the finiteness of the cut $\{t, s\}$ of the dual ${}_2\mathcal{T}^4_{(t,0),(s,s)}$ of $\mathcal{P}$. In fact, for the infinite polytope $\mathcal{P} := {}_2\mathcal{T}^4_{(3,3),(3,0)}$, we have $\mathcal{M}(\mathcal{P}) = \{6, 3\}$ and $\mathcal{M}(\mathcal{P}^*) = \{3, 3\}$; hence, the former correctly predicts that $\mathcal{P}$ is infinite, but the latter points to finiteness. This example also shows that infinite polytopes can have finite cuts.

Next we treat the polytopes $\mathcal{P} := {}_2\mathcal{T}^4_{(s,0),(s,0)}$ with s odd. As in Section 10C, let $\mathcal{Q}$ denote the regular 4-polytope in the class $\langle\{4, 4\}_{(s,0)}, \{4, 4\}_{(s,0)}\rangle$, with group $\Gamma(\mathcal{Q}) = \Gamma = \langle\alpha_0, \ldots, \alpha_3\rangle$, which was constructed from the group ${}_1\Gamma^4_{(s,0)}$ by applying the operation in (10C9). In the notation of Figure 10B1, (10B2) and (10C9), the generators of $\Gamma(\mathcal{M}(\mathcal{Q}))$ are then determined by

$$(\beta_0, \beta_1, \beta_2) = (\alpha_0, \alpha_1\alpha_2\alpha_1, \alpha_3\alpha_2\alpha_3) = (\sigma_2, \sigma_3\sigma_4\sigma_3, \sigma_4),$$

and are related to $\sigma_2, \sigma_3, \sigma_4$ by the facetting operation φ_2 described in Section 7B. Since s is odd, the elements $\sigma_3\sigma_4\sigma_3$ and σ_4 generate $\langle\sigma_3, \sigma_4\rangle$, so that we have

$$\langle\beta_0, \beta_1, \beta_2\rangle = \langle\sigma_2, \sigma_3, \sigma_4\rangle.$$

But the latter group is just a Coxeter group $[3, s]$, and thus belongs to a regular tessellation $\{3, s\}$. Therefore

$$\mathcal{M}(\mathcal{Q}) = \{3, s\}^{\varphi_2},$$

the image of $\{3, s\}$ under φ_2. This map is of type $\{s, s\}$, because the 2-holes of $\{3, s\}$ are s-gons. The only finite maps occur when $s = 3$ or 5. Then $\mathcal{M}(\mathcal{Q})$ is isomorphic to the simplex $\{3, 3\}$ or the Kepler–Poinsot polyhedron $\{5, \frac{5}{2}\}$, respectively; the latter is a finite quotient of the infinite hyperbolic tessellation $\{5, 5\}$ (see Section 7D; abstractly it is $\{5, 5 \mid 3\}$). If $s > 5$, we obtain an infinite map which is a proper quotient of $\{s, s\}$.

For $\mathcal{P}$ itself, we can again use a covering argument. Now the covering of $\mathcal{Q}$ by $\mathcal{P}$ determines a covering of $\mathcal{M}(\mathcal{Q})$ by $\mathcal{M}(\mathcal{P})$. In particular, since the two maps are of the same type, we can conclude that $\mathcal{M}(\mathcal{P})$ is finite if $s = 3$, and infinite if $s > 5$. Indeed, we conjecture here that $\mathcal{M}(\mathcal{P}) = \{s, s\}$ for each odd s.

To conclude, we can provide further evidence for Conjecture 10C7, which is concerned with the exceptional polytopes $\mathcal{P} := {}_2\mathcal{T}^4_{(s,0),(t,0)}$, where s and t are odd and distinct. For now let us assume that these polytopes exist. Then their cuts $\mathcal{M}(\mathcal{P})$ are maps of type $\{s, t\}$, and our previous results suggest that it is indeed not unreasonable to expect that they are regular tessellations $\{s, t\}$. But this would imply that $\mathcal{P}$ can only be finite if $\{s, t\}$ is a spherical tessellation; that is, if $(s, t) = (3, 5)$ or $(5, 3)$. Hence, under some plausible assumptions on the nature of the cuts, we arrive exactly at the finiteness condition for these polytopes stated in Conjecture 10C7, which again expresses just the finiteness of the corresponding cuts.

As we mentioned already in the previous section, the polytope ${}_2\mathcal{T}^4_{(3,0),(5,0)}$ and its dual ${}_2\mathcal{T}^4_{(5,0),(3,0)}$ are indeed finite (see [263, 267]).

The Type {4, 4, 3}

We can proceed similarly with the regular polytopes ${}_1\mathcal{T}^4_{\mathbf{s}} = \{\{4, 4\}_{\mathbf{s}}, \{4, 3\}\}$ and their duals.

For the dual $\mathcal{K} := ({}_1\mathcal{T}^4_{(s,0)})^*$ of ${}_1\mathcal{T}^4_{(s,0)}$, we can appeal to Figure 10B1 and (10B2) to determine the group $\Gamma(\mathcal{M}(\mathcal{K}))$, whose generators are given by

$$(\beta_0, \beta_1, \beta_2) = (\sigma_2, \sigma_3\tau\sigma_3, \sigma_0\tau\sigma_0) = (\sigma_2, \sigma_1\sigma_3\tau, \tau\sigma_0\sigma_4).$$

Indeed, this group is a Coxeter group $[4, s]$, because the related group $\langle\sigma_2, \sigma_1\sigma_3, \sigma_0\sigma_4\rangle$ is also a Coxeter group $[4, s]$. Hence, $\mathcal{M}(\mathcal{K})$ must be a regular tessellation $\{4, s\}$. (We remark that the cut of ${}_1\mathcal{T}^4_{(s,0)}$ itself is a map of type $\{s, 3\}$.)

For the dual $\mathcal{L} := ({}_1\mathcal{T}^4_{(s,s)})^*$ of ${}_1\mathcal{T}^4_{(s,s)}$ we can again work with coverings, now of $\mathcal{K}$ by $\mathcal{L}$, with $\mathcal{K}$ as before. Since $\Gamma(\mathcal{K})$ is a quotient of $\Gamma(\mathcal{L})$, the group $\Gamma(\mathcal{M}(\mathcal{K}))$ is also a quotient of $\Gamma(\mathcal{M}(\mathcal{L}))$; hence $\mathcal{M}(\mathcal{K})$ itself is a quotient of $\mathcal{M}(\mathcal{L})$. But then $\mathcal{M}(\mathcal{L}) = \{4, s\}$, because $\mathcal{M}(\mathcal{K}) = \{4, s\}$ and $\mathcal{M}(\mathcal{L})$ is also a map of type $\{4, s\}$.

For ${}_1\mathcal{T}^4_{(s,s)}$ itself, the cut is a map of type $\{2s, 3\}$. Now we can use Figure 10B2 and (10B4) to obtain the generators

$$(\beta_0, \beta_1, \beta_2) = (\tau_0, \sigma_2\tau_1\sigma_2, \tau_2\tau_1\tau_2) = (\tau_0, \sigma_2\sigma_4\tau_1, \tau_2\tau_1\tau_2).$$

Within the group $\langle\beta_0, \beta_1, \beta_2\rangle$, the generator $\beta_0 = \tau_0$ acts by conjugation on the subgroup

$$\Lambda := \langle\sigma_2\sigma_4\tau_1, \sigma_5\sigma_1\tau_1, \tau_2\tau_1\tau_2\rangle,$$

of index 2, as illustrated by the diagram in Figure 10D2. But Λ really is a Coxeter group with this diagram; this can be checked by considering its rotation subgroup, which is generated by the pairwise products of the generators. Since the full group is obtained from Λ by a simple twisting operation involving τ_0, we must have $\mathcal{M}({}_1\mathcal{T}^4_{(s,s)}) = \{2s, 3\}$.

The following theorem gives a simple criterion for the finiteness of a polytope ${}_1\mathcal{T}^4_{\mathbf{s}}$, which is, in effect, just a finiteness criterion for either its own cut or that of its dual; here we need not include both.

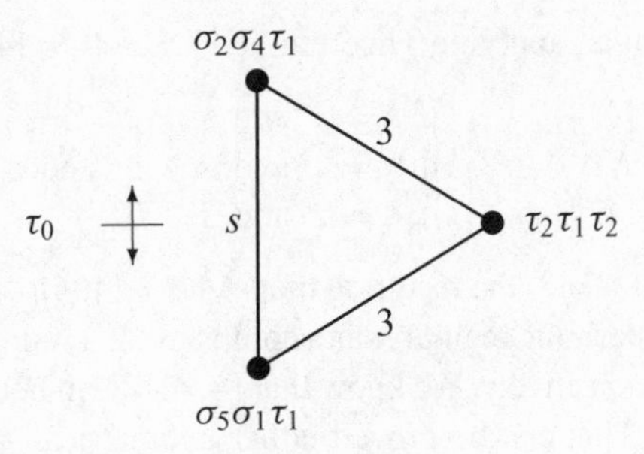

Figure 10D2

10D4 Theorem *Let $s \geqslant 2$. Then the following is a necessary and sufficient condition for the universal polytope $\mathcal{P} = {}_1\mathcal{T}^4_{\mathbf{s}}$ to be finite:*

(a) $\frac{1}{4} + \frac{1}{s} > \frac{1}{2}$, if $\mathbf{s} = (s, 0)$ (that is, $s \leqslant 3$);
(b) $\frac{1}{2s} + \frac{1}{3} > \frac{1}{2}$, if $\mathbf{s} = (s, s)$ (that is, $s = 2$).

Proof. Again, if $\mathcal{P} := {}_1\mathcal{T}^4_{\mathbf{s}}$ is finite, then $\mathcal{M}(\mathcal{P})$ and $\mathcal{M}(\mathcal{P}^*)$ are finite maps. If $\mathbf{s} = (s, 0)$, then $\mathcal{M}(\mathcal{P}^*) = \{4, s\}$. If $\mathbf{s} = (s, s)$, then $\mathcal{M}(\mathcal{P}) = \{2s, 3\}$. Hence the conditions in (a) and (b) are necessary. On the other hand, for the sufficiency we can appeal to the enumeration of the finite polytopes ${}_1\mathcal{T}^4_{\mathbf{s}}$ described in Section 10B, which yields finiteness under the stated conditions. □

We remark that the operation (10D1) which we discussed here is by no means the only way in which cuts of regular 4-polytopes $\mathcal{P}$ of type $\{p, 4, 4\}$ can be defined. For example, whereas (10D1) distinguishes the generator ρ_2 and its "reflexion hyperplane", the following operation relates in a similar way to the generator ρ_3 of $\Gamma(\mathcal{P}) = \langle \rho_0, \ldots, \rho_3 \rangle$:

10D5 $$(\rho_0, \ldots, \rho_3) \mapsto (\rho_0, \rho_1, \rho_2\rho_3\rho_2).$$

If $\mathcal{P}$ is the dual of ${}_1\mathcal{T}^4_{(s,0)}$ or ${}_1\mathcal{T}^4_{(s,s)}$, then this defines a cut which is a regular tessellation $\{3, s\}$ or $\{3, 2s\}$, respectively.

10E Relationships Among Polytopes of Type $\{4, 4, r\}$

In this section, we investigate relationships among the regular polytopes of types $\{3, 4, 4\}$ and $\{4, 4, 4\}$. We shall see that it is preferable to consider $\{3, 4, 4\}$ rather than its dual type $\{4, 4, 3\}$.

A key role is played by the halving operation η described in Section 7B, which only applies to regular polyhedra of type $\{4, q\}$. For a polyhedron $\mathcal{M}$ with group $\Gamma(\mathcal{M}) = \langle \sigma_0, \sigma_1, \sigma_2 \rangle$, this operation is defined by

$$\eta\colon (\sigma_0, \sigma_1, \sigma_2) \mapsto (\sigma_0\sigma_1\sigma_0, \sigma_2, \sigma_1) =: (\rho_0, \rho_1, \rho_2),$$

and usually turns $\mathcal{M}$ into a regular polyhedron $\mathcal{M}^\eta$ with group $\langle \rho_0, \rho_1, \rho_2 \rangle$, which is of type $\{q, q\}$. In particular, applying Theorem 7B17 to $\mathcal{M} := \{4, 4\}_{(s^k, 0^{2-k})}$, we obtain

10E1 $$\mathcal{M}^\eta = \begin{cases} \{4, 4\}_{(m,m)} & \text{if } k = 1 \text{ and } s = 2m \geqslant 2 \text{ is even;} \\ \mathcal{M} & \text{if } k = 1 \text{ and } s \geqslant 3 \text{ is odd;} \\ \{4, 4\}_{(s,0)} & \text{if } k = 2 \text{ and } s \geqslant 2. \end{cases}$$

Note that, for $k = 1$ and $s = 2$, the resulting map $\mathcal{M}^\eta = \{4, 4\}_{(1,1)}$ is not a polyhedron.

Now consider the hyperbolic regular honeycombs $\{3, 4, 4\}$ and $\{4, 4, 4\}$, with groups $[3, 4, 4]$ and $[4, 4, 4]$, respectively. We know that $[4, 4, 4]$ can be viewed as a subgroup of index 3 in $[3, 4, 4]$. This can be proved either geometrically, by dissection of the fundamental simplex of $\{4, 4, 4\}$ [120, §14.8], or algebraically, by considering the mixing operation on $[3, 4, 4] = \langle \rho_0, \ldots, \rho_3 \rangle$ defined by

10E2 $$\kappa\colon (\rho_0, \ldots, \rho_3) \mapsto (\rho_0, \rho_1\rho_2\rho_1, \rho_3, \rho_2) =: (\varphi_0, \ldots, \varphi_3),$$

to obtain its subgroup $[4, 4, 4] = \langle \varphi_0, \ldots, \varphi_3 \rangle$. The corresponding geometric relationship between the honeycombs is then as follows. The facets of $\{3, 4, 4\}$ are octahedra with their vertices on the absolute, and the vertex-figures are horospherical tessellations $\{4, 4\}$. But

10E3 $$\varphi_i = \rho_1\rho_0\rho_{i+1}\rho_0\rho_1 \quad (i = 0, 1, 2),$$

and so the facets of the derived honeycomb $\{4, 4, 4\}$ are vertex-figures of $\{3, 4, 4\}$. On the other hand, the vertex-figure of $\{4, 4, 4\}$ is obtained by halving the vertex-figure of $\{3, 4, 4\}$ (that is, by applying η).

Our next theorem generalizes the relationship between these honeycombs to certain locally toroidal regular polytopes of types $\{3, 4, 4\}$ or $\{4, 4, 4\}$.

10E4 Theorem *Let $\mathcal{P} := \{\{3, 4\}, \{4, 4\}_{\mathbf{s}}\}$ $(= ({}_1\mathcal{T}_{\mathbf{s}}^4)^*)$, where $\mathbf{s} = (s, 0)$ with $s \geqslant 4$ even, or $\mathbf{s} = (s, s)$ with $s \geqslant 2$. Let $\Gamma(\mathcal{P}) = \langle \rho_0, \ldots, \rho_3 \rangle$, where $\rho_0, \ldots, \rho_3$ are the distinguished generators, and let κ and $\varphi_0, \ldots, \varphi_3$ be as in (10E2). Then $\mathcal{P}^\kappa = \{\{4, 4\}_{\mathbf{s}}, \{4, 4\}_{\mathbf{s}}^\eta\}$, and the group $\Gamma(\mathcal{P}^\kappa)$ can be identified with the subgroup $\langle \varphi_0, \ldots, \varphi_3 \rangle$ of $\Gamma(\mathcal{P})$, which is of index 3. In particular, $\mathcal{P}$ and $\mathcal{P}^\kappa$ are either both finite or both infinite.*

Proof. Let $\mathcal{M} := \{4, 4\}_{\mathbf{s}}$. Then $\mathcal{M}^\eta \neq \mathcal{M}$, by our restriction on $\mathbf{s}$. Also, from Section 10C we already know that the universal polytope $\{\mathcal{M}, \mathcal{M}^\eta\}$ exists.

Now consider the subgroup $\Gamma' := \langle \varphi_0, \ldots, \varphi_3 \rangle$ of $\Gamma(\mathcal{P})$, which is the image of the subgroup $[4, 4, 4]$ of $[3, 4, 4]$ under the homomorphism which maps $[3, 4, 4]$ onto $\Gamma(\mathcal{P})$. Then, modulo the intersection property, Γ' is the group of a regular polytope $\mathcal{Q}$ which belongs to the class $\langle \mathcal{M}, \mathcal{M}^\eta \rangle$. Indeed, by (10E3), $\mathcal{Q}$ must have facets isomorphic to $\mathcal{M}$, because the subgroup $\langle \varphi_0, \varphi_1, \varphi_2 \rangle$ of Γ' is the group of a vertex-figure of $\mathcal{P}$, and it must have vertex-figures isomorphic to $\mathcal{M}^\eta$, because $\langle \varphi_1, \varphi_2, \varphi_3 \rangle$ is obtained by halving the vertex-figure of $\mathcal{P}$. Also, Γ' is of index 1 or 3 in $\Gamma(\mathcal{P})$.

Indeed, our arguments in the following will show that $\mathcal{Q} = \{\mathcal{M}, \mathcal{M}^\eta\}$, $|\Gamma(\mathcal{P}) : \Gamma'| = 3$, and $\rho_1 \notin \Gamma'$; in particular, Γ' has the intersection property.

Next, we explain how to reverse the process. Write $\Gamma'' := \langle \varphi_0, \ldots, \varphi_3 \rangle$ for the group of the universal polytope $\{\mathcal{M}, \mathcal{M}^\eta\}$; for simplicity, we are using here the same notation for the generators as before. We now reintroduce ρ_1. We adjoin an element τ to Γ'', such that

10E5 $$\tau^2 = \varepsilon, \quad \tau\varphi_3\tau = \varphi_1, \quad \tau\varphi_2\tau = \varphi_2, \quad (\tau\varphi_0)^3 = \varepsilon.$$

For the resulting group Γ, we define the generators ρ_i by

$$(\rho_0, \ldots, \rho_3) := (\varphi_0, \tau, \varphi_3, \varphi_2).$$

Then it is straightforward to prove that Γ is a quotient of $\Gamma(\mathcal{P})$, if we observe that

$$\varphi_0\tau\varphi_i\tau\varphi_0 = \rho_{i+1} \quad (i = 0, 1, 2).$$

Now we can definitely check that Γ'' has index 3 in Γ. Indeed, if we write $\gamma := \rho_0\rho_1 = \varphi_0\tau$, then

$$\varphi_0\gamma = \gamma^2\varphi_0, \quad \varphi_1\gamma = \gamma\varphi_0\varphi_1\varphi_0, \quad \varphi_2\gamma = \gamma\varphi_2, \quad \varphi_3\gamma = \gamma\varphi_1,$$

and so we can take $\varepsilon, \gamma, \gamma^2$ as the three coset representatives.

In the same fashion, we can also begin with the original group Γ' (or any other quotient of Γ'') and adjoin an involution, such that the new group is again a quotient of $\Gamma(\mathcal{P})$ which contains Γ' as a subgroup of index 3. In particular, this proves then that Γ' has index 3 in $\Gamma(\mathcal{P})$, and that this quotient is $\Gamma(\mathcal{P})$ itself.

In conclusion, we can then also identify Γ with $\Gamma(\mathcal{P})$, and Γ'' with Γ'. In particular, the element τ corresponds to the missing generator ρ_1. Moreover, $\mathcal{Q} = \{\mathcal{M}, \mathcal{M}^\eta\}$. □

The polytopes $\mathcal{P} := \{\{3, 4\}, \{4, 4\}_{(s,0)}\}$, with s odd, were excluded from Theorem 10E4 for the following reason. If $\mathcal{M} := \{4, 4\}_{(s,0)}$ and s is odd, then $\mathcal{M} = \mathcal{M}^\eta$, $\Gamma' = \Gamma(\mathcal{P})$, and the index is 1. All we can say in this case is that κ relates the original polytope $\mathcal{P}$ to a quotient $\mathcal{P}^\kappa$ of the universal polytope $\{\mathcal{M}, \mathcal{M}^\eta\} = \{\{4, 4\}_{(s,0)}, \{4, 4\}_{(s,0)}\}$. Here it is helpful to recall that a similar relation had enabled us in Section 10C to enumerate all the finite universal polytopes of this kind. In particular, we know from this discussion that the only finite instance occurs if $s = 3$, and that then $\mathcal{P}^\kappa$ itself is the universal polytope. However, we conjecture that $\mathcal{P}^\kappa$ is not universal if $s > 5$.

We conclude this section by describing another connexion between pairs of polytopes of type {3, 4, 4} or {4, 4, 4}.

Let $\mathcal{P}$ be a regular 4-polytope in the class $\langle\{4, 4\}_{(s,0)}, \mathcal{M}\rangle$, where $\mathcal{M}$ is a regular polyhedron of type $\{4, r\}$ with $r = 3$ or 4. Consider again the operation κ on $\Gamma(\mathcal{P}) = \langle\rho_0, \ldots, \rho_3\rangle$ defined in (10E2). Then, modulo the intersection property for the new group, we obtain a polytope $\mathcal{P}^\kappa$ of type $\{s, r, r\}$ whose vertex-figures are isomorphic to $\mathcal{M}^\eta$.

The most interesting cases seem to be those with $s = 3$. Here we have

10E6 $$\kappa\colon \begin{cases} {}_2\mathcal{T}^4_{(3,0),(3,0)} \overset{1}{\longmapsto} \left({}_1\mathcal{T}^4_{(3,0)}\right)^*, \\ {}_2\mathcal{T}^4_{(3,0),(2,2)} \overset{12}{\longmapsto} \left({}_1\mathcal{T}^4_{(2,0)}\right)^*, \\ {}_2\mathcal{T}^4_{(3,0),(4,0)} \overset{48}{\longmapsto} \left({}_1\mathcal{T}^4_{(2,2)}\right)^*, \end{cases}$$

where, as before, ${}_2\mathcal{T}^4_{\mathbf{s},\mathbf{t}} := \{\{4, 4\}_{\mathbf{s}}, \{4, 4\}_{\mathbf{t}}\}$ and ${}_1\mathcal{T}^4_{\mathbf{s}} := \{\{4, 4\}_{\mathbf{s}}, \{4, 3\}\}$, and the number above the arrow is the index of the new group $\Gamma(\mathcal{P}^\kappa)$ (on the right) in the old group $\Gamma(\mathcal{P})$ (on the left). These relationships can be verified as follows.

Generally, if $s \geqslant 3$ and $\mathcal{M} = \{4, 4\}_{(t,0)}$ with t odd, then $\mathcal{M}^\eta = \mathcal{M}$, and the same argument as was used in Section 10C in showing that ${}_1\mathcal{T}^4_{(3,0)}$ and ${}_2\mathcal{T}^4_{(3,0),(3,0)}$ have the same group (namely, $S_6 \times C_2$), again works here. Namely, since $(\rho_1\rho_2)^2 = (\rho_2\rho_1)^2$, then with $t = 2k + 1$ we have

$$\begin{aligned}(\varphi_3\varphi_1\varphi_2)^t &= (\rho_2\rho_1\rho_2\rho_1\rho_3)^t \\ &= [(\rho_2\rho_1\rho_2\rho_3\rho_1)(\rho_1\rho_2\rho_1\rho_2\rho_3)]^k(\rho_2\rho_1\rho_2\rho_1)\rho_3 \\ &= (\rho_2\rho_1\rho_2\rho_3)^t\rho_1 = \rho_1.\end{aligned}$$

Hence, $\Gamma(\mathcal{P}) = \Gamma(\mathcal{P}^\kappa)$. In particular, this proves the first assertion of (10E6).

As in the proof of Theorem 10E4, we can recover a quotient of the group of the polytope $\{\{4, 4\}_{(3,0)}, \mathcal{M}\}$ from the group of $\{\{3, 4\}, \mathcal{M}^\eta\}$ by adjoining a suitable involution τ. Thus one universal group is indeed a subgroup of the other. With $\mathcal{M} = \{4, 4\}_{(2,2)}$ or $\{4, 4\}_{(4,0)}$, this then settles the remaining cases of (10E6).

The other finite universal example with $s \geqslant 3$ is

$$\kappa\colon\ {}_2\mathcal{T}^4_{(4,0),(3,0)} \stackrel{1}{\longmapsto} {}_2\mathcal{T}^4_{(4,0),(3,0)}.$$

Here, $\mathcal{P} = \mathcal{P}^\kappa$.

Finally, if the operation κ is applied to a polytope $\mathcal{P}$ in the class $\langle\{4, 4\}_{(s,0)}, \{4, 3\}\rangle$ or $\langle\{4, 4\}_{(s,s)}, \{4, 3\}\rangle$, then we obtain Schläfli types $\{s, 3, 3\}$ or $\{2s, 3, 3\}$, respectively. In particular,

10E7
$$\kappa\colon \begin{cases} {}_1\mathcal{T}^4_{(3,0)} \stackrel{12}{\longmapsto} \{3, 3, 3\}, \\ {}_1\mathcal{T}^4_{(2,2)} \stackrel{12}{\longmapsto} \{4, 3, 3\}_4, \end{cases}$$

where $\{4, 3, 3\}_4$ is the hemi-4-cube (see Section 6C). Here, the first relationship is obvious, because now $s = 3$. To check the second, we can appeal to the construction of ${}_1\mathcal{T}^4_{(2,2)}$ described in Figure 10B2 and (10B4). Then we have

$$\langle\varphi_0, \ldots, \varphi_3\rangle = \langle\tau_0, \sigma_2\sigma_4\tau_1, \tau_2, \tau_1\rangle = \langle\sigma_2\sigma_4, \tau_0, \tau_1, \tau_2\rangle;$$

the latter is just a semi-direct product of C_2^6 by D_6, where each factor C_2 corresponds to one of the 6 "short" diagonals of a hexagon (equivalent under D_6 to the diagonal with end points 2 and 4).

11

Locally Toroidal 4-Polytopes: II

In the previous chapter, we extensively investigated the locally toroidal regular polytopes of types $\{4, 4, 3\}$ and $\{4, 4, 4\}$, and obtained a nearly complete enumeration of those universal regular polytopes which are finite. In this chapter, we move on to the remaining types of locally toroidal 4-polytopes, namely, $\{6, 3, p\}$ with $p = 3, 4, 5, 6$, and $\{3, 6, 3\}$, and carry out a similar classification. For the types $\{6, 3, p\}$ our enumeration is complete, and for $\{3, 6, 3\}$ substantial partial results are known.

It is striking that complex hermitian forms now enter the scene. Indeed, the structure of the universal polytopes is frequently governed by a complex hermitian form. We shall explain our basic enumeration technique in Section 11A, and then, when it applies, enumerate many finite universal polytopes in Sections 11B to 11E by appealing to the results of Chapter 9. In particular, we shall see that such a polytope is finite if and only if the corresponding hermitian form is positive definite. The automorphism group of a finite polytope is then a semi-direct product of a finite unitary reflexion group by a small finite group.

This link between polytopes and hermitian forms generalizes the classical situation, where the structure of a regular tessellation is determined by a real quadratic form which defines the geometry of the ambient space.

For a discussion of the general classification program for locally toroidal regular polytopes, we refer to Section 10A.

11A The Basic Enumeration Technique

We begin the discussion by describing the basic technique which is applied to enumerate those finite universal locally toroidal regular 4-polytopes which are of some type $\{6, 3, p\}$ or $\{p, 3, 6\}$ with $p = 3, 4, 5, 6$, or $\{3, 6, 3\}$. We shall largely concentrate on the types $\{6, 3, p\}$, but our method will also work for certain, but not all, polytopes of type $\{3, 6, 3\}$ (see Section 11E).

In particular, we shall investigate the universal regular polytopes

$$_p\mathcal{T}^4_{\mathbf{s}} := \{\{6, 3\}_{\mathbf{s}}, \{3, p\}\} \qquad (p = 3, 4, 5),$$

with $\mathbf{s} = (s^k, 0^{2-k})$ and $k = 1, 2$, as well as

$${}_6\mathcal{T}^4_{\mathbf{s},\mathbf{t}} := \{\{6, 3\}_{\mathbf{s}}, \{3, 6\}_{\mathbf{t}}\},$$

with $\mathbf{s} = (s^k, 0^{2-k})$, $\mathbf{t} = (t^l, 0^{2-l})$ and $k, l = 1, 2$. Our assumptions on the parameters are as follows. For all the types, $s \geqslant 2$ if $k = 1$, and $s \geqslant 1$ if $k = 2$, and, similarly, for the last type, $t \geqslant 2$ if $l = 1$, and $t \geqslant 1$ if $l = 2$. The notation we have adopted for these polytopes is now more telling than that in the previous chapter; again the superscript denotes the rank, which now is 4, but the subscript p to the left is now the last entry in the Schläfli symbol. (In [199, p. 196], the polytopes with $p = 3$ were denoted by $\mathcal{H}_{s,0}$ or $\mathcal{H}_{s,s}$, respectively.)

The basic construction tool for these polytopes is again twisting, but now the twisting operations are performed on groups which are more complicated than Coxeter groups. In fact, the underlying groups W are of the type discussed in Chapter 9 (where they were denoted by Γ), and are determined by diagrams which contain labelled circuits (usually triangles). These generalize the Coxeter diagrams, in that they also describe the structure of certain 3-generator subgroups (see Section 9G). In particular, we shall require the results of Chapter 9 about the enumeration of the finite groups which belong to certain types of diagrams.

We also need some basic facts about the toroidal polyhedron $\{6, 3\}_{\mathbf{s}}$, which occurs as the facet of ${}_p\mathcal{T}^4_{\mathbf{s}}$ and ${}_6\mathcal{T}^4_{\mathbf{s},\mathbf{t}}$; its group is denoted by $[6, 3]_{\mathbf{s}}$. In particular, $[6, 3]_{(s^k, 0^{2-k})}$ is $[6, 3] = \langle \rho_0, \rho_1, \rho_2 \rangle$, factored out by the single extra relation

11A1
$$\begin{cases} (\rho_0\rho_1\rho_2)^{2s} &= \varepsilon \quad \text{if } k = 1, \\ (\rho_2(\rho_1\rho_0)^2)^{2s} &= \varepsilon \quad \text{if } k = 2 \end{cases}$$

(see [131, §8.4] and the brief introduction in Section 1D). These relations say that, in terms of its Petrie polygons or the 2-zigzags of its dual, the polyhedron can also be represented as $\{6, 3\}_{2s}$ or the dual of $\{3, 6\}_{\cdot,2s}$, respectively (see Section 7B). For the polyhedron $\{3, 6\}_{\cdot,2s}$, the length of its Petrie polygons (1-zigzags) is determined by the length of its 2-zigzags; this explains our notation for the first suffix.

For the polytopes ${}_p\mathcal{T}^4_{\mathbf{s}}$, we can now proceed as in the previous chapter and consider the corresponding abstract group ${}_p\Gamma^4_{\mathbf{s}}$, which is defined as the Coxeter group $[6, 3, p] = \langle \rho_0, \ldots, \rho_3 \rangle$, factored out by the extra relation in (11A1). For either choice of k, this extra relation involves only the first three generators ρ_0, ρ_1, ρ_2, and turns the facet $\{6, 3\}$ of the hyperbolic honeycomb $\{6, 3, p\}$ into its finite quotient $\{6, 3\}_{\mathbf{s}}$. In particular, ${}_p\mathcal{T}^4_{\mathbf{s}}$ exists if and only if ${}_p\Gamma^4_{\mathbf{s}}$ is a C-group, whose subgroups $\langle \rho_0, \rho_1, \rho_2 \rangle$ and $\langle \rho_1, \rho_2, \rho_3 \rangle$ are isomorphic to $[6, 3]_{\mathbf{s}}$ and $[3, p]$, respectively. In this case,

$$\Gamma\big({}_p\mathcal{T}^4_{\mathbf{s}}\big) = {}_p\Gamma^4_{\mathbf{s}}$$

(see Proposition 4A8).

On the other hand, for ${}_6\mathcal{T}^4_{\mathbf{s},\mathbf{t}}$ there are two extra relations, each with three generators. The first involves ρ_0, ρ_1, ρ_2, and is of the same kind as before. The second involves ρ_1, ρ_2, ρ_3, and is dual to a relation of (11A1); thus it is obtained from (11A1) by replacing ρ_i by ρ_{3-i}, k by l, and s by t. These extra relations turn the facet $\{6, 3\}$ and vertex-figure $\{3, 6\}$ of the honeycomb $\{6, 3, 6\}$ into the polyhedra $\{6, 3\}_{\mathbf{s}}$ and $\{3, 6\}_{\mathbf{t}}$, respectively. Now the corresponding group ${}_6\Gamma^4_{\mathbf{s},\mathbf{t}}$ is defined to be the factor group of [6, 3, 6] determined by these two extra relations. In particular, ${}_6\mathcal{T}^4_{\mathbf{s},\mathbf{t}}$ exists if and only if ${}_6\Gamma^4_{\mathbf{s},\mathbf{t}}$ is a C-group, whose subgroups $\langle \rho_0, \rho_1, \rho_2 \rangle$ and $\langle \rho_1, \rho_2, \rho_3 \rangle$ are isomorphic to $[6, 3]_{\mathbf{s}}$ and $[3, 6]_{\mathbf{t}}$, respectively. In this case,

$$\Gamma\left({}_6\mathcal{T}^4_{\mathbf{s},\mathbf{t}}\right) = {}_6\Gamma^4_{\mathbf{s},\mathbf{t}}.$$

We now describe the technique applied to enumerate the finite polytopes. Let $\mathcal{P} := {}_p\mathcal{T}^4_{\mathbf{s}}$ or ${}_6\mathcal{T}^4_{\mathbf{s},\mathbf{t}}$, and let $\Gamma := {}_p\Gamma^4_{\mathbf{s}}$ or ${}_6\Gamma^4_{\mathbf{s},\mathbf{t}}$, respectively. We proceed by the following three steps.

- We find a "suitable" normal subgroup W of finite index in Γ, such that Γ is a semi-direct product of W by a subgroup of its automorphisms.
- For this subgroup W, we construct a complex "locally unitary" representation $r\colon W \to \mathrm{GL}_m(\mathbb{C})$, as in Chapter 9, with m determined by $\mathcal{P}$. In particular, r will preserve a hermitian form $\langle\cdot,\cdot\rangle$ on complex m-space $\mathbb{C}^m$, such that the image group $G := Wr$ consists of isometries with respect to this form.
- We study the hermitian form to analyse the structure of $\mathcal{P}$ and Γ.

For much of the effort required in the second and third steps we can appeal to the results of Chapter 9; in particular, we shall adopt the notation introduced there. The remaining task is then to describe the subgroup W.

Our approach works only under the mild restriction that $\mathbf{s}, \mathbf{t} \neq (1, 1)$, which we shall assume from now on. This leaves the case $\mathbf{s} = (1, 1)$ or $\mathbf{t} = (1, 1)$, which cannot be settled by hermitian forms; we treat this separately in Section 11C.

The construction of W and its representation r is based on the following simple observation, which relates the group of the toroidal polyhedron $\{6, 3\}_{\mathbf{s}}$ to the generalized triangle group $[1\ 1\ 1]^s$ discussed in Section 9D. It shows that the technique can already be applied in rank 3 to obtain $\{6, 3\}_{\mathbf{s}}$.

Recall that the geometric group $[1\ 1\ 1]^s$ (with $s \geqslant 2$) is (isomorphic to that) generated by involutions $\sigma_1, \sigma_2, \sigma_3$, and abstractly defined by the presentation

11A2 $\quad \sigma_1^2 = \sigma_2^2 = \sigma_3^2 = (\sigma_1\sigma_2)^3 = (\sigma_2\sigma_3)^3 = (\sigma_1\sigma_3)^3 = (\sigma_1\sigma_3\sigma_2\sigma_3)^s = \varepsilon$

(see Section 9D). It is represented by the diagram in Figure 11A1, which is a triangular

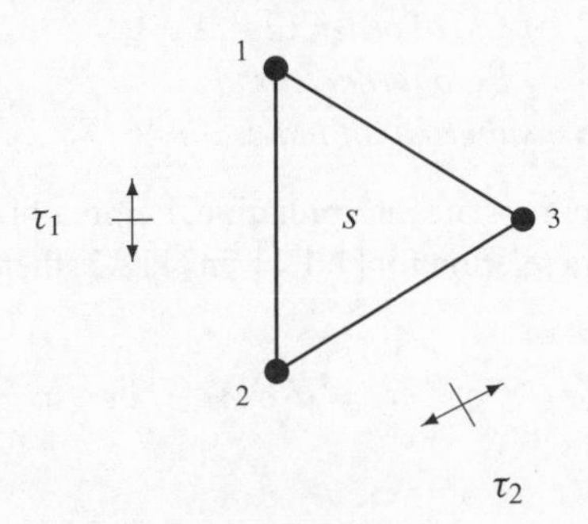

Figure 11A1. The group $[1\ 1\ 1]^s$.

Coxeter diagram with unmarked branches, and a mark s inside the triangle to indicate the rightmost extra relation. Its order is $6s^2$. The last relation in (11A2) can be replaced by any of six equivalent relations of the type $(\sigma_i\sigma_j\sigma_k\sigma_j)^s = \varepsilon$ with $\{i, j, k\} = \{1, 2, 3\}$. Using the group automorphisms τ_1 and τ_2, which act on the generators σ_i as indicated, we can now extend $[1\ 1\ 1]^s$ in two ways; both are simple examples of twisting operations. (We shall comment on the geometric realization of these automorphisms later.)

First, we can extend by τ_1, and consider the operation

11A3 $$(\sigma_1, \sigma_2, \sigma_3; \tau_1) \mapsto (\tau_1, \sigma_2, \sigma_3) =: (\rho_0, \rho_1, \rho_2).$$

Then we obtain

$$[6, 3]_{(s,0)} = [1\ 1\ 1]^s \rtimes C_2,$$

which is a group of order $12s^2$. Indeed, we can recast the presentation (11A2) for $[1\ 1\ 1]^s$ into one for the new group $\langle \rho_0, \rho_1, \rho_2 \rangle$, which is equivalent to a presentation for $[6, 3]_{(s,0)}$. For example, the extra relation in (11A1) for $[6, 3]_{(s,0)}$ follows from

$$(\rho_0\rho_1\rho_2)^2 = \tau_1\sigma_2\sigma_3\tau_1\sigma_2\sigma_3 = \sigma_1\sigma_3\sigma_2\sigma_3$$

and the rightmost relation in (11A2).

Second, if we also extend by τ_2 and apply the operation

11A4 $$(\sigma_1, \sigma_2, \sigma_3; \tau_1, \tau_2) \mapsto (\sigma_1, \tau_1, \tau_2) =: (\rho_0, \rho_1, \rho_2),$$

then we obtain

$$[6, 3]_{(s,s)} = [1\ 1\ 1]^s \rtimes S_3,$$

which is of order $36s^2$. Now the extra relation in (11A1) for $[6, 3]_{(s,s)}$ is a consequence of

$$(\rho_2(\rho_1\rho_0)^2)^2 = (\tau_2\tau_1\sigma_1\tau_1\sigma_1)^2 = \tau_2\sigma_2\sigma_1\tau_2\sigma_2\sigma_1 = \sigma_3\sigma_1\sigma_2\sigma_1 = \sigma_3\sigma_2\sigma_1\sigma_2 \sim \sigma_1\sigma_2\sigma_3\sigma_2$$

and the rightmost relation of (11A2); as usual, $\sim$ denotes conjugacy.

In summary, we have proved the following

11A5 Lemma *Let $s \geqslant 2$. Then*

(a) $[6, 3]_{(s,0)} \cong [1\ 1\ 1]^s \rtimes C_2$, *of order* $12s^2$*;*
(b) $[6, 3]_{(s,s)} \cong [1\ 1\ 1]^s \rtimes S_3$, *of order* $36s^2$*;*
(c) $[6, 3]_{(s,0)}$ *occurs as a subgroup of index* 3 *in* $[6, 3]_{(s,s)}$.

We remark that, with appropriate interpretation, Lemma 11A5(b) also remains true if $s = 1$. In fact, the last extra relation for $[1\ 1\ 1]^s$ in (11A2) then becomes $\sigma_1\sigma_2\sigma_3\sigma_2 = \varepsilon$, or equivalently,

$$\sigma_1 = \sigma_2\sigma_3\sigma_2, \quad \text{or} \quad \sigma_2 = \sigma_1\sigma_3\sigma_1, \quad \text{or} \quad \sigma_3 = \sigma_1\sigma_2\sigma_1.$$

This implies that

11A6 $$[1\ 1\ 1]^1 \cong S_3,$$

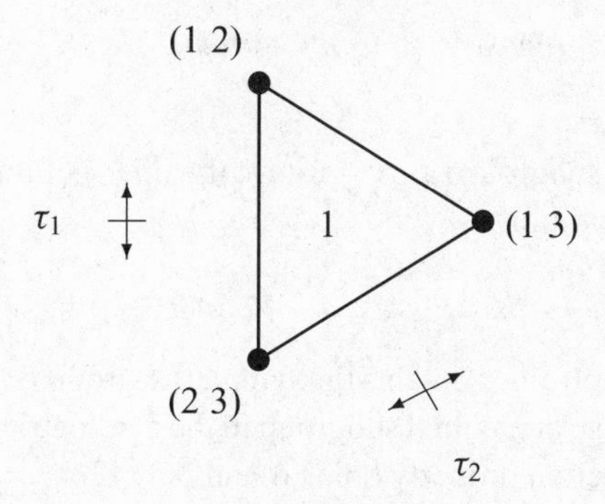

Figure 11A2. The group $[1\ 1\ 1]^1 = S_3$.

as illustrated by the diagram in Figure 11A2. Again we have a semi-direct product $[6, 3]_{(1,1)} \cong S_3 \rtimes S_3$, but now it can be rewritten as a direct product:

11A7 $$[6, 3]_{(1,1)} \cong S_3 \times S_3.$$

Indeed, in the symmetric group $S_{\{1,\ldots,6\}}$ (on $1, \ldots, 6$), we can take $[1\ 1\ 1]^1$ as the subgroup $S_{\{1,2,3\}}$, and realize its group automorphisms τ_1 and τ_2 by conjugation with

$$\tau_1 = (1\ 3)(4\ 6), \quad \tau_2 = (1\ 2)(4\ 5)$$

(for simplicity, the conjugating elements are also denoted by τ_1 and τ_2); then σ_1, τ_1, τ_2 generate the group $S_{\{1,2,3\}} \times S_{\{4,5,6\}}$.

Our approach in subsequent sections is based on reversing this line of argument. So, from the group $[6, 3]_{(s,0)} = \langle \rho_0, \rho_1, \rho_2 \rangle$ we obtain the subgroup (isomorphic to) $[1\ 1\ 1]^s$ by the mixing operation

11A8 $$(\rho_0, \rho_1, \rho_2) \mapsto (\rho_0\rho_1\rho_0, \rho_1, \rho_2) =: (\sigma_1, \sigma_2, \sigma_3),$$

while we obtain the same subgroup from $[6, 3]_{(s,s)} = \langle \rho_0, \rho_1, \rho_2 \rangle$ by the mixing operation

11A9 $$(\rho_0, \rho_1, \rho_2) \mapsto (\rho_0, \rho_1\rho_0\rho_1, \rho_2\rho_1\rho_0\rho_1\rho_2) =: (\sigma_1, \sigma_2, \sigma_3).$$

It is interesting to note what happens if we apply these mixing operations to the "wrong" group. For example, if we apply the operation of (11A8) to $[6, 3]_{(s,s)}$, or that of (11A9) to $[6, 3]_{(3s,0)}$, then the group which we obtain has the presentation of (11A2), with the last relation replaced by

$$(\sigma_1\sigma_2\sigma_3)^{2s} = \varepsilon.$$

This relation still determines a quotient of the infinite euclidean group $[1\ 1\ 1]^\infty$ by a translation, but one which does not fit into the present context.

We conclude with two lemmas which are required in later sections.

11A10 Lemma *Let Γ be a group generated by involutions $\sigma_0, \ldots, \sigma_{n-1}$ (with $n \geqslant 4$), such that $(\sigma_i\sigma_j)^2 = \varepsilon$ if $0 \leqslant i < j - 1 \leqslant n - 2$. Assume that its subgroups $\langle \sigma_0, \ldots, \sigma_{n-2} \rangle$ and $\langle \sigma_1, \ldots, \sigma_{n-1} \rangle$ are C-groups, and that the latter is the group of the*

$(n-1)$-simplex $\{3^{n-2}\}$ (considered as a C-group). Then Γ is a C-group if and only if $\sigma_{n-1} \notin \langle \sigma_0, \ldots, \sigma_{n-2} \rangle$.

Proof. Since the two subgroups are C-groups, the intersection property for Γ is equivalent to the single condition

$$\langle \sigma_0, \ldots, \sigma_{n-2} \rangle \cap \langle \sigma_1, \ldots, \sigma_{n-1} \rangle = \langle \sigma_1, \ldots, \sigma_{n-2} \rangle.$$

But the group on the left side contains the symmetric group $\langle \sigma_1, \ldots, \sigma_{n-2} \rangle = S_{n-1}$ as a subgroup, which is also a maximal subgroup in the symmetric group $\langle \sigma_1, \ldots, \sigma_{n-1} \rangle = S_n$. Hence the intersection property holds if and only if $\sigma_{n-1} \notin \langle \sigma_0, \ldots, \sigma_{n-2} \rangle$. □

11A11 Lemma *Let $\mathcal{K}$ be a regular polyhedron of type $\{3, k\}$ with $k \geqslant 2$. Then the universal polytope $\mathcal{P} := \{\{6, 3\}_{(2,0)}, \mathcal{K}\}$ exists if and only if the universal polytope $\mathcal{Q} := \{\{3, 3\}, \mathcal{K}\}$ exists. In this case, $\Gamma(\mathcal{P}) = \Gamma(\mathcal{Q}) \times C_2$.*

Proof. This is just an application of Theorem 7A8, bearing in mind (7A10). Indeed, we have

$$\mathcal{P} = \mathcal{Q} \diamond \{\,\},$$

and, according to the theorem, the two sides are polytopal (or not) together. □

11B The Polytopes ${}_p\mathcal{T}^4_{(s,0)} := \{\{6, 3\}_{(s,0)}, \{3, p\}\}$

In this section, we enumerate the finite regular 4-polytopes ${}_p\mathcal{T}^4_{(s,0)}$ with $p = 3, 4, 5$ and $s \geqslant 2$. The enumeration of the finite polytopes ${}_p\mathcal{T}^4_{(s,s)}$ uses a different technique, and is treated in the next section.

We begin the discussion with a general remark. Let $\mathcal{P}$ be any regular n-polytope with 3-faces of the form $\{6, 3\}_{(s,0)}$ for some $s \geqslant 2$. If its group $\Gamma := \Gamma(\mathcal{P})$ is, as usual, $\langle \rho_0, \ldots, \rho_{n-1} \rangle$, then the mixing operation

$$(\rho_0, \ldots, \rho_{n-1}) \mapsto (\rho_0\rho_1\rho_0, \rho_1, \ldots, \rho_{n-1}) =: (\sigma_1, \ldots, \sigma_n)$$

yields a subgroup $W := \langle \sigma_1, \ldots, \sigma_n \rangle$ of index at most 2 in Γ (with coset representative ρ_0 if the index is 2 – the notation has been chosen to accord with what we do immediately below). We can recover Γ from W by means of the involutory twisting operation τ $(= \rho_0)$, acting by

$$\tau\sigma_1\tau = \sigma_2, \qquad \tau\sigma_j\tau = \sigma_j \quad (j = 3, \ldots, n).$$

Examples (if they exist) with index 1 are of less interest to us; the twisting operation would then simply be conjugation by ρ_0 within W $(= \Gamma)$.

As we saw in the previous section, the subgroup $\langle \sigma_1, \sigma_2, \sigma_3 \rangle$ of W is isomorphic to $[1\ 1\ 1]^s$. Indeed, we have $\sigma_1\sigma_2 = (\rho_0\rho_1)^2$ and $\sigma_1\sigma_3\sigma_2\sigma_3 = (\rho_0\rho_1\rho_2)^2$; hence

$$(\sigma_1\sigma_2)^3 = \varepsilon, \quad (\sigma_1\sigma_3\sigma_2\sigma_3)^s = \varepsilon.$$

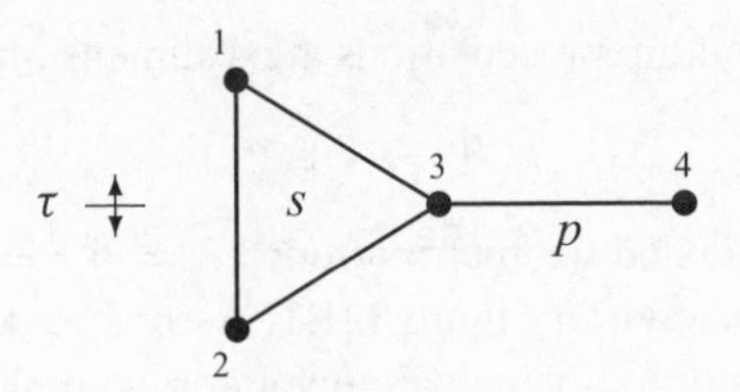

Figure 11B1

In general, this observation is of little value. However, when the vertex-figure of $\mathcal{P}$ is a spherical or toroidal polyhedron, there results a criterion for the finiteness (or otherwise) of the corresponding universal polytope.

The Polytopes ${}_p\mathcal{T}^4_{(s,0)}$

For this class, the observation leads us to begin with the abstract group $W = \langle \sigma_1, \ldots, \sigma_4 \rangle$ corresponding to the diagram in Figure 11B1, which consists of a marked triangle with a tail. Here, τ indicates the group automorphism of W which corresponds to the symmetry of the diagram in its horizontal axis. In more concrete terms, W has a presentation which consists of the relations in (11A2) and

11B1 $$\sigma_4^2 = (\sigma_1\sigma_4)^2 = (\sigma_2\sigma_4)^2 = (\sigma_3\sigma_4)^p = \varepsilon.$$

We now extend W using the operation

11B2 $$(\sigma_1, \ldots, \sigma_4; \tau) \mapsto (\tau, \sigma_2, \sigma_3, \sigma_4) =: (\rho_0, \rho_1, \rho_2, \rho_3).$$

This leads to the new group

$$\Gamma := \langle \rho_0, \ldots, \rho_3 \rangle = W \rtimes C_2,$$

where the factor C_2 is generated by τ. Computing a presentation, we find that Γ is the Coxeter group $[6, 3, p]$, factored out by the extra relation

11B3 $$(\rho_0\rho_1\rho_2)^{2s} = \varepsilon.$$

This extra relation arises from

$$(\rho_0\rho_1\rho_2)^2 = (\tau\sigma_2\sigma_3)^2 = \sigma_1\sigma_3\sigma_2\sigma_3 = \sigma_1\sigma_2\sigma_3\sigma_2.$$

In particular, we have

$$\Gamma = {}_p\Gamma^4_{(s,0)},$$

because the defining relations for Γ are just those for the latter group.

This completes our first step, and produces a subgroup of finite index in ${}_p\Gamma^4_{(s,0)}$, namely, W, which we shall investigate further. In particular, we employ the techniques of Chapter 9, and work with a complex representation r of W; for the general setup of these representations see Sections 9B and 9C. More specifically, we shall require the results of Section 9E, which concern (among others) triangle diagrams with tails.

In the present context, all representations are 4-dimensional, and so we have

$$r: W \to \mathrm{GL}_4(\mathbb{C}).$$

The diagram of the associated geometric group $G := Wr = \langle S_1, \ldots, S_4 \rangle$ belongs to the family of diagrams shown in Figure 11B1. As before, the nodes of the diagram correspond to the generators S_i, which are involutory complex hyperplane reflexions. In the notation of Section 9E, we have $W = \Gamma^4(3, s; 3, 3; p)$ and $G = G^4(3, s; 3, 3; p)$. Abstractly, the diagram admits a twist τ; we shall postpone until later a discussion of the geometry of such a twist.

As we have said, to each of these diagrams corresponds an abstract group with generators $\sigma_1, \sigma_2, \sigma_3, \sigma_4$, whose defining relations are just those implied by the diagram. We also have a geometric group in each case (see Theorem 9E5). When the geometric group is finite, we know that it is abstractly defined by these relations; such groups are enumerated in Theorem 9E4(a), (b) and (c). In all cases, the mapping $\sigma_i \mapsto S_i$ $(i = 1, \ldots, 4)$ indeed induces a representation of the abstract group; when the geometric group (or the abstract group) is finite, it is an isomorphism. Whether the group is finite or not, the restriction of r to the subgroup $\langle \sigma_1, \sigma_2, \sigma_3 \rangle$ of W is one-to-one; hence

$$\langle \sigma_1, \sigma_2, \sigma_3 \rangle \cong \langle S_1, S_2, S_3 \rangle \cong [1\ 1\ 1]^s.$$

Similarly, r is one-to-one on $\langle \sigma_1, \sigma_3, \sigma_4 \rangle$ and $\langle \sigma_2, \sigma_3, \sigma_4 \rangle$, and we have

$$\langle \sigma_1, \sigma_3, \sigma_4 \rangle \cong \langle \sigma_2, \sigma_3, \sigma_4 \rangle \cong [3, p].$$

We have seen that the finite groups W can be obtained from Theorem 9E4. However, before we can translate this into polytope language, we need to verify the intersection property for the corresponding group $\Gamma = W \rtimes \langle \tau \rangle$. This will follow straightforwardly from the intersection property for W with respect to $\sigma_1, \ldots, \sigma_4$, which will imply that Γ has the same property with respect to $\rho_0, \ldots, \rho_3$.

11B4 Lemma *Let $W = \langle \sigma_1, \sigma_2, \sigma_3, \sigma_4 \rangle$ be the abstract group corresponding to the diagram of Figure 11B1. Then W is a C-group.*

Proof. This follows from Theorem 9E6, but instead we can also obtain a simple direct proof. We first establish the lemma in case s is prime. It is easy to see here that $\langle \sigma_2, \sigma_3 \rangle$ is a maximal subgroup of $\langle \sigma_1, \sigma_2, \sigma_3 \rangle$ (we can use the geometry of the group $[1\ 1\ 1]^s$, to which it is isomorphic). Since $\sigma_4 \notin \langle \sigma_1, \sigma_2, \sigma_3 \rangle$, the result follows.

When s is composite, we can vary the proof of the quotient criterion of Theorem 2E17. If $t \mid s$ is prime, then the natural projection map onto the corresponding group with s replaced by t is one-to-one on the subgroups $\langle \sigma_1, \sigma_3, \sigma_4 \rangle$ and $\langle \sigma_2, \sigma_3, \sigma_4 \rangle$; the analogous argument to that of the theorem carries over. □

To fill out the details of the intersection property for Γ, we have, for example,

$$\langle \rho_0, \rho_1, \rho_2 \rangle \cap \langle \rho_1, \rho_2, \rho_3 \rangle = (\langle \sigma_1, \sigma_2, \sigma_3 \rangle \rtimes \langle \tau \rangle) \cap \langle \sigma_2, \sigma_3, \sigma_4 \rangle = \langle \sigma_2, \sigma_3 \rangle = \langle \rho_1, \rho_2 \rangle.$$

It follows that Γ is a C-group, whose subgroups $\langle \rho_0, \rho_1, \rho_2 \rangle$ and $\langle \rho_1, \rho_2, \rho_3 \rangle$ are isomorphic to $[6, 3]_{(s,0)}$ and $[3, p]$, respectively (see Lemma 11A5).

If $p = 3$, then

$$W \cong [1\,1\,2]^s \; (= [1\,1\,2^3]^s)$$

(see Sections 9A and 9E). Recall here our convention that, as in [114, §4], $[k\,l\,m^r]^s$ denotes the geometric group whose presentation it signifies. However, if this group is finite, then its presentation is given by the diagram relations, and so the abstract and the geometric groups are isomorphic.

Summarizing, we now have the following.

11B5 Theorem *The universal regular 4-polytope* ${}_p\mathcal{T}^4_{(s,0)} = \{\{6,3\}_{(s,0)}, \{3,p\}\}$ *exists for all* $s \geqslant 2$ *and* $p = 3, 4, 5$. *Its group* ${}_p\Gamma^4_{(s,0)}$ *is isomorphic to* $W \rtimes C_2$, *where* W *is the group abstractly defined by the diagram in Figure 11B1 (that is,* $W = [1\,1\,2]^s$ *if* $p = 3$*). In particular,* ${}_p\mathcal{T}^4_{(s,0)}$ *is finite precisely in the following cases:*

- *(a)* $p = 3$ *and* $s = 2$, 3 *or* 4, *with group* $S_5 \times C_2$, $[1\,1\,2]^3 \rtimes C_2$ *or* $[1\,1\,2]^4 \rtimes C_2$, *of order* 240, 1296 *or* 15360, *respectively;*
- *(b)* $p = 4$ *and* $s = 2$, *with group* $[3,3,4] \times C_2$, *of order* 768*;*
- *(c)* $p = 5$ *and* $s = 2$, *with group* $[3,3,5] \times C_2$, *of order* 28800.

Proof. For the proof, it only remains to describe the finite polytopes. Clearly, ${}_p\mathcal{T}^4_{(s,0)}$ is finite if and only if W is finite. Therefore, by Theorem 9E4, finiteness can only occur for the parameter values listed in the theorem, and the groups are then as indicated. In particular, if $s = 2$, then Γ is really a direct product of W and C_2, but now the factor C_2 is not generated by τ. Indeed, applying Lemma 11A11 to ${}_p\mathcal{T}^4_{(2,0)}$, we obtain

$${}_p\Gamma^4_{(2,0)} = [3,3,p] \times C_2,$$

which is $S_5 \times C_2$ if $p = 3$. This completes the proof. □

In Table 11B1, we list each finite polytope ${}_p\mathcal{T}^4_{\mathbf{s}}$, along with the numbers v of its vertices and f of its facets, and the order g of its group. For $\mathbf{s} = (s,0)$, these are just the finite polytopes of Theorem 11B5. For $\mathbf{s} = (s,s)$, we refer to Section 11C.

Theorem 11B5 has a striking corollary, which says that the structure of ${}_p\mathcal{T}^4_{(s,0)}$ is governed by the hermitian form which is associated with the underlying geometric diagram (see the proof of Theorem 9E4 for the Gram matrix).

Table 11B1. *The Finite Polytopes* ${}_p\mathcal{T}^4_{\mathbf{s}}$ $(p = 3, 4, 5)$

p	$\mathbf{s}$	v	f	g	Group
3	(2, 0)	10	5	240	$S_5 \times C_2$
	(3, 0)	54	12	1296	$[1\,1\,2]^3 \rtimes C_2$
	(4, 0)	640	80	15360	$[1\,1\,2]^4 \rtimes C_2$
	(2, 2)	120	20	2880	$S_5 \times S_4$
4	(1, 1)	12	8	288	$S_3 \rtimes [3,4]$
	(2, 0)	16	16	768	$[3,3,4] \times C_2$
5	(2, 0)	240	600	28800	$[3,3,5] \times C_2$

11B6 Corollary *Let $s \geqslant 2$, and let $p = 3$, 4 or 5. Then the universal regular 4-polytope ${}_p\mathcal{T}^4_{(s,0)}$ is finite if and only if the corresponding hermitian form $\langle\cdot,\cdot\rangle$ in $\mathbb{C}^4$ is positive definite. In this case, its group ${}_p\Gamma^4_{(s,0)}$ is a semi-direct product of a finite reflexion group in unitary 4-space by C_2.*

Proof. Again we can apply Theorems 9E4 and 9E5, and deduce that W, and thus $\Gamma = {}_p\Gamma^4_{(s,0)}$, is finite if and only if $\langle\cdot,\cdot\rangle$ is positive definite. If indeed the form is positive definite, then it defines a unitary geometry in the ambient space and the representation for W is faithful. In this case, W is conjugate, in the general linear group $GL_4(\mathbb{C})$, to a finite reflexion group in standard unitary 4-space. Now the corollary follows. □

In the traditional theory, the structure of a regular tessellation $\{p_1, \ldots, p_{n-1}\}$ in spherical, euclidean or hyperbolic n-space is determined by a real quadratic form which defines the geometry of the ambient space, and is associated with the canonical bilinear form for the corresponding Coxeter group $[p_1, \ldots, p_{n-1}]$ (see Sections 3A and 3B). We know that this form is positive definite if and only if it belongs to a spherical tessellation (regular convex polytope). Thus a positive definite quadratic form determines a finite object, and vice versa.

Corollary 11B6 (and similar statements in the following) says that this correspondence between tessellations and forms carries over to an important subclass of locally toroidal polytopes, namely, certain of those of types $\{6, 3, p\}$ (including $p = 6$), and their complex hermitian forms. In particular, it remains true that a positive definite form determines a finite object, and vice versa.

Recapitulating, so far we have (in effect) employed the abstract group W to construct the universal polytope ${}_p\mathcal{T}^4_{(s,0)}$. Now the question is whether or not a similar procedure works for the geometric group G. We know that W and G are isomorphic if one of them is finite, but in general this may not be true. The following arguments show that the same twisting operation applied to G gives also a locally toroidal regular 4-polytope in the class $\langle\{6, 3\}_{(s,0)}, \{3, p\}\rangle$.

First, observe that the twist τ carries over to an involutory group automorphism T, which permutes the generators S_i of G in the same way in which τ permutes the generators σ_i of W. (Our labelling of the diagram in Figure 11B1 anticipates this.) Such an automorphism could be obtained from the general principle described in Section 11D. The problem here is the geometric realization of this automorphism.

When $s \geqslant 3$, we can proceed directly; if we coordinatize with respect to the normal vectors v_i of the generating reflexions of G, then we can realize it by conjugation with the semi-linear mapping $T\colon \mathbb{C}^4 \to \mathbb{C}^4$, given by

11B7 $$(\xi_1, \xi_2, \xi_3, \xi_4)T := (\bar{\xi}_2, \bar{\xi}_1, \bar{\xi}_3, \bar{\xi}_4),$$

by which we can now extend the group. More precisely, the operation

11B8 $$(S_1, \ldots, S_4; T) \mapsto (T, S_2, S_3, S_4) =: (R_0, R_1, R_2, R_3)$$

on G determines the new group

$$\bar{G} := \langle R_0, \ldots, R_3\rangle = G \rtimes C_2,$$

where the factor C_2 is generated by T. In particular, $\bar{G}$ is a representation of $\Gamma = {}_p\Gamma^4_{(s,0)}$ in the group of all linear and semi-linear transformations of $\mathbb{C}^4$, which is induced by the representation r of W. Note that

$$\langle xT, yT\rangle = \overline{\langle x, y\rangle} \qquad (x, y \in \mathbb{C}^4).$$

Hence, up to complex conjugation, the hermitian form is preserved by all elements of $\bar{G}$.

Of course, our standard assumptions continue to apply. Specifically, we may suppose that $\langle v_i, v_j\rangle$ is real (actually, minus the cosine of the angle between the vectors), except when $\{i, j\} = \{1, 2\}$; that is, we load the whole turn ϑ (whatever it might be) on the branch $\{1, 2\}$. Some care is necessary in the degenerate case $\Delta = 0$; the mapping T of (11B7) will only work if we take the original group G as a linear group in $\mathbb{C}^4$ (preserving a positive semi-definite form), rather than as an affine (unitary) group in $\mathbb{C}^3$ (see Lemma 9B19).

It is straightforward to verify that $\bar{G}$ is indeed a C-group. In fact, $\bar{G}$ has the intersection property, and so we can argue as in Lemma 11B4 for Γ. Let $\mathcal{Q}$ denote the regular 4-polytope with group $\bar{G}$. Then $\mathcal{Q}$ must have the same types of facets and vertex-figures as the universal polytope ${}_p\mathcal{T}^4_{(s,0)}$, because r is faithful when restricted to the basic 3-generator subgroups of W. In particular, we have a covering

$${}_p\mathcal{T}^4_{(s,0)} \searrow \mathcal{Q},$$

which is an isomorphism if one of the two polytopes is finite.

There is a corresponding 4-dimensional complex model for the polytopes $\mathcal{Q}$, related to the geometric groups G. When considered in real 8-space, this becomes a real model for $\mathcal{Q}$. Technically, however, it is not always a realization in the sense of Chapter 5, because the ambient space will not be euclidean in general; it will be a genuine realization when $\mathcal{Q}$ is finite. Nevertheless, as a geometric model, it captures much of the combinatorics of both $\mathcal{Q}$ and ${}_p\mathcal{T}^4_{(s,0)}$ (the latter via the covering map ${}_p\mathcal{T}^4_{(s,0)} \searrow \mathcal{Q}$). Each model is obtained from its group by Wythoff's construction, in exactly the same way as for ordinary realizations (see Chapter 5).

We shall concentrate on the complex models in $\mathbb{C}^4$; however, we must now think of $\mathbb{C}^4$ as the euclidean space $\mathbb{E}^8$. The hyperplane reflexions S_i now become involutions with 6-dimensional mirrors. The mapping T of (11B7), on the other hand, is an involution with a 4-dimensional mirror. The corresponding Wythoff space for $\bar{G}$ is the complex line (that is, real plane)

$$R_1 \cap R_2 \cap R_3 = S_2 \cap S_3 \cap S_4.$$

(Recall our convention that, in dealing with realizability questions, we identify a reflexion with its mirror in the ambient space.)

Now, if we apply Wythoff's construction to $\bar{G}$, with any point in the Wythoff space as initial vertex, we arrive at a complex model for $\mathcal{Q}$ whose "geometric symmetry group" is $\bar{G}$. If $\bar{G}$ is finite, then this is a finite "unitary model" for $\mathcal{Q}$ (and thus for ${}_p\mathcal{T}^4_{(s,0)}$), in the sense that $\bar{G}$ (and thus Γ) consists of linear and semi-linear transformations which are "unitary" (that is, up to complex conjugation, they leave the positive definite form $\langle\cdot,\cdot\rangle$ invariant). In this case, the corresponding real model in $\mathbb{E}^8$ is a genuine realization of $\mathcal{Q}$ in the sense of Chapter 5, because now the real quadratic form in 8 variables is positive definite and the symmetries are isometries which preserve this form. For infinite groups $\bar{G}$, the complex forms and real forms are generally indefinite, and the models are generally not realizations in the strict sense.

Unfortunately, this approach no longer works when $s = 2$, because the corresponding unitary group is (essentially) real. The automorphism T still exists, but now it is inner (indeed, on the *real* space spanned by $v_1, \ldots, v_4$, the mapping T acts like $S_2S_1S_2 \in G$). In this case, abstractly, as we have noted,

$$\big\{\{6,3\}_{(2,0)},\{3,p\}\big\} = \{3,3,p\} \diamond \{\,\},$$

and so when $\{3,3,p\}$ is finite (that is, when $p = 3, 4$ or 5), we obtain a faithful realization (in the full geometric sense) in $\mathbb{R}^5$ as the blend of $\{3,3,p\}$ in $\mathbb{R}^4$ with the line segment $\{\,\}$ (see Section 5A). (When $p = 3$, there is a different faithful realization in $\mathbb{E}^4$, with $T := -I$.)

In a similar fashion, we can also construct a geometric model for the dual polytope $\mathcal{Q}^*$ of $\mathcal{Q}$, which then also provides a model for $({}_p\mathcal{T}^4_{(s,0)})^*$. Here we take the generators of $\bar{G}$ in reverse order. In particular, the Wythoff space is given by

$$R_0 \cap R_1 \cap R_2 = T \cap S_2 \cap S_3 = S_1 \cap S_2 \cap S_3.$$

Then the complex model for $\mathcal{Q}^*$ in $\mathbb{C}^4$ is found by Wythoff's construction. In the finite cases of Theorem 11B5, the corresponding real model is again a realization in the strict sense. However, when $s = 2$ (the real case), it is not faithful; that is, it is only faithful if $p = 3$ and $s = 3$ or 4.

Generalizations

In the remainder of this section, we shall explore two directions in which the construction of the polytopes ${}_p\mathcal{T}^4_{(s,0)}$ can be generalized.

First, we can begin with the group $W = \langle\sigma_1,\ldots,\sigma_4\rangle$ depicted in Figure 11B2, and apply the same operation (11B2) as before. For each $l \geqslant 2$, this yields a new group $\Gamma = W \rtimes C_2$, which satisfies all the defining relations for the group of the universal regular polytope

$$\mathcal{P} := \{\{2l,3\}_{2s},\{3,p\}\},$$

if the latter should exist. In fact, Γ is just the quotient of the Coxeter group $[2l,3,p]$

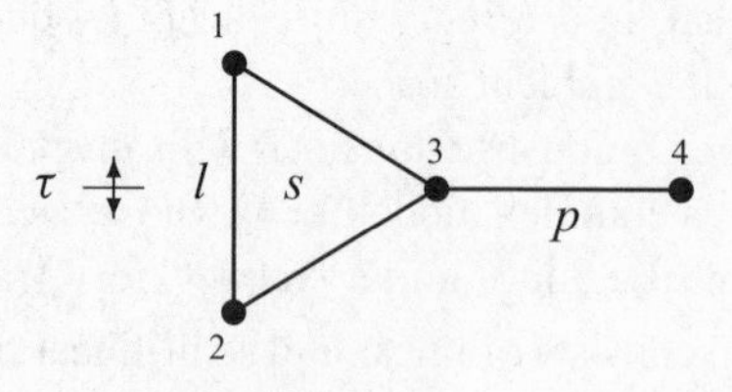

Figure 11B2

defined by the extra relation (11B3). Therefore, if $\mathcal{P}$ indeed exists, then Γ must be its group. However, it seems that little can be said about the existence of $\mathcal{P}$ unless further restrictions on l and s are imposed. Note that, for $l = 3$, we recover the original case.

On the other hand, let us assume that $\mathcal{P}$ exists. Then the subgroup $\langle \sigma_1, \sigma_2, \sigma_3 \rangle$ of W, which accounts for the type of the facets, must necessarily be isomorphic to the triangle group $[1\ 1\ 1^l]^s$; in fact, in $\Gamma = \Gamma(\mathcal{P})$ we must have

$$\langle \rho_0, \rho_1, \rho_2 \rangle = \Gamma(\{2l, 3\}_{2s}) = [1\ 1\ 1^l]^s \ltimes C_2.$$

It follows that $\mathcal{P}$ is finite only if the corresponding triangle group $[1\ 1\ 1^l]^s$ is also finite; that is, if and only if $l = 3$ (and $s < \infty$), or $s = 3$ (and $l < \infty$), or $(l, s) = (4, 4), (4, 5)$ or $(5, 4)$ (see Theorem 9D12 and Table 9A2). But now, if we impose these conditions on l and s, then the methods of Section 9E, become available, and we can produce the required hermitian form on $\mathbb{C}^4$. Theorem 9E6 shows that $W (= \Gamma^4(l, s; 3, 3; p))$ is a C-group, and so we can argue as before that Γ must also be a C-group. For all other properties, we can appeal to Theorems 9E4 and 9E5.

11B9 Theorem *Suppose that the pair (l, s) satisfies the following condition: $l \geqslant 2$, $s = 3$; or $l = 3$, $s \geqslant 2$; or $(l, s) = (4, 4)$, $(4, 5)$ or $(5, 4)$. Let $p = 3$, 4 or 5. Then the universal regular 4-polytope $\{\{2l, 3\}_{2s}, \{3, p\}\}$ exists. Its group is $W \ltimes C_2$, where W is the group abstractly defined by the diagram in Figure 11B2. Except for the finite examples of Theorem 11B5 (obtained for $l = 3$), each other finite instance is a polytope $\{\{2l, 3\}_6, \{3, 3\}\}$ (where $l \geqslant 2$), with group $[1\ 1\ 2^l]^3 \ltimes C_2$ of order $48l^3$.*

In Theorem 11B9, an interesting special case occurs if $(l, s) = (4, 3)$. Then the facet of the polytope $\{\{8, 3\}_6, \{3, p\}\}$ is isomorphic to Dyck's map $\{8, 3\}_6$ of genus 3 (see [131, §8.6; 157, p. 488; 158]). Now the polytope is finite only if $p = 3$, and then its group is $[1\ 1\ 2^4]^3 \ltimes C_2$, of order 3072. Here it is quite remarkable that $\{\{8, 3\}_6, \{3, 3\}\}$ is still finite, while $\{\{8, 3\}_6, \{3, 4\}\}$ is already infinite.

Finally, another direction of generalization employs groups whose diagrams have a longer tail (see Section 9F). This leads to analogues of ${}_p\mathcal{T}^4_{(s,0)}$ of higher ranks.

For example, if $n \geqslant 4$, then we can discover new regular n-polytopes $\mathcal{P} = \mathcal{P}_n$ of type $\{2l, 3^{n-2}\}$ by starting from the abstract group

$$W = [1\,1\,(n-2)^l]^s = \langle \sigma_1, \ldots, \sigma_n \rangle$$

shown in Figure 11B3. Now we can apply the operation

11B10 $$(\sigma_1, \ldots, \sigma_n; \tau) \mapsto (\tau, \sigma_2, \ldots, \sigma_n) =: (\rho_0, \rho_1, \ldots, \rho_{n-1})$$

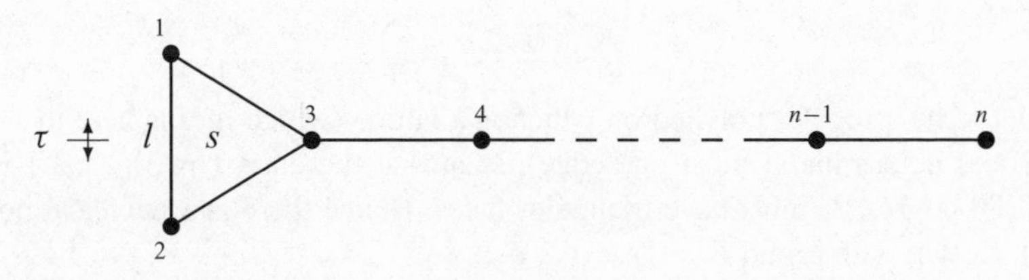

Figure 11B3

and obtain

$$\Gamma(\mathcal{P}) = \langle \rho_0, \ldots, \rho_{n-1} \rangle = W \rtimes C_2.$$

In particular, this construction obeys the recursion formula

$$\mathcal{P}_{n+1} = \{\mathcal{P}_n, \{3^{n-1}\}\} \qquad (n \geqslant 3),$$

beginning with $\mathcal{P}_3 = \{2l, 3\}_{2s}$.

From Sections 9A and 9F, if $n > 4$, then we have finiteness only if either $(l, s) = (3, 2)$ or $l \geqslant 2, s = 3$. But if $(l, s) = (3, 2)$ (and $n \geqslant 3$), then $W = S_{n+1}$ and τ acts on W like an inner automorphism; by Lemma 8A2, we must then have $\Gamma(\mathcal{P}) = S_{n+1} \times C_2$. On the other hand, if $l \geqslant 2$ and $s = 3$, then

$$\Gamma(\mathcal{P}) = [1\,1\,(n-2)^l]^3 \rtimes C_2,$$

of order $2l^{n-1}n!$; in particular, $\mathcal{P} = \{4, 3^{n-2}\}$ if $l = 2$.

If the mark 3 on the last branch of the diagram in Figure 11B3 is changed to 4, then in general we obtain an infinite group W, the only exception being $l = 3,\ s = 2$ (see again Section 9F). Now the operation (11B10) can be applied to produce polytopes of type $\{2l, 3^{n-3}, 4\}$, the finite case being

$$\{3^{n-2}, 4\} \diamond \{\,\}.$$

11C Polytopes with Facets $\{6, 3\}_{(s,s)}$

In this section, we shall investigate the finite regular 4-polytopes whose facets are toroidal polyhedra $\{6, 3\}_{(s,s)}$ with $s \geqslant 1$, and classify those universal polytopes (with any appropriate vertex-figure) which are finite (see [307]). For the most part, we apply the technique described in Section 11A; this will deal with the case $s \geqslant 2$. The case $s = 1$ requires a slightly different (but still fairly similar) treatment. In particular, we shall enumerate all the finite polytopes

$${}_p\mathcal{T}^4_{(s,s)} = \{\{6, 3\}_{(s,s)}, \{3, p\}\},$$

with $p = 3$, 4 or 5, and

$${}_6\mathcal{T}^4_{(s,s),\mathbf{t}} = \big\{\{6, 3\}_{(s,s)}, \{3, 6\}_{\mathbf{t}}\big\},$$

with $\mathbf{t} = (t^k, 0^{2-k})$, $t \geqslant 2$ if $k = 1$ or $t \geqslant 1$ if $k = 2$. Once again, the classification depends on one general result, and an easy observation.

The Case $s \geqslant 2$

First, let $\mathcal{K}$ be a regular polyhedron which is a lattice (which means here that two of its vertices determine at most one edge), such that the class $\langle\{6, 3\}_{(s,s)}, \mathcal{K}\rangle$ is non-empty. Of course, $\mathcal{K}$ must have triangular faces. Hence there is a universal polytope $\{\{6, 3\}_{(s,s)}, \mathcal{K}\}$, with group $\Gamma = \langle \rho_0, \ldots, \rho_3 \rangle$, say.

Writing, as usual, $\Gamma_0 := \langle \rho_1, \rho_2, \rho_3 \rangle$ ($\cong \Gamma(\mathcal{K})$), consider the subset

$$V := \{\gamma^{-1}\rho_0\gamma \mid \gamma \in \Gamma_0\},$$

which consists of involutions (conjugates of ρ_0), and the corresponding subgroup $N_0 := \langle V \rangle$. Then N_0 is the normal closure of ρ_0 in Γ, and $\Gamma = N_0 \cdot \Gamma_0$ (see Lemma 4E7). Now, if $\Gamma_{01} := \langle \rho_2, \rho_3 \rangle$ and $\beta \in \Gamma_{01}\gamma$, then $\beta^{-1}\rho_0\beta = \gamma^{-1}\rho_0\gamma$; it thus follows that there is a natural map $\kappa\colon \mathcal{K}_0 \to V$, with $\mathcal{K}_0$ the vertex-set of $\mathcal{K}$, which takes the vertex associated with $\Gamma_{01}\gamma$ (for $\gamma \in \Gamma_0$) onto the conjugate $\gamma^{-1}\rho_0\gamma$. Notice that κ commutes with the action of Γ_0; that is, if we identify (for a moment) the vertices in $\mathcal{K}$ with the cosets of Γ_{01}, then we have

$$((\Gamma_{01}\gamma)\beta)\kappa = \beta^{-1} \cdot ((\Gamma_{01}\gamma)\kappa) \cdot \beta,$$

for all $\gamma, \beta \in \Gamma_0$.

Suppose that $\lambda, \mu, \nu \in V$ correspond to the vertices of a face of $\mathcal{K}$; without loss of generality, we can take

$$\lambda = \rho_0, \quad \mu = \rho_1\rho_0\rho_1, \quad \nu = \rho_2\rho_1\rho_0\rho_1\rho_2.$$

Observe that conjugation by an element of $\langle \rho_1, \rho_2 \rangle$ merely permutes λ, μ, ν. Then we have

$$\lambda\mu = (\rho_0\rho_1)^2,$$

so that $(\lambda\mu)^3 = \varepsilon$ (and hence $\lambda\nu$ and $\mu\nu$ also have period 3), and

$$\lambda\mu\lambda\nu = \rho_0 \cdot \rho_1\rho_0\rho_1 \cdot \rho_0 \cdot \rho_2\rho_1\rho_0\rho_1\rho_2 = (\rho_0\rho_1\rho_0\rho_1\rho_2)^2,$$

so that (11A1) yields $(\lambda\mu\lambda\nu)^s = \varepsilon$.

Now suppose that $\mathcal{K}$ has v vertices. We can identify $\mathcal{K}_0$ with $\mathcal{N} := \{1, \ldots, v\}$, and then identify the edge-set $\mathcal{K}_1$ and face-set $\mathcal{K}_2$ of $\mathcal{K}$ with pairs $\{i, j\}$ and triples $\{i, j, k\}$ in $\mathcal{N}$. Define the group W by

11C1
$$W := \big\langle \sigma_1, \ldots, \sigma_v \,\big|\, \sigma_i^2 = \varepsilon \ (i \in \mathcal{K}_0),\ (\sigma_i\sigma_j)^3 = \varepsilon \ (\{i, j\} \in \mathcal{K}_1), \\ (\sigma_i\sigma_j\sigma_k\sigma_j)^s = \varepsilon \ (\{i, j, k\} \in \mathcal{K}_2)\big\rangle.$$

Once again, recall that the generators in the last relation can be taken in any order (see Section 9G). Then N_0 is clearly a quotient of W (under the homomorphism which sends σ_i to the conjugate of ρ_0 corresponding to the vertex i).

We can now twist W. We let $\Gamma(\mathcal{K}) = \langle \tau_0, \tau_1, \tau_2 \rangle$ act on $\{\sigma_i \mid i \in \mathcal{K}_0\}$ in the natural way as a group of automorphisms, to obtain a group

$$\bar{\Gamma} := W \rtimes \Gamma(\mathcal{K}),$$

under the operation

11C2
$$(\{\sigma_i \mid i \in \mathcal{K}_0\}; \tau_0, \tau_1, \tau_2) \mapsto (\sigma_1, \tau_0, \tau_1, \tau_2),$$

where we associate σ_1 with the initial vertex of $\mathcal{K}$. Since we recover Γ from N_0 by the exactly analogous operation, we deduce that Γ is a quotient of $\bar{\Gamma}$; in other words, since κ commutes with the action of Γ_0, the homomorphism from W onto N_0 extends to one from $\bar{\Gamma}$ onto Γ. But because Γ was the group of the universal polytope $\{\{6,3\}_{(s,s)}, \mathcal{K}\}$, and since, from the quotient criterion of Theorem 2E17 applied to the group $\Gamma(\mathcal{K})$ of the vertex-figure, $\bar{\Gamma}$ is the group of a polytope in $\langle\{6,3\}_{(s,s)}, \mathcal{K}\rangle$, we conclude that $\bar{\Gamma} = \Gamma$, so that Γ itself is obtained by means of the twisting operation of (11C2). Moreover, $N_0 \cong W$, and $\Gamma \cong N_0 \rtimes \Gamma_0$.

A very similar approach deals with the case when the vertex-figure $\mathcal{K}$ is not a lattice. Then there is a natural quotient $\mathcal{L}$ of $\mathcal{K}$, obtained by identifying edges and faces with the same vertices. Here, we must allow $\mathcal{L} = \{3, 2\}$ (a ditope) and $\mathcal{L} = \{3, 1\}$ (a poset of rank 3 with a single triangular face) as possible quotients. (We slightly abuse the Schläfli symbol notation, and allow 1 as an entry.) More precisely, we have the following

11C3 Lemma *Let $\mathcal{K}$ be a regular polyhedron of type $\{3, k\}$ with $k \geqslant 2$, but let $\mathcal{K} \neq \{3, 4\}/2$. Let $\Gamma(\mathcal{K}) = \langle \tau_0, \tau_1, \tau_2 \rangle$, and let $N := \langle (\tau_2\tau_1)^p \rangle$, where $p \geqslant 1$ is the smallest integer such that $\tau_0(\tau_2\tau_1)^p\tau_0 \in \langle \tau_1, \tau_2 \rangle$. Then $p > 1$, $p \mid k$, and N is a normal subgroup of $\Gamma(\mathcal{K})$ of order k/p contained in $\langle \tau_1, \tau_2 \rangle$. The quotient $\mathcal{L} := \mathcal{K}/N$ is a regular polyhedron of type $\{3, p\}$, whose vertex-set can naturally be identified with that of $\mathcal{K}$. Moreover, $\mathcal{L}$ is a lattice unless $\mathcal{L} = \{3, 2\}$, and $\mathcal{L} \cong \mathcal{K}$ if $\mathcal{K}$ is a lattice.*

Proof. Suppose that we have two vertices in the vertex-figure at the base vertex x (say) which coincide. We know that this must occur when $\mathcal{K}$ is not a lattice. Then the element $(\tau_2\tau_1)^q$, for some q, takes one of these vertices onto the other; hence $\tau_0(\tau_2\tau_1)^q\tau_0$ fixes x. Clearly, $p \neq 1$, $p \mid q$ and $p \mid k$. It follows that the vertex-figure has p distinct vertices through which it cycles k/p times. If $\mathcal{K}$ is a lattice, then $p = k$, and the lemma holds trivially. The same is true if $k = 2$, and so we may assume that $k \geqslant 3$.

The subgroup N is normal in $\Gamma(\mathcal{K})$. Indeed, $\tau := \tau_0(\tau_2\tau_1)^p\tau_0$ fixes the two vertices x and y (say) in the base edge, and so we must have $\tau = \tau_2{}^i(\tau_2\tau_1)^q$, where $i = 0, 1$ and again $p \mid q$. Then $\tau \in N$ if $i = 0$; hence N is invariant under conjugation by the generators in this case. We can rule out $i = 1$ as a possibility as follows. First observe that $i = 1$ forces $k = 2p$ (and $q = p$), because now τ is an involution; in the vertex-figure at y, it acts like a half-turn, and in the vertex-figure at x, it acts like a reflexion. Now, if z (say) is the third vertex of the base face (distinct from x and y), then we have $z\tau = z$, because $z\tau$ is the vertex opposite to z in the vertex-figure at y; on the other hand, being the image of z under the reflexion $\tau_2(\tau_2\tau_1)^p$, the vertex $z\tau$ is $p - 2$ steps away from z on the vertex-figure at x. This contradicts the definition of p unless $p = 2$ (and $k = 4$). But $p = 2$ would imply that $\mathcal{K} = \{3, 4\}/2$, which is excluded by assumption.

Finally, the vertices, edges and faces of $\mathcal{L}$ are the orbits of the vertices, edges and faces of $\mathcal{K}$ under N, respectively. Since N is normal, and contained in $\langle \tau_1, \tau_2 \rangle$, it fixes every vertex of $\mathcal{K}$, and so we can identify the vertices of $\mathcal{L}$ with those of $\mathcal{K}$. Furthermore, N maps every edge and every face of $\mathcal{K}$ onto an edge or a face with the same vertices. Indeed, if two edges of $\mathcal{K}$ share the same vertices, then they are equivalent under N; hence are identified. (This may not be true for the faces.) It follows that two vertices of $\mathcal{L}$ determine at most one edge in $\mathcal{L}$. Hence $\mathcal{L}$ is a lattice unless $\mathcal{L} = \{3, 2\}$. □

Observe that the case $\mathcal{L} = \{3, 2\}$ occurs for the toroidal polyhedron $\mathcal{K} = \{3, 6\}_{(1,1)}$ (and of course for $\{3, 2\}$ itself).

Notice also that the quotient $\mathcal{L}$ in Lemma 11C3 is "minimal", meaning that N is the largest normal subgroup of $\Gamma(\mathcal{K})$ for which polytopality and the vertex-set are preserved for the resulting quotient.

The projective polyhedron $\mathcal{K} = \{3, 4\}/2$ is special and was excluded in the lemma. Here we have $\tau_0(\tau_2\tau_1)^2\tau_0 = \tau_2(\tau_2\tau_1)^2$ (that is, the case $i = 1$ of the proof occurs here), and so τ_2 is a product of conjugates of $(\tau_2\tau_1)^2$. These conjugates generate an abelian normal subgroup N of order 4 which contains τ_2. Now the quotient $\mathcal{L} := \mathcal{K}/N$ is a poset of rank 3 with a single face; that is, $\mathcal{L} = \{3, 1\}$. Note that $\mathcal{K} = \{3, 4\}/2$ is the only polyhedron for which the associated quotient $\mathcal{L}$ is not itself a polyhedron.

At this stage, we have a natural quotient $\mathcal{L}$ for every polyhedron $\mathcal{K}$ with triangular faces. Now, if $\langle\{6, 3\}_{(s,s)}, \mathcal{L}\rangle \neq \emptyset$, then also $\langle\{6, 3\}_{(s,s)}, \mathcal{K}\rangle \neq \emptyset$, by the quotient criterion of Theorem 2E17 applied to the group of the facet $\{6, 3\}_{(s,s)}$. The quotient mapping $\mathcal{K} \searrow \mathcal{L}$ induces a quotient mapping $\{\{6, 3\}_{(s,s)}, \mathcal{K}\} \searrow \{\{6, 3\}_{(s,s)}, \mathcal{L}\}$ determined by the same subgroup N, which remains normal in the larger group. (When $\mathcal{L} = \{3, 1\}$, the quotient $\{\{6, 3\}_{(s,s)}, \mathcal{L}\}$ is not actually a polytope but instead a poset of rank 4 with a single facet $\{6, 3\}_{(s,s)}$.) Moreover, if Γ and Γ' are the groups of $\{\{6, 3\}_{(s,s)}, \mathcal{K}\}$ and $\{\{6, 3\}_{(s,s)}, \mathcal{L}\}$, respectively, then the corresponding homomorphism $\Gamma \to \Gamma'$ takes the product $N_0 \cdot \Gamma_0 = \Gamma$ into the semi-direct product $N_0' \cdot \Gamma_0' = \Gamma'$ (say), where $N_0 \cong N_0' \cong W$, $\Gamma_0 \cong \Gamma(\mathcal{K})$ and $\Gamma_0' \cong \Gamma(\mathcal{L})$. Since the homomorphism is one-to-one on N_0, we must have $N_0 \cap \Gamma_0 = \{\varepsilon\}$, and so $N_0 \cdot \Gamma_0$ is semi-direct as well.

Let us summarize this analysis.

11C4 Lemma *Let $\mathcal{K}$ be a regular polyhedron with triangular faces, and let $\mathcal{L}$ be the associated regular polyhedron (poset) for which the covering map $\mathcal{K} \searrow \mathcal{L}$ preserves vertices. If $s \geqslant 2$, and if $\mathcal{L}$ is such that $\langle\{6, 3\}_{(s,s)}, \mathcal{L}\rangle \neq \emptyset$, then also $\langle\{6, 3\}_{(s,s)}, \mathcal{K}\rangle \neq \emptyset$, and the group of the universal polytope $\{\{6, 3\}_{(s,s)}, \mathcal{K}\}$ is obtained from the group W of (11C1) by means of the twisting operation of (11C2).*

To complete the solution of the classification problem, we must now show that, if we start with an arbitrary regular polyhedron $\mathcal{K}$, then the group W of (11C1) does not collapse to one with fewer than v generators. (Notice that it certainly will if $s = 1$, which is why this case is excluded from the present discussion.) Of course, W depends on the quotient polyhedron $\mathcal{L}$ (if we need to pass to it), rather than on the original vertex-figure $\mathcal{K}$. Indeed, its definition involves $\mathcal{L}_0\,(= \mathcal{K}_0)$, $\mathcal{L}_1$ and $\mathcal{L}_2$, with an appropriate interpretation of $\mathcal{L}_2$ if $\mathcal{L} = \{3, 2\}$ (for $\mathcal{L} = \{3, 2\}$, we can pick any one of the two faces to represent $\mathcal{L}_2$). Once all the properties of W have been established, then we are basically done. The twisting operation in (11C2) will give us a regular polytope in $\langle\{6, 3\}_{(s,s)}, \mathcal{L}\rangle$ to start with, and the rest is taken care of by Lemma 11C4. Therefore it suffices to concentrate on W.

Suppose first that $\mathcal{L}$ (or, equivalently, $\mathcal{K}$) is not neighbourly (recall that every pair of vertices of a neighbourly polytope determines an edge). We now have exactly the situation of Theorem 9G4, with v instead of n. The group we constructed there genuinely has v generators. Since it is a quotient (in the obvious way) of the group W of (11C1), we conclude

11C5 Lemma *The group W defined by (11C1) has v generators. In particular, it does not collapse.*

The last step involves a deeper investigation of W. If the underlying polyhedron $\mathcal{L}$ is non-neighbourly, then the period of a product $\sigma_i\sigma_j$ with $\{i, j\} \notin \mathcal{L}_1$ is infinite, again as in (the proof of) Theorem 9G4; hence W is infinite. There remains the case that $\mathcal{L}$ is neighbourly, and, with a further appeal to Theorem 9G4, we know that, for finiteness, we must have $s = 2$ if $v \geqslant 4$, whereas any finite s will give a finite group if $v = 3$. We need a subsidiary result.

11C6 Lemma *The only neighbourly regular polyhedra with triangular faces which are lattices are the tetrahedron $\{3, 3\}$ and the hemi-icosahedron $\{3, 5\}_5$.*

Proof. Let $\mathcal{L}$ be such a polyhedron, of type $\{3, p\}$, say; then $\mathcal{L}$ has $p + 1$ vertices. We label its initial vertex ∞, and the vertices adjacent to ∞ (that is, those of the vertex-figure) $0, 1, \ldots, p - 1$ in cyclic order. We take $\{\infty, 0\}$ as the initial edge, and $\{\infty, 0, 1\}$ as initial triangular face; further, let τ_0, τ_1, τ_2 be the associated distinguished generators of its group $\Gamma(\mathcal{L})$.

Let r be the third vertex of the other face which contains $\{0, 1\}$. Then r lies in the vertex-figure of $\mathcal{L}$ at ∞ and is fixed by τ_1 (which fixes ∞ and interchanges 0 and 1), and so must be the vertex of the vertex-figure opposite to $\{0, 1\}$. It follows that p is odd, and that $r = \frac{1}{2}(p + 1)$. Now, by symmetry, the edge $\{0, r\}$ of $\{0, 1, r\}$ belongs to the face $\{0, r - 1, r\}$ as well; since the edges like $\{0, r\}$ are the only ones which can be free (in a planar drawing), we see that $\mathcal{L}$ closes up. There are just two possibilities. Either $p = 3$, and $\mathcal{L}$ is closed up by one face not containing ∞, or $p = 5$, and we have 5 extra faces, giving the hemi-icosahedron. □

Note that the groups of the polyhedra in the last lemma are doubly transitive on the vertices. It is shown in [249] that any doubly transitive automorphism group of a triangulated surface is one of the groups S_4, A_5 or $C_7 \times C_6$, and corresponds to the tetrahedron, the hemi-icosahedron or the (chiral) 7-vertex torus, respectively.

We now tie up the loose ends. If $\mathcal{L} = \{3, 3\}$ and $s = 2$, then $W = S_5$, and so W is finite (see again Theorem 9G4). We are thus reduced to the case $\mathcal{L} = \{3, 5\}_5$ (with $s = 2$), with the labelling of its vertices as in Lemma 11C6.

For $\{3, 5\}_5$, we directly construct a representation of W as a reflexion group G in $\mathbb{C}^6$ (using the notation of Chapter 9). We wish to mark the triangles of a diagram (of G) on 6 nodes and all 15 branches unmarked, so that those corresponding to faces in $\mathcal{L}$ get marks 2, while (some at least of) the non-faces are marked ∞. We can obtain the required marking by a specific description of the Gram matrix for the normal vectors of the generating reflexions. Its off-diagonal entries are given by

$$2\alpha_{jk} = \begin{cases} 1, & \text{if } j = 0, \ldots, 4 \text{ and } k = j + 1 \pmod 5, \\ -1, & \text{otherwise.} \end{cases}$$

Here, ∞ is permitted as a suffix. But now we see that a typical triangle such as $\{\infty, 0, 2\}$ is unmarked (its turn is 0); it thus yields an infinite subgroup $[1\ 1\ 1]^\infty$ of G, which is therefore always itself infinite. (In fact, all 10 triangles corresponding to non-faces are unmarked, although we did not initially require this.) It follows that W is infinite as well.

We have now come full circle. Whether finite or not, we can twist W as in (11C2) and obtain a regular polytope in $\langle\{6, 3\}_{(s,s)}, \mathcal{L}\rangle$ (its group is a semi-direct product, and so the intersection property is trivial here). The rest is taken care of by Lemma 11C4.

In summary, we conclude that we have proved the following theorem.

11C7 Theorem *Let $\mathcal{K}$ be a regular polyhedron with triangular faces, and let $s \geqslant 2$. Then the universal regular 4-polytope $\mathcal{P} := \{\{6, 3\}_{(s,s)}, \mathcal{K}\}$ exists. Its group is $\Gamma(\mathcal{P}) = W \rtimes \Gamma(\mathcal{K})$, where W is abstractly defined by (11C1). Moreover, if $\mathcal{L}$ is the regular polyhedron (poset) for which the covering map $\mathcal{K} \searrow \mathcal{L}$ preserves vertices, then $\mathcal{P}$ is finite only when*

(a) $\mathcal{L} = \{3, 1\}$ or $\{3, 2\}$ for any $s \geqslant 2$, with group $[1\ 1\ 1]^s \rtimes \Gamma(\mathcal{K})$;
(b) $\mathcal{L} = \{3, 3\}$ and $s = 2$, with group $S_5 \rtimes \Gamma(\mathcal{K})$.

Note that $\Gamma(\mathcal{K})$ acts on W in the natural way, whether or not $\mathcal{K}$ is a lattice. The projective polytope $\mathcal{K} = \{3, 4\}/2$ is the only instance of a regular polyhedron which fits into Theorem 11C7(a) with $\mathcal{L} = \{3, 1\}$.

We should recall that this chapter is intended to treat locally toroidal regular polytopes. As far as these are concerned, we can extract the following results from Theorem 11C7.

11C8 Corollary *Let $s \geqslant 2$. The only finite universal locally toroidal regular polytopes $\mathcal{P} = \{\{6, 3\}_{(s,s)}, \mathcal{K}\}$ are given by the following:*

(a) $\mathcal{K} = \{3, 6\}_{(1,1)}$ for any s, with group $\Gamma(\mathcal{P}) = [1\ 1\ 1]^s \rtimes (S_3 \times S_3)$;
(b) $\mathcal{K} = \{3, 3\}$ for $s = 2$, with group $\Gamma(\mathcal{P}) = S_5 \times S_4$;
(c) $\mathcal{K} = \{3, 6\}_{(2,0)}$ for $s = 2$, with group $\Gamma(\mathcal{P}) = S_5 \times (S_4 \times C_2)$.

Notice that the products in the second and third part of Corollary 11C8 are direct. The second part covers the polytopes of type ${}_3\mathcal{T}^4_{(s,s)}$ with tetrahedral vertex-figures, whose group is ${}_3\Gamma^4_{(s,s)}$. Here we can embed $\Gamma(\mathcal{P}) = W \rtimes \Gamma(\mathcal{K})$ in S_9 if $s = 2$, taking $\sigma_i = (i\ 5)$ for $i = 1, \ldots, 4$, and $\tau_i = (i\ i+1)(5+i\ 5+i+1)$ for $i = 1, 2, 3$; thus $\Gamma(\mathcal{P}) = S_5 \times S_4$. From Lemma 11A11, we then also have the direct product in the third part. We shall look at the case $\mathcal{K} = \{3, 6\}_{(1,1)}$ from a different point of view later.

For the dual $({}_3\mathcal{T}^4_{(2,2)})^* = \{\{3, 3\}, \{3, 6\}_{(2,2)}\}$ of the polytope in Corollary 11C8(b), a faithful 7-dimensional realization can be obtained from a 9-dimensional real representation of its group $S_5 \times S_4$, consisting of the linear mappings of $\mathbb{E}^9$ which permute the coordinates of a vector $(\xi_1, \ldots, \xi_9)$ according to a permutation in $S_5 \times S_4$ (embedded in S_9). The group acts irreducibly on the 7-dimensional subspace of $\mathbb{E}^9$ defined by

$$\xi_1 + \cdots + \xi_5 = 0, \quad \xi_6 + \cdots + \xi_9 = 0,$$

and the Wythoff space is spanned by the vectors

$$a := (1, 1, 1, -4, 1, 0, \ldots, 0), \quad b := (0, \ldots, 0, 1, 1, 1, -3).$$

(The generators correspond to $\tau_3, \tau_2, \tau_1, \sigma_1$, in this order, and the Wythoff space is the intersection of the mirrors of the last three generators.) With initial vertex $a + b$, Wythoff's construction leads to a realization of $({}_3\mathcal{T}^4_{(2,2)})^*$ whose vertex set consists of the 20 points $(\xi_1, \ldots, \xi_9)$ with

11C9
$$\begin{cases} \xi_i = -4, & \text{for one } i \text{ with } i \leqslant 5, \\ \xi_i = -3, & \text{for one } i \text{ with } i \geqslant 6, \\ \xi_i = 1, & \text{otherwise.} \end{cases}$$

This realization is the blend of the two degenerate realizations of $({}_3\mathcal{T}^4_{(2,2)})^*$ constructed by taking a and b as initial vertices.

The same arguments we have used here will work in a more general context. Let $\mathcal{K}$ be a regular $(n-1)$-polytope with triangular 2-faces. We slightly abuse our usual notation, and write $\langle \{6, 3\}_{(s,s)}, \mathcal{K} \rangle$ for the class of regular n-polytopes with 3-faces isomorphic to $\{6, 3\}_{(s,s)}$ and vertex-figures isomorphic to $\mathcal{K}$. (Strictly speaking, we should set up a recursive definition, and prove first that, if $n > 4$ and $\mathcal{J}$ is the facet of $\mathcal{K}$, then $\langle \{6, 3\}_{(s,s)}, \mathcal{J} \rangle$ contains a universal member $\mathcal{Q} := \{\{6, 3\}_{(s,s)}, \mathcal{J}\}$, say. We then discuss the class $\langle \mathcal{Q}, \mathcal{K} \rangle$.)

The same considerations lead us to the existence of an abstract diagram (for W) on v nodes, with v the number of vertices of $\mathcal{K}$ (see Section 9G). Branches corresponding to edges of $\mathcal{K}$ are unmarked, while those associated with non-edges are marked ∞ (note that two vertices of $\mathcal{K}$ may be joined by more than one edge). Similarly, triangular circuits corresponding to 2-faces of $\mathcal{K}$ are marked s, while all others are unmarked.

Theorem 9G4 leads at once to the following; we shall not give the straightforward proof.

11C10 Theorem *Let $s \geqslant 2$, and for $n \geqslant 5$ let $\mathcal{K}$ be a finite regular $(n-1)$-polytope with triangular 2-faces. Then the class $\langle \{6, 3\}_{(s,s)}, \mathcal{K} \rangle$ is non-empty. Its universal member $\{\{6, 3\}_{(s,s)}, \mathcal{K}\}$ is infinite if $\mathcal{K}$ has at least four vertices and*

(a) $s \geqslant 3$,
(b) $s = 2$ and $\mathcal{K}$ is non-neighbourly.

It is probable that $\{\{6, 3\}_{(s,s)}, \mathcal{K}\}$ is infinite even if $s = 2$ and $\mathcal{K}$ is neighbourly, unless every triangular circuit arises from a 2-face of $\mathcal{K}$, so that the 2-skeleton of $\mathcal{K}$ must collapse onto that of some simplex by natural identification of its edges and faces with the same vertices. However, to prove this would require a stronger version of Theorem 9G4 than we have been able to establish. The condition does give finite universal polytopes with $\mathcal{K} = \{3^{n-2}\}$ (the $(n-1)$-simplex) or $\{3^{n-3}, 4\}/2$ (the hemi-$(n-1)$-cross-polytope).

If $\mathcal{K}$ has only three vertices, then $\langle \{6, 3\}_{(s,s)}, \mathcal{K} \rangle$ is finite for every $s \geqslant 2$, and its group is $[1\ 1\ 1]^s \rtimes \Gamma(\mathcal{K})$. We shall see an example of a regular 4-polytope with only three vertices in Section 11E, namely, $\{\{3, 6\}_{(1,1)}, \{6, 3\}_{(1,1)}\}$.

Before we move on, note that the dual of the polytope ${}_3\mathcal{T}_{\mathbf{s}}^4$ (with $\mathbf{s} \neq (1, 1), (2, 0)$) is actually a 3-dimensional simplicial complex whose vertex links are isomorphic to the toroidal polyhedron $\{3, 6\}_{\mathbf{s}}$. A general result due to Altshuler [2, Theorem 2] says that, given a finite set of abstract polyhedra which are simplicial 2-complexes, there always exists a finite 3-dimensional simplicial complex whose vertex links belong to the set (with each polyhedron in the set actually occurring as a link). This complex will be an abstract 4-polytope, but will generally not be regular.

The Case $s = 1$

The regular polyhedron $\{6, 3\}_{(1,1)}$ (that is, the case $s = 1$) was excluded from the previous discussion, because of the lack of appropriate hermitian forms. We now investigate polytopes whose facets are of this kind. Since the polyhedron is flat, any such polytope will necessarily also be flat, and coincide with the universal member in its class (see Proposition 4E5). Recall from (11A7) that $[6, 3]_{(1,1)} \cong S_3 \times S_3$, where the first factor S_3 is illustrated in Figure 11A2. Here, however, we prefer to consider the group in its alternative form, as a semi-direct product

$$[6, 3]_{(1,1)} \cong S_3 \rtimes S_3$$

(see Section 11A).

Let $\mathcal{K}$ be a regular polyhedron of type $\{3, p\}$ with $p \geqslant 3$ (we are not requiring that $\mathcal{K}$ be a lattice). We shall find a criterion for the existence of the universal regular 4-polytope

$$\mathcal{P} := \big\{\{6, 3\}_{(1,1)}, \mathcal{K}\big\}.$$

First, let us assume that $\mathcal{P}$ exists. We show that the edge-graph of $\mathcal{K}$ must be vertex 3-colourable. To do this, we exploit ideas similar to those described at the beginning of the section.

Let $\Gamma := \Gamma(\mathcal{P}) = \langle \rho_0, \ldots, \rho_3 \rangle$, where $\rho_0, \ldots, \rho_3$ are the distinguished generators of $\Gamma(\mathcal{P})$. Then $\Gamma_0 := \langle \rho_1, \rho_2, \rho_3 \rangle$ is the group of the vertex-figure $\mathcal{K}$, and ρ_1, ρ_2, ρ_3 are the distinguished generators belonging to a base flag $\{F_0, F_1, F_2\}$ (say) of $\mathcal{K}$. Let N_0 denote the normal closure of ρ_0 in $\Gamma(\mathcal{P})$. Then N_0 is generated by the conjugates $\gamma^{-1}\rho_0\gamma$ with $\gamma \in \Gamma_0$, and we have

11C11 $$\Gamma(\mathcal{P}) = N_0 \cdot \Gamma_0$$

(see Lemma 4E7). In effect, N_0 replaces the abstract group W which we employed previously (indeed, we will use the same notation for the generators of the two groups); at this stage, the treatment is just like that for facets $\{6, 3\}_{(s,s)}$ with $s \geqslant 2$. In particular, we shall prove that

$$N_0 = \langle \sigma_1, \sigma_2, \sigma_3 \rangle = S_3,$$

where

$$\sigma_1 := \rho_0, \quad \sigma_2 := \rho_1\rho_0\rho_1, \quad \sigma_3 := \rho_2\rho_1\rho_0\rho_1\rho_2.$$

These σ_i are just the generators of the first factor S_3 in the earlier semi-direct product decomposition of $[6, 3]_{(1,1)}$, and so we have $\sigma_i \sigma_j \sigma_k \sigma_j = \varepsilon$ if i, j, k are distinct.

Let $V(\mathcal{K})$ denote the vertex set of $\mathcal{K}$. As before, we label the vertices of $\mathcal{K}$ by conjugates $\gamma^{-1}\rho_0\gamma$ with $\gamma \in \Gamma_0$; in other words, we define the vertex labelling $\kappa: V(\mathcal{K}) \to N_0$ by

11C12 $$(F_0\gamma)\kappa := \gamma^{-1}\rho_0\gamma$$

for $\gamma \in \Gamma_0$. We show that this gives a vertex 3-colouring. As before, κ commutes with the action of Γ_0; that is, if $\beta \in \Gamma_0$, then

$$((F_0\gamma)\beta)\kappa = \beta^{-1} \cdot ((F_0\gamma)\kappa) \cdot \beta$$

for all $\gamma \in \Gamma_0$.

Now let $\{F, G, H\}$ and $\{G, H, K\}$ be the vertex sets of two adjacent 2-faces of $\mathcal{K}$. Then there exists an involution $\beta \in \Gamma_0$ such that $G\beta = G$, $H\beta = H$, and $F\beta = K$. By the properties of κ,

$$\beta^{-1} \cdot (G\kappa) \cdot \beta = G\kappa, \quad \beta^{-1} \cdot (H\kappa) \cdot \beta = H\kappa, \quad \beta^{-1} \cdot (F\kappa) \cdot \beta = K\kappa;$$

hence, under conjugation by β, the labels of $\{F, G, H\}$ are changed to those of $\{G, H, K\}$, and two of these labels remain invariant.

If $\{F, G, H\}$ is the vertex-set of the base face F_2 of $\mathcal{K}$, chosen so that $F = F_0$ and $F_1 = \{F, G\}$, then the labels are given by

$$F\kappa = \rho_0 = \sigma_1, \quad G\kappa = \rho_1\rho_0\rho_1 = \sigma_2, \quad H\kappa = \rho_2\rho_1\rho_0\rho_1\rho_2 = \sigma_3.$$

But in the group S_3 which these elements generate, any generator σ_i is determined by the two others. Hence, if two generators remain fixed under conjugation with β, then so does the third. Thus the set of labels on any 2-face adjacent to $\{F, G, H\}$ is again $\{\sigma_1, \sigma_2, \sigma_3\}$.

Now we can appeal to the connectedness of $\mathcal{K}$. In particular, the latter remains true for all 2-faces of $\mathcal{K}$. This proves our claim that the edge-graph of $\mathcal{K}$ is vertex 3-colourable. Moreover, we can show that $\Gamma(\mathcal{P})$ is a semi-direct product of S_3 by $\Gamma(\mathcal{K})$. Indeed, since the labels $\gamma^{-1}\rho_0\gamma$ of the vertices of $\mathcal{K}$ are just the elements $\sigma_1, \sigma_2, \sigma_3$, we must have

$$N_0 = \langle \gamma^{-1}\rho_0\gamma \mid \gamma \in \Gamma_0 \rangle = \langle \sigma_1, \sigma_2, \sigma_3 \rangle = S_3.$$

But since $\sigma_i \notin \Gamma_0$ for each i, the product in (11C11) must be semi-direct. This proves our statement about $\Gamma(\mathcal{P})$.

By inspecting the edge-graph of the vertex-figure, we can now conclude that the universal polytope $\mathcal{P} = \{\{6, 3\}_{(1,1)}, \mathcal{K}\}$ cannot exist if $\mathcal{K} = \{3, 3\}, \{3, 5\}$, or $\{3, 6\}_{(s,0)}$ with $s \not\equiv 0 \pmod 3$. On the other hand, we shall see that $\mathcal{P}$ does exist if $\mathcal{K} = \{3, 4\}$, $\{3, 6\}_{(s,0)}$ with $s \equiv 0 \pmod 3$, or $\{3, 6\}_{(s,s)}$ with $s \geqslant 1$.

Indeed, more generally, let $\mathcal{K}$ be a regular polyhedron of type $\{3, p\}$ whose edge-graph is vertex 3-colourable. We shall prove that this condition on $\mathcal{K}$ is also sufficient for the existence of $\mathcal{P}$.

In order to construct $\mathcal{P}$, begin by 3-colouring the vertices of $\mathcal{K}$. Consider the abstract group $W = W(\mathcal{D})$ defined by the following diagram $\mathcal{D}$. The nodes of $\mathcal{D}$ are the vertices of $\mathcal{K}$; any two nodes are connected by an unlabelled branch if and only if they have different colours; finally, each triple of distinctly coloured vertices of $\mathcal{K}$ is represented by a triangle of $\mathcal{D}$ and marked by 1. (In other words, in general the underlying simplicial 2-complex for $\mathcal{D}$ is obtained from $\mathcal{K}$ by adding further edges and 2-faces.) Then similar considerations to those above show that

$$W = \langle \sigma_F \mid F \in V(\mathcal{K}) \rangle = S_3;$$

the isomorphism can be established by "colouring" the vertices of $\mathcal{K}$ by the transpositions (1 2), (2 3) and (1 3). By construction, $\Gamma(\mathcal{K})$ acts on W as a group of group automorphisms permuting the generators σ_F.

Finally, let $\Gamma(\mathcal{K}) = \langle \tau_0, \tau_1, \tau_2 \rangle$, where the distinguished generators τ_0, τ_1, τ_2 are defined with respect to a base flag $\{F_0, F_1, F_2\}$ of $\mathcal{K}$. Now proceed as in (11C2), and construct a new group $\Gamma = \langle \rho_0, \ldots, \rho_3 \rangle$ by the operation

$$(\{\sigma_F \mid F \in V(\mathcal{K})\}; \tau_0, \tau_1, \tau_2) \mapsto (\sigma_{F_0}, \tau_0, \tau_1, \tau_2) =: (\rho_0, \rho_1, \rho_2, \rho_3).$$

Then, again as before, we can prove that Γ is a C-group isomorphic to $W \rtimes \Gamma(\mathcal{K}) = S_3 \rtimes \Gamma(\mathcal{K})$. In particular, the corresponding polytope must have facets $\{6,3\}_{(1,1)}$ and vertex-figures $\mathcal{K}$, because now $\langle \rho_0, \rho_1, \rho_2 \rangle = S_3 \rtimes S_3$ and $\langle \rho_1, \rho_2, \rho_3 \rangle = \Gamma(\mathcal{K})$. Hence this must be the desired polytope $\mathcal{P}$, because $\mathcal{P}$ is the only polytope with facets and vertex-figures of this kind.

We can summarize this discussion as follows.

11C13 Theorem *Let $\mathcal{K}$ be a regular polyhedron of type $\{3, p\}$, with $p \geqslant 3$. Then the universal regular 4-polytope $\mathcal{P} := \{\{6,3\}_{(1,1)}, \mathcal{K}\}$ exists if and only if the edge-graph of $\mathcal{K}$ is vertex 3-colourable. In this case, $\mathcal{P}$ is combinatorially flat, and $\Gamma(\mathcal{P}) = S_3 \rtimes \Gamma(\mathcal{K})$. In particular, $\mathcal{P}$ is finite if and only if $\mathcal{K}$ is finite.*

11C14 Corollary *Let $\mathcal{P} := \{\{6,3\}_{(1,1)}, \mathcal{K}\}$. Then*

(a) $\mathcal{P}$ exists if $\mathcal{K} = \{3,4\}$, $\{3,6\}_{(s,0)}$ with $s \equiv 0 \pmod 3$, or $\{3,6\}_{(s,s)}$ with $s \geqslant 1$;
(b) $\mathcal{P}$ does not exist if $\mathcal{K} = \{3,3\}$, $\{3,5\}$, or $\{3,6\}_{(s,0)}$ with $s \not\equiv 0 \bmod 3$.

The proof of Theorem 11C13 and its corollary is inspired by the results of Sections 4E and 4F about polytopes with the flat amalgamation property (FAP). Indeed, it can be shown that, among the regular polyhedra $\mathcal{K}$ listed in Corollary 11C14, precisely those of part (a) have the FAP with respect to their facets. Since $\{6,3\}_{(1,1)}$ also has the FAP with respect to its vertex-figure, we can appeal to Theorem 4F9 and obtain a regular 4-polytope with facets $\{6,3\}_{(1,1)}$ and vertex-figures $\mathcal{K}$, which then must be $\mathcal{P}$ itself, because $\{6,3\}_{(1,1)}$ is flat. These arguments apply more generally to all regular polyhedra $\mathcal{K}$ of type $\{3, p\}$ whose edge-graph is vertex 3-colourable. Indeed, it can be proved that a regular polyhedron $\mathcal{K}$ of type $\{3, p\}$ has the FAP with respect to its facets if and only if it is vertex 3-colourable.

11D The Polytopes ${}_6\mathcal{T}^4_{(s,0),(t,0)} := \{\{6,3\}_{(s,0)}, \{3,6\}_{(t,0)}\}$

In this section, we complete the enumeration of the finite regular 4-polytopes ${}_6\mathcal{T}^4_{\mathbf{s},\mathbf{t}}$. In the previous section, we investigated those polytopes which have facets $\{6,3\}_{(s,s)}$ or vertex-figures $\{3,6\}_{(t,t)}$. There thus remains to consider the polytopes ${}_6\mathcal{T}^4_{(s,0),(t,0)}$, whose facets are $\{6,3\}_{(s,0)}$ and whose vertex-figures are $\{3,6\}_{(t,0)}$. In fact, as we shall see, our techniques apply to a somewhat wider class of polytopes, namely, those of type $\{\{2p,3\}_{2s}, \{3,2r\}_{2t}\}$.

The Polytopes ${}_6\mathcal{T}^4_{(s,0),(t,0)}$

Our preliminary considerations run much along the lines of those for the previous two sections. Let $\mathcal{P}$ be a (not necessarily universal) regular polytope with facets $\{6,3\}_{(s,0)}$ and vertex-figures $\{3,6\}_{(t,0)}$ for some $s, t \geqslant 2$, and let $\Gamma = \langle \rho_0, \ldots, \rho_3 \rangle$ be its group. Thus Γ has a presentation which includes the relations

11D1 $$(\rho_0\rho_1\rho_2)^{2s} = (\rho_1\rho_2\rho_3)^{2t} = \varepsilon,$$

adjoined to the standard presentation of the Coxeter group [6, 3, 6].

Consider the mixing operation

11D2 $$(\rho_0, \ldots, \rho_3) \mapsto (\rho_0\rho_1\rho_0, \rho_1, \rho_2, \rho_3\rho_2\rho_3) =: (\sigma_1, \ldots, \sigma_4).$$

As in Section 11B, we see that $\langle \sigma_1, \sigma_2, \sigma_3 \rangle \cong [1\ 1\ 1]^s$; moreover, since $\langle \sigma_1, \sigma_2, \sigma_4 \rangle$ is the conjugate of $\langle \sigma_1, \sigma_2, \sigma_3 \rangle$ by ρ_3, we have $\langle \sigma_1, \sigma_2, \sigma_4 \rangle \cong [1\ 1\ 1]^s$ also. In a similar way, $\langle \sigma_1, \sigma_3, \sigma_4 \rangle \cong [1\ 1\ 1]^t \cong \langle \sigma_2, \sigma_3, \sigma_4 \rangle$. Thus, $W = \langle \sigma_1, \ldots, \sigma_4 \rangle$ has a presentation which includes the relations

11D3 $$\begin{cases} \sigma_i^2 = (\sigma_i\sigma_j)^3 = \varepsilon \quad (1 \leqslant i, j \leqslant 4;\ i \neq j), \\ (\sigma_1\sigma_2\sigma_3\sigma_2)^s = (\sigma_1\sigma_2\sigma_4\sigma_2)^s = (\sigma_1\sigma_3\sigma_4\sigma_3)^t = (\sigma_2\sigma_3\sigma_4\sigma_3)^t = \varepsilon. \end{cases}$$

Once again, recall that the generators in the last four relations can be taken in any order (see Section 9E).

Again as in Section 11B, we can recover the original group Γ by means of the two involutory twisting operations (group automorphisms) τ_1 $(= \rho_0)$ and τ_2 $(= \rho_3)$, which act on W by

$$\tau_1\sigma_1\tau_1 = \sigma_2, \qquad \tau_1\sigma_j\tau_1 = \sigma_j \quad (j = 3, 4),$$

and

$$\tau_2\sigma_3\tau_2 = \sigma_4, \qquad \tau_2\sigma_j\tau_2 = \sigma_j \quad (j = 1, 2).$$

Thus Γ is given by the operation

11D4 $$(\sigma_1, \ldots, \sigma_4; \tau_1, \tau_2) \mapsto (\tau_1, \sigma_2, \sigma_3, \tau_2) =: (\rho_0, \ldots, \rho_3).$$

Analogous arguments to those of Section 11C may now be pursued. In order for the original polytope $\mathcal{P}$ to be universal, W must satisfy only the relations of (11D3). We thus consider the abstract group $W = \langle \sigma_1, \ldots, \sigma_4 \rangle$ corresponding to the tetrahedral

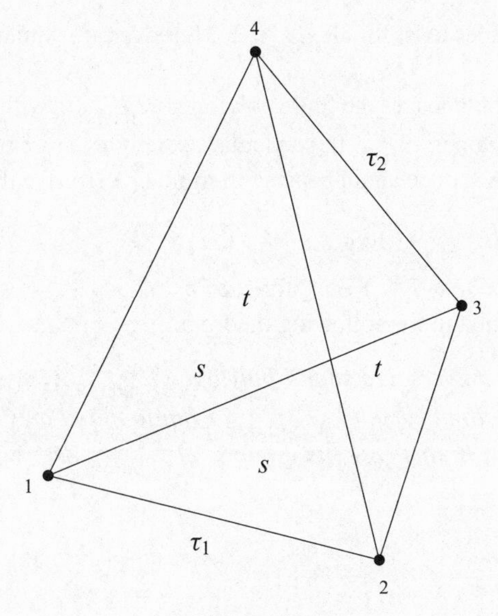

Figure 11D1

diagram of Figure 11D1, with triangles $\{1, 2, 3\}$ and $\{1, 2, 4\}$ marked s and $\{1, 3, 4\}$ and $\{2, 3, 4\}$ marked t (so we have replaced the q and s of the discussion of the diagrams $\mathcal{T}_4(s, s, q, q)$ in Section 9E by s and t, respectively). Then W admits the two automorphisms which are indicated in the diagram, and correspond to the permutations (1 2) and (3 4) of the nodes, respectively. The operation of (11D4) leads back to the group $\Gamma = \langle \rho_0, \ldots, \rho_3 \rangle$, which is thus a semi-direct product of W by $C_2 \times C_2$.

We can now appeal directly to Theorem 9E14 (or Theorem 9E17 with $p = r = 3$). In particular, W is a C-group (and so has the intersection property), and its 3-generator subgroups are isomorphic to $[1\,1\,1]^s$ or $[1\,1\,1]^t$, as appropriate. Moreover, W is finite if and only if $(s, t) = (2, k)$ or $(k, 2)$ with $k = 2, 3, 4$. In this case we have $W = [1\,1\,2]^k = [1\,1\,2^3]^k$, where $[1\,1\,2]^k$ denotes the unitary reflexion group defined by Figure 9A2, whose order is $5!$ if $k = 2$, $3^3 \cdot 4!$ if $k = 3$, or $64 \cdot 5!$ if $k = 4$. The hermitian form associated with the diagram is positive semi-definite if and only if $s = t = 3$ (see the proof of Theorem 9E14 for the Gram matrix); in this case, the corresponding geometric group G can also be regarded as an infinite (discrete) unitary reflexion group in 3 dimensions (see Lemma 9B19).

Now it is immediate that Γ is a C-group. Indeed, we have $\Gamma = W \rtimes (C_2 \times C_2)$, and so its intersection property is inherited directly from that of W. For example, we have

$$\begin{aligned}\langle \rho_0, \rho_1, \rho_2 \rangle \cap \langle \rho_1, \rho_2, \rho_3 \rangle &= (\langle \sigma_1, \sigma_2, \sigma_3 \rangle \rtimes \langle \tau_1 \rangle) \cap (\langle \sigma_2, \sigma_3, \sigma_4 \rangle \rtimes \langle \tau_2 \rangle)\\ &= \langle \sigma_2, \sigma_3 \rangle\\ &= \langle \rho_1, \rho_2 \rangle.\end{aligned}$$

But since $\Gamma = {}_6\Gamma^4_{(s,0),(t,0)}$ is a C-group, its corresponding polytope must be ${}_6\mathcal{T}^4_{(s,0),(t,0)}$.

Therefore ${}_6\mathcal{T}^4_{(s,0),(t,0)}$ does exist for all $s, t \geqslant 2$. Moreover, it is finite if and only if W is finite.

Up to duality, this only leaves the finite polytopes ${}_6\mathcal{T}^4_{(2,0),(t,0)}$ with $t = 2, 3$ or 4. They are related to the polytopes ${}_3\mathcal{T}^4_{(t,0)}$. In particular, exactly as in Lemma 11A11, we can appeal to Theorem 7A8, once again bearing in mind (7A10). We thus see that

$$ {}_6\mathcal{T}^4_{(2,0),(t,0)} = {}_3\mathcal{T}^4_{(t,0)} \lozenge \{\}, $$

so that $\Gamma({}_6\mathcal{T}^4_{(2,0),(t,0)}) = \Gamma({}_3\mathcal{T}^4_{(t,0)}) \times C_2$ for each $t \geqslant 2$.

Summarizing, we obtain the following theorem.

11D5 Theorem *The universal regular 4-polytope* ${}_6\mathcal{T}^4_{(s,0),(t,0)} := \{\{6,3\}_{(s,0)}, \{3,6\}_{(t,0)}\}$ *exists for all* $s, t \geqslant 2$. *In particular,* ${}_6\mathcal{T}^4_{(s,0),(t,0)}$ *is finite if and only if* $(s,t) = (2,k)$ *or* $(k,2)$ *with* $k = 2, 3, 4$. *In this case, its group is* ${}_6\Gamma^4_{(s,0),(t,0)} \cong [1\,1\,2]^k \rtimes (C_2 \times C_2)$, *of order*

(a) 480, *if* $k = 2$,
(b) $108 \cdot 4!$, *if* $k = 3$,
(c) $256 \cdot 5!$, *if* $k = 4$.

Moreover, ${}_6\Gamma^4_{(s,0),(2,0)} = {}_3\Gamma^4_{(s,0)} \times C_2$ *for each* $s \geqslant 2$.

Note that we also have

11D6 $$ {}_6\Gamma^4_{(2,0),(2,0)} = S_5 \times C_2 \times C_2, $$

which can be derived by applying Lemma 11A11 twice, first to $\{3,3,3\}$ and then to ${}_3\mathcal{T}^4_{(2,0)}$.

As in the previous section, we have a finiteness criterion in terms of the corresponding hermitian form.

11D7 Corollary *Let* $s, t \geqslant 2$. *Then the polytope* ${}_6\mathcal{T}^4_{(s,0),(t,0)}$ *is finite if and only if the hermitian form associated with the diagram in Figure 11D1 is positive definite. In this case its group* ${}_6\Gamma^4_{(s,0),(t,0)}$ *is a semi-direct product of a finite reflexion group in unitary 4-space by* $C_2 \times C_2$.

In Table 11D1 we list all the finite polytopes ${}_6\mathcal{T}^4_{\mathbf{s},\mathbf{t}}$, along with the numbers v of their vertices and f of their facets, and the orders g of their groups. Again, we mention

Table 11D1. *The Finite Polytopes* ${}_6\mathcal{T}^4_{\mathbf{s},\mathbf{t}}$

s	**t**	v	f	g	Group
(2, 0)	(2, 0)	10	10	480	$S_5 \times C_2 \times C_2$
(3, 0)	(2, 0)	54	24	2592	$[1\,1\,2]^3 \rtimes (C_2 \times C_2)$
(4, 0)	(2, 0)	640	160	30720	$[1\,1\,2]^4 \rtimes (C_2 \times C_2)$
(1, 1)	$(s, 0)$, $s \equiv 0 \pmod 3$	6	$2s^2$	$72s^2$	$S_3 \rtimes [3,6]_{(s,0)}$
(1, 1)	(s, s), $s \geqslant 1$	6	$6s^2$	$216s^2$	$S_3 \rtimes [3,6]_{(s,s)}$
(2, 2)	(2, 0)	120	40	5760	$S_5 \times S_4 \times C_2$

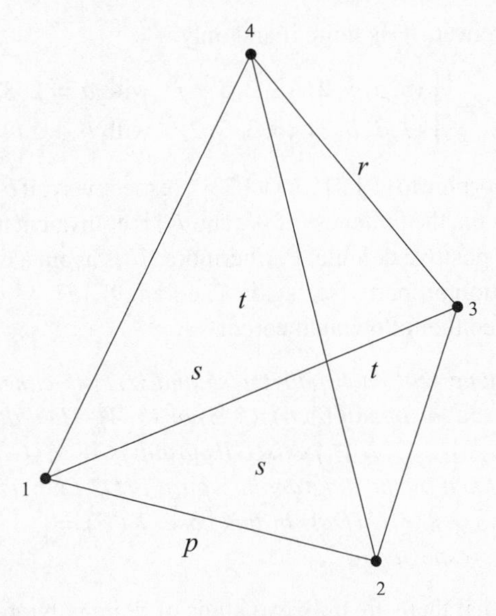

Figure 11D2

only one from a pair of dual polytopes. If $\mathbf{s} = (s, 0)$ and $\mathbf{t} = (t, 0)$, then up to duality these are just the finite polytopes of Theorem 11D5. If $\mathbf{s} = (s, s)$ or $\mathbf{t} = (t, t)$, then, up to duality, we have the finite polytopes of Corollaries 11C8 and 11C14. (Recall that $[3, 6]_{\mathbf{s}}$ denotes the group of the polyhedron $\{3, 6\}_{\mathbf{s}}$.)

The Polytopes $\{\{2p, 3\}_{2s}, \{3, 2r\}_{2t}\}$

The construction of ${}_6\mathcal{T}^4_{(s,0),(t,0)}$ can be generalized to obtain the regular 4-polytopes $\mathcal{P} := \{\{2p, 3\}_{2s}, \{3, 2r\}_{2t}\}$. Once again, we apply the mixing operation of (11D2) to the group $\Gamma = \langle \rho_0, \ldots, \rho_3 \rangle$ of $\mathcal{P}$. This group Γ has a presentation which consists of the standard relations for the Coxeter group $[2p, 3, 2r]$, along with the two extra relations (11D1) which (as with the cases $p = r = 3$) define the facets and vertex-figures by specifying the lengths of their Petrie polygons (see Section 7B). Now we have to consider the abstract group W corresponding to the diagram of Figure 11D2, in which the two branches $\{1, 2\}$ and $\{3, 4\}$ have been marked p and r, respectively. Then the same operation (11D4) applied to $W = \langle \sigma_1, \ldots, \sigma_4 \rangle$ recovers the group Γ, which is thus again a semi-direct product of W by $C_2 \times C_2$.

Hence, if Γ is a C-group, then the polytope $\mathcal{P}$ exists and $\Gamma = \Gamma(\mathcal{P})$. We shall prove this under the assumption that the parameters p, r, s, t are such that the groups $[1\ 1\ 1^p]^s$ and $[1\ 1\ 1^r]^t$ are finite; that is, each pair (p, s) and (r, t) is either of the form $(l, 3)$ or $(3, l)$ with $2 \leqslant l < \infty$, or one of $(4, 4)$, $(4, 5)$ or $(5, 4)$ (see Section 9D). In particular, if $p = r = 3$, then we recover the polytopes ${}_6\mathcal{T}^4_{(s,0),(t,0)}$.

Let p, r, s, t be as specified. Then we can appeal to Theorem 9E17, where W occurs as the group with diagram $\mathcal{S}_4(p, s; r, t)$. In particular, each 3-generator subgroup $\langle \sigma_i, \sigma_j, \sigma_k \rangle$ of W is isomorphic to $[1\ 1\ 1^p]^s$ or $[1\ 1\ 1^r]^t$ according as the triangle $\{i, j, k\}$

is marked s or t. Moreover, W is finite if and only if

11D8 $$(p, s, r, t) = \begin{cases} (3, a, 3, 2), (3, 2, 3, a) & \text{with } a = 2, 3, 4, \text{ or} \\ (3, 2, b, 3), (b, 3, 3, 2) & \text{with } b \geqslant 2. \end{cases}$$

In this case, W is isomorphic to $[1\,1\,2]^a$ or $[1\,1\,2^b]^3$, respectively; if $b = 2$, the latter is the Coxeter group D_4. Again, the finiteness of W (and Γ) is equivalent to the corresponding hermitian form being positive definite. Furthermore, Γ is again a C-group, because W itself has the intersection property (see again Theorem 9E17).

We have thus proved the following theorem.

11D9 Theorem *Assume that each pair* (p, s) *and* (r, t) *is either of the form* $(l, 3)$ *or* $(3, l)$ *with* $2 \leqslant l < \infty$, *or one of* $(4, 4)$, $(4, 5)$ *or* $(5, 4)$. *Then the universal regular 4-polytope* $\mathcal{P} := \{\{2p, 3\}_{2s}, \{3, 2r\}_{2t}\}$ *exists. Its group is* $W \rtimes (C_2 \times C_2)$, *where* W *is the abstract group defined by the diagram in Figure 11D2. In particular,* $\mathcal{P}$ *is finite if and only if* (p, s, r, t) *is as in (11D8). In this case,* $\Gamma(\mathcal{P}) = [1\,1\,2]^a \rtimes (C_2 \times C_2)$ *or* $[1\,1\,2^b]^3 \rtimes (C_2 \times C_2)$, *respectively.*

The theorem says that there are only two kinds of finite polytopes $\mathcal{P}$, namely,

$$\big\{\{2p, 3\}_6, \{3, 6\}_{(2,0)}\big\},$$

with $p \geqslant 2$, and the finite polytopes ${}_6\mathcal{T}^4_{(s,0),(t,0)}$. The first kind is related to the simple polytopes $\mathcal{Q} := \{\{2p, 3\}_6, \{3, 3\}\}$; in particular, $\Gamma(\mathcal{P}) = \Gamma(\mathcal{Q}) \times C_2$ (see Lemma 11A11). See also Theorem 8E10 for another approach to the enumeration for $p = 2$.

Geometric Realizations

In an abstract sense, we have solved our initial problem, in that we have associated with each regular polytope ${}_6\mathcal{T}^4_{(s,0),(t,0)}$ a locally unitary reflexion group G, which is a representation of a subgroup W of index 4 in its group. Whether this geometric group G is finite (and then unitary and isomorphic to W) determines the finiteness of our polytope. In this case, we then run up against the problem of realizing certain involutory automorphisms of such groups geometrically. In some cases, this can be achieved by means of suitable linear or semilinear mappings of the ambient space, but examples show that these cannot always suffice.

Let us look more closely at what we want to do, in slightly more general terms than we actually need. Let

$$W = W(\mathcal{D}) = \langle \sigma_1, \ldots, \sigma_n \rangle$$

be a finite group which is associated with a diagram $\mathcal{D}$ on n nodes. Suppose further that W (as an abstract group) admits group automorphisms τ corresponding to symmetries of the diagram (diagram automorphisms). There is often some ambiguity in associating a geometric group G with W, and we attempt to exploit this in order to find *some* suitable realizations of the τ. These automorphisms τ form a group Λ, and we aim at realizing Λ completely.

We attempt to realize our group as a *sum* (blend) of copies of certain representations of W (see also Section 5A). Let $E_1, \ldots, E_k$ be subspaces of the real or complex space E such that

$$E = E_1 \oplus \cdots \oplus E_k.$$

For $i = 1, \ldots, k$, let $r_i \colon W \to \mathrm{GL}(E_i)$ be a representation of W on E_i as a reflexion group, and let

$$G_i := Wr_i = \langle S_{1,i}, \ldots, S_{n,i} \rangle,$$

with the reflexions $S_{j,i}$ corresponding to σ_j for $j = 1, \ldots, n$. Extend every reflexion $S_{j,i}$ trivially to the ambient space E, such that $G_1, \ldots, G_k$ acts on E. Then

11D10 $$S_j := S_{j,1} S_{j,2} \cdots S_{j,k} \quad (j = 1, \ldots, n)$$

defines a representation

$$G = \langle S_1, \ldots, S_n \rangle$$

of W, which is just the sum of the representations $r_1, \ldots, r_k$ (see [434, §3.2]). The generators of G are reflexions in mirrors of codimension k in E. If the r_i happen to be orthogonal or unitary representations, then we can take $E_1, \ldots, E_k$ to be mutually orthogonal subspaces of the euclidean or unitary space E, so that G also becomes an orthogonal or unitary representation. More generally, if the G_i preserve a hermitian form on E_i, then G preserves the orthogonal sum of these forms; this is the hermitian form which extends each form from E_i to E, such that $E_1, \ldots, E_k$ are mutually orthogonal subspaces.

In the situation of the previous sections, each representation of the underlying abstract group W as a reflexion group in $\mathbb{C}^n$ is associated with its Gram matrix, which defines the corresponding hermitian form $\langle \cdot, \cdot \rangle$ as in Chapter 9. Generally, we have some freedom in choosing the Gram matrix for W, and different admissible choices of Gram matrices A lead to different representations $G(A)$ (say) of W. Such choices usually come from differing signs of the turns associated with circuits, which are for the most part triangular.

We pick a basis $\{v_1, \ldots, v_n\}$ of $\mathbb{C}^n$ and keep it fixed. All Gram matrices A and representations $G(A)$ are taken with respect to this basis; that is, $A = (\alpha_{jk})$ with $\alpha_{jk} = \langle v_j, v_k \rangle$, and the vectors v_j are the normal vectors for the generating reflexions.

Let us consider one fixed Gram matrix A, with representation $G(A)$. Any allowable permutation of the nodes of $\mathcal{D}$, which induces a group automorphism τ of W, corresponds to a permutation of the rows of A, together with the *same* permutation of its columns. We denote the new matrix by A^τ; then $A^\tau = (\beta_{jk})$ with $\beta_{jk} = \alpha_{j\tau^{-1}k\tau^{-1}}$. Under this permutation, the cyclic products of coefficients of (9C10) are preserved; that is, we have $\gamma_{A^\tau}(C^\tau) = \gamma_A(C)$, where $C := (j(1), \ldots, j(m))$ is a cycle of $\mathcal{D}$, $C^\tau := (j(1)\tau, \ldots, j(m)\tau)$ is its image under τ, and $\gamma_{A^\tau}(C^\tau)$ and $\gamma_A(C)$ are the corresponding cyclic products for C^τ and C relative to A^τ and A, respectively. Hence the corresponding turns are also preserved (see (9C12)); that is, $\vartheta_{A^\tau}(C^\tau) = \vartheta_A(C)$, with

$\vartheta_{A^\tau}(C^\tau)$ and $\vartheta_A(C)$ denoting the turns relative to A^τ and A, respectively. Thus the turn of an (undirected) circuit $\mathcal{C}$ in $\mathcal{D}$ is either preserved, or reverses its sign.

The group Λ of automorphisms τ is finite, since the diagram has finitely many nodes. So, take the sum G of all the corresponding representations $G(A^\tau)$ with matrices A^τ. Each automorphism $\tau \in \Lambda$ can now be realized geometrically by a linear mapping T (say) – just permute the coordinates within each component according to τ, and then permute the components by τ. More precisely, if $E = \bigoplus_{\lambda\in\Lambda} E_\lambda$, where $E_\lambda = \langle v_{1,\lambda}, \ldots, v_{n,\lambda}\rangle$ is the ambient space for $G(A^\lambda)$ (with $\{v_{1,\lambda}, \ldots, v_{n,\lambda}\}$ corresponding to $\{v_1, \ldots, v_n\}$), then $v_{j,\lambda}T = v_{j\tau,\lambda\tau}$ for each $j = 1, \ldots, n$ and $\lambda \in \Lambda$; hence $T^{-1}S_jT = S_{j\tau}$ for each $j = 1, \ldots, n$. Finally, since the set of all such mappings T forms a group L (say), we now have a complex representation of $W \rtimes \Lambda$, namely, $G \rtimes L$, with $L \cong \Lambda$. Note that the entire group $G \rtimes L$ preserves the hermitian form on E which is the orthogonal sum of the forms on the E_λ (its Gram matrix has blocks A^λ along the diagonal, one for each $\lambda \in \Lambda$). If the hermitian forms on the E_λ are positive definite (it suffices to know this for one), then so is the sum, and we have a unitary representation of $W \rtimes \Lambda$, and of course a real orthogonal representation of twice the dimension as well.

Notice that among these automorphisms will frequently be one equivalent to complex conjugation of the whole Gram matrix A (in instances discussed earlier, this happens because of the way we have loaded turns on particular edges); this results in changing the signs of *all* the turns, and obviously can be realized geometrically by a semi-linear mapping in the same space (just permute the coordinates according to τ, and then apply complex conjugation). This gives a linear mapping in real space of twice the dimension.

Sometimes, variants of these methods can be employed to reduce the dimension of these representations. For example, we can group representations $G(A^\lambda)$ with the same Gram matrices together, and then pass to a suitable invariant (diagonal) subspace of E (now the new components represent distinct Gram matrices A, and each has a basis in which the ith vector is the sum of all $e_{i,\lambda}$ with $A^\lambda = A$). However, when restricted to this subspace, an element of L may coincide with an element of G or another element of L, so that we no longer have enough distinct mappings to realize all the automorphisms geometrically. Similarly, we can cut the dimension in half by pairing up representations whose Gram matrices are complex conjugates of each other (if this occurs); however, we must then admit semi-linear mappings to realize automorphisms. We can also mix these two constructions.

Notwithstanding what we have said previously, whether or not a regular polytope obtained from W by twisting with Λ admits some suitable realization is actually a more complicated matter than just finding a representation of its group Γ. Indeed, we must still check in each individual case whether the representation of Γ has a non-trivial Wythoff space and, if so, whether it yields a faithful realization (see Chapter 5). Moreover, on the ambient space of a faithful realization (spanned by the orbit of the initial vertex), a map realizing an automorphism of W must not coincide with one realizing an element or another automorphism of W. This generally limits the number of admissible representations (if there are any at all).

The polytope ${}_6\mathcal{T}^4_{(s,0),(t,0)}$ was constructed from the diagram in Figure (11D1) by the operation in (11D4). Unfortunately, for it to be finite, we have seen in Theorem 11D5 that $\{s, t\} = \{2, 2\}, \{2, 3\}$ or $\{2, 4\}$. Suppose that we pick the basis $\{v_1, \dots, v_4\}$ so that the turns $2\pi/s$ and $2\pi/t$ are loaded into the branches $\{1, 2\}$ and $\{3, 4\}$ of the diagram, respectively. Then, in the first case, the two automorphisms of W which interest us preserve the (real) Gram matrix, while in the other two cases only the equivalent of complex conjugation leads to a distinct Gram matrix (because $s = 2$). The first case is an example where grouping of automorphisms by identical Gram matrices will not work. (indeed, the map T_1 realizing τ_1 will act like $S_1S_2S_1 \in G$ on the invariant subspace). On the other hand, the vertex-figure $\{3, 6\}_{(t,0)}$ does not have a faithful realization in any case if $t = 2$, and neither does the polytope itself (see Theorem 5A2).

However, we can make some progress by combining this approach with another. If $t = 3$ or 4, then the semi-linear mapping of (11B7) will realize the diagram automorphism which yields ${}_3\mathcal{T}^4_{(t,0)} = \{\{6, 3\}_{(t,0)}, \{3, 3\}\}$ from the group $[1\ 1\ 2]^t$; hence $({}_3\mathcal{T}^4_{(t,0)})^* = \{\{3, 3\}, \{3, 6\}_{(t,0)}\}$ by duality (this will not work when $t = 2$). As we noticed earlier, we may then construct ${}_6\mathcal{T}^4_{(2,0),(t,0)}$ as a mix

$${}_6\mathcal{T}^4_{(2,0),(t,0)} = \left({}_3\mathcal{T}^4_{(t,0)}\right)^* \diamond \{\,\},$$

in the same way that $\{6, 3\}_{(2,0)} = \{3, 3\} \diamond \{\,\}$. We thus obtain realizations of ${}_6\mathcal{T}^4_{(2,0),(t,0)}$ from realizations of $({}_3\mathcal{T}^4_{(t,0)})^*$ by blending with the line segment (see Section 11B for an example of a realization of $({}_3\mathcal{T}^4_{(t,0)})^*$). The same construction applied with $t = 2$ yields a (non-faithful) realization of ${}_6\mathcal{T}^4_{(2,0),(2,0)}$ from a (non-faithful) realization of $({}_3\mathcal{T}^4_{(2,0)})^*$, with the same vertices as $\{3, 3, 3\} \diamond \{\,\}$, but with twice as many automorphisms.

11E The Type $\{3, 6, 3\}$

In the previous sections, we enumerated all those locally toroidal regular 4-polytopes of type $\{6, 3, p\}$, with $p = 3$, 4, 5 or 6, which are universal and finite. For the type $\{3, 6, 3\}$, the corresponding enumeration is not yet complete, and the general method of Section 11A seems to fail except for certain parameter values. In this section, we describe partial results which follow from this approach, as well as those which can be obtained by alternative methods.

We must investigate the polytopes

11E1 $${}_7\mathcal{T}^4_{\mathbf{s},\mathbf{t}} := \{\{3, 6\}_{\mathbf{s}}, \{6, 3\}_{\mathbf{t}}\},$$

with $\mathbf{s} = (s^k, 0^{2-k})$, $\mathbf{t} = (t^l, 0^{2-l})$. The general assumptions on the parameters are as before: $s \geqslant 2$ if $k = 1$ and $s \geqslant 1$ if $k = 2$; $t \geqslant 2$ if $l = 1$ and $t \geqslant 1$ if $l = 2$. The superscript in the name for the polytope indicates the rank, and the subscript 7 on the left refers to $\{3, 6, 3\}$ as the *7th* (and last) type of Schläfli symbol which requires consideration. Our general results will mainly concern special instances of the case $k = 2$ and $l = 1$, but special techniques will yield results for other cases. As usual, ${}_7\Gamma^4_{\mathbf{s},\mathbf{t}}$ will denote the abstract group corresponding to ${}_7\mathcal{T}^4_{\mathbf{s},\mathbf{t}}$.

Table 11E1. *The Known Finite Polytopes* ${}_7\mathcal{T}^4_{\mathbf{s},\mathbf{t}}$

s	**t**	v	f	g
(1, 1)	(1, 1)	3	3	108
(1, 1)	(3, 0)	3	9	324
(2, 0)	(2, 0)	5	5	240
(2, 0)	(2, 2)	5	15	720
(3, 0)	(3, 0)	27	27	2916
(3, 0)	(2, 2)	288	384	41472
(3, 0)	(4, 0)	1260	2240	241920

To set the scene, we list in Table 11E1 those universal locally toroidal polytopes of type {3, 6, 3} which are known to be finite; some of the entries are taken from [83, Table 1]. As usual, we shall only list one of each dual pair, that with fewer vertices. We do not describe the groups here, since for the most part their structures are not exactly known.

Because we can say a certain amount about it, we begin with the class of abstract regular polytopes of type

$$ {}_7\mathcal{T}^4_{(s,s),(t,0)} := \big\{\{3,6\}_{(s,s)}, \{6,3\}_{(t,0)}\big\}, $$

whose group

$$ \Gamma = {}_7\Gamma^4_{(s,s),(t,0)} = \langle \rho_0, \ldots, \rho_3 \rangle, $$

say, has a presentation consisting of the standard relations for the Coxeter group [3, 6, 3], along with the two extra relations

11E2
$$ (\rho_0(\rho_1\rho_2)^2)^{2s} = (\rho_1\rho_2\rho_3)^{2t} = \varepsilon. $$

Consider the mixing operation

$$ (\rho_0, \ldots, \rho_3) \mapsto (\rho_0\rho_1\rho_2\rho_1\rho_0, \rho_1\rho_2\rho_1, \rho_2, \rho_3) =: (\sigma_1, \ldots, \sigma_4). $$

Then we easily check that the σ_j satisfy the relations

11E3
$$ \begin{cases} \sigma_i^2 = (\sigma_i\sigma_j)^3 = \varepsilon \quad (1 \leqslant i, j \leqslant 4;\ i \neq j), \\ (\sigma_1\sigma_2\sigma_3\sigma_2)^s = (\sigma_1\sigma_2\sigma_4\sigma_2)^t = (\sigma_1\sigma_3\sigma_4\sigma_3)^t = (\sigma_2\sigma_3\sigma_4\sigma_3)^t = \varepsilon. \end{cases} $$

Indeed, we have

$$ (\rho_0(\rho_1\rho_2)^2)^2 = \sigma_1\sigma_3\sigma_2\sigma_3 = \sigma_1\sigma_2\sigma_3\sigma_2, \qquad (\rho_1\rho_2\rho_3)^2 = \sigma_2\sigma_4\sigma_3\sigma_4 = \sigma_2\sigma_3\sigma_4\sigma_3. $$

Recall that we can take the generators in the last four relations of (11E3) in any order (see Section 9E).

As in Section 11C, we first consider the case $s \geqslant 2$ (of course, we also have $t \geqslant 2$). The group $W = \langle \sigma_1, \ldots, \sigma_4 \rangle$ corresponds to the abstract tetrahedral diagram in Figure 11E1, where $s, t \geqslant 2$. Here the triangle $\{1, 2, 3\}$ is marked s, while all the other triangles are marked t. Further, W has two group automorphisms τ_1 $(= \rho_0)$

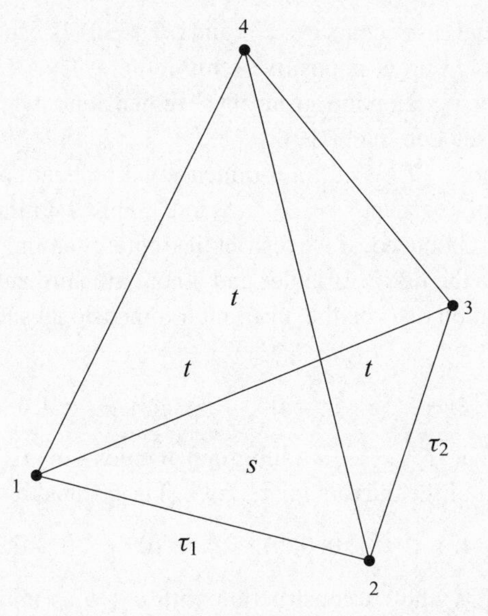

Figure 11E1

and τ_2 ($= \rho_1$), which correspond to the transpositions (1 2) and (2 3) on the nodes, respectively.

We can recover Γ by means of the operation

11E4 $$(\sigma_1, \ldots, \sigma_4; \tau_1, \tau_2) \mapsto (\tau_1, \tau_2, \sigma_3, \sigma_4) =: (\rho_0, \ldots, \rho_3),$$

which shows that

$$\Gamma = W \rtimes S_3.$$

In particular, if W happens to be a C-group, then Γ will be also, and is then the group of the regular polytope ${}_7\mathcal{T}^4_{(s,s),(t,0)}$.

We now appeal to Section 9E, where groups with tetrahedral diagrams are treated. In particular, we can attempt to associate with W a reflexion group G which preserves a complex hermitian form in 4 variables (here we need that $s, t \geqslant 2$). However, the compatibility conditions (9E8) for a geometric diagram only hold if $t = s$ or $t = 3s$. Theorem 9E14 then shows that, in the former case, the only finite group is given by $s = 2$; there are no finite groups in the latter case.

We thus deduce the following.

11E5 Theorem *For $s \geqslant 2$, let $\mathbf{s} := (s, s)$ and $\mathbf{t} := (s, 0)$ or $(3s, 0)$. Then the universal regular 4-polytope ${}_7\mathcal{T}^4_{\mathbf{s},\mathbf{t}} := \{\{3, 6\}_{\mathbf{s}}, \{6, 3\}_{\mathbf{t}}\}$ exists. In particular,*

$$\Gamma\left({}_7\mathcal{T}^4_{\mathbf{s},\mathbf{t}}\right) = {}_7\Gamma^4_{\mathbf{s},\mathbf{t}} = W \rtimes S_3,$$

where W is the group abstractly defined by the diagram in Figure 11E1, with $t = s$ or $3s$, respectively. The only finite instance is ${}_7\mathcal{T}^4_{(2,2),(2,0)}$, with group $S_5 \times S_3$.

For the sole finite example, refer also to Theorem 11E10.

An interesting special case occurs if $s = 3$ and $\mathbf{t} = (3, 0)$, because then the hermitian form of the geometric group G is positive semi-definite. Thus G can be regarded as an infinite locally unitary reflexion group in 3 dimensions, which we have already remarked is discrete (see Lemma 9B19).

For the finite polytope ${}_7\mathcal{T}^4_{(2,2),(2,0)}$, a 6-dimensional realization can be constructed from a representation of its group $\Gamma := S_5 \times S_3$ in $\mathbb{E}^8$. This is similar to what we did with $({}_3\mathcal{T}^4_{(2,2)})^*$ in Section 11C. Indeed, if we restrict the representation of the corresponding group $S_5 \times S_4$ in $\mathbb{E}^9$ to the first 8 variables and ignore the third automorphism τ_3, then we obtain a representation of Γ on the invariant 6-dimensional subspace of $\mathbb{E}^8$ defined by

$$\xi_1 + \cdots + \xi_5 = 0, \quad \xi_6 + \xi_7 + \xi_8 = 0.$$

This exploits the fact that ${}_7\Gamma^4_{(2,2),(2,0)}$ is a subgroup of index 4 in ${}_3\Gamma^4_{(2,2)}$. Now the Wythoff space (the intersection of the mirrors for $\tau_2, \sigma_3, \sigma_4$) is spanned by a and b, with

$$a := (-4, 1, \ldots, 1, 0, 0, 0), \quad b := (0, \ldots, 0, -2, 1, 1),$$

and an application of Wythoff's construction with $a + b$ as initial vertex leads to a 6-dimensional realization for ${}_7\mathcal{T}^4_{(2,2),(2,0)}$, whose vertex-set consists of the 15 points

11E6
$$\begin{cases} \xi_i = -4, & \text{for one } i \text{ with } i \leqslant 5, \\ \xi_i = -2, & \text{for one } i \text{ with } i \geqslant 6, \\ \xi_i = 1, & \text{otherwise.} \end{cases}$$

This realization is the blend of the two degenerate realizations of ${}_7\mathcal{T}^4_{(2,2),(2,0)}$ constructed by taking a and b as initial vertices, and exhibits it as the mix

$${}_7\mathcal{T}^4_{(2,2),(2,0)} = \{3, 3, 3\} \diamond \{3\}$$

(see also Theorem 7A13).

For the enumeration of certain finite polytopes ${}_7\mathcal{T}^4_{(s,0),(s,0)}$, we also refer to Theorem 11H7.

We now turn to polytopes with facets $\{3, 6\}_{(1,1)}$, which are the duals of those with vertex-figures $\{6, 3\}_{(1,1)}$. Recall that our main aim is always to list the locally toroidal polytopes. Here, we only describe the universal polytopes

$$\mathcal{Q} := \big\{\{3, 6\}_{(1,1)}, \{6, 3\}_{(t,0)}\big\} = {}_7\mathcal{T}^4_{(1,1),(t,0)},$$

corresponding to the case $s = 1$.

As with $s = 2$, if $\mathcal{Q}$ exists, then its group $\Gamma(\mathcal{Q})$ can be obtained by applying operation (11E4) to the abstract group $W = \langle \sigma_1, \ldots, \sigma_4 \rangle$ defined by the diagram in Figure 11E1, now with $s = 1$. But the mark $s = 1$ forces $\sigma_3 = \sigma_2\sigma_1\sigma_2$; hence

$$W = \langle \sigma_1, \sigma_2, \sigma_4 \rangle = [1\ 1\ 1]^t.$$

In particular, we have

$$\sigma_3\sigma_1\sigma_3\sigma_4 = \sigma_2\sigma_1\sigma_2\sigma_1\sigma_2\sigma_1\sigma_2\sigma_4 = \sigma_2\sigma_4,$$

so that necessarily $t = 3$.

Conversely, if $t = 3$, then we can produce the group W of Figure 11E1, with $s = 1$ and $t = 3$, by taking $[1\ 1\ 1]^3 = \langle \sigma_1, \sigma_2, \sigma_4 \rangle$, and introducing a new generator $\sigma_3 := \sigma_2\sigma_1\sigma_2$; then $W = [1\ 1\ 1]^3$. This proves the following:

11E7 Theorem *The universal regular 4-polytope* ${}_7\mathcal{T}^4_{(1,1),(t,0)} := \{\{3,6\}_{(1,1)}, \{6,3\}_{(t,0)}\}$ *exists if and only if* $t = 3$. *In particular,* ${}_7\mathcal{T}^4_{(1,1),(3,0)}$ *is (combinatorially) flat and finite, with group* ${}_7\Gamma^4_{(1,1),(3,0)} = [1\ 1\ 1]^3 \rtimes S_3$, *of order* 324.

We then observe that the universal polytope

$$ {}_7\mathcal{T}^4_{(1,1),(1,1)} = \big\{\{3,6\}_{(1,1)}, \{6,3\}_{(1,1)}\big\} $$

also exists, and is finite, since it can be obtained from ${}_7\mathcal{T}^4_{(1,1),(3,0)} = \{\{3,6\}_{(1,1)}, \{6,3\}_{(3,0)}\}$ by identifications. It is flat, so that it has 3 vertices and 3 facets, and a group of order 108.

We also know that the universal regular polytope ${}_7\mathcal{T}^4_{(3,0),(3,0)} = \{\{3,6\}_{(3,0)}, \{6,3\}_{(3,0)}\}$ is finite (see [83, Table 1; 443, §6]). It has 27 vertices and 27 facets, and a group of order $2916 = 2^2 \cdot 3^6$. It is, perhaps, of interest to say a little about this polytope. It has 27 vertices, and clearly covers ${}_7\mathcal{T}^4_{(3,0),(1,1)}$ with 9 vertices and ${}_7\mathcal{T}^4_{(1,1),(3,0)}$ (and ${}_7\mathcal{T}^4_{(1,1),(1,1)}$) with 3 vertices, both of which are flat. We can now appeal to Theorems 5A4 and 5B14. Since ${}_7\mathcal{T}^4_{(1,1),(3,0)}$ has a 2-dimensional realization, while the natural geometric realization of the flat polytope ${}_7\mathcal{T}^4_{(3,0),(1,1)}$ is that of its facet $\{3,6\}_{(3,0)}$, which has dimension 6, we should expect a realization of ${}_7\mathcal{T}^4_{(3,0),(3,0)}$ of dimension $(27-1)-2-6 = 18$. We do indeed have one; it is faithful, and the vertices are those of 9 mutually orthogonal equilateral triangles. In fact, it is more natural to work in $\mathbb{C}^9$, where the generating reflexions of the symmetry group are semi-linear. The actual details are a little complicated, and so we shall not give them here.

There is another general class of these polytopes which we can classify, namely, those with $\mathbf{s} = (2,0)$. Initially, at least, let us consider a general polyhedron $\mathcal{K}$ with facets $\{6\}$, and ask whether it is suited to be a vertex-figure of a polytope of type $\{\{3,6\}_{(2,0)}, \mathcal{K}\}$. Now $\mathcal{Q} := \{3,6\}_{(2,0)}$ has four vertices, and covers the tetrahedron $\{3,3\}$ twice. The natural identification $\mathcal{Q} \searrow \{3,3\}$ then forces an analogous identification of the hexagonal faces of $\mathcal{K}$ onto triangles. (This is even more obvious, if we think of realizing the vertex-figure $\mathcal{K}$, with vertices the other vertices of edges through an initial vertex.)

When $\mathcal{K} = \{6,3\}_{\mathbf{t}}$ (or even $\{6,3\}$), the only possible identification leads to $\mathcal{K} \searrow \{3,3\}$. We conclude

11E8 Lemma *If* $\mathcal{P} := \{\{3,6\}_{(2,0)}, \{6,3\}_{\mathbf{t}}\}$ *exists, then the covering* $\{6\} \searrow \{3\}$ *of its section* $\{6\}$ *induces a covering* $\{6,3\}_{\mathbf{t}} \searrow \{3,3\}$, *and hence a covering* $\mathcal{P} \searrow \{3,3,3\}$ *which preserves vertices.*

Actually, Lemma 11E8 considerably understates what really happens. The census of Colbourn and Weiss [83, Table 1] lists $\{\{3,6\}, \{6,3\}_{(2,0)}\}$ (with the tessellation $\{3,6\}$ as facet!) as a polytope with group order 720. Obviously, the facet here cannot in fact be infinite (the Schläfli symbol notation of [83] is slightly more general than ours). We explain this (in the dual formulation) as follows.

11E9 Lemma *The imposition of the relation* $(\rho_0\rho_1\rho_2)^4 = \varepsilon$ *on the group* $[3, 6, 3]$ *(with standard generators) implies that* $(\rho_1\rho_2\rho_1\rho_2\rho_3)^4 = \varepsilon$.

In other words, $\{\{3, 6\}_{(2,0)}, \{6, 3\}\}$ collapses to $\{\{3, 6\}_{(2,0)}, \{6, 3\}_{(2,2)}\}$.

Proof. Note first that

$$\varepsilon = (\rho_0\rho_1\rho_2)^4 = (\rho_0\rho_1\rho_0\rho_2\rho_1\rho_2)^2 = (\rho_1\rho_0\rho_1\rho_2\rho_1\rho_2)^2 \sim (\rho_0\rho_1\rho_2\rho_1\rho_2\rho_1)^2,$$

hence

$$\rho_0 \leftrightharpoons \rho_1\rho_2\rho_1\rho_2\rho_1$$

(recall that $\leftrightharpoons$ means "commutes with"). Using this fact and the other relations of the groups freely, it then follows that

$$\begin{aligned}
(\rho_1\rho_2\rho_1\rho_2\rho_3)^2 &= \rho_1\rho_2\rho_1\rho_2\rho_1 \cdot \rho_3\rho_2\rho_1\rho_2\rho_3 \\
&\sim \rho_1\rho_2\rho_1\rho_2\rho_1 \cdot \rho_0\rho_3\rho_2\rho_1\rho_2\rho_3\rho_0 \\
&= \rho_1\rho_2\rho_1\rho_2\rho_1 \cdot \rho_3\rho_2\rho_1\rho_0\rho_1\rho_2\rho_3 \\
&\sim \rho_3\rho_2\rho_1\rho_2\rho_3 \cdot \rho_1\rho_2\rho_1\rho_0\rho_1\rho_2\rho_1 \\
&= \rho_3\rho_2\rho_1\rho_2\rho_3\rho_2\rho_1 \cdot \rho_1\rho_2\rho_1\rho_2\rho_1\rho_0\rho_1\rho_2\rho_1 \\
&= \rho_3\rho_2\rho_1\rho_2\rho_3\rho_2\rho_1 \cdot \rho_0\rho_1\rho_2 \\
&\sim \rho_1\rho_2\rho_3\rho_2\rho_1\rho_0\rho_1\rho_2\rho_3\rho_2 \\
&= \rho_1\rho_3\rho_2\rho_3\rho_1\rho_0\rho_1\rho_3\rho_2\rho_3 \\
&\sim \rho_1\rho_2\rho_1\rho_0\rho_1\rho_2.
\end{aligned}$$

We see at once that $(\rho_1\rho_2\rho_1\rho_2\rho_3)^4 \sim (\rho_0\rho_1\rho_2\rho_1\rho_2\rho_1)^2 = \varepsilon$, which establishes our claim. □

As an immediate consequence, we have

11E10 Theorem *The only universal regular 4-polytopes* $\mathcal{P} := \{\{3, 6\}_{(2,0)}, \{6, 3\}_{\mathbf{t}}\}$ *are those with* $\mathbf{t} = (2, 0)$ *and group* $S_5 \times C_2$, *and* $\mathbf{t} = (2, 2)$ *with group* $S_5 \times S_3$.

Proof. Lemma 11E9 implies that $\mathcal{P}$ can only be a quotient of $\{\{3, 6\}_{(2,0)}, \{6, 3\}_{(2,2)}\}$; apart from this itself, the only possibility is $\mathbf{t} = (2, 0)$ (recall that the case $\mathbf{t} = (1, 1)$ was eliminated by the dual form of Theorem 11E7). Because both polytopes have 5 vertices by Lemma 11E8, the group orders are easily calculated. From Theorem 11E5 we know the group to be $S_5 \times S_3$ if $\mathbf{t} = (2, 2)$. Finally, if $\mathbf{t} = (2, 0)$, we can either determine the group as a quotient of $S_5 \times S_3$ using (11E4) or observe that the element $(\rho_1\rho_2)^3$ of $\Gamma(\mathcal{P})$ which defines the covering $\mathcal{P} \searrow \{3, 3, 3\}$ is central. □

An entry in Table 11E1 which has not been mentioned hitherto is $\{\{3, 6\}_{(3,0)}, \{6, 3\}_{(4,0)}\}$ (again see [83, Table 1; 443]). Its group order was found using the Todd–Coxeter coset enumeration method, but its group structure has not been determined. We would guess that, up to duality, there is just one further possibility for a finite polytope of type $\{3, 6, 3\}$, namely, $\{\{3, 6\}_{(3,0)}, \{6, 3\}_{(5,0)}\}$. The reason for this speculation is that, as we have seen, $\{\{3, 6\}_{(3,0)}, \{6, 3\}_{(3,3)}\}$ is infinite, but only just, in that the associated hermitian form is positive semi-definite. Since $\{6, 3\}_{(3,3)}$ has 54 vertices, whereas $\{6, 3\}_{(5,0)}$ has only 50, it seems plausible that the slightly smaller vertex-figure might yield a finite polytope. We shall look at this argument from a different point of view in Section 11F.

In Section 11H, we shall discuss further non-finiteness results obtained by exploiting relationships between regular polytopes of types $\{3, 3, 6\}$ and $\{3, 6, 3\}$.

11F Cuts of Polytopes of Type $\{6, 3, p\}$ or $\{3, 6, 3\}$

In this section, we again employ the cut method to explain why certain parameter values lead to universal polytopes which are finite, while others only give polytopes which are infinite. In Section 10D, we illustrated this method for the types $\{4, 4, 3\}$ and $\{4, 4, 4\}$. We begin with polytopes of type $\{6, 3, p\}$ with $p = 3, 4, 5, 6$, $\{3, 6, 3\}$, or their duals. As before, we write

$$ {}_p\mathcal{T}^4_{\mathbf{s}} := \{\{6, 3\}_{\mathbf{s}}, \{3, p\}\} \quad (p = 3, 4, 5), $$

and

$$ {}_6\mathcal{T}^4_{\mathbf{s},\mathbf{t}} := \{\{6, 3\}_{\mathbf{s}}, \{3, 6\}_{\mathbf{t}}\}, \quad {}_7\mathcal{T}^4_{\mathbf{s},\mathbf{t}} := \{\{3, 6\}_{\mathbf{s}}, \{6, 3\}_{\mathbf{t}}\}, $$

with the usual restrictions on $\mathbf{s}, \mathbf{t}$.

Recall that, for a regular polytope $\mathcal{P}$, a cut $\mathcal{K}$ is another regular polytope of lower rank which "cuts through it". More precisely, $\mathcal{K}$ has the following two properties: $\Gamma(\mathcal{K})$ is a subgroup of $\Gamma(\mathcal{P})$; the vertices of $\mathcal{K}$ are (usually) among the vertices of $\mathcal{P}$ which are invariant under a certain subgroup of $\Gamma(\mathcal{P})$, which is contained in the centralizer of a subset of the distinguished generators (see Sections 7C and 10D).

For the locally toroidal regular polytopes $\mathcal{P} := {}_p\mathcal{T}^4_{\mathbf{s}}, {}_6\mathcal{T}^4_{\mathbf{s},\mathbf{t}}, {}_7\mathcal{T}^4_{\mathbf{s},\mathbf{t}}$, or their duals, these cuts are frequently isomorphic to spherical, euclidean or hyperbolic tessellations. Here, it is remarkable that quite often the polytope $\mathcal{P}$ is finite if and only if its cut is finite, which then means that the tessellation is spherical. However, since this is not always true, the cuts are (in a sense) only "locally inscribed" into $\mathcal{P}$. Therefore, the corresponding finiteness criteria derived from Theorems 11F4, 11F8 and 11F9 can only be viewed as local ones. Moreover, as pointed out before, our considerations in the following will not in general imply the existence of the original polytopes $\mathcal{P}$.

We begin with the types $\{p, 3, 6\}$, which are the duals of $\{6, 3, p\}$. Recall that a 3-hole of a regular polyhedron $\mathcal{K}$ is an edge-path which leaves a vertex by the third edge from which it entered, in the same sense (that is, keeping always to the left, say) in some local orientation.

The Polyhedron $\mathcal{K}(\mathcal{P})$

Let $\mathcal{P}$ be a regular 4-polytope of type $\{p, q, r\}$, and let $\Gamma(\mathcal{P}) = \langle \rho_0, \ldots, \rho_3 \rangle$, where $\rho_0, \ldots, \rho_3$ are the distinguished generators. Assume that the vertex-figures of $\mathcal{P}$ have 3-holes of lengths h (say). Then the (mixing) operation

11F1 $$(\rho_0, \ldots, \rho_3) \mapsto (\rho_0, \rho_1, \rho_2\rho_3\rho_2\rho_3\rho_2) =: (\beta_0, \beta_1, \beta_2)$$

determines a regular polyhedron $\mathcal{K} = \mathcal{K}(\mathcal{P})$ with group $\Gamma(\mathcal{K}) = \langle \beta_0, \beta_1, \beta_2 \rangle$, which is of type $\{p, h\}$. Indeed, we have

$$\beta_1\beta_2 = \rho_1\rho_2(\rho_3\rho_2)^2,$$

so that the last entry in the Schläfli symbol must be h.

Let $\{F_0, \ldots, F_3\}$ be the base flag of $\mathcal{P}$, where F_i denotes an i-face. Then $\mathcal{K}$ can be obtained from $\mathcal{P}$ by (an abstract version of) Wythoff's construction, now applied to the order complex $\mathcal{C}(\mathcal{P})$ of $\mathcal{P}$, with F_0 as initial vertex, which is fixed by β_1 and β_2 (see Section 5A). In particular, $\{F_0, F_1, F_2\}$ becomes the base flag of $\mathcal{K}$, and the transforms of F_i under $\Gamma(\mathcal{K})$ are the i-faces of $\mathcal{K}$. The neighbouring vertices of F_0 in $\mathcal{K}$ are given by

$$F_0\beta_0(\beta_1\beta_2)^j = F_0\rho_0(\rho_1\rho_2(\rho_3\rho_2)^2)^j \quad (j = 0, \ldots, h-1);$$

that is, as we go around F_0 we pick as vertices for its vertex-figure in $\mathcal{K}$ precisely the vertices in a 3-hole of the vertex-figure of $\mathcal{P}$ at F_0.

If $\mathcal{P}$ is of type $\{p, q, 6\}$, then ρ_3 commutes with each β_j, so that $\mathcal{K}$ is invariant under ρ_3. Hence, in a sense, we can think of $\mathcal{K}$ as lying in the reflexion wall of ρ_3. In particular, if the vertex-figures of $\mathcal{P}$ are toroidal polyhedra $\{3, 6\}_{(s,t)}$, then $\mathcal{K}$ is a cut of type $\{p, s\}$ or $\{p, 3s\}$ as $t = 0$ or s.

Polytopes with Vertex-Figures $\{3, 6\}_{(s,0)}$

We now investigate the cut $\mathcal{K} = \mathcal{K}(\mathcal{P})$ for the polytope $\mathcal{P} := \{\{p, 3\}, \{3, 6\}_{(s,0)}\}$, which is the dual of ${}_p\mathcal{T}^4_{(s,0)}$, when $p = 3$, 4 or 5. Recall from Section 11B that $\mathcal{P}$ can be constructed from the abstract group $W = \langle \sigma_1, \ldots, \sigma_4 \rangle$ corresponding to the diagram of Figure 11F1 by the twisting operation

$$(\sigma_1, \ldots, \sigma_4; \tau) \mapsto (\sigma_1, \sigma_2, \sigma_3, \tau) =: (\rho_0, \ldots, \rho_3).$$

(Note that the generators σ_i have been renamed.) In particular, $\Gamma(\mathcal{P}) = W \rtimes C_2$.

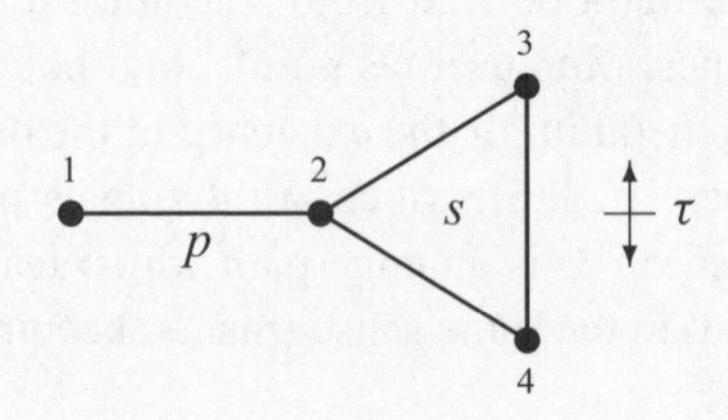

Figure 11F1

But now

11F2 $$(\beta_0, \beta_1, \beta_2) = (\sigma_1, \sigma_2, \sigma_3\sigma_4\sigma_3) = (\sigma_1, \sigma_2, \sigma_4\sigma_3\sigma_4),$$

so that the second entry in the Schläfli symbol $\{p, s\}$ of $\mathcal{K}$ corresponds to the extra relation $(\sigma_2\sigma_3\sigma_4\sigma_3)^s = \varepsilon$ of W. Equivalently, the period s of

$$\beta_1\beta_2 = \rho_1\rho_2(\rho_3\rho_2)^2 = \rho_2(\rho_1\rho_2\rho_3)^2\rho_2 \sim (\rho_1\rho_2\rho_3)^2$$

genuinely does specify the original polytope, where $\sim$ indicates as usual that the elements are conjugates.

The following lemma describes the cut, and is of interest in its own right.

11F3 Lemma *Let $p,\ s \geqslant 2$, and let $W = \langle \sigma_1, \ldots, \sigma_4 \rangle$ be the abstract group depicted in Figure 11F1. Then the subgroup $\langle \sigma_1, \sigma_2, \sigma_4\sigma_3\sigma_4 \rangle$ of W is isomorphic to the Coxeter group $[p, s]$.*

Proof. Indeed, this just comes from the basic change of generators $\sigma_3 \mapsto \sigma_4\sigma_3\sigma_4 =: \bar{\sigma}_3$ of W, which leads to the equivalent diagram of Figure 11F2 for W. In this figure, the label $\bar{3}$ denotes the new generator $\bar{\sigma}_3$. (Here we are using terminology adopted in Section 9C for geometric groups G and their diagrams. The change of generators of W is analogous to the basic change of generators $S_3 \mapsto S_4S_3S_4 =: \bar{S}_3$ of G.) □

The following theorem is now an immediate consequence of (11F2) and Lemma 11F3, and covers the polytopes $({}_p\mathcal{T}^4_{(s,0)})^*$ if $p = 3, 4$ or 5.

11F4 Theorem *Let $p,\ s \geqslant 2$, and let $\mathcal{P} := \{\{p, 3\}, \{3, 6\}_{(s,0)}\}$. Then $\mathcal{K}(\mathcal{P})$ is the regular tessellation $\{p, s\}$, and thus is finite if and only if $\frac{1}{p} + \frac{1}{s} > \frac{1}{2}$.*

It is interesting to note here that an infinite polytope $\mathcal{P}$ can have a cut $\mathcal{K}(\mathcal{P})$ which is a finite tessellation. In Theorem 11F4, if $p \geqslant 3$, this occurs precisely for the polytopes

$$\big\{\{3, 3\}, \{3, 6\}_{(5,0)}\big\}, \quad \big\{\{4, 3\}, \{3, 6\}_{(3,0)}\big\}, \quad \big\{\{5, 3\}, \{3, 6\}_{(3,0)}\big\},$$

which have "locally inscribed" tessellations $\{3, 5\}$, $\{4, 3\}$ and $\{5, 3\}$, respectively. We can see this geometrically from a representation of W as a reflexion group G (see Section 9E and the proof of the previous lemma). If G' (say) is the subgroup of G associated with $\mathcal{K}(\mathcal{P})$, then G' is a finite unitary (indeed, euclidean) reflexion group, while G itself is not a finite unitary group. In other words, the hermitian form restricts to a positive definite form on the 3-dimensional subspace of $\mathbb{C}^4$ spanned by the normals of the generators of G', but is not positive definite on all of $\mathbb{C}^4$.

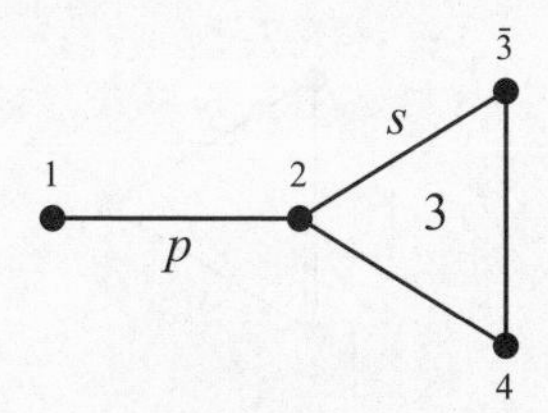

Figure 11F2

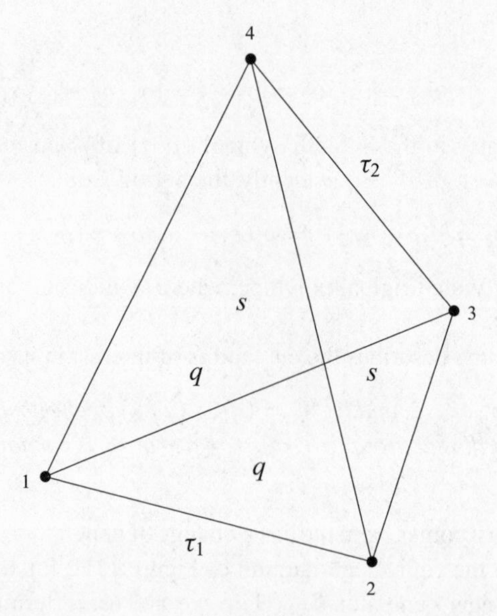

Figure 11F3

We continue our discussion with the polytopes

$$\mathcal{P} := \big\{\{6,3\}_{(q,0)}, \{3,6\}_{(s,0)}\big\} = {}_6\mathcal{T}^4_{(q,0),(s,0)},$$

which can be constructed from the abstract group $W = \langle \sigma_1, \ldots, \sigma_4 \rangle$ corresponding to the diagram of Figure 11F3 by applying the twisting operation

$$(\sigma_1, \ldots, \sigma_4; \tau_1, \tau_2) \mapsto (\tau_1, \sigma_2, \sigma_3, \tau_2) =: (\rho_0, \ldots, \rho_3)$$

(see Section 11D); thus $\Gamma(\mathcal{P}) = W \rtimes (C_2 \times C_2)$. The cut $\mathcal{K}(\mathcal{P})$ is now of type $\{6, s\}$, and the generators of its group are given by

11F5 $$(\beta_0, \beta_1, \beta_2) = (\tau_1, \sigma_2, \sigma_4\sigma_3\sigma_4).$$

Then $\Gamma(\mathcal{K}(\mathcal{P})) = \Lambda \rtimes C_2$, where $C_2 = \langle \tau_1 \rangle$ and $\Lambda := \langle \sigma_1, \sigma_2, \sigma_4\sigma_3\sigma_4 \rangle$ is associated with the diagram of Figure 11F4; here, the generator $\sigma_4\sigma_3\sigma_4$ is represented by the label $\bar{3}$. In particular, $\Gamma(\mathcal{K}(\mathcal{P}))$ can be recovered from Λ by the twisting operation

11F6 $$(\sigma_1, \sigma_2, \sigma_4\sigma_3\sigma_4; \tau_1) \mapsto (\tau_1, \sigma_2, \sigma_4\sigma_3\sigma_4) = (\beta_0, \beta_1, \beta_2).$$

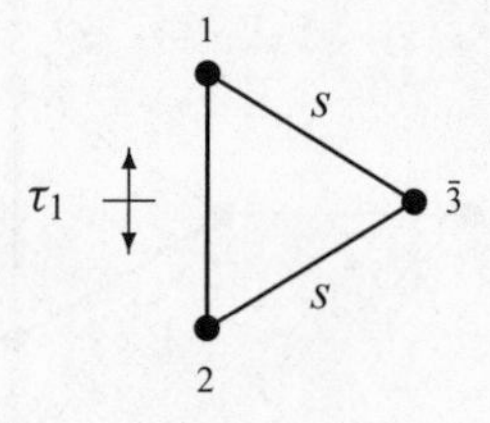

Figure 11F4

From the geometry, we can determine the parameters q and s which yield a finite group Λ. So, once again, consider a geometric group G representing W in $\mathbb{C}^4$, with generators S_i corresponding to σ_i as usual (see Section 9E). The subgroup G' of G associated with Λ has generators $S_1, S_2, S_4S_3S_4$, whose normal vectors must determine a positive definite Gram matrix if G' is finite (see Lemmas 9B7 and 9B19). A direct calculation of the determinant Δ of this matrix shows that

$$8\Delta = 6 - 8\left(1 + 2\cos^2\frac{\pi}{q}\right)\cos^2\frac{\pi}{s}.$$

(To calculate Δ, load the turns $2\pi/q$ and $2\pi/s$ into the branches $\{1, 2\}$ and $\{3, 4\}$, respectively, of the diagram in Figure 11F3.) Now the condition $\Delta > 0$ limits our possibilities to $s = 2$ (and q arbitrary), $s = 3$ and $q < \infty$, and $(s, q) = (4, 2)$ or $(5, 2)$. We will see that these parameters indeed give finite groups G' and that Λ will also be finite except (possibly) when $s = 3$; in all other cases, G' is infinite, and so are Λ and $\mathcal{K}(\mathcal{P})$.

Before we move on, let us consider an alternative approach. For our purposes, the (geometric) type of the triangular diagram of Figure 11F4 is best determined by the period of $\beta_0\beta_1\beta_2$ (compare Theorem 9D3). A direct calculation in G shows that the element $S_1S_4S_3S_4S_2S_4S_3S_4$, corresponding to $(\beta_0\beta_1\beta_2)^2 = \sigma_1\sigma_4\sigma_3\sigma_4\sigma_2\sigma_4\sigma_3\sigma_4$, is a rotation through an angle $2\varphi\,(\leqslant \pi)$ given by

$$\cos^2\varphi = \frac{1}{4}\left(1 + 4\left(1 + \cos\frac{2\pi}{s}\right)\cos\frac{2\pi}{q} + 4\left(1 + \cos\frac{2\pi}{s}\right)^2\right),$$

provided that the term on the right is less than 1. (The other alternative is that the term is 1, but then we have an affine subspace on which the element acts like a translation.)

Strictly speaking, of course, we are only concerned with finite polytopes (then the alternative does not occur). Now Theorem 9E14 tells us that the abstract group corresponding to the diagram of Figure 11F3 is finite precisely when $\{q, s\} = \{2, 2\}$, $\{2, 3\}$ or $\{2, 4\}$; in these cases, it is isomorphic to the geometric group G generated by unitary reflexions in $\mathbb{C}^4$. The case $s = 2$ is clearly of little interest (indeed, then $\mathcal{K}(\mathcal{P}) = \{6, 2\}$); it is thus appropriate to take $q = 2$; hence $s = 2, 3$ or 4. With $q = 2$, we have

$$\cos\varphi = \frac{1}{2}\left|1 + 2\cos\frac{2\pi}{s}\right|,$$

or $\varphi = \pi/3$ if $s = 2$ or 4, and $\varphi = \pi/2$ if $s = 3$. Thus, the triangle in Figure 11F4 belongs to a geometric group $G^3(3, k; s, s)$ with $k = 3$ if $s = 2$ or 4, or $k = 2$ if $s = 3$; in any case, the latter is just the group $[3, s]$ (see Section 9D). More generally, if $q = 2$, then we can identify the diagram of Figure 11F4 as that of the group $[3, s]$ for each $s \geqslant 2$. Hence, when $s = 5$, once again we have a finite cut of an infinite polytope.

We can confirm this as follows. Apply the basic change of generators $\sigma_1 \mapsto \sigma_2\sigma_1\sigma_2$ to Λ. If $q = 2$ and s is arbitrary, then, since

$$(\sigma_2\sigma_1\sigma_2\sigma_4)\sigma_3\sigma_4 = (\sigma_4\sigma_2\sigma_1\sigma_2)\sigma_3\sigma_4 \sim \sigma_2\sigma_1\sigma_2\sigma_3,$$

we have the relation

11F7 $$(\sigma_2\sigma_1\sigma_2\sigma_4\sigma_3\sigma_4)^2 = \varepsilon.$$

Now we see that the basic change of generators directly yields the diagram for the (abstract) group $[3, s]$. Alternatively, we can argue geometrically as follows. We find, again after some calculation in G (with any $p, s \geqslant 2$), that $\sigma_2\sigma_1\sigma_2\sigma_4\sigma_3\sigma_4$ corresponds to a rotation through an angle 2ψ $(\leqslant \pi)$ given by

$$\cos\psi = 2\cos\frac{\pi}{q}\cos\frac{\pi}{s},$$

provided that this term is less than 1. (As before, the other possibility is that the term is 1.) When $q = 2$, we obtain $\psi = \pi/2$, so that the corresponding change of generators directly yields the geometric group $[3, s]$. Moreover, when $s = 3$ and $q \geqslant 2$, we obtain $\psi = \pi/q$, and the geometric group is thus $[1\,1\,1^q]^3 = [1\,1\,1]^q$, yielding $\{6, 3\}_{(q,0)}$ (if $q < \infty$) or $\{6, 3\}$ (if $q = \infty$) by twisting with a single automorphism (see Lemma 11A5).

Finally, if $q = 2$, there is yet another approach, obtained by exploiting the relationship between the polytopes $\mathcal{P} := \{\{6, 3\}_{(2,0)}, \{3, 6\}_{(s,0)}\}$ and $\mathcal{Q} := \{\{3, 3\}, \{6, 3\}_{(s,0)}\}$ (see Lemma 11A11). Indeed, $\mathcal{K}(\mathcal{P})$ is related to $\mathcal{K}(\mathcal{Q})$ in just the same way as $\mathcal{P}$ to $\mathcal{Q}$, and so we can appeal to Theorem 11F4 to find its structure. In particular,

$$\Gamma(\mathcal{K}(\mathcal{P})) \cong \Gamma(\mathcal{K}(\mathcal{Q})) \times C_2 = [3, s] \times C_2;$$

we now have a direct product.

We can summarize these results in the following theorem. See Section 7B for the notation used in part (a), and recall the way that the star notation is used for holes in the dual.

11F8 Theorem *Let $q, s \geqslant 2$, and let $\mathcal{P} := \{\{6, 3\}_{(q,0)}, \{3, 6\}_{(s,0)}\}$. Then the cut $\mathcal{K}(\mathcal{P})$ is of type $\{6, s\}$, and has the following properties.*

(a) If $q = 2$, then $\mathcal{K}(\mathcal{P}) = \{6, s \mid \cdot, *2\} = \{s, 6 \mid \cdot, 2\}^*$, and $\Gamma(\mathcal{K}(\mathcal{P})) = [s, 3] \times C_2$. In particular, $\mathcal{K}(\mathcal{P})$ is finite if and only if $s \leqslant 5$.
(b) If $s = 2$, then $\mathcal{K}(\mathcal{P}) = \{6, 2\}$.
(c) If $s = 3$ and $q < \infty$, then $\mathcal{K}(\mathcal{P})$ covers the toroidal polyhedron $\{6, 3\}_{(q,0)}$ (indeed, most likely, it is isomorphic to this polyhedron). If $s = 3$ and $q = \infty$, then $\mathcal{K}(\mathcal{P}) = \{6, 3\}$.
(d) In all other cases $\mathcal{K}(\mathcal{P})$ is infinite.

Proof. Recall that $\mathcal{K}(\mathcal{P})$ is constructed from Λ by means of the twisting operation in (11F6). For part (a), we use (11F7) to obtain

$$\beta_2\beta_1(\beta_0\beta_1)^2 = \sigma_4\sigma_3\sigma_4\sigma_2\sigma_1\sigma_2;$$

thus $(\beta_2\beta_1(\beta_0\beta_1)^2)^2 = \varepsilon$. Then, since $\Lambda = [s, 3]$, this is the only extra relation required to define $\mathcal{K}(\mathcal{P})$. Now (a) follows from the observation that this is also the defining relation for the dual of $\{s, 6 \mid \cdot, 2\}$.

For part (b), if $s = 2$, then (11F6) directly applied to $\Lambda = S_3 \times C_2$ yields the map $\{6, 2\}$. For part (c), if $s = 3$, the geometric group for Λ is $[1\,1\,1]^q$, and therefore $\mathcal{K}(\mathcal{P})$ covers the polyhedron obtained from $[1\,1\,1]^q$ by twisting with a single automorphism. Finally, as we have mentioned before, part (d) is obvious, because the corresponding groups are infinite. □

In Theorem 11F8, if $(q, s) = (2, 5)$, then we again obtain a finite cut of an infinite polytope $\mathcal{P}$. Indeed, $\mathcal{K}(\mathcal{P}) = \{6, 5 \,|\, \cdot, *2\}$ is finite, with group $[5, 3] \times C_2$, but $\{\{6, 3\}_{(2,0)}, \{3, 6\}_{(5,0)}\}$ is infinite.

For the polytope $\{\{6, 3\}_{(q,q)}, \{3, 6\}_{(s,0)}\}$ the structure of its cut is not known. Limited information is available from the fact that this polytope covers $\{\{6, 3\}_{(q,0)}, \{3, 6\}_{(s,0)}\}$, and so there must be a corresponding covering between the cuts.

Polytopes with Vertex-Figures $\{3, 6\}_{(s,s)}$

For the regular polytope $\mathcal{P} := \{\{p, 3\}, \{3, 6\}_{(s,s)}\}$, which is the dual of ${}_p\mathcal{T}^4_{(s,s)}$, the cut $\mathcal{K}(\mathcal{P})$ is a polyhedron of type $\{p, 3s\}$. But since the length $3s$ of the 3-holes does not uniquely determine $\{3, 6\}_{(s,s)}$ among the toroidal polyhedra of type $\{3, 6\}$, it is now less clear what the structure of $\mathcal{K}(\mathcal{P})$ will be. We shall restrict ourselves mainly to the case $p = 3$. In any event, since we have a complete classification in Section 11C (in the dual formulation) of all the finite universal polytopes with vertex-figures $\{3, 6\}_{(s,s)}$, information given us by cuts may be interesting, but is rather less important.

We summarize our results in the following theorem. We omit the proof, which, once again, employs the geometric group associated with $\mathcal{P}$.

11F9 Theorem *For the polytopes* $\mathcal{P} := \{\{3, 3\}, \{3, 6\}_{(s,s)}\}$ *with* $s \geqslant 2$, *the cut* $\mathcal{K}(\mathcal{P})$ *is of type* $\{3, 3s\}$. *In particular,* $\mathcal{K}(\mathcal{P}) = \{3, 6\}_{(2,2)}$ *if* $s = 2$, *and is infinite if* $s \geqslant 3$.

In Theorem 11F9, the polytope $\mathcal{P}$ is finite if and only if its cut is finite. Under the assumption that $s \geqslant 6$, the following argument gives another explanation why the cut must be infinite.

Since $\mathcal{P}$ covers the polytope $\mathcal{Q} := \{\{3, 3\}, \{3, 6\}_{(s,0)}\}$, its cut $\mathcal{K}(\mathcal{P})$, which is of type $\{3, 3s\}$, must also cover $\mathcal{K}(\mathcal{Q}) = \{3, s\}$ (see Theorem 11F4); indeed, the group homomorphism $\Gamma(\mathcal{P}) \to \Gamma(\mathcal{Q})$ induces a corresponding homomorphism $\Gamma(\mathcal{K}(\mathcal{P})) \to \Gamma(\mathcal{K}(\mathcal{Q}))$. Hence, $\mathcal{K}(\mathcal{P})$ must be infinite if $s \geqslant 6$. Note that we also have a covering $\mathcal{R} := \{\{3, 3\}, \{3, 6\}_{(3s,0)}\} \searrow \mathcal{P}$, and then, in turn, a covering $\{3, 3s\} = \mathcal{K}(\mathcal{R}) \searrow \mathcal{K}(\mathcal{P})$.

Other Cuts

As we explained earlier, the cut $\mathcal{K}(\mathcal{P})$ is naturally associated with the generator ρ_3 of $\Gamma(\mathcal{P})$, in the sense that Wythoff's construction of $\mathcal{K}(\mathcal{P})$ in the order complex $\mathcal{C}(\mathcal{P})$ leads to a polyhedron which is invariant under ρ_3. (For the order complex, see Section 2C. Recall that, combinatorially, $\mathcal{C}(\mathcal{P})$ is constructed by barycentric subdivision of the facets of $\mathcal{P}$ in a consistent way.) We now describe other cuts for regular polytopes of types $\{3, 3, 6\}$ or $\{3, 6, 3\}$.

For a regular polytope $\mathcal{P}$ in the class $\langle\{3,3\},\{3,6\}_{(s,t)}\rangle$, with $t=0$ or s, consider the operation

11F10 $$(\rho_0,\rho_1,\rho_2,\rho_3)\mapsto(\rho_0,\rho_1\rho_2\rho_3\rho_2\rho_1,\rho_3\rho_2\rho_3\rho_2\rho_3)=:(\gamma_0,\gamma_1,\gamma_2)$$

on $\Gamma(\mathcal{P})=\langle\rho_0,\ldots,\rho_3\rangle$. Then

$$\begin{aligned}\gamma_0\gamma_1 &\sim \rho_1\rho_0\rho_1\rho_2\rho_3\rho_2 = \rho_0\rho_1\rho_0\rho_2\rho_3\rho_2 = \rho_0\rho_1\rho_2\rho_3\rho_2\rho_0\\ &\sim \rho_1\rho_2\rho_3\rho_2 \sim \rho_2\rho_1\rho_2\rho_3 = \rho_1\rho_2\rho_1\rho_3 = \rho_1\rho_2\rho_3\rho_1 \sim \rho_2\rho_3,\end{aligned}$$

and

$$\gamma_1\gamma_2 = \rho_1\rho_2\rho_3\rho_2\rho_1\cdot\rho_3\rho_2\rho_3\rho_2\rho_3 = (\rho_1(\rho_2\rho_3)^2)^2.$$

It follows that $\langle\gamma_0,\gamma_1,\gamma_2\rangle$ is the group of a regular polyhedron $\mathcal{L}(\mathcal{P})$ of type $\{6,s\}$. Note that the period of $\gamma_1\gamma_2$ is s in both cases $t=0$ and $s=t$ (see (11A1)).

This polyhedron can be obtained from $\mathcal{P}$ (or, more exactly, from $\mathcal{C}(\mathcal{P})$) by applying Wythoff's construction with the base vertex F_0 of $\mathcal{P}$ as initial vertex. This shows that $\mathcal{L}(\mathcal{P})$ has its vertices and edges among those of $\mathcal{P}$. Now the generator ρ_2 commutes with $\gamma_0,\gamma_1,\gamma_2$, so that the cut $\mathcal{L}(\mathcal{P})$ is now invariant under ρ_2; that is, in a sense, $\mathcal{L}(\mathcal{P})$ lies on the reflexion wall of ρ_2.

Once again, let $\mathcal{P}:=\{\{3,3\},\{3,6\}_{(s,s)}\}$, the dual of ${}_3\mathcal{T}^4_{(s,s)}$. Then, by (11A1), the relation $(\gamma_1\gamma_2)^s=\varepsilon$ genuinely does specify the original polytope. Now, using the construction of ${}_3\mathcal{T}^4_{(s,s)}$ described in Section 11C (see (11C1) and (11C2)), we find that

$$(\gamma_0,\gamma_1,\gamma_2)=(\tau_3,\sigma_3,\sigma_1\sigma_2\sigma_1),$$

so that $\Gamma(\mathcal{L}(\mathcal{P}))=\langle\sigma_1\sigma_2\sigma_1,\sigma_3,\sigma_4\rangle\rtimes\langle\tau_3\rangle$. A similar situation was encountered for the polytope $\mathcal{Q}:=\{\{6,3\}_{(s,0)},\{3,6\}_{(s,0)}\}$ and its cut $\mathcal{K}(\mathcal{Q})$ (see Figure 11F3 and (11F5) with $q=s$). In particular, we have

$$\mathcal{L}(\mathcal{P})=\mathcal{K}(\mathcal{Q}),$$

and so we can obtain further properties of the cut from Theorem 11F8.

If $\mathcal{P}:=\{\{3,3\},\{3,6\}_{(s,0)}\}$, then the relation $(\gamma_1\gamma_2)^s=\varepsilon$ does not specify the group of the original polytope, and so the structure of $\mathcal{L}(\mathcal{P})$ is less clear.

To end this section, we briefly comment on the polytopes $\mathcal{P}$ of type $\{3,6,3\}$. Here the "sections" of $\mathcal{P}$ by its "reflexion walls" do not appear to yield useful information. However, in certain cases, limited information is available by other means.

Consider the operation

$$(\rho_0,\rho_1,\rho_2,\rho_3)\mapsto(\rho_0,(\rho_1\rho_2)^3,\rho_3)=:(\kappa_0,\kappa_1,\kappa_2)$$

on $\Gamma(\mathcal{P})=\langle\rho_0,\ldots,\rho_3\rangle$, which gives the group $\langle\kappa_0,\kappa_1,\kappa_2\rangle$ of a regular map $\mathcal{L}(\mathcal{P})$. Observe that $\mathcal{L}(\mathcal{P}^*)=\mathcal{L}(\mathcal{P})^*$. Now the period of $\kappa_0\kappa_1$ is just the length of the 3-zigzags of the facets of $\mathcal{P}$ (see Section 7B). In particular, if these facets are toroidal polyhedra

$\{3,6\}_{(q,r)}$, then this period is given by

$$\begin{cases} q & \text{if } r = 0 \text{ and } q \text{ is even;} \\ 2q & \text{if } r = 0 \text{ and } q \text{ is odd;} \\ 3q & \text{if } r = q \text{ and } q \text{ is even;} \\ 6q & \text{if } r = q \text{ and } q \text{ is odd.} \end{cases}$$

Indeed, a 3-zigzag of the polyhedron $\{3,6\}_{(q,r)}$ takes all the edges of a 3-hole, but traverses them twice if q is odd (because it uses opposite orientations at consecutive vertices). Moreover, if q is odd, then $\rho_2 \in \langle \kappa_0, \kappa_1 \rangle$. All this is most easily seen geometrically. A similar remark applies to the vertex-figures $\{6,3\}_{(s,t)}$ (say). In particular, if q and s are odd, then $\Gamma(\mathcal{L}(\mathcal{P})) = \Gamma(\mathcal{P})$. It is thus only when $\mathcal{P} = \{\{3,6\}_{(2m,0)}, \{6,3\}_{(2n,0)}\}$ that the periods of $\kappa_0\kappa_1$ and $\kappa_1\kappa_2$ specify the type of the polytope; in this case, $\mathcal{L}(\mathcal{P})$ is of type $\{2m, 2n\}$. We may guess that $\mathcal{L}(\mathcal{P})$ is universal; certainly, $\mathcal{L}(\mathcal{P})$ is finite when $m = 1$, just as $\mathcal{P}$ is, whereas it is likely that $\mathcal{P}$ is infinite if $m, n \geqslant 2$, which would accord with $\mathcal{L}(\mathcal{P})$ being infinite.

The geometry of the cut appears to be a little different from the algebra. Whether or not q is even, applying the analogue of Wythoff's construction to the initial vertex of $\mathcal{P}$ yields a q-gonal face if $r = 0$, and a $3q$-gonal one if $r = q$; when q is odd, the abstract definition (using the group) gives a double covering (now the 3-zigzags of the facets traverse the 3-holes twice). The same argument applies to the dual. However, when s is odd, the vertex-figure is definitely a $2s$- or $6s$-gon (as appropriate), and the natural covering identification of the vertex-figure (to yield one with half as many vertices) seems not to carry over to one of the polytope as a whole. In the known finite case (namely, $\{\{3,6\}_{(3,0)}, \{6,3\}_{(3,0)}\}$), the group $[q, s]$ cannot be a quotient of $\Gamma(\mathcal{P})$ (its order does not divide that of $\Gamma(\mathcal{P})$).

Nevertheless, in the case of $\{\{3,6\}_{(q,0)}, \{6,3\}_{(s,0)}\}$, there remains the feeling that, in some sense, there is an associated polyhedron $\{q, s\}$. When $(q, s) = (3, 5)$, this will be finite, and hence reinforces the notion expressed at the end of Section 11E that $\{\{3,6\}_{(3,0)}, \{6,3\}_{(5,0)}\}$ could be finite.

11G Hyperbolic Honeycombs in $\mathbb{H}^3$

In this section, we shall describe subgroup relationships among the symmetry groups of certain regular honeycombs in hyperbolic 3-space $\mathbb{H}^3$. They translate into corresponding relationships between certain pairs of locally toroidal regular 4-polytopes, which we investigate in the next section.

Here we are concerned with the honeycombs $\{3,3,6\}$, $\{4,3,6\}$, $\{6,3,6\}$ and $\{3,6,3\}$ in $\mathbb{H}^3$, which have all their vertices on the absolute quadric; the two self-dual honeycombs $\{6,3,6\}$ and $\{3,6,3\}$ also have their facets inscribed in the absolute (see Coxeter [113, §3; 136, §4], and Sections 3C and 10A). The subgroup relationships are displayed in Figure 11G1, which also includes a hyperbolic Coxeter group with a branched diagram, which itself is not the symmetry group of a regular honeycomb. Inclusion of the smaller group in the larger group is indicated by an arrow, which is marked by the corresponding subgroup index.

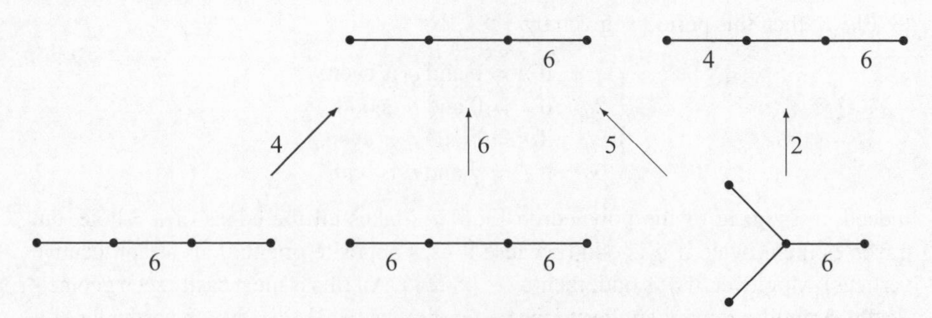

Figure 11G1. Hyperbolic subgroup relationships.

Each subgroup relation of Figure 11G1 can be obtained by simplex dissection of hyperbolic tetrahedra. If the index of the subgroup is k, then its fundamental simplex is decomposed into k congruent copies of the fundamental simplex for the larger group. We shall not discuss the details of these dissections, but refer instead to [300, §3], where they are described in terms of the diagram notation introduced in [120, p. 281] (see also [318, 319]).

Using these simplex dissections, it is easy to obtain generators for the appropriate subgroups. Let $[3, 3, 6] = \langle \sigma_0, \ldots, \sigma_3 \rangle$. Then we get generators $\rho_0, \ldots, \rho_3$ for the subgroups by the following (mixing) operations:

11G1

$$\lambda\colon (\sigma_0, \ldots, \sigma_3) \longrightarrow (\sigma_0, \sigma_1, \sigma_2\sigma_3\sigma_2, \sigma_3) =: (\rho_0, \ldots, \rho_3) \qquad [3, 6, 3]$$

$$\mu\colon (\sigma_0, \ldots, \sigma_3) \longrightarrow (\sigma_0, \sigma_1\sigma_2\sigma_3\sigma_2\sigma_1, \sigma_3, \sigma_2) =: (\rho_0, \ldots, \rho_3) \qquad [6, 3, 6]$$

$$\nu\colon (\sigma_0, \ldots, \sigma_3) \longrightarrow (\sigma_0, \sigma_2, \sigma_1, \sigma_3\sigma_2\sigma_3\sigma_2\sigma_3) =: (\rho_0, \ldots, \rho_3) \qquad \left[\begin{matrix}3, \\ 3,\end{matrix}\, 6\right]$$

The notation for the last group (which corresponds to the last diagram of Figure 11G1), is adapted from [120, §8.1]. For example, for the group [3, 6, 3], a fundamental simplex can be constructed from the fundamental simplex for [3, 3, 6] by keeping all walls but the third (corresponding to σ_2), while replacing the reflexion wall for σ_2 by the image under σ_2 of the reflexion wall for σ_3 (see Figure 11G2).

Finally, the group $[4, 3, 6] = \langle \psi_0, \ldots, \psi_3 \rangle$ is derived from the last group in (11G1) by the twisting operation indicated in Figure 11G3. More exactly, the operation is given by

11G2
$$\kappa\colon (\rho_0, \ldots, \rho_3; \tau) \mapsto (\tau, \rho_1, \rho_2, \rho_3) =: (\psi_0, \ldots, \psi_3).$$

We now investigate the geometric relationships for the honeycombs corresponding to these subgroup relationships. In describing these relationships we require the following two operations on the groups of regular polyhedra (see Section 7B).

Let $\mathcal{L}$ be a regular polyhedron of type $\{p, q\}$, with group $\Gamma(\mathcal{L}) = \langle \psi_0, \psi_1, \psi_2 \rangle$ (say). Recall that the 2nd facetting operation φ_2 on $\Gamma(\mathcal{L})$ is given by

11G3
$$\varphi_2\colon (\psi_0, \psi_1, \psi_2) \mapsto (\psi_0, \psi_1\psi_2\psi_1, \psi_2) =: (\alpha_0, \alpha_1, \alpha_2).$$

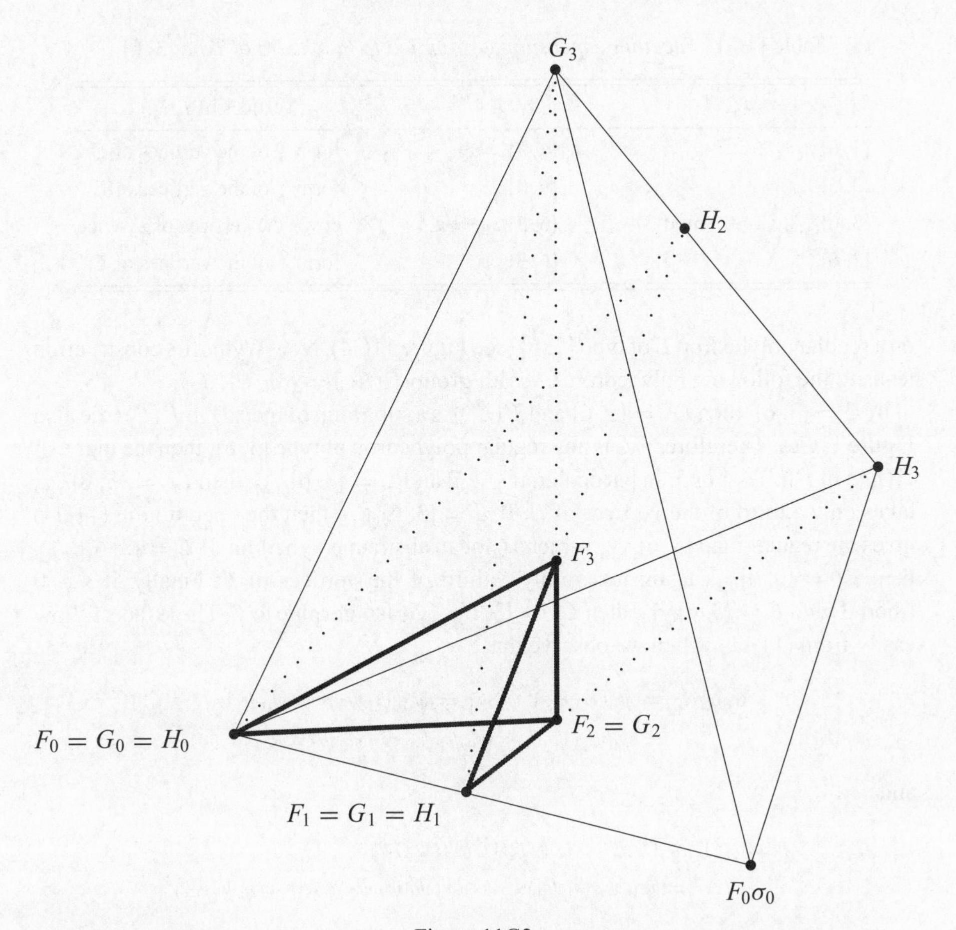

Figure 11G2

If q is odd, then $\Gamma := \langle \alpha_0, \alpha_1, \alpha_2 \rangle = \Gamma(\mathcal{L})$, and Γ is the group of a regular polyhedron $\mathcal{L}^{\varphi_2}$ of type $\{l, q\}$, where l is the length of the 2-holes of $\mathcal{L}$. If q is even, then in general Γ is a proper subgroup of $\Gamma(\mathcal{L})$. The effect of φ_2 is best studied by employing Wythoff's construction to find $\mathcal{L}^{\varphi_2}$. In particular, we obtain the results of Table 11G1.

We also need the operation

11G4 $\chi: (\psi_0, \psi_1, \psi_2) \mapsto (\psi_0\psi_1\psi_2\psi_1\psi_0, \psi_2, \psi_1) =: (\alpha_0, \alpha_1, \alpha_2)$

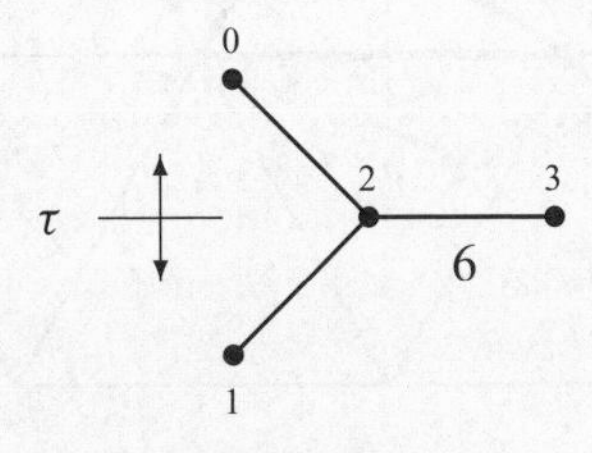

Figure 11G3

Table 11G1. *Facetting Operation φ_2 Applied to Polyhedra of Type* $\{3, 6\}$

$\mathcal{L}$	$\mathcal{L}^{\varphi_2}$	Vertices of $\mathcal{L}^{\varphi_2}$
$\{3, 6\}$	$\{6, 3\}$	form $\frac{2}{3}$ of the vertices of $\mathcal{L}$
$\{3, 6\}_{(3r,0)}$	$\{6, 3\}_{(r,r)}$	form $\frac{2}{3}$ of the vertices of $\mathcal{L}$
$\{3, 6\}_{(s,0)}$, $s \not\equiv 0 \pmod 3$	$\{6, 3\}_{(s,0)} = \mathcal{L}^*$	cover the vertices of $\mathcal{L}$ twice
$\{3, 6\}_{(s,s)}$	$\{6, 3\}_{(s,0)}$	form $\frac{2}{3}$ of the vertices of $\mathcal{L}$

on a regular polyhedron $\mathcal{L}$ of type $\{3, 6\}$ (see Figure 11G4). Now Wythoff's construction leads to the following polyhedron $\mathcal{L}^\chi$ with group $\Gamma(\mathcal{L}^\chi) = \langle \alpha_0, \alpha_1, \alpha_2 \rangle$.

If $\mathcal{L} = \{3, 6\}$, then $\mathcal{L}^\chi = \{3, 6\}$ and $\Gamma(\mathcal{L}^\chi)$ is a subgroup of index 3 in $\Gamma(\mathcal{L})$ (see also Figure 11G6). Therefore, if $\mathcal{L}$ is any regular polyhedron of type $\{3, 6\}$, then the index of $\Gamma(\mathcal{L}^\chi)$ in $\Gamma(\mathcal{L})$ is 1 or 3. In particular, if $s \geqslant 2$ and $\mathcal{L} = \{3, 6\}_{(s,s)}$, then $\mathcal{L}^\chi = \{3, 6\}_{(s,0)}$ takes only a third of the vertices of $\mathcal{L}$. If $\mathcal{L} = \{3, 6\}_{(1,1)}$, then the operation in (11G4) gives the regular map $\{3, 6\}_{(1,0)}$, which is not an abstract polyhedron. If $\mathcal{L} = \{3, 6\}_{(3r,0)}$, then $\mathcal{L}^\chi = \{3, 6\}_{(r,r)}$ again takes only a third of the vertices of $\mathcal{L}$. Finally, if $s \not\equiv 0 \pmod 3$ and $\mathcal{L} = \{3, 6\}_{(s,0)}$, then $\mathcal{L}^\chi = \{3, 6\}_{(s,0)}$ is isomorphic to $\mathcal{L}$. These facts follow easily from (11A1), when we observe that

$$\begin{aligned}\alpha_0\alpha_1\alpha_2 &= \psi_0(\psi_1\psi_2)^2\psi_0\psi_1 = \psi_0(\psi_1\psi_2)^2\psi_1\psi_0\psi_1\psi_0 \\ &= \psi_0\psi_1(\psi_2\psi_1)^2\psi_0\psi_1\psi_0 \sim (\psi_2\psi_1)^2\psi_0,\end{aligned}$$

and

$$\begin{aligned}\alpha_0(\alpha_1\alpha_2)^2 &= (\psi_0\psi_1)\psi_2\psi_1\psi_2(\psi_0\psi_1\psi_2)\psi_1 \\ &= \psi_1(\psi_0\psi_1\psi_0)\psi_2\psi_1\psi_2(\psi_0\psi_1\psi_2)\psi_1 \sim (\psi_0\psi_1\psi_2)^3.\end{aligned}$$

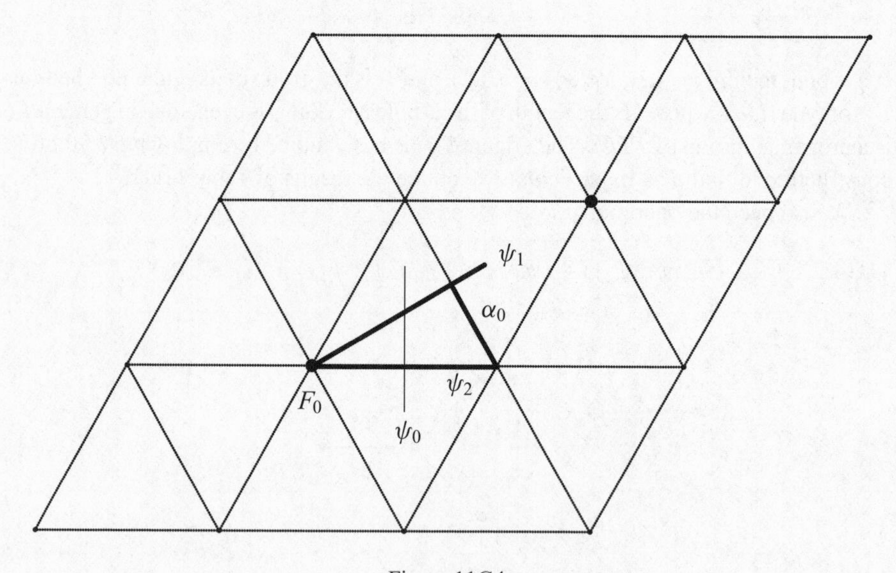

Figure 11G4

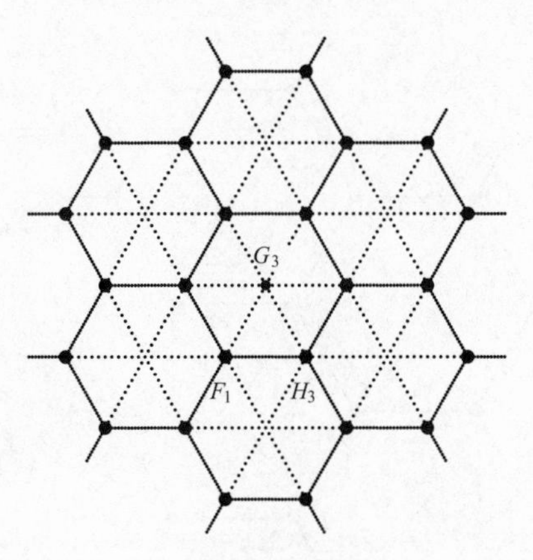

Figure 11G5

The geometric relationships between the honeycombs can now be described as follows. Once again, we employ Wythoff's construction.

For the regular honeycomb $\{3, 6, 3\}$, the facets are (in one-to-one correspondence with) certain vertex-figures of $\{3, 3, 6\}$, while its vertex-figures are tessellations $\{6, 3\}$ whose vertices and edges occur among those of the original vertex-figure $\{3, 6\}$ (see Figure 11G5). Indeed, consider the fundamental simplex T for $[3, 3, 6]$ defined by the base flag $\Phi := \{F_0, \ldots, F_3\}$ of $\{3, 3, 6\}$, whose vertices are again denoted by $F_0, \ldots, F_3$ (see Figure 11G2). Then the operation λ of (11G1) corresponds to changing T to the 4 times larger fundamental simplex T_λ for the subgroup $[3, 6, 3]$ with vertices $G_0 := F_0$, $G_1 := F_1$, $G_2 := F_2$ and G_3. Here our notation is such that G_i corresponds to (the centre of) the i-face in the base flag of $\{3, 6, 3\}$ belonging to the generators $\rho_0, \ldots, \rho_3$. Now, since the vertex G_3 of T_λ is fixed under the subgroup $\langle \rho_0, \rho_1, \rho_2 \rangle$ of $[3, 6, 3]$, Wythoff's construction applied with $G_0 = F_0$ as initial vertex shows that the 3-face G_3 of $\{3, 6, 3\}$ corresponds to the vertex-figure of $\{3, 3, 6\}$ at its vertex $F_0\sigma_0\sigma_1\sigma_2$ ($= G_3$). Similarly, the vertex F_0 of T_λ is invariant under the subgroup $\langle \rho_1, \rho_2, \rho_3 \rangle$ of $[3, 6, 3]$, so that the vertex-figure of $\{3, 6, 3\}$ at F_0 is obtained by clustering triangles in the vertex-figure of $\{3, 3, 6\}$ at F_0 as indicated in Figure 11G5. This clustering corresponds to an application of the facetting operation φ_2 of (11G3).

We can deal with the honeycomb $\{6, 3, 6\}$ in a similar fashion. Now the facets are some of the vertex-figures of $\{3, 6, 3\} = \{3, 3, 6\}^\lambda$, while the vertex-figures are tessellations $\{3, 6\}$ obtained from vertex-figures $\{6, 3\}$ of $\{3, 3, 6\}^\lambda$ by an operation indicated in Figure 11G6. The latter corresponds to an application of the operation χ of (11G4) to the vertex-figure of $\{3, 3, 6\}$. In fact, the operation μ of (11G1) is equivalent to changing T to the 6 times larger fundamental simplex T_μ for $[6, 3, 6]$ with vertices $H_0 := F_0$, $H_1 := F_1$, H_2 and $H_3 := F_0\sigma_0\sigma_1$ (see Figure 11G2). To find the structure of the facets, note that the subgroup $\langle \rho_0, \rho_1, \rho_2 \rangle$ fixes H_3, and that H_3 ($= F_0\sigma_0\sigma_1$) is a vertex of the honeycomb $\{3, 3, 6\}^\lambda$. For the vertex-figures, note that $\langle \rho_1, \rho_2, \rho_3 \rangle$ fixes F_0, while $\langle \rho_2, \rho_3 \rangle$ fixes F_1.

Finally, the operation ν of (11G1) also has a geometric counterpart (see Figure 11G7). If T_ν denotes the simplex with vertices $K_0, \ldots, K_3$, then $\rho_0, \ldots, \rho_3$ are

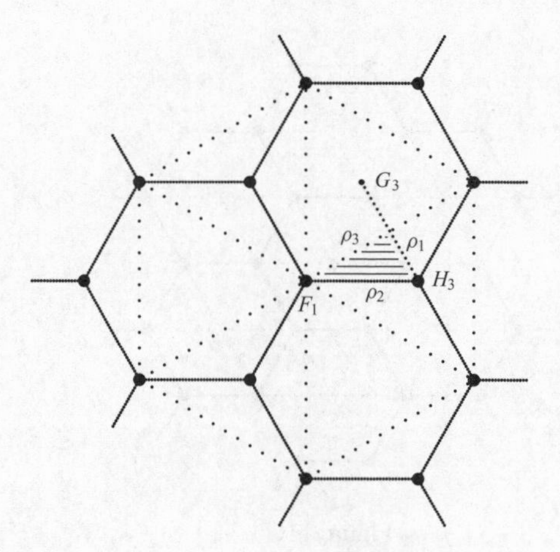

Figure 11G6

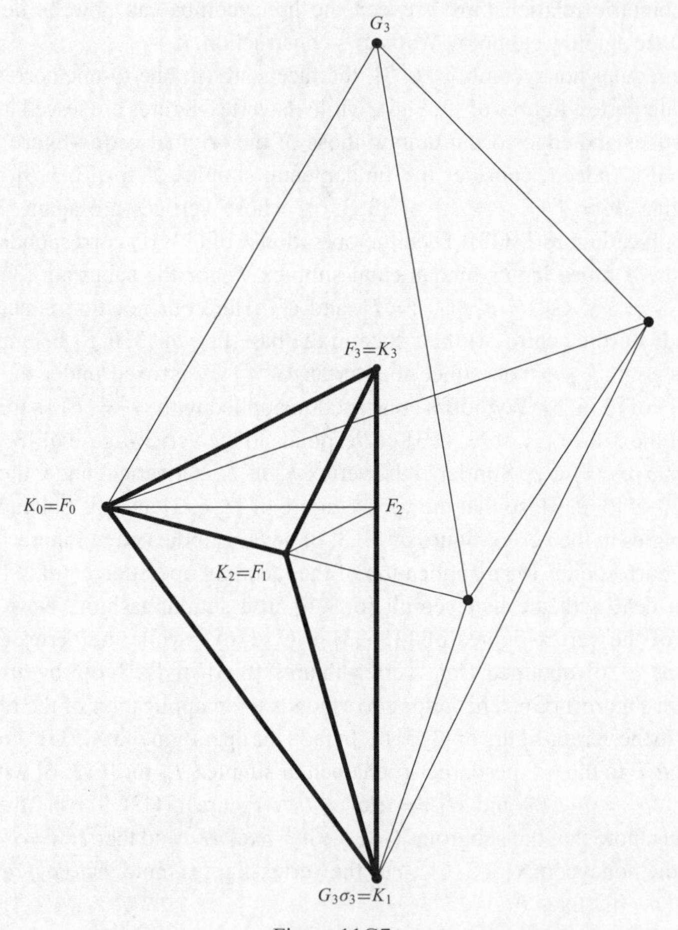

Figure 11G7

just the generating reflexions in the walls of T_ν, such that ρ_i is the reflexion in the wall opposite to K_i. Since $\langle \rho_0, \rho_1, \rho_2 \rangle = \langle \sigma_0, \sigma_1, \sigma_2 \rangle$ is the group of the facet F_3 of {3, 3, 6}, the transforms of T_ν under this subgroup fit together to give five tetrahedral facets of {3, 3, 6}, namely, F_3 and its four adjacent facets. These five tetrahedra form a cube, and this clustering of tetrahedra in fives extends to give a honeycomb {4, 3, 6} inscribed in {3, 3, 6}. In this context, note that the cubical facets really only have the symmetry of tetrahedra. The index of the subgroup $\langle \rho_0, \ldots, \rho_3 \rangle$ in $\langle \sigma_0, \ldots, \sigma_3 \rangle$ is 5, because the fundamental region T_ν for $\langle \rho_0, \ldots, \rho_3 \rangle$ can now be decomposed into 5 copies of T.

11H Relationships Among Polytopes of Types $\{6, 3, p\}$ or $\{3, 6, 3\}$

In this section, we discuss relationships among locally toroidal regular 4-polytopes of types {6, 3, p}, {p, 3, 6} or {3, 6, 3}. Since their groups are quotients of the symmetry group of hyperbolic honeycombs in $\mathbb{H}^3$, we can appeal to the results of the previous section, and obtain certain subgroup relationships. For example, by (11G1), a quotient of [3, 3, 6] must contain as subgroups certain quotients of [3, 6, 3] or [6, 3, 6], whose indices divide 4 or 6, respectively. A similar remark applies to the other groups of Figure 11G1.

The geometric relationships are established by applying the operations in (11G1) to the group of a regular polytope $\mathcal{P}$ of type {3, 3, 6}, with group $\Gamma(\mathcal{P}) = \langle \sigma_0, \ldots, \sigma_3 \rangle$. Since its facets are 3-simplices, such a polytope $\mathcal{P}$ is almost always a simplicial complex in the usual sense. For example, if $\mathcal{P}$ is the polytope

$$\mathcal{P}_{\mathbf{s}} := \{\{3, 3\}, \{3, 6\}_{\mathbf{s}}\},$$

which is the dual of ${}_3\mathcal{T}_{\mathbf{s}}^4$, then this is true unless $\mathbf{s} = (2, 0)$.

We shall also employ the action of $\Gamma(\mathcal{P})$ on the order complex $\mathcal{C}(\mathcal{P})$ of $\mathcal{P}$. If $\Phi = \{F_0, \ldots, F_3\}$ denotes the base flag of $\mathcal{P}$, then the 3-simplex of $\mathcal{C}(\mathcal{P})$ corresponding to Φ is a fundamental region for the action of $\Gamma(\mathcal{P})$ on $|\mathcal{C}(\mathcal{P})|$, the topological space associated with $\mathcal{C}(\mathcal{P})$. Alternatively, $\mathcal{C}(\mathcal{P})$ can also be obtained from the geometric barycentric subdivision of the hyperbolic honeycomb {3, 3, 6} by making identifications corresponding to the extra defining relations for $\Gamma(\mathcal{P})$.

The Types {3, 3, 6} *and* {3, 6, 3}

Let $\mathcal{P}$ be a regular 4-polytope in $\langle \{3, 3\}, \{3, 6\}_{\mathbf{s}} \rangle$. We begin with the construction of the polytope $\mathcal{P}^\lambda$ of type {3, 6, 3}, where, as in (11G1),

11H1 $$\lambda\colon (\sigma_0, \ldots, \sigma_3) \mapsto (\sigma_0, \sigma_1, \sigma_2\sigma_3\sigma_2, \sigma_3) =: (\rho_0, \ldots, \rho_3).$$

We shall impose on the original polytope $\mathcal{P}$ the (weak) condition that its edge-graph (consisting of all vertices and edges of $\mathcal{P}$) has no multiple edges, so that any two vertices of $\mathcal{P}$ are joined by at most one edge of $\mathcal{P}$. This is satisfied, for example, if $\mathcal{P}$ is a simplicial complex. Indeed, we do not know of any polytope which violates this condition; it even holds for $\mathcal{P}_{(2,0)}$.

Now, $\mathcal{P}^\lambda$ can be constructed from $\mathcal{P}$ in a similar fashion as $\{3, 6, 3\}$ was from $\{3, 3, 6\}$. Figure 11G2 illustrates again the relation between the fundamental simplex T for $\Gamma(\mathcal{P})$, with vertices F_0, F_1, F_2, F_3, and the simplex T_λ for $\Gamma(\mathcal{P}^\lambda)$, with vertices G_0, G_1, G_2, G_3; here we use the same notation as in the previous section.

In particular, applying Wythoff's construction shows that each facet of $\mathcal{P}^\lambda$ is a vertex-figure $\{3, 6\}_{\mathbf{s}}$ of the original polytope $\mathcal{P}$. On the other hand, each vertex-figure of $\mathcal{P}^\lambda$ is a transform of a vertex-figure $\{3, 6\}_{\mathbf{s}}$ of $\mathcal{P}$ under the facetting operation φ_2, and thus depends on $\mathbf{s}$ as described in Table 11G1. For the proof that $\Gamma := \langle \rho_0, \ldots, \rho_3 \rangle$ is indeed a C-group, namely, the group of the polytope $\mathcal{P}^\lambda$, we need to verify the intersection property.

Let $\psi \in \langle \rho_0, \rho_1, \rho_2 \rangle \cap \langle \rho_1, \rho_2, \rho_3 \rangle$. Then ψ fixes the two vertices F_0 and G_3 of $\mathcal{P}$. Now, since the graph of $\mathcal{P}$ has no multiple edges, these vertices are joined by just one edge, namely, that edge of F_3 which connects them. It follows that $\psi \in \langle \rho_1, \rho_2 \rangle$, as required.

If $\mathbf{s} = (s, 0)$ and $s \not\equiv 0 \pmod 3$, then $\Gamma = \Gamma(\mathcal{P})$, because $\langle \rho_1, \rho_2, \rho_3 \rangle = \langle \sigma_1, \sigma_2, \sigma_3 \rangle$. More generally, to find the index of Γ in $\Gamma(\mathcal{P})$, observe that the simplex T_λ is dissected into four copies of the simplex T, namely, T, $T\sigma_2$, $T\sigma_1\sigma_2$ and $T\sigma_0\sigma_1\sigma_2$. If the index is not 4, then T_λ is not a fundamental region for Γ, and so two of these copies must be equivalent under Γ. Since $\sigma_0, \sigma_1 \in \Gamma$, it follows in this case that $\sigma_2 \in \Gamma$; thus $\Gamma = \Gamma(\mathcal{P})$. Hence the index must always be 1 or 4. If $\Gamma = \Gamma(\mathcal{P})$, then $\sigma_2 \in \Gamma$, and one is tempted to conclude that $\sigma_2 \in \langle \rho_1, \rho_2, \rho_3 \rangle$ (because σ_2 stabilizes the base vertex $G_0 = F_0$ of $\mathcal{P}^\lambda$). However, we cannot be sure that Wythoff's construction of $\mathcal{P}^\lambda$ in $\mathcal{C}(\mathcal{P})$ will indeed give a faithful "realization".

Bearing in mind that $\Gamma = \Gamma(\mathcal{P}^\lambda)$, we have now completed the proof of part (a) of the following theorem.

11H2 Theorem *Let $\mathcal{P}$ be a regular 4-polytope in the class $\langle \{3, 3\}, \{3, 6\}_{\mathbf{s}} \rangle$, and let λ be the operation in (11H1). Assume that the graph of $\mathcal{P}$ has no multiple edges.*

(a) Then $\mathcal{P}^\lambda$ is a regular 4-polytope in the class

$$\begin{cases} \langle \{3, 6\}_{(3r,0)}, \{6, 3\}_{(r,r)} \rangle, & \text{if } \mathbf{s} = (3r, 0) \text{ with } r \geqslant 1; \\ \langle \{3, 6\}_{(s,0)}, \{6, 3\}_{(s,0)} \rangle, & \text{if } \mathbf{s} = (s, 0) \text{ with } s \geqslant 2,\ s \not\equiv 0 \pmod 3; \\ \langle \{3, 6\}_{(s,s)}, \{6, 3\}_{(s,0)} \rangle, & \text{if } \mathbf{s} = (s, s) \text{ with } s \geqslant 2. \end{cases}$$

Furthermore, $|\Gamma(\mathcal{P}) : \Gamma(\mathcal{P}^\lambda)| = 1$ or 4, and $\Gamma(\mathcal{P}^\lambda) = \Gamma(\mathcal{P})$ in the second case.

(b) If $\mathcal{P} = \mathcal{P}_{\mathbf{s}} := \{\{3, 3\}, \{3, 6\}_{\mathbf{s}}\}$, then in the three cases of part (a) the index is given by $|\Gamma(\mathcal{P}) : \Gamma(\mathcal{P}^\lambda)| = 4$, 1 and 4, respectively. In particular, $\mathcal{P}^\lambda_{(s,s)} = \{\{3, 6\}_{(s,s)}, \{6, 3\}_{(s,0)}\}$ $(= {}_7\mathcal{T}^4_{(s,s),(s,0)})$ for each $s \geqslant 2$.

Proof. By construction, if $\mathcal{P}$ is universal in its class, then $\mathcal{P}^\lambda$ is universal among all polytopes obtained by applying λ to a polytope which is in the same class as $\mathcal{P}$. Now Theorem 11H2 says that, at least for $\mathcal{P}_{(s,s)}$, this new polytope is indeed universal in its class. In this context, recall that ${}_7\mathcal{T}^4_{\mathbf{s},\mathbf{t}} := \{\{3, 6\}_{\mathbf{s}}, \{6, 3\}_{\mathbf{t}}\}$ (see Section 11E).

The proof of part (b) of Theorem 11H2 proceeds in several steps. We begin by reviewing the construction of the universal polytopes $\mathcal{P}_{\mathbf{s}}$.

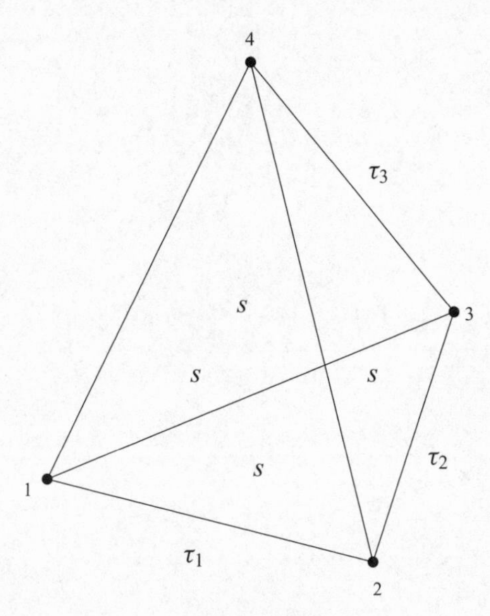

Figure 11H1

Note that the second case of (b) has already been settled. We next deal with the third case. Recall that the polytope $\mathcal{P}_{(s,s)}$, which is the dual of ${}_3\mathcal{T}^4_{(s,s)}$, was obtained from the abstract group $W_1 = \langle \alpha_1, \ldots, \alpha_4 \rangle$ (say) of Figure 11H1 by applying the twisting operation

11H3 $$(\alpha_1, \ldots, \alpha_4; \tau_1, \tau_2, \tau_3) \mapsto (\tau_3, \tau_2, \tau_1, \alpha_1) =: (\sigma_0, \ldots, \sigma_3)$$

(see Section 11C). Then we have the semi-direct product

$$\Gamma(\mathcal{P}_{(s,s)}) = \langle \alpha_1, \ldots, \alpha_4 \rangle \rtimes \langle \tau_1, \tau_2, \tau_3 \rangle = W_1 \rtimes S_4.$$

In particular, $\mathcal{P}_{(s,s)}$ is finite if and only if $s = 2$. In this case, $W_1 = S_5$, and $\Gamma(\mathcal{P}_{(2,2)}) = S_5 \times S_4$.

Now the generators $\rho_0, \ldots, \rho_3$ of $\Gamma = \Gamma(\mathcal{P}^\lambda_{(s,s)})$ are given by

11H4 $$(\rho_1, \rho_1, \rho_2, \rho_3) = (\sigma_0, \sigma_1, \sigma_2\sigma_3\sigma_2, \sigma_3) = (\tau_3, \tau_2, \alpha_2, \alpha_1),$$

so that $\Gamma = W_1 \rtimes S_3$. In particular, the index is $|\Gamma(\mathcal{P}_{(s,s)}) : \Gamma| = 4$.

In order to identify $\mathcal{P}^\lambda_{(s,s)}$, recall that the polytope ${}_7\mathcal{T}^4_{(s,s),(t,0)} = \{\{3, 6\}_{(s,s)}, \{6, 3\}_{(t,0)}\}$, with $t = s$ or $t = 3s$, was constructed from the abstract group $W_2 = \langle \beta_1, \ldots, \beta_4 \rangle$ (say) of Figure 11H2 by means of the twisting operation

11H5 $$(\beta_1, \ldots, \beta_4; \tau_1, \tau_2) \mapsto (\tau_1, \tau_2, \beta_3, \beta_4)$$

(see Section 11E). Then $\Gamma({}_7\mathcal{T}^4_{(s,s),(t,0)}) = W_2 \rtimes S_3$, which is a finite group if and only if $s = t = 2$. In particular, $\Gamma({}_7\mathcal{T}^4_{(2,2),(2,0)}) = S_5 \times S_3$.

If $t = s$, then the generators in (11H4) are (up to renaming) just those of (11H5). It follows that $\mathcal{P}^\lambda_{(s,s)} = {}_7\mathcal{T}^4_{(s,s),(s,0)}$, as claimed. In particular, this completes the proof of the third case of Theorem 11H2(b). Note that the only finite instance occurs for $s = 2$.

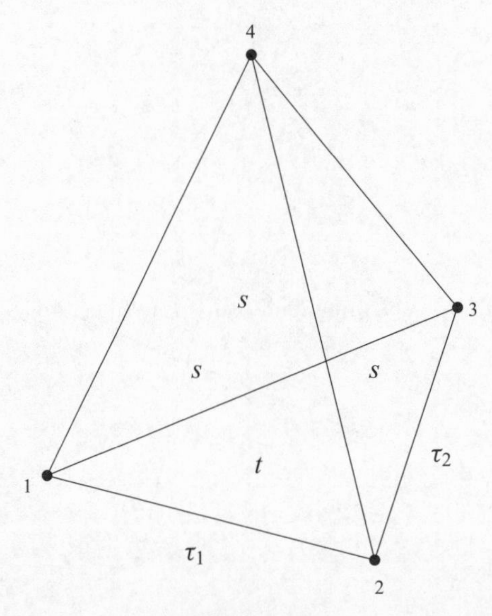

Figure 11H2

Finally we consider the first case of Theorem 11H2(b), in which $\mathcal{P}^{\lambda}_{(3r,0)}$ belongs to the class $\langle\{3,6\}_{(3r,0)},\{6,3\}_{(r,r)}\rangle$. Now recall that the polytope $\mathcal{P}_{(s,0)}$ (with $s \geqslant 2$), which is the dual of ${}_3\mathcal{T}^4_{(s,0)}$, was constructed from the abstract group $W_3 := [1\ 1\ 2]^s$ in Figure 11H3, with generators $\gamma_1,\ldots,\gamma_4$ (say), by means of the twisting operation

11H6 $$(\gamma_1,\ldots,\gamma_4;\tau) \mapsto (\gamma_1,\gamma_2,\gamma_3,\tau) =: (\sigma_0,\ldots,\sigma_3)$$

(see Section 11B). In particular, $\Gamma(\mathcal{P}_{(s,0)}) = W_3 \rtimes C_2$. This group is finite if and only if $s \leqslant 4$.

Now we can argue as follows. For each r, we have a covering map of $\mathcal{P}_{(3r,0)}$ onto $\mathcal{P}_{(3,0)}$, because the group of the latter is a quotient of the former. Hence, there is also a covering map of $\mathcal{P}^{\lambda}_{(3r,0)}$ onto $\mathcal{P}^{\lambda}_{(3,0)}$. Therefore, in order to prove that $|\Gamma(\mathcal{P}_{(3r,0)}) : \Gamma(\mathcal{P}^{\lambda}_{(3r,0)})| = 4$ for each r, it suffices to verify this for $r = 1$. But for $r = 1$ we obtain $\Gamma(\mathcal{P}_{(3,0)}) = [1\ 1\ 2]^3 \rtimes C_2$, of order 1296, and $\Gamma(\mathcal{P}^{\lambda}_{(3,0)}) = [1\ 1\ 1]^3 \rtimes S_3$, of order 324. Indeed, in this case we must have

$$\mathcal{P}^{\lambda}_{(3,0)} = \big\{\{3,6\}_{(3,0)},\{6,3\}_{(1,1)}\big\} = {}_7\mathcal{T}^4_{(3,0),(1,1)},$$

because now the universal polytope is combinatorially flat (see Proposition 4E5 and Theorem 11E7). This completes the proof of Theorem 11H2. □

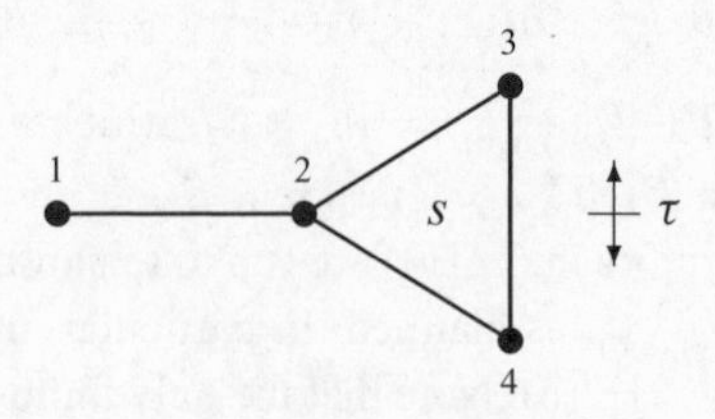

Figure 11H3

The following is a consequence of Theorem 11H2, which complements our results of Section 11E about the universal polytopes ${}_7\mathcal{T}^4_{\mathbf{s},\mathbf{t}}$.

11H7 Theorem *The universal regular 4-polytope* ${}_7\mathcal{T}^4_{(s,0),(s,0)} := \{\{3,6\}_{(s,0)}, \{6,3\}_{(s,0)}\}$ *exists (at least) for all s with* $s \not\equiv 0 \bmod 3$. *For such s, it is infinite if* $s \geqslant 5$ *(and most likely also if* $s = 4$*). If* $s = 2$*, it is finite, and its group is* $S_5 \times C_2$.

Proof. Indeed, if $s \not\equiv 0 \pmod 3$, then $\mathcal{P}^\lambda_{(s,0)} \in \langle\{3,6\}_{(s,0)}, \{6,3\}_{(s,0)}\rangle$; this proves the existence statement. But $\mathcal{P}^\lambda_{(s,0)}$ is finite if and only if $\mathcal{P}_{(s,0)}$ is finite. Hence, the universal polytope ${}_7\mathcal{T}^4_{(s,0),(s,0)}$ must be infinite if $\mathcal{P}_{(s,0)}$ is infinite; this occurs if $s \geqslant 5$. Finally, ${}_7\mathcal{T}^4_{(2,0),(2,0)}$ is combinatorially flat, because $\{6,3\}_{(2,0)}$ is flat (see Proposition 4E5). Hence ${}_7\mathcal{T}^4_{(2,0),(2,0)} = \mathcal{P}^\lambda_{(2,0)}$, with group $\Gamma(\mathcal{P}^\lambda_{(2,0)}) = \Gamma(\mathcal{P}_{(2,0)}) = S_5 \times C_2$ (see Theorems 11B5 and 11H2). □

Compared to the six other types of locally toroidal regular 4-polytopes we know only little about ${}_7\mathcal{T}^4_{\mathbf{s},\mathbf{t}}$. Observe that Theorem 11H7 implies that, for any s with $s \neq 2^k 3^l$, the polytope ${}_7\mathcal{T}^4_{(s,0),(s,0)}$ either does not exist or, if it exists, must be infinite. Indeed, if $s \neq 2^k 3^l$, then choose a prime divisor p with $p \geqslant 5$, and apply Theorem 11H7 with s replaced by p. Since the group for p is a quotient of that for s, the claim follows.

Indeed, we actually know quite a bit more. If $s = 2^k 3^l \geqslant 5$, then 6, 8 or 9 divides s, and so ${}_7\mathcal{T}^4_{(s,0),(s,0)}$ (if it exists) must cover ${}_7\mathcal{T}^4_{(3,0),(3,3)}$, ${}_7\mathcal{T}^4_{(8,0),(8,0)}$ or ${}_7\mathcal{T}^4_{(6,0),(2,2)}$, respectively. Since the latter polytopes are infinite (see Theorems 11E5 and 11H7), so must be the former. In particular, the polytope ${}_7\mathcal{T}^4_{(s,0),(s,0)}$ (if it exists) must be infinite whenever $s \geqslant 5$, without further restrictions on the congruence class of s.

For further properties of the polytopes $\mathcal{P}^\lambda_{\mathbf{s}}$, we refer to [300, pp. 95–97].

The Types {3, 3, 6} *and* {6, 3, 6}

We next discuss the polytopes $\mathcal{P}^\mu$ of type {6, 3, 6}, where $\mathcal{P}$ is again a regular 4-polytope in $\langle\{3,3\}, \{3,6\}_{\mathbf{s}}\rangle$ and μ is the operation on $\Gamma(\mathcal{P}) = \langle\sigma_0, \ldots, \sigma_3\rangle$ given by

11H8 $$\mu\colon (\sigma_0, \ldots, \sigma_3) \mapsto (\sigma_0, \sigma_1\sigma_2\sigma_3\sigma_2\sigma_1, \sigma_3, \sigma_2) =: (\rho_0, \ldots, \rho_3),$$

as in (11G1). We assume as before that the graph of $\mathcal{P}$ has no multiple edges; $\mathcal{P}^\mu$ is then constructed from $\mathcal{P}$ in the same way as {6, 3, 6} was from {3, 3, 6} (see Figure 11G2).

Applying Wythoff's construction to the order complex $\mathcal{C}(\mathcal{P})$, we find that the facets of $\mathcal{P}^\mu$ are just some of the vertex-figures of $\mathcal{P}^\lambda$, while its vertex-figures are transforms of the original vertex-figure $\{3,6\}_{\mathbf{s}}$ under the operation χ of (11G4). To check that $\Gamma := \langle\rho_0, \ldots, \rho_3\rangle$ is indeed a C-group with $\mathcal{P}^\mu$ as the corresponding polytope, we must verify the intersection property. To this end, let $\psi \in \langle\rho_0, \rho_1, \rho_2\rangle \cap \langle\rho_1, \rho_2, \rho_3\rangle$. Then ψ fixes the vertices F_0 $(= H_0)$ and H_3 of $\mathcal{P}$, and thus the unique edge (of F_3) connecting F_0 and H_3; here we have used our assumption that the graph of $\mathcal{P}$ has no multiple edges. It follows that $\psi \in \langle\rho_1, \rho_2\rangle$, as required.

To find the index of $\Gamma = \Gamma(\mathcal{P}^\mu)$ in $\Gamma(\mathcal{P})$, observe that, in $\mathcal{C}(\mathcal{P})$, the simplex T_μ with vertices $H_0, \ldots, H_3$ is dissected into six copies of the fundamental simplex T for

$\Gamma(\mathcal{P})$ with vertices $F_0, \ldots, F_3$. These copies are

$$T,\ T\sigma_1,\ T\sigma_0\sigma_1,\ T\sigma_2\sigma_1,\ T\sigma_0\sigma_2\sigma_1,\ T(\sigma_0\sigma_1\sigma_2)^2.$$

We shall see later that, in the most interesting case, we have $|\Gamma(\mathcal{P}) : \Gamma| = 1$ or 6, proving that either T or T_μ, respectively, is a fundamental simplex for Γ. In particular, if the index is 6, then we have $(\sigma_0\sigma_1\sigma_2)^2 \notin \Gamma$; note that, in Figure 11G2, the element $(\sigma_0\sigma_1\sigma_2)^2$ corresponds to the "half-turn" of F_3 about the edge connecting F_0 and H_3, and thus maps T_μ onto itself.

Note that Γ can only be normal in $\Gamma(\mathcal{P})$ if $\Gamma = \Gamma(\mathcal{P})$. In fact, if Γ is normal, then $\sigma_0\sigma_1\sigma_0 = \sigma_1\sigma_0\sigma_1 \in \sigma_1\Gamma\sigma_1 = \Gamma$; thus $\Gamma = \Gamma(\mathcal{P})$. It follows that the index cannot be 2.

Further, note that, by our remarks on (11G4), we must have $\Gamma = \Gamma(\mathcal{P})$ if $\mathbf{s} = (s, 0)$ with $s \not\equiv 0 \pmod 3$. It is likely that this is the only case where $\Gamma = \Gamma(\mathcal{P})$. In fact, if $\Gamma = \Gamma(\mathcal{P})$, then $\sigma_1 \in \Gamma$, and we are tempted to conclude that $\sigma_1 \in \langle \rho_1, \rho_2, \rho_3 \rangle$ (since σ_1 stabilizes the base vertex $F_0 = H_0$ of $\mathcal{P}^\mu$). However, as in the case of the operation λ, we cannot be sure that the realization of $\mathcal{P}^\mu$ on $\mathcal{C}(\mathcal{P})$ is faithful. Indeed, if $\sigma_1 \in \langle \rho_1, \rho_2, \rho_3 \rangle$, then our remarks on (11G4) imply that $\mathbf{s} = (s, 0)$ with $s \not\equiv 0 \pmod 3$.

Now let $\mathcal{P} = \mathcal{P}_{(3r,0)}$ with $r \geqslant 1$, where again $\mathcal{P}_\mathbf{s} := \{\{3, 3\}, \{3, 6\}_\mathbf{s}\}$ (see (11H6) and Figure 11H3). Then $\Gamma(\mathcal{P}_{(3,0)})$ is a quotient of $\Gamma(\mathcal{P})$. Since the index of a subgroup can only get smaller under a homomorphism, it suffices to prove that $|\Gamma(\mathcal{P}) : \Gamma| = 6$ if $r = 1$. If $r = 1$, then $\mathcal{P}^\mu \in \langle \{6, 3\}_{(1,1)}, \{3, 6\}_{(1,1)} \rangle$. But the universal polytope $\{\{6, 3\}_{(1,1)}, \{3, 6\}_{(1,1)}\}$ is combinatorially flat, and so it is the only member in its class. It follows that $\mathcal{P}^\mu = \{\{6, 3\}_{(1,1)}, \{3, 6\}_{(1,1)}\}$ and $\Gamma(\mathcal{P}^\mu) = S_3 \ltimes \Gamma(\{3, 6\}_{(1,1)})$, of order 216 (see Theorem 11C13). On the other hand, $\Gamma(\mathcal{P}) = [1\ 1\ 2]^3 \ltimes C_2$, of order 1296. Hence $|\Gamma(\mathcal{P}) : \Gamma| = 6$, as required.

If $\mathcal{P}$ is a non-universal member of $\langle \{3, 3\}, \{3, 6\}_{(3r,0)} \rangle$, we do not know whether $\Gamma(\mathcal{P}_{(3,0)})$ is always a quotient of $\Gamma(\mathcal{P})$. Here we cannot completely rule out the possibility that the index is not 6, though very likely it is.

Now let $\mathcal{P} = \mathcal{P}_{(s,s)}$ with $s \geqslant 2$ (see (11H3) and Figure 11H1). Then, by (11H8), the generators of Γ are given by

11H9 $$(\rho_0, \rho_1, \rho_2, \rho_3) = (\tau_3, \alpha_3, \alpha_1, \tau_1).$$

It follows that

$$\Gamma = W_1 \ltimes \langle \tau_1, \tau_3 \rangle = W_1 \ltimes (C_2 \times C_2).$$

But $\Gamma(\mathcal{P}) = W_1 \ltimes S_4$, so that $|\Gamma(\mathcal{P}) : \Gamma| = 6$, as required. Note further that (11H9) itself defines a twisting operation on $W_1 = \langle \alpha_1, \ldots, \alpha_4 \rangle$. Up to renaming the generators, this is in fact the same operation which was used to construct the universal polytope $\{\{6, 3\}_{(s,0)}, \{3, 6\}_{(s,0)}\}$ (see Section 11D). Hence $\mathcal{P}^\mu = \{\{6, 3\}_{(s,0)}, \{3, 6\}_{(s,0)}\}$.

Finally, if $\mathcal{P}$ is any member of $\langle \{3, 3\}, \{3, 6\}_{(s,s)} \rangle$, then $\Gamma(\mathcal{P})$ is the image of $\Gamma(\mathcal{P}_{(s,s)}) = W_1 \ltimes S_4$ under a homomorphism f (say) mapping distinguished generators to distinguished generators. But then we have $\Gamma(\mathcal{P}) = (W_1 f) \cdot (\langle \tau_1, \tau_2, \tau_3 \rangle f)$, with the second factor isomorphic to S_4, and $\Gamma(\mathcal{P}^\mu) = (W_1 f) \cdot (\langle \tau_1, \tau_3 \rangle f)$. If f is such that $(W_1 f) \cap (\langle \tau_1, \tau_2, \tau_3 \rangle f) = \{\varepsilon\}$, then the two products are semi-direct products and the index is 6. However, we are not sure whether this is always true.

Summarizing, we have proved the following theorem.

11H10 Theorem *Let $\mathcal{P}$ be a regular 4-polytope in the class $\langle \{3, 3\}, \{3, 6\}_{\mathbf{s}} \rangle$, and let μ be the operation in (11H8). Assume that the edge graph of $\mathcal{P}$ has no multiple edges.*

(a) Then $\mathcal{P}^\mu$ is a regular 4-polytope in the class

$$\begin{cases} \langle \{6, 3\}_{(r,r)}, \{3, 6\}_{(r,r)} \rangle, & \text{if } \mathbf{s} = (3r, 0) \text{ with } r \geqslant 1; \\ \langle \{6, 3\}_{(s,0)}, \{3, 6\}_{(s,0)} \rangle, & \text{if } \mathbf{s} = (s, 0) \text{ with } s \geqslant 2,\ s \not\equiv 0 \bmod 3; \\ \langle \{6, 3\}_{(s,0)}, \{3, 6\}_{(s,0)} \rangle, & \text{if } \mathbf{s} = (s, s) \text{ with } s \geqslant 2. \end{cases}$$

Furthermore, $|\Gamma(\mathcal{P}) : \Gamma(\mathcal{P}^\mu)| = 1$, 3 or 6, and $\Gamma(\mathcal{P}^\mu) = \Gamma(\mathcal{P})$ in the second case.

(b) If $\mathcal{P} = \mathcal{P}_{\mathbf{s}} := \{\{3, 3\}, \{3, 6\}_{\mathbf{s}}\}$, then in the three cases of part (a) the index is given by $|\Gamma(\mathcal{P}) : \Gamma(\mathcal{P}^\mu)| = 6$, 1 and 6, respectively. In particular, $\mathcal{P}^\mu_{(s,s)} = \{\{6, 3\}_{(s,0)}, \{3, 6\}_{(s,0)}\}$ $(= {}_6\mathcal{T}^4_{(s,0),(s,0)})$ for each $s \geqslant 2$.

We briefly discuss applications to finite universal polytopes. In the second case of Theorem 11H10(a), if $s = 2$ or 4, then $\Gamma(\mathcal{P}^\mu_{(s,0)}) = S_5 \times C_2$ or $[1\ 1\ 2]^4 \rtimes C_2$, respectively. Hence $\mathcal{P}^\mu_{(s,0)}$ is not universal in its class, and $\Gamma(\mathcal{P}^\mu)$ has index 2 or infinite index in the group of the universal polytope, respectively (see Theorem 11D5). The remaining finite polytopes $\mathcal{P}_{\mathbf{s}}$ are obtained for $\mathbf{s} = (2, 2)$ or $(3, 0)$ (see Theorem 11B5 and Corollary 11C8). Here we have

$$\mathcal{P}^\mu_{(2,2)} = \{\{6, 3\}_{(2,0)}, \{3, 6\}_{(2,0)}\}, \qquad \mathcal{P}^\mu_{(3,0)} = \{\{6, 3\}_{(1,1)}, \{3, 6\}_{(1,1)}\},$$

and the index is 6 in each case.

The Types $\{3, 3, 6\}$ *and* $\{4, 3, 6\}$

In Section 11G, the hyperbolic honeycomb $\{4, 3, 6\}$ was constructed from $\{3, 3, 6\}$ by clustering tetrahedra in fives. For the groups, this implied an application of the operation ν in (11G1), followed by a twisting operation κ as in (11G2).

Now let $\mathcal{P}$ be a regular polytope in $\langle \{3, 3\}, \{3, 6\}_{\mathbf{s}} \rangle$ with group $\Gamma(\mathcal{P}) = \langle \sigma_0, \ldots, \sigma_3 \rangle$. We cannot generally expect to obtain a new regular polytope of type $\{4, 3, 6\}$ from $\mathcal{P}$ by applying the corresponding operations ν and κ. Clearly ν can be defined as before by

11H11 $\qquad \nu\colon (\sigma_0, \ldots, \sigma_3) \mapsto (\sigma_0, \sigma_2, \sigma_1, \sigma_3\sigma_2\sigma_3\sigma_2\sigma_3) =: (\rho_0, \ldots, \rho_3),$

but in general the resulting group will not admit a suitable group automorphism τ to allow a twisting operation. Equivalently, in general the clustering of the tetrahedral facets of $\mathcal{P}$ in fives will only give a polytope with cubical facets which is not regular. The following explains why the construction fails.

First, note that the construction gives new vertex-figures of two kinds: old vertex-figures $\{3, 6\}_{\mathbf{s}}$, and new vertex-figures obtained from the old vertex-figure $\{3, 6\}_{\mathbf{s}}$ by clustering faces $\{3\}$ in fours. If $\mathbf{s} = (s, t)$ with s is even (and $t = 0$ or s), this latter vertex-figure is of type $\{3, 6\}_{(s/2,t/2)}$; otherwise it is again of type $\{3, 6\}_{\mathbf{s}}$ and takes every face of the old vertex-figure four times (but with roles switched). The two kinds of vertex-figures correspond to the subgroups $\langle \rho_0, \rho_2, \rho_3 \rangle = \sigma_3 \langle \sigma_0, \sigma_1, \sigma_2\sigma_3\sigma_2 \rangle \sigma_3$ and $\langle \rho_1, \rho_2, \rho_3 \rangle$ of $\langle \rho_0, \ldots, \rho_3 \rangle$, respectively (see also (11H1)).

Hence, for the construction to give a polytope which is regular, s must be odd. But then $\sigma_3 \in \langle \rho_1, \rho_2, \rho_3 \rangle$; hence $\langle \rho_0, \ldots, \rho_3 \rangle = \Gamma(\mathcal{P})$. But now, since σ_3 maps the facet F_3 of $\mathcal{P}$ to its neighbour, the clustering process covers each facet of $\mathcal{P}$ five times (but with roles switched). It follows that each edge of $\mathcal{P}$ is also a new edge. However, only in case $\mathbf{s} = (1, 1)$ or $\mathbf{s} = (3, 0)$ are the faces $\{3\}$ of the (second kind of) new vertex-figures equivalent under the group to the faces $\{3\}$ of the old vertex-figures, so only here can there be a twisting operation. It follows that the construction gives a regular polytope only if $\mathbf{s} = (1, 1)$ or $(3, 0)$.

However, there are other interesting cases where the clustering process gives a non-regular polytope of type $\{4, 3, 6\}$. For example, if $\mathcal{P} = \{\{3, 3\}, \{3, 6\}_{(4,0)}\}$ and thus $\Gamma(\mathcal{P}) = [1\ 1\ 2]^4 \rtimes C_2$, then we obtain a polytope with 80 vertices (at 16 of which the vertex-figure is $\{3, 6\}_{(4,0)}$, while at the remaining 64 it is $\{3, 6\}_{(2,0)}$), 256 edges, 384 square faces and 128 cubical facets.

12

Higher Toroidal Polytopes

Just as in the previous two chapters we almost completed the classification of the locally toroidal regular polytopes of rank 4, we shall now give an almost complete description of those regular polytopes of higher rank whose vertex-figures and facets are either spherical or toroidal, with at least one of each. We shall briefly refer to these as *higher toroidal polytopes*.

As a necessary preliminary, in Section 12A we look at certain of the regular hyperbolic honeycombs, and the relationships among them. We then consider the higher toroidal polytopes of rank 5 in Section 12B, and those of rank 6 in Sections 12C–12E.

The techniques which we bring to bear on these classification problems are various. On occasions, a more or less direct geometric construction will suffice to settle a problem. More frequently, though, we shall rely heavily on twisting methods; we shall thus make many references to Chapter 8. A stock in trade is then to pass from a twisted group to a suitable quotient group, which will be that in which we are ultimately interested.

12A Hyperbolic Honeycombs in $\mathbb{H}^4$ and $\mathbb{H}^5$

In Sections 6D and 6E we have already described the regular toroids which are the potential candidates for facets or vertex-figures of higher toroidal polytopes. (We shall soon see that it is only the cases of rank 4 or 5 that are of interest in the present context.) In further preparation for our investigations, we now consider the regular $(n + 1)$-honeycombs in hyperbolic space $\mathbb{H}^n$ of dimensions $n = 4$ and 5. Bear in mind that we need only consider those honeycombs which have an euclidean facet or vertex-figure (or both), and that we need not count dual pairs as essentially different.

In $\mathbb{H}^4$, the only suitable candidate is $\{3, 4, 3, 4\}$. Our concern here is just with a certain cut of the honeycomb; we refer to Section 7C for a general discussion of cuts. Since we are working with the geometric groups here, we shall use the same notational conventions as we did in Chapter 6 for geometric groups.

12A1 Theorem *Let* $[3, 4, 3, 4] = \langle R_0, \ldots, R_4 \rangle$. *Then the operation*

$$(R_0, \ldots, R_4) \mapsto (R_0, R_1, R_2, R_3 R_4 R_3) =: (S_0, \ldots, S_3)$$

yields the group $\langle S_0, \ldots, S_3 \rangle \cong [3, 4, 4]$ *of a cut* $\{3, 4, 4\}$ *of* $\{3, 4, 3, 4\}$ *by the hyperplane spanned by an octahedral face* $\{3, 4\}$.

Proof. It is easy to verify that

$$R_2 R_3 R_4 R_3 \sim R_3 R_4,$$

where $\sim$ denotes conjugacy; the remaining group relations are also straightforward. □

We may remark at this point that the structure of a cut can also be recognized by using the canonical representations of Coxeter groups. To give an example, for the honeycomb $\{3, 4, 3, 4\}$, we can find the Gram matrix for the generating reflexions $S_0, \ldots, S_3$ of the group of the cut from the Gram matrix for $R_0, \ldots, R_4$. But $S_0, \ldots, S_3$ commute with the reflexion R_4, so that $\langle S_0, \ldots, S_3 \rangle$ is determined by its action on the mirror of R_4. The Gram matrix for $S_0, \ldots, S_3$ can now be identified as the Gram matrix of $[3, 4, 4]$.

There are more relationships involving the honeycombs in $\mathbb{H}^5$. We first look at those between pairs of honeycombs of the same dimension.

12A2 Theorem *The Coxeter group* $[3, 3, 3, 4, 3]$ *has subgroups* $[3, 3, 4, 3, 3]$ *of index 5 and* $[3, 4, 3, 3, 4]$ *of index 10.*

Proof. Writing $[3, 3, 3, 4, 3] = \langle R_0, \ldots, R_5 \rangle$, the operations which yield these subgroups are

$$(R_0, \ldots, R_5) \mapsto (R_0, R_1, R_2, R_3 R_4 R_3, R_5, R_4) =: (S_0, \ldots, S_5)$$

for $[3, 3, 4, 3, 3]$, and

$$(R_0, \ldots, R_5) \mapsto (R_0, R_1, R_2 R_3 R_4 R_3 R_2, R_5, R_4, R_3) =: (T_0, \ldots, T_5)$$

for $[3, 4, 3, 3, 4]$. It is again relatively easy to check the relations (or this may be done geometrically), and the indices can be verified by simplex dissection arguments (see [146]), or directly (see Corollary 8B12 or [303, §4]). Geometrically, the meaning is that both $\{3, 3, 4, 3, 3\}$ and $\{3, 4, 3, 3, 4\}$ can be inscribed in $\{3, 3, 3, 4, 3\}$. □

We now describe various cuts of these honeycombs. We first consider the polytopes of type $\{3, 3, 3, 4, 3\}$. We recall the relations of Theorem 6E6 for the toroids of type $\{3, 3, 4, 3\}$, and the fact that Theorem 6E8 shows that these relations are induced by those on a cut $\{4, 4\}$ of $\{3, 3, 4, 3\}$. Throughout, we use the notation for the group generators introduced in the last proof. We have

12A3 Theorem *Let* $[3, 3, 3, 4, 3] = \langle R_0, \ldots, R_5 \rangle$. *Then the operation*

$$(R_0, \ldots, R_5) \mapsto (R_0, R_1, R_2 R_3 R_4 R_3 R_2, R_5 R_4 R_3 R_4 R_5) =: (U_0, \ldots, U_3)$$

yields the group $\langle U_0, \ldots, U_3 \rangle \cong [3, 4, 4]$ *of a cut* $\{3, 4, 4\}$ *of* $\{3, 3, 3, 4, 3\}$ *by the 3-dimensional plane containing a diametral octahedron* $\{3, 4\}$ *of a facet* $\{3, 3, 3, 4\}$.

Next, we have the type $\{3, 3, 4, 3, 3\}$.

12A4 Theorem *Let* $[3, 3, 4, 3, 3] = \langle S_0, \ldots, S_5 \rangle$. *Then the operation*

$$(S_0, \ldots, S_5) \mapsto (S_0, S_1S_2S_3S_2S_1, S_4S_3S_2S_3S_4, S_5) =: (V_0, \ldots, V_3)$$

yields the group $\langle V_0, \ldots, V_3 \rangle \cong [4, 4, 4]$ *of a cut* $\{4, 4, 4\}$ *of* $\{3, 3, 4, 3, 3\}$ *by a 3-dimensional plane which is induced by a cut* $\{4, 4\}$ *of a facet* $\{3, 3, 4, 3\}$.

The cut $\{4, 4\}$ is as in Theorems 6E6 and 6E8.

Finally, we have the type $\{3, 4, 3, 3, 4\}$, or, rather, its dual.

12A5 Theorem *Let* $[3, 4, 3, 3, 4] = \langle T_0, \ldots, T_5 \rangle$. *Then the operation*

$$(T_0, \ldots, T_5) \mapsto (T_5, T_4, T_3T_2T_1T_2T_3, T_0T_1T_2T_1T_0) =: (W_0, \ldots, W_3)$$

yields the group $\langle W_0, \ldots, W_3 \rangle \cong [4, 4, 4]$ *of a cut* $\{4, 4, 4\}$ *of* $\{4, 3, 3, 4, 3\}$ *by a 3-dimensional plane which is induced by a cut* $\{4, 4\}$ *of a facet* $\{4, 3, 3, 4\}$.

In each of these three cases, the relations for the subgroups are easily verified.

We conclude this section by describing how representations over finite rings can be constructed for the symmetry group of a regular honeycomb. These representations are useful in deciding the structure of subgroups for certain abstractly defined groups. We illustrate the method for the group $[3, 3, 3, 4, 3]$, though it generalizes to other groups as well.

We employ the real 6-dimensional canonical representation of (the Coxeter group) $[3, 3, 3, 4, 3]$. The Gram matrix for the generating reflexions $R_0, \ldots, R_5$ of $[3, 3, 3, 4, 3]$ only has entries $0, 1, -\frac{1}{2}$ $(= -\cos\frac{\pi}{3})$ and $-\frac{1}{2}\sqrt{2}$ $(= -\cos\frac{\pi}{4})$. It follows that $R_0, \ldots, R_5$ are representated by matrices over $\mathbb{Q}(\sqrt{2})$. Now, if m is odd, then we can regard all the entries of such a matrix as lying in the ring $\mathbb{Z}_m$ if 2 is a quadratic residue molulo m (in which we can fix one solution of $x^2 \equiv 2 \pmod{m}$), or in $\mathbb{Z}_m[\sqrt{2}]$ if 2 is a quadratic non-residue modulo m. This gives us a representation of $[3, 3, 3, 4, 3]$ over the ring $\mathbb{K} := \mathbb{Z}_m$ or $\mathbb{Z}_m[\sqrt{2}]$ respectively, which also supports the quadratic form defined by the corresponding Gram matrix M over $\mathbb{K}$.

Specifically, we are interested in finding the order of the element R_1STS, with $S := R_2R_3R_4R_3R_2$ and $T := R_5R_4R_3R_4R_5$; compare Theorem 6E6 with $k = 1$ and the indices shifted by 1. We shall show that R_1STS has order m for either choice of the ring $\mathbb{K}$.

Let $\{e_0, \ldots, e_5\}$ be the canonical basis of $\mathbb{K}^6$ and let $\langle \cdot, \cdot \rangle = \langle \cdot, \cdot \rangle_M$ be the quadratic form over $\mathbb{K}$ whose Gram matrix with respect to $\{e_0, \ldots, e_5\}$ is M. Then, over $\mathbb{K}$,

$$xR_i = x - 2\langle x, e_i \rangle e_1 \quad \text{for } i = 0, \ldots, 5,$$

just as in the real case. Now, T and STS are "reflexions" with "normal vectors" $e_3R_4R_5$ and $e_3R_4R_5S$, respectively. To find the order of R_1STS, we need to compute $\langle e_1, e_3R_4R_5S \rangle = \langle e_1S, e_3R_4R_5 \rangle$. This is straightforward, since all products $\langle e_i, e_j \rangle$ are $0, 1, -\frac{1}{2}$ and $-\frac{1}{2}\sqrt{2}$. In particular, $\langle e_1, e_3R_4R_5S \rangle = -1$. But R_1STS fixes the intersection of the "reflexion planes" of R_1 and STS, and thus is completely determined by its effect on the linear span E of $e_1, e_3R_4R_5S$. With respect to these two base vectors, the restriction of R_1STS to E is represented by the matrix

$$B := \begin{bmatrix} 3 & -2 \\ 2 & -1 \end{bmatrix}$$

over $\mathbb{K}$. An easy induction argument yields

$$B^k = \begin{bmatrix} 2k+1 & -2k \\ 2k & -(2k-1) \end{bmatrix}$$

for $k \geqslant 0$, so that B has order m (recall that m is odd). It follows that R_1STS also has order m, as claimed. All this is most easily checked by first working over the real numbers $\mathbb{R}$ and then reducing modulo m; in fact, R_1STS is a translation over $\mathbb{R}$, as we should expect.

Note that, if m is an odd prime, then we obtain representations over fields $GF(p)$ if $p \equiv \pm 1 \pmod 8$, or $GF(p^2)$ if $p \equiv \pm 3 \pmod 8$. Similar such representations were also used in [284, 292, 322–324] for the construction of certain regular 3- and 4-polytopes; see Chapter 13.

As we said when introducing the chapter, we shall employ twisting techniques in our investigations. The following examples illustrate twisting in the present context; we met them earlier in Chapter 8. In all cases, $\mathcal{G}$ will be the trivial diagram on $\mathcal{K}$, so that $\mathcal{L}^{\mathcal{K},\mathcal{G}} = \mathcal{L}^{\mathcal{K}}$. In particular, if $\mathcal{L}$ is the triangle $\{3\}$ and $\mathcal{K}$ is the regular 4-simplex $\{3, 3, 3\}$ (so that $\mathcal{G}$ has 5 vertices), then we have

$$\{3, 4, 3, 3, 3\} = \{3\}^{\{3,3,3\}}, \quad [3, 4, 3, 3, 3] = W(\mathcal{D}) \rtimes S_5,$$

with $\mathcal{D}$ the Coxeter diagram

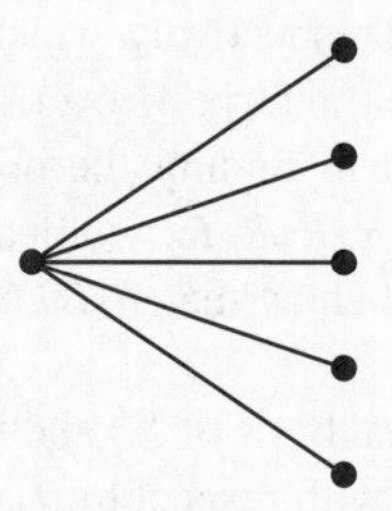

Similarly, if $\mathcal{L}$ and $\mathcal{K}$ are both copies of the tetrahedron $\{3, 3\}$ (with 4 vertices), then we obtain

$$\{3, 3, 4, 3, 3\} = \{3, 3\}^{\{3,3\}}, \quad [3, 3, 4, 3, 3] = W(\mathcal{D}) \rtimes S_4,$$

and now the same diagram $\mathcal{D}$ occurs in the form

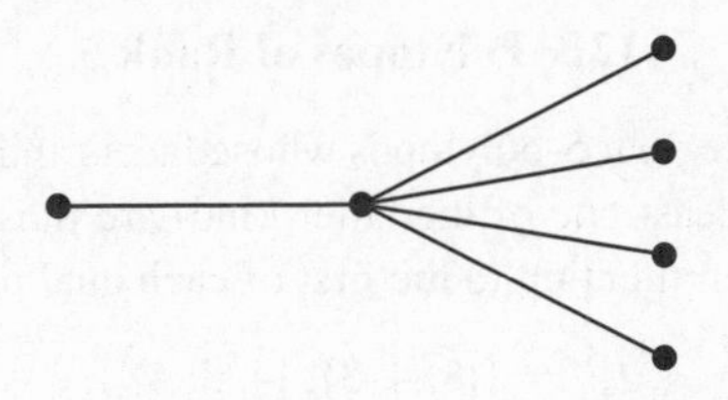

Finally, taking this diagram $\mathcal{D}$ with $W(\mathcal{D}) = \langle \sigma_0, \ldots, \sigma_5 \rangle$ in the form

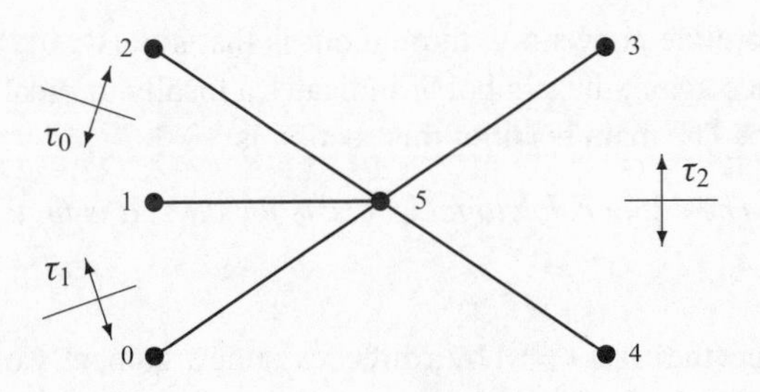

then the twisting operation

$$(\sigma_0, \ldots, \sigma_5; \tau_0, \tau_1, \tau_2) \mapsto (\tau_0, \tau_1, \sigma_0, \sigma_5, \sigma_3, \tau_2) =: (\rho_0, \ldots, \rho_5)$$

on $W(\mathcal{D})$ defines the polytope $\{3, 4, 3, 3, 4\}$ with group

$$\langle \rho_0, \ldots, \rho_5 \rangle = [3, 4, 3, 3, 4] = W(\mathcal{D}) \rtimes (S_3 \times C_2).$$

As we previously remarked, these observations provide an alternative proof of Theorem 12A2, that the Coxeter group $[3, 4, 3, 3, 3]$ has subgroups $[3, 3, 4, 3, 3]$ and $[3, 4, 3, 3, 4]$ of indices 5 and 10, respectively; these facts can also be proved by simplex dissection of hyperbolic simplices.

Note that, with $\mathcal{L} = \{3, 3, 3\}$ and $\mathcal{K} = \{3\}$ (on 3 vertices), we also have

$$\{3, 3, 3, 4, 3\} = \{3, 3, 3\}^{\{3\}}, \quad [3, 3, 3, 4, 3] = W(\mathcal{D}) \rtimes S_3,$$

with diagram

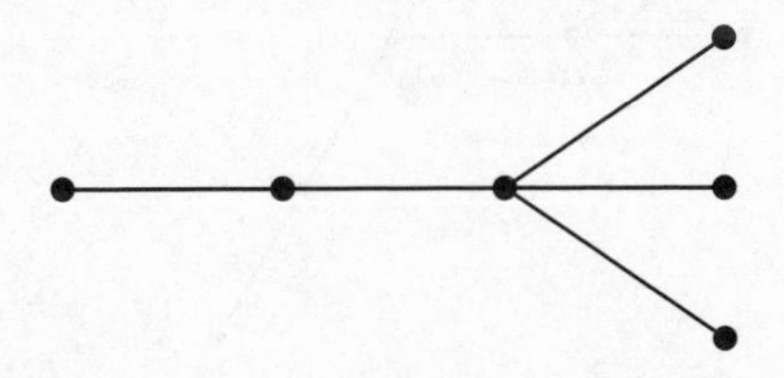

and the same group as in the first example is now expressed in a different way.

12B Polytopes of Rank 5

The only candidates for regular 5-polytopes whose facets and vertex-figures are spherical or toroidal (with at least one of the latter kind) are those of type {3, 4, 3, 4} and their duals. Confining our attention to the first of each dual pair, we shall write here

$$\mathcal{T}_{\mathbf{s}}^5 := \{\{3, 4, 3\}, \{4, 3, 4\}_{\mathbf{s}}\},$$

and employ the abbreviation

$$\Gamma_{\mathbf{s}}^5 := \Gamma(\mathcal{T}_{\mathbf{s}}^5)$$

for its group, where the convention throughout is that $\mathbf{s} = (s^k, 0^{3-k})$ with $s \geqslant 2$ and $k = 1,\ 2$ or 3. Once again, the symbol $\mathcal{T}$ indicates a locally toroidal regular polytope, this time of rank 5. The main result of this section is

12B1 Theorem *The regular polytope $\mathcal{T}_{\mathbf{s}}^5$ exists for all* **s**. *It is finite when $s = 2$, and infinite when $s \geqslant 3$.*

Proof. We construct the cases $k = 1$ by a direct twisting argument. Consider the Coxeter diagram of Figure 12B1. The involutions ρ_0, ρ_1 and ρ_4 are in the original Coxeter group G (say), while ρ_2 and ρ_3 are involutory automorphisms of the diagram. The presentation of the resulting group is easily checked, as is the intersection property (the same group occurs as the case $q = 3$ of [296, Figure (85)]). Only if $s = 2$ (when the branches marked s in the diagram will be absent) is G finite (see [120, §11.5]); it then has order $2^3.192 = 1536$. The group of outer automorphisms is the dihedral group D_3 of order 6; we conclude that $\Gamma(\mathcal{T}_{(2,0,0)}^5)$ has order $6{\cdot}1536 = 9216$, while $\Gamma(\mathcal{T}_{(s,0,0)}^5)$ is infinite for $s \geqslant 3$.

For $k = 2$ or 3, if $\mathcal{T}_{\mathbf{s}}^5$ exists, then it covers $\mathcal{T}_{(s,0,0)}^5$, so that it is infinite when $s \geqslant 3$. To prove that $\mathcal{T}_{(s,0,0)}^5$ does indeed exist, let $\Gamma_{\mathbf{s}}^5$ be the group abstractly defined by the presentation belonging to $\mathcal{T}_{(s,0,0)}^5$; if $\mathcal{T}_{(s,0,0)}^5$ exists, then $\Gamma_{\mathbf{s}}^5$ is its group. We shall also allow $s = 1$ in our discussion, even though $\Gamma_{\mathbf{s}}^5$ cannot then be the group of a polytope. The intersection property for $\Gamma_{\mathbf{s}}^5$ with $s \geqslant 2$ is guaranteed by the quotient criterion of

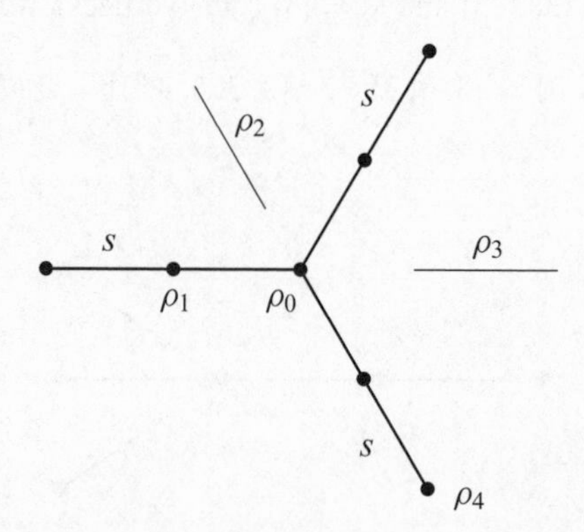

Figure 12B1. The group of $\{\{3, 4, 3\}, \{4, 3, 4\}_{(s,0,0)}\}$.

Table 12B1. *The Finite Polytopes* $\{\{3, 4, 3\}, \{4, 3, 4\}_{\mathbf{s}}\}$

$\mathbf{s}$	v	f	g
(2, 0, 0)	24	8	9216
(2, 2, 0)	48	32	36864
(2, 2, 2)	1536	2048	23 59296

Theorem 2E17, since the quotient is one-to-one on (the group of) the facet $\{3, 4, 3\}$. It remains to prove that the vertex-figure group in $\Gamma^5_{\mathbf{s}}$ is indeed that of the torus $\{4, 3, 4\}_{\mathbf{s}}$. This can be checked using the homomorphisms $\Gamma^5_{(2s,0,0)} \mapsto \Gamma^5_{(s^k,0^{3-k})} \mapsto \Gamma^5_{(s,0,0)}$ and $\Gamma^5_{(s^k,0^{3-k})} \mapsto \Gamma_{(1^k,0^{3-k})}$. For $k = 2$, if the group is not that of $\{4, 3, 4\}_{(s,s,0)}$, then because it has the intersection property it must be that of $\{4, 3, 4\}_{(s,0,0)}$; hence $\Gamma^5_{(s,s,0)}$ and $\Gamma^5_{(s,0,0)}$ are the same. But then $\Gamma^5_{(1,1,0)}$ and $\Gamma^5_{(1,0,0)}$ must also be the same, which can easily be disproved by identifying both groups as quotients of $\Gamma^5_{(2,0,0)}$ and using Figure 12B1. For $k = 3$, the proof is exactly similar.

There remain the cases $\mathbf{s} = (2, 2, 0)$ and $(2, 2, 2)$. The existence (intersection property) follows by the same quotient argument as that immediately preceding, so only the finiteness remains to be established. This was done using the Todd–Coxeter coset enumeration method (however, see also the following), with the results listed in Table 12B1. This completes the proof. □

There is an alternative common approach to the cases $\mathbf{s} = (2^k, 0^{3-k})$, which identifies the groups involved. The idea is to represent all these cases as quotients of the infinite polytope $\mathcal{T}^5_{(4,0,0)}$; this is constructed by a different twisting argument from that before. We have met this approach already in Section 8F.

Let us also note that, when $k = 1$, we have a cut $\{\{3, 4\}, \{4, 4\}_{\tilde{\mathbf{s}}}\}$ of $\{\{3, 4, 3\}, \{4, 3, 4\}_{\mathbf{s}}\}$, where we write $\tilde{\mathbf{s}} := (s^k, 0^{2-k})$ with $s \geqslant 2$, induced by the corresponding cut of $\{3, 4, 3, 4\}$. In fact, the corresponding operation (of Theorem 12A1) on the group of Figure 12B1 is

$$(\rho_0, \ldots, \rho_4) \mapsto (\rho_0, \rho_1, \rho_2, \rho_3\rho_4\rho_3),$$

and this employs only the generators of the left and upper right branches of the diagram. But this operation is known to give $\{\{3, 4\}, \{4, 4\}_{(s,0)}\}$ (see Section 10B or [299, §3]), so that the cut is universal. We conjecture that the corresponding cut for $k = 2$ is also universal, because the relations on the latter cut which yield $\{\{3, 4\}, \{4, 4\}_{\tilde{\mathbf{s}}}\}$ are again just those which need to be imposed on $\{3, 4, 3, 4\}$ to yield $\{\{3, 4, 3\}, \{4, 3, 4\}_{\mathbf{s}}\}$. While there is an analogous cut in case $k = 3$, it cannot be universal; in fact, it is a finite quotient of the infinite polytope $\{\{3, 4\}, \{4, 4\}_{(4,0)}\}$.

While on the subject of cuts, let us make a final observation. It is known (again see Section 10B or [299, §3]) that the polytope $\{\{3, 4\}, \{4, 4\}_{(3,0)}\}$ is finite, and, as before, it is a universal cut of $\{\{3, 4, 3\}, \{4, 3, 4\}_{(3,0,0)}\}$, which is infinite. Thus an infinite regular polytope can have finite universal cuts.

The main result of the rest of this section, Theorem 12B22, uses the geometry of the octahedron to obtain an infinite polytope in $\langle\{3, 4, 3\}, \{4, 3, 4\}_{\mathbf{s}}\rangle$, for each $k = 1, 2, 3$ and each odd $\mathbf{s} = (s^k, 0^{3-k})$ (meaning that s is odd). This is based on Corollary 8E18 and gives a more constructive way of proving the non-finiteness of the corresponding polytope $\mathcal{T}^5_{\mathbf{s}}$. If $\mathbf{s} = (s, 0, 0)$, then the construction gives $\mathcal{T}^5_{\mathbf{s}}$ itself.

It is interesting to note that the groups of these polytopes are residually finite, because they admit faithful representations as linear groups over the real field [272, 439, Chapter 4]. Recall that a group U is *residually finite* if, for each finite subset $\{\varphi_1, \ldots, \varphi_m\}$ of $U \setminus \{\varepsilon\}$, there exists a homomorphism f of U onto some finite group, such that $f(\varphi_j) \neq \varepsilon$ for each $j = 1, \ldots, m$. It was proved in Theorem 4C4 (and [298, Theorem 1]) that, if a class $\langle\mathcal{T}_1, \mathcal{T}_2\rangle$ contains an infinite polytope $\mathcal{T}$ with a residually finite group, then it contains infinitely many regular polytopes which are finite and are covered by $\mathcal{T}$. In particular, the construction of such infinite polytopes implies that all classes $\langle\{3, 4, 3\}, \{4, 3, 4\}_{\mathbf{s}}\rangle$ with $\mathbf{s}$ odd contain infinitely many finite regular polytopes. We conjecture this result to be true for even $\mathbf{s}$ with $s > 2$ as well. Note that there are also similar results for each class which contains an infinite polytope $2^{\mathcal{K},\mathcal{G}}$ or $\mathcal{L}^{\mathcal{K},\mathcal{G}}$ with $\mathcal{K}$ finite.

In our construction, we shall frequently use the following quotient relations among the polytopes $\mathcal{T}^5_{\mathbf{s}}$; these follow directly from Theorem 6F1.

12B2

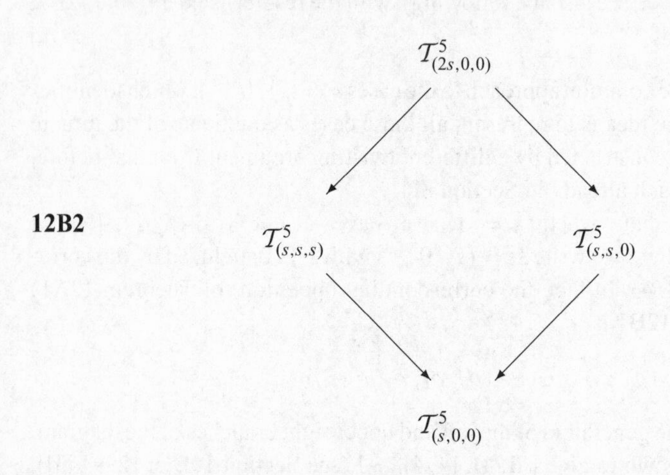

Now let $\mathcal{K} := \{3, 4\}$ be the octahedron with vertices $1, \ldots, 6$, where antipodal vertices are labelled i and $i + 3 \pmod 6$. Then $\Gamma(\mathcal{K}) = \langle\tau_0, \tau_1, \tau_2\rangle$, with

12B3 $$\tau_0 := (1\ 2)(4\ 5), \quad \tau_1 = (2\ 3)(5\ 6), \quad \tau_2 = (3\ 6).$$

The construction of Corollary 8E18 amounts to applying the twisting operation

12B4 $$(\sigma_0, \ldots, \sigma_6; \tau_0, \tau_1, \tau_2) \mapsto (\sigma_0, \sigma_1, \tau_0, \tau_1, \tau_2) =: (\rho_0, \ldots, \rho_4).$$

to the Coxeter group $W_s = \langle \sigma_0, \ldots, \sigma_6 \rangle$ with diagram

12B5 $\mathcal{D}_s$:

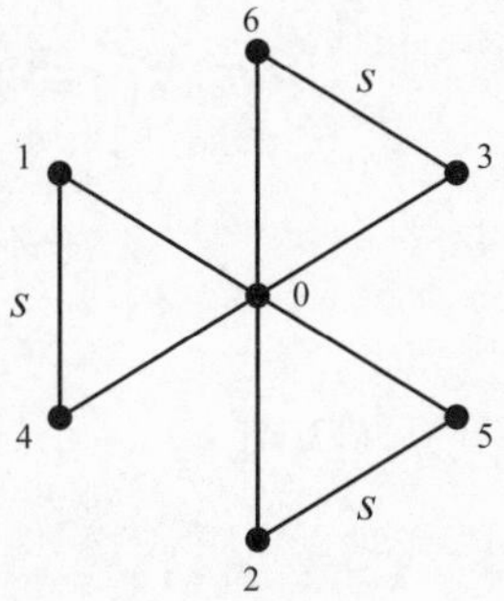

(The generator σ_j corresponds to the node marked j.) The resulting polytope is $\mathcal{T}_{(2s,0,0)}$, with group $\Gamma(\mathcal{T}_{(2s,0,0)}) = \langle \rho_0, \ldots, \rho_4 \rangle = W_s \rtimes [3, 4]$.

We use the quotient relations (12B2) to relate the structure of the polytopes $\mathcal{T}_{(s,0,0)}$, $\mathcal{T}^5_{(s,s,0)}$ and $\mathcal{T}^5_{(s,s,s)}$ to $\mathcal{T}_{(2s,0,0)}$. As we shall see, this works particularly well if s is odd; the geometry of the octahedron then comes into play. However, we begin with the case where $s = 2t$ is even, when we shall recognize the group as a certain semi-direct product.

Let us take the generators of $\Gamma(\mathcal{T}_{(2s,0,0)}) = W_{2t} \rtimes [3, 4]$ as in (12B4). If we write $\chi := (1\,2\,3\,4\,5\,6)$ and $\omega := (1\,2\,4\,5)$, then we have $\rho_1\rho_2\rho_3\rho_4 = \sigma_1\chi$ and $\rho_1\rho_2\rho_3\rho_4\rho_3 = \sigma_1\omega$; these elements have orders $12t$ and $8t$ in $\Gamma(\mathcal{T}_{(2s,0,0)})$, respectively. By Theorem 6D4, the groups of $\mathcal{T}^5_{(s,s,s)}$ and $\mathcal{T}^5_{(s,s,0)}$ are the Coxeter group $[3, 4, 3, 4]$, factored out by the single extra relations of (6D5), namely, $(\rho_1\rho_2\rho_3\rho_4)^{6t} = \varepsilon$ and $(\rho_1\rho_2\rho_3\rho_4\rho_3)^{4t} = \varepsilon$, respectively. But in $\Gamma(\mathcal{T}_{(2s,0,0)})$, we have

$$(\rho_1\rho_2\rho_3\rho_4)^{6t} = (\sigma_1\sigma_4)^t(\sigma_2\sigma_5)^t(\sigma_3\sigma_6)^t$$

and

$$(\rho_1\rho_2\rho_3\rho_4\rho_3)^{4t} = (\sigma_1\sigma_4)^t(\sigma_2\sigma_5)^t.$$

Therefore, in $\Gamma^5_{(s,s,s)}$ and $\Gamma^5_{(s,s,0)}$, these relations impose extra relations on the generators of W_{2t}, namely,

12B6 $$(\sigma_1\sigma_4)^t(\sigma_2\sigma_5)^t(\sigma_3\sigma_6)^t = \varepsilon$$

and

12B7 $$(\sigma_1\sigma_4)^t(\sigma_2\sigma_5)^t = (\sigma_1\sigma_4)^t(\sigma_3\sigma_6)^t = (\sigma_2\sigma_5)^t(\sigma_3\sigma_6)^t = \varepsilon,$$

respectively; because of the action of $\Gamma(\mathcal{K})$, the two latter relations of (12B7) are equivalent to the first.

Let $\hat{W}_{2t}$ and $\hat{\hat{W}}_{2t}$ denote the quotients of W_{2t} defined by the extra relations (12B6) and (12B7), respectively. Then we have

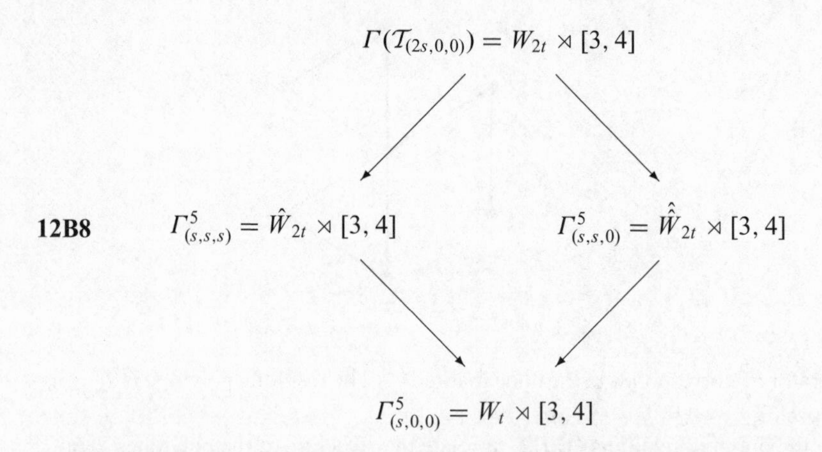

with $s = 2t$ and $t \geqslant 2$. Note that we do indeed have semi-direct products here, because the surjective homomorphisms onto $\Gamma^5_{(s,0,0)} = W_t \rtimes [3, 4]$ are one-to-one on the $[3, 4]$-subgroup.

The more interesting case is when $s = 2t + 1$ ($\geqslant 3$) is odd. First, we describe a new construction of the polytopes $\mathcal{T}^5_{(s,0,0)}$. We begin with two lemmas.

12B9 Lemma *Let $s = 2t + 1 \geqslant 3$, and let $\langle \varphi_0, \varphi_1, \varphi_2 \rangle$ be the Coxeter group with diagram*

12B10 φ_0, φ_1, φ_2, s

factored out by the extra relation

12B11 $$(\varphi_0\varphi_1(\varphi_2\varphi_1)^t)^2 = \varepsilon.$$

Then $\langle \varphi_0, \varphi_1, \varphi_2 \rangle \cong [3, s]$, with the ψ_i's in the diagram

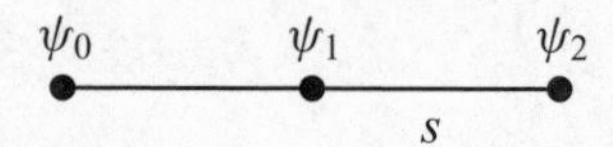

for $[3, s]$ related to the φ_j's by

$$(\psi_0, \psi_1, \psi_2) = (\varphi_0, \varphi_1, \varphi_1(\varphi_2\varphi_1)^t),$$
$$(\varphi_0, \varphi_1, \varphi_2) = (\psi_0, \psi_1, \psi_2\psi_1\psi_2).$$

Proof. It is easy to check that the relations of (12B10) and (12B11) in terms of φ_0, φ_1 and φ_2 are equivalent to those for $[3, s]$ in terms of ψ_0, ψ_1 and ψ_2. □

12B12 Lemma *Let $s = 2t + 1 \geqslant 3$, and let $U_s := \langle \varphi_0, \varphi_1, \ldots, \varphi_6 \rangle$ be the group abstractly defined by the standard relations given by the diagram*

12B13

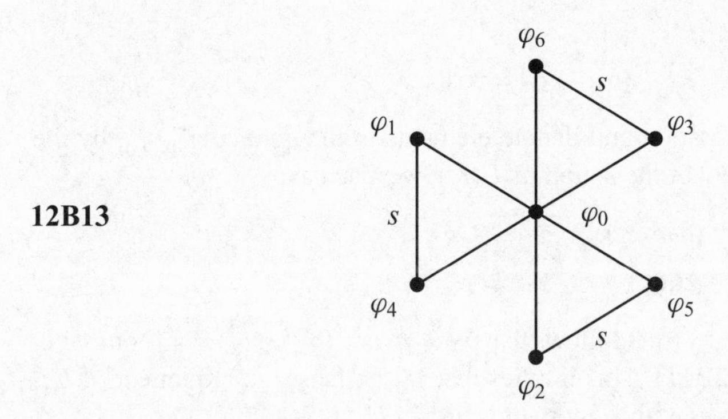

and the three extra relations

12B14 $$(\varphi_0 \varphi_i (\varphi_{i+3} \varphi_i)^t)^2 = \varepsilon \quad \textit{for } i = 1, 2, 3.$$

Then U_s is isomorphic to the Coxeter group with diagram

12B15

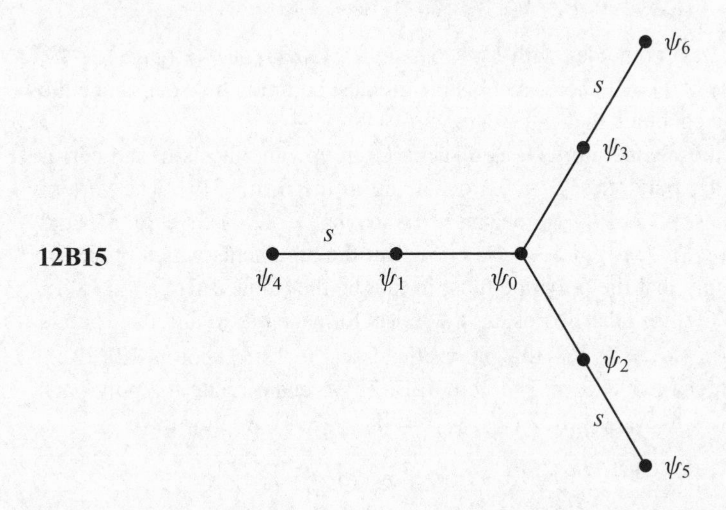

Proof. Start from the group with diagram (12B15), and apply Lemma 12B9 to each of its subgroups $\langle \psi_0, \psi_i, \psi_{i+3} \rangle$ for $i = 1, 2, 3$. Then the change of generators to φ_0, φ_i, φ_{i+3} in one subgroup does not effect the changes in the other subgroups. It is now straightforward to prove the isomorphism. □

The results of the remainder of this section are summarized in Theorem 12B22. First, we consider the case $\mathbf{s} = (s, 0, 0)$. To construct the universal polytope $\mathcal{T}^5_{(s,0,0)}$

with $s = 2t + 1$, we begin with the following observation. Relating $\mathcal{T}^5_{(s,0,0)}$ to $\mathcal{T}_{(2s,0,0)}$ as in (12B2), we see that, in $\Gamma(\mathcal{T}_{(2s,0,0)}) = \langle \rho_0, \ldots, \rho_4 \rangle = W_s \rtimes [3,4]$, we have

$$(\rho_1\rho_2\rho_3\rho_4\rho_3\rho_2)^s = (\sigma_1\sigma_4)^t\sigma_1 \cdot \tau_{14},$$

with $\tau_{14} := (1\ 4)$, so that the imposition of the defining relation $(\rho_1\rho_2\rho_3\rho_4\rho_3\rho_2)^s = \varepsilon$ forces

$$(\sigma_1\sigma_4)^t\sigma_1 = \tau_{14}$$

in $\Gamma^5_{(s,0,0)}$; here we abuse notation and denote elements in the quotient $\Gamma^5_{(s,0,0)}$ by the same symbols. Then, in view of the action of $\Gamma(\mathcal{K})$, we also have

$$(\sigma_2\sigma_5)^t\sigma_2 = \tau_{25},$$
$$(\sigma_3\sigma_6)^t\sigma_3 = \tau_{36} = \tau_2 = \rho_4,$$

with $\tau_{25} := (2\ 5)$ and $\tau_{36} := (3\ 6)$. It follows that $\sigma_0 = \rho_0$ and $(\sigma_3\sigma_6)^t\sigma_3 = \rho_4$ commute, leading to relations like (12B11). This implies that W_s collapses to a quotient of U_s. As we shall see in the following, this quotient is in fact U_s itself.

We can now construct $\mathcal{T}^5_{(s,0,0)}$ using the group $U_s = \langle \varphi_0, \varphi_1, \ldots, \varphi_6 \rangle$ in the form (12B13) and applying an operation like (12B4), taking into account the identifications which have to occur. More precisely, consider the group $U_s \rtimes S_3 = U_s \rtimes \langle \tau_0, \tau_1 \rangle$ with generators

12B16 $$(\alpha_0, \ldots, \alpha_4) := (\varphi_0, \varphi_1, \tau_0, \tau_1, (\varphi_3\varphi_6)^t\varphi_3).$$

Now $\Gamma := \langle \alpha_0, \ldots, \alpha_4 \rangle$ coincides with $U_s \rtimes S_3$, since $\tau_0\tau_1\alpha_4\tau_1\tau_0\varphi_1 = (\varphi_1\varphi_4)^t \in \Gamma$; thus $\varphi_4 \in \Gamma$ because $(s,t) = 1$ (here, (s,t) is the greatest common divisor). Since the φ_j are conjugate, we conclude that $\varphi_1, \ldots, \varphi_6 \in \Gamma$.

With respect to these generators, Γ is a string C-group. The facets of the corresponding polytope are polytopes $\{3,4,3\}$, occurring in the form $\{3\}^{(3)}$. The vertex-figures are polytopes $\{4,3,4\}_{(s,0,0)}$, because $\langle \alpha_2, \alpha_3, \alpha_4 \rangle \cong C_2^3 \rtimes S_3 \cong [3,4]$, and $\langle \alpha_1, \alpha_2, \alpha_3, \alpha_4 \rangle \cong \langle \varphi_1, \ldots, \varphi_6 \rangle \rtimes S_3 = D_s^3 \rtimes S_3$. But the arguments which preceded this construction imply that the polytope must in fact be the (universal) $\mathcal{T}_{(s,0,0)}$ itself, because it covers $\mathcal{T}_{(s,0,0)}$. In other words, $\mathcal{T}_{(s,0,0)}$ exists for each odd s, and its group is $U_s \rtimes S_3$. If we set $\hat{\mathcal{T}}_{(s,0,0)} := \mathcal{T}_{(s,0,0)}$, this proves the case $k = 1$ of Theorem 12B22.

Next, we treat the case $k = 3$ of Theorem 12B22. We again relate the polytopes $\mathcal{T}^5_{(s,s,s)}$ with $s = 2t + 1$ to $\mathcal{T}_{(2s,0,0)}$. In $\Gamma(\mathcal{T}_{(2s,0,0)}) = W_s \rtimes [3,4]$, we now have

$$(\rho_1\rho_2\rho_3\rho_4)^{3s} = (\sigma_1\sigma_4)^t\sigma_1 \cdot (\sigma_2\sigma_5)^t\sigma_2 \cdot (\sigma_3\sigma_6)^t\sigma_3 \cdot \chi^3,$$

where $\chi = (1\ 2\ 3\ 4\ 5\ 6)$ as before, so that $\chi^3 = (1\ 4)(2\ 5)(3\ 6)$ is the central involution of $[3,4]$. In $\Gamma(\mathcal{T}_{(2s,0,0)})$, the elements $\sigma_0 = \rho_0$ and χ^3 $(\in \langle \rho_2, \rho_3, \rho_4 \rangle)$ commute, so that the imposition of the relation $(\rho_1\rho_2\rho_3\rho_4)^{3s} = \varepsilon$ leads to

12B17 $$(\sigma_0 \cdot (\sigma_1\sigma_4)^t\sigma_1 \cdot (\sigma_2\sigma_5)^t\sigma_2 \cdot (\sigma_3\sigma_6)^t\sigma_3)^2 = \varepsilon$$

in $\Gamma^5_{(s,s,s)}$. While it seems difficult to get a handle on the quotient of W_s defined by (12B17), we can still collapse this quotient further onto U_s, corresponding to passing from $\mathcal{T}^5_{(s,s,s)}$ to $\mathcal{T}^5_{(s,0,0)}$. This suggests that we construct from U_s a polytope $\hat{\mathcal{T}}^5_{(s,s,s)}$ in the class $\langle\{3,4,3\},\{4,3,4\}_{(s,s,s)}\rangle$; however, this will not coincide with the universal $\mathcal{T}^5_{(s,s,s)}$.

To do so, consider the regular 3-simplex $\mathcal{T}$ whose vertices lie at the centres of alternate 2-faces of $\{3,4\}$, with one of its vertices at the centre of that 2-face with vertices $1, 2, 3$. Now the group of the polytope $\hat{\mathcal{T}}^5_{(s,s,s)}$ will be $U_s \rtimes S_4$, with S_4 realized as the subgroup $S(\mathcal{T})$ of $[3,4]$ which preserves $\mathcal{T}$. Note that $S(\mathcal{T}) \cong [3,4]/\langle\chi^3\rangle$, and $S(\mathcal{T}) = \langle\tau_0, \tau_1, \omega^2\rangle$, with ω as before, so that $\omega^2 = (1\,4)(2\,5)$. As generators, we take

12B18 $(\beta_0, \ldots, \beta_4) := (\varphi_0, \varphi_1, \tau_0, \tau_1, (\varphi_1\varphi_4)^t\varphi_1 \cdot (\varphi_2\varphi_5)^t\varphi_2 \cdot (\varphi_3\varphi_6)^t\varphi_3 \cdot \omega^2).$

In this context, one should think of β_4 as $\chi^3\omega^2 = \tau_2$. First note that $\Gamma = \langle\beta_0, \ldots, \beta_4\rangle$ coincides with $U_s \rtimes S(\mathcal{T})$. In fact, $\varphi_1, \varphi_2, \varphi_3 \in \Gamma$ and $\varphi_6 = (\varphi_1\varphi_2\varphi_3\beta_4)^2\varphi_3 \in \Gamma$; hence $\varphi_4, \varphi_5 \in \Gamma$, and thus $\Gamma = U_s \rtimes S_4$. Again, Γ is a string C-group, and the facets of the corresponding polytope $\hat{\mathcal{T}}^5_{(s,s,s)}$ are isomorphic to $\{3,4,3\}$. The group of the vertex-figure is

$$\langle\beta_1, \ldots, \beta_4\rangle = \langle\varphi_1, \ldots, \varphi_6\rangle \rtimes S(\mathcal{T}) \cong D_s^3 \rtimes S_4 \cong [4,3,4]_{(s,s,s)},$$

so that the vertex-figures of $\hat{\mathcal{T}}^5_{(s,s,s)}$ are isomorphic to $\{4,3,4\}_{(s,s,s)}$. In particular, $\hat{\mathcal{T}}^5_{(s,s,s)}$ is an infinite regular polytope in $\langle\{3,4,3\},\{4,3,4\}_{(s,s,s)}\rangle$ whose group is the semi-direct product $U_s \rtimes S_4$. This proves the case $k = 3$ of Theorem 12B22.

We treat the case $k = 2$ in a similar fashion, and construct a regular polytope $\hat{\mathcal{T}}^5_{(s,s,0)}$ in $\langle\{3,4,3\},\{4,3,4\}_{(s,s,0)}\rangle$ with group $U_s \rtimes D_6$, where $s = 2t + 1$ is odd. In $\Gamma(\mathcal{T}_{(2s,0,0)}) = W_s \rtimes [3,4]$, we have

$$(\rho_1\rho_2\rho_3\rho_4\rho_3)^{2s} = (\sigma_1\sigma_4)^t\sigma_1 \cdot (\sigma_2\sigma_5)^t\sigma_2 \cdot \hat{\kappa}_3,$$

with $\hat{\kappa}_3 := \omega^2 = (1\,4)(2\,5)$. In $\Gamma(\mathcal{T}_{(2s,0,0)})$, the elements $\sigma_0 = \rho_0$ and ω^2 commute, so that the imposition of the relation $(\rho_1\rho_2\rho_3\rho_4\rho_3)^{2s} = \varepsilon$ leads to

12B19 $$(\sigma_0 \cdot (\sigma_1\sigma_4)^t\sigma_1 \cdot (\sigma_2\sigma_5)^t\sigma_2)^2 = \varepsilon,$$

or equivalent relations modulo $[3,4]$ in $\Gamma(\mathcal{T}_{(s,0,0)})$. Let $\hat{\kappa}_1 := (2\,5)(3\,6)$ and $\hat{\kappa}_2 := (1\,4)(3\,6)$; then $\langle\hat{\kappa}_1, \hat{\kappa}_2, \hat{\kappa}_3\rangle \cong C_2^2$. But in $\Gamma^5_{(s,s,0)}$ we have $\hat{\kappa}_3 = (\sigma_1\sigma_4)^t\sigma_1 \cdot (\sigma_2\sigma_5)^t\sigma_2$ and, by conjugation, $\hat{\kappa}_1 = (\sigma_2\sigma_5)^t\sigma_2 \cdot (\sigma_3\sigma_6)^t\sigma_3$ and $\hat{\kappa}_2 = (\sigma_1\sigma_4)^t\sigma_1 \cdot (\sigma_3\sigma_6)^t\sigma_3$, so that $\langle\hat{\kappa}_1, \hat{\kappa}_2, \hat{\kappa}_3\rangle$ becomes a subgroup of the corresponding quotient of W_s. Finally, for our construction of $\hat{\mathcal{T}}^5_{(s,s,0)}$, we note that in $[3,4]$ the groups $\langle\hat{\kappa}_1, \hat{\kappa}_2, \hat{\kappa}_3\rangle$ and $\langle\tau_0\tau_2, \tau_1\rangle = \langle(1\,2)(4\,5)(3\,6), (2\,3)(5\,6)\rangle$ ($\cong D_6$) are complementary, with the latter occurring as the group of the Petrie polygon $\mathcal{L}$ of $\{3,4\}$ with successive vertices $1, 2, 6, 4, 5, 3$ (see Section 1D or [120, §2.6] for the definition, and Figure 12B2). Again, since it is difficult to identify the quotient of W_s defined by (12B19) and equivalent relations, we collapse the group further onto U_s, corresponding to passing from $\mathcal{T}^5_{(s,s,0)}$ to $\mathcal{T}^5_{(s,0,0)}$.

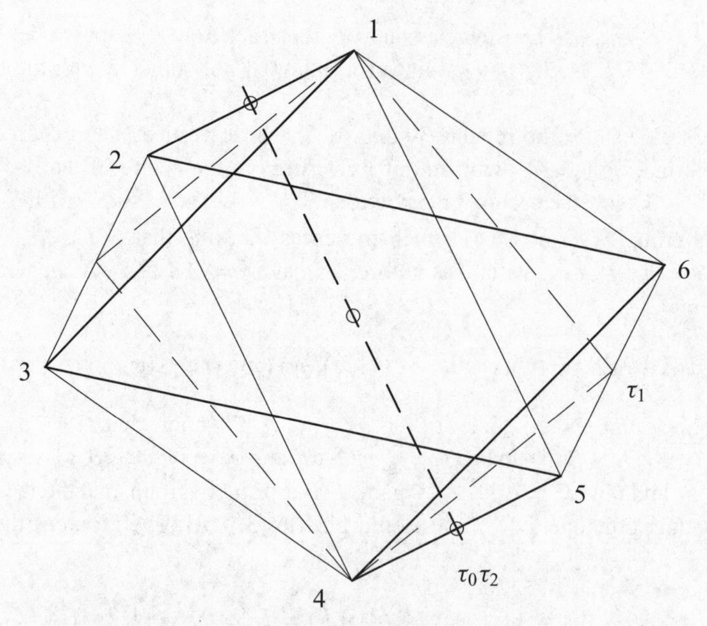

Figure 12B2. The Petrie polygon $\mathcal{L}$ of $\{3, 4\}$.

As a subgroup of $[3, 4]$, the symmetry group of $\mathcal{L}$ is given by $S(\mathcal{L}) = \langle \tau_0\tau_2, \tau_1 \rangle$. Define the elements $\kappa_1, \kappa_2, \kappa_3$ of $U_s = \langle \varphi_0, \ldots, \varphi_6 \rangle$ by

12B20
$$\begin{aligned} \kappa_1 &:= (\varphi_2\varphi_5)^t \varphi_2 \cdot (\varphi_3\varphi_6)^t \varphi_3, \\ \kappa_2 &:= (\varphi_1\varphi_4)^t \varphi_1 \cdot (\varphi_3\varphi_6)^t \varphi_3, \\ \kappa_3 &:= (\varphi_1\varphi_4)^t \varphi_1 \cdot (\varphi_2\varphi_5)^t \varphi_2. \end{aligned}$$

Then $\langle \kappa_1, \kappa_2, \kappa_3 \rangle \cong C_2^2$ and $\langle \kappa_1, \kappa_2, \kappa_3, \tau_0\tau_2, \tau_1 \rangle \cong C_2^2 \rtimes D_6 \cong [3, 4]$. Now the group of $\hat{\mathcal{T}}^5_{(s,s,0)}$ will be $U_s \rtimes D_6$, with D_6 realized as $S(\mathcal{L})$. As generators, we take

12B21
$$(\gamma_0, \ldots, \gamma_4) := (\varphi_0, \varphi_1, \tau_0\tau_2\chi^3\kappa_3, \tau_1, \chi^3\kappa_3),$$

with $\chi^3 = (1\,4)(2\,5)(3\,6)$, again as before. Here, $\chi^3 = (\tau_0\tau_2\tau_1)^3 \in S(\mathcal{L})$ and $\tau_0\tau_2\chi^3 = (1\,5)(2\,4) \in S(\mathcal{L})$. Keeping in mind the previous identifications, one should think of γ_2 as τ_0 $(= (1\,5)(2\,4) \cdot (1\,4)(2\,5) = \tau_0\tau_2\chi^3 \cdot \kappa_3)$ and of γ_4 as τ_2 $(= (1\,4)(2\,5)(3\,6) \cdot (1\,4)(2\,5) = \chi^3 \cdot \kappa_3)$. To see that $\Gamma := \langle \gamma_0, \ldots, \gamma_4 \rangle$ coincides with $U_s \rtimes S(\mathcal{L})$, note that $\tau_0\tau_2 = \gamma_2\gamma_4 \in \Gamma$, and thus $S(\mathcal{L}) \leqslant \Gamma$; then the conjugates of φ_1 by elements in $S(\mathcal{L})$ are also in Γ, proving that $\Gamma = U_s \rtimes S(\mathcal{L})$. Again, Γ is a string C-group with respect to its generators $\gamma_0, \ldots, \gamma_4$. To check that the facets of the corresponding polytope $\hat{\mathcal{T}}^5_{(s,s,0)}$ are isomorphic to $\{3, 4, 3\}$, we note that $\langle \gamma_0, \ldots, \gamma_3 \rangle = \langle \varphi_0, \ldots, \varphi_3 \rangle \rtimes S_3$, with S_3 given by $\langle \gamma_2, \gamma_3 \rangle$. By construction, $\langle \gamma_2, \gamma_3, \gamma_4 \rangle \cong \langle \kappa_1, \kappa_2, \kappa_3 \rangle \rtimes S(\mathcal{L}) \cong C_2^2 \rtimes D_6 = [3, 4]$. The vertex-figures are now isomorphic to $\{4, 3, 4\}_{(s,s,0)}$, because

$$\langle \gamma_1, \ldots, \gamma_4 \rangle = \langle \varphi_1, \ldots, \varphi_6 \rangle \rtimes S(\mathcal{L}) = D_s^3 \rtimes D_6 = [4, 3, 4]_{(s,s,0)}.$$

It follows that $\hat{\mathcal{T}}^5_{(s,s,0)}$ is in $\langle\{3,4,3\},\{4,3,4\}_{(s,s,0)}\rangle$. This completes the proof of the following result.

12B22 Theorem *Let $k = 1, 2$ or 3, let $s = 2t + 1 \geqslant 3$, and let U_s denote the Coxeter group with diagram (12B15). Then there exists an infinite regular 5-polytope $\hat{\mathcal{T}}_{(s^k,0^{3-k})}$ in $\langle\{3,4,3\},\{4,3,4\}_{(s^k,0^{3-k})}\rangle$; its group is $U_s \rtimes S_3$ (for $k = 1$), $U_s \rtimes D_6$ (for $k = 2$), or $U_s \rtimes S_4$ (for $k = 3$). In particular, $\hat{\mathcal{T}}^5_{(s,0,0)} = \mathcal{T}^5_{(s,0,0)}$, the universal member in its class.*

Remarks

(a) In the proof of Theorem 12B22, we have identified the groups $[4,3,4]_{(s^k,0^{3-k})}$ for odd s as $D_s^3 \rtimes S_3$ if $k = 1$, $D_s^3 \rtimes D_6$ if $k = 2$, or $D_s^3 \rtimes S_4$ if $k = 3$.

(b) We can also construct a regular 5-polytope $\hat{\mathcal{T}}_{(2s,0,0)}$ in the class $\langle\{3,4,3\},\{4,3,4\}_{(2s,0,0)}\rangle$ with group $U_s \rtimes [3,4]$, by applying the exact analogue of (12B4) directly to U_s. Then (12B2) remains true with $\mathcal{T}_{(s^k,0^{3-k})}$ replaced by $\hat{\mathcal{T}}_{(s^k,0^{3-k})}$, and the corresponding quotient constructions are equivalent to our constructions.

To conclude this section, we observe that our techniques already work in rank 4. For example, if $\mathcal{F} = \{4\}$ and $\mathcal{H}_s(\mathcal{F})$ is the diagram on the vertices of a square connecting antipodes by a branch labelled s ($\geqslant 2$), then, by Theorem 8E16,

$$\{3\}^{\mathcal{F},\mathcal{H}_s(\mathcal{F})} = \{\{3,4\},\{4,4\}_{(2s,0)}\}$$

for all s. This amounts to an operation like (12B4) on a diagram like (12B5) with only two triangular subdiagrams. For odd s, we have an analogue of Lemma 12B12 involving only two tails of (12B15). The construction of the corresponding polytope is equivalent to that of Section 10B (and [299, §3]) for $\{\{3,4\},\{4,4\}_{(s,0)}\}$.

12C Polytopes of Rank 6: Type $\{3,3,3,4,3\}$

For locally toroidal regular polytopes of rank 6, there is as yet no case completely settled. We begin with the type $\{3,3,3,4,3\}$, and explain why we conjecture that there are precisely three finite examples.

Let us adopt the notation of Section 6E, except that we replace ρ_j by ρ_{j+1} in the definitions of σ and τ for $j = 1, \ldots, 4$. In other words,

$$\sigma := \rho_2\rho_3\rho_4\rho_3\rho_2, \quad \tau := \rho_5\rho_4\rho_3\rho_4\rho_5.$$

Then, as we saw in Theorem 12A3, the operation

$$(\rho_0, \ldots, \rho_5) \mapsto (\rho_0, \rho_1, \sigma, \tau)$$

defines a cut of type $\{3,4,4\}$. Moreover, the defining relations, added to those of the (Coxeter) group of the cut, which yield $\{\{3,4\},\{4,4\}_{\tilde{\mathbf{s}}}\}$, with $\tilde{\mathbf{s}} := (s^k, 0^{2-k})$ when $\mathbf{s} = (s^k, 0^{4-k})$, are just those which give the quotient ${}_1\mathcal{T}^6_{\mathbf{s}} := \{\{3,3,3,4\},\{3,3,4,3\}_{\mathbf{s}}\}$ in which we are interested. This leads us to conjecture that this cut is indeed universal.

Table 12C1. *The Known Finite Polytopes* $\{\{3,3,3,4\},\{3,3,4,3\}_{\mathbf{s}}\}$

$\mathbf{s}$	v	f	g
(2, 0, 0, 0)	20	960	3 68640
(2, 2, 0, 0)	160	30720	117 96480
(3, 0, 0, 0)	780	189540	727 83360

Now, we know from Theorems 10B3 and 10B5 (see also [299, §3]) that the only finite cases of regular polytopes $\{\{3,4\},\{4,4\}_{\tilde{\mathbf{s}}}\}$ are those with $\tilde{\mathbf{s}} = (2,0), (2,2)$ or $(3,0)$. Thus we have

12C1 Conjecture *The regular polytope* ${}_1\mathcal{T}_{\mathbf{s}}^6 = \{\{3,3,3,4\},\{3,3,4,3\}_{\mathbf{s}}\}$ *is finite only for* $\mathbf{s} = (2,0,0,0), (2,2,0,0)$ *and* $(3,0,0,0)$.

The finiteness has been checked in each case by the Todd-Coxeter coset enumeration method. The resulting details of the polytopes are listed in Table 12C1.

We shall denote by ${}_1\Gamma_{\mathbf{s}}^6$ the group abstractly defined by the presentation belonging to ${}_1\mathcal{T}_{\mathbf{s}}^6$; if ${}_1\mathcal{T}_{\mathbf{s}}^6$ exists, then ${}_1\Gamma_{\mathbf{s}}^6$ is its group. We have only little to offer at present on the groups, although relationships between these groups and those of other regular 6-polytopes will appear in the next two sections. Here we just mention that the group ${}_1\Gamma_{(3,0,0,0)}^6$ is closely related to a certain finite orthogonal group (see [193, 321, 468]).

In the remainder of this section, we shall provide arguments to support the following

12C2 Conjecture *The regular polytope* ${}_1\mathcal{T}_{\mathbf{s}}^6 = \{\{3,3,3,4\},\{3,3,4,3\}_{\mathbf{s}}\}$ *exists for each* $\mathbf{s} = (s^k, 0^{4-k})$ *with* $s \geqslant 2$ *and* $k = 1, 2$.

We shall explain how the case $k = 2$ can be deduced from that with $k = 1$, and how the latter case can be established under an additional assumption on a subgroup of ${}_1\Gamma_{\mathbf{s}}^6$. Let us introduce some notation, which we shall use here and in the next two sections. We write $\Gamma := \langle \rho_0, \ldots, \rho_5 \rangle$ for a group like ${}_1\Gamma_{\mathbf{s}}^6$, or the group of a regular 6-polytope, $\Gamma_0 := \langle \rho_1, \ldots, \rho_5 \rangle$ (corresponding to the group of its vertex-figure), $\Gamma_5 := \langle \rho_0, \ldots, \rho_4 \rangle$ (for the group of its facet), and $\Gamma_{05} := \langle \rho_1, \ldots, \rho_4 \rangle$.

We begin with an obvious remark about the subgroup Γ_5.

12C3 Lemma *For all* $\mathbf{s}$, *the subgroup* Γ_5 *is isomorphic to* $[3,3,3,4]$.

Proof. Whatever $\mathbf{s}$ may be, ${}_1\Gamma_{\mathbf{s}}^6$ has a quotient with $\mathbf{s} = (2,0,0,0)$ or $(p,0,0,0)$ with p an odd prime. In each case, we know that then the corresponding quotient subgroup is indeed isomorphic to $[3,3,3,4]$ (the case $(p,0,0,0)$ was dealt with in Section 12A, and the case $(2,0,0,0)$ will be treated in the following). Thus Γ_5, being a quotient of $[3,3,3,4]$, must itself be isomorphic to $[3,3,3,4]$. □

We shall now prove Conjecture 12C2 under the assumption that the subgroup Γ_0 of $\Gamma = {}_1\Gamma_{(s,0,0,0)}^6$ really is isomorphic to $[3,3,4,3]_{(s,0,0,0)}$. By the considerations at the end of Section 12A, we know this to be true if s is odd. Indeed, in the representation of $[3,3,3,4,3]$ over $\mathbb{Z}_s$ or $\mathbb{Z}_s[\sqrt{2}]$ (as appropriate), the element R_1STS has order s;

thus we also have a representation of Γ in which the subgroup corresponding to Γ_0 is isomorphic to $[3, 3, 4, 3]_{(s,0,0,0)}$; by Lemma 12C3, the corresponding statement for Γ_5 also holds. It follows that, if s is odd, then the subgroups Γ_0 and Γ_5 have the required structure. The same also remains true if $s = 2t$ with t odd. In this case, we can work with the natural homomorphisms $\Gamma \mapsto {}_1\Gamma^6_{(t,0,0,0)}$ and $\Gamma \mapsto {}_1\Gamma^6_{(2,0,0,0)}$ to identify Γ_0 and Γ_5. In fact, the element of Γ corresponding to R_1STS must now have an order divisible by t and 2, which must hence be $s = 2t$. On the other hand, Γ_0 cannot be isomorphic to $[3, 3, 4, 3]_{(t,t,0,0)}$, since otherwise $\Gamma = {}_1\Gamma^6_{(t,t,0,0)}$; the latter can be disproved by using the quotients onto the groups with $t = 1$. Note that the general case could be derived in a similar way if the structure of Γ_0 in Γ with $s = 2^m$ were known to be of the required type.

Using our assumption, we first prove the intersection property for both $k = 1$ and $k = 2$. Because of the quotient criterion of Theorem 2E17, we need only check the cases $k = 1$ with s a prime; every other example covers one of these, with the quotient mapping one-to-one on the group $\Gamma_5 \cong [3, 3, 3, 4]$ of the facet. With Γ now the group ${}_1\Gamma^6_{\mathbf{s}}$, we have

12C4 Lemma *The only subgroup H with $\Gamma_{05} < H < \Gamma_5$ has $[H : \Gamma_{05}] = 2$.*

Proof. We can see this geometrically; Γ_5 is the group of the regular 5-cross-polytope X^5, and Γ_{05} is the group of its vertex-figure. Some coset representative of Γ_{05} in H must move the base vertex of X^5, and if this new vertex is not antipodal to the base vertex, then conjugation by Γ_{05} will move it to any vertex adjacent to the base vertex, and there follows $H = \Gamma_5$, contrary to assumption. Thus the index is 2, as claimed. □

We now prove the intersection property. The easy case is $s = p$, an odd prime. We have $|\Gamma_5| = 2^5 \cdot 5! = 10 \cdot 384$, and $|\Gamma_0| = p^4 \cdot 1152 = 3p^4 \cdot 384$. By Lemma 12C4, the only possible indices for Γ_{05}, of order 384, in $\Gamma_0 \cap \Gamma_5$ are $1, 2$ or 10. But the latter two are impossible, and the result follows.

For $s = 2$, we argue quite differently. In $\{3, 3, 4, 3\}_{(2,0,0,0)}$, we see that opposite vertices of the vertex-figure $\{3, 4, 3\}$ are identified; in other words, the polytope has the same vertices as $\mathcal{L}^5_0 := \{\{3, 3, 4\}, \{3, 4, 3\}_6\}$ (we make a forward reference here to Section 14A – see (14A1) for the definition). It follows that ${}_1\mathcal{T}^6_{(2,0,0,0)}$ (if it exists) has the same vertices as $\mathcal{L}^6_0 := \{\{3, 3, 3, 4\}, \mathcal{L}^5_0\}$. However, we shall show in Theorems 14A2 and 14A3 and the discussion before the latter (see also [287]) that $\mathcal{L}^6_0$ does indeed exist, has 20 vertices, and has a faithful realization in $\mathbb{E}^{15}$. We now apply the quotient criterion of Theorem 2E17 to show that ${}_1\mathcal{T}^6_{(2,0,0,0)}$ also exists, since the covering map is one-to-one on the facet. This completes the proof of the intersection property.

Under the previous assumption on Γ_0 in ${}_1\Gamma^6_{(s,0,0,0)}$, it now follows that the polytope ${}_1\mathcal{T}^6_{(s,0,0,0)}$ does indeed exist. But the same assumptions also imply the existence for $k = 2$. In fact, if the subgroup Γ_0 of ${}_1\Gamma^6_{(s,s,0,0)}$ is not isomorphic to $[3, 3, 4, 3]_{(s,s,0,0)}$, then it can only be $[3, 3, 4, 3]_{(s,0,0,0)}$, by the homomorphism ${}_1\Gamma^6_{(s,s,0,0)} \mapsto {}_1\Gamma^6_{(s,0,0,0)}$, by our assumptions on ${}_1\Gamma^6_{(s,0,0,0)}$, and the fact that it is known to be a C-group. Were that true, then ${}_1\Gamma^6_{(s,s,0,0)} \cong {}_1\Gamma^6_{(s,0,0,0)}$, which can easily be disproved by using quotients onto the groups with $s = 1$. It follows that Γ_0 and Γ_5 are the groups we are interested in; hence ${}_1\mathcal{T}^6_{(s,s,0,0)}$ also exists.

Note that, since our assumptions on ${}_1\Gamma^6_{(s,0,0,0)}$ are known to hold if s is odd, or $s = 2t$ with t odd, our arguments actually prove Conjecture 12C2 for all such s with $k = 1$ or 2.

12D Polytopes of Rank 6: Type $\{3, 3, 4, 3, 3\}$

The situation for the remaining two types of locally toroidal regular polytopes of rank 6 is somewhat similar. We can appeal in each case to known results about which of the regular 4-polytopes of type $\{4, 4, 4\}$ exists and is finite, but since these do not, at present, cover all possibilities, our knowledge of the polytopes of rank 6 is correspondingly incomplete. We shall not quote here the complete results known about these 4-polytopes, but instead refer the reader to Section 10C (see also [294, 299, 300]).

Let us begin by giving the list of those polytopes of type $\{3, 3, 4, 3, 3\}$ which are known to exist and be finite; this is done in Table 12D1 (we have only listed one of each dual pair).

Let us write throughout

12D1
$$ {}_2\mathcal{T}^6_{\mathbf{s},\mathbf{t}} := \{\{3, 3, 4, 3\}_{\mathbf{s}}, \{3, 4, 3, 3\}_{\mathbf{t}}\}, $$

and denote the corresponding abstract group by ${}_2\Gamma^6_{\mathbf{s},\mathbf{t}} := \Gamma({}_2\mathcal{T}^6_{\mathbf{s},\mathbf{t}})$. If we define σ and τ exactly as in Section 6E (that is, without the shift of indices in Section 12C), then the operation

$$ (\rho_0, \ldots, \rho_5) \mapsto (\rho_0, \sigma, \tau, \rho_5) $$

yields a cut of ${}_2\mathcal{T}^6_{\mathbf{s},\mathbf{t}}$ in the class $\langle \{4, 4\}_{\tilde{\mathbf{s}}}, \{4, 4\}_{\tilde{\mathbf{t}}} \rangle$ (as usual in such contexts, $\tilde{\mathbf{s}} = (s^k, 0^{2-k})$ when $\mathbf{s} = (s^k, 0^{4-k})$, and so on). Evidence indicates that this cut is indeed $\{\{4, 4\}_{\tilde{\mathbf{s}}}, \{4, 4\}_{\tilde{\mathbf{t}}}\}$ (that is, the cut is universal), but so far we have not been able to prove this. Now several cases of the regular polytopes of type $\{4, 4, 4\}$ are completely settled; as a consequence, we have

12D2 Theorem *Under the assumption that the cut above is universal, if the polytope ${}_2\mathcal{T}^6_{\mathbf{s},\mathbf{t}}$ exists, then it is infinite in at least the following cases:*

Table 12D1. *The Known Finite Polytopes* $\{\{3, 3, 4, 3\}_{\mathbf{s}}, \{3, 4, 3, 3\}_{\mathbf{t}}\}$

$\mathbf{s}$	$\mathbf{t}$	v	f	g
$(2, 0, 0, 0)$	$(t, 0, 0, 0)$ (t even)	32	$2t^4$	$36864t^4$
$(2, 0, 0, 0)$	$(t, t, 0, 0)$ (t even)	32	$8t^4$	$147476t^4$
$(2, 2, 0, 0)$	$(2, 2, 0, 0)$	2048	2048	1509 94944
$(3, 0, 0, 0)$	$(3, 0, 0, 0)$	2340	2340	2183 50080

(a) $\mathbf{s} = (s, 0, 0, 0)$, $\mathbf{t} = (t, t, 0, 0)$ *and*

$$\frac{1}{s} + \frac{1}{2t} \leqslant \frac{1}{2};$$

(b) $\mathbf{s} = (s, s, 0, 0)$, $\mathbf{t} = (t, t, 0, 0)$ *and*

$$\frac{1}{s} + \frac{1}{2t} \leqslant \frac{1}{2} \quad \textit{or} \quad \frac{1}{2s} + \frac{1}{t} \leqslant \frac{1}{2};$$

(c) $\mathbf{s} = (s, 0, 0, 0)$, $\mathbf{t} = (t, 0, 0, 0)$, *with s or t even, or $s = t$ odd, and*

$$\frac{1}{s} + \frac{1}{t} \leqslant \frac{1}{2}.$$

The cases (a) and (b) of Theorem 12D2 exactly describe the infinite cases of the corresponding 4-polytopes; however, case (c) is incomplete (see Section 10C and [299, §4]). In particular, in case (a), this applies to the cases $s = 4, t = 2$ and $s = t = 3$, where we have equality; these immediately cover the two sporadic examples of Table 12D1. Computer application of the Todd–Coxeter coset enumeration process suggests that no other regular polytopes of type ${}_2\mathcal{T}^6_{\mathbf{s},\mathbf{t}}$ can be finite, except for those in the two infinite sequences in Table 12D1; several cases are, in any event, eliminated by the known results about the class {4, 4, 4}. Three of the probably infinite examples are given by $\mathbf{s} = (3, 0, 0, 0)$ and $\mathbf{t} = (2, 2, 0, 0)$, $(4, 0, 0, 0)$ or $(5, 0, 0, 0)$, for which the corresponding 4-polytope is finite.

There remain the two infinite sequences ${}_2\mathcal{T}_{\mathbf{s},\mathbf{t}}$ with $\mathbf{s} = (2, 0, 0, 0)$. The first sequence, with $\mathbf{t} = (t, 0, 0, 0)$, is easily dealt with, regardless of our assumptions about universality. The corresponding 4-polytope collapses if t is odd, and so the only possibilities are with t even. In fact, the same is also true for the class $\mathbf{t} = (t, t, 0, 0)$, as the following argument shows. In $\mathcal{Q} := \{3, 3, 4, 3\}_{(2,0,0,0)}$, opposite vertices of the vertex-figure {3, 4, 3} coincide, so that $\mathcal{Q}$ covers the locally projective regular polytope $\{\{3, 3, 4\}, \{3, 4, 3\}_6\}$ with the same number of vertices (see Sections 12C and 14A, and [287]). The existence of this covering forces the opposite vertices of the facets of the vertex-figure $\{3, 4, 3, 3\}_{\mathbf{t}}$ to coincide. Now the product of the reflexions in the centres of adjacent facets of {3, 4, 3, 3} is conjugate to the translation by $(2, 0, 0, 0)$, and so any group of translations by which we identify must be a subgroup of the group $\Lambda_{(2,0,0,0)}$. Thus t must be even in both cases.

Actually, we can go further than this. In the vertex-figure $\{3, 4, 3, 3\}_{\mathbf{t}}$, whatever t might be, vertices equivalent under the translations by $\Lambda_{(2,0,0,0)}$ coincide. That is, all the polytopes $\mathcal{T}^6_{(2,0,0,0),\mathbf{t}}$ have the same number of vertices, namely, that of the case $\mathbf{t} = (2, 0, 0, 0)$. Direct calculation (using the Todd–Coxeter coset enumeration process; see also the following) yields 32 vertices for this, and then the remainder of the details in Table 12D1 can be filled in easily.

We now discuss the general question of existence of the polytopes ${}_2\mathcal{T}^6_{\mathbf{s},\mathbf{t}}$. As in the previous sections, we provide arguments which support the following

12D3 Conjecture *The regular polytope* ${}_2\mathcal{T}^6_{\mathbf{s},\mathbf{t}} = \{\{3, 3, 4, 3\}_{\mathbf{s}}, \{3, 4, 3, 3\}_{\mathbf{t}}\}$ *exists for each* $\mathbf{s} = (s^k, 0^{4-k})$, $\mathbf{t} = (t^m, 0^{4-m})$ *with* $s, t \geqslant 2$ *and* $k, m = 1, 2$, *except when* $\mathbf{s} = (2, 0, 0, 0)$ *and t is odd, or* $\mathbf{t} = (t, 0, 0, 0)$ *and s is odd.*

We have already discussed the first exception; the second is its dual analogue. For the remaining cases, we shall (as before) prove the conjecture under the additional assumptions that the subgroups Γ_0 and Γ_5 of $\Gamma := {}_2\Gamma^6_{(s,0,0,0),(t,0,0,0)}$ are distinct from Γ and are actually isomorphic to $[3,3,4,3]_{(s,0,0,0)}$ and $[3,4,3,3]_{(t,0,0,0)}$, respectively. We must observe that, in contrast to the previous section, our assumptions cannot be reduced to the study of representations of Γ over finite rings. In fact, in the corresponding representations of $[3,3,4,3,3]$ over $\mathbb{Z}_s$ or $\mathbb{Z}_s[\sqrt{2}]$, the two subgroups corresponding to Γ_0 and Γ_5 are related, with parameter vectors **s**, **t** as in Theorem 12D5. In general, of course, **s** and **t** must be allowed to vary independently.

We first discuss the intersection property for ${}_2\Gamma^6_{\mathbf{s},\mathbf{t}}$ with arbitrary **s** and **t**. The case $\mathbf{s} = \mathbf{t} = (2,0,0,0)$ can be checked directly, if tediously. All the polytopes with $\mathbf{s} = (2,0,0,0)$ (and t even) cover ${}_2\mathcal{T}^6_{(2,0,0,0),(2,0,0,0)}$ while preserving the facet, and so the intersection property follows from the quotient criterion of Theorem 2E17. The general polytope with $\mathbf{s} = (s^k, 0^{4-k})$, $\mathbf{t} = (t^m, 0^{4-m})$, where s and t are both even, covers one of the latter while preserving the group of the vertex-figure, and so this group also has the intersection property. In particular, this proves polytopality if $k = m = 1$.

All remaining cases can be dealt with by covering arguments reducing to a case **s** (or **t**) of the form $(p,0,0,0)$ with p an odd prime. Adopting the notation of Section 12C, but with $\Gamma = {}_2\Gamma^6_{\mathbf{s},\mathbf{t}}$, we now have

12D4 Lemma *If* $\mathbf{s} = (p,0,0,0)$ *with p an odd prime, then Γ_{05} is a maximal subgroup of Γ_5.*

Proof. In fact, by our assumption, $\Gamma_5 = [3,3,4,3]_{\mathbf{s}}$, even if $m = 2$. Now let H be a subgroup such that $\Gamma_{05} < H < \Gamma_5$. The cosets of Γ_{05} in H are represented by translations (we think here of the original construction of Γ_5), so since $H \neq \Gamma_{05}$, then H contains some translation $(a_1,\ldots,a_4) \not\equiv o \bmod p$. If $a_k \not\equiv 0 \bmod p$, then (as in Section 6E) $2a_k e_k \in H$, so that $e_k \in H$ also. But this means that $H = \Gamma_5$, and so H cannot be proper, contrary to assumption. □

An immediate consequence is that, if p is an odd prime, then the intersection property holds for the group ${}_2\Gamma_{(p,0,0,0),\mathbf{t}} =: \langle \rho_0,\ldots,\rho_5 \rangle$ in the following cases, to which the remaining ones can be reduced by applications of the quotient criterion. First, if $\mathbf{t} = (2,2,0,0)$, then $|\Gamma_0 \cap \Gamma_5|$ is a common divisor of $1152 \cdot p^4$ and $1152 \cdot 64$, and thus of 1152. Hence, $\Gamma_0 \cap \Gamma_5 = \Gamma_{05}$, as required. Otherwise, we can suppose that $\mathbf{t} = (q,0,0,0)$, with q an odd prime. If $q \neq p$, then $|\Gamma_0 \cap \Gamma_5|$ is a common divisor of $1152 \cdot p^4$ and $1152 \cdot q^4$, and thus again of 1152; we reach the same conclusion. If $q = p$, the situation is only slightly different; this time the order of the intersection could be $1152 \cdot p^4$, but this would mean that the whole group collapsed to that of the vertex-figure, so that $\Gamma_0 = \Gamma_5$, contrary to assumption. Thus the intersection cannot have order $1152 \cdot p^4$, and again we have reached the desired conclusion. This completes the proof of the intersection property for all **s** and **t**.

It now follows that ${}_2\mathcal{T}^6_{\mathbf{s},\mathbf{t}}$ exists at least for $\mathbf{s} = (s,0,0,0)$ and $\mathbf{t} = (t,0,0,0)$. To settle the existence in the remaining cases, we proceed as in the previous section, and work with the homomorphisms

$$_2\Gamma^6_{(s,s,0,0),(t,t,0,0)} \to {}_2\Gamma^6_{(s,s,0,0),(t,0,0,0)} \to {}_2\Gamma^6_{(s,0,0,0),(t,0,0,0)},$$

and the corresponding coverings of polytopes. For example, if in $\Gamma = {}_2\Gamma^6_{(s,s,0,0),(t,0,0,0)}$ the subgroup Γ_5 is not $[3, 3, 4, 3]_{(s,s,0,0)}$, then it can only be $[3, 3, 4, 3]_{(s,0,0,0)}$, because it has the intersection property. But this forces $\Gamma = {}_2\Gamma^6_{(s,0,0,0),(t,0,0,0)}$, which can easily be disproved by using the quotients onto the corresponding groups with $s = t = 1$. The case $\Gamma = {}_2\Gamma^6_{(s,s,0,0),(t,t,0,0)}$ can be shown in a similar way. Modulo our initial assumptions, our proof of Conjecture 12D3 is now complete.

As we saw in Section 12A, the group $[3, 3, 4, 3, 3]$ is a subgroup of index 5 in $[3, 3, 3, 4, 3]$. Under our assumptions that Conjectures 12C2 and 12D3 hold, the corresponding relationship between the quotient groups of the locally toroidal polytopes is

12D5 Theorem *For each $s \geqslant 2$, ${}_2\Gamma^6_{(s,s,0,0),(s,0,0,0)}$ is a subgroup of index 5 in ${}_1\Gamma^6_{(s,s,0,0)}$, while ${}_2\Gamma^6_{(2s,0,0,0),(s,s,0,0)}$ is a subgroup of index 5 in ${}_1\Gamma^6_{(2s,0,0,0)}$.*

Actually, rather weaker assumptions than the ones we have made would suffice. To identify the subgroups ${}_2\Gamma^6_{\mathbf{s},\mathbf{t}}$ in Theorem 12D5, we note that the structure of the group of the vertex-figure is given by Theorem 6F5, while that of the facet is conjugate (under $\rho_3\rho_2\rho_1\rho_0$) to the group of the original vertex-figure. The index is indeed 5 (rather than the 1 it could theoretically become under the quotient); in fact, the index remains 5, even if we take the quotient onto the groups with $s = 1$. In the case $s = 2$, we observe that five copies of ${}_2\mathcal{T}^6_{(2,2,0,0),(2,0,0,0)}$ (each with 128 vertices) can be inscribed in ${}_1\mathcal{T}^6_{(2,2,0,0)}$ (which has 160 vertices), to form a regular compound; each vertex of the latter belongs to four copies of the former. The complementary set of 32 vertices to the 128 vertices of a ${}_2\mathcal{T}^6_{(2,2,0,0),(2,0,0,0)}$ in the 160 vertices of ${}_1\mathcal{T}^6_{(2,2,0,0)}$ belong to those of a ${}_2\mathcal{T}^6_{(2,0,0,0),(2,0,0,0)}$, so that another regular compound of five of the latter is also formed.

If we try to apply the same construction to ${}_1\Gamma_{(s,0,0,0)}$ with s odd, then something curious happens. We obtain a group isomorphic to the original (because the vertex-figure essentially remains the same) which satisfies the relations of ${}_2\Gamma_{(s,0,0,0),(s,0,0,0)}$, as well as some others. In the interesting case $s = 3$, the order of the resulting group is a third of that of the universal polytope (whose existence can be proved directly), as we see by comparing group orders. We deduce

12D6 Theorem *The group ${}_2\Gamma_{(3,0,0,0),(3,0,0,0)}$ has a normal subgroup of order 3; taking the quotient by this subgroup identifies the vertices of ${}_2\mathcal{T}_{(3,0,0,0),(3,0,0,0)}$ by threes.*

In fact, the subgroup is probably central (we have not yet identified it), and also identifies the facets by threes. For more details about ${}_2\Gamma_{(3,0,0,0),(3,0,0,0)}$, see [321].

12E Polytopes of Rank 6: Type $\{3, 4, 3, 3, 4\}$

As in Section 12D, we begin the section by giving a list of those locally toroidal regular polytopes of type $\{3, 4, 3, 3, 4\}$ which are known to exist and be finite. This is done in Table 12E1.

Table 12E1. *The Known Finite Polytopes* $\{\{3, 4, 3, 3\}_{\mathbf{s}}, \{4, 3, 3, 4\}_{\mathbf{t}}\}$

$\mathbf{s}$	$\mathbf{t}$	v	f	g
$(s, 0, 0, 0)$ (s even)	$(2, 0, 0, 0)$	$3s^4$	16	$18432s^4$
$(s, s, 0, 0)$	$(2, 0, 0, 0)$	$12s^4$	16	$73728s^4$
$(s, 0, 0, 0)$ (s even)	$(2, 2, 0, 0)$	$6s^4$	64	$73728s^4$
$(s, s, 0, 0)$ (s even)	$(2, 2, 0, 0)$	$24s^4$	64	$294912s^4$
$(2, 0, 0, 0)$	$(2, 2, 2, 2)$	384	1024	188 74368
$(2, 0, 0, 0)$	$(4, 0, 0, 0)$	12288	65536	12079 59552
$(3, 0, 0, 0)$	$(3, 0, 0, 0)$	2340	780	727 83360

Let us write

12E1 $$_3\mathcal{T}^6_{\mathbf{s},\mathbf{t}} := \{\{3, 4, 3, 3\}_{\mathbf{s}}, \{4, 3, 3, 4\}_{\mathbf{t}}\}.$$

As in Section 12D, one of the main tools in our investigation is a certain cut, actually (in geometric terms) of the dual

$$\left({}_3\mathcal{T}^6_{\mathbf{s},\mathbf{t}}\right)^* = \{\{4, 3, 3, 4\}_{\mathbf{t}}, \{3, 3, 4, 3\}_{\mathbf{s}}\}.$$

The cut is induced by the mixing operation

$$(\rho_0, \ldots, \rho_5) \mapsto (\rho_5, \rho_4, \sigma, \tau),$$

with

$$\sigma := \rho_3\rho_2\rho_1\rho_2\rho_3,$$
$$\tau := \rho_0\rho_1\rho_2\rho_1\rho_0,$$

and is thus of type $\{4, 4, 4\}$. More specifically, if $\mathbf{t} = (t^k, 0^{4-k})$ with $k = 1$ or 2, the cut is in the class $\langle\{4, 4\}_{\tilde{\mathbf{t}}}, \{4, 4\}_{\tilde{\mathbf{s}}}\rangle$, where the notation for the suffixes is that introduced earlier. In fact, we conjecture that the cut is universal, so that it is isomorphic to $\{\{4, 4\}_{\tilde{\mathbf{t}}}, \{4, 4\}_{\tilde{\mathbf{s}}}\}$. On the other hand, if $k = 4$, this cut cannot be universal.

We can immediately deduce (independently of the universality of the cut) that collapse of the polytope occurs if $\mathbf{s} = (2, 0, 0, 0)$ and $\mathbf{t} = (t, 0, 0, 0)$ with t odd, or if $\mathbf{t} = (2, 0, 0, 0)$ and $\mathbf{s} = (s, 0, 0, 0)$ with s odd. In fact, when $\mathbf{t} = (2, 0, 0, 0)$, this is the only restriction on $\mathbf{s}$. To see this, it is better to consider the dual polytope. Since $\{4, 3, 3, 4\}_{(2,0,0,0)}$ is flat with 16 facets, we see that $({}_3\mathcal{T}^6_{\mathbf{s},(2,0,0,0)})^*$ is also flat with 16 vertices, when it does not degenerate. Now the whole polytope collapses onto the facet $\{4, 3, 3\}$ of $\{4, 3, 3, 4\}_{(2,0,0,0)}$. This must induce a corresponding collapse of the vertex-figure $\{3, 3, 4, 3\}_{(s,0,0,0)}$ onto the vertex-figure $\{3, 3\}$ of $\{4, 3, 3\}$. In the universal polytope, with group

$$[3, 3, 4, 3] = \langle\rho_4, \ldots, \rho_0\rangle,$$

this is given by imposing the relation $(\rho_1\rho_2)^2 = \varepsilon$, that is, by making the generators ρ_1 and ρ_2 commute. Therefore this must be compatible with the extra relation imposed on $[3, 3, 4, 3]$ to get the group of $\{3, 3, 4, 3\}_{\mathbf{s}}$.

The implications of this commutativity for the elements σ and τ are

$$\sigma = \rho_3\rho_2\rho_1\rho_2\rho_3 = \rho_3\rho_1\rho_3 = \rho_1,$$
$$\tau = \rho_0\rho_1\rho_2\rho_1\rho_0 = \rho_0\rho_2\rho_0 = \rho_2.$$

In case $\mathbf{s} = (s, 0, 0, 0)$, we have

$$\varepsilon = (\rho_4\sigma\tau\sigma)^s = (\rho_4\rho_1\rho_2\rho_1)^s = (\rho_4\rho_2)^s,$$

from which it follows that s must be even. In case $\mathbf{s} = (s, s, 0, 0)$, we have

$$\varepsilon = (\rho_4\sigma\tau)^{2s} = (\rho_4\rho_1\rho_2)^{2s},$$

which holds automatically, imposing no additional condition. This gives the corresponding entries of Table 12E1. (See also Sections 8E and 8F, and [303, §7] for an explicit construction of these polytopes.)

For a deeper investigation of the cases $\mathbf{s} = (2, 0, 0, 0)$ and $(2, 2, 0, 0)$, it is again appropriate to consider the dual polytope. However, we shall set these cases in a much more general context in Section 14B, and so here we merely summarize the results. Note that, in $\{3, 3, 4, 3\}_{(2,0,0,0)}$, opposite vertices of its vertex-figure $\{3, 4, 3\}$ are identified, so that it has the same vertices as those of the locally projective regular polytope $\mathcal{L}_0^5 := \{\{3, 3, 4\}, \{3, 4, 3\}_6\}$ (see (14A1) and [287]).

12E2 Corollary *The regular polytope* ${}_3\mathcal{T}^6_{\mathbf{s},(2,2,0,0)}$ *exists when* $\mathbf{s} = (s^k, 0^{4-k})$ *with* $k = 1, 2$ *and* s *even, and then has 64 facets; otherwise it degenerates.*

Proof. The corollary follows immediately from Theorem 14B11. Dualizing, the finite polytope $({}_3\mathcal{T}^6_{\mathbf{s},(2,2,0,0)})^*$ will degenerate unless it covers $\{\{4, 3, 3, 4\}_{(2,2,0,0)}, \mathcal{L}_0^5\}$, because both this and the universal polytope have 64 vertices. As far as toroidal vertex-figures are concerned, since $\{3, 3, 4, 3\}_{(2,0,0,0)}$ and $\mathcal{L}_0^5$ have the same vertices, this implies that the vertex-figure must cover $\{3, 3, 4, 3\}_{(2,0,0,0)}$; thus s is even, as required. □

We are left with the sporadic example $\mathbf{s} = \mathbf{t} = (3, 0, 0, 0)$. There is little to say about this, other than that its group order was found using the Todd–Coxeter coset enumeration process. We have no corresponding description of the group; however, see the remark at the end of the section.

We next discuss the general question of existence of the polytope ${}_3\mathcal{T}^6_{\mathbf{s},\mathbf{t}}$. Again, we provide arguments to support

12E3 Conjecture *The regular polytope* ${}_3\mathcal{T}^6_{\mathbf{s},\mathbf{t}} = \{\{3, 4, 3, 3\}_{\mathbf{s}}, \{4, 3, 3, 4\}_{\mathbf{t}}\}$ *exists for* $\mathbf{s} = (s^k, 0^{4-k})$, $\mathbf{t} = (t^m, 0^{4-m})$ *with* $s, t \geqslant 2$, $k = 1, 2$ *and* $m = 1, 2, 4$, *except when*

(a) $\mathbf{s} = (2, 0, 0, 0$ *and* $\mathbf{t} = (t, 0, 0, 0)$ *with* t *odd,*
(b) $\mathbf{t} = (2, 0, 0, 0)$ *and* $\mathbf{s} = (s, 0, 0, 0)$ *with* s *odd,*
(c) $\mathbf{t} = (2, 2, 0, 0)$ *and* s *odd.*

This conjecture was confirmed in Corollary 8E15 (and [303, Corollary 5.6]) for all $\mathbf{s}$ and $\mathbf{t}$ with $m = 1$ and t even. To explain the general case, we again write $\Gamma = {}_3\Gamma^6_{\mathbf{s},\mathbf{t}}$ for the group corresponding to ${}_3\mathcal{T}^6_{\mathbf{s},\mathbf{t}}$, Γ_0 for the group of the vertex-figure, and so on. We also allow $s = 1$ or $t = 1$, to act as targets for quotient maps. We shall prove the conjecture under the assumption that the groups ${}_3\Gamma^6_{\mathbf{s},\mathbf{t}}$ are distinct for $\mathbf{s} = (s, 0, 0, 0$ and $\mathbf{t} = (t, t, 0, 0)$ or (t, t, t, t), and $\mathbf{t} = (t, 0, 0, 0)$ and $\mathbf{s} = (s, 0, 0, 0)$ or $(s, s, 0, 0)$, and that the subgroups Γ_0 and Γ_5 of $\Gamma = {}_3\Gamma^6_{(s,0,0,0),(t,0,0,0)}$ are distinct from Γ, and are actually isomorphic to $[4, 3, 3, 4]_{(t,0,0,0)}$ and $[3, 3, 4, 3]_{(s,0,0,0)}$, respectively. We may observe that the quotients of the groups ${}_3\Gamma^6_{\mathbf{s},\mathbf{t}}$ onto the corresponding groups obtained for $s = 1$, $t = 1$ do not distinguish between those pairs of groups where equality is excluded by assumption. Further, concerning representations of $[3, 4, 3, 3, 4]$ over finite rings, the same remark applies as in the previous section. We begin with some lemmas.

12E4 Lemma *If* $\mathbf{t} = (p, 0, 0, 0)$ *with* p *an odd prime, then* Γ_{05} *is a maximal subgroup of* Γ_0.

Proof. To prove this, we work with the original group $[4, 3, 3, 4]$. The cosets of Γ_{05} in a subgroup H with $\Gamma_{05} < H < \Gamma_0$ are represented by translations. Since we can conjugate by Γ_{05}, we can use the same analysis as that in Section 6E, and conclude that these translations themselves form a subgroup generated by a vector $(q^k, 0^{4-k})$, for some q and k. But this subgroup must contain $p\mathbb{Z}^4$, which is impossible, since this latter subgroup is itself maximal.

Note that, by our assumptions, the group of the vertex-figure is indeed $[4, 3, 3, 4]_{(p,0,0,0)}$. The lemma thus follows. □

With a very similar proof, we also have

12E5 Lemma *If* $\mathbf{s} = (p, 0, 0, 0)$ *with* p *an odd prime, then* Γ_{05} *is a maximal subgroup of* Γ_5.

We now consider the intersection property for $\Gamma = {}_3\Gamma^6_{\mathbf{s},\mathbf{t}}$, with $\mathbf{s}, \mathbf{t}$ arbitrary. First, any case $(\mathbf{s}, \mathbf{t})$ with s, t both even covers the case $\mathbf{s} = \mathbf{t} = (2, 0, 0, 0)$; the corresponding flat polytope can be checked directly, and we can then apply the quotient criterion. Eliminating the degenerate cases, bearing in mind that we must also check the infinite examples for polytopality, and again employing the quotient lemma, we are reduced to considering the following.

$$\mathbf{s} = (p, p, 0, 0), \mathbf{t} = (2, 0, 0, 0), \ p \text{ an odd prime}.$$

Note that the polytope (if it exists) is flat. Now, $\Gamma_5 \cong [3, 4, 3, 3]_{(p,p,0,0)}$; in fact, otherwise we would have $\Gamma_5 \cong [3, 4, 3, 3]_{(p,0,0,0)}$, and thus ${}_3\Gamma^6_{(p,p,0,0),(2,0,0,0)} \cong {}_3\Gamma^6_{(p,0,0,0),(2,0,0,0)}$, contrary to our assumptions on the groups ${}_3\Gamma^6_{\mathbf{s},\mathbf{t}}$. But then $|\Gamma_5| = 4p^4 \cdot 1152$ and $|\Gamma_0| = 16 \cdot 384$. Thus $|\Gamma_5 \cap \Gamma_0|$ is a divisor of $4 \cdot 384$. Thus the index of Γ_{05} in $\Gamma_5 \cap \Gamma_0$ is 1, 2 or 4. The cosets are represented by translations (in both Γ_0 and Γ_5), and, as in Lemma 12E4, if one translation occurs, then so do its conjugates under Γ_{05}. Considering the various cases shows that the only non-zero translation we can have in Γ_0 is that by $(1, 1, 1, 1)$, when the index would be 2. If this translation, τ say, occurs in $\Gamma_0 \cap \Gamma_5$, then it is a central involution (because $\langle \tau \rangle$ is normal) and $\Gamma_0 \cap \Gamma_5 \cong \Gamma_{05} \times C_2$. Since p is odd, as an element of the group Γ_5 of the facet, the translation τ', say, in $\Gamma_{05}\tau$ must be that by $(p, 0, 0, 0)$, since only then do we also obtain index 2. We now have the required contradiction. As a subset of Γ_0, we see that $\Gamma_{05}\tau$ must leave fixed the initial vertex F_0, say, of the whole polytope, but as a subset of Γ_5 it must move F_0 (recall again that the polytope is flat). We conclude that the index was, in fact, just 1, and the proof is complete.

$\mathbf{s} = (2, 2, 0, 0)$, $\mathbf{t} = (p, 0, 0, 0)$, p **an odd prime.**

Now $|\Gamma_5| = 64 \cdot 1152 = 192 \cdot 384$ and $|\Gamma_0| = p^4 \cdot 384$, and it follows at once that $\Gamma_5 \cap \Gamma_0 = \Gamma_{05}$ (the possible exception $p = 3$ is covered by Lemma 12E4).

$\mathbf{s} = (p, 0, 0, 0)$, $\mathbf{t} = (q, 0, 0, 0)$, p, q **odd primes.**

The maximality of Γ_{05} in Γ_0 shows that the order of $\Gamma_5 \cap \Gamma_0$ is either 384 or $q^4 \cdot 384$. If $p \neq q$, the latter is immediately excluded. But even if $p = q$, the latter cannot occur, because $\Gamma \neq \Gamma_5$ by assumption.

This completes the discussion of the intersection property. The existence of ${}_3\mathcal{T}^6_{\mathbf{s},\mathbf{t}}$ now follows for $\mathbf{s} = (s, 0, 0, 0)$ and $\mathbf{t} = (t, 0, 0, 0)$ (and for some other cases). For the remaining cases, we again work with the homomorphisms

$$ {}_3\Gamma^6_{(s,0,0,0),(t,t,t,t)} \to {}_3\Gamma^6_{(s,0,0,0),(t,t,0,0)} \to {}_3\Gamma^6_{(s,0,0,0),(t,0,0,0)} $$

and

$$ {}_3\Gamma^6_{(s,s,0,0),\mathbf{t}} \to {}_3\Gamma^6_{(s,0,0,0),\mathbf{t}}. $$

For example, if in $\Gamma = {}_3\Gamma^6_{(s,0,0,0),(t,t,0,0)}$ the subgroup Γ_0 is not $[4, 3, 3, 4]_{(t,t,0,0)}$, then it must be $[4, 3, 3, 4]_{(t,0,0,0)}$, because it has the intersection property. But this implies that $\Gamma = {}_3\Gamma^6_{(s,0,0,0),(t,0,0,0)}$, which can be disproved using the quotients onto the groups with $s = t = 1$. The other groups are dealt with similarly, except that occasionally we draw on our assumption that certain of the groups are distinct. Modulo all our assumptions, the proof of Conjecture 12E3 is now complete.

As we saw in Section 12A, the group [3, 4, 3, 3, 4] is a subgroup of index 10 in [3, 3, 3, 4, 3]. Under the assumption that Conjectures 12C2 and 12E3 hold (in fact, somewhat weaker assumptions suffice), the corresponding relationship between groups of locally toroidal regular polytopes is

12E6 Theorem *If $s \geqslant 2$, then ${}_3\Gamma^6_{(s,0,0,0),(s,s,0,0)}$ is a subgroup of index 10 in ${}_1\Gamma^6_{(s,s,0,0)}$, while ${}_3\Gamma^6_{(s,s,0,0),(s,s,s,s)}$ is a subgroup of index 10 in ${}_1\Gamma^6_{(2s,0,0,0)}$.*

The verification of Theorem 12E6 is similar to that of Theorem 12D5 and is omitted, because there is again only one finite example, namely, the first case with $s = 2$. However, compare also Theorem 6F5 for the corresponding subgroups relationships for the facets and vertex-figures. Observe that the second case corrects a mistaken assertion in [302, Theorem 8.6]. We have an associated regular compound; ten copies of ${}_3\mathcal{T}^6_{(2,0,0,0),(2,2,0,0)}$ (each with 96 vertices) can be inscribed in the 160 vertices of ${}_1\mathcal{T}^6_{(2,2,0,0)}$, and each vertex of the latter belongs to six copies of the former.

If we apply the same construction to the group ${}_1\Gamma^6_{(s,0,0,0)}$ with s odd, we obtain a group satisfying the relations of ${}_3\Gamma^6_{(s,0,0,0),(s,0,0,0)}$, but possibly some others as well. However, comparison of the group orders shows

12E7 Theorem *The groups* ${}_1\Gamma^6_{(3,0,0,0)}$ *and* ${}_3\Gamma^6_{(3,0,0,0),(3,0,0,0)}$ *are isomorphic, and the polytopes* ${}_1\mathcal{T}^6_{(3,0,0,0)}$ *and* ${}_3\mathcal{T}^6_{(3,0,0,0),(3,0,0,0)}$ *have the same vertices.*

We refer to [321] for more details about the structure of the two groups ${}_1\Gamma^6_{(3,0,0,0)}$ and ${}_3\Gamma^6_{(3,0,0,0),(3,0,0,0)}$.

13

Regular Polytopes Related to Linear Groups

Since the automorphism group of each finite regular polytope can be thought of as a permutation group, it can also be regarded as a linear group. However, there are families of regular polytopes, of ranks 3 and 4, whose groups are intimately related to certain linear groups, in that they are, for example, central quotients of special linear groups or the like. In this chapter, we shall discuss these families. While it might be natural to include some chiral polytopes in the discussion, since they arise by the same or similar constructions, in fact we shall not do so.

The chapter is arranged as follows. In Section 13A, we shall describe families of regular polyhedra related to projective linear groups. In Section 13B, we consider various connexions among these polyhedra; in particular, in certain cases two of them mix to form another. We discuss realizations of some of the polyhedra in Section 13C; in principle, we could describe the realization spaces of all the polyhedra, but in practice the general case is rather complicated, and so we confine our attention to primes. In Section 13D, we investigate analogous families of 4-polytopes, whose vertex-figures fall among those of the previous polyhedra. Finally, in Section 13E, we describe various connexions among these 4-polytopes.

13A Regular Polyhedra

In this section, we shall discuss a family of regular polyhedra whose groups are intimately related to certain projective linear groups. Except in some matters of notation, we shall follow fairly closely the treatment of [292]; this paper extended the earlier work of [284], which only dealt with the prime case.

Let $\mathcal{G}$ be any subgroup of $\mathbb{Z}_m^*$, the group of units in the ring $\mathbb{Z}_m := \mathbb{Z}/m\mathbb{Z}$ of residues modulo m; in practice, we shall have $m \geqslant 3$, but occasionally it is useful to allow $m = 2$ also. We denote by $\mathrm{L}_2^{\mathcal{G}}(\mathbb{Z}_m)$ the group of all 2×2 matrices over $\mathbb{Z}_m$ with determinants in $\mathcal{G}$. Thus, if $I := I_2$ is the identity matrix, then $C^{\mathcal{G}}(m) = \{\lambda I \mid \lambda^2 \in \mathcal{G}\}$ is the centre of $\mathrm{L}_2^{\mathcal{G}}(\mathbb{Z}_m)$. For example, when $\mathcal{G} = \{1\}$, we write $\mathrm{SL}_2(\mathbb{Z}_m) = \mathrm{L}_2^{\mathcal{G}}(\mathbb{Z}_m)$, whose centre is written $C(m)$; this is the usual *special linear group*. Similarly, when $\mathcal{G} = \mathbb{Z}_m^*$, then $\mathrm{L}_2^{\mathcal{G}}(\mathbb{Z}_m) = \mathrm{GL}_2(\mathbb{Z}_m)$, the *general linear group*.

The case which will concern us most is that of $\mathcal{G} = \mathcal{I} := \{\pm 1\}$ (we adopt the abbreviation $\mathcal{I}$, because this subgroup is nearly ubiquitous). We shall then write $\mathcal{C}(m) := \{\lambda \in \mathbb{Z}_m \mid \lambda^2 = \pm 1\}$ for the subgroup of $\mathbb{Z}^*$ which corresponds to the central elements of $\mathrm{L}_2^{\mathcal{I}}(\mathbb{Z}_m)$. More generally, if $\mathcal{I} \leqslant \mathcal{H} \leqslant \mathcal{C}(m)$, then we write $H = \mathcal{H}I := \{\lambda I \mid \lambda \in \mathcal{H}\}$ for the corresponding subgroup of such central elements. We shall need the special case $\mathcal{C}(m)^+ := \{\lambda \in \mathbb{Z}_m \mid \lambda^2 = 1\}$, which is such that $C(m) = \mathcal{C}(m)^+ I$.

In harmony with the notation which we shall employ in Section 13D, we could also write $\mathrm{SL}_2^{\#}(\mathbb{Z}_m) := \mathrm{SL}_2^{\mathcal{I}}(\mathbb{Z}_m)$.

Whenever H is a subgroup of $C^{\mathcal{G}}(m)$, we have a corresponding quotient $\mathrm{L}_2^{\mathcal{G}}(\mathbb{Z}_m)/H$. In particular, we define

$$\mathrm{PL}_2^{\mathcal{G}}(\mathbb{Z}_m) := \mathrm{L}_2^{\mathcal{G}}(\mathbb{Z}_m)/C^{\mathcal{G}}(m),$$

$$\hat{\mathrm{P}}\mathrm{L}_2^{\mathcal{G}}(\mathbb{Z}_m) := \mathrm{L}_2^{\mathcal{G}}(\mathbb{Z}_m)/\{\pm I\}.$$

When $\mathcal{G} = \{1\}$, we also write

$$\mathrm{PSL}_2(\mathbb{Z}_m) := \mathrm{PL}_2^{\mathcal{G}}(\mathbb{Z}_m) = \mathrm{SL}_2(\mathbb{Z}_m)/C(m);$$

this is the *projective special linear group*. Observe that, if m is not prime, then we can have $C(m) \neq \{\pm I\}$. It will also be useful in the following to recall that

$$\hat{\mathrm{P}}\mathrm{SL}_2(\mathbb{Z}_m) := \mathrm{SL}_2(\mathbb{Z}_m)/\{\pm I\}.$$

Finally, when $\mathcal{G} = \mathbb{Z}_m^*$, we write

$$\mathrm{PGL}_2(\mathbb{Z}_m) := \mathrm{L}_2^{\mathcal{G}}(\mathbb{Z}_m)/C^{\mathcal{G}}(m);$$

this is the *projective general linear group*.

Now consider the matrices

13A1 $$R_0 := \begin{bmatrix} 0 & 1 \\ 1 & 0 \end{bmatrix}, \quad R_1 := \begin{bmatrix} 1 & 0 \\ -1 & -1 \end{bmatrix}, \quad R_2 := \begin{bmatrix} 1 & 0 \\ 0 & -1 \end{bmatrix}$$

in the group $\mathrm{SL}_2^{\mathcal{I}}(\mathbb{Z}_m)$. (In [284, p. 162], the corresponding matrix R_1 was defined incorrectly, and later used in certain calculations; however, the results of these calculations remain valid.) (Unless otherwise stated, in what follows we shall take $\mathcal{G} = \mathcal{I} = \{\pm 1\}$.) It is a straightforward exercise to show that the two matrices

$$R_1 R_2 = \begin{bmatrix} -1 & 0 \\ 1 & -1 \end{bmatrix}, \quad R_0 R_2 = \begin{bmatrix} 0 & -1 \\ 1 & 0 \end{bmatrix}$$

generate $\mathrm{SL}_2(\mathbb{Z}_m)$; thus the matrices in (13A1) generate $\mathrm{SL}_2^{\mathcal{I}}(\mathbb{Z}_m)$. It follows that the $\rho_j := R_j H$ generate $\mathrm{SL}_2^{\mathcal{I}}(\mathbb{Z}_m)/H$, and satisfy

$$\rho_0^2 = \rho_1^2 = \rho_2^2 = (\rho_0\rho_1)^3 = (\rho_1\rho_2)^m = (\rho_0\rho_2)^2 = \varepsilon$$

precisely when $\{\pm I\} \leqslant H \leqslant C^{\mathcal{I}}(m)$. Thus we always assume henceforth that $(R_0 R_1)^3 = -I \in H$; this is why we demand that, if $H = \mathcal{H}I$, then $\mathcal{I} \leqslant \mathcal{H}$. Since obviously $\rho_0 \notin \langle \rho_1, \rho_2 \rangle$, the intersection property also holds (so that $\langle \rho_0, \rho_1 \rangle \cap \langle \rho_1, \rho_2 \rangle = \langle \rho_1 \rangle$), which is the remaining property required of a string C-group. We shall often represent an element $\alpha \in \mathrm{SL}_2^{\mathcal{I}}(\mathbb{Z}_m)/H$ by a suitable matrix A in the coset of H which contains α, that is, an A such that $\alpha = AH$; observe that A and $-A$ may always be identified.

Thus, when $m \geqslant 3$, the string C-group $\Gamma_{m;\mathcal{H}} := \mathrm{SL}_2^{\mathcal{I}}(\mathbb{Z}_m)/H$ is the automorphism group of an (abstract) regular polyhedron $\mathcal{L}^3_{m;\mathcal{H}}$ of type $\{3, m\}$, where we shall always write $\mathcal{H} := \{\lambda \in \mathbb{Z}_m \mid \lambda I \in H\}$. (The notation "$\mathcal{L}^3$" here indicates a 3-polytope derived from a linear group.) For example, $\mathcal{H} = \mathcal{I}$ gives the maps $\mathcal{L}^3_{m;\mathcal{I}}$, which were those considered in [284] (particularly when m is a prime). As we shall see in the following, of particular interest is the map $\mathcal{L}^3_{m;\mathcal{C}(m)} := \mathcal{P}(\mathrm{PL}_2^{\mathcal{I}}(\mathbb{Z}_m))$, obtained when $\mathcal{H} = \mathcal{C}(m)$, so that H is the full centre $C^{\mathcal{I}}(m)$. Summarizing, we have

13A2 Theorem *For each $m \geqslant 3$ and each subgroup $\mathcal{H}$ such that $\mathcal{I} \leqslant \mathcal{H} \leqslant \mathcal{C}(m)$, there is a regular polyhedron $\mathcal{L}^3_{m;\mathcal{H}}$ of type $\{3, m\}$, with group isomorphic to $\mathrm{SL}_2^{\mathcal{I}}(\mathbb{Z}_m)/\mathcal{H}I$.*

In the degenerate case $m = 2$, we have $\rho_2 = \varepsilon$, $H = \{I\}$ (because $\mathcal{C}(m) = \{1\}$) and $\mathrm{L}_2(\mathbb{Z}_2) = \mathrm{SL}_2(\mathbb{Z}_2) \cong S_3$. The resulting group is then that of the triangle $\mathcal{L}^3_2 = \{3\}$.

The structure of $\Gamma_{m;\mathcal{H}} = \mathrm{SL}_2^{\mathcal{I}}(\mathbb{Z}_m)/H$ depends on the prime factorization of m, say $m = p_0^{e_0} p_1^{e_1} \cdots p_k^{e_k}$, where $p_0 = 2$ and $p_1, \ldots, p_k$ are distinct odd primes. Now let $r_m^{\pm}$ be the number of distinct solutions to the congruence $x^2 \equiv \pm 1 \pmod{m}$. It follows from elementary number theory (for example, [213, p. 74]) that

13A3
$$r_m^+ = \begin{cases} 2^k, & \text{if } e_0 < 2, \\ 2^{k+1}, & \text{if } e_0 = 2, \\ 2^{k+2}, & \text{if } e_0 > 2, \end{cases}$$

and

13A4
$$r_m^- = \begin{cases} 2^k, & \text{if } e_0 < 2 \text{ and } p_j \equiv 1 \pmod{4} \text{ for } 1 \leqslant j \leqslant k, \\ 0, & \text{otherwise.} \end{cases}$$

It follows, for example, that the original map of [284], now denoted $\mathcal{L}^3_{m;\mathcal{I}}$, coincides with $\mathcal{L}^3_{m;\mathcal{C}(m)}$ if and only if $m = 4$, p^e or $2p^e$ with $p \equiv -1 \pmod{4}$. Note that, throughout this chapter, p will denote a prime. More generally, $\mathcal{L}^3_{m;\mathcal{I}} = \mathcal{L}^3_{m;\mathcal{C}(m)^+}$ when $m = 4$, p^e or $2p^e$ with p odd. In all other cases, $\mathcal{L}^3_{m;\mathcal{I}}$ covers $\mathcal{L}^3_{m;\mathcal{H}}$, as described in the following.

Now any subgroup $\mathcal{H}$ with $\mathcal{I} \leqslant \mathcal{H} \leqslant \mathcal{C}(m)$ must yield an abelian subgroup $\mathcal{H}I$ of matrices; as usual, we avoid the trivial case by assuming that $m \geqslant 3$. Each matrix $M \in C(m) = \mathcal{C}(m)^+ I$ satisfies $M^2 = I$, so that $C(m) = C_2 \times \cdots \times C_2$, a direct product of groups of order 2. Further, if $e_0 \geqslant 2$ or $p_j \equiv -1 \pmod{4}$ for at least one j, then $\mathcal{C}(m) = C(m)$ has order $2^{s(m)}$, where $s(m) = k, k+1$ or $k+2$ according as $e_0 < 2$, $e_0 = 2$ or $e_0 > 2$. Thus there is some h, with $1 \leqslant h \leqslant s(m)$, such that $H \cong \langle \lambda_1 \rangle \times \cdots \times \langle \lambda_h \rangle$ has order 2^h, with $\lambda_1 \equiv -1$ and each $\lambda_j^2 \equiv 1 \pmod{m}$.

Similarly, if $e_0 < 2$ and $p_j \equiv 1 \pmod{4}$ for $1 \leqslant j \leqslant k$, then $\mathcal{C}(m) \cong C_4 \times C_2 \times \cdots \times C_2$ of order $4 \cdot 2^{k-1} = 2^{k+1}$. Here $C_4 \cong \langle \lambda \rangle$, where $\lambda^2 \equiv -1 \pmod{m}$. Again, for some h with $1 \leqslant h \leqslant k$, we have $H \cong \langle \lambda_1 \rangle \times \cdots \times \langle \lambda_h \rangle$, where $\lambda_1 \equiv -1$ or $\lambda \pmod{m}$ and $\lambda_j^2 \equiv 1 \pmod{m}$ for $2 \leqslant j \leqslant h$.

Using [206, p. 9], we find that, for $|H| = 2^n$ (with $n = h$ or $h + 1$),

13A5
$$\left|\mathrm{SL}_2^{\mathcal{I}}(\mathbb{Z}_m)/H\right| = 2^{1-n}m^3 \prod_{p|m}(1 - p^{-2}).$$

We may now easily calculate the number of faces of each rank in $\mathcal{L}^3_{m;\mathcal{H}}$.

The following table contains several familiar examples. In it, v, e, f and g denote the numbers of vertices, edges, and faces, and the order of the group, respectively.

m	$\mathcal{H}$	v	e	f	$\mathcal{L}^3_{m;\mathcal{H}}$	g
3	$\mathcal{I}$	4	6	4	$\{3, 3\}$	24
4	$\mathcal{I}$	6	12	8	$\{3, 4\}$	48
5	$\mathcal{I}$	12	30	20	$\{3, 5\}$	120
5	$\{\pm 1, \pm 2\}$	6	15	10	$\{3, 5\}_5$	60
6	$\mathcal{I}$	12	36	24	$\{3, 6\}_{(2,2)}$	144
7	$\mathcal{I}$	24	84	56	$\{3, 7\}_8$	336
8	$\{\pm 1, \pm 3\}$	12	48	32	$\{3, 8\}_6$	192
13	$\{\pm 1, \pm 5\}$	42	273	182	$\{3, 13\}_7$	1092

Before investigating the structure of the maps $\mathcal{L}^3_{m;\mathcal{H}}$ further, it will first be useful to describe certain groups $\mathrm{SL}_2^{\mathcal{I}}(\mathbb{Z}_m)/H$ in more familiar terms.

13A6 Proposition $\mathrm{PL}_2^{\mathcal{I}}(\mathbb{Z}_m)$ *is isomorphic to*

(a) $\mathrm{PSL}_2(\mathbb{Z}_m)$*, if* $e_0 < 2$ *and* $p_j \equiv 1 \pmod 4$ *for* $1 \leqslant j \leqslant k$,
(b) $\mathrm{PSL}_2(\mathbb{Z}_m) \times C_2$ *otherwise.*

Proof. The canonical map

$$\chi\colon \mathrm{SL}_2(\mathbb{Z}_m)/C(m) \to \mathrm{SL}_2^{\mathcal{I}}(\mathbb{Z}_m)/C^{\mathcal{I}}(m)$$

is a well-defined homomorphism since $C(m) \leqslant C^{\mathcal{I}}(m)$, and is injective since $C(m)I \cap \mathrm{SL}_2(\mathbb{Z}_m) = C(m)$. Since χ is surjective if and only if some element of $C^{\mathcal{I}}(m)$ has determinant -1, part (a) follows from (13A4). For part (b), note that ρ_0 is an involution in $\mathrm{SL}_2^{\mathcal{I}}(\mathbb{Z}_m)/C^{\mathcal{I}}(m)$ which is not in $\chi(\mathrm{PSL}_2(\mathbb{Z}_m))$. □

13A7 Proposition *If* $m = 4$*,* p^e *or* $2p^e$ *with* p *odd and* $e \geqslant 1$*, then* $\mathrm{PGL}_2(\mathbb{Z}_m) \cong \mathrm{PSL}_2(\mathbb{Z}_m) \times C_2$.

Proof. The image of the canonical monomorphism $\mathrm{SL}_2(\mathbb{Z}_m)/C(m) \to \mathrm{PGL}_2(\mathbb{Z}_m)$ is a subgroup whose elements are represented by matrices with determinants in the set R of quadratic residues. The conditions on m imply that m has a primitive root (see [224, p. 44]), so that R has index 2 in $\mathbb{Z}_m^*$. □

These propositions provide the key ideas needed to prove

13A8 Proposition *The following group isomorphisms hold.*

$$\mathrm{PL}_2^{\mathcal{I}}(\mathbb{Z}_{p^e}) \cong \begin{cases} \mathrm{PSL}_2(\mathbb{Z}_{p^e}), & \text{if } p \equiv 1 \pmod 4, \\ \mathrm{PGL}_2(\mathbb{Z}_{p^e}), & \text{if } p \equiv -1 \pmod 4. \end{cases}$$

$$\mathrm{PL}_2^{\mathcal{I}}(\mathbb{Z}_{2^e}) \cong \begin{cases} \mathrm{PGL}_2(\mathbb{Z}_{2^e}), & \text{if } e \leqslant 2, \\ \mathrm{PSL}_2(\mathbb{Z}_{2^e}) \times C_2, & \text{if } e > 2. \end{cases}$$

$$\hat{\mathrm{P}}\mathrm{L}_2^{\mathcal{I}}(\mathbb{Z}_{p^e}) \cong \begin{cases} \hat{\mathrm{P}}\mathrm{SL}_2(\mathbb{Z}_{p^e}) \times C_2, & \text{if } p \equiv 1 \pmod 4, \\ \mathrm{PGL}_2(\mathbb{Z}_{p^e}), & \text{if } p \equiv -1 \pmod 4. \end{cases}$$

$$\hat{\mathrm{P}}\mathrm{L}_2^{\mathcal{I}}(\mathbb{Z}_{2^e}) \cong \begin{cases} \mathrm{PGL}_2(\mathbb{Z}_{2^e}), & \text{if } e \leqslant 2, \\ \hat{\mathrm{P}}\mathrm{SL}_2(\mathbb{Z}_{2^e}) \times C_2, & \text{if } e > 2. \end{cases}$$

We now examine presentations for many of these groups. In Theorem 13A10, we explain how to derive a presentation for $\mathrm{SL}_2^{\mathcal{I}}(\mathbb{Z}_m)/H$ from a given one for $\mathrm{SL}_2^{\mathcal{I}}(\mathbb{Z}_m)/\{\pm I\}$ in terms of the distinguished generators ρ_0, ρ_1, ρ_2. The major difficulty is obtaining an efficient presentation for $\hat{\mathrm{P}}\mathrm{SL}_2(\mathbb{Z}_m)$. When m is odd, Sunday [406] showed that $\hat{\mathrm{P}}\mathrm{SL}_2(\mathbb{Z}_m)$ has presentation

$$\sigma^m = (\sigma\tau)^3 = \tau^2 = \left(\sigma^{(m+1)/2}\tau\sigma^4\tau\right)^2 = \varepsilon$$

in terms of the generators

$$\sigma := \rho_2\rho_1 = \begin{bmatrix} 1 & 0 \\ 1 & 1 \end{bmatrix}, \quad \tau := \rho_0\rho_2 = \begin{bmatrix} 0 & -1 \\ 1 & 0 \end{bmatrix}.$$

Thus, when m is odd, $\hat{\mathrm{P}}\mathrm{L}_2^{\mathcal{I}}(\mathbb{Z}_m)$ has presentation

$$\begin{aligned} \rho_0^2 = \rho_1^2 = \rho_2^2 = (\rho_0\rho_1)^3 = (\rho_1\rho_2)^m = (\rho_0\rho_2)^2 \\ = ((\rho_1\rho_2)^{(m+1)/2}\rho_0(\rho_2\rho_1)^4\rho_0)^2 = \varepsilon. \end{aligned} \tag{13A9}$$

When m is even, the situation is yet more complicated, and some of the available presentations involve a large number of relations (see [348, 424]).

Now let $m \geqslant 3$, and express H as a direct product, as described previously.

13A10 Theorem *For $m \geqslant 3$, consider any presentation for $\Gamma_{m;\mathcal{I}}$ in terms of the distinguished generators ρ_0, ρ_1, ρ_2. Suppose that $\mathcal{H} = \langle\lambda_1\rangle \times \cdots \times \langle\lambda_h\rangle$, where $\lambda_j^2 \equiv 1 \pmod m$ for $2 \leqslant j \leqslant h$, and either $\lambda_1 \equiv -1$ or $\lambda_1^2 \equiv -1 \pmod m$. Then a presentation for $\Gamma_{m;\mathcal{H}} = \mathrm{SL}_2^{\mathcal{I}}(\mathbb{Z}_m)/\mathcal{H}I$ is obtained by adjoining to the standard presentation of $[3, m]$ the relations*

$$(\rho_0(\rho_1\rho_2)^{\lambda_j}\rho_2)^3 = \varepsilon,$$

for $2 \leqslant j \leqslant h$, together with

$$(\rho_0(\rho_1\rho_2)^{\lambda_1})^3 = \varepsilon$$

if $\lambda_1^2 \equiv -1 \pmod m$.

The latter alternative involving λ_1 can occur only if $e_0 < 2$ and $p_j \equiv 1 \pmod 4$ for $1 \leqslant j \leqslant k$.

Proof. Taking as generators the matrices of (13A1), we easily check that $(\rho_0(\rho_1\rho_2)^{\lambda_j}\rho_2)^3 = \varepsilon$ when $\lambda_j^2 \equiv 1 \pmod m$, and that $(\rho_0(\rho_1\rho_2)^{\lambda_1})^3 = \varepsilon$ when $\lambda_1^2 \equiv -1 \pmod m$. The proof is completed by elementary group theory. □

Note that $\mathcal{L}^3_{m;\mathcal{I}}$ is directly regular (that is, it does not coincide with its rotation subgroup, consisting of products of even numbers of the generators ρ_j) for each $m \geqslant 3$, since $\hat{\mathrm{P}}\mathrm{SL}_2(\mathbb{Z}_m)$ has index 2 in $\hat{\mathrm{P}}\mathrm{L}^{\mathcal{I}}(\mathbb{Z}_m)$. It follows from Theorem 13A10 that $\mathcal{L}^3_{m;\mathcal{H}}$ is directly regular if and only if $\mathcal{H}$ contains no element λ with $\lambda^2 \equiv -1 \pmod m$. For instance, when $m = 65$, define $\mathcal{H}_1 := \langle 8 \rangle \cong C_4$ and $\mathcal{H}_2 := \langle -1, 14 \rangle \cong C_2 \times C_2$. Then $\mathcal{L}^3_{65;\mathcal{H}_1}$ and $\mathcal{L}^3_{65;\mathcal{H}_2}$ are not isomorphic, since only the second map is directly regular. Of course, the corresponding groups are not isomorphic either; however, the two polyhedra do have the same numbers of vertices, edges and faces, and the same group orders.

We can interpret the relations of Theorem 13A10 in more geometric terms, as follows. The length of a λ-hole in $\mathcal{L}^3_{m;\mathcal{I}}$ is the period of $\rho_0(\rho_1\rho_2)^{\lambda}\rho_2$ in $\mathrm{SL}_2^{\mathcal{I}}(\mathbb{Z}_m)/H$. Thus, if $\lambda^2 \equiv 1 \pmod m$, then λ-holes have length 3 or 6 according as λI does or does not belong to H. Similarly, a λ-zigzag has length the period of $\rho_0(\rho_1\rho_2)^{\lambda}$. (Recall that the Petrie polygon corresponds to the case $\lambda = 1$.) If $\lambda^2 \equiv -1 \pmod m$, then a λ-zigzag has length 3 or 6 according as λ does or does not belong to $\mathcal{H}$ (see also the following).

As another example, take $m = 24$, $\mathcal{H}_1 = \langle -1, 5 \rangle$ and $\mathcal{H}_2 = \langle -1, 7 \rangle$. Then $\mathcal{L}^3_{24;\mathcal{H}_1} \ncong \mathcal{L}^3_{24;\mathcal{H}_2}$, since in these maps 5-holes have lengths 3 and 6, respectively.

We now come to the promised models of these polyhedra. (Remember that we use the term *model* in preference to *realization*, when the concrete representation of the automorphism group is other than a subgroup of isometries of some euclidean space.) First, we remark that the group $\mathrm{SL}_2^{\mathcal{I}}(\mathbb{Z}_m)$ acts transitively on

$$V_m := \{(\alpha, \beta) \in \mathbb{Z}_m \times \mathbb{Z}_m \mid (\alpha, \beta, m) = 1\}.$$

To see this, note that $(1, 0) \in V_m$, and that, if

$$M := \begin{bmatrix} \alpha & \beta \\ \gamma & \delta \end{bmatrix},$$

then $(1, 0)M = (\alpha, \beta)$. But $M \in \mathrm{SL}_2^{\mathcal{I}}(\mathbb{Z}_m)$ has $\alpha\delta - \beta\gamma \equiv \pm 1 \pmod m$. Since this congruence is soluble if and only if $(\alpha, \beta, m) = 1$, transitivity then follows easily.

Now fix $\mathcal{H}$ with $\mathcal{I} \leqslant \mathcal{H} \leqslant \mathcal{C}(m)$, and again let $H := \mathcal{H}I$. We employ the exact analogue of Wythoff's construction to give an explicit description of $\mathcal{L}^3_{m;\mathcal{H}}$. Each $\lambda \in \mathcal{H}$ satisfies $\lambda^2 \equiv \pm 1 \pmod m$, so that λ is a unit. We may thus define an equivalence relation on V_m by $(\alpha, \beta) \sim (\lambda\alpha, \lambda\beta)$ for $\lambda \in \mathcal{H}$. Let $V_{m;\mathcal{H}}$ be the resulting set of equivalence classes, that of (α, β) being denoted by $[\alpha, \beta]$. Clearly, $\mathrm{SL}_2^{\mathcal{I}}(\mathbb{Z}_m)/H$ acts transitively on $V_{m;\mathcal{H}}$.

We begin the construction by defining a base flag in $V_{m;\mathcal{H}}$, namely,

$$\begin{aligned} F_{-1} &:= \emptyset, \\ F_0 &:= \{v_0\}, \\ F_1 &:= \{v_0, v_1\}, \\ F_2 &:= \{v_0, v_1, v_2\}, \\ F_3 &:= V_{m;\mathcal{H}}, \end{aligned}$$

where $v_0 := [1, 0]$, $v_1 := v_0\rho_0 = [0, 1]$ and $v_2 := v_1\rho_1 = [1, 1]$. (Thus, we here identify a face with its vertex-set, as we did when defining realizations.) Finally, let

$$\mathcal{Q} := \left\{F_i\sigma \mid -1 \leqslant i \leqslant 3 \text{ and } \sigma \in \mathrm{SL}_2^{\mathcal{I}}(\mathbb{Z}_m)/H\right\},$$

which is a partially ordered set with respect to set inclusion. In fact, the following result shows that $\mathcal{Q}$ is a polyhedron.

13A11 Theorem *Let m and $\mathcal{H}$ be as before. Then*

(a) $\mathcal{L}^3_{m;\mathcal{H}}$ and $\mathcal{Q}$ are isomorphic partially ordered sets, so that $\mathcal{Q}$ is a regular polyhedron with group $\Gamma_{m;\mathcal{H}}$;

(b) $V_{m;\mathcal{H}}$ is the vertex-set of $\mathcal{Q}$, and

$$|V_{m;\mathcal{H}}| = \left(m^2 \prod_{p|m} (1 - p^{-2})\right) \Big/ |H|;$$

(c) two vertices $x = [\alpha, \beta]$ and $y = [\gamma, \delta]$ of $\mathcal{Q}$ determine an edge if and only if $\mu := \alpha\delta - \beta\gamma \in \mathcal{H}$. In this case, the third vertices of the two triangular faces on this edge are $\mu x \pm y = [\mu\alpha \pm \gamma, \mu\beta \pm \delta]$.

Proof. We make use of the distinguished subgroups $\Gamma_j := \langle \rho_i \mid i \neq j \rangle$ of $\Gamma := \mathrm{SL}_2^{\mathcal{I}}(\mathbb{Z}_m)/H$, for $j = 0, 1, 2$. Define a map

$$f : \mathcal{L}^3_{m;\mathcal{H}} \to \mathcal{Q}$$

by $f(\Gamma_j\sigma) = F_j\sigma$, for $-1 \leqslant j \leqslant 3$ and $\sigma \in \Gamma$. Noting that each subgroup Γ_j acts transitively on the elements of F_j, it is easy to prove that f is well defined, surjective and order preserving. Next, observe that if σ fixes each element of $F_j = \{v_0, \ldots, v_j\}$, then $\sigma \in \langle \rho_i \mid i > j \rangle$. For example, suppose that

$$\sigma = \begin{bmatrix} \alpha & \beta \\ \gamma & \delta \end{bmatrix}$$

fixes $v_0 = [1, 0]$, so that $\beta \equiv 0$ and $\alpha \equiv \lambda$ with $\lambda^2 \equiv \pm 1$. Thus,

$$\sigma = \begin{bmatrix} 1 & 0 \\ \kappa & \pm 1 \end{bmatrix} \begin{bmatrix} \alpha & 0 \\ 0 & \alpha \end{bmatrix},$$

where $\alpha^2 \equiv \pm 1$ and $\kappa \equiv \alpha\gamma$. The first matrix in this product is of the form $(\rho_1\rho_2)^{-\kappa}$ or $\rho_2(\rho_1\rho_2)^{\kappa}$, so that $\sigma \in \langle \rho_1, \rho_2 \rangle$. After verifying similar statements for $j = 1$ and 2, we can easily prove that f is injective, and that f^{-1} is order preserving. This completes the proof of (a).

By transitivity, $V_{m;\mathcal{H}}$ must be the vertex set of $\mathcal{Q}$, and

$$|V_{m;\mathcal{H}}| = \left|\mathrm{SL}_2^{\mathcal{I}}(\mathbb{Z}_m)/H\right|/|\langle\rho_1,\rho_2\rangle|.$$

Part (b) now follows from (13A5).

Now consider two vertices $x = [\alpha, \beta]$ and $y = [\gamma, \delta]$ with $\mu = \alpha\delta - \beta\gamma \in \mathcal{H}$. We wish to show that these vertices determine an edge in $\mathcal{Q}$. Since the group acts transitively on $V_{m;\mathcal{H}}$, and also preserves the condition that the determinant $\mu \in \mathcal{H}$, we may assume that $[\alpha, \beta] = [1, 0]$. Thus, $\delta = \mu \in \mathcal{H}$, so that $[\gamma, \delta] = [\kappa, 1] = v_1(\rho_1\rho_2)^{-k}$, where $\kappa \equiv \pm\mu\gamma$. On the other hand, it is clear that if $x = [\alpha, \beta]$ and $y = [\gamma, \delta]$ determine an edge of $\mathcal{Q}$, then $\mu \in \mathcal{H}$. Finally, suppose that z is the third vertex of a triangular face on such an edge. Then, again by transitivity, we may assume that $[\alpha, \beta] = [\kappa, 0] = v_0$ and $[c, d] = [0, \lambda] = v_1$, where $\kappa\lambda = \pm\mu$; the third vertices for the two faces on this edge are $\kappa[1, \pm 1] = \mu v_0 \pm v_1$. The characterization of these third vertices as $z = \mu x \pm y$ now follows from the linearity of the group. □

13B Connexions Among the Polyhedra

In this section, we shall look at two kinds of connexions among the regular polyhedra $\mathcal{L}^3_{m;\mathcal{H}}$. The first kind concerns covering relationships between two polyhedra with either the same or different m. The second concerns isomorphic polyhedra with the same vertices, but possibly different edges or faces.

Let $k > 1$ be a divisor of m. Then the natural ring epimorphism $\mathbb{Z}_m \to \mathbb{Z}_k$ induces a group homomorphism

$$\eta\colon \mathrm{L}_2^{\mathcal{I}}(\mathbb{Z}_m) \to \mathrm{L}_2^{\mathcal{I}}(\mathbb{Z}_k).$$

Since the canonical homomorphism $\mathrm{SL}_2(\mathbb{Z}) \to \mathrm{SL}_2(\mathbb{Z}_m)$ is surjective (compare the remarks in Section 13A), η is also surjective.

Let $\mathcal{K}$ be any subgroup of $\mathcal{C}(k)$ which contains $\mathcal{I}$; similarly, take $\mathcal{I} \leqslant \mathcal{M} \leqslant \mathcal{C}(m)$, where additionally $\mathcal{M}\eta \leqslant \mathcal{K}$. Further, write $H(k) := \mathcal{K}I$, and so on. Then η induces an epimorphism of C-groups

$$\bar{\eta}\colon \mathrm{L}_2^{\mathcal{I}}(\mathbb{Z}_m)/H(m) \to \mathrm{L}_2^{\mathcal{I}}(\mathbb{Z}_k)/H(k).$$

Since these are just the groups $\Gamma_{m;\mathcal{M}}$ and $\Gamma_{k;\mathcal{K}}$, this $\bar{\eta}$ in turn induces an order-preserving surjection

$$\pi\colon \mathcal{L}^3_{m;\mathcal{M}} \to \mathcal{L}^3_{k;\mathcal{K}}$$

of the corresponding polytopes. In fact, if $k > 2$, then π is a *covering* of polyhedra in the sense of Section 2D, meaning that π is a rank-preserving homomorphism which takes adjacent flags to adjacent flags. (The term *ordinary collapse* was used in [284, p. 41].) When $k = 2$, we have a collapse of $\mathcal{L}^3_{m;\mathcal{M}}$ onto the triangle $\mathcal{L}^3_{2;\mathcal{C}(2)} = \{3\}$. These facts are easily verified when we recall the construction of the polytope $\mathcal{P}$ from a given string C-group Γ. When $\Gamma = \Gamma_{m;\mathcal{M}}$ and $\mathcal{P} = \mathcal{L}^3_{m;\mathcal{M}}$, then every automorphism of $\mathcal{L}^3_{m;\mathcal{M}}$ induces (by means of $\bar{\eta}$) an automorphism of $\mathcal{L}^3_{k;\mathcal{K}}$.

Taking $m = k$, we note that any $\mathcal{L}^3_{m;\mathcal{H}}$ covers $\mathcal{L}^3_{m;\mathcal{C}(m)}$. Indeed, we take $\eta = \iota$, noting that $\mathcal{H}_1 = \mathcal{M}$ is a subgroup of $\mathcal{H}_2 = \mathcal{K}$, with index 2^ℓ, say, for some $\ell \geqslant 0$. In this case it makes sense to assert that the covering π has *multiplicity* 2^ℓ; each face of $\mathcal{L}^3_{m;\mathcal{H}_2}$ is the image under π of 2^ℓ faces in $\mathcal{L}^3_{m;\mathcal{H}_1}$ of the same rank. However, in general we shall have $m > d$, and then the multiplicity of the covering for edges and faces is m/k times that for vertices (we simply count using (13A5)).

We now recall from Section 7A the definition of the mix of two abstract regular polytopes $\mathcal{P}$ and $\mathcal{Q}$. Let $\Gamma(\mathcal{P}) = \langle \sigma_0, \ldots, \sigma_{m-1} \rangle$ and $\Gamma(\mathcal{Q}) = \langle \tau_0, \ldots, \tau_{n-1} \rangle$, and define $\rho_j := (\sigma_j, \tau_j) \in \Gamma(\mathcal{P}) \times \Gamma(\mathcal{Q})$, where $\sigma_j = \varepsilon$ if $j \geqslant m$ or $\tau_j = \varepsilon$ if $j \geqslant n$. Let $\Gamma := \langle \rho_j \mid 0 \leqslant j \leqslant \max(n, r) - 1 \rangle$. Whenever Γ satisfies the intersection property, it must be a C-group, and the corresponding polytope $\mathcal{P} \diamondsuit \mathcal{Q}$ is called the *mix* of $\mathcal{P}$ and $\mathcal{Q}$. In a similar way, we may define the mix $\mathcal{P}_1 \diamondsuit \cdots \diamondsuit \mathcal{P}_r$ of several polytopes. Note that the corresponding group Γ is a subdirect product of the groups $\Gamma(\mathcal{P}_j)$; in other words, Γ is a subgroup of the direct product $\prod_{j=1}^r \Gamma(\mathcal{P}_j)$, and each induced projection $\Gamma \to \Gamma(\mathcal{P}_j)$ is surjective.

Now suppose that $m = k_1 k_2$, where each $k_j > 1$ and $(k_1, k_2) = 1$. Then we define as before $\eta_j \colon \mathrm{L}_2^{\mathcal{I}}(\mathbb{Z}_m) \to \mathrm{L}_2^{\mathcal{I}}(\mathbb{Z}_{k_j})$, and take $\mathcal{I} \leqslant \mathcal{H}(k_j) \leqslant \mathcal{C}(k_j)$. Further, let $H(m) \leqslant \mathrm{L}_2^{\mathcal{I}}(\mathbb{Z}_m)$ be defined by

$$H(m) = (\mathcal{H}(k_1)I)\eta_1^{-1} \cap (\mathcal{H}(k_2)I)\eta_2^{-1}.$$

With this convention, we may now state

13B1 Theorem *Suppose that $m = k_1 k_2$, where each $k_j > 1$ and $(k_1, k_2) = 1$. Then $H(m) = \mathcal{M}I$ for some $\mathcal{I} \leqslant \mathcal{M} \leqslant \mathcal{C}(m)$, and*

$$\mathcal{L}^3_{k_1;\mathcal{H}(k_1)} \diamondsuit \mathcal{L}^3_{k_2;\mathcal{H}(k_2)} \cong \mathcal{L}^3_{m;\mathcal{M}}.$$

Proof. We first note that indeed $H(m) = \mathcal{M}I$ with a suitable $\mathcal{I} \leqslant \mathcal{M} \leqslant \mathcal{C}(m)$. To see this, let

$$M = \begin{bmatrix} \alpha & \beta \\ \gamma & \delta \end{bmatrix} \in H(m),$$

so that $\alpha \equiv \xi_j \equiv \delta \pmod{k_j}$, with $\xi_j^2 \equiv \pm 1 \pmod{k_j}$, and $\beta \equiv 0 \equiv \gamma \pmod{k_j}$, for each $j = 1, 2$. Since $(k_1, k_2) = 1$, we have $\beta \equiv \gamma \equiv 0 \pmod{m}$, $\alpha \equiv \delta \pmod{m}$; hence $M \in \mathcal{C}(m)I$.

Now let Γ be the group of the mix $\mathcal{L}^3_{k_1;\mathcal{H}(k_1)} \diamondsuit \mathcal{L}^3_{k_2;\mathcal{H}(k_2)}$. Define $\vartheta \colon \Gamma_{m;\mathcal{M}} \to \Gamma$ by $\sigma\vartheta = (\sigma\bar{\eta}_1, \sigma\bar{\eta}_2)$. By the previous remarks, ϑ is a well-defined epimorphism. To show that ϑ is injective, suppose that $\sigma = \mathcal{M}M \in \ker(\vartheta)$, with

$$M = \begin{bmatrix} \alpha & \beta \\ \gamma & \delta \end{bmatrix} \in \mathrm{L}_2^{\mathcal{I}}(\mathbb{Z}_m).$$

Since $\sigma\vartheta = (\mathcal{H}(k_1)M\eta_1, \mathcal{H}(k_2)M\eta_2)$, we may argue as before that

$$M \equiv \begin{bmatrix} \alpha & 0 \\ 0 & \alpha \end{bmatrix} \pmod{m},$$

and finally that $M \in \mathcal{M}I$. This isomorphism ϑ of C-groups thus induces the desired isomorphism of polytopes. □

13B2 Corollary $\mathcal{L}^3_{m;\mathcal{C}(m)} \cong \Diamond_{p^e \| m} \mathcal{L}^3_{p^e;\mathcal{C}(p^e)}$, *where the mix is taken over all exact prime power divisors* p^e *of* m.

Proof. Let $m = p_0^{e_0} p_1^{e_1} \cdots p_r^{e_r}$ be the prime decomposition of m. The proof is by induction on r. Take $k_1 = p_0^{e_0} p_1^{e_1} \cdots p_{r-1}^{e_{r-1}}$ and $k_2 = p_k^{e_k}$. It follows from the proof of Theorem 13A10 that

$$\mathcal{C}(m) = \mathcal{C}(k_1)\eta_1^{-1} \cap \mathcal{C}(k_2)\eta_2^{-1}.$$

This establishes both the initial case $k = 1$ and the inductive step. □

It is worth remarking that Corollary 13B2 does not extend to the original maps of [284], with each $\mathcal{L}^3_{m;\mathcal{C}(m)}$ replaced by $\mathcal{L}^3_{m;\mathcal{I}}$). For example, with $k_1 = 3$, $k_2 = 5$ and $m = 15$, we find that

$$\mathcal{L}^3_{3;\mathcal{I}} \Diamond \mathcal{L}^3_{5;\mathcal{I}} \cong \mathcal{L}^3_{15;\mathcal{C}(15)} \cong \mathcal{L}^3_{3;\mathcal{C}(3)} \Diamond \mathcal{L}^3_{5;\mathcal{C}(5)}$$

is doubly covered by $\mathcal{L}^3_{15;\mathcal{I}}$. This is because $\mathcal{C}(15) = \{\pm 1, \pm 4\}$, whereas $4 \equiv 1 \pmod 3$ and $4 \equiv -1 \pmod 5$.

We next consider the structure of the groups $\mathrm{PL}_2^{\mathcal{I}}(\mathbb{Z}_m)$ and $\mathrm{PSL}_2(\mathbb{Z}_m)$ in terms of the prime factorization of m. It follows from the proof of Theorem 13B1 that $\mathrm{PL}_2^{\mathcal{I}}(\mathbb{Z}_m)$ is always isomorphic to some subdirect product of the groups $\mathrm{PL}_2^{\mathcal{I}}(\mathbb{Z}_{p^e})$ with $p^e \| m$, although not usually to the full direct product. For example, when $m = 12$,

$$\left(\mathcal{C}(4)\begin{bmatrix} 1 & 0 \\ 0 & -1 \end{bmatrix}, \mathcal{C}(3)I\right) \notin (\mathrm{PL}_2^{\mathcal{I}}(\mathbb{Z}_{12}))\,\vartheta.$$

In certain cases, we can say a little more. Recall that $p_0 = 2$ in the prime decomposition of $m = p_0^{e_0} p_1^{e_1} \cdots p_r^{e_r}$.

13B3 Corollary $\mathrm{PL}_2^{\mathcal{I}}(\mathbb{Z}_m) \cong \prod_{p^e \| m} \mathrm{PL}_2^{\mathcal{I}}(\mathbb{Z}_{p^e})$ *if and only if either*

(a) $e_0 < 2$ *and* $p_j \equiv -1 \pmod 4$ *for at most one* $j \geqslant 1$, *or*
(b) $e_0 \geqslant 2$ *and* $p_j \equiv 1 \pmod 4$ *for each* $1 \leqslant j \leqslant k$.

Proof. We must show that the natural monomorphism

$$\vartheta \colon \mathrm{PL}_2^{\mathcal{I}}(\mathbb{Z}_m) \to \prod_{p^e \| m} \mathrm{PL}_2^{\mathcal{I}}(\mathbb{Z}_{p^e})$$

is surjective. However, using (13A3), (13A4) and the fact that $\mathrm{SL}_2(\mathbb{Z}_m) \cong \prod_{p^e \| m} \times \mathrm{SL}_2(\mathbb{Z}_{p^e})$ (see [270, p. 117]), we can easily calculate the orders of all groups involved. □

A more constructive proof of surjectivity is possible using the Chinese Remainder Theorem.

If we confine our attention to matrices of determinant 1 only, we can similarly and more easily prove

13B4 Corollary $\mathrm{PSL}_2(\mathbb{Z}_m) \cong \prod_{p^e \| m} \mathrm{PSL}_2(\mathbb{Z}_{p^e})$.

Here we recall that $\mathrm{PSL}_2(\mathbb{Z}_m)$ is obtained from $\mathrm{SL}_2(\mathbb{Z}_m)$ by factoring out the full centre. It is well known, however, that $\hat{\mathrm{P}}\mathrm{SL}_2(\mathbb{Z}_m) = \mathrm{SL}_2(\mathbb{Z}_m)/\{\pm I\}$ is not generally isomorphic to $\prod_{p^e \| m} \hat{\mathrm{P}}\mathrm{SL}_2(\mathbb{Z}_{p^e})$.

We now move on to a different kind of relationship between pairs of our polyhedra. The example of the icosahedron $\{3, 5\}$ and great icosahedron $\{3, \frac{5}{2}\}$ shows that we can have distinct isomorphic regular polyhedra which share the same vertices. These are particular instances of polyhedra $\mathcal{L}^3_{m;\mathcal{H}}$ (in this case, $m = 5$). However, the relationship between their groups (considered as $\mathrm{L}_2^{\mathcal{I}}(\mathbb{Z}_5)/\{\pm I\}$) is that of outer rather than inner automorphism. We shall see that this is a very general phenomenon.

We begin with the following result, assuming once more that $m \geqslant 3$.

13B5 Theorem *Let $\mathcal{Q}$ be a polyhedron isomorphic to $\mathcal{L}^3_{m;\mathcal{H}}$, with vertex-set $V_{m;\mathcal{H}}$ and group $\Gamma(\mathcal{Q}) \leqslant \Gamma_{m;\mathcal{H}}$. Then the distinguished generators τ_0, τ_1, τ_2 of $\Gamma(\mathcal{Q})$ can be chosen to be of the form $\tau_j = \Lambda^{-1}\rho_j\Lambda$ for $j = 0, 1, 2$, where ρ_0, ρ_1, ρ_2 are the distinguished generators of $\Gamma(\mathcal{L}^3_{m;\mathcal{H}})$, and*

$$\Lambda := \begin{bmatrix} \lambda & 0 \\ 0 & 1 \end{bmatrix}$$

for some $\lambda \in \mathbb{Z}_m^*$.

Proof. Since $V_{m;\mathcal{H}}$ is the vertex-set of $\mathcal{Q}$, we can pick $[1, 0]$ as our initial vertex. The stabilizer of $[1, 0]$ in $\Gamma := \Gamma_{m;\mathcal{H}}$ is just $\langle\rho_1, \rho_2\rangle$ (of order $2m$), so this must be the group of the vertex-figure of $\mathcal{Q}$. Let $[\alpha, \beta]$ be one vertex of $\mathcal{Q}$ adjacent to $[1, 0]$. Applying $\langle\rho_1, \rho_2\rangle$, the adjacent vertices are all those of the form $[\alpha + \kappa\beta, \pm\beta]$, with $\kappa \in \mathbb{Z}_m$. Let $k = (\beta, m)$. Then we obtain at most $2m/k$ such adjacent vertices, so that $k = 1$ or 2.

Suppose, if possible, that $k = 2 = \gamma\beta + \mu m$ (say). Then we can check that the two non-indentity matrices

$$\begin{bmatrix} 1 & 0 \\ \frac{1}{2}m & 1 \end{bmatrix}, \quad \begin{bmatrix} -1 & 0 \\ \alpha\gamma & 1 \end{bmatrix}$$

both stabilize $[1, 0]$ and $[\alpha, \beta]$. But this is not permitted; hence $k = 1$.

It now follows that $[0, \beta]$ is also adjacent to $[1, 0]$; since ρ_2 is the only non-trivial element of $\mathrm{L}_2^{\mathcal{I}}(\mathbb{Z}_m)/H$ which stabilizes $[1, 0]$ and $[0, \beta]$, we must therefore have $\tau_2 = \rho_2$.

The vertices adjacent to $[1, 0]$ are all the $[\alpha, \beta]$ with $\alpha \in \mathbb{Z}_m$. Let $[\alpha, \beta]$ be adjacent to $[0, \beta]$ also, and let τ_1 interchange $[0, \beta]$ and $[\alpha, \beta]$. Since τ_1 is an involution in $\langle\rho_1, \rho_2\rangle$, it is of the form

$$\begin{bmatrix} 1 & 0 \\ \frac{1}{2}m & 1 \end{bmatrix} \quad \text{or} \quad \begin{bmatrix} -1 & 0 \\ \delta & 1 \end{bmatrix}$$

for some $\delta \in \mathbb{Z}_m$. The former case is excluded (since $\langle \tau_1, \tau_2 \rangle = \langle \rho_1, \rho_2 \rangle$), and so we must have the latter, with $\delta \equiv \alpha\beta^{-1}$.

Finally, τ_0 is an element of $\Gamma_{m;\mathcal{H}}$ which interchanges $[1, 0]$ and $[0, \beta]$, and so must be of the form

$$\begin{bmatrix} 0 & \beta \\ \eta\beta^{-1} & 0 \end{bmatrix}$$

with $\eta \equiv \pm 1$. Since τ_0 stabilizes $[\alpha, \beta]$, we also have

$$[\eta, \alpha\beta] = \xi[\alpha, \beta]$$

for some $\xi \in \mathcal{H}$, so that $\alpha \equiv \xi$ and $\eta \equiv \xi^2$ (recall that $\beta \in \mathbb{Z}_m^*$). We can now rewrite τ_0 in the equivalent form

$$\xi \begin{bmatrix} 0 & \beta \\ \eta\beta^{-1} & 0 \end{bmatrix} = \begin{bmatrix} 0 & \xi\beta \\ (\xi\beta)^{-1} & 0 \end{bmatrix};$$

we can thus replace $\xi\beta$ by β, and take $\eta \equiv 1$ above. This also has the effect of setting $\alpha \equiv 1$.

Our three generators are now

$$\tau_0 = \begin{bmatrix} 0 & \beta \\ \beta^{-1} & 0 \end{bmatrix}, \quad \tau_1 = \begin{bmatrix} -1 & 0 \\ \beta^{-1} & 1 \end{bmatrix}, \quad \tau_2 = \begin{bmatrix} 1 & 0 \\ 0 & -1 \end{bmatrix},$$

for some $\beta \in \mathbb{Z}_m^*$. Since, for $j = 0, 1, 2$, we have $\tau_j = \Lambda^{-1}\rho_j\Lambda$, with

$$\Lambda = \begin{bmatrix} \lambda & 0 \\ 0 & 1 \end{bmatrix}$$

and $\lambda \equiv \beta^{-1} \in \mathbb{Z}_m^*$, we see that we have proved the theorem. □

Let us write $\mathcal{L}^3_{m;\mathcal{H}}(\lambda)$ for the polyhedron $\mathcal{Q}$ of Theorem 13A11. An alternative view of the theorem is that $\mathcal{L}^3_{m;\mathcal{H}}(\lambda)$ is obtained from $\mathcal{L}^3_{m;\mathcal{H}}$ by the mapping $v \mapsto v\Lambda$, with $v \in V_{m;\mathcal{H}}$. By a minor abuse of notation, we can then write $\mathcal{L}^3_{m;\mathcal{H}}\Lambda := \mathcal{L}^3_{m;\mathcal{H}}(\lambda)$, and further extend this abuse to

$$\mathcal{L}^3_{m;\mathcal{H}}(\lambda_1\lambda_2) = \mathcal{L}^3_{m;\mathcal{H}}\Lambda_1\Lambda_2 = \left(\mathcal{L}^3_{m;\mathcal{H}}\Lambda_1\right)\Lambda_2 = \mathcal{L}^3_{m;\mathcal{H}}(\lambda_1)\Lambda_2,$$

and so on, when

$$\Lambda_j := \begin{bmatrix} \lambda_j & 0 \\ 0 & 1 \end{bmatrix}$$

for $j = 1, 2$. It follows that $\mathcal{L}^3_{m;\mathcal{H}}(\lambda_1) = \mathcal{L}^3_{m;\mathcal{H}}(\lambda_2)$ if and only if $\Lambda_1\Lambda_2^{-1} \in D_{m;\mathcal{H}}$; that is, $\lambda_1\lambda_2^{-1} \equiv \pm 1$, or $\lambda_1 \equiv \pm\lambda_2$. Thus, there are exactly $\varphi(m)/2$ essentially distinct polyhedra $\mathcal{L}^3_{m;\mathcal{H}}(\lambda)$, where φ is Euler's totient function.

Naturally, we now have new relationships for the edges and faces of $\mathcal{L}^3_{m;\mathcal{H}}(\lambda)$, but these are easily obtained from the originals. For instance, the condition that $x := x'\Lambda$ and $y := y'\Lambda$ yield an edge is now $\det(x, y) := \delta \in \lambda\mathcal{H}$. Then, writing $\delta := \lambda\eta$, with $\eta \in \mathcal{H}$, the third vertices which, together with such x and y, form a face, are

$$(\eta x' \pm y')\Lambda = \delta\lambda^{-1}x \pm y.$$

Let us note that, since we identify the vertices x and ηx with $\eta \in \mathcal{H}$, it might be thought that λ and $\eta\lambda$ give rise to the same automorphism. However, this is not in fact the case, as we can see by considering the determining condition for triangular faces. We shall return to this question later.

Indeed, a particularly interesting case is when $\lambda \in \mathcal{H}$ itself; we shall consider this case separately. The condition for adjacent vertices x and y now reads $\det(x, y) = \mu \in \lambda\mathcal{H} = \mathcal{H}$, so that the edges of $\mathcal{L}^3_{m;\mathcal{H}}(\lambda)$ coincide with those of the original $\mathcal{L}^3_{m;\mathcal{H}}$.

Since $\lambda^2 \equiv \pm 1$, we have $\lambda^{-1} \equiv \pm\lambda$, so that the condition for triangular faces now gives the third vertices as $\mu\lambda x \pm y$, bearing in mind that we always indentify by ± 1. These faces are not, of course, faces of $\mathcal{L}^3_{m;\mathcal{H}}$; however, we can easily identify the resulting polyhedron $\mathcal{L}^3_{m;\mathcal{H}}(\lambda)$ as a particular transform of $\mathcal{L}^3_{m;\mathcal{H}}$.

In fact, let use use the notation of Section 7B, and write φ_λ for the λth facetting operation and π for the Petrie operation. We recall that φ_λ corresponds to the operation on generators of the group of a polyhedron $\mathcal{Q}$:

$$(\rho_0, \rho_1, \rho_2) \mapsto (\rho_0, (\rho_1\rho_2)^{\lambda-1}\rho_1, \rho_2),$$

and leads to the polyhedron $\mathcal{Q}^{\varphi_\lambda}$, whose vertices and edges are those of $\mathcal{Q}$, and whose faces are the λ-holes of $\mathcal{Q}$. Further, π corresponds to the operation

$$(\rho_0, \rho_1, \rho_2) \mapsto (\rho_0\rho_2, \rho_1, \rho_2),$$

and leads to the polyhedron $\mathcal{Q}^\pi$, whose edges are again those of $\mathcal{Q}$ and whose faces are the Petrie polygons of $\mathcal{Q}$. We recall as well that π commutes with each of the φ_λ.

If we now calculate the generators $\Lambda^{-1}\rho_j\Lambda$ of $\Gamma(\mathcal{L}^3_{m;\mathcal{H}}(\lambda))$, we have

$$\Lambda^{-1}\rho_0\Lambda = \begin{bmatrix} 0 & \lambda^{-1} \\ \lambda & 0 \end{bmatrix} = \begin{cases} \rho_0, & \text{if } \lambda^2 \equiv 1 \pmod{m}, \\ \rho_0\rho_2, & \text{if } \lambda^2 \equiv -1 \pmod{m}; \end{cases}$$

$$\Lambda^{-1}\rho_1\Lambda = \begin{bmatrix} -1 & 0 \\ \lambda & 1 \end{bmatrix} = (\rho_1\rho_2)^\lambda\rho_2;$$

$$\Lambda^{-1}\rho_2\Lambda = \rho_2.$$

Hence the new polyhedron is

$$\mathcal{L}^3_{m;\mathcal{H}}(\lambda) = \begin{cases} (\mathcal{L}^3_{m;\mathcal{H}})^{\varphi_\lambda}, & \text{if } \lambda^2 \equiv 1 \pmod{m}, \\ (\mathcal{L}^3_{m;\mathcal{H}})^{\varphi_\lambda\pi}, & \text{if } \lambda^2 \equiv -1 \pmod{m}. \end{cases}$$

As a further remark, it is now clear why the relations of Theorem 13A10 hold for the group of $\mathcal{L}^3_{m;\mathcal{H}}$, since they just correspond to the triangular faces of the various $\mathcal{L}^3_{m;\mathcal{H}}(\lambda)$.

Now let us return to the case of general λ. We shall now observe that certain of the $\mathcal{L}^3_{m;\mathcal{H}}(\lambda)$ can be obtained by dilatation, which (almost) means starting the analogue of Wythoff's construction with a different vertex, but using the same generating reflexions ρ_j. (The situation is actually a little more complicated than this, as we shall shortly see.)

Specifically, a *dilatate* of $\mathcal{L}^3_{m;\mathcal{H}}$ is a polyhedron $\mu\mathcal{L}^3_{m;\mathcal{H}}$ for some $\mu \in \mathbb{Z}^*_m$; its vertex corresponding to the vertex v of $\mathcal{L}^3_{m;\mathcal{H}}$ is μv. Since

$$M = \begin{bmatrix} \pm\mu & 0 \\ 0 & \mu^{-1} \end{bmatrix} \in \Gamma_{m;\mathcal{H}},$$

we can equivalently replace v by $\mu v M = v\Lambda$, with an appropriate $\lambda \equiv \pm\mu^2$. It follows at once that $\mathcal{L}^3_{m;\mathcal{H}}(\lambda_1)$ and $\mathcal{L}^3_{m;\mathcal{H}}(\lambda_2)$ can be identified under dilatation, so that $\mathcal{L}^3_{m;\mathcal{H}}(\lambda_2) = \mu\mathcal{L}^3_{m;\mathcal{H}}(\lambda_1)$ for some $\mu \in \mathbb{Z}^*_m$, precisely when $\lambda_2 \equiv \pm\mu^2\lambda_1$. In particular, we recover $\mathcal{L}^3_{m;\mathcal{H}}(\lambda_1)$ if and only if

$$\mu \in \mathcal{C}(m) = \{\gamma \in \mathbb{Z}^*_m \mid \gamma^2 \equiv \pm 1\}.$$

Now there are $\varphi(m)$ dilatations, so that $\varphi(m)/|\mathcal{C}(m)|$ of the $\varphi(m)/2$ distinct polyhedra $\mathcal{L}^3_{m;\mathcal{H}}(\lambda)$ are equivalent under dilatation. We have thus proved

13B6 Theorem *Let $m \geqslant 3$. Then the number of classes modulo dilatation of polyhedra $\mathcal{L}^3_{m;\mathcal{H}}(\lambda)$ (with vertex-set $V_{m;\mathcal{H}}$ and group $\Gamma_{m;\mathcal{H}}$) is*

$$\tfrac{1}{2}|\mathcal{C}| = \begin{cases} r^+_m, & \text{if } e_0 < 2 \text{ and } p_j \equiv 1 \pmod 4 \text{ for } 1 \leqslant j \leqslant k, \\ \tfrac{1}{2}r_m, & \text{otherwise.} \end{cases}$$

The notation r^+_m is as in (13A3).

Let us give one illustration of this result; we take $m = 5$. If $\mathcal{H} = \mathcal{I}$, then $\mathcal{L}^3_{5;\mathcal{H}} = \{3, 5\}$, and, since dilatation by 2 is a symmetry (actually, the central symmetry), the only distinct automorph occurs with $\lambda = 2$, when we obtain the great icosahedron $\{3, \frac{5}{2}\}$. This, of course, has different edges from those of $\{3, 5\}$. However, if $\mathcal{H} = \mathcal{C}(5) = \{\pm 1, \pm 2\}$, when the polyhedron is the hemi-icosahedron $\mathcal{L}^3_{m;\mathcal{H}} = \{3, 5\}_5$, the automorph (again with $\lambda = 2$) has the same edges, but different triangular faces, and is obtained by applying the operation $\varphi_2\pi$.

Finally, let us return to our earlier remark about the Wythoff space. If $4 \mid m$, we can see that more points that those of the form $(\mu, 0)$ can be invariant under ρ_1 and ρ_2. Indeed, let $\eta := \frac{1}{2}m - 1$, so that $\eta^2 \equiv 1$, and suppose that $\eta \in \mathcal{H}$. Then the point $v := (1, \frac{1}{2}m)$ is such that $v\rho_1 = (-1 + \frac{1}{2}m, \frac{1}{2}m) = \eta v$ and $v\rho_2 = (1, -\frac{1}{2}m) = v$; hence v also lies in the Wythoff space, since vertices are identified under $\mathcal{H}$.

Actually, it is not hard to see that, if (λ, μ) does lie in the Wythoff space with $\mu \not\equiv 0$, then $4 \mid m$, and, with the previous notation, the point is of the form λv for some $\lambda \in \mathbb{Z}^*_m$, and $\eta \in \mathcal{H}$. Thus what we have just described is essentially the only exceptional situation.

However, in fact we have found nothing new, and investigation of the incidence relations which must result from the choice of $(1, \frac{1}{2}m)$ as initial vertex shows that we have merely obtained $\mathcal{L}^3_{m;\mathcal{H}}(\eta) = (\mathcal{L}^3_{m;\mathcal{H}})^{\varphi_\eta}$ from a different point of view.

13C Realizations of the Polyhedra

We shall now describe realizations of certain of the regular polyhedra $\mathcal{L}^3_{m;\mathcal{H}}$. In view of Section 13A, in this context the most interesting case is $\mathcal{H} = \mathcal{I} = \{\pm 1\}$, since all the other polyhedra $\mathcal{L}^3_{m;\mathcal{H}}$ are covered by this one; hence all their realizations occur among those of $\mathcal{L}^3_{m;\mathcal{I}}$.

We shall see that the pure realizations of $\mathcal{L}^3_{m;\mathcal{I}}$ depend upon the structure of the group $\mathbb{Z}^*_m/\mathcal{I}$. This is a product of cyclic groups, and is in general rather complicated to handle. Indeed, even in the rather more special case when $m = p^e$ is a prime power, the situation has still not been completely resolved. For that reason, we shall concentrate on the case when $m = p$ is a prime; here, things are much more straightforward.

So, let $p \geqslant 3$ be prime. We write $q := \frac{1}{2}\varphi(p) = \frac{1}{2}(p-1)$. Now, as we know, if (α, β) is a vertex of $\mathcal{L}^3_{p;\mathcal{I}}$ (in the original model), then $\gamma(\alpha, \beta) = (\gamma\alpha, \gamma\beta)$ is also a vertex for each $\gamma \in \mathbb{Z}^*_p$. Moreover, we identify the vertices $\pm(\alpha, \beta)$. In other words, since $\mathbb{Z}^*_p/\mathcal{I} \cong C_q$ is a cyclic group, there is a natural action of C_q on (the vertices of) $\mathcal{L}^3_{p;\mathcal{I}}$, which permutes cyclically the points on any line of the model. Observe that

$$(\gamma, 0) = (1, 0)\begin{bmatrix} \gamma & 0 \\ 0 & \gamma^{-1} \end{bmatrix},$$

so that, on each individual line, these permutations are inducible by automorphisms of $\mathcal{L}^3_{p;\mathcal{I}}$.

In any realization of $\mathcal{L}^3_{p;\mathcal{I}}$, the vertices corresponding to those on a line must therefore form a regular q-gon, which may be degenerate, or even trivial. This suggests immediately how we may construct some realizations. The q vertices on each of the $p+1$ lines in the model are mapped to congruent (possibly degenerate) q-gons lying in $p+1$ mutually orthogonal planes in $\mathbb{E}^{2(p+1)}$ with their centres at the origin o. (Even the trivial case is covered here, if we interpret the congruent 1-gons as points in orthogonal lines at the same distance from o. Similarly, 2-gons are just congruent line segments centred at o in orthogonal lines.) It is clear that this will yield a realization.

It is convenient at first to identify $\mathbb{E}^{2(p+1)}$ with $\mathbb{C}^{p+1}$. However, we must constantly bear in mind that, for example, a complex line in $\mathbb{C}^{p+1}$ is ultimately to be thought of as a real (2-dimensional) plane. Let $\omega\colon \mathbb{Z}^*_p/\mathcal{I} \to \mathbb{C}^*$ be a homomorphism; we write ω_γ for the image of $\gamma \in \mathbb{Z}^*_p$ under ω, so that $\omega_{-\gamma} = \omega_\gamma$. Thus ω_γ is a qth root of 1. We label basis vectors of $\mathbb{C}^{p+1}$ by $e_\infty, e_0, \ldots, e_{p-1}$, define the realization of $\mathcal{L}^3_{p;\mathcal{I}}$ by

$$\begin{aligned} &(1, 0) \mapsto e_\infty, \\ &(\alpha, 1) \mapsto e_\alpha, \quad \text{for } \alpha = 0, \ldots, p-1, \end{aligned}$$

and extend to the remaining vertices by letting $\gamma x \mapsto \omega_\gamma v$ if $x \mapsto v$ for any $x = (1, 0)$ or $(\alpha, 1)$ and $\gamma \neq 0$.

If ω is trivial, then the vertices of $\mathcal{L}^3_{p;\mathcal{I}}$ are mapped to those of a regular p-simplex; each vertex corresponds to a line of the model. This realization must be pure, since the p vertices adjacent to a given one are images of a vertex-figure.

So, let us suppose that ω is non-trivial, although the calculations which follow apply to the trivial case as well. The reflexions R_j corresponding to the generators ρ_j of $\Gamma(\mathcal{L}^3_{p;\mathcal{I}})$ are as follows. First, $(1, 0)\rho_0 = (0, 1)$ gives $e_\infty R_0 = e_0$, and since $(\alpha, 1)\rho_0 = (1, \alpha) = \alpha(\alpha^{-1}, 1)$ for $\alpha \neq 0$, we have $e_\alpha = \omega_\alpha e_{\alpha^{-1}}$ for $\alpha = 1, \ldots, p-1$. Further, $(1, 0)$ is left fixed by ρ_1 and ρ_2, while $(\alpha, 1)\rho_1 = (1-\alpha, 1)$ and $(\alpha, 1)\rho_2 = (-\alpha, 1)$ yield

$e_\infty R_j = e_\infty$ for $j = 1$ and 2, and $e_\alpha R_1 = e_{1-\alpha}$ and $e_\alpha R_2 = e_{-\alpha}$ for $\alpha = 0, \ldots, p-1$ (with all suffixes taken modulo p, of course).

Now this realization is not pure. To see this, we first note that the Wythoff space of the realization is spanned by e_∞ and $\sum_{\alpha=0}^{p-1} e_\alpha$. Thus, f_∞, defined by

$$f_\infty := \sqrt{p} e_\infty + \sum_{\alpha=0}^{p-1} e_\alpha,$$

is also suitable to be an initial vertex of a realization. We check that

$$f_0 := f_\infty R_0 = \sqrt{p} e_0 + e_\infty + \sum_{\gamma=1}^{p-1} \bar{\omega}_\gamma e_\gamma,$$

noting that $\omega_{\gamma^{-1}} = \bar{\omega}_\gamma$, and then, applying powers of $R_2 R_1$, the vertex corresponding to e_α for $\alpha = 1, \ldots, p-1$ is

$$f_\alpha = \sqrt{p} e_\alpha + e_\infty + \sum_{\gamma=1}^{p-1} \bar{\omega}_\gamma e_{\alpha+\gamma}.$$

The remaining vertices of the realization are obtained by multiplying by the qth roots ω_β for $\beta \neq 0$.

Now define g_∞ and $g_0, \ldots, g_{p-1}$ similarly, by changing the sign of $\sqrt{p}$. It is easy to check that each f_α is orthogonal to g_∞. Indeed,

$$\langle f_\infty, g_\infty \rangle = -(\sqrt{p})^2 + \sum_{\alpha=0}^{p-1} 1 = 0,$$

while for $\alpha = 0, \ldots, p-1$, we have

$$\langle f_\alpha, g_\infty \rangle = \sqrt{p} - \sqrt{p} + \sum_{\gamma \neq 0} \omega_\gamma = 0.$$

Thus every vertex of the first realization is orthogonal to g_∞, and applying the operations of the groups shows immediately that each f_α is orthogonal to each g_β. In other words, the realization has two components, each of the same dimension. (Strictly speaking, we can only claim as yet that there are at least two components, but we shall soon see that these are actually pure.)

Let us also observe that, except when the q-gon degenerates to a segment, when the Wythoff space is a line, these two realizations are associated with the same group, although they are non-congruent (as they must be). The Wythoff space of each realization is 2-dimensional (any complex multiple of f_0 can serve as initial vertex). When the q-gon is a segment, then the components must be associated with different representations of the group; if the groups were the same, then the components would have to be similar, and their blend could then not have double their dimension.

Summarizing the previous discussion, we see that we have proved most of

13C1 Theorem *Let $p \geqslant 3$ be prime. Then the pure realizations of $\mathcal{L}^3_{p;\mathcal{I}}$ consist of a regular p-simplex, and*

(a) $\frac{1}{4}(p-3)$ *with* $d = p+1$ *and* $w = 2$ *if* $p \equiv 3 \pmod 4$*;*

(b) $\frac{1}{4}(p-5)$ *with* $d = p+1$ *and* $w = 2$ *and* 2 *with* $d = \frac{1}{2}(p+1)$ *and* $w = 1$ *if* $p \equiv 1 \pmod 4$.

Proof. To complete the proof, we first count diagonal classes of $\mathcal{L}^3_{p;\mathcal{I}}$, both symmetric and asymmetric. For each $\beta \neq 0$, there is a single diagonal class from $(1, 0)$ to the set of vertices (α, β); this class is symmetric, since (for example)

$$B := \begin{bmatrix} 0 & \beta \\ \beta^{-1} & 0 \end{bmatrix}$$

is an involution in $\Gamma(\mathcal{L}^3_{p;\mathcal{I}})$ which interchanges $(1, 0)$ and $(0, \beta)$. Bearing in mind identification of the vertices $\pm x$, we see that there are $\frac{1}{2}\varphi(p) = \frac{1}{2}(p-1)$ such classes.

On the other hand, for $\gamma \neq 0$, the diagonal from $(1, 0)$ to $(\gamma, 0)$ is equivalent only to that from $(\gamma^{-1}, 0)$ to $(1, 0)$; thus we have an asymmetric class unless $\gamma^{-1} = \pm\gamma$, which can only happen if $p \equiv 1 \pmod 4$. Here, there are $\lfloor \frac{1}{4}\varphi(p) \rfloor = \lfloor \frac{1}{4}(p-1) \rfloor$ classes, all asymmetric except for 1 when $p \equiv 1 \pmod 4$.

The count for r and $\bar{w}$ (in the notation of Section 5B) is thus

$$r = \tfrac{1}{2}(p-1) + \left\lfloor \tfrac{1}{4}(p-1) \right\rfloor = \left\lfloor \tfrac{3}{4}(p-1) \right\rfloor,$$

$$\bar{w} = \tfrac{1}{2}(p-1) + \left(\tfrac{1}{2}(p-1) - 1\right) = p - 2.$$

On the other hand, the corresponding count from our realizations is

$$r = \sum \tfrac{1}{2}w(w+1) = \begin{cases} 1 + \frac{1}{4}(p-3) \cdot 3, & \text{if } p \equiv 3 \pmod 4, \\ 1 + \frac{1}{4}(p-5) \cdot 3 + 2 \cdot 1, & \text{if } p \equiv 1 \pmod 4, \end{cases}$$

$$= \left\lfloor \tfrac{3}{4}(p-1) \right\rfloor,$$

$$\bar{w} = \sum w^2 = \begin{cases} 1 + \frac{1}{4}(p-3) \cdot 4, & \text{if } p \equiv 3 \pmod 4, \\ 1 + \frac{1}{4}(p-5) \cdot 4 + 2 \cdot 1, & \text{if } p \equiv 1 \pmod 4, \end{cases}$$

$$= p - 2,$$

giving the same answers.

Notice, by the way, that our assertion that the groups of the two components corresponding to a homomorphism ω whose image is not contained in $\mathcal{I}$ must be the same follows from this count. The contrary would give larger numbers for r and $\bar{w}$ than those obtained from enumerating diagonal classes.

The corresponding count for $v - 1$ is

$$v - 1 = \sum dw = p \cdot 1$$

$$+ \begin{cases} \frac{1}{4}(p-3) \cdot (p+1) \cdot 2, & \text{if } p \equiv 3 \pmod 4, \\ \frac{1}{4}(p-5) \cdot (p+1) \cdot 2 + 2 \cdot \frac{1}{2}(p+1) \cdot 1, & \text{if } p \equiv 1 \pmod 4, \end{cases}$$

$$= \tfrac{1}{2}(p^2 - 3),$$

as required. ☐

We now make some remarks about the more general case. It must be emphasized that we have so far failed to resolve this case completely, even for a power $m = p^e$, with p a prime and an exponent $e \geqslant 2$.

A first step in the decomposition of the realization cone depends on the group $Q_m := \mathbb{Z}_m^*/\mathcal{I}$. Just as in the prime case, Q_m acts on each line of vertices in the model as a group of permutations; for $\lambda \in \mathbb{Z}_m$ with $(\lambda, m) = 1$, the action on the points $(\alpha, 0)$ is by

$$x \mapsto x \begin{bmatrix} \lambda & 0 \\ 0 & \lambda^{-1} \end{bmatrix}.$$

Write $Q_m^* := \hom(Q_m, \mathbb{C}^*)$ for the group of homomorphisms from Q_m to the non-zero complex numbers. With each $\omega \in Q_m^*$ we can associate a realization of $\mathcal{L}_{m;\mathcal{I}}^3$. We take a complex line for each of the

$$t := \tfrac{1}{2}m^2 \prod_{p|m}\Big(1 - \frac{1}{p^2}\Big) \Big/ \tfrac{1}{2}m \prod_{p|m}\Big(1 - \frac{1}{p}\Big) = m \prod_{p|m}\Big(1 + \frac{1}{p}\Big)$$

lines in the model, all through o in $\mathbb{C}^t$ and mutually orthogonal. The vertex $(\alpha, 0)$ on the initial line is mapped to $\omega(\alpha)e_\infty$ in the corresponding line of the realization; the other lines in the model are then mapped to their corresponding lines in the realization.

We observe that ω and $\bar{\omega}$ give congruent realizations, and so are to be identified. Moreover, if $\operatorname{im}(\omega) \in \mathcal{I}$, then the complex lines reduce to real ones (or points, in fact, if ω is trivial). Counting total real dimensions of these realizations gives

$$|Q_m^*|t - 1 = |Q_m|t - 1 = v - 1,$$

the dimension of the realization cone. Thus to find the complete decomposition, we look to decompose further the realizations corresponding to the different ω.

However, the straightforward analysis in the prime case does not carry over. The Wythoff spaces now have (in general, complex) dimension greater than 2, since $\mathcal{L}_{m;\mathcal{I}}^3$ has diagonals from $(1, 0)$ to other vertices than those in the vertex-figure. We merely remark on the problems which thus arise.

In one of the previous realizations, we identified the points lying on each line of the model; this gave a regular $(t-1)$-simplex. For convenience, we shall refer to this as the *small simplex* realization. However, unlike what occurs in the prime case, the small simplex is no longer pure. Indeed, there are now more lines in the model than those through the initial vertex $(1, 0)$ and those of the vertex-figure, and so at least two different diagonal classes survive in the small simplex. Of course, we should expect this, because the further identifications induced by $\alpha \mapsto \alpha \pmod{k}$, with k a proper divisor of m, must provide non-trivial components.

We end with a few comments on the prime power case.

13C2 Lemma *If $m = p^e \geqslant 4$ is a prime power, then $\mathbb{Z}_m^*/\mathcal{I}$ is a cyclic group.*

Proof. Whether or not m is a prime power, $\mathbb{Z}_m^*/\mathcal{I}$ has order $\frac{1}{2}\varphi(m)$ for $m \geqslant 3$. When $m = p^e$, this order is $\frac{1}{2}\varphi(p^e) = \frac{1}{2}(p-1)p^{e-1}$.

We first take $p = 2$. We show that $\mathbb{Z}_{2^e}^*/\mathcal{I} = \langle 3 \rangle$ for each $e \geqslant 3$. (In fact, it is trivially true if $e = 2$ also.) Here we have $\frac{1}{2}\varphi(2^e) = 2^{e-2}$. We make the inductive assumption that $3^{2^{e-2}} \equiv 2^{e-1} + 1 \pmod{2^e}$, which is certainly true for $e = 3$. For $e \geqslant 4$, if the result holds with $e - 1$ in place of e, then we have, for some odd number k,

$$3^{2^{e-2}} = (3^{2^{e-3}})^2 = (2^{e-2}k + 1)^2 = 2^{2e-4}k^2 + 2^{e-1}k + 1 \equiv 2^{e-1} + 1 \pmod{2^e},$$

which is the inductive step, and this establishes the claim. Now it is obvious that there is no solution to $\alpha^2 \equiv -1 \pmod{2^e}$ for $e \geqslant 2$. Hence, factoring out by $\langle -1 \rangle$ yields the cyclic group $\mathbb{Z}_{2^e}^*/\mathcal{I} = \langle 3 \rangle$.

Now suppose that p is an odd prime. We know that $\mathbb{Z}_p^*$ is cyclic; the argument is familiar – we use the fact that $\mathbb{Z}_p$ is a field, and just count the numbers of elements of $\mathbb{Z}_p^*$ of each possible order. (We used this fact without comment earlier.) This deals with the case $e = 1$. When $e \geqslant 2$, let α be a generator of $\mathbb{Z}_p^*$. If $\alpha^{p-1} \equiv 1 \pmod{p^2}$, then replace α by $\alpha + p$; in any event, we may assume that $\alpha^{p-1} \equiv kp + 1 \pmod{p^2}$, for some k not divisible by p. It is now an easy induction (similar to that before) to show that $p^{(p-1)p^{e-1}} \equiv kp^e \pmod{p^{e+1}}$ for each e. It is again an immediate consequence that α generates $\mathbb{Z}_{p^e}^*$ for each $e \geqslant 1$. Factoring out by $\mathcal{I}$ then shows that $\mathbb{Z}_{p^e}^*$ is also cyclic. □

It might be thought that counting diagonals would point to what the realization space looks like. However, even in the prime power case, the count of diagonals is tedious and complicated, particularly if we wish to distinguish between the symmetric and the asymmetric. For the latter, however, the following result is useful.

13C3 Lemma *Let $1 \leqslant r < e$, and let $p^r \parallel \beta$. Then the diagonal class containing $\{(1, 0), (\alpha, \beta)\}$ is symmetric if and only if $\alpha^2 \equiv \pm 1 \pmod{p^r}$.*

Proof. The condition $r \geqslant 1$ ensures that $(\alpha, p) = 1$, so that $\alpha^{-1} \pmod{m}$ exists. Suppose that the matrix

$$M := \begin{bmatrix} \kappa & \lambda \\ \mu & \nu \end{bmatrix},$$

with determinant $\det M = 1$ (without loss of generality), is such that $(\alpha, \beta)M = (1, 0)$. Then (with all calculations modulo m)

$$\kappa\alpha + \mu\beta = 1$$
$$\lambda\alpha + \nu\beta = 0,$$

from which we deduce that $\kappa = \alpha^{-1}(1 - \mu\beta)$ and $\lambda = -\nu\alpha^{-1}\beta$, and hence

$$1 = \kappa\nu - \lambda\mu = \alpha^{-1}\nu,$$

or $\nu = \alpha$; thus $\lambda = -\beta$. Hence

$$(1, 0)M = (\kappa, \lambda) = (\alpha^{-1} - \mu\alpha^{-1}\beta, -\beta) \sim (-\alpha^{-1} + \mu\alpha^{-1}\beta, \beta).$$

Since the images of (α, β) under $\langle \rho_1, \rho_2 \rangle$ are all the $(\pm\alpha + \gamma\beta, \beta)$, it follows that the condition for symmetry of the diagonal class is

$$-\alpha^{-1} + \mu\alpha^{-1}\beta = \pm\alpha + \gamma\beta,$$

for some $\gamma \in \mathbb{Z}_m$, or (since $p^r \parallel \beta$),

$$\alpha^2 \equiv \pm 1 \pmod{p^r},$$

as claimed. □

Notice, by the way, that the diagonal classes with $(\beta, p) = 1$ are automatically symmetric, since they contain all the diagonals $\{(1, 0), (\alpha, \beta)\}$ with $\alpha \in \mathbb{Z}_m$.

13D The 4-Polytopes

In two papers [322, 323], Monson and Weiss have described families of regular (and chiral) 4-polytopes, with triangular 2-faces and vertex-figures among the polyhedra $\mathcal{L}^3_{m;\mathcal{H}}$ we have been looking at in the previous sections. We now reproduce those of their results which pertain to regular polytopes; for the most part we shall follow [323], although in some places we have used alternative simpler arguments. However, we have changed some definitions, so that the elements which generate the groups of the vertex-figures coincide with those of Section 13A.

We shall work within the ring $\mathbb{D} := \mathbb{Z}[i]$ of Gaussian integers and certain of its quotients. Note that $\mathbb{D}$ is an euclidean domain. If $m \in \mathbb{D}$, then (as usual) $m\mathbb{D} := \{mz \mid z \in \mathbb{D}\}$ is the principal ideal generated by m; we denote by $\mathbb{D}_m := \mathbb{D}/m\mathbb{D}$ the corresponding residue class ring. We may usually safely identify m with its *associates* $\pm m$ and $\pm im$; we write $m \sim m'$ for associated elements of $\mathbb{D}$.

For later reference, we observe here that we shall adopt the same conventions as in Section 13A, namely, that $\mathcal{I} := \{\pm 1\} \leqslant \mathbb{D}^*_m$ (the multiplicative subgroup of units of $\mathbb{D}_m$), that $\mathcal{C}(m) := \{\lambda \in \mathbb{D}_m \mid \lambda^2 = \pm 1\}$, and $\mathcal{H} \leqslant \mathbb{D}^*_m$ is such that $\mathcal{I} \leqslant \mathcal{H} \leqslant \mathcal{C}(m)$. For the time being, we shall only work with the case $\mathcal{H} = \mathcal{I}$.

Recall that there are three kinds of prime number p in $\mathbb{D}$.

- $p = a + ib$, where $q = p\bar{p} = a^2 + b^2 \equiv 1 \pmod 4$ is a prime in $\mathbb{Z}$.
- $p = 1 + i$; here, $p \sim \bar{p}$.
- $p \equiv 3 \pmod 4$, an ordinary prime in $\mathbb{Z}$.

In [322], the case where $m = p$, an ordinary prime number, is considered. In [323], using a different construction, the case of general m is described. However, unless m is of the form k or $k + ki$, with k an ordinary (positive) integer, the resulting polytopes are chiral rather than regular; observe that these are precisely the cases in which $m \sim \bar{m}$. We shall therefore not treat the remaining cases.

Our method is much the same as that in Section 13A, except that now we use complex 2×2 matrices. Complex conjugation plays a rôle; if M is a 2×2 matrix, we define

13D1 $$z^{\#}M := \bar{z}M,$$

where $\bar{z} := (\bar{\xi}, \bar{\eta})$ is the conjugate vector of $z = (\xi, \eta)$. (It is not absolutely necessary to work with complex vectors and matrices; all our mappings induce linear mappings on $\mathbb{E}^4$, which preserve a certain indefinite quadratic form of signature $(-1, 1, 1, 1)$.)

As far as taking products of such mappings is concerned, all we need note is that

$$({}^{\#})^2 = E, \qquad M({}^{\#}N) = {}^{\#}\bar{M}N.$$

(This latter equation is misdescribed in [323, (5)].)

With the convention of (13D1), we define the following four mappings:

$$R_0 := {}^{\#}\begin{bmatrix} i & 0 \\ 1 & i \end{bmatrix}, \quad R_1 := {}^{\#}\begin{bmatrix} 0 & 1 \\ 1 & 0 \end{bmatrix}, \quad R_2 := {}^{\#}\begin{bmatrix} 1 & 0 \\ -1 & -1 \end{bmatrix}, \quad R_3 := {}^{\#}\begin{bmatrix} 1 & 0 \\ 0 & -1 \end{bmatrix}.$$

13D2

It is not too hard to check that $\langle R_0, \ldots, R_3 \rangle$ is a group which contains $-I$, where $I := I_2$ is the identity matrix; indeed, $(R_1R_2)^3 = -I$. Since $-I$ is clearly central, we may take the quotient by $\mathcal{I}I = \{\pm I\}$; we then define $\rho_j := \mathcal{I}R_j$ for $j = 0, \ldots, 3$. While it is not immediately obvious, in fact $\Gamma := \langle \rho_0, \ldots, \rho_3 \rangle \cong [3, 3, \infty]$.

If we adjoin

$$R_{-1} := {}^{\#}\begin{bmatrix} i & 0 \\ 0 & i \end{bmatrix},$$

and define $\rho_{-1} := R_{-1}\{\pm 1\}$, then $\bar{\Gamma} := \langle \rho_{-1}, \ldots, \rho_3 \rangle$ contains Γ as a subgroup of infinite index. This group is denoted $\mathrm{PSL}_2^{\#}(\mathbb{D})$ in [323, p. 189]; its subgroup of index 2 generated by the $\sigma_j := \rho_{j-1}\rho_j$ for $j = 0, \ldots, 3$ is the *Picard group* $\mathrm{PSL}_2(\mathbb{D})$. We may note that $\mathrm{PSL}_2^{\#}(\mathbb{D}) \cong [\infty, 3, 3, \infty]$ is isomorphic to a Coxeter group; we shall not use that fact directly, but we shall consider some quotients of the group.

Just as in Section 13A, we shall usually represent ρ_j by R_j, tacitly identifying by $\pm I$ where necessary.

Let $m \in \mathbb{D}$ be of the form $m = k$ or $m = k(1 + i)$, with k a positive integer; we write $\hat{m} := k$ or $2k$ respectively – in other words, $\hat{m} = m\bar{m}/k$. We now pass from $\mathbb{D}$ to its quotient domain $\mathbb{D}_m := \mathbb{D}/m\mathbb{D}$. This induces a quotient mapping $\Gamma/\mathcal{I}I \mapsto \Gamma_{m;\mathcal{I}}$, say; notice that the condition $m \mid \bar{m}$ is needed to ensure that conjugation ${}^{\#}$ is compatible with taking the quotient. We emphasize here that we always factor out by $\pm I$. At the risk of some confusion, we find it convenient to use the same symbols $\sigma_j := \mathcal{I}S_j$ and $\rho_j = \mathcal{I}R_j$ for the corresponding elements of $\Gamma_{m;\mathcal{I}}$; moreover, our calculations will invariably be with the original matrices, rather than with their equivalence classes. We trivially find that

$$R_2R_3 = \begin{bmatrix} 1 & 0 \\ -1 & 1 \end{bmatrix},$$

so that

13D3 Lemma *The period of $\rho_2\rho_3$ in $\Gamma_{m;\mathcal{I}}$ is $\hat{m}$.*

Thus the quotient $\Gamma_{m;\mathcal{I}}$ is a group of type $[3, 3, \hat{m}]$. Next, we have

13D4 Lemma *The group $\Gamma_{m;\mathcal{I}}$ is a C-group.*

Proof. Since $\Gamma_{m;\mathcal{I}}$ is obviously a sggi, we must verify that it has the intersection property. However, this is a purely mechanical task, since we must have $\langle \rho_0, \rho_1, \rho_2 \rangle \cong [3, 3]$ (it cannot be a proper quotient), and then showing that $\langle \rho_1, \rho_2, \rho_3 \rangle \cap \langle \rho_0, \rho_1, \rho_2 \rangle = \langle \rho_1, \rho_2 \rangle$ is easy. □

Subsequently, we shall wish to relate the groups $\Gamma_{m;\mathcal{I}}$ for different $m \in \mathbb{D}$. We observe that, whenever $k \mid m$, there is a natural surjective homomorphism

13D5
$$\varphi_{m,k} \colon \Gamma_{m;\mathcal{I}} \to \Gamma_{k;\mathcal{I}}.$$

Let us return for the moment to the subgroup $\langle R_{-1}, \ldots, R_3 \rangle$ of the general semilinear group $\Gamma\mathrm{GL}_2(\mathbb{D})$. As before, we write $S_j = R_{j-1}R_j$ for $j = 0, \ldots, 3$, so that

13D6
$$S_0 = \begin{bmatrix} 1 & 0 \\ -i & 1 \end{bmatrix}, \quad S_1 = \begin{bmatrix} 0 & -i \\ -i & 1 \end{bmatrix}, \quad S_2 = \begin{bmatrix} -1 & -1 \\ 1 & 0 \end{bmatrix}, \quad S_3 = \begin{bmatrix} 1 & 0 \\ -1 & 1 \end{bmatrix}.$$

Since $S_2 S_j^{-1} S_2^{-1} = S_j^{\mathsf{T}}$ for $j = 0$ and 3, we see that $\mathrm{SL}_2(\mathbb{Z}) = \langle S_2, S_3 \rangle$ and $\mathrm{SL}_2(\mathbb{D}) = \langle S_0, S_2, S_3 \rangle$. Compare here [270, Lemma 3.1], where it is shown that $\mathrm{SL}_2(R)$ is generated by transvections for any euclidean domain R; recall that a *transvection* is a matrix of the form

$$\begin{bmatrix} 1 & \zeta \\ 0 & 1 \end{bmatrix} \quad \text{or} \quad \begin{bmatrix} 1 & 0 \\ \zeta & 1 \end{bmatrix},$$

with $\zeta \in R$. Notice that, in our context, if $\zeta = \xi + i\eta$ with $\xi, \eta \in \mathbb{Z}$, then

$$\begin{bmatrix} 1 & 0 \\ \zeta & 1 \end{bmatrix} = S_3^{-\xi} S_0^{-\eta},$$

with an analogous expression for the transpose in terms of S_3^{T} and S_0^{T}.

Because $\mathbb{D}$ is invariant under conjugation, $\mathrm{SL}_2(\mathbb{D})$ has a natural extension within $\Gamma\mathrm{GL}_2(\mathbb{D})$. In [453, §10], Wilker introduced the *extended special linear group*

13D7
$$\mathrm{SL}_2^{\#}(\mathbb{D}) := \mathrm{SL}_2(\mathbb{D}) \cup \{{}^{\#}N \mid N \in \mathrm{GL}_2(\mathbb{D}) \text{ and } \det N = -1\}.$$

To explain what this new group is, recall that $\mathrm{SL}_2(\mathbb{D})$ consists of the orientation-preserving Möbius transformations. On the other hand, $^{\#}$ (that is, $z \mapsto \bar{z}$) is a (complex) reflexion, so that a general orientation-reversing Möbius transformation is of the form ${}^{\#}N$, with N as in (13D7). We may thus identify $\mathrm{SL}_2^{\#}(\mathbb{D})$, or, rather, its quotient $\mathrm{PSL}_2^{\#}(\mathbb{D})$ by $\pm I$, with the Möbius group. The previous remark about $\mathrm{SL}_2(\mathbb{D})$ shows that $\mathrm{SL}_2^{\#}(\mathbb{D}) = \langle R_{-1}, \ldots, R_3 \rangle$.

We find the order of $\Gamma_{m;\mathcal{I}}$ by determining the index $r_m := [\mathrm{PSL}_2^{\#}(\mathbb{D}_m) : \Gamma_{m;\mathcal{I}}] = [\mathrm{SL}_2(\mathbb{D}_m) : \Delta_m]$, where $\Delta_m := \langle S_1, S_2, S_3 \rangle \leqslant \mathrm{SL}_2^{\#}(\mathbb{D}_m)$. We shall see that r_m is multiplicative (that is, that $r_{km} = r_k r_m$ when $(k, m) = 1$); the main difficulties in the calculation of r_m arise from the primes $1 + i$ and 3.

13D8 Lemma *Let $m \in \mathbb{D}$. Then*

(a) *the matrix S_3 has period $\hat{m}$ in Δ_m;*
(b) *if $\mathbb{Z}_k := \mathbb{Z}/k\mathbb{Z}$ with $k \in \mathbb{Z}$, then $\mathrm{SL}_2(\mathbb{Z}_{\hat{m}}) \leqslant \Delta_m$;*
(c) *if $m \neq 0$, then*

$$|\,\mathrm{SL}_2(\mathbb{D}_m)| = (m\bar{m})^3 \prod_{p|m} (1 - (p\bar{p})^{-2}),$$

where the product extends over all (non-associated) prime divisors p of m in $\mathbb{D}$; moreover,

$$|\,\mathrm{SL}_2(\mathbb{Z}_{\hat{m}})| = \hat{m}^3 \prod_{q|\hat{m}} (1 - q^{-2}),$$

where the product extends over all prime divisors q of $\hat{m}$ in $\mathbb{Z}$.

Proof. We have already in effect observed (a), and (b) follows from the fact that $\mathrm{SL}_2(\mathbb{Z}) = \langle S_2, S_3 \rangle$. For the proof of (c), we can refer to [206, p. 9]; alternatively, use Lemmas 13D9 and 13D17, and induction on e when $m = p^e$. □

It should be noted that a complex prime divisor $p = a + ib$ of a real prime $p\bar{p} \equiv 1 \pmod 4$ will, in effect, occur twice in the expression of part (c), since its conjugate $\bar{p} = a - ib$ will also make a contribution.

Our first step in determining r_m is

13D9 Lemma *Let $k, m \in \mathbb{D}$ with $(k, m) = 1$. Then there is an injective homomorphism*

$$\psi\colon \mathrm{SL}_2(\mathbb{D}_{km}) \to \mathrm{SL}_2(\mathbb{D}_k) \times \mathrm{SL}_2(\mathbb{D}_m),$$

and $r_{km} = r_k r_m [\Delta_k \times \Delta_m : \Delta_{km}\psi]$.

Proof. We appeal to the Chinese remainder theorem, from which it follows that the homomorphism ψ given by

$$A\psi := (A\varphi_{km,k}, A\varphi_{km,m})$$

is an injection. We now count indices in the chain of subgroups

$$\mathrm{SL}_2(\mathbb{D}_k) \times \mathrm{SL}_2(\mathbb{D}_m) \geqslant \Delta_k \times \mathrm{SL}_2(\mathbb{D}_m) \geqslant \Delta_k \times \Delta_m \geqslant \Delta_{km}\psi,$$

and the claim of the lemma results. □

We now see how to apply Lemma 13D9.

13D10 Lemma *If $m \in \mathbb{D}$, then $S_0^{24} \in \Delta_m$. More particularly,*

(a) *if $1 + i \nmid m$, then $S_0^3 \in \Delta_m$;*
(b) *if $3 \nmid m$, then $S_0^8 \in \Delta_m$;*
(c) *if $(m, 6) = 1$, then $S_0 \in \Delta_m$.*

Proof. For any $\alpha \in \mathbb{Z}$ such that $(\alpha, m) = 1$, since $(\alpha, \hat{m}) = 1$ also, there is a $\beta \in \mathbb{Z}$ such that $\alpha\beta \equiv -1 \pmod{m}$. Then

13D11 $$U := \begin{bmatrix} 0 & \alpha \\ \beta & 0 \end{bmatrix}, \quad V := \begin{bmatrix} 0 & \beta \\ \alpha & 0 \end{bmatrix}$$

are elements of $\mathrm{SL}_2(\mathbb{Z}_{\hat{m}}) \leqslant \Delta_m$. By direct calculation, we find that

13D12 $$S_0^{a^2-1} = \begin{bmatrix} 1 & 0 \\ -(a^2-1)i & 1 \end{bmatrix} = S_1 U S_1 V \in \Delta_m.$$

We now apply this in the various particular cases.

For (a), we can take $a = 2$, giving $S_0^3 \in \Delta_m$. For (b), we can take $a = 3$, giving $S_0^8 \in \Delta_m$. For (c), if $(m, 6) = 1$, then cases (a) and (b) both apply, so that $S_0 = (S_0^3)^3(S_0^8)^{-1} \in \Delta_m$. In any event, we have $S_0^{24} \in \Delta_m$, since 24 is the least common multiple of 3 and 8. □

The previous lemma settles many cases immediately.

13D13 Proposition *Let $k, m \in \mathbb{D}$.*

(a) If $(m, 6k) = 1$, then $r_{km} = r_k r_m$.
(b) If $(m, 6) = 1$, then $r_m = 1$ and $\Delta_m = \mathrm{SL}_2(\mathbb{D}_m)$.
(c) If $k = 1 + i$ and $m = 3^f$, then $r_{km} = r_k r_m = r_m$.

Proof. Using Lemma 13D9, for part (a) we need only show that

$$\{I\} \times \Delta_m \leqslant \Delta_{km}\psi_{k,m}.$$

But Lemma 13D10(b) shows that $S_0 \in \Delta_m$. Since $(m, k) = 1$, and hence $(m, \hat{k}) = 1$, we can solve the equation $x\hat{k} \equiv 1 \pmod{m}$. It follows that $(I, S_0) = S_0^{x\hat{k}}\psi_{k,m} \in \Delta_{km}\psi_{k,m}$. Moreover, we clearly have $(I, S_3) = S_3^{x\hat{k}}\psi_{k,m} \in \Delta_{km}\psi_{k,m}$. Using the earlier remark that $S_2 S_j^{-1} S_2^{-1} = S_j^{\mathsf{T}}$ for $j = 0, 3$, we see that we now have enough transvections to generate $\{I\} \times \Delta_m$.

Part (b) is just the special case of part (a) with $k = 1$. For part (c), all we have to note is that, in Δ_{1+i},

$$S_0 S_3 = \begin{bmatrix} 1 & 0 \\ -(1+i) & 1 \end{bmatrix} = I,$$

so that $S_0 = S_3^{-1}$ can be disregarded. □

We have now established that r_m depends only on the exponents e and f in the expression of m as $m = (1+i)^e 3^f k$, with $(k, 6) = 1$. We shall therefore employ the temporary notation $r(e, f) := r_m$. (It is worth recalling at this stage that k here is now an ordinary integer, in view of our initial restriction on m. However, our arguments often apply to more general cases.) To focus the discussion, we list the values of $r(e, 0)$ and $r(0, f)$ which we shall subsequent determine; as we said earlier, we shall also show that $r(e,f) = r(e,0)r(0,f)$.

13D14
$$r(e,0) = \begin{cases} 1, & \text{if } e = 0, 1, \\ 2, & \text{if } e = 2, 3, \\ 8, & \text{if } e = 4, 5, \\ 32, & \text{if } e \geqslant 6. \end{cases}$$

13D15
$$r(0,f) = \begin{cases} 1, & \text{if } f = 0, \\ 6, & \text{if } f \geqslant 1. \end{cases}$$

For $k \mid m$, we define

$$K_{m,k} := \ker \varphi_{m,k} \trianglelefteq \mathrm{SL}_2(\mathbb{D}_m).$$

13D16 Lemma *If $k, m \in \mathbb{D}$ are such that $k \mid m$, then $r_m = r_k[K_{m,k} : \Delta_m \cap K_{m,k}]$.*

Proof. This result follows from restricting $\varphi_{m,k}$ to Δ_m, noticing that $\Delta_m \varphi_{m,k} = \Delta_k$, and using

$$|G| = |G\varphi_{m,k}| \cdot |G \cap K_{m,k}|,$$

for the groups $G \leqslant \mathrm{SL}_2(\mathbb{D}_m)$ which occur. □

13D17 Lemma *Let $k, m \in \mathbb{D}$ be such that $k \mid m \mid k^2$. Then*

$$K_{m,k} \cong \mathbb{D}^3_{m/k} = \{(\xi, \eta, \zeta) \mid \xi, \eta, \zeta \in \mathbb{D}_{m/k}\},$$

is an abelian group (and, indeed, a $\mathbb{D}_k$-module). Moreover, the automorphisms $A \mapsto S_j^{-1} A S_j$ acting on $K_{m,k}$ induce linear mappings on $\mathbb{D}^3_{m/k}$, with corresponding matrices

$$\tilde{S}_1 = \begin{bmatrix} -1 & -2i & 0 \\ -i & 1 & 1 \\ 0 & 1 & 0 \end{bmatrix}, \quad \tilde{S}_2 = \begin{bmatrix} -1 & 0 & 2 \\ 0 & 0 & -1 \\ -1 & -1 & 1 \end{bmatrix}, \quad \tilde{S}_3 = \begin{bmatrix} 1 & 0 & 2 \\ -1 & 1 & -1 \\ 0 & 0 & 1 \end{bmatrix}.$$

Proof. To prove this, we observe that an element of $\ker \varphi_{k^{s+1}m, k^s m}$ must be of the form

$$A := \begin{bmatrix} 1 + \xi k & \eta k \\ \zeta k & 1 + \omega k \end{bmatrix},$$

with $\xi, \eta, \zeta, \omega \in \mathbb{D}_{m/k}$. (If, say, $\xi' \equiv \xi \pmod{m/k}$, then ξ can clearly be replaced by ξ'.) Since $\det A = 1$ in $\mathbb{D}_m$, we must have $\omega \equiv -\xi \pmod{m/k}$. Moreover, if we define

$$A(\xi, \eta, \zeta) := \begin{bmatrix} 1 + \xi k & \eta k \\ \zeta k & 1 - \xi k \end{bmatrix},$$

then we see directly that $A(\xi, \eta, \zeta) A(\xi', \eta', \zeta') = A(\xi + \xi', \eta + \eta', \zeta + \zeta')$ modulo m. The verification of the second half of the lemma is purely mechanical. □

We now apply these notions to the remaining primes $1 + i$ and 3. The latter is easier, and so we begin there.

13D18 Lemma *Let $k, m \in \mathbb{D}$. If $m/k \sim 3$, and $3 \mid k$, then $K_{m,k} \leqslant \Delta_m$.*

Proof. The conditions of Lemma 13D17 hold, and (reducing modulo 3) the automorphisms of $\mathbb{D}_3^3$ become

$$\tilde{S}_1 = \begin{bmatrix} -1 & i & 0 \\ -i & 1 & 1 \\ 0 & 1 & 0 \end{bmatrix}, \quad \tilde{S}_2 = \begin{bmatrix} -1 & 0 & -1 \\ 0 & 0 & -1 \\ -1 & -1 & 1 \end{bmatrix}, \quad \tilde{S}_3 = \begin{bmatrix} 1 & 0 & -1 \\ -1 & 1 & -1 \\ 0 & 0 & 1 \end{bmatrix}.$$

Now we always have $S_3^{-\hat{k}} \in K_{m,k}$, since $k \mid \hat{k}$, so that $A(0, 0, \zeta) \in \Delta_m$, with $\zeta = 1$ or $1 - i$ (as $1 + i$ divides k to an even or odd power). But from $(0, 0, \zeta)$ we obtain (by applying $\tilde{S}_1$ twice) $(0, \zeta, 0)$ and then $(-i\zeta, \zeta, \zeta)$; the latter then yields

$$(-i\zeta, 0, 0) = (-i\zeta, \zeta, \zeta) - (0, \zeta, 0) - (0, 0, \zeta).$$

Applying $\tilde{S}_2$ to this yields $(i\zeta, 0, i\zeta)$; thus

$$(0, 0, i\zeta) = (i\zeta, 0, i\zeta) + (-i\zeta, 0, 0).$$

Repeating the argument, we conclude that Δ_m contains all matrices $A(z)$ with $z \in \zeta\mathbb{D}_3^3$, and since $\mathbb{D}_3 \cong \mathrm{GF}(3^2)$ is a field, we thus have all of $K_{m,k}$, as required. □

For the prime $1 + i$, the situation is more complicated, and depends on whether $(1 + i)^e \parallel k$ with e even or odd.

13D19 Lemma *Let $k, m \in \mathbb{D}$. If $m/k \sim 1 + i$, and $(1 + i)^e \parallel k$ with $e = 2$, $e = 4$ or $e \geqslant 6$, then $K_{m,k} \leqslant \Delta_m$.*

Proof. We now have $\mathbb{D}_{1+i} \cong \mathrm{GF}(2)$; in particular, $i \equiv 1 \pmod{1 + i}$. Reducing modulo $1 + i$, the automorphisms $\mathbb{D}_{1+i}^3$ take the form

$$\tilde{S}_1 = \begin{bmatrix} 1 & 0 & 0 \\ 1 & 1 & 1 \\ 0 & 1 & 0 \end{bmatrix}, \quad \tilde{S}_2 = \begin{bmatrix} 1 & 0 & 0 \\ 0 & 0 & 1 \\ 1 & 1 & 1 \end{bmatrix}, \quad \tilde{S}_3 = \begin{bmatrix} 1 & 0 & 0 \\ 1 & 1 & 1 \\ 0 & 0 & 1 \end{bmatrix}.$$

We see (even more easily than in Lemma 13D18) that from $(0, 0, 1)$ we obtain vectors which span $\mathbb{D}_{1+i}^3$. When $e = 2t$ is even, then (modulo powers of i) we have $A(0, 0, 1) = S_3^{-|k|} \in \Delta_m$. When $e = 2t + 1$ is odd, then (again modulo powers of i)

$$A(0, 0, 1) = S_3^{-k'} S_0^{-k'} \in \Delta_m,$$

where $k' = |k/(1 + i)|$, since $8 \mid k'$ if $3 \nmid k'$ or $24 \mid k'$ otherwise. Thus $K_{m,k} \leqslant \Delta_m$ in these cases, as required. □

Consequently, we have established

13D20 Corollary *The following reduction formulae hold.*

(a) *If $f \geqslant 0$ and either $e = 2t + 1$ for some $t \geqslant 0$ or $e = 2t$ for some $t \geqslant 3$, then $r(e, f) = r(e - 1, f)$.*
(b) *If $f \geqslant 2$ and either $e = 2t$ for some $t \geqslant 0$ or $e = 2t + 1$ for some $t \geqslant 3$, then $r(e, f) = r(e, f - 1)$.*

We now find bounds for the remaining ratios $r(e, f)/r(e-1, f)$ and $r(e, f)/r(e, f-1)$. Bear in mind that we have reduced the problem to the cases $e \leqslant 6$ even and $f \leqslant 1$. It is at this point that our treatment departs from that of [323]. We first look at the prime 3.

13D21 Lemma $r(0, 1) = 6$.

Proof. We do this directly. The whole group $SL_2(\mathbb{D}_3)$ is a quotient of the rotation group $[3, 3, 3, 3]^+ \cong A_6$ of the 6-simplex, while the subgroup of Δ_3 of index $r(0, 1)$ is similarly a quotient of the rotation group $[3, 3, 3]^+ \cong A_5$ of the 5-simplex. But both these groups are simple, and so have no proper quotients. We conclude that $r(0, 1) = [SL_2(\mathbb{D}_3) : \Delta_3] = [A_6 : A_5] = 6$, as claimed. □

We next look at the prime $1 + i$; observe that we are only concerned with the ratio $r(e, f)/r(e-1, f)$ in case e is even, so that the conditions $k \mid m \mid k^2$ of Lemma 13D17 hold.

13D22 Lemma

$$r(e, 0) = \begin{cases} 2, & \text{if } e = 2, \\ 8, & \text{if } e = 4, \\ 32, & \text{if } e = 6. \end{cases}$$

Proof. Our treatment is different in the two cases. First, let $e = 2$. Then the whole group $SL_2(\mathbb{D}_2)$ is a quotient of $[2, 3, 3, 2]^+ \cong [2, 3, 3]$, while the subgroup Δ_2 is quotient of $[3, 3, 2]^+ \cong [3, 3]$. Hence $r(1, 0) = 2$.

Now let $e = 4$ or 6. Then the family of all matrices of the form

$$\begin{bmatrix} 1 + 2^t\xi & 2^t\eta \\ 2^t\zeta & 1 - 2^t\xi \end{bmatrix},$$

with $t = \frac{1}{2}e - 1$ and $\xi, \eta, \zeta \in \mathbb{D}$, forms an abelian subgroup J_m of Δ_m (with $m = 2^{t+1}$), and the automorphisms $\tilde{S}_1$, $\tilde{S}_2$ and $\tilde{S}_3$ act on this subgroup, and not just on the usual kernel. Since $S_3^{2^t} \in J_m$, it follows by direct calculation that J_m contains the subgroup corresponding to the subspace $\langle e_1, ie_1, e_2, e_3 \rangle$; in particular, J_m contains the element corresponding to $(1 + i)e_1$, which lies in the kernel. Now $(S_0S_3)^{2^t} = S_0^{2^t} S_0^{2^t}$, which corresponds to $(1 + i)e_3$, also lies in the kernel, and similarly for its transpose, which corresponds to $(1 + i)e_2$. It then easily follows that the (abelian) subgroup $\langle S_0^{2^t}, (S_0^{\mathsf{T}})^{2^t} \rangle$ comprises the coset representatives for $K_{m,k}$ in $K_{m,k} \cap \Delta_m$ (with the appropriate values of k and m). We conclude that $r(4, 0) = 4 \cdot 2 = 8$, and $r(6, 0) = 4 \cdot 8 = 32$. □

We shall finish off the problem by returning to the approach of [323], and considering the Petrie element $T := S_1S_3\ (= R_0R_1R_2R_3)$. However, we are enabled to avoid the detailed calculations which were employed there. Recall that $\langle S_2, S_3 \rangle = SL_2(\mathbb{Z}_{\hat{m}})$, so that to find $|\Delta_m|$, it suffices to find $[\Delta_m : SL_2(\mathbb{Z}_{\hat{m}})]$. Since T satisfies its characteristic equation, so that $T^2 - (1 + i)T + I = O$, the zero matrix, it follows that $T^r = \alpha_r T - \alpha_{r-1} I$, where α_r is given recursively by

$$\alpha_0 = 0,\ \alpha_1 = 1,\ \alpha_r = (1 + i)\alpha_{r-1} - \alpha_{r-2} \text{ for } r \geqslant 2.$$

Thus the first few terms of the sequence $(\alpha_r \mid r \geqslant 0)$ are

$$0,\ 1,\ 1+i,\ -1+2i,\ -4,\ -3-6i,\ 7-9i,\ 19+4i,\ 8+32i.$$

More generally, we have the following useful recursions, which are easily established by induction.

13D23 Lemma *For* $s \geqslant 1$,

$$\alpha_{2s-1} = \alpha_s^2 - \alpha_{s-1}^2,$$
$$\alpha_{2s} = \alpha_s(\alpha_{s+1} - \alpha_{s-1}).$$

We then have

13D24 Lemma *The period of* T *in* Δ_m *is*

(a) 5 *if* $m = 3$,
(b) 2^{t+1} *if* $m = 2^t$ *with* $t \geqslant 1$,
(c) $5 \cdot 2^{t+1}$ *if* $m = 3 \cdot 2^t$ *with* $t \geqslant 1$.

Proof. That T has period 5 in Δ_3 is given by the fact that $\alpha_5 \equiv 0 \pmod 3$, which proves (a). Alternatively, we can argue that $\Delta_3 \cong [3, 3, 3]^+$, and we know the period of T in this group already.

For (b), we use Lemma 13D23 and induction to show that, if $m = 2^t$, then $\alpha_{m-1} \equiv \pm 1 + (1+i)^{2t-1}$ and $\alpha_m \equiv 0 \pmod m$. Hence $T^m = (\mp 1 + (1+i)^{2t-1})I$ in Δ_m, so that $T^{2m} = I$, and T has period $2m = 2^{t+1}$, as asserted.

Now (c) follows at once, since $(3, 2^t) = 1$. □

The final step is given by

13D25 Lemma *For* $m = 2^t 3^f$, *with* $0 \leqslant t \leqslant 3$ *and* $0 \leqslant f \leqslant 1$, *the right coset representatives of* $\mathrm{SL}_2(\mathbb{Z}_{\hat{m}})$ *in* Δ_m *are the distinct powers of* T.

Proof. Since $(3, 8) = 1$ and, by good fortune, $(5, 16) = 1$ (the latter coming from the order of T in Δ_m with $m = 3$ or 8), it is enough to verify the lemma for $m = 3$ and $m = 8$. In each case, it is easy to verify that, if $T^r \neq I$, then T^r is not a real matrix modulo m, and so does not lie in $\mathrm{SL}_2(\mathbb{Z}_{\hat{m}})$. Hence the distinct powers of T give different cosets.

However, we can now appeal to the known orders of $\mathrm{SL}_2(\mathbb{D}_m)$ and $\mathrm{SL}_2(\mathbb{Z}_{\hat{m}})$ (with $m = 3$ or 8) from Lemma 13D8, and to the known indices of Δ_m in $\mathrm{SL}_2(\mathbb{D}_m)$. It follows that the cosets of $\mathrm{SL}_2(\mathbb{Z}_{\hat{m}})$ represented by the powers of T exhaust the group Δ_m.

Finally, we must have $r(e, f) \geqslant r(e, 0)r(0, f)$ for each e and f, since indices of subgroups never increase on taking quotients, and also $r(e, f) \leqslant r(e, 0)r(0, f)$ because the powers of T give distinct coset representatives. This concludes the argument. □

Let us now consider the polytopes corresponding to these groups, when they exist. It is now convenient to return to the full groups $\Gamma_{m;\mathcal{I}}$, rather than working with the rotation subgroups Δ_m.

13D26 Theorem *Let $m = k$ or $m = k(1+i)$, with $k \geqslant 1$ an integer. Except when $m = 1$ or $1+i$, there exists a regular 4-polytope $\mathcal{L}^4_{m;\mathcal{I}} := \mathcal{P}(\Gamma_{m;\mathcal{I}})$.*

Proof. First, we have already in effect dealt with the case $m = 2$, which gives the (degenerate) polytope $\mathcal{P}_2 = \{3, 3, 2\}$. Thus there are no corresponding polytopes for $m = 1$ or $1 + i$, since $\mathcal{P}_2$ has no polytopal quotients.

Next, let $m = p$ be an odd (real) prime. Again, we have dealt with the case $p = 3$, where we have $\mathcal{P}_3 = \{3, 3, 3\}$. For $p \geqslant 5$, we can look at the rotation subgroups, and show that $\langle \sigma_1, \sigma_2 \rangle \cap \langle \sigma_2, \sigma_3 \rangle = \langle \sigma_2 \rangle$. But this is a routine calculation, since we can specifically list the real matrices in the group $\langle S_1, S_2 \rangle$.

Finally, for the remaining cases, we can appeal to the quotient criterion, because the natural quotient maps (arising from divisors of m) enable us to reduce the problem to the prime case, since these are one-to-one on the group $\langle \rho_0, \rho_1, \rho_2 \rangle$ of the facet. $\square$

At this point, it is appropriate to pose an important question. We know that the polytope $\mathcal{L}^4_{m;\mathcal{I}}$ is of type $\{\{3, 3\}, \mathcal{L}^3_{\hat{m};\mathcal{I}}\}$, with $\mathcal{L}^3_{\hat{m};\mathcal{I}}$ the polyhedron initially described in Section 13A. When m is real, then $\hat{m} = m$, and so it is natural to ask

13D27 Question *If $m \geqslant 3$ is an integer, is the 4-polytope $\mathcal{L}^4_{m;\mathcal{I}}$ universal of its type?*

For small integers m, this property can be checked directly; it has also been verified for prime m up to 37 (private communication from Barry Monson and Asia Weiss).

We shall consider relationships among these polytopes which arise from quotients (by subgroups of the centre) and mixing in the next section.

We saw earlier that $\mathrm{SL}(\mathbb{D}_3) \cong [3, 3, 3, 3]^+$, from which we deduce that $\mathrm{SL}^{\#}_2(\mathbb{D}_3) \cong [3, 3, 3, 3]$ is the group of a regular 5-polytope. It is natural to ask whether any of the other groups $\mathrm{SL}^{\#}_2(\mathbb{D}_m)$ are also the groups of polytopes; in other words, which of them also possess the intersection property?

Let us eliminate many of the possibilities from the outset. In most cases, we can violate the intersection property in quite a trivial way. For example, we know that $S_0^{24} \in \Delta_m$ whatever m might be. Since S_0 has period $\hat{m}$, the only chance we have of polytopality is that $\hat{m} \mid 24$. Indeed, unless $1 + i \mid m$, we actually have $S_0^3 \in \Delta_m$, yielding $\hat{m} = 3$, so that $m = 3$.

We may thus suppose that $1 + i \mid m$; hence (by the remark we have just made) m is of the form $(1 + i)^e 3^f$, with $0 \leqslant e \leqslant 6$ and $f = 0$ or 1. We first consider the powers of $1 + i$. Now $m = 1 + i$ is not big enough (because not even the facet and vertex-figure are polytopal), but $m = 2$ does give the degenerate polytope $\{2, 3, 3, 2\}$. We cannot allow $m = (1 + i)^3 \sim 2 + 2i$, because $\mathcal{P}_{2+2i} \cong \{3, 3, 4\}/2 = \{3, 3, 4\}_4$, and, as we saw in Chapter 6, the only quotients of $\{4, 3, 3, 4\}$ are toroids.

The following more general argument disposes of all the cases $e \geqslant 3$. First, we have

$$|\,\mathrm{SL}_2(\mathbb{D}_{(1+i)^e})| = 3 \cdot 2^{3e-2}.$$

Hence we have

$$|\Delta_{(1+i)^e}| = 3 \cdot 2^{3e-2}/r(e, 0),$$

and so the number of vertices of the dual $(\mathcal{P}_{(1+i)^e})^*$, namely, the number of facets of $\mathcal{P}_{(1+i)^e}$, is

$$|\Delta_{(1+i)^e}|/12 = 2^{3e-4}/r(e,0).$$

However, the number of vertices of the supposed regular 5-polytope is just the index $r(e,0)$, and thus to avoid the facet (or vertex-figure) degenerating, we must perforce have $2^{3e-4}/r(e,0) \leqslant r(e,0)$, or

$$2^{3e-4} \leqslant r(e,0)^2.$$

This condition is violated for $e \geqslant 3$, and so the corresponding 5-polytope cannot exist. The criterion does allow the case $m = 2$, which we have already admitted.

In the case $m = 3(1+i)^e$, similar calculations for $e \geqslant 1$ lead to the possibility of existence only when

$$3 \cdot 2^{3e-1} \leqslant r(e,1)^2,$$

which excludes all but $e = 1$ and 2.

13E Connexions Among 4-Polytopes

We end the chapter by considering briefly various connexions among the regular 4-polytopes $\mathcal{P}_m$ which we constructed in Section 13D. These are exactly analogous to those among the polyhedra constructed from linear groups, which we discussed in Section 13B.

There are two main kinds of relationship between pairs of the 4-polytopes – those arising from quotients, and those coming from mixing. We look at the former first.

In fact, we have already implicitly discussed the main quotient relationships. At this point, we must modify our earlier notation. In analogy to the 3-polytopes $\mathcal{L}^3_{m;\mathcal{H}}$, we write $\mathcal{L}^4_{m;\mathcal{H}} := \mathcal{P}(\Gamma_{m;\mathcal{H}})$ for the polytope (if it exists) whose group $\Gamma_{m;\mathcal{H}}$ is the quotient of Γ_m by a subgroup $H := \mathcal{H}I$ of its centre $\mathcal{C}(m)I$ which contains $\{\pm I\}$. The basic case $\mathcal{H} = \mathcal{I}$ (with $\mathcal{I} = \{\pm 1\}$) is, of course, what we have discussed hitherto.

We begin with a trivial remark.

13E1 Lemma *If $\mathcal{I} \leqslant \mathcal{H} \leqslant \mathcal{J} \leqslant \mathcal{C}(m)$, then $\mathcal{L}^4_{m;\mathcal{J}}$ is a quotient of $\mathcal{L}^4_{m;\mathcal{H}}$.*

As before, if $k \mid m$, then we denote by $\varphi_{m,k}$ the natural quotient map from $\Gamma_{m;\mathcal{I}}$ to $\Gamma_{k;\mathcal{I}}$ which is induced by the corresponding natural quotient map $\eta_{m,k}\colon \mathbb{D}_m \to \mathbb{D}_k$. Our main quotient result follows immediately from the definition and Lemma 13E1.

13E2 Theorem *Let $k \mid m$, let $\mathcal{I} \leqslant \mathcal{J} \leqslant \mathcal{C}(k)$, and let $\mathcal{I} \leqslant \mathcal{H} \leqslant \mathcal{C}(m)$ be such that $\mathcal{H}\eta_{m,k} \leqslant \mathcal{J}$. Then $\mathcal{L}^4_{k;\mathcal{J}}$ is a quotient of $\mathcal{L}^4_{m;\mathcal{H}}$.*

One might expect that the mixing properties of the 4-polytopes $\mathcal{L}^4_{m;\mathcal{H}}$ should be similar to those of the polyhedra $\mathcal{L}^3_{m;\mathcal{H}}$. For the polyhedra, we saw that the ones which mix most naturally are those for which $\mathcal{H} = \mathcal{C}(m)$ corresponds to the whole centre. For the 4-polytopes we have the exact analogue.

13E3 Theorem *If* $n := (k, m)$, $\mathcal{I} \leqslant \mathcal{K} \leqslant \mathcal{C}(k)$ *and* $\mathcal{I} \leqslant \mathcal{M} \leqslant \mathcal{C}(m)$, *then*

$$\mathcal{L}^4_{k;\mathcal{K}} \diamond \mathcal{L}^4_{m;\mathcal{M}} = \mathcal{L}^4_{n;\mathcal{H}},$$

where $\mathcal{H} := \mathcal{K}\eta^{-1}_{n,k} \cap \mathcal{M}\eta^{-1}_{n,m}$.

Proof. It suffices to consider the rotation subgroups $\Gamma^+_{m;\mathcal{I}}$, and so on. There is a natural homomorphism $\Gamma^+_{n;\mathcal{H}} \mapsto \Gamma^+_{k;\mathcal{K}} \times \Gamma^+_{m;\mathcal{M}}$; to complete the proof, we need only observe that this mapping is injective. □

Once again, therefore, if the subgroup $\mathcal{H}$ is as large as possible (that is, $\mathcal{H} = \mathcal{C}(n)$), then $\mathcal{L}^4_{n;\mathcal{H}}$ is the natural mix of corresponding prime power constituents. For instance, we have

$$\mathcal{L}^4_{21;\mathcal{C}} = \mathcal{L}^4_{3;\mathcal{I}} \diamond \mathcal{L}^4_{7;\mathcal{I}},$$

where $\mathcal{C} = \{\pm 1, \pm 8\}$. In contrast, $\mathcal{L}^4_{21;\mathcal{I}}$ cannot be decomposed in such a simple way.

14

Miscellaneous Classes of Regular Polytopes

In the final chapter of the book, we shall discuss several classes of regular polytopes which do not fit naturally into any of the earlier general categories. We begin in Section 14A with locally projective polytopes. The philosophy here is a little different from that of Chapter 12, in that, in the most important class, the sections which are projective regular polytopes have rank 4. In Section 14B, we then consider classes in which the minimal non-spherical sections will be toroidal, projective, or of some other topological type, but at least two different of these types.

14A Locally Projective Regular Polytopes

A few *locally projective* regular polytopes have been described hitherto; here, the minimal non-spherical sections are projective spaces. For example, Grünbaum [199, p. 197] and Coxeter [122] independently found $\{\{3,5\}_5,\{5,3\}_5\}$ which has 11 vertices and 11 hemi-icosahedral facets, while Coxeter [121] found $\{\{5,3\}_5,\{3,5\}_5\}$, which has 57 vertices and 57 hemi-dodecahedral facets. However, we can recall mention of no others which are not themselves just projectives spaces. (A nice speculation would be that these two last polytopes could be concatenated, to yield

$$\{\{3,5\}_5,\{5,3\}_5,\{3,5\}_5\},$$

with a fairly obvious meaning, but unfortunately a computer check using the Todd–Coxeter coset enumeration method has shown that this "polytope" degenerates.) It is interesting to note here that the Petersen graph, which is the edge-graph of the hemi-dodecahedron $\{5,3\}_5$, occurs in the two examples as the graph of the vertex-figure or facet. We shall further comment on this at the end of the section.

As an outcome of the original investigation of locally toroidal regular polytopes, we found new classes of locally projective regular n-polytopes, for each $n \geqslant 5$. As we said previously, the minimal non-spherical sections all have rank 4, and the family begins with the three projective polytopes $\{3,4,3\}_6$, $\{4,3,3\}_4$ and $\{3,3,4\}_4$ of rank 4. It turns out that all are finite and universal of their types.

We begin by describing the basic polytope $\mathcal{L}_0^n$ of each rank $n \geqslant 4$. The automorphism groups of all the other polytopes $\mathcal{L}_k^n$ of the same rank are subgroups of the automorphism group $\Gamma_0^n := \Gamma(\mathcal{L}_0^n)$ of $\mathcal{L}_0^n$. We shall obtain $\mathcal{L}_0^n$ in two ways – by means of a modified twisting contruction, and directly through a geometric realization (see Chapter 5 and [282]).

It turns out to be notationally more convenient to describe the group of the dual $\mathcal{L}_{n-1}^n$ of $\mathcal{L}_0^n$. Actually, we shall start by defining an abstract group Γ_{n-1}^n; only later will it become apparent that the geometric group Γ which we next construct is a faithful representation of Γ_{n-1}^n.

The group Γ_{n-1}^n is the Coxeter group $[3, 4, 3^{n-3}]$, with the single extra relation

$$(\rho_0\rho_1\rho_2\rho_3)^6 = \varepsilon.$$

(As usual, we use 3^{n-3} as a shorthand for a string of $n-3$ consecutive 3s.) We shall show that Γ_{n-1}^n is the group of a regular polytope

$$\mathcal{L}_{n-1}^n := \{3, 4, 3^{n-3}\}/\langle\langle (0123)^6 \rangle\rangle,$$

in the notation we introduced in Section 2F. Thus the original polytope (to which we have already made reference in Chapter 12) is

14A1 $$\mathcal{L}_0^n = \{3^{n-3}, 4, 3\}/\langle\langle ((n-4)(n-3)(n-2)(n-1))^6 \rangle\rangle.$$

We now describe the representation Γ of Γ_{n-1}^n. For each $n \geqslant 4$, we take $n-1$ copies of the symmetric group S_4, which we think of as the symmetry group of the regular tetrahedron. The generators of the ith copy are then ω_{i0}, ω_{i1} and ω_{i2}, satisfying the relations

$$\omega_{ij}^2 = (\omega_{i0}\omega_{i1})^3 = (\omega_{i0}\omega_{i2})^2 = (\omega_{i1}\omega_{i2})^3 = \varepsilon,$$

for $j = 0, 1, 2$. We now take the wreath product of S_4 with S_{n-1}: the $n-2$ generators of S_{n-1} are $\chi_i = (i-1\ i)$, for $i = 2, \ldots, n-1$, which permute the copies of S_4 accordingly, so that

$$\chi_i\omega_{i-1,j}\chi_i = \omega_{ij},$$

for each i and j.

The group Γ is then $\langle \rho_0, \ldots, \rho_{n-1} \rangle$, where

$$\begin{aligned}
\rho_0 &:= \omega_{11}\omega_{21}\cdots\omega_{n-1,1},\\
\rho_1 &:= \omega_{10}\omega_{22}\omega_{32}\cdots\omega_{n-1,2},\\
\rho_i &:= \chi_i, \quad \text{for } i = 2, \ldots, n-1.
\end{aligned}$$

(Our use of the same symbols to denote the generators of Γ as for those of the group Γ_{n-1}^n should cause no confusion.)

We first check the relations satisfied by $\rho_0, \ldots, \rho_{n-1}$. Those satisfied by $\rho_2, \ldots, \rho_{n-1}$ are just those of S_{n-1}, namely,

$$\rho_i^2 = (\rho_i\rho_{i+1})^3 = (\rho_i\rho_j)^2 = \varepsilon,$$

for each appropriate i, j with $i \leqslant j-2$. Next,

$$\rho_i^2 = \varepsilon, \quad \text{for } i = 0, 1.$$

Since

$$\begin{aligned} \rho_0\rho_1 &= \omega_{11}\omega_{10} \cdot \omega_{21}\omega_{22} \cdots \omega_{n-1,1}\omega_{n-1,2}, \\ \rho_i\rho_0\rho_i &= \rho_0, \quad \text{for } i = 2, \ldots, n-1, \\ (\rho_1\rho_2)^2 &= \omega_{10}\omega_{12} \cdot \omega_{20}\omega_{22}, \end{aligned}$$

we conclude that

$$(\rho_0\rho_1)^3 = (\rho_1\rho_2)^4 = (\rho_0\rho_i)^2 = \varepsilon,$$

for $i = 2, \ldots, n-1$. That is, Γ is a quotient of the group $[3, 4, 3^{n-3}]$.

There is one last relation to verify. We have

$$\begin{aligned} (\rho_0\rho_1\rho_2\rho_3)^2 &= \rho_0\rho_1\rho_0\rho_2\rho_1\rho_3\rho_2\rho_3 \\ &= \omega_{11}\omega_{10}\omega_{11}\omega_{12} \cdot \omega_{21}\omega_{22}\omega_{21}\omega_{20} \cdot \prod_{i=3}^{n-1} (\omega_{i1}\omega_{i2})^2 \cdot (\rho_2\rho_3)^2, \end{aligned}$$

and so

$$(\rho_0\rho_1\rho_2\rho_3)^6 = \varepsilon.$$

We are now ready to state the main result of the section.

14A2 Theorem *The group Γ is isomorphic to Γ_{n-1}^n.*

So far, of course, we have only shown Γ to be a quotient of Γ_{n-1}^n. As we have already suggested, we shall leave the proof incomplete, until we draw a number of strands together towards the end of the section.

We next produce a realization of the polytope $\mathcal{L}_0^n$, in the sense of Chapter 5. This is obtained directly from the abstract formulation of the group Γ described previously.

We have an obvious representation of the group Γ_0^n as $\Gamma = \langle \rho_0, \ldots, \rho_{n-1} \rangle$ in $\mathbb{E}^{3(n-1)}$, where we continue to employ the same notation. (It is obviously a faithful representation of the group Γ, but for the moment Theorem 14A2 remains open.) Let us take the generators of S_4, as the symmetry group of the regular tetrahedron, in the form

$$\begin{aligned} \omega_0 &: x \mapsto (\xi_1, -\xi_3, -\xi_2), \\ \omega_1 &: x \mapsto (\xi_2, \xi_1, \xi_3), \\ \omega_2 &: x \mapsto (\xi_1, \xi_3, \xi_2), \end{aligned}$$

where $x := (\xi_1, \xi_2, \xi_3)$. The ith copy of S_4 operates on the space with coordinate vectors $x_i := (\xi_{i1}, \xi_{i2}, \xi_{i3})$ for $i = 1, \ldots, n-1$. We regard this copy of $\mathbb{E}^3$ as a coordinate subspace of $\mathbb{E}^{3(n-1)}$, of which a general vector is thus written $x = (x_1, \ldots, x_{n-1})$. The elements of Γ will act *coherently* on $\mathbb{E}^{3(n-1)}$, in that they respect its decomposition as an orthogonal direct sum of $n-1$ copies of $\mathbb{E}^3$. In fact, they do more. If we think of the coordinates of vectors in $\mathbb{E}^{3(n-1)}$ as split into *blocks* of 3, not only do elements of Γ permute the blocks, but also induce the same permutations of coordinates within the blocks, if we ignore changes of sign. (Note that ω_0 and ω_2 occur together in ρ_1.) Hence we can now write the generators ρ_j as

$$\begin{aligned} x\rho_0 &:= (x_1\omega_1, \ldots, x_{n-1}\omega_1), \\ x\rho_1 &:= (x_1\omega_0, x_2\omega_2, \ldots, x_{n-1}\omega_2), \\ x\rho_j &:= (x_1, \ldots, x_{j-2}, x_j, x_{j-1}, x_{j+1}, \ldots, x_{n-1}) \quad \text{for } j = 2, \ldots, n-1. \end{aligned}$$

We now apply Wythoff's construction. The initial vertex of $\mathcal{L}_0^n$ lies on all the mirrors ρ_i except ρ_{n-1}, and so is $(0^{3n-6}, 1, 1, 1)$ (up to scalar multiple). It is now easy to check that the vertices are all those of the form

$$(0^{3k}, \pm 1, \pm 1, \pm 1, 0^{3n-3k-6}),$$

with an even number of minus signs, for some $k = 0, \ldots, n-2$. Since $\mathcal{L}_0^n$ has $4(n-1)$ vertices, and the initial group Γ_0^4 is that of the projective polytope $\{3, 4, 3\}_6 = \{3, 4, 3\}/2$, and so has order $1152/2 = 576 = 3 \cdot 2^5 \cdot 3!$, we obtain by induction

14A3 Theorem *The group* Γ_0^n *has order* $3 \cdot 2^{2n-3} \cdot (n-1)!$.

Once again, we emphasize that our proof of Theorem 14A3 is incomplete, because we do not know as yet that Γ is a faithful representation of the abstract group Γ_{n-1}^n. All we are entitled to say for now is that the group of Theorem 14A3 is a quotient of the abstract group.

We now move back up to the Coxeter group $[3, 4, 3^{n-3}]$, without any additionally imposed relations. We appeal to a simplex dissection result of Debrunner [146], which gives a general version of a folkloristic result in group theory. We denote by $S(p_1, \ldots, p_{n-1})$ the $(n-1)$-dimensional simplex (in some space of constant curvature) bounded by the facet hyperplanes $H_0, \ldots, H_{n-1}$, such that the dihedral angle between H_i and H_j is given by

$$\begin{aligned} &\frac{\pi}{p_j}, \quad \text{if } i = j-1 \quad \text{for } j = 1, \ldots, n-1, \\ &\frac{\pi}{2}, \quad \text{otherwise.} \end{aligned}$$

14A4 Theorem *For each* $k = 0, \ldots, n-1$, *the simplex* $S(p, 4, 3^{n-3})$ *can be dissected into* $\binom{n-1}{k}$ *copies of the simplex* $S(3^{k-2}, 4, p, p, 4, 3^{n-k-3})$.

The conventions for small (or large) k are obvious; for example, when $k = 1$, the simplex is $S(p, p, 4, 3^{n-4})$.

In fact, we can rephrase this result in terms of Coxeter groups.

14A5 Theorem *Let* $p \geqslant 3$ *be an integer. Then, for* $k = 0, \ldots, n-1$, *the Coxeter group*

$$[p, 4, 3^{n-3}] := \langle \rho_0, \ldots, \rho_{n-1} \rangle$$

has a subgroup

$$[3^{k-2}, 4, p, p, 4, 3^{n-k-3}] := \langle \sigma_0, \ldots, \sigma_{n-1} \rangle,$$

of index $\binom{n-1}{k}$*, where*

$$\sigma_i = \begin{cases} \rho_{k-i}, & \text{if } 0 \leqslant i \leqslant k, \\ \rho_{k+1}\rho_k \cdots \rho_2\rho_1\rho_2 \cdots \rho_k\rho_{k+1}, & \text{if } i = k+1, \\ \rho_i, & \text{if } k+2 \leqslant i \leqslant n-1. \end{cases}$$

The generators of the subgroup can be found by inspection of the geometric construction of Debrunner [146]. Our interest will naturally be in the case $p = 3$ of Theorem 14A5.

We now pass to the subgroup of the group Γ^n_{n-1}, whose generators are obtained in the same way as in Theorem 14A5. That is, we consider the operation

$$\begin{aligned} (\rho_0, \ldots, \rho_{n-1}) \mapsto (&\rho_k, \ldots, \rho_0, \rho_{k+1}\rho_k \cdots \rho_2\rho_1\rho_2 \cdots \rho_k\rho_{k+1}, \rho_{k+2}, \ldots, \rho_{n-1}) \\ &=: (\sigma_0, \ldots, \sigma_{n-1}); \end{aligned} \quad \textbf{14A6}$$

we write $\Gamma^n_{n-k-1} := \langle \sigma_0, \ldots, \sigma_{n-1} \rangle$. Then we have

14A7 Theorem *For each n and k with* $0 \leqslant k \leqslant n-1$*, the subgroup* Γ^n_k *has index* $\binom{n-1}{k}$ *in* Γ^n_{n-1}*.*

Again, all we can really claim at the moment is that this index is a divisor of $\binom{n-1}{k}$. What we shall do next is find relations which the generators of the subgroup satisfy.

The relations for the products $\sigma_i\sigma_j$ are easily (if tediously) checked; as we have said, they are a feature of the relationship between the original Coxeter groups. So, we check the periods of certain products of four consecutive generators.

First, $\sigma_{k-3}\sigma_{k-2}\sigma_{k-1}\sigma_k = \rho_3\rho_2\rho_1\rho_0$, so that

$$(\sigma_{k-3}\sigma_{k-2}\sigma_{k-1}\sigma_k)^6 = \varepsilon.$$

Second,

$$\begin{aligned} (\sigma_{k-2}\sigma_{k-1}\sigma_k\sigma_{k+1})^2 &= (\rho_2\rho_1\rho_0\rho_{k+1}\rho_k \cdots \rho_2\rho_1\rho_2 \cdots \rho_k\rho_{k+1})^2 \\ &= (\rho_{k+1}\rho_k \cdots \rho_4\rho_2\rho_3\rho_1\rho_2\rho_0\rho_1\rho_2 \cdots \rho_k\rho_{k+1})^2 \\ &\sim (\rho_2\rho_3\rho_1\rho_2\rho_0\rho_1\rho_2\rho_3)^2 \\ &\sim (\rho_3\rho_2\rho_1\rho_2\rho_0\rho_1)^2 \\ &\sim (\rho_3\rho_1\rho_2\rho_1\rho_2\rho_0)^2 \\ &= \rho_1\rho_3\rho_2\rho_1\rho_0\rho_2\rho_3\rho_2\rho_1\rho_2\rho_1\rho_0 \\ &\sim \rho_3\rho_2\rho_1\rho_0 \cdot \rho_3\rho_2\rho_3\rho_1\rho_2\rho_0\rho_1\rho_0 \\ &= (\rho_3\rho_2\rho_1\rho_0)^3, \end{aligned}$$

and so

$$(\sigma_{k-2}\sigma_{k-1}\sigma_k\sigma_{k+1})^4 = \varepsilon.$$

Here, as usual, $\sim$ designates conjugacy. Similarly,

$$\begin{aligned}(\sigma_{k-1}\sigma_k\sigma_{k+1}\sigma_{k+2})^2 &= (\rho_1\rho_0\rho_{k+1}\rho_k\cdots\rho_2\rho_1\rho_2\cdots\rho_k\rho_{k+1}\rho_{k+2})^2\\ &= (\rho_{k+1}\rho_k\cdots\rho_3\rho_1\rho_2\rho_0\rho_1\rho_2\rho_3\cdots\rho_k\rho_{k+1}\rho_{k+2})^2\\ &\sim (\rho_1\rho_2\rho_0\rho_1\rho_2\rho_3)^2\\ &\sim (\rho_0\rho_1\rho_2\rho_3)^3,\end{aligned}$$

so that

$$(\sigma_{k-1}\sigma_k\sigma_{k+1}\sigma_{k+2})^4 = \varepsilon.$$

Here, we have used successively relations of the form

$$\rho_m\gamma\rho_m\rho_{m+1} \sim \gamma\rho_m,$$

when $(\rho_m\rho_{m+1})^3 = \varepsilon$ and ρ_{m+1} commutes with γ. Finally,

$$\begin{aligned}\sigma_k\sigma_{k+1}\sigma_{k+2}\sigma_{k+3} &= \rho_0\rho_{k+1}\rho_k\cdots\rho_2\rho_1\rho_2\cdots\rho_k\rho_{k+1}\rho_{k+2}\rho_{k+3}\\ &\sim \rho_{k+1}\rho_k\cdots\rho_2\rho_0\rho_1\rho_2\cdots\rho_k\rho_{k+1}\rho_{k+2}\rho_{k+3}\\ &\sim \rho_0\rho_1\rho_2\rho_3,\end{aligned}$$

so that

$$(\sigma_k\sigma_{k+1}\sigma_{k+2}\sigma_{k+3})^6 = \varepsilon.$$

Here, we have used successively relations of the form

$$\begin{aligned}\rho_m\gamma\rho_m\rho_{m+1}\rho_{m+2} &\sim \gamma\rho_m\rho_{m+1}\rho_m\rho_{m+2}\\ &= \gamma\rho_{m+1}\rho_m\rho_{m+1}\rho_{m+2}\\ &\sim \gamma\rho_m\rho_{m+1}\rho_{m+2}\rho_{m+1}\\ &= \gamma\rho_m\rho_{m+2}\rho_{m+1}\rho_{m+2}\\ &\sim \gamma\rho_m\rho_{m+1},\end{aligned}$$

when $(\rho_m\rho_{m+1})^3 = \varepsilon = (\rho_{m+1}\rho_{m+2})^3$ and ρ_{m+1} and ρ_{m+2} commute with γ.

As usual, we write $[p, q, r]_s$ for the Coxeter group $[p, q, r]$ with the imposition of the single extra relation $(\rho_0\rho_1\rho_2\rho_3)^s = \varepsilon$. In each case, $[p, q, r]$ will be finite, and thus the group of a regular convex 4-polytope. We shall have also $s = h/2$, where h is the period of $\rho_0\rho_1\rho_2\rho_3$ in $[p, q, r]$, so the group is the automorphism group of the projective polytope $\{p, q, r\}_s = \{p, q, r\}/2$. We now have

14A8 Theorem *For each n and k with $0 \leqslant k \leqslant n-1$, the group Γ^n_{n-k-1} is the Coxeter group $[3^{k-2}, 4, 3, 3, 4, 3^{n-k-3}] = \langle\sigma_0, \ldots, \sigma_{n-1}\rangle$, with the imposition of the extra relations*

$$(\sigma_{k-3}\sigma_{k-2}\sigma_{k-1}\sigma_k)^6 = \varepsilon,$$

$$(\sigma_{k-2}\sigma_{k-1}\sigma_k\sigma_{k+1})^4 = \varepsilon,$$

$$(\sigma_{k-1}\sigma_k\sigma_{k+1}\sigma_{k+2})^4 = \varepsilon,$$

$$(\sigma_k\sigma_{k+1}\sigma_{k+2}\sigma_{k+3})^6 = \varepsilon,$$

whenever appropriate. It is a subgroup of index $\binom{n-1}{k}$ *in* Γ_0^n, *and hence has order*

$$|\Gamma_{n-k-1}^n| = 3 \cdot 2^{2n-3} k!(n-k-1)!.$$

Once again, the proof is incomplete; all that we have shown so far is that Γ_{n-k-1}^n is a quotient group of the group specified in Theorem 14A8, and that the index is at most $\binom{n-1}{k}$.

We recall that a regular polytope $\mathcal{P}$ is *flat* if each of its vertices is incident with each of its facets. By Lemmas 4E2 and 4E3, if a regular polytope $\mathcal{P}$ has flat facets or flat vertex-figures, then $\mathcal{P}$ is itself flat.

If we identify opposite vertices of a regular cube (of whatever rank), then we obtain a flat regular polytope. It follows that, if we construct a regular polytope $\mathcal{L}_{n-k-1}^n$ from the group Γ_{n-k-1}^n defined by the relations of Theorem 14A8, then the sections of $\mathcal{L}_{n-k-1}^n$ determined by the subgroups $\langle \sigma_{k-2}, \sigma_{k-1}, \sigma_k, \sigma_{k+1} \rangle$ and $\langle \sigma_{k-1}, \sigma_k, \sigma_{k+1}, \sigma_{k+2} \rangle$ are flat, and at least one occurs if $1 \leqslant k \leqslant n-2$. Hence, $\mathcal{L}_{n-k-1}^n$ must be flat for these k.

It is clear that the dual of $\mathcal{L}_{n-k-1}^n$ is $\mathcal{L}_k^n$, and so this latter polytope is also flat for the same k. Given that $\mathcal{L}_k^n$ is flat if $1 \leqslant k \leqslant n-2$, and assuming that it does not degenerate, we can immediately calculate the numbers of its vertices and facets, and the order of its symmetry group. Indeed, $\mathcal{L}_k^n$ must have the same number of vertices as its facet $\mathcal{L}_{k-1}^{n-1}$, or the same number of facets as its vertex-figure $\mathcal{L}_k^{n-1}$. Proceeding by induction, starting from the basic cases $\mathcal{L}_1^4 = \{4, 3, 3\}_4$ and $\mathcal{L}_2^4 = \{3, 3, 4\}_4$, with common group of order $192 = 3 \cdot 2^5 \cdot 1! \cdot 2!$, we easily deduce

14A9 Theorem *If* $1 \leqslant k \leqslant n-1$, *then* $\mathcal{L}_k^n$ *has* $4(n-k-1)$ *vertices,* $4k$ *facets, and group of order* $3 \cdot 2^{2n-3} k!(n-k-1)!$.

If we had degeneracy, then, of course, these numbers would be smaller. But we can now draw the strands together, as we promised near the beginning of the section. We consider the abstract groups Γ_k^n alone. First, Γ_0^n has order at least $3 \cdot 2^{2n-3}(n-1)!$, since, as we have shown, its quotient group Γ has a faithful representation of that order. Next, by construction, if $1 \leqslant k \leqslant n-2$, then Γ_k^n is a subgroup of Γ_0^n of index at most $\binom{n-1}{k}$. Finally, again by construction of the corresponding polytopes, Γ_k^n can have order at most $3 \cdot 2^{2n-3} k!(n-k-1)!$ for these intermediate values of k. Hence, we must have equality throughout, and the qualifying remarks after each of the theorems can now be removed.

There is one property of a regular polytope, or, rather, of its automorphism group, which we have hitherto neglected – this is the intersection property.

However, for our groups, this will be fairly easy to prove, because we have the isomorphism $\Gamma_0^n \cong \Gamma$ of Theorem 14A2.

When $1 \leqslant k \leqslant n-2$, we see that Γ_{n-k-1}^n lacks the permutation $\rho_{k+1} = \chi_{k+1} = (k\ k+1)$. Nevertheless, Γ_{n-k-1}^n still contains all the conjugates of ρ_1 under $\langle \chi_2, \ldots, \chi_{n-1} \rangle$, namely, the elements

$$\prod_{i \neq j} \omega_{i2} \omega_{j0}$$

for $j = 1, \ldots, n-1$, since ρ_1 and its conjugate ρ_{k+1} both belong to the group. We may now prove independently that the index satisfies $[\Gamma_0^n : \Gamma_k^n] = \binom{n-1}{k}$. The intersection property can then be verified directly, first because we have the geometric representation of the group itself, and second because we can use Theorem 2E16(a), which means that (making the natural inductive assumptions) it need only be checked for $I = \{0, \ldots, n-2\}$ and $J = \{1, \ldots, n-1\}$. Given the description of the group Γ, this is straightforward, if somewhat tedious, and we omit the details. This concludes the proof.

Finally, it is of course desirable to enumerate all the locally projective regular polytopes. As we have hinted at the beginning of this section, the famous Petersen graph occurs naturally when the Schläfli symbol consists of 3's and 5's. Petersen graphs have been extensively studied in graph theory and diagram geometries, and this work may be relevant in the present context. There are many examples of diagram geometries related to finite simple groups which are "locally Petersen" (see [65, §5; 207, 225–230, 231, §1.10; 393, 427]). For example, from the locally projective polytope $\{\{3, 5\}_5, \{5, 3\}_5\}$ with group PSL(2, 11), we obtain a locally Petersen diagram geometry of rank 3 by omitting the facets from the face poset. We do not know whether there are other interesting regular polytopes which correspond to locally Petersen diagram geometries. Similarly, the edge-graph of $\{\{3, 5\}_5, \{5, 3\}_5\}$ is a graph which is locally Petersen (that is, the induced subgraph on the neighbours of each vertex is a Petersen graph). The finite graphs which are locally Petersen have been completely described (see [53, p. 37]).

14B Mixed Topological Types

We next consider regular polytopes which are defined by non-spherical sections of different topological types. We shall largely confine our attention here to sections which are toroidal or projective (with at least one of each). Indeed, we shall concentrate almost exclusively on n-polytopes of type $\{4, 3^{n-4}, 4, 3\}$ for $n \geqslant 4$; there will be two complementary strands to our approach.

We begin geometrically. As we saw in Section 14A, the locally projective regular polytope $\mathcal{L}^{n-1} := \mathcal{L}_0^{n-1}$ (in the previous notation) has a faithful realization in $\mathbb{E}^{3(n-2)}$ (we change the rank, because $\mathcal{L}^{n-1}$ will soon be the vertex-figure of n-polytopes); its $4(n-2)$ vertices are those of $n-2$ congruent regular tetrahedra centred at the origin lying in orthogonal 3-dimensional subspaces, and its edges join vertices of different tetrahedra. It follows that this $\mathcal{L}^{n-1}$ is suitable to be the vertex-figure of a regular n-apeirotope $\mathcal{Q}^n$ in the class $\langle\{4, 3^{n-4}, 4\}, \mathcal{L}^{n-1}\rangle$. This apeirotope is not universal, and we shall discover the extra relations satisfied by its group below; however, it turns out to be useful in the investigation of certain locally toroidal polytopes (see Section 12E).

There is a representation $\Delta^n := \langle \rho_0, \ldots, \rho_{n-1}\rangle$ of the group of $\{\{4, 3, 3, 4\}, \mathcal{L}^{n-1}\}$, which is determined as follows. To facilitate the description, we recall from Section 14A that the group $\Delta_0^n := \Gamma(\mathcal{L}^{n-1})$ of $\mathcal{L}^{n-1}$ acts coherently on $\mathbb{E}^{3(n-2)}$, in that it respects its decomposition as an orthogonal direct sum of $n-2$ copies of $\mathbb{E}^3$. We therefore adapt the notation of Section 14A in the obvious way (and bear in mind that there things were set up in the dual context); we further define $u := (1, 1, 1)$. We then obtain for the generators:

$$\begin{aligned} x\rho_0 &:= (u - x_1, x_2, \ldots, x_{n-2}), \\ x\rho_j &:= (x_1, \ldots, x_{j-1}, x_{j+1}, x_j, x_{j+2}, \ldots, x_{n-2}) \quad \text{for } j = 1, \ldots, n-3, \\ x\rho_{n-2} &:= (x_1\omega_2, \ldots, x_{n-3}\omega_2, x_{n-2}\omega_0), \\ x\rho_{n-1} &:= (x_1\omega_1, \ldots, x_{n-2}\omega_1), \end{aligned}$$

with ω_0, ω_1 and ω_2 as before the generators of the symmetry group S_4 of the tetrahedron. It is tedious, but easy, to verify that these generators satisfy the required relations: for the subgroup $\langle \rho_1, \ldots, \rho_{n-1} \rangle$, see Section 14A, and for $\langle \rho_0, \ldots, \rho_{n-2} \rangle$ observe that it is the group of an infinite quotient of $\{4, 3^{n-4}, 4\}$, which must therefore be isomorphic to it. Thus a realization $\mathcal{Q}^n$ (not faithful) of $\{\{4, 3^{n-4}, 4\}, \mathcal{L}^{n-1}\}$ can be obtained by applying Wythoff's construction to the group Δ^n with initial vertex o. The polytopality of the realization may be checked directly, if tediously. (We can also obtain it as a consequence of the polytopality of various of its quotients, which have the same vertex-figure $\mathcal{L}^{n-1}$.)

We can observe at this stage that the case $n = 4$ is also covered by our discussion, even though the vertex-figure is the ordinary (non-projective) cube $\{4, 3\}$, in a 6-dimensional realization as $\{4, 3\} \# \{4, 3\}_3$. Indeed, there is a corresponding 6-dimensional cut $\mathcal{Q}^4$ (of type $\{4, 4, 3\}$) of the general $\mathcal{Q}^n$ for $n \geqslant 5$.

In a sense we can reduce even further to an apeirohedron $\mathcal{Q}^3$; we employ the mixing operation

$$(\rho_0, \ldots, \rho_{n-1}) \mapsto (\rho_{n-3}\rho_{n-4} \cdots \rho_0\rho_1 \cdots \rho_{n-3}, \rho_{n-2}, \rho_{n-1}) =: (\sigma_0, \sigma_1, \sigma_2).$$

The resulting $\mathcal{Q}^3$ is indeed of type $\{\infty, 3\}$, as we might expect; however, it does not (quite) fall into the general pattern. Instead, we see that the generators σ_j act as

$$\begin{aligned} x\sigma_0 &= (x_1, \ldots, x_{n-3}, u - x_{n-2}), \\ x\sigma_1 &= (x_1\omega_2, \ldots, x_{n-3}\omega_2, x_{n-2}\omega_0), \\ x\sigma_2 &= (x_1\omega_1, \ldots, x_{n-3}\omega_1, x_{n-2}\omega_1). \end{aligned}$$

With initial vertex o as before, it follows that the vertices of $\mathcal{Q}^3$ lie in the last component $\mathbb{E}^3$. They do not form all the vertices of $\mathcal{Q}^n$ in this component – the restriction of the graph of $\mathcal{Q}^n$ to a component coincides with the graph $\mathcal{G}$ (say) of the 4-apeirotope $\{\{\infty, 3\}_6, \{3, 3\}\}$ which we described in Section 7F – but this clearly points to

14B1 Lemma *The 3-apeirohedron $\mathcal{Q}^3$ is isomorphic to $\{\infty, 3\}_6$.*

Proof. All we need to verify is that, in the notation we have introduced, $\sigma_0\sigma_1\sigma_2$ has period 6. This is straightforward, since

$$x\sigma_0\sigma_1\sigma_2 = (x_1\omega_2\omega_1, \ldots, x_{n-3}\omega_2\omega_1, (u - x_{n-2})\omega_0\omega_1),$$

and each of $\omega_2\omega_1$ and $\omega_0\omega_1$ has period 3. It then follows that the group of $\mathcal{Q}^3$ is an infinite quotient of $[6, 3]$, and so must be isomorphic to $[6, 3]$. □

We shall conclude from Lemma 14B1 that there is one extra relation which must be imposed on the abstract group $\langle \rho_0, \ldots, \rho_{n-1} \rangle$ of the universal polytope $\{\{4, 3, 3, 4\}, \mathcal{L}^{n-1}\}$ to yield Δ^n. We obtain this by using the circuit criterion of Theorem 2F4. The apeirohedron $\mathcal{Q}^n$ has two kinds of minimal circuits. A typical circuit of the first kind is a square face $\{4\}$, which goes between different components $\mathbb{E}^3$, and that of the second kind is a skew hexagon within a component. The latter is defined by the relation $(\sigma_0\sigma_1\sigma_2)^6 = \varepsilon$, as we saw in Lemma 14B1, or, in view of the previous definition of the σ_j,

14B2 $$(\rho_{n-4}\rho_{n-5} \cdots \rho_0\rho_1 \cdots \rho_{n-2})^6 = \varepsilon.$$

Bearing in mind the already known presentation of the group of the vertex-figure $\mathcal{L}^{n-1}$, we have thus nearly proved

14B3 Theorem *For each $n \geqslant 4$, the group of the n-apeirotope $\mathcal{Q}^n$ is the Coxeter group $[4, 3^{n-4}, 4, 3] = \langle \rho_0, \ldots, \rho_{n-1} \rangle$, with the extra relations $(\rho_{n-4}\rho_{n-3}\rho_{n-2}\rho_{n-1})^6 = \varepsilon$ (except when $n = 4$), and (14B2).*

Proof. To complete the proof, we must show that a general edge-circuit in $\mathcal{Q}^n$ can be obtained by concatenating copies of the basic minimal circuits. The first observation is that ρ_0 acts in exactly the same way on each translate of the first component $\mathbb{E}^3$ of $\mathbb{E}^{3(n-2)}$, and in particular on each such translate through a vertex of $\mathcal{Q}^n$ in the sum of the remaining components. It then follows that the graph of $\mathcal{Q}^n$ is $\mathcal{G}^{n-2}$, where (as before) $\mathcal{G}$ is the subgraph of $\mathcal{Q}^n$ lying in one component. The notation here means that the vertex-set is

$$V(\mathcal{G}^{n-2}) := V(\mathcal{G}) \times \cdots \times V(\mathcal{G}),$$

with $n - 2$ terms in the product, while the edges consist of sums of an edge of one component and vertices of the remaining $n - 3$.

There are now two steps in the proof. The first is inductive. The edges in $\mathcal{G}^{n-2}$ are of two kinds: those within a single translate of the first component, and those between two different translates of it. Let C be an edge-circuit, and let C' be its image under projection on the sum of the last $n - 3$ components. Then C' is a circuit in $\mathcal{G}^{n-3}$. Consider an edge E of C of the second kind, and its corresponding edge E' in $\mathcal{G}^{n-3}$. Then E and E' can be joined by a strip of square faces, meeting in edges parallel to E and E'. We conclude that, modulo circuits in translates of the first component, C is equivalent to C'.

We are now done. We contract circuits in the first component (and its translates) using hexagons alone, while contractability of circuits in the $\mathcal{G}^{n-3}$ lying in the sum of the other components by means of squares and hexagons is our inductive assumption. □

The relation (14B2) admits a useful reformulation. We define $\sigma_0, \sigma_1, \sigma_2$ as before, so that

$$\sigma_0 := \rho_{n-3}\rho_{n-4} \cdots \rho_0\rho_1 \cdots \rho_{n-3},$$

$\sigma_1 = \rho_{n-2}$ and $\sigma_2 = \rho_{n-1}$. Then (14B2) says that $(\sigma_0\sigma_1\sigma_2)^6 = \varepsilon$, which was how it was derived. Now

$$\begin{aligned}(\sigma_0\sigma_1\sigma_2)^6 &\sim (\sigma_2\sigma_1\sigma_2 \cdot \sigma_0\sigma_1\sigma_2\sigma_0\sigma_1\sigma_0)^2 \\ &= (\sigma_1\sigma_2\sigma_1 \cdot \sigma_0\sigma_1\sigma_2\sigma_0\sigma_1\sigma_0)^2 \\ &\sim ((\sigma_0\sigma_1)^2\sigma_2)^2 \cdot ((\sigma_1\sigma_0)^2\sigma_2)^2.\end{aligned}$$

Thus (14B2) is equivalent to $(\sigma_0\sigma_1)^2 \rightleftharpoons \sigma_2(\sigma_0\sigma_1)^2\sigma_2$ (with $\rightleftharpoons$ meaning "commutes with"). Since

$$\tau := (\sigma_0\sigma_1)^2 = (\rho_{n-3}\rho_{n-4}\cdots\rho_0\rho_1\cdots\rho_{n-3}\rho_{n-2})^2$$

is just the translation by $(0^{3n-8}, -2, -2)$ (we leave the reader to check this), the relation (14B2) expresses the fact that two certain translations in the group Δ do indeed commute. (It was conjectured in [302, §8], in the particular case $n = 6$, that this relation would suffice to determine the group.) In summary, we thus have

14B4 Proposition *The relation (14B2) is equivalent to* $\tau \rightleftharpoons \rho_{n-1}\tau\rho_{n-1}$, *where*

$$\tau := (\rho_{n-3}\rho_{n-4}\cdots\rho_0\rho_1\cdots\rho_{n-3}\rho_{n-2})^2.$$

In the course of the subsequent discussion, we construct three sublattices Λ_1, Λ_2 and Λ_{n-2} of $\mathbb{Z}^{3(n-2)}$; these correspond to the the lattices $\Lambda_\mathbf{s}$ in the group of the facet $\{4, 3^{n-4}, 4\}$ generated by $(2^k, 0^{n-k-2})$ (with $k = 1, 2, n-2$) and their images under the subgroup of the vertex-figure (compare Section 6D). It will be helpful to notice that $4\mathbb{Z}^{3(n-2)} \subseteq \Lambda_k$ for each $k = 1, 2, n-2$. In view of Theorem 14B3, we can argue geometrically, in the knowledge that the group Δ^n of the abstract apeirotope $\mathcal{Q}^n$ defined by the relations of the theorem possesses an abelian subgroup of "translations" acting upon it, which is generated by $\tau = (\rho_{n-3}\rho_{n-4}\cdots\rho_0\rho_1\cdots\rho_{n-3}\rho_{n-2})^2$ and its conjugates.

First, we recall that τ acts on $\mathcal{Q}^n$ as the translation by $((0,0,0)^{n-3}, (0,-2,-2))$. The conjugates of τ under Δ^n are the translations by the vectors of the form

$$((0, \pm 2, \pm 2), (0,0,0)^{n-3}),$$

with arbitrary signs, and those obtained from them by coherent permutations (of the blocks, and within them); these translations generate the lattice Λ_1. We shall count various equivalence classes of vertices of $\mathcal{Q}^n$; our first count is

14B5 Lemma *The apeirotope* $\mathcal{Q}^n$ *has* 2^{n-2} *vertex classes modulo* Λ_1.

Proof. As before, we set $u := (1,1,1)$; we also write $o := (0,0,0)$ for convenience. Since $((1,-1,-1), o^{n-3}) \equiv (u, o^{n-3})$ modulo Λ_1, and so on, it easily follows that the equivalence classes of vertices modulo Λ_1 are represented by the (block) permutations of (u^k, o^{n-k-2}) for $k = 0, \ldots, n-2$. This gives 2^{n-2} classes, as claimed. □

A little later, we shall calculate this number in a different way.

We now consider Λ_2. Again comparing with Section 6D (particularly Proposition 6D6), or by direct calculation, we see that Λ_2 is generated by the conjugates of

$$\tau \cdot (\rho_{n-3}\tau\rho_{n-3}) = (\tau\rho_{n-3})^2,$$

which is the translation by $(o^{n-4}, (0, -2, -2)^2)$. These conjugates are the vectors of the form

$$((0, \pm 2, \pm 2)^2, o^{n-4}),$$

with arbitrary signs, and those obtained from them by coherent permutations (as before, both of the blocks and within them). Our next count is

14B6 Lemma *The index of Λ_2 in Λ_1 is $[\Lambda_1 : \Lambda_2] = 4$.*

Proof. Using the observation that $4\mathbb{Z}^{3(n-2)} \subset \Lambda_2$, it is easy to see that coset representatives of Λ_2 in Λ_1 may be taken to be (o^{n-2}), and $((0, 2, 2), o^{n-3})$ and the two vectors obtained from it by permuting the first three coordinates. This gives index 4, as claimed. □

Finally, we look at Λ_{n-2}. Again, either using Proposition 6D6 or directly, we find that Λ_{n-2} is generated by $((0, -2, -2)^{n-2})$ and its conjugates, which are the coherent permutations of the vectors of the form $((0, \pm 2, \pm 2)^{n-2})$, with arbitrary signs. Our last count is

14B7 Lemma *The index of Λ_{n-2} in Λ_1 is $[\Lambda_1 : \Lambda_{n-2}] = 2^{2n-6}$.*

Proof. This time, the coset representatives are the vectors of the form $((0, 2, 2)^k, o^{n-k-2})$ for $k = 0, \ldots, n-3$, and those obtained from them by permutations of the first $n-3$ blocks (that is, the last block is always o), and (not necessarily coherent) permutations within the blocks. Again, we appeal to the fact that $4\mathbb{Z}^{3(n-2)} \subset \Lambda_{n-2}$. This yields the index $4^{n-3} = 2^{2n-6}$, as claimed. □

When n is even, we have the additional subgroup relationship $\Lambda_{n-2} \subset \Lambda_2$, with index 2^{2n-8}.

Since the lattice Λ_1 has full rank in $\mathbb{E}^{3(n-2)}$, it follows that $2\Lambda_1$ has index 2^{3n-6} in Λ_1. As a check, we may verify that $2\Lambda_1$ has index 2^{3n-8} in Λ_2, and 2^n in Λ_{n-2}. The easiest way to see the latter is to notice that $[2\mathbb{Z}^{3(n-2)} : \Lambda_1] = 2^{n-2}$, with coset representatives all block permutations of $((2, 0, 0)^k, o^{n-k-2})$ for each $k = 0, \ldots, n-2$, and that $[\Lambda_{n-2} : 4\mathbb{Z}^{3(n-2)}] = 4$, with coset representatives (o^{n-2}), and the coherent permutations of $((0, 2, 2)^{n-2})$ with no changes of sign.

We now pass to certain quotients of $\mathcal{Q}^n$. If we recall what relationship the lattices Λ_k have to the facets of $\mathcal{Q}^n$, we see that the group of $\{\{4, 3^{n-4}, 4\}_{(4,0^{n-3})}, \mathcal{L}^{n-1}\}$ is obtained by imposing on the abstract group of $\{\{4, 3^{n-4}, 4\}, \mathcal{L}^{n-1}\}$ the extra relation $\tau^2 = \varepsilon$. The relation of Proposition 14B4 says that, in the group Δ^n of $\mathcal{Q}^n$, we have

$$(\tau\rho_{n-1})^2(\tau^{-1}\rho_{n-1})^2 = \varepsilon.$$

Table 14B1. *Finite Polytopes* $\{\{4, 3^{n-4}, 4\}_{\mathbf{s}}, \mathcal{L}^{n-1}\}$

$\mathbf{s}$	v	f	g
$(2, 0^{n-3})$	2^{n-2}	$3 \cdot 2^{n-3}$	$3 \cdot 2^{3n-7} \cdot (n-2)!$
$(2, 2, 0^{n-4})$	2^{n}	$3 \cdot 2^{n-1}$	$3 \cdot 2^{3n-5} \cdot (n-2)!$
(2^{n-2})	2^{3n-8}	$3 \cdot 2^{2n-6}$	$3 \cdot 2^{5n-13} \cdot (n-2)!$
$(4, 0^{n-3})$	2^{4n-8}	$3 \cdot 2^{3n-7}$	$3 \cdot 2^{6n-13} \cdot (n-2)!$

Now $\tau^2 = \varepsilon$ implies that $\tau^{-1} = \tau$, and so we are interested in

$$(\tau\rho_{n-1})^2(\tau^{-1}\rho_{n-1})^2 = (\tau\rho_{n-1}\tau^{-1}\rho_{n-1})^2.$$

If we write $\psi := \rho_{n-3}\rho_{n-4}\cdots\rho_0\rho_1\cdots\rho_{n-3}$, so that $\tau = (\psi\rho_{n-2})^2$, and bear in mind that ψ is an involution which commutes with ρ_{n-1}, this becomes

$$\begin{aligned}((\psi\rho_{n-2})^2\rho_{n-1})^2((\rho_{n-2}\psi)^2\rho_{n-1})^2 &\sim (\psi\rho_{n-2}\rho_{n-1}\rho_{n-2}\psi\rho_{n-1}\rho_{n-2}\rho_{n-1})^2\\ &= (\psi\rho_{n-1}\rho_{n-2}\rho_{n-1})^4\\ &\sim (\psi\rho_{n-2})^4\\ &= \tau^2.\end{aligned}$$

It follows that $\tau^2 = \varepsilon$ implies (14B2). In other words,

14B8 Theorem *The quotient mapping*

$$\{\{4, 3^{n-4}, 4\}, \mathcal{L}^{n-1}\} \mapsto \left\{\{4, 3^{n-4}, 4\}_{(4,0^{n-3})}, \mathcal{L}^{n-1}\right\}$$

can be factored through $\mathcal{Q}^n$.

There follows immediately

14B9 Corollary *The quotient mappings*

$$\{\{4, 3^{n-4}, 4\}, \mathcal{L}^{n-1}\} \mapsto \{\{4, 3^{n-4}, 4\}_{\mathbf{s}}, \mathcal{L}^{n-1}\},$$

with $\mathbf{s} = (2, 0^{n-3})$, $(2, 2, 0^{n-4})$ *and* (2^{n-2}), *can be factored through $\mathcal{Q}^n$.*

We are assuming here that these quotient mappings genuinely yield polytopes. In fact, we need only check the case $\mathbf{s} = (2, 0^{n-3})$; the others will follow from the quotient criterion of Theorem 2E17. Before we proceed further, we give in Table 14B1 the numerical data about these polytopes. We make no claim for the completeness of our enumeration, although we may conjecture that the list is complete for large enough n. It is worth repeating that the previous arguments still work in the case $n = 4$, so that these data remain valid for the polytopes of type $\{4, 4, 3\}$.

Let us now discuss two of these polytopes geometrically. We write

$$\mathcal{K}^n_{\mathbf{s}} := \{\{4, 3^{n-4}, 4\}_{\mathbf{s}}, \mathcal{L}^{n-1}\},$$

if the polytope exists. In case $\mathbf{s} = (2, 0^{n-3})$, the facet $\{4, 3^{n-4}, 4\}_{(2,0^{n-3})}$ is flat, so that $\mathcal{K}^n_{(2,0^{n-3})}$ should also be flat; this accords with it having 2^{n-2} vertices. In fact, we can appeal to the results of Section 4F.

14B10 Theorem *If $n \geqslant 4$, then $\mathcal{K}^n_{(2,0^{n-3})}$ is a polytope.*

Proof. This follows from Theorem 4F9, once we have verified the free amalgamation property (FAP) for appropriate component polytopes. First, the $(n-2)$-cube $\{4, 3^{n-4}\}$ has the FAP with respect to its vertex-figures. Second $\mathcal{L}^{n-1}$ has the FAP with respect to its $(n-3)$-faces. (Both these facts are easily verified using Lemma 4E10.) It follows that the 1-mix

$$\{4, 3^{n-4}\} \Diamond_1 \mathcal{L}^{n-1}$$

exists; it is a $(0, n-2)$-flat polytope, and hence in particular is flat. It must therefore be $\mathcal{K}^n_{(2,0^{n-3})}$, as required. □

We now consider the case $\mathbf{s} = (2, 2, 0^{n-4})$. Here we have a very general result.

14B11 Theorem *For each $n \geqslant 4$, the regular polytope*

$$\big\{\{4, 3^{n-4}, 4\}_{(2,2,0^{n-4})}, \{3^{n-4}, 4, 3\}\big\}$$

exists, and has 2^n vertices, which coincide with those where the vertex-figure $\{3^{n-4}, 4, 3\}$ is replaced by its locally projective quotient $\mathcal{L}^{n-1}$.

Proof. In case $n = 4$, as before, we interpret $\mathcal{L}^3$ as the cube $\{4, 3\}$. Our proof proceeds in two stages: first, we give a direct construction for the universal polytope; second, we present a model for the polytope with the locally projective vertex-figure, and show that it too has 2^n vertices.

For the universal polytope, we begin with the facet $\mathcal{T}^{n-1} := \{4, 3^{n-4}, 4\}_{(2,2,0^{n-4})}$, which has 2^{n-1} vertices and facets. All its vertices belong to a (closed) *ring*, which is composed of four $(n-2)$-cubes, meeting on opposite $(n-3)$-faces, to form a subcomplex $C^{n-3} \times C^2_1$, where C^k is a k-cube and C^k_1 is its 1-skeleton (graph). An $(n-2)$-cube adjacent to one in the ring meets it again two steps along, on the opposite side, so that the new ring which they determine shares two cubes with the old ring. The whole facet $\mathcal{T}^{n-1}$ can be built up step by step in this fashion, moving from one ring to another related to it in this way.

We now consider the subcomplex of the universal polytope, which we obtain just from sticking facets $\mathcal{T}^{n-1}$ together along the $(n-2)$-cubes C^{n-2} in a single ring of each. Start with one ring. Since three facets fit together around each face C^{n-3}, we see that the four new facets meeting the initial one on the ring use up three out of the four faces of their rings.

We readily conclude that the remaining faces C^{n-3} in the new rings, one from each new facet, form a ring of a sixth facet. Thus the cubes C^{n-3} in these six rings actually close up to form a subcomplex $C^{n-3} \times C^3_1$, thereby involving six facets, and giving 2^n vertices in all. (Perhaps the picture the reader should have in mind is that of an ordinary cube with hollow square faces corresponding to the rings; its vertices and edges are now thickened to $(n-3)$- and $(n-2)$-cubes.) We claim that these 2^n are all the vertices of the polytope.

To see this, we just continue to add new facets. At each stage, we fit a third facet, F say, around a face C^3 which already belongs to two existing facets F' and F''; further, we may suppose that F' and F'' meet on a 4-cube which already belongs to a subcomplex $\mathcal{C} = C^3 \times C_1^3$ as before, whose vertices are just the previous 2^n. Then F meets F' and F'' on $(n-2)$-cubes C' and C'', respectively, which contain the given $(n-3)$-cube C^{n-3}, and are adjacent to the rings of F' and F'' in $\mathcal{C}$. Now complete C' to the ring in F' which shares two 4-cubes (one of which is that in F'' also) with the existing one. These two $(n-2)$-cubes belong to $\mathcal{C}$; it follows that the ring in F which contains C' and C'' already has all its vertices in $\mathcal{C}$, since the only vertices yet unaccounted for belong to the facet in $\mathcal{C}$ which meets the ring of F' on the second of the $(n-2)$-cubes in the existing ring. We finish the stage by completing any subcomplex $C^{n-3} \times C_1^3$ formed by rings, in the manner just described. Summarizing the procedure, in adding a new facet $\mathcal{T}^{n-1}$, we see that three out of four subfacets in one of its rings will already have all their vertices in the existing subcomplex; hence all the vertices of this facet are among those already found. In other words, after the first six facets, the introduction of a new facet does not require any new vertices. This establishes the first claim.

The model is constructed as follows. Let $A^n(2)$ be the n-dimensional affine space over the field $GF(2)$ with 2 elements. We define the following n reflexions (involutions), where $x := (\xi_1, \ldots, \xi_n)$.

$$\begin{aligned} xR_0 &:= (\xi_1 + 1, \xi_2, \ldots, \xi_n), \\ xR_j &:= (\xi_1, \ldots, \xi_{j-1}, \xi_{j+1}, \xi_j, \xi_{j+2}, \ldots, \xi_n), \quad \text{for } j = 1, \ldots, n-4, n-2, n-1, \\ xR_{n-3} &:= (\xi_1, \ldots, \xi_{n-3}, \xi_{n-2} + \xi_{n-1} + \xi_n, \xi_{n-3} + \xi_{n-1} + \xi_n, \xi_{n-1}, \xi_n). \end{aligned}$$

Then we may verify that $\langle R_0, \ldots, R_{n-1} \rangle$ is the group $[4, 3^{n-4}, 4, 3]$, with the additional relations

$$(R_0 R_1 \cdots R_{n-2} R_{n-3} \cdots R_2)^4 = E = (R_{n-4} R_{n-3} R_{n-2} R_{n-1})^6.$$

That is, the corresponding polytope has facet $\{4, 3^{n-4}, 4\}_{(2,2,0^{n-4})}$, and vertex-figure the projective polytope $\mathcal{L}^5$. It is also easy to check that the group is transitive on the 2^n points of $A^n(2)$, so that each is a vertex. This latter fact shows that the facet cannot collapse onto the flat polytope $\{4, 3^{n-4}, 4\}_{(2,0^{n-3})}$, so that the first of the previous extra relations cannot be replaced by the stronger one $(R_0 R_1 \cdots R_{n-2} R_{n-3} \cdots R_1)^2 = E$.

Finally, since the universal polytope and its quotient with the locally projective vertex-figure both have 2^n vertices, the vertex-sets of the two polytopes can be thought of as coinciding in an obvious way. □

There is an immediate consequence of Theorem 14B11.

14B12 Corollary *Let $\mathcal{P}_1$ and $\mathcal{P}_2$ be regular $(n-1)$-polytopes, with $\mathcal{P}_1$ isomorphic to $\{4, 3^{n-4}, 4\}_{(2,2,0^{n-4})}$, and $\mathcal{P}_2$ of Schläfli type $\{3^{n-4}, 4, 3\}$. Then the (universal) n-polytope $\{\mathcal{P}_1, \mathcal{P}_2\}$ exists if and only if $\mathcal{P}_2$ has the locally projective polytope $\mathcal{L}^{n-1}$ as a quotient.*

As a consequence of results of Chapter 8, we can characterize the groups of these polytopes.

14B13 Theorem *Let $n \geqslant 4$, and let $\mathcal{Q}$ be a regular $(n-1)$-polytope of type $\{3^{n-4}, 4, 3\}$ which covers the locally projective regular polytope $\mathcal{L}^{n-1}$. Then the regular n-polytope $\{\{4, 3^{n-4}, 4\}_{(2,2,0^{n-4})}, \mathcal{Q}\}$ exists, and has group $C_2^n \rtimes \Gamma(\mathcal{Q})$.*

Bibliography

[1] A. Altshuler, Construction and enumeration of regular maps on the torus. *Discrete Math.** **4** (1973), 201–217. [6J]†

[2] A. Altshuler, 3-pseudomanifolds with preassigned links. *Trans. Amer. Math. Soc.* **241** (1978), 213–237. [11C]

[3] A. Altshuler, Construction and representation of neighborly manifolds. *J. Combin. Theory Ser. A* **77** (1997), 246–267.

[4] A. Altshuler, J. Bokowski, and P. Schuchert, Spatial polyhedra without diagonals. *Israel J. Math.* **86** (1994), 373–396. [8E]

[5] A. Altshuler, J. Bokowski, and P. Schuchert, Sphere systems and neighborly spatial polyhedra with 10 vertices. *Rend. Circ. Mat. Palermo (2) Suppl.* **35** (1994), 15–28.

[6] A. Altshuler, J. Bokowski, and P. Schuchert, Neighborly 2-manifolds with 12 vertices. *J. Combin. Theory Ser. A* **75** (1996), 148–162. [8E]

[7] A. Altshuler and U. Brehm, A non-schlegelian polyhedral map on the torus. *Mathematika* **31** (1984), 83–88.

[8] A. Altshuler and U. Brehm, Neighborly maps with few vertices. *Discrete Comput. Geom.* **8** (1992), 93–104. [8E]

[9] D. S. Archdeacon, P. Gvozdjak, and J. Širáň, Constructing and forbidding automorphisms in lifted maps. *Math. Slovaca* **47** (1997), 113–129. [4C]

[10] D. S. Archdeacon and R. B. Richter, The construction and classification of self-dual polyhedra. *J. Combin. Theory Ser. A* **54** (1992), 37–63. [2A]

[11] J. L. Arocha, J. Bracho, and L. Montejano, Regular projective polyhedra with planar faces, I. *Aequationes Math.* **59** (2000), 55-73. [7E]

[12] E. Artin, *Geometric Algebra* (2nd edition). Wiley & Sons (New York, 1988). [7A, 8F]

[13] M. Aschbacher, Flag structures on Tits geometries. *Geom. Dedicata* **14** (1983), 21–32. [7A]

[14] M. Aschbacher, *Sporadic Groups*. Cambridge University Press (Cambridge, 1994). [1A, 7A]

[15] J. Ashley, B. Grünbaum, G. C. Shephard, and W. Stromquist, Self-duality groups and ranks of self-dualities. In *Applied Geometry and Discrete Mathematics (The Victor Klee Festschrift)* (eds. P. Gritzmann and B. Sturmfels), DIMACS Series in Discrete Mathematics and Theoretical Computer Science **4**, Amer. Math. Soc. and Assoc. Computing Machinery (1991), 11–50. [2A]

* Serials are, for the most part, listed using the abbreviated names of *Mathematical Reviews*.

† At the end of a reference we list the sections in which the reference occurs. A number without a letter refers to the opening paragraph of the corresponding chapter. Some references are not quoted in the text.

[16] L. Balke and A. Valverde, Σ-chamber systems, coloured graphs and orbifolds. *Beiträge Algebra Geom.* **37** (1996), 17–29.

[17] W. Ballmann and M. Brin, Polygonal complexes and combinatorial group theory. *Geom. Dedicata* **50** (1994), 165–191. [6B]

[18] T. F. Banchoff, Tightly embedded 2-dimensional polyhedral manifolds. *Amer. J. Math.* **87** (1965), 462–472. [8D]

[19] T. F. Banchoff, Torus decompositions of regular polytopes in 4-space. In *Shaping Space – A polyhedral approach* (eds. M. Senechal and G. Fleck), Birkhäuser (Boston–Basel, 1988), 221–230. [6B]

[20] T. F. Banchoff, *Beyond the Third Dimension*. Scientific American Library (New York, 1996). [1A]

[21] D. Barnette, Graph theorems for manifolds. *Israel J. Math.* **16** (1973), 62–72. [2A]

[22] M. Bayer and C. W. Lee, Combinatorial aspects of convex polytopes. In *Handbook of Convex Geometry* (eds. P. M. Gruber and J. M. Wills), Elsevier Publishers (Amsterdam, 1993), 485–534.

[23] A. F. Beardon, *The Geometry of Discrete Groups*. Springer-Verlag (New York–Heidelberg–Berlin, 1983). [6A]

[24] L. W. Beineke and F. Harary, The genus of the n-cube. *Canad. J. Math.* **17** (1965), 494–496. [8D]

[25] L. Bieberbach, Über die Bewegungsgruppen der euklidischen Räume: erste Abhandlung. *Math. Ann.* **70** (1910), 297–336. [5C, 7E]

[26] N. L. Biggs, *Finite Groups of Automorphisms*. London Math. Soc. Lecture Notes Ser. **6**, Cambridge University Press (London, 1971). [6J]

[27] N. L. Biggs and A. T. White, *Permutation Groups and Combinatorial Structures*. London Math. Soc. Lecture Notes Ser. **33**, Cambridge University Press (London, 1979). [6J]

[28] A. Björner, Posets, regular CW complexes and Bruhat order. *European J. Combin.* **5** (1984), 7–16. [6B]

[29] A. Björner, Some combinatorial and algebraic properties of Coxeter complexes and Tits buildings. *Adv. Math.* **52** (1984), 173–212. [6B]

[30] A. Björner, Topological methods. In *Handbook of Combinatorics* (eds. R. L. Graham, M. Grötschel and L. Lovász), Elsevier Science B. V. (Amsterdam, 1995), 1819–1872. [2C, 6B]

[31] A. Björner, M. Las Vergnas, B. Sturmfels, N. White, and G. Ziegler, *Oriented Matroids*. Encyclopedia of Mathematics and Its Applications **46**, Cambridge University Press (Cambridge, 1993). [3B]

[32] G. Blind and R. Blind, The semi-regular polytopes. *Comment. Math. Helv.* **66** (1991), 150–154. [1B]

[33] A. Blokhuis, A. E. Brouwer, D. Buset, and A. M. Cohen, The locally icosahedral graphs. In *Finite Geometries* (eds. C. A. Baker and L. M. Batten), Lecture Notes Pure Appl. Math. **103**, Marcel Dekker (New York, 1985), 19–22. [6C]

[34] J. Bokowski, A geometric realization without self-intersection does exist for Dyck's regular map. *Discrete Comput. Geom.* **4** (1989), 583–589. [5B]

[35] J. Bokowski and A. Eggert, All realizations of Möbius' torus with seven vertices. *Structural Topology* **17** (1991), 59–76. [8E]

[36] J. Bokowski and A. G. de Oliviera, On the generation of oriented matroids. *Discrete Comput. Geom.* **24** (2000), 197–208. [5B]

[37] J. Bokowski and B. Sturmfels, *Computational Synthetic Geometry*. Lecture Notes in Mathematics **1355**, Springer-Verlag (New York–Heidelberg–Berlin, 1989). [5B]

[38] J. Bokowski and J. M. Wills, Regular polyhedra with hidden symmetries. *Math. Intelligencer* **10** (1988), 27–32. [5B]

[39] D. Bonheure, F. Buekenhout, and D. Leemans, On the Petrials of thin rank 3 geometries. *J. Geom.* **71** (2001), 19–25. [7B]

[40] N. Bourbaki, *Groupes et Algébres de Lie, Ch. IV–VI*. Herman (Paris, 1968). [3B]

[41] S. Bouzette, F. Buekenhout, E. Dony, and A. Gottcheiner, A theory of nets for polyhedra

and polytopes related to incidence geometries. *Des. Codes Cryptogr.* **10** (1997), 115–136.

[42] J. Bracho, Regular projective polyhedra with planar faces, II. *Aequationes Math.* **59** (2000), 160–176. [7E]

[43] H. R. Brahana, Regular maps on an anchor ring. *Amer. J. Math.* **48** (1926), 225–240. [6J]

[44] H. R. Brahana, Regular maps and their groups. *Amer. J. Math.* **49** (1927), 268–284. [6J]

[45] H. R. Brahana and A. B. Coble, Maps of twelve countries with five sides with a group of order 120 containing an icosahedral subgroup. *Amer. J. Math.* **48** (1926), 1–20. [6J]

[46] U. Brehm, Maximally symmetric polyhedral realizations of Dyck's regular map. *Mathematika* **34** (1987), 229–236. [5B]

[47] U. Brehm, A maximally symmetric polyhedron of genus 3 with 10 vertices. *Mathematika* **34** (1987), 237–242. [5B]

[48] U. Brehm, The homology of the power-complex of a simplicial complex (in preparation). [8D]

[49] U. Brehm, W. Kühnel and E. Schulte, Manifold structures on abstract regular polytopes. *Aequationes Math.* **49** (1995), 12–35. [6B, 8D, 10A, 10B]

[50] U. Brehm and E. Schulte, Polyhedral maps. In *Handbook of Discrete and Computational Geometry* (eds. J. E. Goodman and J. O'Rourke), CRC Press (Boca Raton–New York, 1997), 345–358. [1D]

[51] U. Brehm and J. M. Wills, Polyhedral manifolds. In *Handbook of Convex Geometry* (eds. P. M. Gruber and J. M. Wills), Elsevier Publishers (Amsterdam, 1993), 535–554. [1D, 7E]

[52] T. Breuer, *Characters and Automorphism Groups of Compact Riemann Surfaces*. London Math. Soc. Lecture Note Series **280**, Cambridge University Press (Cambridge, 2000). [6J]

[53] A. E. Brouwer, A. M. Cohen, and A. Neumaier, *Distance-Regular Graphs*. Springer-Verlag (New York–Heidelberg–Berlin, 1989). [6C, 14A]

[54] K. S. Brown, *Buildings*. Springer-Verlag (New York–Heidelberg–Berlin, 1989). [2, 2C, 4D]

[55] F. Buekenhout, Diagrams for geometries and groups. *J. Combin. Theory Ser. A* **27** (1979), 121–151. [1A, 2, 7A]

[56] F. Buekenhout, Diagram geometries for sporadic groups. *Contemp. Math.* **45** (1985), 1–32. [7A]

[57] F. Buekenhout, The geometry of the finite simple groups. In *Buildings and the Geometry of Diagrams* (ed. L. A. Rosati), Lecture Notes in Mathematics **1181**, Springer-Verlag (New York–Heidelberg–Berlin, 1986), 1–78. [7A]

[58] F. Buekenhout (ed.), *Handbook of Incidence Geometry*. Elsevier Science B. V. (1995). [1A, 2]

[59] F. Buekenhout and A. M. Cohen, *Diagram Geometry* (in preparation). [2]

[60] F. Buekenhout, M. Dehon, and P. Cara, Inductively minimal flag-transitive geometries. In *Mostly Finite Geometries* (ed. N. L. Johnson), Lecture Notes Pure Appl. Math. **190**, Dekker (New York, 1997), 185–190.

[61] F. Buekenhout, M. Dehon, and P. Cara, Geometries of small almost simple groups based on maximal subgroups. *Bull. Belg. Math. Soc. Simon Stevin* (1998), suppl. (128 pages). [7A]

[62] F. Buekenhout, M. Dehon, and D. Leemans, All geometries of the Mathieu group M_{11} based on maximal subgroups. *Experiment. Math.* **5** (1996), 101–110. [7A]

[63] F. Buekenhout, M. Dehon, and D. Leemans, An Atlas of residually weakly primitive geometries for small groups. *Acad. Roy. Belgique, Mém. Cl. Sci., Collect.* 8° *(3)* Tome XIV (1999). [7A]

[64] F. Buekenhout and E. Dony, The regular polyhedra whose symmetry group is $Z_2 \times SYM(5)$. *Bull. Soc. Math. Belg. Ser. A* **42** (1990), 471–476. [5B]

[65] F. Buekenhout and A. Pasini, Finite diagram geometries extending buildings. In *Handbook of Incidence Geometry* (ed. F. Buekenhout), Elsevier Publishers (Amsterdam, 1995), 1143–1254. [2E, 7A, 14A]

[66] H. Burgiel, *Realizations of regular maps*. Ph. D. Thesis, University of Washington (Seattle, 1995). [5B]

[67] H. Burgiel, Toroidal skew polyhedra. In *Symmetry: Culture and Science*, Vol. 6 (1995), No. 1, 96–99. [5B]

[68] H. Burgiel and D. Stanton, Realizations of regular abstract polyhedra of types {3, 6} and {6, 3}. *Discrete Comput. Geom.* **24** (2000), 241–255. [5B]

[69] M. Burt, *Spatial Arrangement and Polyhedra with Curved Surfaces and their Architectural Applications*. M. Sc. Thesis, Technion-Israel Institute of Technology (Haifa, 1966). [7E]

[70] D. Buset, Graphs which are locally a cube. *Discrete Math.* **46** (1983), 221–226. [6C]

[71] D. Buset, Construction of a locally icosahedral graph. *Mitt. Math. Sem. Giessen* **163** (1984), 221–227. [6C]

[72] D. Buset, Locally polyhedral graphs. In *Finite Geometries* (eds. C. A. Baker and L. M. Batten), Lecture Notes Pure Appl. Math. **103**, Marcel Dekker (New York, 1985), 23–25. [6C]

[73] P. J. Cameron, Permutation groups. In *Handbook of Combinatorics* (eds. R. L. Graham, M. Grötschel, and L. Lovász), Elsevier Science B. V. (Amsterdam, 1995), 611–645. [8E]

[74] P. J. Cameron, J. J. Seidel, and S. V. Tsaranov, Signed graphs, root lattices, and Coxeter groups. *J. Algebra* **164** (1994), 173–209. [9G]

[75] R. W. Carter, *Simple Groups of Lie Type*. Wiley (London–New York, 1972). [8]

[76] L. Cauchy, Recherches sur les polyèdres. *Journal de l'École Polytechnique* **16** (1813), 68–86. (In *Ostwalds Klassiker der exakten Wissenschaften* **151** (ed. R. Haussner), Engelmann (Leipzig, 1906), 49–72.) [1A]

[77] A. Cayley, On Poinsot's four new regular solids. *Philos. Mag.* (4) **17** (1859), 123–128. (Reprinted in *Collected Math. Papers* **IV**, Cambridge University Press (Cambridge, 1891).)

[78] A. M. Cohen, Finite complex reflection groups. *Ann. Sci. École Norm. Sup.* (4) **9** (1976), 379–436. [9]

[79] A. M. Cohen, Finite quaternionic reflection groups. *J. Algebra* **64** (1980), 293–324. [9A]

[80] A. M. Cohen, Local recognition of graphs, buildings, and related geometries. In *Finite Geometries, Buildings and Related Topics* (eds. W. M. Kantor, R. A. Liebler, S. E. Payne, and E. E. Shult), Clarendon Press (Oxford, 1990), 85–94. [6C]

[81] A. M. Cohen, Coxeter groups and three related topics. In *Generators and Relations in Groups and Geometries* (eds. A. Barlotti, E. W. Ellers, P. Plaumanu and K. Strambach), NATO ASI Series C **333**, Kluwer (Dordrecht etc., 1991), 235–278. [3]

[82] A. M. Cohen, Recent results on Coxeter groups. In *Polytopes: Abstract, Convex and Computational* (eds. T. Bisztriczky, P. McMullen, R. Schneider, and A. Ivić Weiss), NATO ASI Series C **440**, Kluwer (Dordrecht etc., 1994), 1–19. [3]

[83] C. J. Colbourn and A. I. Weiss, A census of regular 3-polystroma arising from honeycombs. *Discrete Math.* **50** (1984), 29–36. [6H, 6J, 10A, 10C, 11E]

[84] M. Conder, The genus of compact Riemann surfaces with maximal automorphism group. *J. Algebra* **108** (1987), 204–247. [6J]

[85] M. Conder, Maximal automorphism groups of symmetric Riemann surfaces with small genus. *J. Algebra* **114** (1988), 16–28. [6J]

[86] M. Conder, Hurwitz groups: A brief survey. *Bull. Amer. Math. Soc.* **23** (1990), 359–370. [6J]

[87] M. Conder, Regular maps with small parameters. *J. Austral. Math. Soc.* A **57** (1994), 103–112. [6J]

[88] M. Conder, Regular maps on non-orientable surfaces. *Geom. Dedicata* **56** (1995), 209–219. [6J]

[89] M. Conder, Asymmetric combinatorially-regular maps. *J. Algebraic Combin.* **5** (1996), 323–328. [6J]

[90] J. H. Conway, R. T. Curtis, S. P. Norton, R. A. Parker, and R. A. Wilson, *Atlas of Finite Groups*. Clarendon Press (Oxford, 1985). [7A, 10C]

[91] J. H. Conway and N. J. A. Sloane, *Sphere Packings, Lattices and Groups*. Springer-Verlag (New York, 1988). [3B, 6D, 6E]

[92] R. Cordovil and K. Fukuda, Oriented matroids and combinatorial manifolds. *European J. Combin.* **14** (1993), 9–15. [2A]

[93] R. Cori and A. Machì, Maps, hypermaps, and their automorphisms – A survey I,II,III. *Exposition. Math.* **10** (1992), 403–427, 429–447, 449–467. [6J]

[94] A. F. Costa, Locally regular coloured graphs. *J. Geom.* **43** (1992), 57–74. [6B]

[95] H. S. M. Coxeter, The densities of the regular polytopes. *Proc. Cambridge Philos. Soc.* **27** (1930–31), 201–211. (In *Kaleidoscopes: Selected Writings of H. S. M. Coxeter* (eds. F. A. Sherk, P. McMullen, A. C. Thompson, and A. I. Weiss), Wiley-Interscience (New York, 1995), 35–45.) [3E, 7D]

[96] H. S. M. Coxeter, Groups whose fundamental regions are simplexes. *J. London Math. Soc.* **6** (1931), 132–136. (In *Kaleidoscopes: Selected Writings of H. S. M. Coxeter* (eds. F. A. Sherk, P. McMullen, A. C. Thompson, and A. I. Weiss), Wiley-Interscience (New York, 1995), 133–137.) [3B]

[97] H. S. M. Coxeter, The densities of the regular polytopes, II. *Proc. Cambridge Philos. Soc.* **28** (1931–1932), 509–521. (In *Kaleidoscopes: Selected Writings of H. S. M. Coxeter* (eds. F. A. Sherk, P. McMullen, A. C. Thompson, and A. I. Weiss), Wiley-Interscience (New York, 1995), 47–59.) [7D]

[98] H. S. M. Coxeter, The densities of the regular polytopes, III. *Proc. Cambridge Philos. Soc.* **29** (1932–1933), 1–22. (In *Kaleidoscopes: Selected Writings of H. S. M. Coxeter* (eds. F. A. Sherk, P. McMullen, A. C. Thompson, and A. I. Weiss), Wiley-Interscience (New York, 1995), 61–82.) [7D]

[99] H. S. M. Coxeter, Discrete groups generated by reflections. *Ann. of Math.* **35** (1934), 588–621. (In *Kaleidoscopes: Selected Writings of H. S. M. Coxeter* (eds. F. A. Sherk, P. McMullen, A. C. Thompson, and A. I. Weiss), Wiley-Interscience (New York, 1995), 145–178.) [3B]

[100] H. S. M. Coxeter, Finite groups generated by reflections, and their subgroups generated by reflections. *Proc. Cambridge Philos. Soc.* **30** (1933–1934), 466–482. (In *Kaleidoscopes: Selected Writings of H. S. M. Coxeter* (eds. F. A. Sherk, P. McMullen, A. C. Thompson, and A. I. Weiss), Wiley-Interscience (New York, 1995), 179–195.)

[101] H. S. M. Coxeter, The functions of Schläfli and Lobatschefsky. *Quart. J. Math.* **6** (1935), 13–29. (In *Twelve Geometric Essays*, Southern Illinois University Press (Carbondale, 1968), 3–20.) [6B]

[102] H. S. M. Coxeter, The complete enumeration of finite groups of the form $R_i^2 = (R_i R_j)^{k_{ij}} = 1$. *J. London Math. Soc.* **10** (1935), 21–35. (In *Kaleidoscopes: Selected Writings of H. S. M. Coxeter* (eds. F. A. Sherk, P. McMullen, A. C. Thompson, and A. I. Weiss), Wiley-Interscience (New York, 1995), 139–143.) [3B, 3E]

[103] H. S. M. Coxeter, Wythoff's construction for uniform polytopes. *Proc. London Math. Soc. (2)* **38** (1935), 327–339. [1B]

[104] H. S. M. Coxeter, The abstract groups $R^m = S^m = (R^j S^j)^{p_j} = 1, S^m = T^2 = (S^j T)^{2p_j} = 1$, and $S^m = T^2 = (S^{-j} T S^j T)^{p_j} = 1$. *Proc. London Math. Soc. (2)* **41** (1936), 278–301. [7B]

[105] H. S. M. Coxeter, Regular skew polyhedra in 3 and 4 dimensions and their topological analogues. *Proc. London Math. Soc. (2)* **43** (1937), 33–62. (Reprinted with amendments in *Twelve Geometric Essays*, Southern Illinois University Press (Carbondale, 1968), 76–105.) [1A, 5B, 7A, 7B, 7D, 7E, 8C, 8D]

[106] H. S. M. Coxeter, The abstract groups $G^{m,n,p}$. *Trans. Amer. Math. Soc.* **45** (1939), 73–150. [7B]

[107] H. S. M. Coxeter, The regular sponges, or skew polyhedra. *Scripta Math.* **6** (1939), 240–244. (In *Kaleidoscopes: Selected Writings of H. S. M. Coxeter* (eds. F. A. Sherk, P. McMullen, A. C. Thompson, and A. I. Weiss), Wiley-Interscience (New York, 1995), 19–23.) [7B]

[108] H. S. M. Coxeter, Regular and semi-regular polytopes, I. *Math. Z.* **46** (1940), 380–407. (In *Kaleidoscopes: Selected Writings of H. S. M. Coxeter* (eds. F. A. Sherk, P. McMullen, A. C. Thompson, and A. I. Weiss), Wiley-Interscience (New York, 1995), 251–278.) [1B]

[109] H. S. M. Coxeter, Configurations amd maps. *Rep. Math. Colloq. (2)* **8** (1948), 18–38. [6J]

[110] H. S. M. Coxeter, Self-dual configurations and regular graphs. *Bull. Amer. Math. Soc.* **56** (1950), 413–455. (Reprinted in *Twelve Geometric Essays*, Southern Illinois University Press (Carbondale, 1968), 106–149.) [6J]

[111] H. S. M. Coxeter, The product of the generators of a finite group generated by reflections. *Duke Math. J.* **18** (1951), 765–782. (In *Kaleidoscopes: Selected Writings of H. S. M. Coxeter* (eds. F. A. Sherk, P. McMullen, A. C. Thompson, and A. I. Weiss), Wiley-Interscience (New York, 1995), 207–224.) [6C]

[112] H. S. M. Coxeter, Regular honeycombs in elliptic space. *Proc. London Math. Soc. (3)* **4** (1954), 471–501. [6C]

[113] H. S. M. Coxeter, Regular honeycombs in hyperbolic space. *Proc. Internat. Congress Math. Amsterdam (1954)*, Vol. 3, North-Holland (Amsterdam, 1956), 155–169. (In *Twelve Geometric Essays,* Southern Illinois University Press (Carbondale, 1968), 199–214.) [3C, 4C, 6J, 8B, 10A, 11G]

[114] H. S. M. Coxeter, Groups generated by unitary reflections of period two. *Canad. J. Math.* **9** (1957), 243–272. (In *Kaleidoscopes: Selected Writings of H. S. M. Coxeter* (eds. F. A. Sherk, P. McMullen, A. C. Thompson, and A. I. Weiss), Wiley-Interscience (New York, 1995), 385–414.) [9, 9A, 9D, 9E, 9F, 9G, 11B]

[115] H. S. M. Coxeter, Regular compound tessellations of the hyperbolic plane. *Proc. Roy. Soc. Lond. Ser. A* **278** (1964), 147–167. [6J]

[116] H. S. M. Coxeter, Finite groups generated by unitary reflections. *Abh. Math. Sem. Univ. Hamburg* 31 (1967), 125–135. (In *Kaleidoscopes: Selected Writings of H. S. M. Coxeter* (eds. F. A. Sherk, P. McMullen, A. C. Thompson, and A. I. Weiss), Wiley-Interscience (New York, 1995), 415–425.) [9, 9A, 9E, 9G]

[117] H. S. M. Coxeter, *Twelve Geometric Essays*. Southern Illinois University Press (Carbondale, 1968). (Reprinted as *The Beauty of Geometry: Twelve Essays.* Dover (Mineola, NY, 1999).)

[118] H. S. M. Coxeter, *Non-Euclidean Geometry* (5th edition). University of Toronto Press (Toronto, 1968). [10A]

[119] H. S. M. Coxeter, *Twisted Honeycombs*. Regional Conf. Ser. in Math., Amer. Math. Soc. (Providence, 1970). [2B, 6J]

[120] H. S. M. Coxeter, *Regular Polytopes* (3rd edition). Dover (New York, 1973). [1A, 1B, 2, 2A, 3, 3A, 3B, 3E, 5B, 5C, 6B, 6C, 6J, 7, 7D, 7E, 7F, 8B, 9E, 10A, 10E, 11G, 12B]

[121] H. S. M. Coxeter, Ten toroids and fifty-seven hemi-dodecahedra. *Geom. Dedicata* **13** (1982), 87–99. [7A, 14A]

[122] H. S. M. Coxeter, A symmetrical arrangement of eleven hemi-icosahedra. In *Convexity and Graph Theory (Jerusalem 1981)*, North-Holland Math. Stud. 87, North-Holland (Amsterdam, 1984), 103–114. [14A]

[123] H. S. M. Coxeter, Regular and semi-regular polytopes, II. *Math. Z.* **188** (1985), 559–591. (In *Kaleidoscopes: Selected Writings of H. S. M. Coxeter* (eds. F. A. Sherk, P. McMullen, A. C. Thompson, and A. I. Weiss), Wiley-Interscience (New York, 1995), 279–311.) [1B, 2B]

[124] H. S. M. Coxeter, Regular and semi-regular polytopes, III. *Math. Z.* **200** (1988), 3–45.

(In *Kaleidoscopes: Selected Writings of H. S. M. Coxeter* (eds. F. A. Sherk, P. McMullen, A. C. Thompson, and A. I. Weiss), Wiley-Interscience (New York, 1995), 313–355.) [1B]

[125] H. S. M. Coxeter, Star polytopes and the Schläfli function $f(\alpha, \beta, \gamma)$. *Elem. Math.* **44** (2) (1989), 25–36. (In *Kaleidoscopes: Selected Writings of H. S. M. Coxeter* (eds. F. A. Sherk, P. McMullen, A. C. Thompson, and A. I. Weiss), Wiley-Interscience (New York, 1995), 121–132.) [3E]

[126] H. S. M. Coxeter, *Regular Complex Polytopes* (2nd edition). Cambridge University Press (Cambridge, 1991). [5B, 8D, 9, 9A, 9E, 9F]

[127] H. S. M. Coxeter, The evolution of Coxeter-Dynkin diagrams. In *Polytopes: Abstract, Convex and Computational* (eds. T. Bisztriczky, P. McMullen, R. Schneider, and A. Ivić Weiss), NATO ASI Series C **440**, Kluwer (Dordrecht, 1994), 21–42. [3A]

[128] H. S. M. Coxeter, R. Frucht, and D. L. Powers, *Zero-Symmetric Graphs*. Academic Press (New York, 1981). [6J]

[129] H. S. M. Coxeter and B. Grünbaum, Face-transitive polyhedra with rectangular faces. *C. R. Math. Acad. Sci. Soc. R. Can.* **20** (1998), 16–21. [7E]

[130] H. S. M. Coxeter and B. Grünbaum, Face-transitive polyhedra with rectangular faces and icosahedral symmetry. *Discrete Comput. Geom.* **25** (2001), 163–172. [7E]

[131] H. S. M. Coxeter and W. O. J. Moser, *Generators and Relations for Discrete Groups* (4th edition). Springer-Verlag (Berlin–New York, 1980). [1A, 2B, 6B, 6G, 6H, 6J, 7B, 8F, 9D, 10B, 11A, 11B]

[132] H. S. M. Coxeter and G. C. Shephard, Regular 3-complexes with toroidal cells. *J. Combin. Theory Ser. B* **22** (1977), 131–138. [6B, 10A, 10B]

[133] H. S. M. Coxeter and G. C. Shephard, Some regular maps and their polyhedral realizations. In *Applied Geometry and Discrete Mathematics (The Victor Klee Festschrift)* (eds. P. Gritzmann and B. Sturmfels), DIMACS Series in Discrete Mathematics and Theoretical Computer Science **4**, Amer. Math. Soc. and Assoc. Computing Machinery (1991), 157–174. [5B]

[134] H. S. M. Coxeter and J. A. Todd, A practical method of enumerating cosets of a finite abstract group. *Proc. Edinburgh Math. Soc.* **5** (2) (1937), 26–34. [8F]

[135] H. S. M. Coxeter and A. I. Weiss, Twisted honeycombs $\{3, 5, 3\}_t$ and their groups. *Geom. Dedicata* **17** (1984), 169–179. [6J]

[136] H. S. M. Coxeter and G. J. Whitrow, World-structure and non-Euclidean honeycombs, *Proc. Roy. Soc. Lond. Ser. A* **201** (1950), 417–437. [6J, 11G]

[137] K. Critchlow, *Time Stands Still: New Light on Megalithic Science.* Gordon Fraser (London, 1979). [1A]

[138] A. Császár, A polyhedron without diagonals. *Acta Sci. Math. Szeged* **13** (1949), 140–142. [8E]

[139] H. Cuypers, Regular quaternionic polytopes. *Linear Algebra Appl.* **226–228** (1995), 311–329. [9A]

[140] L. Danzer, Regular incidence-complexes and dimensionally unbounded sequences of such, I. In *Convexity and Graph Theory (Jerusalem 1981)*, North-Holland Math. Stud. 87, North-Holland (Amsterdam, 1984), 115–127. [8C, 8D]

[141] L. Danzer and E. Schulte, Reguläre Inzidenzkomplexe, I. *Geom. Dedicata* **13** (1982), 295–308. [2, 10A]

[142] B. Datta and N. Nilakantan, Equivelar polyhedra with few vertices. *Discrete Comput. Geom.* **26** (2001), 429–461. [6J]

[143] L. Di Martini, W. E. Kantor, G. Lunardon, and A. Pasini (eds.), *Groups and Geometries*, Birkhäuser (Boston–Basel–Berlin, 1998). [1A]

[144] J. D. Dixon and B. Mortimer, *Permutation Groups.* Springer-Verlag (New York–Heidelberg–Berlin, 1996). [8E]

[145] M. W. Davis, Regular convex cell complexes. In *Geometry and Topology* (eds. C. McCrory and T. Shifrin), Lecture Notes Pure Appl. Math. **105**, Marcel Dekker (New York, 1987), 53–88. [2E, 6B]

[146] H. E. Debrunner, Dissecting orthoschemes into orthoschemes. *Geom. Dedicata* **33** (1990), 123–152. [7D, 12A, 14A]

[147] M. Dehon, Classifying geometries with CAYLEY, *J. Symbolic Comput.* **17** (1994), 259–276. [7A]

[148] A. W. M. Dress, A combinatorial theory of Grünbaum's new regular polyhedra, I: Grünbaum's new regular polyhedra and their automorphism group. *Aequationes Math.* **23** (1981), 252–265. [1A, 5C, 7E]

[149] A. W. M. Dress, Regular polytopes and equivariant tessellations from a combinatorial point of view. In *Algebraic Topology, Göttingen 1984*, Springer Lecture Notes **1172**, Springer (New York, 1985), 56–72. [2]

[150] A. W. M. Dress, A combinatorial theory of Grünbaum's new regular polyhedra, II: complete enumeration. *Aequationes Math.* **29** (1985), 222–243. [1A, 5C, 7B, 7E]

[151] A. W. M. Dress, Presentations of discrete group actions on simply connected manifolds in terms of parametrized systems of Coxeter matrices. *Adv. Math.* **63** (1987), 196–212. [2, 6B]

[152] A. W. M. Dress and D. Huson, On tilings of the plane. *Geom. Dedicata* **24** (1987), 295–310. [2]

[153] A. W. M. Dress, D. Huson, and E. Molnar, The classification of face-transitive 3D-tilings, *Acta Cryst. Sect. A* **49** (1993), 806–817. [2]

[154] A. W. M. Dress and R. Scharlau, Zur Klassifikation äquivarianter Pflasterungen. *Mitt. Math. Sem. Giessen* **164** (1984), 83–136. [2]

[155] A. W. M. Dress and E. Schulte, On a theorem of McMullen about combinatorially regular polytopes. *Simon Stevin* **61** (1987), 265–273. [1B, 6B]

[156] P. Du Val, *Homographies, Quaternions and Rotations*. Oxford University Press (Oxford, 1964). [1B, 2B]

[157] W. Dyck, Über Aufstellung und Untersuchung von Gruppe und Irrationalität regulärer Riemannscher Flächen. *Math. Ann.* **17** (1880), 473–508. [1D, 6J, 11B]

[158] W. Dyck, Gruppentheoretische Studien. *Math. Ann.* **20** (1882), 1–45. [1D, 6J, 11B]

[159] A. L. Edmonds, J. H. Ewing, and R. S. Kulkarni, Regular tessellations of surfaces and $(p, q, 2)$-triangle groups. *Ann. of Math.* **116** (1982), 113–132. [6B]

[160] A. L. Edmonds, J. H. Ewing, and R. S. Kulkarni, Torsion free subgroups of Fuchsian groups and tessellations of surfaces. *Invent. Math.* **69** (1982), 331–346. [6B]

[161] V. A. Efremovič and Ju. S. Il'jašenko, Regular polygons in E^n. (In Russian). *Vest. Mosk. Univ. Ser. I, Mat. Mekhan.* **5** (1962), 18–24. [5B]

[162] E. L. Elte, *The Semiregular Polytopes of the Hyperspaces*. Hoitsema (Groningen), 1912.

[163] P. Engel, *Geometric Crystallography*. Reidel (Dordrecht, 1986). [5C, 6A]

[164] A. Erréra, Sur les polyèdres réguliers de l'Analysis Situs. *Acad. Roy. Belg. Cl. Sci. Mém. Collect.* 8° *(2)* **7** (1922), 1–17. [6J]

[165] S. L. Farris, Completely classifying all vertex-transitive and edge-transitive polyhedra, Part I: Necessary class conditions. *Geom. Dedicata* **26** (1988), 111–124. [7E]

[166] S. L. Farris, Completely classifying all vertex-transitive and edge-transitive polyhedra, Part II: Finite, fully transitive polyhedra. *Preprint announced in [165]*. [7E]

[167] L. Fejes Tóth, *Reguläre Figuren*. Akadémiai Kiadó (Budapest, 1965). (English translation: *Regular Figures*, Pergamon Press (Oxford, 1964).) [1B]

[168] M. Ferri, C. Gagliardi, and L. Grasselli, A graph theoretical representation of PL manifolds. A survey on crystallization. *Aequationes Math.* **31** (1986), 121–141. [6B]

[169] B. Fine, *Algebraic Theory of the Bianchi Groups*. Dekker (New York, 1989). [4D]

[170] R. H. Fox, On Fenchel's conjecture about F-groups. *Mat. Tidskrift B* (1952), 61–65. [4C]

[171] R. Franz and D. Huson, The classification of quasi-regular polyhedra of genus 2. *Discrete Comput. Geom.* **7** (1992), 347–357. [6J]

[172] R. Frucht, J. E. Graver, and M. E. Watkins, The groups of the generalized Petersen graphs. *Proc. Cambridge Philos. Soc.* **70** (1971), 211–218. [6C]

[173] W. Fulton and J. Harris, *Representation Theory*. Graduate Texts in Mathematics, Springer (New York, 1991). [5B]

[174] D. Garbe, Über die regulären Zerlegungen geschlossener orientierbarer Flächen. *J. Reine Angew. Math.* **237** (1969), 39–55. [6J]
[175] D. Garbe, A generalization of the regular maps of type $\{4, 4\}_{b,c}$ and $\{3, 6\}_{b,c}$. *Canad. Math. Bull.* **12** (1969), 293–297. [6J]
[176] D. Garbe, A remark on nonsymmetric compact Riemann surfaces. *Arch. Math.* **30** (1978), 435–437. [6J]
[177] A. Gardiner, R. Nedela, J. Širáň, and M. Škoviera, Characterization of graphs which underlie regular maps on closed surfaces, *J. London Math. Soc. (2)* **59** (1999), 100–108. [6J]
[178] C. W. L. Garner, *Polyhedra and Honeycombs in Hyperbolic Space*. Ph.D. Thesis (University of Toronto, 1964). [6J, 7B]
[179] C. W. L. Garner, Coordinates for the vertices of some regular honeycombs in hyperbolic space. *Proc. Roy. Soc. Lond. Ser. A* **293** (1966), 94–107. [6J]
[180] C. W. L. Garner, Regular skew polyhedra in hyperbolic three-space. *J. Canad. Math. Soc.* **19** (1967), 1179–1186. [7B]
[181] C. W. L. Garner, Compound honeycombs in hyperbolic space. *Proc. Roy. Soc. Lond. Ser. A* **316** (1970), 441–448. [6J]
[182] C. W. L. Garner, Compound honeycombs in hyperbolic four-space. *Preprint* (1998). [6J]
[183] P. Gordan, Über die Auflösung der Gleichungen vom fünften Grade. *Math. Ann.* **13** (1878), 375–404. [6J]
[184] P. Goossens, Incidence complexes and r-connectedness. *Portugal. Math.* **46** (1989), 259–268. [2A]
[185] P. Goossens, Combinatorial and homological properties of some partially ordered sets arising in geometry. *Bull. Soc. Roy. Sci. Liège* **56** (1987), 193–294.
[186] T. Gosset, On the regular and semi-regular figures in space of n dimension. *Messenger of Mathematics* **29** (1900), 43–48.
[187] E. Goursat, Sur les substitutions othogonales et les divisions réguliéres des l'espace. *Ann. Sci. École Norm. Sup.* (3) **6** (1889), 9–102.
[188] A. Gray and S. E. Wilson, A more elementary proof of Grünbaum's conjecture. *Congr. Numer.* **72** (1990), 25–32. [4C]
[189] A. S. Grek, Regular polyhedra on a closed surface with $\chi = -1$. *Trudy Tbliss. Mat. Inst.* **27** (1960), 103–112. [6J]
[190] A. S. Grek, Regular polyhedra of the simplest hyperbolic type. *Ivanov. Gos. Ped. Inst. Uchebn. Zap.* **34** (1963), 27–30. [6J]
[191] A. S. Grek, Regular polyhedra on surfaces with Euler characteristic $\chi = -4$. *Soobshch. Akad. Nauk Gruzin. SSR* **42** (1966), 11–15. [6J]
[192] A. S. Grek, Regular polyhedra on a closed surface whose Euler characteristic is $\chi = -3$. *Izv. Vyssh. Uchebn. Zaved. Mat.* (6) **55** (1966), 50–53. [6J]
[193] R. L. Griess, Quotients of infinite reflection groups, *Math. Ann.* **263** (1983), 267–278. [12C]
[194] M. Gromov, Volume and bounded cohomology. *Inst. Hautes Études Sci. Publ. Math.* **56** (1982), 5–99. [6B]
[195] L. C. Grove and C. T. Benson, *Finite Reflection Groups* (2nd edition). Graduate Texts in Mathematics, Springer-Verlag (New York–Heidelberg–Berlin–Tokyo, 1985). [3]
[196] J. L. Gross and T. W. Tucker, *Topological Graph Theory*. Wiley (New York, 1987). [6J]
[197] B. Grünbaum, *Convex Polytopes*. Wiley-Interscience (New York, 1967). [1B, 3E, 8B]
[198] B. Grünbaum, Regular polyhedra – Old and new. *Aequationes Math.* **16** (1977), 1–20. [1A, 5B, 7B, 7E]
[199] B. Grünbaum, Regularity of graphs, complexes and designs. In *Problèmes combinatoires et théorie des graphes*, Coll. Int. C.N.R.S. **260**, Orsey (1977), 191–197. [2, 4A, 6B, 10, 10A, 10B, 11A, 14A]
[200] B. Grünbaum, Regular polyhedra. In *Companion Encyclopedia of the History and Philosophy of the Mathematical Sciences, Vol. 2* (ed. I. Grattan-Guinness), Routledge (London, 1994), 866–876.

[201] B. Grünbaum, Polyhedra with hollow faces. In *Polytopes: Abstract, Convex and Computational* (eds. T. Bisztriczky, P. McMullen, R. Schneider, and A. Ivić Weiss), NATO ASI Series C **440**, Kluwer (Dordrecht, 1994), 43–70. [5B, 7E]

[202] B. Grünbaum, Realizations of symmetric maps by symmetric polyhedra. *Discrete Comput. Geom.* **20** (1998), 19–33. [5B]

[203] B. Grünbaum and G. C. Shephard, Polyhedra with transitivity properties. *C. R. Math. Acad. Sci. Soc. R. Can.* **6** (1984), 61–66. [7E]

[204] B. Grünbaum and G. C. Shephard, *Tilings and Patterns*. Freeman (New York, 1987).

[205] B. Grünbaum and G. C. Shephard, Is self-duality always involutory? *Amer. Math. Monthly* **95** (1988), 729–733. [2A]

[206] R. C. Gunning, *Lectures on Modular Forms*. Princeton University Press (Princeton, NJ, 1962). [13A, 13D]

[207] J. I. Hall, Locally Petersen graphs. *J. Graph Theory* **4** (1980), 173–187. [14A]

[208] M. I. Hartley, *Combinatorially Regular Euler Polytopes.* Ph.D. Thesis, University of Western Australia (Nedlands, Perth, 1995). [2E]

[209] M. I. Hartley, All polytopes are quotients, and isomorphic polytopes are quotients by conjugate subgroups. *Discrete Comput. Geom.* **21** (1999), 289–298. [2E]

[210] M. I. Hartley, More on quotient polytopes. *Aequationes Math.* **57** (1999), 108–120. [2E]

[211] M. I. Hartley, Polytopes of finite type. *Discrete Math.* **218** (2000), 97–108. [6C]

[212] M. I. Hartley, P. McMullen, and E. Schulte, Symmetric tessellations on euclidean space-forms. *Canad. J. Math.* (6) **51** (1999), 1230–1239. [6G, 6H]

[213] H. Hasse, *Number Theory.* Springer-Verlag (Berlin, 1980). [13A]

[214] P. J. Heawood, Map-colour theorem, *Quart. J. Math.* **24** (1890), 332–338. [6J]

[215] E. Hess, Über die regulären Polytope höherer Art. *Sitz. Ges. Bef. Ges. Naturwiss. Marburg* (1885), 31–57. [1A, 7D]

[216] M. Hermand, *Géométries, Langage CAYLEY et Groupe de Hall-Janko*, Thesis, Université de Bruxelles (1991). [7A]

[217] D. Hilbert and S. Cohn-Vossen, *Anschauliche Geometrie*. Springer-Verlag (Berlin, 1932). (English translation: *Geometry and the Imagination*, Chelsea (New York, 1952).) [1B]

[218] G. Hillebrandt, *Über dünne Geometrien vom sphärischen Typ*. Dissertation, Technische Universität Braunschweig (1986).

[219] H. Hiller, *Geometry of Coxeter Groups*. Pitman Books Limited (London, 1982). [3]

[220] S. G. Hoggar, Two quaternionic 4-polytopes. In *The Geometric Vein: the Coxeter Festschrift* (eds. C. Davis, B. Grünbaum, and F. A. Sherk), Springer-Verlag (New York–Heidelberg–Berlin, 1981), 219–230. [9A]

[221] R. Hoppe, Regelmässige linear begrenzte Figuren von vier Dimensionen. *Arch. Math. Phys.* **67** (1881), 29–44.

[222] J. E. Humphreys, *Reflection Groups and Coxeter Groups*. Cambridge University Press (Cambridge, 1990). [3, 3B, 3C, 6C, 9B]

[223] A. Hurwitz, Über algebraische Gebilde mit eindeutigen Transformationen in sich, *Math. Ann.* **41** (1893), 403–442. [6J]

[224] K. Ireland and M. Rosen, *A Classical Introduction to Modern Number Theory*. Academic Press (1974). [13A]

[225] A. A. Ivanov and S. V. Shpectorov, Geometries for sporadic groups related to the Petersen graph, I. *Comm. Algebra* **16** (1988), 925–954. [14A]

[226] A. A. Ivanov and S. V. Shpectorov, Geometries for sporadic groups related to the Petersen graph, II. *European J. Combin.* **10** (1989), 347–362. [14A]

[227] A. A. Ivanov and S. V. Shpectorov, The P-geometry for M_{23} has no non-trivial 2-coverings. *European J. Combin.* **11** (1989), 373–379. [14A]

[228] A. A. Ivanov and S. V. Shpectorov, The last flag-transitive P-geometry. *Israel J. Math.* **82** (1993), 341–362. [14A]

[229] A. A. Ivanov and S. V. Shpectorov, The flag-transitive tilde and Petersen type geometries are all known. *Bull. Amer. Math. Soc.* **31** (1994), 173–184. [14A]

[230] A. A. Ivanov and C. E. Praeger, On locally projective graphs of girth 5. *J. Algebraic Combin.* **7** (1998), 259–283. [6C, 14A]

[231] A. A. Ivanov, *Geometry of Sporadic Groups I*. Cambridge University Press (Cambridge, 1999). [7A, 14A]

[232] L. D. James, *Maps and Hypermaps – Operations and Symmetry* Ph.D. Thesis, Southampton University (1985). [6J, 7B]

[233] L. D. James, Complexes and Coxeter groups – Operations and outer automorphisms. *J. Algebra* **113** (1988), 339–345. [6J, 7B]

[234] S. Jendrol, R. Nedela, and M. Škoviera, Constructing regular maps and graphs from planar quotients. *Math. Slovaca* **47** (1997), 155–170. [4C]

[235] N. W. Johnson, *Uniform Polytopes*. Cambridge University Press (Cambridge, to appear). [1B, 6J, 7B, 9E]

[236] N. W. Johnson, R. Kellerhals, J. G. Ratcliffe, and S. T. Tschantz, The size of a hyperbolic Coxeter simplex. *Transform. Groups* **3** (1999), 329–353. [6J]

[237] N. W. Johnson and A. I. Weiss, Quadratic integers and Coxeter groups. *Canad. J. Math.* (6) **51** (1999), 1307–1336. [2B, 10A]

[238] N. W. Johnson and A. I. Weiss, Quaternionic modular groups. *Linear Algebra Appl.* **295** (1999), 159–189.

[239] G. A. Jones and D. Singerman, Theory of maps on orientable surfaces. *Proc. London Math. Soc. (3)* **37** (1978), 273–307. [6J]

[240] G. A. Jones and D. Singerman, *Complex Functions*. Cambridge University Press (Cambridge, 1987). [6J]

[241] G. A. Jones and J. S. Thornton, Operations on maps, and outer automorphisms. *J. Combin Theory Ser. B* **35** (1983), 93–103. [6J, 7B]

[242] G. Kalai, Polytope skeletons and paths. In *Handbook of Discrete and Computational Geometry* (eds. J. E. Goodman and J. O'Rourke), CRC Press (Boca Raton–New York, 1997), 331–344. [1B]

[243] W. M. Kantor, Generalized polygons, SCABs and GABs, In *Buildings and the Geometry of Diagrams*, Springer Lecture Notes **1181**, Springer (New York, 1986), 79–158. [2]

[244] W. M. Kantor, R. A. Liebler, S. E. Payne, and E. E. Shult, *Finite Geometries, Buildings, and Related Topics*. Clarendon Press (Oxford, 1990). [1A]

[245] M. Kato, On combinatorial space-forms. *Scientific Papers of the College of General Education, University of Tokyo* **30** (1980), 107–146. [2E, 6B]

[246] R. Kellerhals, Shape and size through hyperbolic eyes. *Math. Intelligencer* **17** (1995), 21–30. [6B, 6J]

[247] R. Kellerhals, *Volumina von hyperbolischen Raumformen*. Habilitationsschrift, Universität Bonn (1995). [6B]

[248] J. Kepler, *Harmonices Mundi*. J. Planck (Linz, 1619). (Reprinted in *Opera omnia, Vol. V*, Heyder & Zimmer (Frankfurt, 1864), 75–334; also in *Gesammelte Werke, Bd. VI*, Beck (Munich, 1940), 3–377.) [1A, 6J]

[249] W. Kimmerle and E. Kouzoudi, Doubly transitive automorphism groups of combinatorial surfaces (preprint). [8E, 11C]

[250] F. Klein, Über die Transformationen siebenter Ordnung der elliptischen Functionen. *Math. Ann.* **14** (1879), 428–471. (Revised version in *Gesammelte Mathematische Abhandlungen, Vol. 3*, Springer (Berlin, 1923).) [1D, 6J]

[251] F. Klein, *Vorlesungen über das Ikosaeder und die Auflösung der Gleichungen fünften Grades*. (Leipzig, 1884). [1D, 6J]

[252] F. Klein and R. Fricke, *Vorlesungen über die Theorie der elliptischen Modulfunktionen*. (Leipzig, 1890). [1D, 6J]

[253] E. A. Komissartschik and S. V. Tsaranov, Construction of finite group amalgams and geometries – Geometries of the group $U_4(2)$. *Comm. Algebra* **18** (1990), 1071–1117. [7A]

[254] J. L. Koszul, *Lectures on Hyperbolic Coxeter Groups* (Notes by T. Ochiai). University of Notre Dame (Indiana, 1967). [3]

[255] W. Kühnel, *Tight Polyhedral Submanifolds and Tight Triangulations*. Lecture Notes in Mathematics Vol. 1612, Springer-Verlag (Berlin–Heidelberg–New York, 1995). [8D]

[256] W. Kühnel, Equilibrium decompositions of 4-manifolds, and abstract regular 5-polytopes. *Mathematika* **44** (1997), 100–112. [6B]

[257] W. Kühnel, Topological aspects of twofold triple systems. *Exposition. Math.* **16** (1998), 289–332. [6J]

[258] W. Kühnel and Ch. Schulz, Submanifolds of the cube. In *Applied Geometry and Discrete Mathematics (The Victor Klee Festschrift)* (eds. P. Gritzmann and B. Sturmfels), DIMACS Series in Discrete Mathematics and Theoretical Computer Science **4**, Amer. Math. Soc. & Assoc. Computing Machinery (1991), 423–432. [8D]

[259] F. Lannér, On complexes with transitive groups of automorphisms. *Medd. Lunds Univ. Math. Sem.* **11** (1950), 1–71. [2C]

[260] D. Leemans, Thin geometries for the Suzuki simple group $Sz(8)$. *Bull. Belg. Math. Soc. Simon Stevin* **5** (1998), 373–387. [7A]

[261] D. Leemans, An atlas of regular thin geometries for small groups. *Math. Comput.* **68** (1999), 1631–1647. [7A]

[262] C. Lefevre-Percsy, N. Percsy, and D. Leemans, New geometries for finite groups and polytopes. *Bull. Belg. Math. Soc. Simon Stevin* **7** (2000), 583–610. [7A]

[263] Ch. Leytem, The automorphism group of the abstract polytope $\{\{4,4\}_{3,0},\{4,4\}_{5,0}\}$. *Unpublished manuscript* (1996). [10C, 10D]

[264] Ch. Leytem, Pseudo-Petrie operators on Grünbaum polyhedra. *Math. Slovaca* **47** (1997), 175–188. [7E]

[265] Ch. Leytem, Regular polyhedra with translational symmetries. *Period. Math. Hungar.* **34** (1997), 111–122. [7E]

[266] Ch. Leytem, Regular coloured rank 3 polyhedra with tetragonal vertex figure. *Studia Sci. Math. Hungar.* **35** (1999), 17–38. [7E]

[267] Ch. Leytem, A presentation of $S_4 \times C_2 \times C_2$ (preprint). [10C, 10D]

[268] S. Lins, Graph-encoded maps. *J. Combin Theory Ser. B* **32** (1982), 171–181. [6J, 7B]

[269] R. C. Lyndon and P. E. Schupp, *Combinatorial Group Theory*. Springer (Berlin, 1977). [4A]

[270] W. Magnus, *Noneuclidean Tessellations and their Groups.* Academic Press (1974). [6J, 13B, 13D]

[271] W. Magnus, A. Karrass, and D. Solitar, *Combinatorial Group Theory* (2nd edition). Dover (New York, 1976). [4C]

[272] A. I. Malcev, On faithful representations of infinite groups of matrices (Russian). *Mat. Sb.* **8** (1940), 405–422. (English translation: *Amer. Math. Soc. Transl. (2)* **45** (1965), 1–18.) [4C, 12B]

[273] H. Martini, Hierarchical classification of euclidean polytopes with regularity properties. In *Polytopes: Abstract, Convex and Computational* (eds. T. Bisztriczky, P. McMullen, R. Schneider, and A. Ivić Weiss), NATO ASI Series C **440**, Kluwer (Dordrecht, 1994), 71–96. [1B, 7E]

[274] B. Maskit, *Kleinian Groups*. Grundlehren Math. Wiss. **287**, Springer-Verlag (New York–Heidelberg–Berlin, 1988). [6A]

[275] G. Maxwell, Wythoff's construction for Coxeter groups. *J. Algebra* **123** (1989), 351–377. [3D]

[276] P. McMullen, *A Classification of the Regular Polytopes and Honeycombs in Unitary Space.* Qualifying M.Sc. Thesis, University of Birmingham (1966). [9E, 9F]

[277] P. McMullen, Combinatorially regular polytopes. *Mathematika* **14** (1967), 142–150. [1B, 2]

[278] P. McMullen, Affinely and projectively regular polytopes. *J. London Math. Soc.* **43** (1968), 755–757. [1B]

[279] P. McMullen, *On the Combinatorial Structure of Convex Polytopes.* Ph.D. Thesis, University of Birmingham (1968). [1B, 6B]

[280] P. McMullen, Regular star-polytopes, and a theorem of Hess. *Proc. London Math. Soc. (3)* **18** (1968), 577–596. [1A, 1C, 2, 3E, 7D]

[281] P. McMullen, Angle-sum relations for polyhedral sets. *Mathematika* **33** (1986), 173–188. [3E]

[282] P. McMullen, Realizations of regular polytopes. *Aequationes Math.* **37** (1989), 38–56. [5A, 14A]

[283] P. McMullen, Nondiscrete regular honeycombs. Chapter 10 in *Quasicrystals, Networks, and Molecules of Fivefold Symmetry* (ed. I. Hargittai), VCH Publishers (New York, 1990), 159–179. [5C]

[284] P. McMullen, Regular polyhedra related to projective linear groups. *Discrete Math.* **91** (1991), 161–170. [12A, 13A, 13B]

[285] P. McMullen, The order of a finite Coxeter group. *Elem. Math.* **46** (1991), 121–130. [3E]

[286] P. McMullen, The regular polyhedra of type $\{p, 3\}$ with $2p$ vertices. *Geom. Dedicata* **43** (1992), 285–289. [6C]

[287] P. McMullen, Locally projective regular polytopes. *J. Combin. Theory Ser. A* **65** (1994), 1–10. [8F, 12C, 12D, 12E]

[288] P. McMullen, Realizations of regular apeirotopes. *Aequationes Math.* **47** (1994), 223–239. [5A, 5C]

[289] P. McMullen, Modern developments in regular polytopes. In *Polytopes: Abstract, Convex and Computational* (eds. T. Bisztriczky, P. McMullen, R. Schneider, and A. Ivić Weiss), NATO ASI Series C **440**, Kluwer (Dordrecht, 1994), 97–124.

[290] P. McMullen, The groups of the regular star-polytopes. *Canad. J. Math.* (2) **50** (1998), 426–448. [7D]

[291] P. McMullen, and B. R. Monson, Realizations of regular polytopes, II *Aequationes Math.* (to appear). [5B]

[292] P. McMullen, B. R. Monson, and A. I. Weiss, Regular maps constructed from linear groups. *European J. Combin.* **14** (1993), 541–552. [5A, 12A, 13A]

[293] P. McMullen and E. Schulte, Self-dual regular polytopes and their Petrie-Coxeter-polyhedra. *Results Math.* **12** (1987), 366–375. [7B]

[294] P. McMullen and E. Schulte, Regular polytopes from twisted Coxeter groups. *Math. Z.* **201** (1989), 209–226. [5C, 8B, 10A, 12D]

[295] P. McMullen and E. Schulte, Constructions for regular polytopes. *J. Combin. Theory Ser. A* **53** (1990), 1–28. [5C, 7B]

[296] P. McMullen and E. Schulte, Regular polytopes from twisted Coxeter groups and unitary reflexion groups. *Adv. Math.* **82** (1990), 35–87. [5C, 10A, 12B]

[297] P. McMullen and E. Schulte, Hermitian forms and locally toroidal regular polytopes. *Adv. Math.* **82** (1990), 88–125. [5C, 10A]

[298] P. McMullen and E. Schulte, Finite quotients of infinite universal polytopes. In *Discrete and Computational Geometry* (eds. J. Goodman, R. Pollack, and W. Steiger), DIMACS Series in Discrete Mathematics and Theoretical Computer Science **6**, Amer. Math. Soc. & Ass. Comput. Mach. (1991), 231–236. [4C, 12B]

[299] P. McMullen and E. Schulte, Regular polytopes of type $\{4, 4, 3\}$ and $\{4, 4, 4\}$. *Combinatorica* **12** (1992), 203–220. [10A, 12B, 12C, 12D]

[300] P. McMullen and E. Schulte, Locally toroidal regular polytopes of rank 4. *Comment. Math. Helv.* **67** (1992), 77–118. [4E, 10A, 11G, 11H, 12D]

[301] P. McMullen and E. Schulte, Quotients of polytopes and C-groups. *Discrete Comput. Geom.* **11** (1994), 453–464. [2D, 2E]

[302] P. McMullen and E. Schulte, Higher toroidal regular polytopes. *Adv. Math.* **117** (1996), 17–51. [6G, 10A, 12E, 14B]

[303] P. McMullen and E. Schulte, Twisted groups and locally toroidal regular polytopes. *Trans. Amer. Math. Soc.* **348** (1996), 1373–1410. [8B, 8F, 10A, 12A, 12E]

[304] P. McMullen and E. Schulte, Regular polytopes in ordinary space. *Discrete Comput. Geom.* **17** (1997), 449–478. [5C, 7B, 7E]

[305] P. McMullen and E. Schulte, Flat regular polytopes. *Ann. Comb.* **1** (1997), 261–278. [4E, 4F]

[306] P. McMullen and E. Schulte, Locally unitary groups generated by involutory reflexions (preprint). [9]

[307] P. McMullen and E. Schulte, Locally unitary groups and regular polytopes. *Adv. in Appl. Math.* (to appear). [9, 11C]

[308] P. McMullen and E. Schulte, The mix of a regular polytope with a face *Ann. Comb.* (to appear). [7A]

[309] P. McMullen, E. Schulte, and J. M. Wills, Infinite series of combinatorially regular maps in three-space. *Geom. Dedicata* **26** (1988), 299–307. [1D, 5B, 8D]

[310] P. McMullen, Ch. Schulz, and J. M. Wills, Equivelar polyhedral manifolds in E^3. *Israel J. Math.* **41** (1982), 331–346. [1B, 1D, 2A, 5B]

[311] P. McMullen, Ch. Schulz, and J. M. Wills, Polyhedral manifolds in E^3 with unusually large genus. *Israel J. Math.* **46** (1983), 127–144. [1D, 5B, 8D]

[312] P. McMullen, Ch. Schulz, and J. M. Wills, Two remarks on equivelar manifolds. *Israel J. Math.* **52** (1985), 28–32. [5B]

[313] T. Meixner, Groups acting transitively on locally finite classical Tits chamber systems. In *Finite Geometries, Buildings and Related Topics* (eds. W. M. Kantor, R. A. Liebler, S. E. Payne, and E. E. Shult), Clarendon Press (Oxford, 1990), 45–65.

[314] G. A. Miller, Groups defined by the orders of two generators and the order of their product. *Amer. J. Math.* **24** (1902), 96–106. [4C]

[315] J. Milnor, On the 3-dimensional Brieskorn manifolds M(p,q,r). In *Knots, Groups, and 3-manifolds* (ed. L. P. Neuwirth), Annals of Mathematics Studies **84**, Princeton University Press (Princeton, 1975), 175–225.

[316] A. F. Möbius, Theorie der symmetrischen Figuren. In *Gesammelte Werke* **2**, Hirzel Verlag (Leipzig, 1886), 561–708.

[317] E. Molnar, Discontinuous groups in homogeneous Riemannian spaces by classification of D-symbols. *Publ. Math. Debrecen* **49** (1996), 265–294. [2]

[318] B. R. Monson, The densities of certain regular star-polytopes. *C. R. Math. Acad. Sci. R. Can.* **2** (1980), 73–78. [11G]

[319] B. R. Monson, Simplicial quadratic forms. *Canad. J. Math.* **35** (1983), 101–116. [11G]

[320] B. R. Monson, A family of uniform polytopes with symmetric shadows. *Geom. Dedicata* **23** (1987), 355–363. [7A]

[321] B. R. Monson and E. Schulte, Reflection groups and polytopes over finite fields (in preparation). [12C, 12D, 12E]

[322] B. R. Monson and A. I. Weiss, Regular 4-polytopes related to general orthogonal groups. *Mathematika* **37** (1990), 106–118. [2B, 10A, 12A, 13D]

[323] B. R. Monson and A. I. Weiss, Polytopes related to the Picard group. *Linear Algebra Appl.* **218** (1995), 185–204. [2B, 10A, 12A, 13D]

[324] B. R. Monson and A. I. Weiss, Eisenstein integers and related C-groups. *Geom. Dedicata* **66** (1997), 99–117. [2B, 10A, 12A]

[325] B. R. Monson and A. I. Weiss, Realizations of regular toroidal maps of type $\{4, 4\}$. *Discrete Comput. Geom.* **24** (2000), 453–465. [5B]

[326] B. R. Monson and A. I. Weiss, Realizations of regular toroidal maps. *Canad. J. Math.* (6) **51** (1999), 1240–1257. [5B]

[327] J. M. Montesinos, *Classical Tessellations and Three-Manifolds*. Springer-Verlag (New York–Heidelberg–Berlin, 1987). [6B]

[328] B. Mühlherr, Coxeter groups in Coxeter groups. In *Finite Geometry and Combinatorics* (eds. A. Beutelspacher, F. Buekenhout, F. DeClerck, J. Doyen, J. W. P. Hirschfeld, and J. A. Thas), Cambridge University Press (Cambridge, 1993), 277–287.

[329] T. Muir, *A Treatise on the Theory of Determinants.* Dover (New York, 1960). [5B]

[330] J. R. Munkres, *Elements of Algebraic Topology*. Addison–Wesley (New York, 1984).

[331] Q. Mushtaq and H. Servatius, Permutation representations of the symmetry groups of regular hyperbolic tessellations. *J. London Math. Soc. (2)* **48** (1993), 77–86. [4C]

[332] R. Nedela and M. Škoviera, Regular maps on surfaces with large planar width. *European J. Combin.* **22** (2001), 243–261. [4C]

[333] B. Nostrand, *Chiral Honeycombs*. Ph.D. Thesis, Northeastern University (Boston, 1993). [2B, 6H, 10A]

[334] B. Nostrand and E. Schulte, Chiral polytopes from hyperbolic honeycombs. *Discrete Comput. Geom.* **13** (1995), 17–39. [2B, 6H, 10A]

[335] B. Nostrand, E. Schulte, and A. I. Weiss, Constructions of chiral polytopes. *Congr. Numer.* **97** (1993), 165–170. [2B, 6H, 10A]

[336] P. Orlik and H. Terao, *Arrangements of Hyperplanes*, Grundlehren Serie **300**, Springer-Verlag (New York–Heidelberg–Berlin, 1992). [3B, 9A]

[337] S. L. van Oss, Die regelmässigen vierdimensionalen Polytope höherer Art. *Verh. Konink. Akad. Weten. Amsterdam* (Eerste sectie) **12.1** (1915). [1A, 7D]

[338] J. Oxley, *Regular Combinatorial Polyhedra*. Master's Thesis, Australian National University (1975). [6J]

[339] P. Pearce, *Synestructures*. A report to the Graham Foundation. [7E]

[340] A. Pasini, *Diagram Geometries*, Oxford University Press (Oxford, 1994). [2, 2B]

[341] S. E. Payne and J. A. Thas, *Finite Generalized Quadrangles*. Pitman Books Limited (London, 1984). [2A]

[342] L. Poinsot, Mémoire sur les polygones et les polyèdres. *J. École Polytechnique* **10** (1810), 16–48. (In *Ostwalds Klassiker der exakten Wissenschaften* **151** (ed. R. Haussner), Engelmann (Leipzig, 1906), 3–48.) [1A]

[343] V. L. Popov, Discrete complex reflection groups. *Notes* (circulated at the Workshop on Reflection Groups and Their Applications, Trieste, January 1998). [9A, 9C]

[344] J. G. Ratcliffe, *Foundations of Hyperbolic Manifolds.* Graduate Texts in Mathematics, Springer-Verlag (New York–Berlin–Heidelberg, 1994). [3C, 5C, 6A, 6B, 6C, 6J, 7E, 9C]

[345] C. Reinhardt, Einleitung in die Theorie der Polyeder. *Jahresbericht d. Fürsten- und Landesschule St. Afra in Meissen* (Meissen, 1890), 1–31. [1D]

[346] G. Ringel, Über drei kombinatorische Probleme am n-dimensionalen Würfel und Würfelgitter. *Abh. Math. Sem. Univ. Hamburg* **20** (1956), 10–19. [8D]

[347] G. Ringel, *Map Color Theorem*. Springer-Verlag (New York, 1974). [6J]

[348] E. F. Robertson and P. D. Williams, Efficient presentations of the groups $PSL(2, 2p)$ and $SL(2, 2p)$. *Canad. Math. Bull.* **32** (1989), 3–10. [13A]

[349] S. A. Robertson, *Polytopes and Symmetry*. London Math. Soc. Lecture Notes Ser. **90**, Cambridge University Press (Cambridge, 1984). [7E]

[350] M. A. Ronan, Coverings and automorphisms of chamber systems. *European J. Combin.* **1** (1980), 259–269. [4A]

[351] M. A. Ronan, *Lectures on Buildings*. Academic Press (San Diego, 1989). [2]

[352] M. A. Ronan and G. Stroth, Minimal parabolic geometries for the sporadic groups. *European J. Combin.* **5** (1984), 59–91. [7A]

[353] R. Scharlau, Geometrical realizations of shadow geometries. *Proc. London Math. Soc.* (3) **61** (1990), 615–656. [3D]

[354] R. Scharlau, Buildings. In *Handbook of Incidence Geometry* (ed. F. Buekenhout), Elsevier Science B. V. (1995), Chapter 11. [2]

[355] L. Schläfli, Theorie der vielfachen Kontinuität. *Denkschriften der Schweizerlichen Naturforschenden Gesellschaft* **38** (1901), 1–237. [1A]

[356] V. Schlegel, Theorie der homogen zusammengesetzten Raumgebilde. *Nova Acta Leop. Carol.* **44** (1883), 343–459.

[357] R. Schneider, *Convex Bodies: the Brunn-Minkowski Theory*. Cambridge University Press (Cambridge, 1993). [1B, 5B]

[358] A. H. Schoen, Infinite regular warped polyhedra (IRWP) and infinite periodic minimal surfaces (IPHS). *Notices Amer. Math. Soc.* **15** (1968), 727 (Abstract 658–30). [7E]

[359] A. H. Schoen, Regular saddle polyhedra. *Notices Amer. Math. Soc.* **15** (1968), 929–930 (Abstract 68T–D6). [7E]

[360] M. Schönert et al., *GAP: Groups, Algorithms, and Programming*. Lehrstuhl D für Mathematik RWTH, Aachen, Germany fifth edition (1995). [10C]

[361] P. H. Schoute, *Mehrdimensionale Geometrie* Bd. 2 (*Die Polytope*). Sammlung Göschen, de Gruyter (Leipzig, 1905).

[362] E. Schulte, *Reguläre Inzidenzkomplexe*. Dissertation, University of Dortmund (1980). [2, 3D]

[363] E. Schulte, Reguläre Inzidenzkomplexe, II. *Geom. Dedicata* **14** (1983), 33–56. [2, 3D]

[364] E. Schulte, Reguläre Inzidenzkomplexe, III. *Geom. Dedicata* **14** (1983), 57–79. [2]

[365] E. Schulte, On arranging regular incidence-complexes as faces of higher-dimensional ones. *European J. Combin.* **4** (1983), 375–384. [4D]

[366] E. Schulte, Regular incidence-polytopes with euclidean or toroidal faces and vertex-figures. *J. Combin. Theory Ser. A* **40** (1985), 305–330. [10A]

[367] E. Schulte, Extensions of regular complexes. In *Finite Geometries* (eds. C. A. Baker and L. M. Batten), Lecture Notes Pure Applied Mathematics **103**, Dekker (New York, 1985), 289–305. [8C, 8D]

[368] E. Schulte, Amalgamations of regular incidence-polytopes. *Proc. London Math. Soc. (3)* **56** (1988), 303–328. [4A, 4E, 4F, 10A]

[369] E. Schulte, On a class of abstract polytopes constructed from binary codes. *Discrete Math.* **84** (1990), 295–301. [8F]

[370] E. Schulte, Classification of locally toroidal regular polytopes. In *Polytopes: Abstract, Convex and Computational* (eds. T. Bisztriczky, P. McMullen, R. Schneider, and A. Ivić Weiss), NATO ASI Series C **440**, Kluwer (Dordrecht, 1994), 125–154. [10A]

[371] E. Schulte, Symmetry of polytopes and polyhedra. In *Handbook of Discrete and Computational Geometry* (eds. J. E. Goodman and J. O'Rourke), CRC Press (Boca Raton–New York, 1997), 311–330. [1B]

[372] E. Schulte and A. I. Weiss, Chiral polytopes. In *Applied Geometry and Discrete Mathematics (The Victor Klee Festschrift)* (eds. P. Gritzmann and B. Sturmfels), DIMACS Series in Discrete Mathematics and Theoretical Computer Science **4**, Amer. Math. Soc. and Assoc. Computing Machinery (1991), 493–516. [2B, 6H]

[373] E. Schulte and A. I. Weiss, Chirality and projective linear groups. *Discrete Math.* **131** (1994), 221–261. [2B, 6H, 10A]

[374] E. Schulte and A. I. Weiss, Free extensions of chiral polytopes. *Canad. J. Math.* **47** (1995), 641–654. [2B, 4D]

[375] E. Schulte and J. M. Wills, A polyhedral realization of Felix Klein's map $\{3,7\}_8$ on a Riemann surface of genus 3. *J. London Math. Soc. (2)* **32** (1985), 539–547. [5B]

[376] E. Schulte and J. M. Wills, On Coxeter's regular skew polyhedra. *Discrete Math.* **60** (1986), 253–262. [5B]

[377] E. Schulte and J. M. Wills, Geometric realization for Dyck's regular map on a surface of genus 3. *Discrete Comput. Geom.* **1** (1986), 141–153. [5B]

[378] E. Schulte and J. M. Wills, Kepler-Poinsot-type realizations of regular maps of Klein, Fricke, Gordan and Sherk. *Canad. Math. Bull.* **30** (1987), 155–164. [5B]

[379] E. Schulte and J. M. Wills, Combinatorially regular polyhedra in three-space. In *Symmetry of Discrete Mathematical Structures and Their Symmetry Groups* (eds. K. H. Hofmann and R. Wille), Research and Exposition in Mathematics **15**, Heldermann (Berlin, 1991), 49–88. [5B]

[380] J. Schwörbel, *Die kombinatorisch regulären Tori.* Diplom-Arbeit, Universität Siegen (1988). [5B]

[381] J. Schwörbel and J. M. Wills, The two Pappus-tori. *Geom. Dedicata* **28** (1988), 359–362. [5B]

[382] J. J. Seidel and S. V. Tsaranov, Two-graphs, related groups, and root systems. *Bull. Soc. Math. Belg. Sér. A* **42** (1990), 695–711. [9G]
[383] H. Seifert and W. Threlfall, *Lehrbuch der Topologie*. Teubner (Leipzig, 1934). (English translation in *A Textbook of Topology*, Academic Press (1980).) [6B]
[384] M. Senechal and G. Fleck (eds.), *Shaping Space – A Polyhedral Approach*. Birkhäuser (Boston–Basel, 1988). [1A]
[385] B. Servatius and H. Servatius, Self-dual maps on the sphere. *Discrete Math.* **134** (1994), 139–150. [2A]
[386] G. C. Shephard, An elementary proof of Gram's theorem for convex polytopes. *Canad. J. Math.* **19** (1967), 1214–1217. [3E]
[387] G. C. Shephard, Regular complex polytopes. *Proc. London Math. Soc. (3)* **2** (1952), 82–97. [8D, 9A]
[388] G. C. Shephard, Unitary groups generated by reflections. *Canad. J. Math.* **5** (1953), 364–383. [9A, 9F]
[389] G. C. Shephard, Abstract definitions for reflection groups. *Canad. J. Math.* **9** (1957), 273–276. [9A]
[390] G. C. Shephard and J. A. Todd, Finite unitary reflection groups. *Canad. J. Math.* **6** (1954), 274–304. [9, 9A]
[391] F. A. Sherk, The regular maps on a surface of genus three. *Canad. J. Math.* **11** (1959), 452–480. [6J]
[392] F. A. Sherk, A family of regular maps of type {6, 6}. *Canad. Math. Bull.* **5** (1962), 13–20. [6J]
[393] S. V. Shpectorov, *On Geometries with Diagram P^n*. Thesis, University of Moscow (1989). [14A]
[394] D. Singerman and P. D. Watson, Weierstrass points on regular maps. *Geom. Dedicata* **66** (1997), 69–88. [6J]
[395] J. Širáň, M. Škoviera and H. J. Voss, Sachs triangulations and regular maps. *Discrete Math.* **134** (1994), 161–175. [6J]
[396] T. A. Springer, *Invariant Theory*. Lecture Notes in Mathematics **585**, Springer-Verlag (New York–Heidelberg–Berlin, 1977). [6C]
[397] R. P. Stanley, *Combinatorics and Commutative Algebra*. Birkhäuser (Boston–Basel–Stuttgart, 1983). [6B]
[398] R. P. Stanley, *Enumerative Combinatorics* **1**. Wadsworth & Brooks/Cole (Monterey, CA, 1986). [2C, 4E, 6B]
[399] R. Steinberg, Finite reflection groups. *Trans. Amer. Math. Soc.* **91** (1959), 493–504. [6C]
[400] A. Stephanides, Regular incidence quasi-polytopes and regular maps. *Geom. Dedicata* **30** (1989), 211–221. [2A]
[401] A. Stephanides, On regular incidence quasi-polytopes. *Geom. Dedicata* **38** (1991), 59–65. [2A]
[402] A. Stephanides, Constructing infinite families of regular incidence (quasi-) polytopes. *Geom. Dedicata* **53** (1994), 263–270. [2A]
[403] J. Stillwell, The story of the regular 120-cell. *Notices Amer. Math. Soc.* **48** (2001), 17–24. [1B]
[404] A. B. Stott, Geometrical deduction of semiregular from regular polytopes and space fillings. *Verhandel. Koninkl. Akad. Wetenschap.* (Eerste sectie) **11.1** (1910).
[405] W. I. Stringham, Regular figures in n-dimensional space. *Amer. J. Math.* **3** (1880), 1–14. [1A]
[406] J. G. Sunday, Presentations of the groups $SL(2, m)$ and $PSL(2, m)$. *Canad. J. Math.* **24** (1972), 1129–1131. [13A]
[407] S. Szabó, Polyhedra without diagonals. *Period. Math. Hungar.* **15** (1984), 44–49.
[408] J. M. Szucs, The asymptotic numbers of certain kinds of regular toroidal maps. *Discrete Math.* **215** (2000), 225–244. [6J]

[409] J. M. Szucs and D. J. Klein, Regular affine tilings and regular maps on a flat torus. *Discrete Appl. Math.* **105** (2000), 225–237. [6J]

[410] W. Threlfall, Gruppenbilder. *Abh. Sächs. Akad. Wiss. Math.-Phys. Kl.* **41** (1932), 1–59. [6J]

[411] W. P. Thurston, *The Geometry and Topology of Three-manifolds*. Lecture Notes, Princeton University Press (Princeton, NJ, 1979). [2E, 6B, 6J]

[412] W. P. Thurston, Three dimensional manifolds, Kleinian groups, and hyperbolic geometry. *Bull. Amer. Math. Soc.* **6** (1982), 357–381. [6J]

[413] W. P. Thurston, *Three-dimensional Geometry and Topology, Volume 1* (ed. S. Levy). Princeton University Press (Princeton, NJ, 1997). [6A]

[414] F. G. Timmesfeld, Tits geometries and parabolic systems in finitely generated groups, I, II. *Math. Z.* **184** (1983), 377–396, 449–487.

[415] J. Tits, Groupes et géométries de Coxeter. Institut des Hautes Études Scientifiques (notes polycopiées), Paris (1961). [1A, 3D]

[416] J. Tits, Géométries polyédriques et groupes simples. *Atti 2a Riunione Groupem. Math. Express. lat. Firenze* (1962), 66–88. [1A, 2B]

[417] J. Tits, *Buildings of Spherical type and finite BN-Pairs*. Lecture Notes in Mathematics Vol. 386, Springer-Verlag (New York–Heidelberg–Berlin, 1974). [1A, 2, 2A, 2B, 2C, 3, 3D]

[418] J. Tits, A local approach to buildings. In *The Geometric Vein: the Coxeter Festschrift* (eds. C. Davis, B. Grünbaum, and F. A. Sherk), Springer-Verlag (New York–Heidelberg–Berlin, 1981), 519–547. [4A]

[419] J. Tits, Simple groups and Buekenhout geometries. In *Finite Simple Groups II* (ed. M. Collins), Academic Press (New York, 1981), 309–320. [7A]

[420] J. Tits, Immeubles de type affine. In *Buildings and the Geometry of Diagrams* (ed. L. A. Rosati), Lecture Notes in Mathematics **1181**, Springer-Verlag (New York–Heidelberg–Berlin, 1986), 157–190.

[421] J. Tits, Sur le groupe des automorphismes de certains groupes de Coxeter. *J. Algebra* **113** (1988), 346–357. [7B]

[422] J. Tits and R. Weiss, *Moufang Polygons* (in preparation). [2A]

[423] J. A. Todd, The groups of symmetries of the regular polytopes. *Proc. Cambridge Philos. Soc.* **27** (1931), 212–231.

[424] J. A. Todd, A second note on the linear fractional group. *J. London Math. Soc.* **11** (1936), 103–107. [13A]

[425] S. V. Tsaranov, On a generalization of Coxeter groups. *Algebras Groups Geom.* **6** (1989), 281–318. [9G]

[426] S. V. Tsaranov, Finite generalized Coxeter groups. *Algebras Groups Geom.* **6** (1989), 421–452. [9G]

[427] S. V. Tsaranov, Geometries and amalgams of J_1. *Comm. Algebra* **18** (1990), 1119–1135. [7A, 14A]

[428] A. Valverde, 3-Dimensional euclidean manifolds represented by locally regular coloured graphs. *Math. Slovaca* **47** (1997), 99–110. [6B]

[429] P. Vanden Cruyce, Geometries related to PSL(2,19). *European J. Combin.* **6** (1985), 163–173. [6J]

[430] P. Vanden Cruyce, A finite graph which is locally a dodecahedron. *Discrete Math.* **54** (1985), 343–346. [6C]

[431] E. B. Vinberg, Discrete groups in Lobachevskiĭ spaces generated by reflections. *Mat. Sb.* **72** (1967), 471–488 (= *Math. USSR-Sb.* **1** (1967), 429–444). [3C]

[432] E. B. Vinberg, Discrete linear groups generated by reflections. *Izv. Akad. Nauk. SSSR Ser. Mat.* **35** (1971) 1072–1112 (= *Math. USSR-Izv.* **5** (1971), 1083–1119). [3C]

[433] E. B. Vinberg, Hyperbolic reflection groups. *Uspekhi Mat. Nauk* **40** (1985), 29–66 (= *Russian Math. Surveys* **40** (1985), 31–75). [3C]

[434] E. B. Vinberg, *Linear Representations of Groups*. Birkhäuser (Basel, 1989). [5B, 11D]

[435] A. Vince, Combinatorial maps. *J. Combin. Theory Ser. B* **34** (1983), 1–21. [4C, 6J]

[436] A. Vince, Regular combinatorial maps. *J. Combin. Theory Ser. B* **35** (1983), 256–277. [4C, 6J]

[437] H. J. Voss, Sachs triangulations, generated by dessin d'enfant, and regular maps. *Math. Slovaca* **47** (1997), 193–210. [6J]

[438] W. C. Waterhouse, The discovery of the regular solids. *Arch. Hist. Exacts Sciences* **9** (1972/73), 212–221. [1A]

[439] B. A. F. Wehrfritz: *Infinite Linear Groups*. Springer (New York, 1973). [4C, 12B]

[440] G. Weetman, A construction of locally homogeneous graphs. *J. London Math. Soc. (2)* **50** (1994), 68–86. [6C]

[441] A. I. Weiss, On trivalent graphs embedded in twisted honeycombs. In *Combinatorics '81 (Rome, 1981)*, North-Holland Math. Stud. 78, North-Holland (Amsterdam, 1983), 781–787. [6J]

[442] A. I. Weiss, Twisted honeycombs $\{3, 5, 3\}_t$. *C. R. Math. Acad. Sci. Soc. R. Can.* **5** (1983), 211–215. [6J]

[443] A. I. Weiss, An infinite graph of girth 12. *Trans. Amer. Math. Soc.* **283** (1984), 575–588. [10A, 11E]

[444] A. I. Weiss, Incidence-polytopes of type $\{6, 3, 3\}$. *Geom. Dedicata* **20** (1986), 147–155. [10A]

[445] A. I. Weiss, Incidence-polytopes with toroidal cells. *Discrete Comput. Geom.* **4** (1989), 55–73. [10A]

[446] A. I. Weiss, Some infinite families of finite incidence-polytopes. *J. Combin. Theory Ser. A* **55** (1990), 60–73. [6J]

[447] A. I. Weiss and Z. Lučić, Regular polyhedra in hyperbolic three-space. *Mitt. Math. Sem. Giessen*. **165** (1984), 237–252. [7B]

[448] A. F. Wells, The geometrical basis of crystal chemistry, I. *Acta Cryst.* **7** (1954), 535–544. [7E]

[449] A. F. Wells, The geometrical basis of crystal chemistry, II. *Acta Cryst.* **7** (1954), 545–554. [7E]

[450] A. F. Wells, The geometrical basis of crystal chemistry, X. Further study of three-dimensional polyhedra. *Acta Cryst. Sect. B* **25** (1969), 1711–1719. [7E]

[451] A. F. Wells, *Three-dimensional Nets and Polyhedra*. Wiley-Interscience (New York, 1977). [7E, 7F]

[452] M. Wester, Endliche fahnentransitive Tits-Geometrien und ihre universellen Überlagerungen. *Mitt. Math. Sem. Giessen* **170** (1985), 1–143.

[453] J. B. Wilker, Inversive geometry. In *The Geometric Vein: The Coxeter Festschrift* (ed. C. Davis, B. Grünbaum, F. A. Sherk), Springer (New York, 1981), 379–442. [13D]

[454] R. Williams, *Natural Structure*. Eudaemon Press (Moorpark, CA, 1972). [7E]

[455] J. M. Wills, Combinatorially regular polyhedra of index 2. *Aequationes Math.* **34** (1987), 206–220. [5B]

[456] J. M. Wills, On regular polyhedra with hidden symmetries. *Results Math.* **12** (1987), 450–458. [5B]

[457] J. M. Wills, A combinatorially regular dodecahedron of genus 3, *Discrete Math.* **67** (1987), 199–204. [5B]

[458] S. E. Wilson, *New Techniques for the Construction of Regular Maps*. Ph.D. Thesis, University of Washington (1976). [6J]

[459] S. E. Wilson, Non-orientable regular maps. *Ars Combin.* **5** (1978), 213–218. [6J]

[460] S. E. Wilson, The smallest nontoroidal chiral maps. *J. Graph Theory* **2** (1978), 315–318. [6J]

[461] S. E. Wilson, Operators over regular maps. *Pacific J. Math.* **81** (1979), 559–568. [6J, 7B]

[462] S. E. Wilson, A construction from an existence proof. *Congr. Numer.* **50** (1985), 25–30. [4C]

[463] S. E. Wilson, Smooth coverings of regular maps. In *Groups, Combinatorics & Geometry*

(eds. M. Liebeck and J. Saxl), London Math. Soc. Lecture Notes Series **165**, Cambridge University Press (Cambridge, 1992), 480–489. [6J]

[464] S. E. Wilson, Applications and refinements of Vince's construction. *Geom. Dedicata* **48** (1993), 231–242. [4C]

[465] E. Witt, Spiegelungsgruppen und Aufzählung halbeinfacher Liescher Ringe. *Abh. Math. Sem. Univ. Hamburg* **14** (1941), 289–322. [3B]

[466] W. A. Wythoff, A relation between the polytopes of the C_{600}-family. *Konink. Akad. Weten. Amsterdam, Proc. Sect. Sciences* **20** (1918), 966–970. [1B]

[467] J. A. Wolf, *Spaces of Constant Curvature* (5th edition). Publish or Perish (Wilmington, DE, 1984). [6A, 6B, 6C]

[468] A. E. Zalesskiǐ and V. N. Serežkin, Finite linear groups generated by reflections. *Math. USSR Izvestija* **17** (1981), 477–503. [12C]

[469] G. M. Ziegler, *Lectures on Polytopes*. Graduate Texts in Mathematics, Springer (New York, 1995)

[470] H. Zieschang, On Heegaard diagrams of 3-manifolds. In *Geometry of Differentiable Manifolds* (Workshop, Rome, 1986) [Astérique **163–164**], Soc. Math. France (Paris, 1989), 7, 247–280, 283. [6B]

List of Symbols

Author Index

Subject Index